REVIEWS in MINERALOGY
Volume 13

MICAS

S. W. BAILEY
Editor

The Authors:

S.W. BAILEY
  Dept. of Geology & Geophysics
  University of Wisconsin - Madison
  Madison, Wisconsin 53706

DONALD M. BURT
  Dept. of Geology
  Arizona State University
  Tempe, Arizona 85287

PETR ČERNÝ
  Dept. of Earth Sciences
  University of Manitoba
  Winnipeg, Manitoba, Canada R3T 2N2

DENNIS D. EBERL & RAY E. WILCOX
  U.S. Geological Survey
  Denver Federal Center
  Denver, Colorado 80225

R.F. GIESE, Jr.
  Dept. of Geological Sciences
  State University of New York
  4240 Ridge Lea Road
  Buffalo, New York 14226

STEPHEN GUGGENHEIM
  Dept. of Geology
  University of Illinois at Chicago
  Chicago, Illinois 60680

CHARLES V. GUIDOTTI
  Dept. of Geology
  University of Maine
  Orono, Maine 04469

D.A. HEWITT, J.A. SPEER, D.R. WONES
  Dept. of Geological Sciences
  Virginia Polytechnic Insitute and
    State University
  Blacksburg, Virginia 24061

J.L. MUNOZ
  Dept. of Geological Sciences
  University of Colorado
  Boulder, Colorado 80309

I. EDGAR ODOM
  American Colloid Company
  5100 Suffield Court
  Skokie, Illinois 60025

GEORGE R. ROSSMAN
  Div. of Geological & Planetary Science
  California Institute of Technology
  Pasadena, California 91125

JAN ŚRODOŃ
  Institute of Geological Sciences PAN
  31-002 Krakow
  Senacka 3, Poland

Series Editor: PAUL H. RIBBE
Department of Geological Sciences
Virginia Polytechnic Institute & State University
Blacksburg, Virginia 24061

MINERALOGICAL SOCIETY OF AMERICA

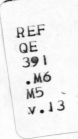

COPYRIGHT 1984
MINERALOGICAL SOCIETY of AMERICA

PRINTED BY

BookCrafters, Inc.
Chelsea, Michigan 48118

REVIEWS in MINERALOGY

(Formerly: SHORT COURSE NOTES)

ISSN 0275-0279

Volume 13: MICAS
ISBN 0-939950-17-0

ADDITIONAL COPIES of this volume as well as those listed below may be obtained from:

Mineralogical Society of America

2000 FLORIDA AVENUE, N.W. WASHINGTON, D. C. 20009

Volume		Pages
1	SULFIDE MINERALOGY, P.H. Ribbe, editor (1974)	284
2	FELDSPAR MINERALOGY, P.H. Ribbe, editor (1975; revised 1983)	362
3	OXIDE MINERALS, Douglas Rumble, III, editor (1976)	502
4	MINERALOGY and GEOLOGY of NATURAL ZEOLITES, F.A. Mumpton, editor (1977)	232
5	ORTHOSILICATES, P.H. Ribbe, editor (1980; revised 1982)	450
6	MARINE MINERALS, R.G. Burns, editor (1979)	380
7	PYROXENES, C.T. Prewitt, editor (1980)	525
8	KINETICS of GEOCHEMICAL PROCESSES, A.C. Lasaga and R.J. Kirkpatrick, editors (1981)	391
9A	AMPHIBOLES and OTHER HYDROUS PYRIBOLES - MINERALOGY, D.R. Veblen, editor (1981)	372
9B	AMPHIBOLES: PETROLOGY and EXPERIMENTAL PHASE RELATIONS, D.R. Veblen and P.H. Ribbe, editors (1982)	390
10	CHARACTERIZATION of METAMORPHISM through MINERAL EQUILIBRIA, J.M. Ferry, editor (1982)	397
11	CARBONATES: MINERALOGY and CHEMISTRY, R.J. Reeder, editor (1983)	394
12	FLUID INCLUSIONS, Edwin Roedder, author (1984)	644
13	MICAS, S.W. Bailey, editor (1984)	

REVIEWS in
MINERALOGY   **MICAS**   Volume 13

FOREWORD

1984 marks the tenth anniversary of the Mineralogical Society of America's short courses, the first of which was held in Miami, Florida on the topic of sulfide mineralogy. This year, S.W. Bailey organized our eleventh short course for presentation prior to the annual meetings of the Society in Reno, Nevada.

A total of fourteen volumes of REVIEWS in MINERALOGY have now been published, two of which are in second editions and one of which — our first single-author monograph — appeared earlier this year without an accompanying short course. See page ii (opposite) for a list of publications.

The present volume on micas incorporates the efforts of fifteen authors in thirteen chapters. The reader with a critical eye will notice a few editorial inconsistencies, including two major departures from our traditional format. Chapters 2 and 7 were prepared on word processors, but only the latter benefitted from proportional spacing to produce full left and right justification of typed lines. To reduce printing costs and thus keep the selling price down, the exceptionally long Chapter 10 was singled out to be single-spaced. Other chapters — or for that matter, all of them — might have been similarly formatted, but given the narrow time frame within which this volume was prepared for publication, there is cause to be thankful that it could be completed at all, let alone perfectly! This book, like others produced in conjunction with M.S.A. short courses, was put together in less than two months from manuscripts in various states of disarray by numerous authors with as many writing styles and a wide range of attitudes toward editorial deadlines. So there are the usual elements of suspense and drama as the technical editor and the copy editor try to guess who might not finish in time, what to do if he/she doesn't, how many pages of text there ultimately will be, and — among other things — how much tolerance our readership has for editorial eccentricities. In this volume, perhaps more so than previously, the last item has been tested. But I am confident that the value of this compendium on micas will more than compensate for a slight unevenness in style, for which I accept sole responsibility.

Paul H. Ribbe
Series Editor
September 1, 1984

PREFACE and ACKNOWLEDGMENTS

Although phyllosilicates are common in almost all types of rocks, their detailed study has not advanced in proportion to their importance. Books and reviews on this subject have been restricted primarily to the areas of clay mineralogy and soils. Such treatments understandably restrict coverage of the occurrences of the macroscopic-size species as well as much of their mineralogical and petrological nature. It was decided at the outset that not all phyllosilicates could be covered in a single book, and the size of this volume addressed only to the micas justifies the original decision. Kaolins, serpentines, chlorites, etc. will have to wait until some later date.

This volume attempts to gather together much of our knowledge of micas, the most abundant phyllosilicate, and to indicate promising areas of future research. Chapters 1-3 lay the foundations of the classification, structures, and crystal chemistry of micas. Chapter 4 treats bonding and electrostatic modeling of micas. Chapters 5 and 6 cover spectroscopic and optical properties. Chapters 7-13, the bulk of the volume, are devoted to geochemistry and petrology. These include phase equilibria and the occurrences, chemistry, and petrology of micas in igneous, metamorphic, and sedimentary rocks, pegmatites, and certain ore deposits. Some treatments are exhaustive. All are at the forefront of our present knowledge, and indicate clearly the practical applications of the study of micas to ascertaining various parameters of origin and crystallization history, as well as the many problems that still exist. The aim of this type of treatment is twofold -- to provide a handy reference volume for teachers and students and to enable researchers to pick more easily those directions and problems for which future research is most needed or is apt to be most productive or most challenging. X-ray powder patterns of micas in the literature are of surprisingly poor quality. The best are collated and supplemented with additional new patterns in the Appendix as an aid to identification.

I wish to thank all of the authors for agreeing enthusiastically to participate in this volume and then for keeping on schedule. I thank the Series Editor, Paul Ribbe, for starting us out on the right foot and then letting us have our head -- in the racing sense. The Department of Geology and Geophysics at the University of Wisconsin-Madison provided me with some free time through the Lewis G. Weeks Endowment Fund plus generous secretarial and logistical support. Our typists were Ada Simmons, Margie Strickler, and Donna Williams.

S.W. Bailey
Madison, Wisconsin
September 1, 1984

MICAS
S. W. BAILEY
Editor

TABLE of CONTENTS

	Page
COPYRIGHT; LIST OF PUBLICATIONS	ii
FOREWORD	iii
PREFACE; ACKNOWLEDGMENTS	iv

1. CLASSIFICATION and STRUCTURES of the MICAS
S.W. Bailey

INTRODUCTION	1
STANDARD MICA POLYTYPES	5
NATURAL MICA STRUCTURES	8
REFERENCES	12

2. CRYSTAL CHEMISTRY of the TRUE MICAS
S.W. Bailey

SHEET CONFIGURATIONS	13
Dioctahedral distortions	13
Trioctahedral distortions	16
TETRAHEDRAL-OCTAHEDRAL LATERAL FIT	16
Ideal lateral dimensions	16
Tetrahedral rotation	17
Changes in sheet thicknesses	19
Tetrahedral tilt and basal surface corrugation	28
Limits of strain	29
INTERLAYER REGION	31
Size of cavity	31
Layer offsets in K-micas	32
Intralayer overshifts in relation to β angle	33
Interlayer cation coordination	34
Layer offset in paragonite; Location of H^+ proton	35
CALCULATED STRUCTURAL MODELS	37
CATION ORDER AND DISORDER	38
Ordering of tetrahedral cations	40
Ordering of octahedral cations	44
Ordering of interlayer cations	50
Local charge balance	51
Standardized cation notation	52
SHORT RANGE ORDERING	54
ACKNOWLEDGMENTS	56
REFERENCES	57

3. The BRITTLE MICAS
Stephen Guggenheim

INTRODUCTION	61
NOMENCLATURE AND GENERAL CHEMICAL CLASSIFICATION	61
CRYSTAL CHEMISTRY OF THE BRITTLE MICAS	64

	Page
Introduction	64
Charge considerations; Sheet size considerations	65
Cation ordering	66
Octahedral ordering	70
Tetrahedral ordering	75
Interlayer region	82
Orientation of the O--H vector	84
PHASE RELATIONS OF MARGARITE AND CLINTONITE	85
Introduction	85
Margarite	86
Margarite/paragonite ± muscovite	94
Clintonite	94
RELATIONSHIP OF CRYSTAL STRUCTURE TO STABILITY	97
ACKNOWLEDGMENTS	100
REFERENCES	101

4.   ELECTROSTATIC ENERGY MODELS of MICAS
R.F. Giese, Jr.

INTRODUCTION	105
METHOD OF CALCULATION	107
APPLICATIONS TO MICAS	110
STRUCTURAL DISTORTIONS	111
Octahedral distortions; Tetrahedral distortion	112
HYDROXYL ORIENTATIONS	114
Ionic model	115
Possible errors	117
OH orientations in micas	118
Mixed tri- and dioctahedral micas	121
INTERLAYER BONDING	123
Ionic model	123
Calculation of interlayer bonding	124
Interlayer bonding of micas	126
Substitution of F for OH	128
CATION ORDERING	129
Theoretical considerations	131
Order/disorder in trioctahedral M sites	132
Order/disorder in dioctahedral T sites	136
SUMMARY	139
ACKNOWLEDGMENTS	140
REFERENCES	141

5.   SPECTROSCOPY of MICAS
George R. Rossman

INTRODUCTION	145
OPTICAL SPECTRA AND COLOR	145
Individual ions; Fe^{2+} in phlogopite-biotite	147
Fe^{2+} in muscovite	148
Fe^{3+} in muscovite; Fe^{3+} in biotite	149
The role of iron in the color of muscovite	149
Cr^{3+} in micas	150
Lepidolites: the role of Mn and Fe; Ti in micas	151
Interactions between cations	152
Mn in phlogopite; Other micas	154
Reverse pleochroism -- tetrahedral ferric iron	156
Reverse pleochroism -- pink muscovite; The ultraviolet region	157
Future needs	158

	Page
MÖSSBAUER SPECTRA	158
Biotite	158
Muscovite	160
Glauconite	162
Clintonite; Other micas; Commentary	163
ELECTRON SPIN RESONANCE SPECTRA	163
NUCLEAR MAGNETIC RESONANCE SPECTRA	164
X-RAY SPECTROSCOPY	168
X-ray photoelectron spectroscopy	168
INFRARED SPECTRA	169
OH groups and their orientation	169
Cation ordering	171
The far-infrared spectral region	172
RAMAN SPECTRA	173
NEAR INFRARED SPECTRA	173
RADIATION HALOES AND OTHER RADIATION EFFECTS	174
OXIDATION AND DEHYDROXYLATION OF BIOTITE	174
QUANTITATIVE INTENSITIES AND ANALYTICAL DETERMINATIONS	175
CONCLUDING REMARKS	176
REFERENCES	177

6.   OPTICAL PROPERTIES of MICAS under the POLARIZING MICROSCOPE
Ray E. Wilcox

	Page
INTRODUCTION	183
OPTICAL PROPERTIES OF INDIVIDUAL MINERALS	183
Explanation of tables	184
Variation of optical properties with chemical composition	192
METHODS FOR DETERMINATION OF OPTICAL PROPERTIES	196
Determination in immersion liquids	196
Coated slide technique	197
Spindle stage technique	197
Determinations in thin sections	198
SUMMARY	199
REFERENCES	200

7.   EXPERIMENTAL PHASE RELATIONS of the MICAS
David A. Hewitt & David R. Wones

	Page
INTRODUCTION	201
SUBSOLIDUS PHASE RELATIONS OF THE DIOCTAHEDRAL MICAS	202
Stability of muscovite and muscovite + quartz	202
Stability of paragonite and paragonite + quartz	204
The system muscovite-paragonite-quartz	206
Stability of margarite and margarite + quartz	207
Stability of pyrophyllite	211
SUBSOLIDUS PHASE RELATIONS FOR THE TRIOCTAHEDRAL MICAS	213
Stability of end-member biotites	213
Stability of biotite solid solutions	219
Stability of talc	226
Stability of other micas and mica reactions	228
MELTING EQUILIBRIA OF MICAS	228

	Page
Melting of phlogopite	228
Micas and siliceous melts	238
Muscovites in granitic plutons -- an attempt at geobarometry	244
REFERENCES	247

8. PARAGENESIS, CRYSTALLOCHEMICAL CHARACTERISTICS, and GEOCHEMICAL EVOLUTION of MICAS in GRANITE PEGMATITES
Petr Černý & Donald M. Burt

INTRODUCTION	257
SOME DEFINITIONS AND OBJECTIVES	258
Classification of granitic pegmatites	258
Internal structure of granitic pegmatites	259
Mica species in granitic pegmatites	259
Graphic representation	260
Limitations of scope	261
MICA ASSEMBLAGES IN OROGENIC PEGMATITES	261
Muscovite pegmatites	261
Rare-element (and miarolitic) pegmatites	262
Gadolinite type; Beryl-columbite type; Complex type	262
Spodumene type; Lepidolite type	263
Late muscovite alteration; Exomorphic micas	263
MICA ASSEMBLAGES IN ANDROGENIC PEGMATITES	264
Rare-element pegmatites; Miarolitic pegmatites	264
CRYSTALLOCHEMICAL CHARACTERISTICS AND POLYTYPISM	265
Phlogopite - biotite	265
Muscovite - lithian muscovite - lepidolite	266
Zinnwaldite - masutomilite	269
Exomorphic micas	270
VECTOR PRESENTATION OF LITHIUM MICA COMPOSITIONS	270
The concept; Capacity and limitations	271
Planar subsystems	272
The 3-dimensional polyhedron	276
GEOCHEMICAL EVOLUTION OF PEGMATITE MICAS	279
Micas of orogenic pegmatites	279
Muscovite pegmatites; Rare-element pegmatites	279
Micas of anorogenic subalkalic pegmatites	283
GENETIC ASPECTS OF MICA CRYSTALLIZATION IN GRANITIC PEGMATITES	284
Muscovite pegmatite class	284
Rare-element (and miarolitic) class of orogenic suites	285
Biotite; Muscovite; Lepidolite	285
Zinnwaldite	287
Late muscovite alteration; Exomorphic micas	288
Rare-element (and miarolitic) class of anorogenic suites	290
CONCLUDING NOTE	290
REFERENCES	292

9. MICAS in IGNEOUS ROCKS
J. Alexander Speer

INTRODUCTION	299
OXYGEN AND WATER FUGACITIES	299
Oxygen fugacity	300
Water fugacity	305
BIOTITES COEXISTING WITH OTHER MINERALS	307
Amphibole	307
Aluminous minerals	313
Apatite	315

	Page
Feldspars	317
Magnetite; Muscovite	318
ISOTOPIC COMPOSITION	319
Geochronology	319
Wall-rock/igneous rock interaction	320
Geothermometry	323
MUSCOVITE IN IGNEOUS ROCKS	325
Texture; Chemistry	326
Mineral assemblage; Occurrence	329
IGNEOUS MICAS AS METALLOGENIC INDICATORS	330
SUBSOLIDUS ALTERATION	332
HALOGENS IN BIOTITE	335
Fluorine	335
Chlorine	338
INTERLAYER SITES	340
Ammonium	340
Barium	341
Calcium; Sodium	342
RARE EARTH ELEMENTS IN BIOTITES	342
PETROLOGY	343
Occurrence	343
Indicators of magmatic evolution	347
Hybrid rocks	348
REFERENCES	349

10. MICAS in METAMORPHIC ROCKS
Charles V. Guidotti

INTRODUCTION	357
DIOCTAHEDRAL WHITE MICAS	360
White micas from a mineralogic perspective	360
Observed chemical variation of the white micas	360
The system $NaAlO_2$-$KAlO_2$-$CaAl_2O_3$-SiO_2-H_2O	360
Theoretical and experimental studies bearing upon the solvi of the ideal white mica plane: Comparison with natural data	363
White mica compositional deviation from the ideal system	363
Details of the deviation of muscovite from the ideal white mica plane	366
Phengite = celadonite = Tschermak substitution	367
The substitution of Fe^{3+} for Al^{vi}	371
The substitution of Ti into the octahedral sites	371
Substitution of F and Cl for (OH)	372
Deviation of ideal Mu from dioctahedral to trioctahedral	373
Substitutions involving 12-coordinated (XII) sites	374
Miscellaneous compositional variations in muscovite	375
White mica lattice spacings and polytypes	375
Composition versus lattice spacings and cell volume	376
General summary of the composition versus lattice spacing relationships	379
White mica polymorphs in metamorphic rocks	379
Crystal chemistry of the white micas	380
Chemical variation within the ideal white mica systems	380
Deviations from the ideal mica plane	381
Effects of celadonite substitution on lattice dimensions	382
Interrelationships between substitutions in XII and substitutions in the octahedral and tetrahedral sites	382
Other substitutions	383

	Page
Petrologic aspects of white micas	384
White micas in the context of gross lithologic types	384
Margarite occurrence	385
Normal, rock-forming occurrences	385
Pseudomorph-forming occurrences	386
Paragonite occurrence	386
Muscovite occurrence	387
White mica composition in the context of gross lithologic differences	388
Metamorphic controls on white mica occurrence and composition	388
Theoretical framework for treating the phase relations of white micas	389
Margarite phase relations	391
Paragonite phase relations	393
Theoretical models of Pg phase relations	394
Pg phase relations based on natural parageneses	394
Experimental and theoretical approaches to Pg stability conditions	395
Miscellaneous aspects of Pg phase relations	396
Muscovite phase relations	401
Occurrence	401
Compositional variation of Mu in response to metamorphic conditions	401
Variation of celadonite content	402
Variation of Na/(Na+K) ratio	406
Summary; Other compositional variations of Mu in response to metamorphism	406
Polymorphs of Mu -- metamorphic aspects	406
White micas as petrogenetic indicators	408
Use of white mica composition data for petrogenetic purposes	408
Geothermometry using white micas	409
The Mu-Pg solvus	409
The assemblage Mu + plagioclase + Al-silicate	410
The assemblage Mu + plagioclase + K-feldspar + Al-silicate	410
Other geothermometers involving the composition of Mu	411
Geobarometry using white micas	411
Information on the fluid phase via white micas	412
White micas: identification and determination of composition	413
Optical methods	413
X-ray methods	414
Electron microprobe analysis of white micas	416
Problems of Fe^{2+} versus Fe^{3+} in white micas	417
Staining techniques	418
TRIOCTAHEDRAL "DARK MICAS"	418
Biotite from mineralogic perspective	419
Observed chemical variation of biotite	419
Compositional variation within the biotite plane	419
Titanium substitution in octahedral sites	423
Fe^{3+} substitution into octahedral sites	425
Substitution of F and Cl for (OH) in biotite	427
Vacancies in octahedral sites and other substitutions	430
Substitutions in the XII sites	431
Crystallochemical aspects of biotite compositional variation	431
Compositional variation within the biotite plane	433
Compositional variation deviating from the biotite plane	434
Experimental data	434
Substitution of Ti	435
Substitution of F for (OH) and its relation to $Mg/(Mg+Fe_T)$	436
Al^{iv} in excess of that required for the Tschermak exchange	437
Substitution of Fe^{3+}; Substitutions in the XII site	437
Petrologic aspects of biotite	437
Biotite occurrence and composition in the context of gross lithology	438
Biotite in pelitic schists; metagreywacke and impure quartzites; calc-silicates in marbles; metabasites; metamorphosed granitic rocks	438
Miscellaneous	439

	Page
Metamorphic controls on the occurrence and composition of biotite	439
Biotite in marbles and calc-silicates	439
Biotite in metabasites	442
Biotite in pelitic schists	443
Petrogenetic aspects of biotite	447
Isogradic reactions; Continuous variation of biotite composition in response to continuous reactions; Relationships with volatile constituents	448
Exchange-reactions and geothermometry	450
Biotite identification and determination of composition	452
Optical methods; Chemical analyses of biotite	452
SUGGESTIONS FOR FUTURE RESEARCH	453
ACKNOWLEDGMENTS	455
REFERENCES	456

11. F-OH and Cl-OH EXCHANGE in MICAS with APPLICATIONS to HYDROTHERMAL ORE DEPOSITS
J.L. Munoz

INTRODUCTION	469
THERMODYNAMICS OF HALOGEN=HYDROXYL EXCHANGE	469
SYSTEMATICS OF F=OH EXCHANGE	470
Effect on physical properties	470
Fe-F avoidance	472
Stability relations	472
F-OH exchange experiments	473
Thermodynamic models; Crystal field theory; Cation ordering	474
Thermodynamics of (OH,F) biotite solutions	475
Fluorine intercept value and fluorine index	477
Uncertainties relating to anion occupancy	477
SYSTEMATICS OF Cl=OH EXCHANGE	479
Mg-Cl avoidance	479
Chlorine intercept value; F/Cl intercept value	480
HALOGEN COMPOSITIONS OF HYDROTHERMAL MICAS RELATED TO ORE DEPOSITS	481
Fluorine	483
Chlorine; Fluorine/chlorine ratio	485
INTERPRETATION OF HALOGEN INTERCEPT DATA IN TERMS OF FLUID COMPOSITIONS	486
CONCLUSIONS	490
REFERENCES	491

12. ILLITE
Jan Środoń & Dennis D. Eberl

INTRODUCTION	495
X-RAY DIFFRACTION ANALYSIS	496
XRD identification of illitic materials by basal reflections	496
Characterization of "standard illites" by basal reflections	500
Illite crystallinity index	502
Interparticle diffraction	503
Polytypes	506
Quantitative analysis of illitic materials in multicomponent systems	510
Analysis of oriented preparations	510
Analysis of random preparations	512
CHEMICAL COMPOSITION	512
Composition of illite in relation to other 2:1 phyllosilicates	512
Chemistry of illite and illite layers	515

	Page
ILLITE IN NATURAL, SYNTHETIC, AND THEORETICAL SYSTEMS	518
Weathering environment	518
Opening of illite layers	518
Neoformation of illite in weathering profiles	518
Illitization of smectite by wetting and drying	520
Sedimentary environment	521
Formation of illite in sea water	521
Formation of illite in lakes	523
Diagenetic/metamorphic environment	524
Illitization of smectite	524
Illitization of kaolinite	527
Illitization of feldspar	529
Illitization of muscovite	529
Neoformation of illite in sandstone pores (Hairy illite)	529
Hydrothermal environment	532
Synthesis	532
Stability diagrams	536
ACKNOWLEDGMENTS	538
REFERENCES	539

13.   GLAUCONITE and CELADONITE MINERALS
I. Edgar Odom

NOMENCLATURE	545
LITERATURE	547
GEOLOGIC OCCURRENCES OF GLAUCONITES AND CELADONITES	547
MORPHOLOGICAL FORMS OF GLAUCONITES AND CELADONITES	549
UNIT STRUCTURES OF GLAUCONITES AND CELADONITES	551
Structural characteristics	551
Mineralogy of glauconites	552
Mineralogy of celadonites	557
CHEMICAL CHARACTERISTICS	557
Glauconites	557
Celadonites	561
MISCELLANEOUS PHYSICAL AND CHEMICAL PROPERTIES OF GLAUCONITES AND CELADONITES	564
ORIGIN OF GLAUCONITE MINERALS	566
Layer lattice theory	567
Neoformation theory	568
ORIGIN OF CELADONITE MINERALS	570
FUTURE RESEARCH	570
REFERENCES	571

APPENDIX   X-RAY POWDER PATTERNS of MICAS

REFERENCES	573
DIOCTAHEDRAL TRUE MICAS	574
TRIOCTAHEDRAL TRUE MICAS	577
BRITTLE MICAS	583

1. CLASSIFICATION and STRUCTURES of the MICAS
S.W. Bailey

INTRODUCTION

In its simplest terms the crystal structure of a mica consists of negatively charged *2:1 layers* that are compensated and bonded together by large, positively charged, *interlayer cations*. A 2:1 layer contains two *tetrahedral sheets* and one *octahedral sheet*. Each tetrahedral sheet is of composition T_2O_5 (T = tetrahedral cation), and within each sheet individual tetrahedra are linked with neighboring tetrahedra by sharing three corners each (the basal oxygens) to form an hexagonal mesh pattern of the sort illustrated in Figure 1a. The fourth tetrahedral corner (the apical oxygen) points in a direction normal to the sheet and at the same time forms part of an immediately adjacent octahedral sheet in which individual octahedra are linked laterally by sharing octahedral edges (Fig. 1b). In a 2:1 layer the upper and lower planes of anions that comprise an octahedral sheet are also the common planes of junction with two oppositely directed tetrahedral sheets, as in Figure 1c. These planes of junction consist of the shared apical

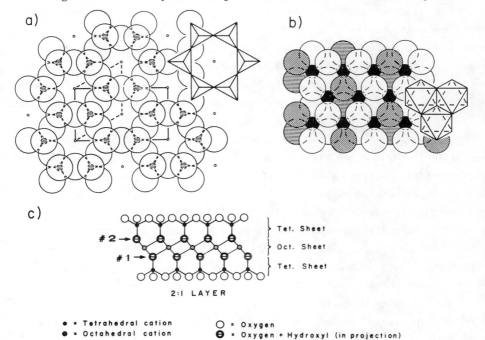

Figure 1. (a) Stylized tetrahedral sheet, fully extended laterally, with possible hexagonal (dashed line) and orthohexagonal (full line) cells shown. (b) Octahedral sheet with OH,F groups shaded. (c) 2:1 layer with #1 and #2 common planes of junction labeled. Modified from Bailey (1980).

oxygens plus unshared OH groups that lie at the center of each tetrahedral six-fold ring at the same z-level as the apical oxygens. Fluorine may substitute for OH in most species, even to the exclusion of OH; sulfur substitutes for OH in the species anandite.

The micas can be classified at the group level by the magnitude of the layer charge (x) per formula unit. This charge ideally is $x = -1.0$ for the true (or flexible) micas with compensation primarily by monovalent interlayer cations, and ideally is $x = -2.0$ for the brittle micas with compensation primarily by divalent interlayer cations. In practice considerable deviation from these ideal charges is realized. The two groups are divided into subgroups on the basis of their *trioctahedral* or *dioctahedral* nature. The smallest structural unit contains three octahedra. If all three octahedra are occupied, i.e., have octahedral cations at their centers, the sheet is classified as trioctahedral. If only two octahedra are occupied and the third octahedron is vacant, the sheet is classified as dioctahedral. In practice most micas do not contain exactly 3.0 or 2.0 octahedral cations, but 2.5 forms a convenient boundary between the two subgroups because few homogeneous micas have octahedral cation totals near that value.

Classification at the species level is done on the basis of ideal end-member compositions, as in Table 1. In naturally occurring micas the tetrahedral cations are normally Si, Al, or Fe^{3+} (rarely Be^{2+}). The interlayer cations are normally K or Na in the true micas (rarely Rb, Cs, NH_4, or H_3O) and Ca or Ba in the brittle micas (rarely Sr). The octahedral cations normally are Mg, Al, Fe^{2+}, and Fe^{3+}, but other medium-sized cations such as Li, Ti, V, Cr, Mn, Co, Ni, Cu, and Zn also occur in some species. Varietal names (fuchsite, mariposite, alurgite, manganophyllite, etc.) also exist in the older literature, and are based on color or lesser compositional differences. Usage of varietal names is discouraged in the scientific literature today. Significant compositional variations from those of the ideal end members are described better with the use of adjectival modifiers (Schaller, 1930), such as chromian muscovite, manganoan phlogopite, or sodian margarite. Lepidolite is used here as a group name to include several trioctahedral Li,Al-rich species and illite both as a species and as a group name to include the micaceous components found in fine-grained sediments and soils.

Generalized formulas for trioctahedral or dioctahedral micas can usually be formulated within a solid solution series, even if several substitution mechanisms are active simultaneously. It is much more difficult to group several such series into a single formula or to group trioctahedral and

Table 1. Classification of the Micas

Group	Subgroup	Species	Ideal Formula Units
True micas ($x \cong -1.0$)	trioctahedral	phlogopite	$KMg_3(Si_3Al)O_{10}(OH,F)_2$
		biotite	$K(Mg_{0.6-1.8}Fe_{2.4-1.2})(Si_3Al)O_{10}(OH,F)_2$
		annite	$KFe_3^{2+}(Si_3Al)O_{10}(OH,F)_2$
		ferri-annite	$KFe_3^{2+}(Si_3Fe^{3+})O_{10}(OH,F)_2$
		polylithionite	$K(Li_2Al)Si_4O_{10}(F,OH)_2$
		trilithionite	$K(Li_{1.5}Al_{1.5})(Si_3Al)O_{10}(F,OH)_2$
		taeniolite	$K(Mg_2Li)Si_4O_{10}(F,OH)_2$
		zinnwaldite	$K[Fe_{1.5-0.5}^{2+}Li_{0.5-1.5}(Al,Fe^{3+})](Si_{3.5-2.5}Al_{0.5-1.5})O_{10}(F,OH)_2$
		masutomilite	$K(Mn_{1.0-0.5}^{2+}Li_{1.0-1.5}Al)(Si_{3.5-3.0}Al_{0.5-1.0})O_{10}(F,OH)_2$
		hendricksite	$K(Zn,Mn)_3(Si_3Al)O_{10}(OH,F)_2$
		Na-phlogopite	$NaMg_3(Si_3Al)O_{10}(OH)_2$
		wonesite	$[(Na,K)_{0.5}\square_{0.5}][(Mg,Fe)_{2.5}Al_{0.5}](Si_3Al)O_{10}(OH,F)_2$
		siderophyllite	$K(Fe_2^{2+}Al)(Si_2Al_2)O_{10}(OH,F)_2$
		ephesite	$Na(LiAl_2)(Si_2Al_2)O_{10}(OH,F)_2$
		preiswerkite	$Na(Mg_2Al)(Si_2Al_2)O_{10}(OH)_2$
		montdorite	$K(Fe^{2+},Mn,Mg)_{2.5}Si_4O_{10}(F,OH)_2$
	dioctahedral	muscovite	$KAl_2(Si_3Al)O_{10}(OH,F)_2$
		paragonite	$NaAl_2(Si_3Al)O_{10}(OH,F)_2$
		phengite	$K[Al_{1.5}(Mg,Fe^{2+})_{0.5}](Si_{3.5}Al_{0.5})O_{10}(OH,F)_2$
		illite	$\sim K_{0.75}(Al_{1.75}R_{0.25}^{2+})(Si_{3.50}Al_{0.50})O_{10}(OH,F)_2$
		chernykhite	$(Ba,Na,NH_4)(V^{3+},Al)_2(Si,Al)_4O_{10}(OH)_2$
		roscoelite	$K(V^{3+},Al,Mg)_2(Si_3Al)O_{10}(OH)_2$
		celadonite	$K(Mg,Fe^{2+})(Fe^{3+},Al)Si_4O_{10}(OH)_2$
		glauconite	$\sim K(R_{1.33}^{3+}R_{0.67}^{2+})(Si_{3.67}Al_{0.33})O_{10}(OH)_2$
		tobelite	$(NH_4,K)Al_2(Si_3Al)O_{10}(OH)_2$
Brittle micas ($x \cong -2.0$)	trioctahedral	clintonite	$Ca(Mg_2Al)(SiAl_3)O_{10}(OH,F)_2$
		kinoshitalite	$Ba(Mg,Mn,Al)_3(Si_2Al_2)O_{10}(OH,F)_2$
		anandite	$BaFe_3^{2+}(Si_3Fe^{3+})O_{10}(OH)S$
		bityite	$Ca(Al_2Li)(Si_2AlBe)O_{10}(OH,F)_2$
	dioctahedral	margarite	$CaAl_2(Si_2Al_2)O_{10}(OH,F)_2$

dioctahedral species together. No attempt will be made to give general formulas in this chapter, whose emphasis is intended to be structural. More detail on the possible substitutions within the micas will be given in succeeding chapters.

The layer charge in a mica arises by some combination of four primary mechanisms: (1) substitution of R^{3+} (usually Al or Fe^{3+}) or R^{2+} (Be) for Si^{4+} in tetrahedral positions; (2) substitution of R^{1+} or R^{2+} for R^{2+} or R^{3+} in octahedral positions; (3) vacancies in octahedral positions; or (4) dehydroxylation of OH to O. The resultant layer charge is due to an excess of negative anion charges relative to positive cation charges. It may originate entirely within the tetrahedral sheet or entirely within the octahedral sheet in some species, or may come partly from both sheets. When the excess tetrahedral charge is greater than -1.0 for a true mica or -2.0 for a brittle mica the octahedral sheet then must have an offsetting excess positive charge (usually due to substitution of R^{3+} for R^{2+}). In the brittle micas the excess tetrahedral charge is known to go as high as -3.0 in the species clintonite where the tetrahedral cation composition is $SiAl_3$. The overall charge is decreased to -2.0 in clintonite by an octahedral cation composition of Mg_2Al (octahedral sheet charge of $+1.0$). The interlayer charge must balance the overall layer charge, but the interlayer cation occupancy need not total 1.0. Vacancies or neutral entities, e.g., Ar or H_2O, are permissible if the charges balance otherwise. The validity of $(H_3O)^+$ as an interlayer entity is not yet certain.

The interlayer cation that compensates for the layer charge lies in the cavity of the six-fold rings where two tetrahedral sheets occur back to back. The 2:1 layers must superimpose in that region so that the hexagonal rings of adjacent basal surfaces line up and enclose the interlayer cations. In most species adjacent layers are propped slightly apart by the interlayer cation. This is because tetrahedral rotation (to be discussed later) moves every other basal oxygen towards the center of each ring and thereby reduces the size of the opening. The interlayer cation then has only six nearest oxygen neighbors, rather than twelve for the case of ideal hexagonal geometry. The resultant basal spacing of 9.5 to 10.2 Å is slightly larger than the thickness of the layer itself (approximately 9.2-9.4 Å), and the basal oxygen surfaces are not believed to be in contact with one another. The interlocking of adjacent layers by the interlayer cation limits the structural diversity in layer arrangements of micas to the variations possible at the octahedral interface.

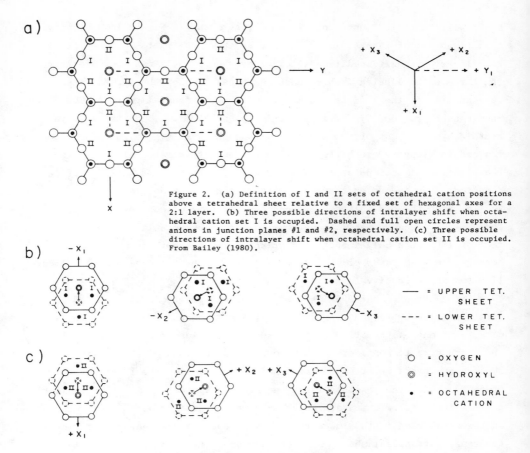

Figure 2. (a) Definition of I and II sets of octahedral cation positions above a tetrahedral sheet relative to a fixed set of hexagonal axes for a 2:1 layer. (b) Three possible directions of intralayer shift when octahedral cation set I is occupied. Dashed and full open circles represent anions in junction planes #1 and #2, respectively. (c) Three possible directions of intralayer shift when octahedral cation set II is occupied. From Bailey (1980).

STANDARD MICA POLYTYPES

Within each 2:1 layer the upper tetrahedral sheet necessarily is staggered by $a/3$ relative to the lower tetrahedral sheet in order to provide octahedral coordination around the medium-sized cations. It is possible to direct this stagger along the positive or negative directions of any of the three crystallographic axes X_1, X_2, or X_3 of a pseudohexagonal network (Fig. 2). If the direction of stagger differs from layer to layer in some regular fashion, even though the layers themselves are keyed together by the interlayer cations, then the periodicity along the Z direction will be some multiple of 10 Å and the mica is said to exhibit *polytypism*. Layer structures of essentially similar composition that differ only in the layer sequences are called *polytypes* and usually have different symmetries and c-periodicities.

For ease of reference the two anion planes composed of apical oxygens and OH groups that are common to both a tetrahedral sheet and an octahedral sheet will be called junctions #1 and #2, as shown in Figure 1c. Relative to a fixed set of lateral X and Y axes, as shown in Figure 2a, the octahedral cations positioned immediately above junction #1 can occupy either the three sites labeled I above each hexagonal ring or the three sites labeled II. If the I sites are occupied, then octahedral coordination about them can be provided by fitting junction plane #2 (i.e., the upper tetrahedral sheet) in any of the three positions shown in Figure 2b. These correspond to staggers or shifts of $a/3$ of the upper tetrahedral sheet along the negative directions of X_1, X_2, and X_3. Conversely, if the II sites are occupied, then the upper tetrahedral sheet can fit on only as shown in Figure 2c, in which the $a/3$ shifts are along the positive directions of these same three axes. Because set I transforms to set II by rotation of the layer by ±60° or 180°, the only important distinction is whether the same sites are occupied in each layer or whether there is an alternation between the I and II sites in successive layers.

Let us assume that the I set of octahedral positions is occupied in the first layer of every polytype and that the tetrahedral stagger of $a/3$ is directed along $-X_1$ relative to the fixed set of axes of Figure 2. It is then necessary to consider the six possible directions of shift within the succeeding layers relative to the fixed orientation of the first layer. In order to depict the possible layer sequences graphically, it is convenient to show the shift within each layer as a vector directed from the center of an hexagonal ring in the lower tetrahedral sheet to the center of a ring in the upper tetrahedral sheet, as viewed in projection on (001). The interlayer stacking angle will be defined as the angle between two such projected vectors in adjacent layers, as measured in a clockwise direction. The six possible interlayer stacking angles are 0°, 60°, 120°, 180°, 240°, and 300°. Because of the plane of symmetry that is parallel to each vector in a given layer, angles 60° and 300° (±60°) are equivalent, as are angles 120° and 240° (±120°).

Smith and Yoder (1956) have shown that there are only six standard or ideal mica polytypes that can be formed with periodicities between one and six layers if only one interlayer stacking angle, positive or negative, is permitted in a given crystal of trioctahedral composition. These six possibilities are illustrated graphically in Figure 3. The structure with only 0° stacking angles, corresponding to successive $-X_1$ shifts, is the one-layer monoclinic polytype. In the shorthand notation of Ramsdell (1947) it would be designated $1M$. The two-layer orthorhombic, or $2Or$, polytype results from

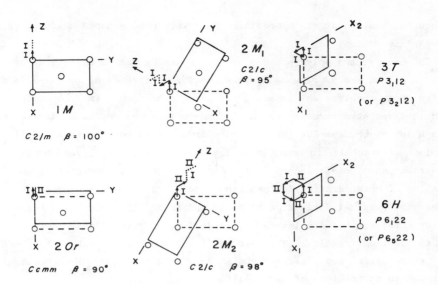

Figure 3. Derivation of six standard mica polytypes (modified from Smith and Yoder, 1956). The orientation of Figure 2a is assumed for the first layer of each polytype with the I set of octahedral cation positions occupied and the intralayer shift of $-a_1/3$. The bases of the resultant unit cell (solid lines) may or may not coincide with the fixed cell (dashed lines) used for the first layer. The origin of each resultant unit cell has been placed as close to the center (circles) of an hexagonal ring in the first layer as the symmetry permits.

180° stacking angles, equivalent to alternate $-X_1$ and $+X_1$ shifts. These are the only polytypes that involve shifts along just one axis. If shifts are considered along two different X axes, then a continuous alternation of ±120° stacking angles gives rise to the two-layer monoclinic polytype designated $2M_1$ and a continuous alternation of ±60° angles to the $2M_2$ polytype. Note that the fixed axes and unit cell (dashed lines) used to derive these two theoretical structures are not the best axes to use for the resultant structure (solid lines). Also, in the $2M_2$ polytype the X and Y axes must be reversed because the symmetry plane of the structure, which by convention must be designated (010), is found to be normal to the short 5.2 Å axis. Although only the X_1 and X_2 axes were used in this derivation, any two X axes could be used. The same two structures would be derived, differing only in the orientation of the resultant unit cells relative to the fixed starting axes. Finally, if all three X axes are used for shift directions a continuous sequence of -120° angles along $-X_1$, $-X_2$, and $-X_3$ gives rise to the three-layer trigonal, or $3T$, polytype, or to its mirror image if +120° angles are used. In the same way continuous stacking angles of -60° or +60° along all three axes give two mirror image structures that are six-layer hexagonal, or $6H$. Hexagonal-shaped cells can be used for the resultant $3T$ and $6H$

structures, instead of the monoclinic- or orthorhombic-shaped cells necessary for the others.

In summary, two polytypes can be derived using shifts along just one X axis. A second group of two polytypes involves shifts along two axes, and a third group involves shifts along all three axes. One member of each group involves structures in which the same octahedral positions are occupied in each layer, whereas for the other member the octahedral sets I and II are occupied alternately in successive layers (equivalent to layer rotations). The derivation of the standard mica polytypes presented here emphasizes the variables of octahedral cation positions and intralayer shift directions, whereas the original derivation by Smith and Yoder (1956) emphasizes the angular rotations between layers having different orientations. Both methods lead to identical results. The more atomistic approach used here has been found to be useful pedagogically and in extending the study of theoretical polytypes to other kinds of layer silicates (Bailey, 1980).

NATURAL MICA STRUCTURES

It is of interest next to see which of the theoretically possible standard mica structures actually exist in nature and whether or not the assumptions made in the derivation (trioctahedral compositions and a single interlaying stacking angle) were justified. It is possible to calculate the diffraction pattern for any theoretical stacking sequence of layers and to compare this calculated pattern with those observed for natural specimens. It is found by this approach that the single crystal patterns are different for each of the six standard polytypes and thus can serve to identify each uniquely. The powder patterns are less diagnostic for trioctahedral compositions because the $1M$ and $3T$ patterns are identical, as are the $2M_2$ and $6H$ patterns. As will be discussed later, the vacant octahedral position in dioctahedral micas distorts these structures to the extent that the powder patterns alone are sufficient for identification. Figure 4 and Table 2 illustrate some of these patterns for muscovite.

Five of the six standard polytypes have been found in natural micas (Table 3). The $6H$ structure has not been found to date, and the $2Or$ structure has been found only at one locality in the only known occurrence of anandite, the Ba-Fe brittle mica. There is a noticeable tendency for compositional control of the layer sequences actually adopted by natural micas. For example, the $2M_1$ structure is predominant for dioctahedral muscovite, with the $1M$, $3T$, and $2M_1$ structures less abundant. The $1M$ structure is most

Table 2. MUSCOVITE MICAS --
Diagnostic diffraction lines. See Fig. 4.

	2M₁		
	d(Å)	I	hkl
1.	4.29	1	111
2.	4.09	1	022
3.	3.88	3	11$\bar{3}$
4.	3.72	3	023
5.	3.49	3	11$\bar{4}$
6.	3.20	3	114
7.	2.98	3½	025
8.	2.86	3	115
9.	2.79	2½	11$\bar{6}$
10.	2.15	1	20$\bar{6}$,222
	2.12	3	135,043

$a = 5.201$ Å
$b = 8.967$
$c = 20.058$
$\beta = 95.8°$

	1M		
	d(Å)	I	hkl
1.	4.35	1½	11$\bar{1}$
2.	4.12	1	02$\bar{1}$
3.	3.66	5	11$\bar{2}$
4.	3.07	5	112
5.	2.93	1	11$\bar{3}$
6.	2.69	2	023

$a = 5.208$ Å
$b = 8.995$
$c = 10.275$
$\beta = 101.6°$

	3T		
	d(Å)	I	hkl
1.	3.87	3½	10$\bar{1}$4
2.	3.60	3	10$\bar{1}$5
3.	3.11	3	10$\bar{1}$7
4.	2.88	4	10$\bar{1}$8
5.	2.68	1	10$\bar{1}$9

$a = 5.196$ Å
$b = 5.196$
$c = 29.970$
$\beta = 90°$
$\gamma = 120°$

	2M₂		
	d(Å)	I	hkl
1.	4.34	2	20$\bar{2}$,111
2.	3.90	3	11$\bar{3}$
3.	3.68	4	20$\bar{4}$
4.	3.52	4	11$\bar{4}$
5.	3.21	4	114
6.	3.07	4	204
7.	2.87	2	115
8.	2.81	1	116

$a = 9.017$ Å
$b = 5.210$
$c = 20.437$
$\beta = 100.4°$

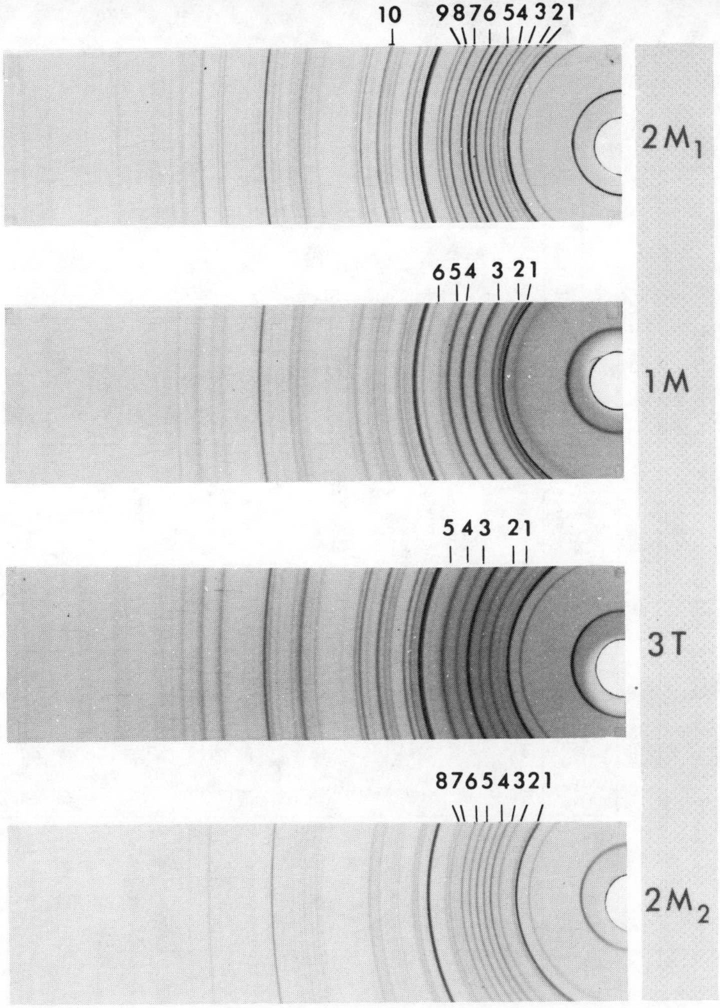

Figure 4. X-ray patterns of muscovites (monochromatic Cu Kα radiation, camera diameter 114.6 mm). Characteristic identifying lines are numbered according to Table 2. From Bailey (1980).

Table 3. Observed natural mica structures

	Trioctahedral				Dioctahedral		
	Abundance				Abundance		
Species	High	Medium	Low	Species	High	Medium	Low
phlogopite	$1M$, $1Md$	$3T$	$2M_1$	muscovite	$2M_1$	$1M$, $1Md$	$3T$
manganoan phlogopite	$1M$	$2M_1$		paragonite	$2M_1$	$3T$	
Na-phlogopite	$1M$			lithian muscovite	$2M_1$		
biotite	$1M$, $1Md$	$3T$	$2M_1$	manganian muscovite	$2M_1$, $3T$	$1M$	
annite	$1M$			chromian muscovite	$2M_1$, $1M$		
ferri-annite	$1M$						
"lepidolite"	$1M$, $2M_2$	$3T$	$2M_1$	phengite	$1M$, $1Md$, $3T$	$2M_1$	$2M_2$
taeniolite	$1M$, $3T$	$2M_1$		roscoelite	$1M$		
zinnwaldite	$1M$, $1Md$	$2M_1$, $3T$		chernykhite	$2M_1$		
masutomilite	$1M$	$2M_1$		tobelite	$1M$, $2M_1$		
hendricksite	$1M$	$2M_1$	$3T$	illite	$1M$, $1Md$		$3T$
wonesite	$1Md$			glauconite	$1M$, $1Md$		
preiswerkite	$2M_1$			celadonite	$1M$, $1Md$		
siderophyllite	$1M$			margarite	$2M_1$		
ephesite	$2M_1$	$1M$					
clintonite	$1M$		$3T$, $2M_1$				
anandite	$2Or$	$2M_1$, $1M$					
kinoshitalite	$1M$, $2M_1$						
bityite	$2M_1$						

abundant for trioctahedral micas, with $3T$, $2M_1$, and $2M_2$ less abundant. Disordered stacking sequences, termed $1Md$, are observed in addition to the regular sequences for both dioctahedral and trioctahedral compositions, but are most abundant for trioctahedral species. Because of the compositional control of the stacking it is evident that many of the natural mica structures are neither truly polytypic nor truly polymorphic with one another. They may be described instead as *polytypoids* of different compositions. Another point to be noted in Table 3 is that the stacking sequences based on 0°, 120° and 240° stacking angles (occupation of the same set of octahedral positions in every layer) occur more frequently than those based on 60°, 180°, and 300° angles (involving alternation between I and II octahedral sets in successive layers). That is, the $1M$, $2M_1$, and $3T$ structures occur more abundantly than the $2M_2$, $2Or$, and $6H$ structures.

Ross et al. (1966) developed a computer program to generate all possible layer sequences with periodicities from 1-layer to 4-layers without restriction on the values of the interlayer stacking angles, plus all sequences from 1-layer to 8-layers with 0° and 120° angles as well as a number of 10-, 11-, and 14-layer forms. The complexity of the possible sequences is indicated by the fact that there are 402 different 6-layer and 9,212 8-layer forms that are theoretically possible. Nevertheless, by making use of the generated cell dimensions and angles plus calculated unmodulated periodic intensity distributions they were able to identify the stacking sequences uniquely of ten natural biotites with layer periodicities between 3 and 23 layers. They concluded that these complex sequences are to be found primarily in the trioctahedral micas and often are the result of periodic stacking faults within the $1M$ and $3T$ basic structures. Non-periodic stacking faults, on the other hand, lead to twins or to tabular domain structures.

Takeda (1971) showed that the 11 complex polytypes known to him at that time all had lower symmetry than the standard polytypes, and belonged to space groups $C\bar{1}$, $C1$, or $C2$. More recently Rieder (1970) has described $5M$, $9Tc$, and $14M$ polytypes in natural zinnwaldite and siderophyllite, and Baronnet et al. (1976) have described $3Tc$ and $5Tc$ forms in synthetic muscovite, Bailey and Christie (1978) have described $3M_2$ crystals in natural lepidolite, and Baronnet (1980) has described $3Tc$ crystals in both synthetic phengite and paragonite. Pandey et al. (1982) showed that all 18 of the complex stacking sequences known at that time could be explained on the basis of spiral growth around single screw dislocations created in a faulted matrix structure of $1M$, $2M_1$, or $3T$ forms. They generated all possible polytypes that can be created in this manner and considered 11 of these higher-layer forms to be most probably on the basis of their relative stacking fault energies. These most probable forms are designated $3Tc$, $4Tc$, $8Tc_1$, $8Tc_2$, $9Tc$, $11Tc$, $23Tc$, $5M$, $8M$, $11M$, and $14M$.

The shorthand notation of Ramsdell (1947) for polytypes used in this chapter are convenient, but do not specify the precise interlayer stacking angles involved. For more detailed information, and especially useful for complex polytypes, the notation scheme of Ross et al. (1966) can be used. In this system the possible interlayer stacking angles or layer rotations are indicated by numbers, as follows:

$0 = 0°$ angle
$\pm 1 = \pm 60°$ angles
$\pm 2 = \pm 120°$ angles
$3 = 180°$ angle

where a positive sign for an angle indicates it is to be measured in a clockwise direction. Any sequence of layers can then be indicated by a sequence of numbers. Zvyagin (1967) has used a somewhat similar system, in which a Greek symbol σ_i designates the relative displacement of tetrahedral and octahedral sheets within each layer and where the subscript i defines the azimuthal orientation of the displacements in 60° clockwise increments about the cleavage normal. The equivalence between the Ramsdell, Ross et al., and Zvyagin symbols for the six standard mica structures of Smith and Yoder (1956) are listed below.

Ramsdell	Ross et al.	Zvyagin
$1M$	[0]	σ_3
$2M_1$	[2$\bar{2}$]	$\sigma_2\sigma_4$
$2M_2$	[1$\bar{1}$]	$\sigma_5\sigma_4$
$2Or$	[33]	$\sigma_3\sigma_6$
$3T$	[222] or [$\bar{2}\bar{2}\bar{2}$]	$\sigma_3\sigma_5\sigma_1$ or $\sigma_3\sigma_1\sigma_5$
$6H$	[111111] or [$\bar{1}\bar{1}\bar{1}\bar{1}\bar{1}\bar{1}$]	$\sigma_3\sigma_4\sigma_5\sigma_6\sigma_1\sigma_2$ or $\sigma_3\sigma_2\sigma_1\sigma_6\sigma_5\sigma_4$

REFERENCES

Bailey, S.W. (1980) Structures of layer silicates. Ch. 1 in: Crystal Structures of Clay Minerals and their X-ray Identification, G.W. Brindley and G. Brown, eds. Mineralogical Society, London.

———— and Christie, O.H.J. (1978) Three-layer monoclinic lepidolite from Tordal, Norway. Am. Mineral. 63, 203-204.

Baronnet, A. (1980) Polytypism in micas: a survey with emphasis on the crystal growth aspect. Current Topics Material Sci. 5, 447-548.

————, Amouric, M. and Chabot, B. (1976) Mécanismes de croissance, polytypisme et polymorphisme de la muscovite hydroxylée synthetique. J. Crystal Growth 32, 37-59.

Pandey, D., Baronnet, A., and Krishna, P. (1982) Influence of the stacking faults on the spiral growth of polytype structures in mica. Phys. Chem. Minerals 8, 268-278.

Ramsdell, L.S. (1947) Studies on silicon carbide. Am. Mineral. 32, 64-82.

Rieder, M. (1970) Lithium-iron micas from the Krusné hory Mountains (Erzgebirge): Twins, epitactic overgrowths and polytypes. Z. Kristallogr. 132, 161-184.

Ross, M. and Wones, D.R. (1965) Polytypism in biotites. Am. Mineral. 50, 291 (abstr.).

————, Takeda, H. and Wones, D.R. (1966) Mica polytypes: systematic description and identification. Science 151, 191-193.

Schaller, W.T. (1930) Adjectival ending of chemical elements used as modifiers to mineral names. Am. Mineral. 15, 566-574.

Smith, J.V. and Yoder, H.S. (1956) Experimental and theoretical studies of the mica polymorphs. Mineral. Mag. 31, 209-235.

Takeda, H. (1971) Distribution of mica polytypes among space groups. Am. Mineral. 56, 1042-1056.

Zvyagin, B.B. (1967) Electron-Diffraction Analysis of Clay Mineral Structures. Plenum Press, New York, 364 p.

2. CRYSTAL CHEMISTRY of the TRUE MICAS
S.W. Bailey

SHEET CONFIGURATIONS

Structural refinements of micas and other layer silicates have shown that the stylized hexagonal geometry of the tetrahedral and octahedral sheets, as illustrated in Figure 1 of Chapter 1 and assumed for the derivation of the standard polytypes, is considerably distorted -- much more so in dioctahedral than in trioctahedral species. These distortions and the stresses that cause them affect the ease of junction of the octahedral and tetrahedral sheets with one another and are believed to influence the layer stacking sequences, crystal morphologies, compositions, and stabilities of the species involved. As a result, it is important to understand the nature and origin of these distortions.

Dioctahedral distortions

Let us consider the dioctahedral case first. Figure 1 shows an octahedral sheet that is part of a 2:1 layer. Thus the upper and lower planes of oxygens (open circles) are also the apices of tetrahedral sheets

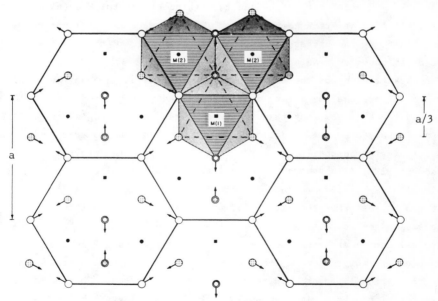

Figure 1. Idealized dioctahedral sheet within 2:1 layer. Open circles represent apical oxygens of attached upper tetrahedral sheet, with its hexagonal rings indicated by bold lines. Stippled circles represent apical oxygens of attached lower tetrahedral sheet. Double circles represent central OH,F groups. Arrows indicate directions of movement of anions to shorten diagonal shared edges between occupied M2 sites (solid circles).

above and below. The apical oxygens of the hexagonal rings in the upper tetrahedral sheet are connected by bold solid lines. OH-groups (double circles), which project in the centers of these tetrahedral rings, lie at the same z-heights as the apical oxygens and complete the octahedral anion junction planes. The lowermost plane of anions is stippled. Each vacant octahedral site ($\square$) is surrounded by six occupied sites (small solid circles). Because of the high charge on the octahedral Al ions, adjacent cations tend to repel one another and would like to move apart but cannot do so easily because each is being repelled in three directions by three neighboring Al cations. Instead, the anions that lie on shared edges between the occupied octahedra tend to shorten these shared edges by moving toward one another along the arrows in Figure 1 so that they can partly shield the cations. Because each anion is being pulled in only one direction in dioctahedral micas extreme shortening of shared edges results. Observed O...O and OH...OH shared edge lengths for Al-rich dioctahedral species are in the range from 2.3 to 2.5 Å in contrast to values from 2.7 to 2.9 Å for unshared edges. Because the shortening of shared edges requires the anions to move in directions diagonal to the sheet surface, the entire octahedral sheet is thinned along the sheet normal and is expanded along X and Y within the sheet. Typical sheet thicknesses are 2.04 to 2.14 Å for Al-rich compositions. The value of 2.04 Å is probably a lower limit, as the oxygens then will be in contact and there will be an effect of anion-anion repulsion.

The lateral edges of the horizontal triads of anions around the vacant sites are all expanded as a result of the shortening of shared edges. The arrows in Figure 1 show that the oxygens of these lateral edges are pulling in approximately opposite directions, and such edges tend to measure 3.2 to 3.4 Å for Al-rich composiitions. Figure 2, taken from the crystal structure of muscovite-$2M_1$, illustrates the much larger resultant size of the vacant octahedral site relative to the six occupied sites (shaded) around it.

It can be seen from the pattern of arrows in Figure 1 that two of the six lateral edges connecting apices of the hexagonal rings in the attached tetrahedral sheets are expanded but the other four edges are twisted. The twisting causes a rotation of the upper and lower anion triads around each octahedral cation in opposite directions by a few degrees. This octahedral counter-rotation occurs simultaneously with the thinning of the octahedral sheet, and both result from the shortening of octahedral shared edges. A

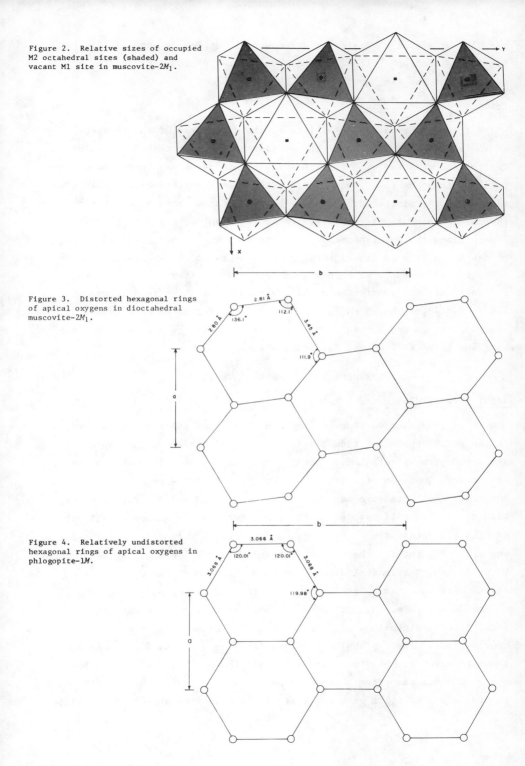

Figure 2. Relative sizes of occupied M2 octahedral sites (shaded) and vacant M1 site in muscovite-$2M_1$.

Figure 3. Distorted hexagonal rings of apical oxygens in dioctahedral muscovite-$2M_1$.

Figure 4. Relatively undistorted hexagonal rings of apical oxygens in phlogopite-$1M$.

top view of a typical distorted dioctahedral network is shown in Figure 3 for the case of muscovite-$2M_1$. Note the values of the edge lengths and the deviation of the ring angles from 120°.

Trioctahedral distortions

For trioctahedral layer silicates, the distortion of the octahedral sheet is much less because of the smaller charge per octahedral cation, averaging 2+ instead of 3+. Each anion now is being pulled in three directions at the same time along three shared edges, instead of only along one edge as in the dioctahedral case. The anion cannot move directly along any shared edge as a result, only along the resultant of the forces, which is inward along the sheet normal. Trioctahedral sheets nevertheless tend to be nearly as thin as dioctahedral sheets. Because of the limitations on direction of movement of anions in trioctahedral micas, the shared edges are only slightly shorter than the unshared edges. Observed values are 2.6 to 2.9 Å for shared edges and 3.0 to 3.2 Å for unshared edges. The hexagonal rings are now much more regular than for the dioctahedral case, as illustrated in Figure 4 for phlogopite-$1M$.

TETRAHEDRAL-OCTAHEDRAL LATERAL FIT

Ideal lateral dimensions

It is important to remember that the upper and lower planes of anions in the octahedral sheet (junctions #1 and #2 of Figure 1c in Chapter 1) are also parts of tetrahedral sheets above and below. In order to have a comfortable junction at these interfaces the lateral dimensions of the octahedra and tetrahedra must be similar. Sheets that fit together only with difficulty will have strained interfaces that influence greatly the resultant crystal size, morphology, and structure. An idea of the relative degree of fit between the tetrahedral and octahedral sheets can be obtained by estimating the lateral dimensions such sheets would have in the free state. Free octahedral sheets occur in the structures of some of the hydroxide minerals, which thus can be used as analogues to estimate the ideal lateral dimensions for the octahedral sheets in micas of similar compositions. Thus, gibbsite and bayerite have b = 8.64 Å and 8.67 Å, respectively, for the dioctahedral composition $Al(OH)_3$ and brucite has b = 9.36 Å for the trioctahedral composition $Mg(OH)_2$. Alhough tetrahedral sheets do not exist in an unconstrained state in minerals, it is possible to estimate their lateral dimensions in the free state by assuming an average T-O bond length and ideal tetrahedral angles:

$$a = (4\sqrt{6}/3)(\text{T-O}) \quad \text{and} \quad b = a\sqrt{3} = (4\sqrt{2})(\text{T-O}) \quad . \quad (1)$$

Thus, for hexagonal geometry and Si-O = 1.618 ± 0.01 Å the resulting lateral dimensions are a = 5.28 ± 0.03 Å and b = 9.15 ± 0.06 Å. Substitution of the larger Al for Si in the tetrahedral sites increases these values. Assuming Al^{iv}--O = 1.748 ± 0.01 Å, then b = 9.89 ± 0.6 Å or

$$b\ (Si_{1-x}Al_x) = 9.15\ \text{Å} + 0.74x \quad . \quad (2)$$

This relationship should be considered as only approximate, not only because of the uncertainties in the assumed bond lengths but also because the tetrahedral angles are known to vary somewhat from structure to structure.

Despite the uncertainties noted above it is evident that a tetrahedral sheet containing only Si has a lateral b dimension that is intermediate between those of octahedral sheets containing only Mg or only Al. The magnitude of the misfit is approximately 3-5% for these compositions. Structural adjustments of some type are necessary in order to eliminate this misfit and to articulate the tetrahedral and octahedral sheets by common planes of junction into a stable 2:1 layer.

Tetrahedral rotation

In the great majority of micas the lateral dimensions of the tetrahedral sheet are larger than those of the octahedral sheet as a result of tetrahedral substitution of Al or Fe^{3+} for Si. Mathieson and Walker (1954), Newnham and Brindley (1956), Zvyagin (1957), Bradley (1959), and Radoslovich (1961) have shown that it is relatively easy to reduce the lateral dimensions of a tetrahedral sheet by rotating adjacent tetrahedra in opposite directions in the (001) plane. The amount of rotation necessary to relieve the misfit is given by

$$\cos \alpha = b(\text{obs})/b(\text{ideal}) \quad (3)$$

where α is the rotation angle and b(ideal) is defined as the calculated value of b for an unconstrained tetrahedral sheet. According to this expression, which is only an approximation in practical applications because of the difficulty in estimating b(ideal), the lateral reduction achieved varies from 1.5% for α = 10° to 6.0% for α = 20° and 13.4% for a maximum possible rotation of 30° that would bring basal oxygens into contact at the center of each hexagonal ring. This is a non-linear change. Toraya (1981) proposed a linear relation between the rotation angle (α) and the dimensional misfit (Δ) between the tetrahedral and octahedral sheets

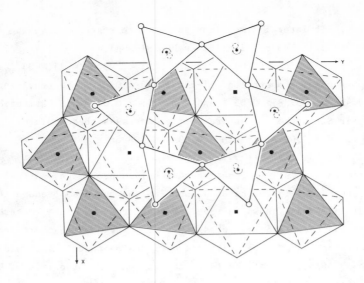

Figure 5. Octahedral sheet and part of attached upper tetrahedral sheet of muscovite-$2M_1$. Adjacent tetrahedra rotate in opposite directions (arrows) so that basal oxygens move towards nearest octahedral site.

$$\alpha = 35.44\Delta - 11.09 \quad , \tag{4}$$

where $\Delta = [2\sqrt{3}e_b - 3\sqrt{2}d_o]$, e_b = the mean of all basal tetrahedral edge lengths, and d_o = the mean of all M-O,OH,F octahedral bond lengths. Rotation angles actually observed in the structures of the true micas range from 1.4° to 11.0° for trioctahedral species and from 6.0° to 19.1° for dioctahedral species. Peterson et al. (1979) calculated that a 6-fold ring existing by itself would have minimum energy for $\alpha = 16°$ rather than 0°. By extrapolation it may be presumed that even the "ideal" tetrahedral sheet for a layer silicate should be shown as rotated to some degree.

The direction of tetrahedral rotation in micas is governed by the attraction of the basal oxygens to the octahedral cations in the same 2:1 layer (Fig. 5). This causes every other basal oxygen to move toward the center of a given 6-fold ring, thus creating a ditrigonal rather than hexagonal symmetry for the tetrahedral sheet. The space available for the interlayer cation within the ring is correspondingly reduced. The interlayer cation therefore props apart adjacent basal oxygen surfaces so they are no longer in contact. The number of nearest basal oxygen neighbors to the interlayer cation is reduced because of tetrahedral rotation from 12 for $\alpha = 0°$ to the six nearest the centers of the ditrigonal rings above and below the cation. Another group of six oxygens forms a second-nearest-neighbor coordination group. McCauley and Newnham (1971) determined a linear relation between α and the difference between the distances from the interlayer cation to the inner and outer oxygen groups. They further showed by multiple regression analysis that the magnitude of α for micas depends, on the average, 90% on the misfit of the lateral dimensions of the tetrahedral and octahedral sheets and 10% on the effect of the interlayer cation, as measured by its field strength (ratio of valence to ionic radius). These average values do not apply necessarily to specific cases, and the interlayer cation may have a large influence on α in some micas.

Changes in sheet thicknesses

Because the sheets are semi-elastic some adjustment for lateral misfit can be accomplished also by thickening the tetrahedral sheet and thinning the octahedral sheet, thereby decreasing and increasing, respectively, their lateral dimensions. These adjustments can be expressed either in terms of the measured sheet thicknesses (t_o) or of the τ_{tet} and ψ_{oct} angles. An ideal tetrahedron has an angle $\tau = (O_{apical}-T-O_{basal}) = 109°28'$, which is increased by thickening and decreased by thinning. An ideal octahedron has an angle ψ, which is defined as the angle between a body diagonal and the vertical and for a sheet of linked octahedra will be calculated here as $\cos\psi$ = octahedral thickness divided by $2(\overline{M-O,OH})$. The ideal ψ angle is 54°44', which increases with octahedral thinning. Observed values of τ and ψ for true micas are listed in Tables 1 and 2, along with other structural details. It can be seen that in all micas the tetrahedral sheet is thickened and the octahedral sheet is thinned. The small value of τ in Table 1 for phlogopite #1 is believed to be an artifact due to incomplete refinement. If one looks at the three individual $\tau = O_{apical}-T-O_{basal}$ angles in a tetrahedron, rather than just their average, it is evident that the size and charge of the adjacent octahedral cations or vacancies also influence the detailed shape of the tetrahedron. Lee and Guggenheim (1981) point out that in dioctahedral micas one of the three individual τ values is about two degrees larger than the other two. The effect of the one large angle is to shift the apical oxygen approximately in the (001) plane toward the other coordinating anion on its shared octahedral edge (Fig. 5) without affecting the positions of the other ions of the tetrahedron. This shift is in the same direction as that required by tetrahedral tilt (discussed in the next section), but its magnitude is only about 20% of the magnitude of the latter shift.

Table 1. Structural details of trioctahedral micas.

	Species	Composition	# Refl.	Final R	α_{tet}	τ_{tet}	ψ_{oct}
1. Zvyagin & Mishchenko (1962)	phlogopite-1$\underline{M}$ (locality ?)	~ $KMg_3(Si_3Al)O_{10}(OH)_2$	109	18.8%	5.9%	108.9°	58.1°, 58.1
2. Steinfink (1962)	ferriphlogopite-1$\underline{M}$ (Langban, Sweden)	$(K_{0.9}Mn_{0.1})Mg_3[Si_3(Fe^{3+},Mn)]O_{10}(OH)_2$	392	13.1	11.0	110.0	58.1, 58.2
3. Donnay et al. (1964b)	synthetic ferriannite-1$\underline{M}$	$KFe_3^{2+}(Si_3Fe^{3+})O_{10}(OH)_2$	433	9.3	6.4	110.3	59.3, 59.3
4. Takeda & Donnay (1966)	synthetic lithian fluorophlogopite-1$\underline{M}$	$K_{0.95}(Mg_{2.8}Li_{0.2})(Si_{3.25}Al_{0.75})O_{10}F_2$	472	7.3	6.2	110.6	59.4, 59.3
5. Tepikin et al. (1969)	biotite-1$\underline{M}$ (locality ?)	$(K_{0.88}Ca_{0.02}Na_{0.02}H3O_{0.04})(Si_{2.71}Al_{1.29})$ $(Fe_{2.31}^{2+}Mg_{0.28}Al_{0.18}Ti_{0.1}Li_{0.04}Fe_{0.01}^{3+})$ $O_{10}(OH)_{1.86}F_{0.14}$	420	13.5	1.4	110.3	59.7, 59.6
6. Takeda & Burnham (1969)	synthetic polylithionite-1$\underline{M}$	$K(Li_2Al)Si_4O_{10}F_2$	328	5.1	3.0	113.8	60.2, 58.1
7. Takeda et al. (1971)	lepidolite-2$\underline{M}_2$ (Rozna, Czecho.)	$(K_{0.87}Na_{0.12}Rb_{0.06}Ca_{0.02})(Li_{1.05}Al_{1.4}R_{0.15}^{2+})$ $(Si_{3.39}Al_{0.61})O_{10}(OH)_{0.8}F_{1.2}$	471	7.2	5.4	112.2 x2	61.1, 58.2
8. Joswig (1972)	(F,OH)-phlogopite-1$\underline{M}$ (Madagascar)	$(K_{0.90}Na_{0.02})(Mg_{2.7}Fe_{0.16}^{2+}Al_{0.07}Ti_{0.03})$ $(Si_{2.91}Al_{1.09})O_{9.9}F_{1.13}(OH)_{0.97}$	468	2.0	7.7	110.6	59.2, 59.1
9. McCauley et al. (1973)	synthetic fluorphlogopite-1$\underline{M}$	$(K_{0.98}Na_{0.04})Mg_{2.97}(Si_{2.98}Al_{1.02})$ $O_{9.9}F_{1.94}(OH)_{0.16}$	535	6.1	5.9	110.1	59.0, 59.0
10. Sartori et al. (1973)	lepidolite-2$\underline{M}_2$ (Elba, Italy)	$(K_{0.87}Na_{0.12}Rb_{0.06}Ca_{0.02})(Al_{1.4}Li_{1.05}R_{0.15}^{2+})$ $(Si_{3.39}Al_{0.61})O_{10}(OH)_{0.8}F_{1.2}$	525	9.6	6.5	111.9, 112.0	60.8, 58.4
11. Hazen & Burnham (1973)	(F,OH)-phlogopite-1$\underline{M}$ (Franklin, N.J.)	$(K_{0.77}Na_{0.16}Ba_{0.05})Mg_3(Si_{2.95}Al_{1.05})$ $O_{10}F_{1.3}(OH)_{0.7}$	1691	4.1	7.5	110.5	59.0, 59.0
12. Hazen & Burnham (1973)	annite-1$\underline{M}$ (Pikes Peak, Colo.)	$(K_{0.88}Na_{0.07}Ca_{0.03})(Si_{2.81}Al_{1.19})$ $(Fe_{2.22}^{2+}Mg_{0.12}Mn_{0.05}Fe_{0.19}^{3+}Al_{0.09}Ti_{0.22})$ $O_{10.35}(OH)_{1.38}F_{0.22}Cl_{0.05}$	1517	4.5	1.6	110.3	58.6, 58.3
13. Rayner (1974)	phlogopite-1$\underline{M}$ (locality ?)	$(K_{0.93}Na_{0.04}Ca_{0.03})(Mg_{2.77}Fe_{0.10}Ti_{0.11})$ $(Si_{2.88}Al_{1.12})O_{10}(OH)_{1.49}F_{0.51}$	293	6.6	8.7	110.4	59.2, 59.0
14. Tateyama et al. (1974)	synthetic Mg^{IV} phlogopite-1$\underline{M}$	$(K_{0.96}Na_{0.04})Mg_{2.84}(Si_{3.63}Mg_{0.31}Fe_{0.03}$ $Al_{0.03})O_{10}(OH)_2$	385	10.4	7.4	110.9	58.0, 58.2
15. Takeda & Ross (1975)	hydrogenated biotite-1$\underline{M}$ (Ruiz Peak, N.M.)	$(K_{0.78}Na_{0.16}Ba_{0.02})(Mg_{1.68}Fe_{0.71}^{2+}Fe_{0.19}^{3+}$ $Ti_{0.34}Al_{0.19}Mn_{0.01})(Si_{2.86}Al_{1.14})$ $O_{11.12}(OH)_{0.71}F_{0.17}$	649	4.4	7.6	110.4	59.2, 58.9
16. Takeda & Ross (1975)	hydrogenated biotite-2$\underline{M}_1$ (Ruiz Peak, N.M.)	$(K_{0.78}Na_{0.16}Ba_{0.02})(Mg_{1.68}Fe_{0.71}^{2+}Fe_{0.19}^{3+}$ $Ti_{0.34}Al_{0.19}Mn_{0.01})(Si_{2.86}Al_{1.14})$ $O_{11.12}(OH)_{0.71}F_{0.17}$	875	5.6	7.7	110.3, 110.2	59.2, 58.9
17. Takeda & Morosin (1975)	synthetic fluorphlogopite-1$\underline{M}$	$(K_{0.98}Na_{0.04})Mg_{2.97}(Si_{2.98}Al_{1.02})$ $O_{9.9}F_{1.94}(OH)_{0.16}$	458	4.3	6.5	110.7	59.4, 59.4
18. Toraya et al. (1976)	synthetic fluormica-1$\underline{M}$	~ $K_{0.88}(Mg_{2.56} \square_{0.44})Si_4O_{10}F_2$	1057	3.8	1.4	111.8	58.0, 58.0
19. Sartori (1976)	lepidolite-1$\underline{M}$ (Elba, Italy)	$(K_{0.88}Na_{0.06}Rb_{0.05}Ca_{0.01})(Li_{1.69}Al_{1.31})$ $(Si_{3.36}Al_{0.64})O_{10}(OH)_{0.47}F_{1.53}$	400	6.7	7.4	112.2	60.8, 58.5
20. Brown (1978)	lepidolite-3$\underline{T}$ (Kalgoorlie, Australia)	$(K_{0.85}Na_{0.11}Rb_{0.05})(Al_{1.25}Li_{1.62}Mn_{0.09}$ $Fe_{0.02}Mg_{0.01})(Si_{3.48}Al_{0.52})O_{10}$ $(OH)_{0.44}F_{1.54}$	705	4.7	7.6	112.6, 111.8	59.6, 60.8, 57.6

Table 1, continued.

	Sheet Thickness		Interlayer Separation	Basal Oxygen Δz	Mean Bond Lengths				Intralayer Shift	Layer Offset	Resultant Shift
	Tet.	Oct.			T--O	M--O,OH	R^{1+}--O Inner	Outer			
1.	2.168 Å	2.192 Å	3.620 Å	0.10 Å	1.627x2 Å	2.075, 2.075x2 Å	3.075 Å	3.334 Å	$-0.319a_1$	$-0.014a_1$	$-0.333a_1$
2.	2.280	2.218	3.475	0.01	1.681x2	2.101, 2.105x2	2.945	3.452	$-0.338a_1$	$+0.001a_1$	$-0.337a_1$
3.	2.318	2.152	3.394	0.01	1.687x2	2.107, 2.106x2	3.055	3.351	$-0.334a_1$	$+0.001a_1$	$-0.333a_1$
4.	2.273	2.101	3.328	0.01	1.651x2	2.061, 2.060x2	2.995	3.278	$-0.332a_1$	0	$-0.332a_1$
5.	2.274	2.100	3.353	0.06	1.667x2	2.078, 2.076x2	3.134	3.197	$-0.330a_1$	$-0.004a_1$	$-0.334a_1$
6.	2.247	2.095	3.274	0.04	1.619x2	2.106, 1.981x2	3.000	3.132	$-0.351a_1$	0	$-0.351a_1$
7.	2.247	2.076	3.360	0.09	1.620, 1.632	2.144, 1.967x2	2.980	3.220	$+0.364b_2$ $-0.364b_3$	$-0.007a_1$ $-0.007a_1$	$-0.379a_1$
8.	2.268	2.116	3.356	0.00	1.654x2	2.066, 2.063x2	2.970	3.319	$-0.334a_1$	0	$-0.334a_1$
9.	2.251	2.124	3.356	0.00	1.642x2	2.063, 2.064x2	3.006	3.273	$-0.334a_1$	0	$-0.334a_1$
10.	2.241	2.074	3.409	0.09	1.630, 1.627	2.123, 1.980x2	2.976	3.261	$-0.361b_3$ $+0.361b_2$	$-0.005a_1$ $-0.005a_1$	$-0.372a_1$
11.	2.261	2.125	3.352	0.00	1.649x2	2.063, 2.065x2	2.969	3.312	$-0.334a_1$	$-0.001a_1$	$-0.335a_1$
12.	2.252	2.208	3.380	0.01	1.659x2	2.121, 2.101x2	3.143	3.215	$-0.334a_1$	$-0.017a_1$	$-0.352a_1$
13.	2.270	2.126	3.418	0.01	1.659x2	2.076, 2.064x2	2.967	3.360	$-0.332a_1$	$-0.003a_1$	$-0.335a_1$
14.	2.231	2.205	3.461	0.03	1.650x2	2.083, 2.094x2	3.010	3.347	$-0.335a_1$	$-0.003a_1$	$-0.338a_1$
15.	2.271	2.138	3.334	0.01	1.659x2	2.086, 2.068x2	2.972	3.318	$-0.335a_1$	$-0.002a_1$	$-0.337a_1$
16.	2.269	2.135	3.336	0.02	1.657, 1.662	2.086, 2.068x2	2.970	3.323	$+0.334a_{2,3}$	$-0.001a_{1,1}$	$-0.335a_1$
17.	2.277	2.095	3.329	0.01	1.650x2	2.056, 2.058x2	2.987	3.282	$-0.334a_1$	0	$-0.334a_1$
18.	2.243	2.186	3.337	0.00	1.625x2	2.062, 2.064x2	3.079	3.142	$-0.333a_1$	$+0.001a_1$	$-0.332a_1$
19.	2.255	2.060	3.387	0.06	1.630x2	2.113, 1.972x2	2.942	3.269	$-0.330a_1$	$+0.017a_1$	$-0.313a_1$
20.	2.257	2.059	3.347	0.13	1.652, 1.617	2.036, 2.113	2.925	3.265	$-0.354a_{2,3,1}$	$-0.007a_{1,2,3}$	0

Table 1, continued.

	Reference	Mineral/Location	Formula						
21.	Guggenheim & Bailey (1977)	zinnwaldite-1$\underline{M}$ (Erzgebirge, D.D.R.)	$(K_{0.90}Na_{0.05})(Al_{1.05}Fe^{3+}_{0.16}Ti_{0.01}Fe^{2+}_{0.77}$ $Mg_{0.01}Mn^{2+}_{0.05}Li_{0.67})(Si_{3.09}Al_{0.91})O_{10}$ $(OH)_{0.79}F_{1.21}$	1493	5.7	5.8	111.0, 111.1	60.8, 56.5, 60.8
22.	Toraya et al. (1977)	synthetic taenolite-1$\underline{M}$	$K(LiMg_2)Si_4O_{10}F_2$	1303	2.4	1.1	112.7	57.8, 57.9
23.	Semenova et al. (1977, 1983)	tetraferri-phlogopite-1$\underline{M}$ (Kovdor massif, U.S.S.R.)	$(K_{0.9}Na_{0.1})(Mg_{2.85}Fe^{2+}_{0.15})$ $(Si_{2.98}Fe^{3+}_{0.45}Al_{0.55}Ti_{0.02})O_{10}(OH)_2$	1059	5.7	9.2	110.4	58.7, 58.7
24.	Toraya et al. (1978a)	synthetic Ge-taeniolite-1$\underline{M}$	$K(LiMg_2)Ge_4O_{10}F_2$	1451	3.8	13.5	114.3	59.3, 59.3
25.	Toraya et al. (1978a)	synthetic Ge-fluorphlogopite-1$\underline{M}$	$K(Mg_{2.5\ 0.5})Ge_4O_{10}F_2$	1413	5.5	12.5	114.3	60.1, 58.4
26.	Toraya et al. (1978b)	synthetic Ge-fluorphlogopite-1$\underline{M}$	$KMg_3(Ge_3Al)O_{10}F_2$	1365	3.7	15.9	111.5	60.2, 60.2
27.	Toraya et al. (1978c)	synthetic fluor-phlogopite-1$\underline{M}$	$K(Mg_{2.75\ 0.25})(Si_{3.5}Al_{0.5})O_{10}F_2$	1228	2.9	3.6	111.1	58.8, 58.8
28.	Sokolova et al. (1979)	ephesite-1$\underline{M}$ (Ilimaussak massif, Greenland)	$(Na_{0.92}Ca_{0.06}K_{0.02})(Al_{1.97}Li_{0.67}Na_{0.10}Fe_{0.02})$ $(Si_{2.20}Al_{1.80})O_{10}(OH)_2$	284	11.5	20.8	108.7, 110.8	61.3, 57.9, 57.9
29	Kato et al. (1979)	Mn-biotite-1$\underline{M}$	$(K_{0.78}Na_{0.11}Ca_{0.02})(Mg_{1.20}Mn^{2+}_{1.36}Fe^{3+}_{0.33}$ $Fe^{2+}_{0.02}Al_{0.22})(Si_{2.69}Al_{1.29}Ti_{0.02})O_{10}$ $(OH)_{1.54}$	426	12.1	7.4	110.8	57.6, 57.7
30.	Kato et al. (1979) and Guggenheim & Kato (1984)	Mn-phlogopite-1$\underline{M}$ (Iwate, Japan)	$(K_{0.85}Na_{0.19}Ba_{0.06})(Mg_{1.74}Mn^{2+}_{0.95}Mn^{3+}_{0.18}$ $Fe^{3+}_{0.06})(Si_{2.75}Al_{1.15}Ti_{0.03}Fe^{3+}_{0.07})O_{10}$ $(OH)_{1.90}F_{0.09}$	529	5.4	6.8	110.8	58.1, 57.9
31.	Kato et al. (1979)	Mn-phlogopite-1$\underline{M}$ (Iwate, Japan)	$(K_{0.73}Na_{0.32}Ba_{0.11})(Mg_{2.27}Mn^{2+}_{0.49}Fe^{3+}_{0.03}$ $Mn^{3+}_{0.20})(Si_{2.86}Al_{1.07}Ti_{0.02}Fe^{3+}_{0.05})O_{10}$ $(OH)_{1.57}F_{0.09}$	559	10.6	6.2	110.6	58.8, 58.7
32.	Bohlen et al. (1980)	biotite-2$\underline{M_1}$ (Au Sable Forks, New York)	$(K_{0.985}Na_{0.02}Ca_{0.002})(Fe^{2+}_{1.393}Mg_{1.161}Al_{0.121}$ $Mn_{0.007}Ti_{0.318})(Si_{2.793}Al_{1.207})O_{10.562}$ $(OH)_{1.218}F_{0.082}Cl_{0.138}$	787	4.2	5.3	110.4, 110.2	58.7, 58.4
33.	Swanson & Bailey (1981)	lepidolite-2$\underline{M_1}$ (Biskupice, Czecho.)	$(K_{0.80}Rb_{0.06}Cs_{0.02})(Li_{1.65}Al_{1.24}Mn_{0.04})$ $(Si_{3.62}Al_{0.38})O_{10}(F,OH)_2$	971	9.1	6.2	112.2 x2	60.7, 58.6
34.	Guggenheim (1981)	lepidolite-1$\underline{M}$ (Radkovice, Czecho.)	$(K_{0.79}Na_{0.03}Rb_{0.07}Ca_{0.01}Cs_{0.03})(Li_{1.48}$ $Al_{1.30}Mg_{0.05}Fe^{3+}_{0.01}Mn_{0.03})(Si_{3.49}$ $Al_{0.51})O_{10}(F,OH)_2$	1164	3.5	7.3	112.1	61.3, 59.0
35.	Guggenheim (1981)	lepidolite-2$\underline{M_2}$ (Radkovice, Czecho.)	$(K_{0.79}Na_{0.03}Rb_{0.07}Ca_{0.01}Cs_{0.03})(Li_{1.48}$ $Al_{1.30}Mg_{0.05}Fe^{3+}_{0.01}Mn_{0.03})(Si_{3.49}$ $Al_{0.51})O_{10}(F,OH)_2$	2764	4.8	6.6	112.1 x2	61.1, 58.5
36.	Guggenheim (1981)	lepidolite-1$\underline{M}$ (Tanakamiyama, Japan)	$(K_{1.01}Na_{0.01}Rb_{0.03})(Li_{1.41}Al_{1.13}Fe^{2+}_{0.07}$ $Mn_{0.05})(Si_{3.87}Al_{0.13})O_{10}F(OH)_2$	807	6.2	3.4	112.6, 112.5	60.4, 56.1, 60.5
37.	Toraya & Marumo (1981)	synthetic Ge,Mn-fluorphlogopite-1$\underline{M}$	$K(Mg_{2.32}Mn_{0.68})(Ge_3Al)O_{10}F_2$	1528	3.9	14.9	111.5	59.9, 59.9
38.	Toraya & Marumo (1981)	synthetic Ge,Mn-fluorphlogopite-1$\underline{M}$	$K(Mg_{1.04}Mn_{1.96})(Ge_3Al)O_{10}F_2$	1413	5.0	13.2	111.4	59.5, 59.9
39.	Hazen et al. (1981)	lithian fluorferri-phlogopite-1$\underline{M}$ (Cupaello, Italy)	$(K_{0.97}Ba_{0.02}Na_{0.01})(Mg_{2.46}Li_{0.23}Na_{0.11}$ $Fe^{2+}_{0.03}Ti^{3+}_{0.09})(Si_{3.31}Fe^{3+}_{0.65}Al_{0.04})O_{10}F_2$	625	3.0	5.7	110.8	58.8, 58.8

Table 1, continued.

No.											
21.	2.252	2.078	3.333	0.12	1.646, 1.639	2.132, 1.882, 2.131	2.990	3.251	$-0.354a_1$	$-0.004a_1$	$-0.358a_1$
22.	2.251	2.192	3.297	0.00	1.625x2	2.058, 2.061x2	3.069	3.116	$-0.332a_1$	0	$-0.332a_1$
23.	2.262	2.167	3.451	0.01	1.665x2	2.085, 2.085x2	2.973	3.393	$-0.333a_1$	0	$-0.333a_1$
24.	2.458	2.138	3.338	0.01	1.744x2	2.092, 2.092x2	2.861	3.494	$-0.333a_1$	$-0.002a_1$	$-0.335a_1$
25.	2.446	2.170	3.306	0.05	1.744x2	2.178, 2.070x2	2.872	3.480	$-0.339a_1$	$-0.003a_1$	$-0.342a_1$
26.	2.425	2.063	3.395	0.01	1.744x2	2.076, 2.078x2	2.824	3.577	$-0.335a_1$	$-0.002a_1$	$-0.337a_1$
27.	2.258	2.137	3.334	0.00	1.638x2	2.062, 2.063x2	3.045	3.209	$-0.335a_1$	0	$-0.335a_1$
28.	2.276	2.047	2.937	0.19	1.764, 1.609	2.128, 1.927, 1.927	2.477	3.454	$-0.371a_1$	$+0.035a_1$	$-0.336a_1$
29.	2.270	2.278	3.380	0.01	1.675x2	2.125, 2.129x2	3.013	3.356	$-0.340a_1$	$-0.004a_1$	$-0.343a_1$
30.	2.257	2.245	3.404	0.01	1.663x2	2.122, 2.110x2	3.027	3.337	$-0.331a_1$	$-0.001a_1$	$-0.332a_1$
31.	2.265	2.202	3.395	0.02	1.657x2	2.101, 2.091x2	3.025	3.307	$-0.331a_1$	$-0.002a_1$	$-0.332a_1$
32.	2.250	2.185	3.393	0.01	1.656, 1.661	2.106, 2.086x2	3.046	3.289	$+0.334a_{2,3}$	$+0.001a_1$	$-0.328a_1$
33.	2.256	2.061	3.365	0.07	1.628, 1.631	2.107, 1.977x2	2.964	3.237	$+0.357a_{2,3}$	$-0.003a_{1,1}$	$-0.362a_1$
34.	2.259	2.056	3.396	0.07	1.632x2	2.118, 1.970x2	2.950	3.270	$-0.357a_1$	$-0.007a_1$	$-0.364a_1$
35.	2.254	2.052	3.387	0.08	1.629, 1.629	2.121, 1.966x2	2.961	3.251	$-0.358b_3$ $+0.358b_2$	$-0.006a_1$ $-0.006a_1$	$-0.368a_1$
36.	2.254	2.092	3.318	0.08	1.639, 1.626	2.120, 1.878, 2.126	3.024	3.178	$-0.356a_1$	$-0.004a_1$	$-0.360a_1$
37.	2.421	2.101	3.355	0.00	1.746x2	2.094, 2.095x2	2.846	3.557	$-0.335a_1$	0	$-0.335a_1$
38.	2.417	2.158	3.308	0.01	1.749x2	2.128, 2.123x2	2.892	3.527	$-0.335a_1$	0	$-0.335a_1$
39.	2.275	2.151	3.364	0.00	1.655x2	2.077, 2.077x2	3.021	3.282	$-0.332a_1$	0	$-0.332a_1$

Table 2, continued.

40. Pavlishin et al. (1981)	protolithionite-3T (Ukraine, U.S.S.R.)	$(K_{0.90}Na_{0.05}Rb_{0.03}Ca_{0.02})(Fe^{2+}_{1.06}Fe^{3+}_{0.26}$ $Al_{0.83}Li_{0.52}Ti_{0.02}Mn_{0.02}Mg_{0.02})(Si_{2.87}$ $Al_{1.13})O_{10}(OH)_{1.99}F_{0.01}$	702	4.7	5.6	110.5, 111.0	60.9, 57.4, 61.4
41. Ohta et al. (1982)	oxybiotite-1M (Ruiz Peak, N.M.)	$(K_{0.77}Na_{0.16}Ba_{0.02})(Mg_{1.67}Al_{0.16}Fe^{2+}_{0.01}$ $Fe^{3+}_{0.86}Ti_{0.34}Mn_{0.01})(Si_{2.84}Al_{1.16})$ $O_{11.62}(OH)_{0.21}F_{0.17}$	1125	5.0	7.3	110.3	59.4, 59.1
42. Ohta et al. (1983)	oxybiotite-$2M_1$ (Ruiz Peak, N.M.)	$(K_{0.77}Na_{0.16}Ba_{0.02})(Mg_{1.67}Al_{0.16}Fe^{2+}_{0.01}$ $Fe^{3+}_{0.86}Ti_{0.34}Mn_{0.01})(Si_{2.84}Al_{1.16})O_{11.62}$ $(OH)_{0.21}F_{0.17}$	1676	4.5	7.4	110.2 x2	59.4, 59.1
43. Toraya & Marumo (1983)	synthetic Ge-fluorphlogopite-1M	$K Mg_{2.64}(Ge_{3.72}Al_{0.28})O_{10}F_2$	1568	4.2	13.9	113.5	60.1, 59.9
44. Toraya et al. (1983)	synthetic Mn-fluorphlogopite-1M	$K(Mg_{2.44}Mn^{2+}_{0.24})(Si_{3.82}Mn^{2+}_{0.18})O_{10}F_2$	1131	4.3	1.7	111.2	60.0, 60.0
45. Semenova et al. (1983)	tetraferri-biotite-1M (Krivoy Rog, U.S.S.R.)	$(K_{0.89}Na_{0.08}Ca_{0.03})(Mg_{1.64}Fe^{2+}_{1.37}Mn_{0.01})$ $(Si_{3.06}Fe^{3+}_{0.87}Al_{0.03})O_{10}(OH)_2$	543	7.9	9.0	110.9	58.6, 58.4
46. Backhaus (1983)	lepidolite-1M (Wakefield, Que., Canada)	$(K_{0.90}Na_{0.08}Rb_{0.04})(Al_{1.10}Li_{1.50}Mn_{0.16}Fe^{2+}_{0.15}$ $Fe^{3+}_{0.03}Ti_{0.01})(Si_{3.48}Al_{0.53})O_{10}(OH)_{0.4}F_{1.7}$	724	7.3	6.0	112.3, 112.3	60.5, 59.9, 57.3
47. Knurr et al. (1984)	manganoan phlogopite-1M (Langban, Sweden)	$(K_{0.84}Na_{0.07}Ba_{0.02})(Mg_{2.51}Al_{0.11}Fe^{3+}_{0.17}Mn^{2+}_{0.13}$ $Ti_{0.01})(Si_{2.79}Al_{1.21})O_{10}(OH)_{1.9}F_{0.1}$	669	3.1	9.9	110.3	58.9, 58.9

For ordered layer silicates Alexandrova et al. (1972) have stated that the Al-rich tetrahedral cation tends to move closer to its apical oxygen in order to compensate for the latter's unsatisfied valence. The Si-rich cation simultaneously moves closer to the basal oxygens, and the bases of these Si-rich tetrahedra contract while those of the Al-rich tetrahedra expand. Analysis of ordered micas (see later section) shows that this trend is followed in general but may be obscured where the degree of ordering is small and the determinative errors large.

Toraya (1981) correlates the octahedral flattening angle (ψ) with the magnitude of the dimensional misfit (Δ) between the tetrahedral and octahedral sheets. Thus, ψ can be estimated from Equations 4 and 5 for given values of basal tetrahedral edge lengths and octahedral cation--anion bond lengths,

$$\psi = \frac{2\,e_b}{3\,d_o} \cos\alpha \;. \tag{5}$$

His graph of ψ versus Δ for a constant e_b value shows a gradual decrease of ψ with decreasing misfit in the vicinity of $\psi = 59°$ and a more rapid decline for ψ values between 59° and 58°. This agrees with earlier calculations by Hazen and Wones (1972). Sheet thicknesses (t_o) are affected also, because

Table 2, continued.

40. 2.251	2.091	3.325	0.10	1.665, 1.633	2.121, 1.909, 2.149	3.005	3.260	$-0.355a_{2,3,1}$	$-0.051a_{1,2,3}$	0
41. 2.275	2.111	3.287	0.02	1.655x2	2.077, 2.059x2	2.962	3.294	$-0.331a_1$	$-0.002a_1$	$-0.333a_1$
42. 2.270	2.114	3.295	0.02	1.656x2	2.076, 2.060x2	2.960	3.300	$+0.331a_{2,3}$	$-0.001a_{1,1}$	$-0.333a_1$
43. 2.455	2.130	3.336	0.01	1.746x2	2.103, 2.091x2	2.857	3.513	$-0.333a_1$	$-0.001a_1$	$-0.334a_1$
44. 2.246	2.172	3.372	0.00	1.632x2	2.071, 2.069x2	3.097	3.174	$-0.333a_1$	0	$-0.334a_1$
45. 2.278	2.99	3.392	0.02	1.676x2	2.111, 2.098x2	2.973	3.387	$-0.336a_1$	0	$-0.336a_1$
46. 2.257	2.065	3.328	0.10	1.631, 1.633	2.098, 2.058 1.914	2.958	3.226	$-0.356a_1$	$-0.004a_1$	$-0.360a_1$
47. 2.266	2.143	3.454	0.01	1.663x2	2.073, 2.072x2	2.950	3.401	$-0.332a_1$	$-0.001a_1$	$-0.333a_1$

$$t_o = 2d_o \cos\psi \quad . \tag{6}$$

It follows that variations in dimensions of the octahedral sheet along directions perpendicular and parallel to the sheet are related to each other, and that the rate of change in thickness of the octahedral sheet is largely affected by the variation of the lateral dimensions of the tetrahedral sheet. Thus, the degree of octahedral flattening possible is limited by the amount of tetrahedral-octahedral misfit and consequent tetrahedral rotation.

Lin and Guggenheim (1983) made a multiple regression analysis of the magnitude of ψ in 26 refined mica structures to demonstrate the relative effects on ψ by octahedral cation size, cation field strength (ratio of cation valence to radius), and tetrahedral-octahedral sheet lateral misfit. They found for trioctahedral silicate and germanate micas that tan ψ depends significantly on the field strengths of neighboring octahedra and on the tetrahedral-octahedral misfit

$$\tan \psi = -0.0803 + 0.0201 \Sigma FS + 1.3181 \text{ TM} \quad , \tag{7}$$

where ΣFS is the sum of neighboring field strengths and the misfit parameter TM = $4D_t/3D_o$ where D_t and D_o are refined values of the tetrahedral and octahedral bond lengths. For dioctahedral micas tan ψ depends only on the field strengths

Table 2. Structural details of dioctahedral micas.

Reference	Species	Composition	# Refl.	Final R	α_{tet}	τ_{tet}	ψ_{oct}
1. Zvyagin (1957, 1967, 1979)	celadonite-1M (Bug River, U.S.S.R.)	~ $(K_{0.86}Ca_{0.04})(Fe^{3+}_{1.20}Fe^{2+}_{0.36}Mg_{0.41}$ $Al_{0.05})(Si_{3.94}Al_{0.06})O_{10}(OH)_2$	400	?	7.5°	108.1°	58.0°, 56.8x2
2. Gatineau (1963)	muscovite-$2M_1$ (Harts Range, Australia)	$(K_{0.94}Na_{0.06})(Al_{1.83}Fe^{3+}_{0.12}Mg_{0.06})$ $(Si_{3.11}Al_{0.89})O_{10}(OH)_2$	550	8.0%	11.7	110.6, 111.1	62.3, 57.1x2
3. Birle & Tettenhorst (1968)	muscovite-$2M_1$ (Harts Range, Australia)	$(K_{0.94}Na_{0.06})(Al_{1.83}Fe^{3+}_{0.12}Mg_{0.06})$ $(Si_{3.11}Al_{0.89})O_{10}(OH)_2$	550	12.0	12.0	110.6, 111.2	62.3, 57.0x2
4. Burnham & Radoslovich (1964)	sodian muscovite-$2M_1$ (Alpe Sponda, Switz.)	~ $(K_{0.66}Na_{0.34})Al_2(Si_3Al)O_{10}(OH)_2$	619	3.8	12.8	110.9 x2	62.2, 57.1x2
5. Burnham & Radoslovich (1964)	potassian paragonite-$2M_1$ (Alpe Sponda, Switz.)	~ $(Na_{0.85}K_{0.15})Al_2(Si_3Al)O_{10}(OH)_2$	557	3.4	15.9	110.5 x2	62.0, 56.9x2
6. Güven & Burnham (1967)	muscovite-3T (Sultan Basin, Wash.)	$(K_{0.90}Na_{0.06}R^{2+}_{0.02})(Al_{1.83}Fe^{3+}_{0.04}Ti_{0.01}$ $Fe^{2+}_{0.04}Mg_{0.09})(Si_{3.11}Al_{0.89})$ $O_{10}(OH)_{1.98}F_{0.03}$	280	2.4	11.8	110.0, 111.3	61.8, 57.6, 56.5
7. Rothbauer (1971)	muscovite-$2M_1$ (Black Hills, S.D.)	$(K_{0.85}Na_{0.09})(Al_{1.81}Fe^{2+}_{0.14}Mg_{0.12})$ $(Si_{3.09}Al_{0.91})O_{9.81}(OH)_2F_{0.19}$	625	2.7	11.3	110.9 x2	62.2, 57.2x2
8. Güven (1971a)	muscovite-$2M_1$ (Georgia)	$[K_{0.86}Na_{0.10}(H_3O)_{0.01}](Al_{1.90}Mg_{0.06}$ $Fe^{2+}_{0.05}Fe^{3+}_{0.02}Ti_{0.01})(Si_{3.02}Al_{0.98})$ $O_{10}(OH)_{1.99}F_{0.01}$	567	3.5	11.4	111.0, 111.1	62.1, 57.0x2
9. Güven (1971a)	phengite-$2M_1$ (Tiburon Penin., California)	$(K_{0.87}Na_{0.07}R^{2+}_{0.03})(Al_{1.43}Fe^{3+}_{0.05}Mg_{0.50}$ $Fe^{2+}_{0.09}Ti_{0.01})(Si_{3.39}Al_{0.61})$ $O_{10.08}(OH)_{1.92}$	557	4.5	6.0	111.5, 111.4	61.4, 57.1x2
10. Zhoukhlistov et al. (1973)	phengite-$2M_2$ (N. Armenia, U.S.S.R.)	~ $(K_{0.68}Na_{0.09})Al_{1.93}(Si_{3.5}Al_{0.05})$ $O_{10.06}(OH)_{1.94}$	504	11.7	11.2	111.4, 111.6	60.9, 57.0x2
11. Rozhdestvenskaya & Frank-Kamenetskii (1974)	chernykhite-$2M_1$ (N. W. Karatau, U.S.S.R.)	$[Ba_{0.28}Na_{0.20}K_{0.07}(NH_4)_{0.1}][V^{4+}_{0.30}V^{3+}_{1.13}$ $Al_{0.65}(Mg,Fe)_{0.20}](Si_{2.30}Al_{1.70})$ $O_{10}(OH)_2$	984	12.0	7.0	111.7 x2	60.4, 56.6x2
12. Sidorenko, et al. (1975)	phengite-1M (Transbaikal, U.S.S.R.)	$(K_{0.65}Na_{0.03})Al_{1.83}Fe^{3+}_{0.25}Fe^{2+}_{0.04}Mn_{0.04}$ $Mg_{0.10})(Si_{3.51}Al_{0.49})O_{10.13}(OH)_{1.80}F_{0.07}$	588	10.9	9.3	110.1, 111.1	61.6, 56.6, 57.3
13. Zhoukhlistov et al. (1977)	celadonite-1M (Krivoi Rog, U.S.S.R.)	$(K_{0.83}Na_{0.01}Ca_{0.04})(Fe_{1.15}Fe^{3+}_{0.36}Al_{0.05}Mg_{0.41}$ $Ti_{0.01})(Si_{3.94}Al_{0.06})O_{10}(OH)_{1.99}F_{0.01}$	163	10.8	1.3	112.6	58.3, 56.6x2
14. Soboleva et al. (1977)	synthetic paragonite-1M	$Na_{0.91}Al_{1.88}(Si_{3.45}Al_{0.55})O_{10}(OH)_2$	196	12.1	19.1	110.4	59.9, 57.8x2
15. Sidorenko et al. (1977a)	paragonite-$2M_1$ (S.Urals, U.S.S.R.)	$(Na_{0.60}K_{0.10}Ca_{0.03})(Al_{1.93}Mg_{0.10}Fe^{2+}_{0.02})$ $(Si_{2.98}Al_{1.02})O_{9.78}(OH)_{2.22}$	430	11.1	17.4	111.0, 112.3	60.6, 57.0x2
16. Sidorenko et al. (1977b)	paragonite-3T	$(Na_{0.71}K_{0.16}Ca_{0.03})(Al_{2.02}Fe^{3+}_{0.01}Mg_{0.01})$ $(Si_{2.96}Al_{1.04})O_{9.98}(OH)_{2.02}$	208	13.0	15.7	111.6, 110.4	59.8, 58.1, 58.4
17. Tsipurskii & Drits (1977)	phengite-1M (locality ?)	$(K_{0.80}Na_{0.02}Ca_{0.01})(Al_{1.66}Mg_{0.28}Fe^{2+}_{0.02}Fe^{3+}_{0.06})$ $(Si_{3.41}Al_{0.59})O_{10}(OH)_2$	211	7.0	8.6	111.5	61.5, 56.4
18. Tsipurskii & Drits (1977)	muscovite-$2M_1$ (locality ?)	$(K_{0.79}Na_{0.04}Ca_{0.03})(Al_{1.84}Mg_{0.08}Fe^{2+}_{0.01}Fe^{3+}_{0.06})$ $(Si_{3.16}Al_{0.84})O_{10}(OH)_2$	184	5.0	11.2	110.9, 110.8	62.2, 57.1
19. Richardson & Richardson (1982)	muscovite-$2M_1$ (Archer's Post, Kenya)	$(K_{0.88}Na_{0.03}Ca_{0.01})(Al_{1.87}Ti_{0.03}Mg_{0.06}Mn^{3+}_{0.03})$ $(Si_{3.01}Al_{0.85}Fe^{3+}_{0.14})O_{10}(OH)_2$	879	12.9	11.0	110.9, 111.1	51.7, 56.6x2
20. Lin & Bailey (1984)	paragonite-$2M_1$ (Zermatt-Saas Fee, Switz.)	$(Na_{0.92}K_{0.04}Ca_{0.02})(Al_{1.99}Fe_{0.03}Mg_{0.01})$ $(Si_{2.94}Al_{1.06})O_{10}(OH)_2$	698	4.5	16.2	110.4, 110.3	62.1, 57.0x2
21. Knurr et al. (1984)	muscovite-$2M_1$ (Minas Gerais, Brazil)	$(K_{0.93}Na_{0.05}Ba_{0.01})(Al_{1.72}Fe^{3+}_{0.15}Mn^{3+}_{0.02}Ti_{0.02}$ $Mg_{0.10})(Si_{3.06}Al_{0.94})O_{10}(OH)_2$	1261	2.7	10.8	111.1, 111.0	62.1, 57.0
22. Rule & Bailey (1984)	phengite-$2M_1$ (Bulautad, Spanish Sahara)	$(K_{0.95}Na_{0.03}Ba_{0.01})(Al_{1.51}Mg_{0.27}Fe^{2+}_{0.14}Cr_{0.09}$ $Ti_{0.01})(Si_{3.25}Al_{0.75})O_{10}(OH)_2$	1276	3.3	7.9	111.6 x2	61.6, 57.1

Table 2, continued.

	Sheet Thickness		Interlayer Separation	Basal Oxygen Δz	Mean Bond Lengths					Intralayer Shift	Layer Offset	Resultant Shift
	Tet.	Oct.			T—O	M—O,OH	R^{1+}—O Inner	Outer				
1.	2.258 Å	2.238 Å	3.357 Å	0.09 Å	1.643x2 Å	2.112 Å, 2.044x2	2.933 Å	3.268 Å		$-0.33a_1$	$-0.013a_1$	$-0.346a_1$
2.	2.244	2.092	3.428	0.23	1.653, 1.639	2.254, 1.924x2	2.857	3.378		$+0.367a_{2,3}$	$+0.002a_{1,1}$	$-0.350a_1$
3.	2.242	2.096	3.426	0.23	1.660, 1.636	2.258, 1.924x2	2.851	3.382		$+0.368a_{2,3}$	$+0.002a_{1,1}$	$-0.350a_1$
4.	2.243	2.091	3.312	0.23	1.645, 1.646	2.241, 1.923x2	2.793	3.369		$+0.376a_{2,3}$	$+0.001a_{1,1}$	$-0.374a_1$
5.	2.244	2.092	3.078	0.24	1.654, 1.650	2.226, 1.913x2	2.640	3.371		$+0.373a_{2,3}$	$+0.035a_{1,1}$	$-0.304a_1$
6.	2.207	2.112	3.465	0.14	1.670, 1.603	2.231, 1.973, 1.913	2.868	3.389		$+0.379a_{2,3,1}$	$-0.004a_{1,2,3}$	0
7.	2.245	2.089	3.393	0.21	1.645, 1.644	2.241, 1.930x2	2.857	3.362		$+0.376a_{2,3}$	$-0.005a_{1,1}$	$-0.386a_1$
8.	2.239	2.103	3.392	0.22	1.643, 1.642	2.245, 1.933x2	2.855	3.362		$+0.377a_{2,3}$	$-0.005a_{1,1}$	$-0.387a_1$
9.	2.219	2.126	3.359	0.16	1.621, 1.633	2.222, 1.956x2	2.971	3.236		$+0.376a_{2,3}$	$-0.004a_{1,1}$	$-0.385a_1$
10.	2.216	2.133	3.414	0.22	1.619, 1.653	2.195, 1.956x2	2.860	3.354		$-0.380b_3$ $+0.380b_2$	$-0.018a_1$ $-0.018a_1$	$-0.419a_1$
11.	2.269	2.217	3.207	0.15	1.657, 1.660	2.243, 2.013x2	2.937	3.258		$+0.361a_{2,3}$	$-0.006a_{1,1}$	$-0.375a_1$
12.	2.196	2.113	3.399	0.18	1.614, 1.633	2.221, 1.920, 1.957	2.897	3.306		$-0.376a_1$	$-0.024a_1$	$-0.400a_1$
13.	2.248	2.249	3.233	0.00	1.636x2	2.141, 2.043x2	3.044	3.103		$-0.354a_1$	$-0.002a_1$	$-0.356a_1$
14.	2.222	2.099	3.059	0.10	1.659x2	2.091, 1.971x2	2.561	3.441		$-0.338a_1$	$+0.020a_1$	$-0.319a_1$
15.	2.264	2.125	3.005	0.14	1.671, 1.661	2.161, 1.950x2	2.586	3.386		$+0.355a_{2,3}$	$+0.053a_{1,1}$	$-0.274a_1$
16.	2.249	2.075	3.001	0.09	1.609, 1.684	2.061, 1.965, 1.981	2.621	3.341		$-0.349a_{2,3,1}$	$-0.024a_{1,2,3}$	0
17.	2.207	2.158	3.395	0.15	1.626x2	2.261, 1.947x2	2.918	3.298		$-0.373a_1$	$-0.010a_1$	$-0.383a_1$
18.	2.241	2.093	3.399	0.22	1.640, 1.644	2.247, 1.927x2	2.859	3.359		$+0.374a_{2,3}$	$-0.008a_{1,1}$	$-0.386a_1$
19.	2.229	2.138	3.405	0.02	1.646, 1.639	2.253, 1.940x2	2.871	3.328		$+0.378a_{2,3}$	$-0.005a_{1,1}$	$-0.390a_1$
20.	2.243	2.078	3.053	0.23	1.653, 1.652	2.221, 1.908x2	2.624	3.370		$+0.374a_{2,3}$	$+0.045a_{1,1}$	$-0.285a_1$
21.	2.243	2.106	3.393	0.22	1.644, 1.644	2.252, 1.934x2	2.872	3.353		$+0.378a_{2,3}$	$-0.006a_{1,1}$	$-0.391a_1$
22.	2.236	2.121	3.342	0.18	1.637, 1.636	2.233, 1.952x2	2.925	3.279		$+0.376a_{2,3}$	$-0.005a_1$	$-0.386a_1$

$$\tan \psi = 1.213 + 0.020 \ \Sigma FS \ , \tag{8}$$

so that Lin and Guggenheim (1983) conclude that the field strength of neighboring octahedra may be more important than either octahedral cation size or the tetrahedral-octahedral sheet misfit in determining the magnitude of ψ. They point out also that octahedral flattening acts in opposition to counter-rotation of upper and lower octahedral anion triads (due to shortening of shared edges) in affecting the lateral dimensions of the octahedral sheet.

Weiss et al. (1984) used the data from 66 refined mica structures in a statistical analysis of octahedral distortions. They concluded that the octahedral sheet is the most rigid element of the mica structure and is under only minor stresses from the tetrahedral sheets and the interlayer region. They found all octahedra to be flattened, those around larger cations usually more so than around smaller cations. Flattening dominates over counter rotation for large octahedra and vice versa for small octahedra, and the difference between the flattening and counter rotation functions ($\psi - \delta$) is linear in a plot versus cation size. These relationships are a direct consequence of the need for a uniform sheet thickness. The distortion of a particular octahedron is a result of interaction with all neighboring octahedra, including their ordering patterns. They give regression equations for predicting the amount of flattening (ψ) and counter rotation (δ) using either bond lengths or cation radii appropriate for any assumed octahedral ordering pattern.

Tetrahedral tilt and basal surface corrugation

For dioctahedral species tetrahedral rotation will serve nicely to reduce the lateral tetrahedral dimensions to provide a good fit with the four short and twisted octahedral edges illustrated in Figure 3 for muscovite. But what about the two elongate edges shown in each ring? These edges are now longer than those of the rotated tetrahedral sheet, and articulation can be accomplished only by tilting the tetrahedra that must join to these edges. For the lower tetrahedral sheet, with apices pointing up, the apices of adjacent tetrahedra across the elongate edge will now tilt slightly away from one another and the bridging basal oxygen between them will be elevated by $\Delta z = 0.1$ to 0.2 Å above the other two basal oxygens of each tetrahedron. The elevated oxygen forms part of a groove on the bottom surface of the layer that transects the structure normal to the elongate octahedral edges and parallel to the direction of intralayer shift of the 2:1 layer. A similar corrugation exists in the upper surface, where tilting with apices down depresses the corresponding bridging basal oxygen relative to its neighbors. All dioctahedral layer silicates show tetrahedral corrugations of this type. The effect of tetrahedral tilting in the case of muscovite can be seen in Figure 5 by noting that the apical oxygens are offset from the tetrahedral cations in a normal projection. A more detailed analysis of the effect of out-of-plane tilting is given by Lee and Guggenheim (1981) for the structure of pyrophyllite-$1Tc$.

One consequence of the nearly automatic structural distortions that occur in dioctahedral micas is the appearance on X-ray single crystal patterns of weak 06ℓ reflections with ℓ odd, which would be "structural absences" in an undistorted structure. Hendricks and Jefferson (1939), who first noted these 06ℓ reflections for muscovite-$2M_1$, postulated that the forces responsible for this structural distortion also cause adoption of a unique stacking sequence of layers for muscovite-$2M_1$. Bailey (1980) noted that other 2-layer dioctahedral structures exhibit similar 06ℓ reflections (e.g., dickite, paragonite-$2M_1$, and cookeite), but so do certain ordered trioctahedral layer silicates (e.g., amesite and chlorite). Radoslovich and Norrish (1962) have shown that for equal tetrahedral sizes and scattering powers (i.e., tetrahedral cation disorder) the effect can be caused by altering the positions of the tetrahedral cations so they no longer have y-coordinates that are exact multiples of $b/12$.

Limits of strain

For trioctahedral species the octahedral edges are usually similar enough (Fig. 4) that articulation of a larger tetrahedral sheet with a smaller octahedral sheet can be accomplished readily by tetrahedral rotation plus thickening and thinning of sheets. It is much more difficult in the opposite case, however, to stretch a smaller tetrahedral sheet to fit onto a larger octahedral sheet. This would be the situation, for example, where there is little or no tetrahedral substitution for Si and where trioctahedral sheets are populated with the larger medium-sized cations such as Cu^{2+}, Mg, Ni, Co, Zn, Fe^{2+}, or Mn^{2+}. In 1:1 layers this type of lateral mismatch may lead to tetrahedral tilting so extreme that the layers curl into elongate scrolls, as in chrysotile asbestos and pecoraite. In 2:1 layers the mismatch can be alleviated by inverting certain tetrahedra and

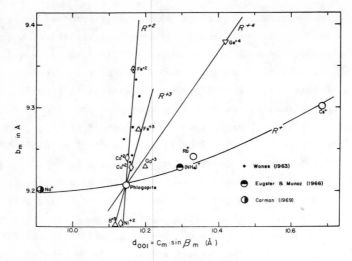

Figure 6 (above). Effects of different synthetic substitutions into the phlogopite-$1M$ structure on the b and $c \sin \beta$ cell dimensions. From Hazen and Wones (1972).

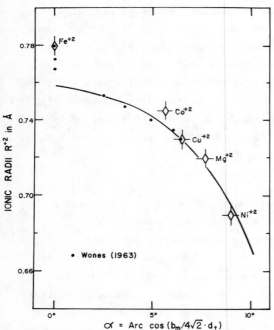

Figure 7 (to the left). Plot of tetrahedral rotation angle α versus ionic radii of octahedral R^{2+} cations in synthetic $1M$ micas of composition $KR_3^{2+}(Si_3Al)O_{10}(OH)_2$. Homogeneous micas could not be synthesized with R^{2+} cations having ionic radii greater than 0.76 Å, where α becomes 0° and the tetrahedral sheet cannot be extended further. From Hazen and Wones (1972).

rebonding the apices in some different but regular pattern, as in stilpnomelane and other Fe,Mn silicates. But the 2:1 layer within a mica by definition is continuous and planar, and thickening and thinning of sheets is necessarily the main method of adjusting a misfit of this type. The mica simply will not crystallize if the limit of strain is exceeded for a particular combination of tetrahedral and octahedral compositions. Hazen and Wones (1972) demonstrated this for a series of synthetic micas in which several different elements were substituted into the tetrahedral, octahedral, and interlayer sites. Figure 6 shows the effect of their multiple substitutions on the b and $c\sin\beta$ cell parameters. Note that octahedral substitutions of different size R^{2+} cations affect only the b parameter, whereas incorporation of a higher charge cation affects $c\sin\beta$ as well. Interlayer cations affect $c\sin\beta$ primarily, and b to a lesser extent. Tetrahedral cation substitutions affect both parameters. For a composition of $KR_3^{2+}(Si_3Al)O_{10}(OH)_2$ they found they could not crystallize a completely trioctahedral mica for any octahedral cation of effective ionic radius greater than 0.76 Å [slightly smaller than that of Fe^{2+} for the radii listed by Shannon (1976)]. This is the value of the radius for which the tetrahedral sheet is predicted to be fully extended with $\alpha = 0°$ (Fig. 7). A nomogram constructed by Hazen and Wones suggests an extrapolated octahedral cation radius limit of 0.70-0.72 Å (near the ionic radii values for Ni, Mg, Cu^{2+}, and Co^{2+}) for a tetrahedral composition of Si_4O_{10}. Toraya (1981) plotted mean octahedral bond length (d_o) values for M1 and M2 octahedra in a number of $1M$ micas versus values of shared octahedral edge lengths (e_s) for given mean tetrahedral basal edge length (e_b) values. He also superimposed onto the graph the predicted locus of the curve for $\alpha = 0°$ and concluded that $\alpha = 0°$ is not always the limit of stability. Large values of shared octahedral edges (e_s) also lead to instability due to increased cation-cation repulsion across these edges.

INTERLAYER REGION

Size of cavity

Potassium is the common interlayer cation in micas. Radoslovich and Norrish (1962) presented evidence that in dioctahedral micas the interlayer K is too large for its six-coordinated hole after full tetrahedral rotation has taken place. This has the result of reducing the amount of tetrahedral rotation that actually occurs, of propping apart the basal oxygen surfaces, and of stretching the octahedral sheet laterally. For a smaller interlayer cation, and the same composition otherwise, one would expect a larger tetrahedral rotation, less interlayer separation, and a smaller b repeat distance. All of these expectations are fulfilled in the $1M$, $2M_1$, and $3T$ structures of paragonite, the dioctahedral Na-mica (Table 2). Burns and

White (1963) also have shown for muscovite that the b value decreases with time as K is leached from the structure. The compression of K within its cavity in muscovite makes it less susceptible to leaching than would be the case for other micas in which tetrahedral rotation can go to completion either because of a smaller amount of lateral misfit present, as in trioctahedral phlogopite, or for a smaller interlayer cation, as in dioctahedral paragonite. The orientation of the OH dipole moment (discussed in a later section) accentuates this difference in weathering tendency between dioctahedral and trioctahedral micas.

In K-micas blocking of tetrahedral rotation by the larger interlayer cation has been credited by Hewitt and Wones (1975) as the reason why Al-substitutions in the synthetic phlogopite-biotite system cease at a tetrahedral composition of $Si_{2.25}Al_{1.75}$ and an octahedral composition of $(Mg,Fe)_{2.25}Al_{0.75}$. Further Al-substitution in both sheets would decrease the lateral dimensions of the octahedral sheet and increase those of the tetrahedral sheet, and thereby require further tetrahedral rotation for relief of misfit than is physically possible. A rather similar compositional limit was found by Robert (1976) in synthetic Ti-rich micas at $Si_{2.3}Al_{1.7}$ and $Ti^{vi}_{0.35}$.

Layer offsets in K-micas

Radoslovich (1960) pointed out that in muscovite-$2M_1$ the interlayer K is not in the exact center of the hexagonal ring, but is displaced along the two-fold axis toward the vacant octahedral sites above and below. This displacement also is in the direction that would partly equalize the lengths of the K-O bonds. Analysis of recent muscovite-$2M_1$ refinements indicates that tetrahedral tilt and resultant basal oxygen surface corrugation create a slightly irregular interlayer cavity. At the same time the $T(1)...T(2)$ distance across the vacant octahedral site is increased by about 0.1 Å relative to T...T distances across occupied octahedral sites. These distortions have the net effect of offsetting adjacent 2:1 layers across the interlayer spaces by about 0.02 Å in muscovite-$2M_1$ (maximum of 0.12 Å in phengite-$1M$), as measured either between the geometric centers of opposing rings or by the displacement of tetrahedral cations in adjacent layers (Fig. 8), so that the layers are not exactly superimposed. This offset is negligible in most trioctahedral species, where tetrahedral tilting and basal oxygen corrugation are minimal (Table 1). Larger offsets of about

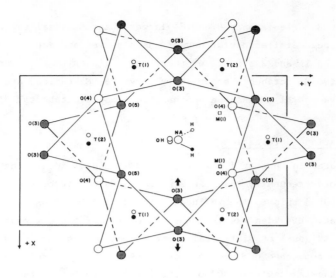

Figure 8. Interlayer region of paragonite-$2M_1$ illustrating repulsion between O(3) basal oxygens (arrows) that approach each other most closely across the interlayer gap as a result of tetrahedral tilt and consequent corrugation of the basal oxygen surfaces. From Lin and Bailey (1984).

0.23 Å are observed in paragonite, but this is due to introduction of a new influence (discussed in a later section) caused by the small size of the interlayer Na.

Intralayer overshifts in relation to β angle

Another distortional effect, observed in dioctahedral but not in trioctahedral micas to date, is the intralayer stagger or "overshift" of the upper tetrahedral sheet relative to the lower tetrahedral sheet of $0.36a$ to $0.38a$ instead of the ideal $0.33a$ (Tables 1,2). Bailey (1975) has shown that the overshift is due to the M1 vacant octahedron being larger than the M2 occupied octahedra. Because apical oxygens of opposed tetrahedral sheets of the 2:1 layer are linked to diagonal edges of the *trans* M1 octahedron (see Fig. 1), increasing the length of these diagonal edges increases the effective intralayer shift. These large overshifts plus the smaller layer offsets combine to create β angles one to two degrees larger than the ideal values, especially evident in the $1M$ structures. Because the intralayer shifts in the $2M_1$ structure are oriented at 120° to each other in adjacent layers, the displacements partly cancel one another so the resultant β angle is only 0.6-0.8° larger than the ideal value in this structural type. Similar deviations of the observed and ideal β values are observed in the $1M$ and $2M_2$ forms of phengite. Celadonite-$1M$ is an exception because the two M2 octahedra occupied by large Fe, Mg cations are the same size as the vacant M1 octahedron so that the usual dioctahedral distortions do not occur.

Interlayer cation coordination

It was noted in Chapter 1 that the $1M$, $2M_1$, and $3T$ ideal mica structures occur much more frequently than the $2Or$, $2M_2$, and $6H$ structures. Radoslovich (1959) attributed this to the influence of adjacent basal oxygen surfaces on one another. His idea was that the distortion of these surfaces from hexagonal to trigonal or ditrigonal symmetry would favor arrangements where the distortions matched one another most closely, as in structures having interlayer stacking angles of 0° or of ± 120°. Interlayer stacking angles differing by ± 60° or 180° would occur only where tetrahedral rotation is small and the basal oxygen surfaces nearly hexagonal as a result. Although this view may have some validity for layer silicates in which opposing layer surfaces are in contact and can influence one another, as in 1:1 type layers and chlorites, it has been pointed out above that in micas adjacent layers are propped apart by the interlayer cation and are not in contact. Detailed refinements of the $2M_2$ structure (having ± 60° stacking angles) have shown also that the oxygen surfaces in these cases are not necessarily hexagonal, with α up to 11.2°. The important feature appears to be that interlayer stacking angles of 0° or ± 120° cause the six nearest oxygen neighbors to form an energetically stable octahedral coordination group around the interlayer cation. Stacking angles ± 60° or 180° require occupation of different sets of octahedral positions, I and II (Chapter 1), in adjacent layers. Lateral rotation of basal oxygens in each layer toward the octahedral sites as a result of tetrahedral-octahedral sheet misfit then creates a trigonal prism coordination around the interlayer cation, in which three basal oxygens from one layer directly superimpose on three basal oxygens from the layer below. The latter coordination is energetically unstable in most cases because tetrahedral substitution leads to unsatisfied negative charges on these basal oxygens so they will tend to repel one another (Takeda et al., 1971). This condition is minimized to the extent that ± 60° or 180° stacking angles may be adopted if tetrahedral substitution and resultant charge are small. See the brittle mica anandite-$2Or$, however, for an exception.

Layer offset in paragonite

Adjacent basal oxygen surfaces are closer to one another in paragonite than in the K-micas, and the influence of the surface corrugations can be noted in the reduced magnitude of the β angle. In paragonite-$2M_1$ the intralayer over-shifts of $+ 0.374a$ along the pseudohexagonal axes X_2 and X_3 of the $2M_1$ cell are quite similar to those in the $2M_1$ K-micas but adjacent 2:1 layers are offset around the interlayer Na by $+ 0.045a_1 = 0.23$ Å (Lin and Bailey, 1984). This offset is five times the magnitude of the offset reported by Rothbauer (1971) for muscovite-$2M_1$, and is opposite in direction so that β is reduced in size rather than increased. Figure 8 shows that the offset of adjacent ditrigonal rings (best noted for the tetrahedral cations) is caused in this case by mutual repulsion between the O(3) anions (arrows) that approach one another most closely across the interlayer gap as a result of tetrahedral tilting and consequent corrugation of the basal oxygen surfaces. In paragonite-$1M$ the effect is less pronounced because the different geometry of layer stacking causes mutual oxygen repulsion along both $- X_2$ and $- X_3$ of the $1M$ cell, with a smaller resultant along $+ X_1$.

Figure 8 also illustrates why the interlayer Na lies in a very irregular cavity in paragonite-$2M_1$. The O(3) atoms that approach each other most closely across the interlayer gap to create the offset of adjacent layers also coordinate most closely with the Na (2.496 Å). The O(4) atoms (unshaded) that have been depressed or elevated away from the mean plane of their neighbors by tetrahedral tilting in turn are farthest from the Na (2.718 Å), whereas the O(5) atoms are intermediate in approach. The closest observed approach of Na-O(3) = 2.496 Å is in accord with the observation that in micas the bond lengths to the interlayer cation tend to be at least 0.1 Å greater than the sum of the ionic radii for six-fold coordination (2.38 Å for Na and O). This deviation increases with decreasing angle of tetrahedral twist and increasing effective coordination number for the cation. An irregular interlayer cavity due to the corrugation effect also exists around K in muscovite-$2M_1$, but the difference between individual K-O bond lengths is much smaller ($\Delta = 0.04$ Å) than in paragonite-$2M_1$ ($\Delta = 0.22$ Å).

Location of H^+ proton

The hydrogen proton associated with the central OH group also exerts an important influence on the interlayer region. The position of the hydrogen proton can be determined directly by neutron diffraction study or by careful X-ray diffraction study, and the O-H transition moment, which should be close to the direction of the O-H bond, can be calculated from infrared measurements. It has been found for micas that the nucleus of the H^+ is located at 0.90 to 0.95 Å from the nucleus of the associated oxygen atom and is pointed away from the octahedral sheet and into the tetrahedral sheet in most cases. For most trioctahedral mica species where cation ordering is not extreme the H^+ is repelled about equally by its three octahedral cation neighbors so that it points quasi-normal to the sheets toward the interlayer cation and thereby presumably weakens the strength of the interlayer bonding. But in dioctahedral species the interlayer cation is able to repulse the proton in the direction of least resistance, namely toward the vacant octahedron M1 (as in Fig. 8). In muscovite-$2M_1$, for example, Rothbauer (1971) found by neutron diffraction that the O-H bond was inclined from the horizontal (001) cleavage by only 12°. In this location the H^+ is at a distance of 2.47 Å from a basal oxygen and is equi-distant from two apical oxygens at 2.68 Å. For micas that are intermediate between dioctahedral and trioctahedral with incomplete occupancy of M1, some unit cells necessarily are dioctahedral in nature and some are trioctahedral. Electron density maps of two such specimens in fact have shown a splitting of the hydrogen proton maximum such that both dioctahedral and trioctahedral orientations of the O-H^+ vector are present (see Chapter 3 for details).

With increasing transfer of cationic positive charge from the octahedral sheet to the tetrahedral sheet, as in going from muscovite to celadonite, Bookin and Drits (1982) calculated from minimization of electrostatic energy considerations that the O...H vectors should first become parallel to (001) and then point into the octahedral sheet so that the two H^+ protons of the vacant *trans* M1 octahedron actually lie inside M1 and act as a divalent octahedral cation. Similar electrostatic calculations for micas have been made by Giese (1971, 1979, and unpubl.), Soboleva and Mineeva (1981), and Bookin et al. (1982). These results form the basis of Chapter 4 of this volume.

Interest in the location of the H^+ protons is not restricted to correlation of structural data with infrared measurements. It also figures prominently in theories regarding the ease of mica alteration (Bassett, 1960), the origin of the different polytypic forms of mica (Takéuchi, 1965, and others), and the stability of regular interstratifications of a mica with some different kind of layer silicate (Norrish, 1973).

CALCULATED STRUCTURAL MODELS

The mica structure is confined within fairly narrow cell dimension and unit-layer volume limits by its ditrigonal to hexagonal geometry and the bonding characteristics of its constituent tetrahedral and octahedral sheets. Comfortable junction planes must be maintained between the sheets, but these can still be maintained by quite a variety of complementary isomorphous substitutions to be described in later chapters.

The geometry of the networks plus our knowledge of the variation of bond lengths and angles with composition have allowed calculation of quite precise structural models, including all atomic positions for a given composition. The pioneering work in this area was that of E.W. Radoslovich and his coworkers at C.S.I.R.O. in Australia, who published a very useful series of papers in 1962 and 1963. These were followed by the study of Donnay et al. (1964a) on prediction of the detailed shapes and sizes of the tetrahedral and octahedral sheets in $1M$ micas from their known compositions and cell dimensions, later modified by Drits (1969) to take into account additional data on observed distortions of the polyhedra. As our knowledge of the interrelationships of the structural elements has increased, so has the intricacy of the calculated models. Important contributions relative to the micas have been made by Hazen and Wones (1972), Zvyagin et al. (1972), Appelo (1978, 1979), Zvyagin and Soboleva (1979), Toraya (1981), and Weiss et al. (1984). The role played by computer minimization of electrostatic energy is covered by Giese in Chapter 4.

Synthesis studies of micas of varying compositons usually have included plots of one or more of the cell dimensions or of cell volume as a function of composition or of the P-T conditions of synthesis (Chapter 7). These plots usually have shown the tetrahedral sheet as being sufficiently flexible to conform to the lateral dimensions dictated by a smaller octahedral sheet. Natural micas normally have more substitutions than attempted in the synthetic systems, but multiple regression analyses have been used to pinpoint the relative contributions of each substitution to a given cell parameter. Some of these specific plots and equations will be mentioned in later chapters, as appropriate. It should be noted also that graphs and tables exist for the variation of the refractive indices and of birefringence as a function of mica composition (Chapter 6) and of the 00ℓ basal reflection intensities as a function of heavy elements present in the octahedral or interlayer sites (e.g., Brown, 1955; Profi et al., 1973; Yoshii et al., 1973).

CATION ORDER AND DISORDER

Where isomorphous substitutions occur in a mica the possibility exists that the different cations that occupy a given type of structural site may do so in either a regular or irregular manner. A truly random distribution over many unit cells is required to meet the definition of substitutional solid solution in which on the average, each cation site in the unit cell is represented by a hybrid atom that is statistically part atom A, part atom B, etc. This is the disordered state. But there may be a tendency instead for complete or partial ordering of the constituent cations over the available positions either as a result of size or bonding differences of the cations involved or of some inherent structural difference between the positions. The presence or absence of such ordering is important in evaluating the overall energy and stability of these structures.

One of the problems in studying ordering in layer silicates is that the ideal space group symmetry that is conferred by a particular stacking sequence of layers with disordered cation distributions may not be the true resultant symmetry. The pattern of cation ordering that has been adopted may lower the true symmetry relative to that of the disordered state. For example, the symmetries of the ideal space groups require all interlayer cations to be equivalent (i.e., disordered) for all six standard mica polytypes of Smith and Yoder (1956) and the tetrahedral cations to be equivalent for the $1M$, $2Or$, and $6H$ micas. Any ordering that takes place for these cations necessarily lowers the symmetry to a subgroup of that of the disordered state symmetry. The lower subgroup symmetry may be very difficult to detect in view of the large influence of the stacking sequence of layers on the diffraction intensities, an unfortunate consequence of a high degree of pseudosymmetry. The ordering must be investigated in each possible subgroup symmetry by some method that negates the pseudosymmetry.

Bailey (1975) summarized the results of detection of cation ordering in layer silicates in both ideal and subgroup symmetries. The results of tetrahedral and octahedral ordering for the micas have been updated by Bailey (1984) and are shown here as Tables 3 and 5. No ordering of

Table 3. Examples of tetrahedral cation ordering in micas.

Reference	Species	Space Group	#Refl.	Final R	Mean T--O	Tet. Comp. from Eq.*	Total from Eq.*	Al^{IV} from Anal.
Güven & Burnham (1967)	muscovite-3T (Sultan Basin Wash.)	$P3_112$	280	2.4%	T(1)=1.670Å T(2)=1.603	$Si_{0.62}Al_{0.38}$ $Si_{1.0}$	0.76	0.89
Güven (1971a)	phengite-2M_1 (Tiburon penin., Calif.)	$C2/c$	557	4.5	T(1)=1.621 T(2)=1.633	$Si_{0.92}Al_{0.08}$ $Si_{0.85}Al_{0.15}$	0.46	0.61
Zhoukhlistov et al. (1973)	phengite-2M_2 (N. Armenia)	$C2/c$	504	11.7	T(1)=1.619 T(2)=1.653	$Si_{0.93}Al_{0.07}$ $Si_{0.72}Al_{0.28}$	0.70	0.50
Sidorenko et al. (1975)	phengite-1M (Transbaikal, U.S.S.R.)	$C2$	588	10.9	T(1)=1.614 T(2)=1.633	$Si_{0.96}Al_{0.04}$ $Si_{0.85}Al_{0.15}$	0.38	0.49
Guggenheim & Bailey (1977)	zinnwaldite-1M (Erzgebirge, D.D.R.)	$C2$	1493	5.7	T(1)=1.646 T(2)=1.639	$Si_{0.77}Al_{0.23}$ $Si_{0.81}Al_{0.19}$	0.84	0.91
Sidorenko et al. (1977b)	paragonite-3T (locality ?)	$P3_112$	208	13.0	T(1)=1.609 T(2)=1.684	$Si_{0.99}Al_{0.01}$ $Si_{0.53}Al_{0.47}$	0.96	1.04
Brown (1978)	lepidolite-3T (Kalgoorlie, Australia)	$P3_112$	705	4.7	T(1)=1.652 T(2)=1.617	$Si_{0.73}Al_{0.27}$ $Si_{0.94}Al_{0.06}$	0.66	0.52
Guggenheim & Bailey (1975, 1978)	margarite-2M_1 (Chester, Pa.)	Cc	1071	4.0	T(1)=1.747 T(2)=1.633 T(22)=1.736 T(11)=1.623	$Si_{0.15}Al_{0.85}$ $Si_{0.85}Al_{0.15}$ $Si_{0.21}Al_{0.79}$ $Si_{0.91}Al_{0.09}$	1.88	1.89
Sokolova et al. (1979)	ephesite-1M (Ilimaussak, Greenland)	$C2$	284	11.5	T(1)=1.764 T(2)=1.609	$Si_{0.04}Al_{0.96}$ $Si_{0.99}Al_{0.01}$	1.94	1.80
Sokolova et al. (1979)	intermediate margarite-bityite-2M_1 (U.S.S.R.)	Cc	450	11.5	T(1)=1.717 T(2)=1.642 T(3)=1.622 T(4)=1.710	(Be^{IV} also present)		
Pavlishin et al. (1981)	protolithionite-3T (Ukraine, U.S.S.R.)	$P3_112$	702	4.7	T(1)=1.665 T(2)=1.633	$Si_{0.65}Al_{0.35}$ $Si_{0.85}Al_{0.15}$	1.00	1.13
Lin & Guggenheim (1983)	intermediate margarite-bityite-2M_1 (Zimbabwe)	Cc	1927	3.0	T(1)=1.723 T(2)=1.628 T(22)=1.721 T(11)=1.632	(Be^{IV} also present)		
Joswig et al. (1983)	margarite-2M_1 (Greiner, Zillerthal, Austria)	Cc	1003	1.7	T(11)=1.748 T(12)=1.628 T(21)=1.626 T(22)=1.748	$Si_{0.14}Al_{0.86}$ $Si_{0.88}Al_{0.12}$ $Si_{0.89}Al_{0.11}$ $Si_{0.14}Al_{0.86}$	1.95	2.06
Filut et al. (1984)	anandite-2Or (Wilagedera, Sri Lanka)	$Pnmn$	1074	6.4	T(1)=1.618 T(2)=1.797	(Fe^{IV} also present)		

* Indicated Si,Al and Al^{IV} contents are as given by the regression equation of Hazen & Burnham (1973)

interlayer cations has been reported. Because of the small number of specimens involved, it is convenient to consider both true micas and brittle micas together. The following text is updated from Bailey (1984).

Ordering of tetrahedral cations

Ordering of tetrahedral cations is relatively rare in micas. Of the large number of mica refinements listed in Chapters 2 and 3, only thirteen examples of Si,Al ordering and one example of Si,Fe^{3+} ordering merit consideration here. One of the major mysteries regarding micas has been why the most common species are disordered (namely muscovite-$2M_1$, phlogopite-$1M$, and biotite-$1M$), even though these species may occur in their host rocks immediately adjacent to other silicates that are completely ordered, such as maximum microcline and low albite.

In Table 3 only those specimens are listed for which the authors have stated that tetrahedral cation ordering occurs and on which reasonable structural refinement has been performed. Less certain examples are not listed here. Even in these best examples the final residual values (R) between observed and calculated spectral amplitudes and the deviations between analyzed and calculated tetrahedral compositions are not as small as desired for some specimens. Assuming the validity of these examples, however, the obvious trends emerge that tetrahedral ordering is favored in the $3T$ structure, for phengitic compositions, and for Si:Al^{iv} ratios near 1:1. The $3T$ structures of muscovite, paragonite, lepidolite, and protolithionite all have been stated to be ordered as have phengites in the $1M$, $2M_1$, and $2M_2$ stacking arrangements. Muscovite-$3T$ is slightly phengitic also. Sidorenko et al. (1975) commented that the correlation between phengitic composition and ordering is probably not coincidental. The accuracy of the phengite-$1M$ refinement by Sidorenko et al. (1975) by electron diffraction in subgroup symmetry $C2$ is questionable, as will be documented by statistical analysis later in this section, and a refinement of a different phengite-$1M$ by Tsipurskii and Drits (1977) concluded that the true symmetry of that specimen is $C2/m$ and therefore that the tetrahedral cations must be disordered. In contrast to the order found in dioctahedral phengite-$2M_2$, three determinations of the structure of trioctahedral lepidolite-$2M_2$ detected no tetrahedral ordering. No significant tetrahedral ordering has been found either for any of the abundant trioctahedral $1M$ micas, although the possibility of ordering to subgroup symmetry has seldom been investigated. Because of similarity in their scattering powers, relative to X-rays, ordered patterns of Si and Al in these studies have been detected by the differences noted in the mean T-O bond lengths between nonequivalent tetrahedra. Hazen and Burnham (1973) found a linear regression relation T-O = 1.608 Å + 0.163[$x_{Al}/(x_{Al} + x_{Si})$] between tetrahedral composition and the mean tetrahedral bond length in the micas, and this is the equation used in Table 3 to calculate the tetrahedral compositions. Alternation of the smaller Si and the larger Al around the six-fold tetrahedral rings has been the only ordering pattern thus far found in the true micas (see Chapter 3 for an exception in the brittle mica anandite).

Although muscovite-$2M_1$ has two independent tetrahedra in its ideal symmetry, all well-refined structures show relatively small differences in the mean T-O bond lengths of the two tetrahedra. In discussing the relationship between the disordered $2M_1$ and the ordered $3T$ forms of muscovite Güven (1971b) pointed out that the two tetrahedral sheets within a single mica layer are related to each other by a center of symmetry in the disordered $2M_1$ structure but by a lateral two-fold rotation axis in the ordered $3T$ structure. An apical oxygen attached to one Si^{4+} tetrahedral cation and to two Al^{3+} octahedral cations has its negative charge exactly balanced, but an excess negative charge exists if Al^{3+} substitutes for Si^{4+}. Güven points out that if tetrahedral cation ordering were present in the $2M_1$ structure there would be two apical oxygens with excess negative charges arrayed along a diagonal octahedral edge that is shared between two Al octahedral cations. Such an arrangement maintains the inversion center that lies on the shared edge, but is potentially unstable because the shortening of shared edges inherent in these dioctahedral sheets will lead to repulsion between apical oxygens with excess negative charges. In the ordered $3T$ structure the octahedral edge in question lies between one occupied M2 and one vacant M1 octahedral site so that edge-shortening is not required (see Fig. 1).

It is also possible to describe an ordered $2M_1$ structure in subgroup Cc so that compositionally similar tetrahedra of adjacent sheets are not related by the inversion center but instead by a lateral two-fold rotation axis that does not hold for the structure as a whole. This latter ordered structure has been found in the $2M_1$ brittle mica margarite (Guggenheim and Bailey, 1975, 1978) and in two specimens that are chemically intermediate between margarite and bityite (Sokolova et al., 1979; Lin and Guggenheim, 1983). The Cc subgroup structure allows complete ordering of the four tetrahedral cations into four nonequivalent sites (in contrast to two sites in $C2/c$). Nevertheless, the greater driving force for ordering inherent in these brittle micas due to their greater tetrahedral substitutions [(2Si + 2Al) in margarite and (2Si + 2Al,Be) in bityite] is required to realize the ordering, because the ordering pattern is not adopted by either muscovite-$2M_1$ (Guggenheim and Bailey, 1975) or paragonite-$2M_1$ (Lin and Bailey, 1984) that have tetrahedral compositions of (3Si + 1Al).

The structures listed in Table 3 have been derived by both X-ray diffraction and electron diffraction methods, and represent differing degrees of accuracy. The quality of the data as compared to the structural model can be judged by the final agreement between observed and calculated spectral amplitudes (R values) and by the agreement between the total Al^{iv} contents as determined by the size differences of the tetrahedra and by chemical analysis (Table 3). The significance of the tetrahedral size differences also can be judged statistically by consideration of the determinative errors involved. If σ_ℓ is the error (standard deviation) of an individual bond length, the error of the mean of n values is $\sigma_n = \sigma_\ell/\sqrt{n}$, where $n = 4$ for a tetrahedron. For the difference Δ between the two mean values of the same accuracy $\sigma_\Delta = \sqrt{2}\sigma_n$, and in order for an observed bond length difference Δ to be statistically significant at the 1% level it should be equal to or greater than $2.33\sigma_\Delta$ or at the 0.1% highly significant level should be equal to or greater than $3.09\sigma_\Delta$ according to the criteria of Cruickshank (1949). Because of the difficulty in deriving true values of the determinative errors, many crystallographers would prefer an observed bond length difference to be at least 3.0 standard deviations, rather than 2.33, for significance at the 1% level. Significance at the 1% level as used here means that there is a 1% probability that by chance a bond length A could be observed as greater than bond length B by at least Δ, although really equal to B. In this section bond length differences between $2.3\sigma_\Delta$ and $3.1\sigma_\Delta$ are treated as being in the borderline area of significance at this level. Table 4 lists the results of the application of this statistical approach to the structures of Table 3. The published σ_ℓ or σ_n values have been used in all cases except for phengite-$2M_2$, where the published errors in atomic coordinates have been averaged for each atom in order to calculate the standard deviations of the bond lengths.

Table 4. Statistical analysis of tetrahedral size differences.

Polytype	Species	σ_n	σ_Δ	Δ	Δ/σ_Δ
1M	phengite	0.009 A	0.0127 A	0.019 A	1.5
	zinnwaldite	0.002	0.0028	0.007	2.5
	ephesite	0.010	0.0141	0.155	11.0
2M_1	phengite	0.003	0.0042	0.012	2.9
	margarite (Penn.)	0.0035	0.0049	0.1135	23.2
	margarite-bityite (Zimbabwe)	0.003	0.0042	0.092	21.9
	margarite-bityite (U.S.S.R)	0.010	0.0141	0.081	5.7
	margarite (Austria)	0.002	0.0028	0.121	43.2
2M_2	phengite	0.005	0.0071	0.034	4.8
3T	muscovite	0.0085	0.0120	0.067	5.6
	paragonite	0.020	0.0283	0.075	2.7
	lepidolite	0.008	0.0113	0.035	3.1
	protolithionite	0.005	0.0071	0.032	4.5
2Or	anandite	0.002	0.0028	0.179	63.9

According to the results in Table 4 the differences between the tetrahedral bond lengths are highly significant for anandite-2Or, margarite-2M_1, margarite-bityite-2M_1, ephesite-1M, muscovite-3T, phengite-2M_2, and protolithionite-3T (Δ/σ_Δ = 63.9 to 4.5). The difference in mean bond lengths reported for phengite-1M is not statistically significant at the 1% level (Δ/σ_Δ = 1.5). All of the other structures group together in the border-line range with Δ/σ_Δ values between 2.5 and 3.1. The reasons for the lower significance levels in this latter group are different for different specimens. For paragonite-3T it is due to the larger determinative error σ_n. For phengite-2M_1 and lepidolite-3T, where the errors are smaller, it appears to be due to a combination of a small amount of tetrahedral substitution and incomplete ordering, and for zinnwaldite 1M to a small degree of ordering. Clearly, only a small number of refinements of high accuracy have detected any substantial degree of tetrahedral ordering in micas. It also should be noted at this point that it is likely that the statistical significance levels used here are not equally valid for refinements carried out in ideal versus subgroup symmetry. The latter probably require a considerably greater observed bond length difference than the former for the same significance level.

Ordering of octahedral cations

Octahedral cation ordering, as judged either by observed differences in mean M–O,OH,F bond lengths or by refinement of octahedral occupancies based on differences in scattering powers of the cations, is more common than tetrahedral ordering in micas. Table 5 gives the bond lengths and reported octahedral compositions for the best documented examples of octahedral ordering in the micas. Note in some cases that verification of ordering depends entirely on refinement of the octahedral occupancies because the mean bond lengths of all of the octahedra are similar. Statistical analysis of the bond lengths is not helpful in such cases. It should be noted also that many specimens selected for structural refinement have been end member compositions with only one element present in octahedral coordination. Although ordering between Mg and Fe^{2+} is not common, Table 5 shows that ordering occurs frequently between other octahedral cations of different sizes and charges. It may be concluded that ordering is to be expected if the octahedral compositions are conducive.

In all dioctahedral micas the vacant octahedral site has been found to be located on the mirror plane of each 2:1 layer, i.e., in site M1 that has its OH,F groups on opposite octahedral corners in the *trans* orientation. This can be considered as a form of ordering and is in accord with the pattern of ordering usually found in trioctahedral micas: the *trans* octahedron M1 tends to be larger than the mean of the M2 octahedra, which have their OH,F groups on adjacent corners in the *cis* orientation. An exception to this generalization has been reported to be the structure of the brittle mica clintonite-1*M* (xanthophyllite) for which Takeúchi and Sadanaga (1966) found the smaller Al in M1 and the larger Mg in the two M2 sites. A redetermination of the clintonite-1*M* structure by neutron diffraction (W. Joswig, pers. comm. 1983), however, showed that one Mg is in M1 and that the other Mg plus the Al are disordered over the two M2 sites. Because the two refinements were done on specimens from different localities, however, it is still possible that the first refinement is valid. Some evidence to this effect is given by the observation that the intralayer shift (ideally $-0.333a_1$) is an *under*shift of $-0.331a_1$ in the Takeúchi and Sadanaga clintonite but an *over*shift of $-0.336a_1$ in the Joswig clintonite, as would be expected for M1 sizes smaller and larger,

Table 5. Examples of octahedral cation ordering in micas.

Reference	Species	Space Group	Refl.	Final R	Mean M-O,OH	Oct. Composition
Takéuchi & Sadanaga (1966)	clintonite-1M (Chichibu mine, Japan)	$C2/m$	384	10.4%	$M(1)=2.019\text{Å}$ $M(2)=2.050\times 2$	$Al_{0.72}Mg_{0.18}\square_{0.10}$ $Mg_{1.0}\times 2$
Güven & Burnham (1967)	muscovite-3T (Sultan Basin, Wash.)	$P3_112$	280	2.4	$M(2)=1.973$ $M(3)=1.913$	$Al_{0.83}Mg_{0.08}Fe_{0.09}^*$ $Al_{1.0}$
Takeda & Burnham (1969)	synthetic lepidolite-1M	$C2/m$	328	5.1	$M(1)=2.106$ $M(2)=1.981\times 2$	$Li_{0.89}Al_{0.11}$ $Li_{0.55}Al_{0.45}$
Takeda et al. (1971)	lepidolite-2M$_2$ (Rozna, Czecho.)	$C2/c$	471	7.2	$M(1)=2.144$ $M(2)=1.967\times 2$	$Li_{0.35}Al_{0.10}\square_{0.55}$ $Li_{0.35}Al_{0.65}$
Sartori et al. (1973)	lepidolite-2M$_2$ (Elba, Italy)	$C2/c$	525	9.6	$M(1)=2.123$ $M(2)=1.980\times 2$	$Li_{0.95}Al_{0.05}$ $Li_{0.37}Al_{0.63}$
Sidorenko et al. (1975)	phengite-1M (Transbaikal, U.S.S.R.)	$C2$	588	10.9	$M(2)=1.920$ $M(3)=1.957$	$Al_{1.0}$ $Al_{0.8}Mg_{0.1}Fe_{0.1}^*$
Guggenheim & Bailey (1975, 1978)	margarite-2M$_1$ (Chester, Pa.)	Cc	1071	3.0	$M(2)=1.903$ $M(3)=1.915$	$Al_{1.0}$ $Al_{0.96}Mg_{0.03}Fe_{0.01}^{2+}$
Sartori (1976)	lepidolite-1M (Elba, Italy)	$C2/m$	400	6.7	$M(1)=2.113$ $M(2)=1.972\times 2$	$Li_{0.95}Al_{0.05}$ $Li_{0.37}Al_{0.63}$
Guggenheim & Bailey (1977)	zinnwaldite-1M (Erzgebirge, D.D.R.)	$C2$	1493	5.7	$M(1)=2.132$ $M(2)=1.882$ $M(3)=2.131$	$Fe_{0.42}^{2+}Li_{0.34}Fe_{0.10}^*\square_{0.14}$ $Al_{1.0}$ $Al_{0.05}Fe_{0.36}^{2+}Li_{0.33}Fe_{0.12}^*\square_{0.14}$
Sidorenko et al. (1977b)	paragonite-3T (locality ?)	$P3_112$	208	13.0	$M(1)=2.061$ $M(2)=1.965$ $M(3)=1.981$	$Al_{0.3}\square_{0.7}$ $Al_{0.9}\square_{0.1}$ $Al_{0.8}\square_{0.2}$
Toraya et al. (1977b)	synthetic taeniolite-1M	$C2/m$	1303	2.4	$M(1)=2.058$ $M(2)=2.061\times 2$	$Mg_{0.71}Li_{0.29}$ $Mg_{0.66}Li_{0.34}$

45

Table 5, continued.

Reference	Mineral (Locality)	Space group	Reflections	R (%)	M-O distances	Site occupancies
Brown (1978)	lepidolite-3T (Kalgoorlie, Australia)	$P3_112$	705	4.7	M(1)=2.036 M(2)=2.113 M(3)=1.920	$Li_{0.71}Al_{0.29}$ $Li_{0.96}R_{0.04}$ $Li_{0.18}Al_{0.82}$
Toraya et al. (1978b)	synthetic 1M $K(Mg_{2.5}\square_{0.5})Ge_4O_{10}F_2$ mica	$C2/m$	1413	5.5	M(1)=2.178 M(2)=2.070x2	$Mg_{0.6}\square_{0.40}$ $Mg_{0.95}\square_{0.05}$
Toraya et al. (1978b)	synthetic Ge-taeniolite-1M	$C2/m$	1451	3.8	M(1)=2.092 M(2)=2.092x2	$Mg_{0.64}Li_{0.36}$ $Mg_{0.68}Li_{0.32}$
Sokolova et al. (1979)	ephesite-1M (Ilímaussak massif, Greenland)	$C2$	284	11.5	M(1)=2.128 M(2)=1.927 M(3)=1.927	$Li_{0.67}Na_{0.10}Fe_{0.02}\square_{0.21}$ $Al_{1.0}$ $Al_{1.0}$
Sokolova et al. (1979)	intermediate margarite-bityite-2M_1 (U.S.S.R.)	Cc	450	11.5	M(1)=2.184 M(2)=1.900 M(3)=1.897	$Li_{0.48}Mg_{0.10}Fe_{0.02}Al_{0.03}\square_{0.37}$ $Al_{1.0}$ $Al_{1.0}$
Swanson & Bailey (1981)	lepidolite-2M_1 (Biskupice, Czecho.)	$C2/c$	971	9.1	M(1)=2.107 M(2)=1.977x2	$Li_{0.93}(Fe^*,Mg)_{0.07}$ $Li_{0.35}Al_{0.58}\square_{0.07}$
Guggenheim (1981)	lepidolite-1M (Radkovice, Czecho.)	$C2/m$	1164	3.5	M(1)=2.118 M(2)=1.970x2	$Li_{0.91}Fe_{0.04}Mg_{0.05}$ $Li_{0.28}Al_{0.65}\square_{0.07}Fe^{3+}_{0.005}$
Guggenheim (1981)	lepidolite-1M (Tanakamiyama, Japan)	$C2$	807	6.2	M(1)=2.120 M(2)=1.878 M(3)=2.126	$Li_{0.70}Al_{0.07}Fe_{0.06}\square_{0.18}$ $Al_{1.0}$ $Li_{0.71}Al_{0.07}Fe_{0.06}\square_{0.16}$
Guggenheim (1981)	lepidolite-2M_2 (Radkovice, Czecho.)	$C2/c$	2764	4.8	M(1)=2.121 M(2)=1.966x2	$Li_{1.0}$ $Li_{0.24}Al_{0.65}(Mn,Mg,Fe^{3+})_{0.05}\square_{0.07}$
Pavlishin et al. (1981)	protolithionite-3T (Ukraine, U.S.S.R.)	$P3_112$	702	4.7	M(1)=2.121 M(2)=1.909 M(3)=2.149	$Li_{0.53}Li_{0.26}Mg_{0.01}Fe^{2+}_{0.06}\square_{0.14}$ $Al_{0.83}Fe^{3+}_{0.17}$ $Fe_{0.53}Li_{0.26}Mg_{0.01}Fe^{2+}_{0.06}\square_{0.14}$
Ohta et al. (1982)	oxybiotite-1M (Ruiz Peak, N.M.)	$C2/m$	1125	5.0	M(1)=2.077 M(2)=2.059x2	$Mg_{0.62}Fe^*_{0.19}Al_{0.19}$ $Mg_{0.51}Fe^*_{0.29}Al_{0.03}Ti_{0.17}$

46

Table 5, concluded.

Reference	Mineral (locality)	Space group	Reflections	R (%)	Bond lengths (Å)	Octahedral composition
Ohta et al. (1982)	oxybiotite-2M₁ (Ruiz Peak, N.M.)	$C2/c$	1676	4.5	M(1)=2.076 M(2)=2.060×2	$Mg_{0.61}Al_{0.19}Fe^{*}_{0.20}$ $Mg_{0.51}Al_{0.02}Fe^{*}_{0.30}Ti_{0.17}$
Lin & Guggenheim (1983)	intermediate margarite-bityite-2M₁ (Zimbabwe)	Cc	1927	3.0	M(1)=2.140 M(2)=1.902 M(3)=1.903	$Li_{0.5}\square_{0.5}$ $Al_{1.0}$ $Al_{1.0}$
Joswig et al. (1983)	margarite-2M₁ (Austria)	Cc	1003	1.7	M(1)=2.193 M(2)=1.909 M(3)=1.912	$Li_{0.11}Mg_{0.10}\square_{0.79}$ $Al_{1.0}$ $Al_{1.0}$
Backhaus (1983)	lepidolite-1M (Wakefield, Canada)	$C2$	724	7.3	M(1)=2.098 M(2)=2.058 M(3)=1.914	$Li_{0.87}Al_{0.13}$ $Li_{0.61}Fe^{*}_{0.39}$ $Al_{1.0}$
Filut et al. (1984)	anandite-2Or (Wilagedera, Ceylon)	$Pnmn$	1074	6.4	M(1)=2.095(OH) M(2)=2.233(S) M(3)=2.225(S) M(4)=2.118(OH)	$Mg_{0.59}Fe_{0.41}$ $Fe_{1.0}$ $Fe_{1.0}$ $Fe_{0.89}Mg_{0.11}$
Joswig (1984)	clintonite-1M (Adamello, Italy)	$C2/m$	558	2.0	M(1)=2.046 M(2)=2.021	$Mg_{1.0}$ $Mg_{0.64}Al_{0.27}Fe^{*}_{0.09}$

Indicated octahedral compositions are as given by the authors. $Fe^{*} = Fe^{3+} + Fe^{2+} + Mn + Ti$ except where these elements are listed separately.

respectively, than M2. This leads to observed β angles slightly smaller and larger, respectively, than the ideal values. Levillain et al. (1981) cited Mossbauer data to suggest that an exception to the generalization may exist also in the structure of a synthetic siderophyllite of composition $K(Fe_2^{2+}Al)(Si_2Al_2)O_{10}(OH)_2$.

The relative ratio of large to small octahedra is not always in accord with the ratio of large to small octahedral cations present. For example, in synthetic lepidolite-1M of the polylithionite composition (Takeda and Burnham, 1969) there are two large octahedral Li ions and one smaller Al ion by chemical analysis. Yet the ordering pattern creates only one large octahedral site at M1 on the mirror plane but two smaller symmetry-related M2 sites. The compositions inferred from the refinement of scattering powers in these sites are M1 = $Li_{0.89}Al_{0.11}$ and M2 = $(Li_{0.55}Al_{0.45}) \times 2$. Similar ordering patterns have been observed in both the 1M and 2M_2 forms of natural lepidolites from Elba, Italy, and Radkovice, Czechoslovakia (Table 5).

Hybrid atoms, such as those cited above that are part atom A and part atom B, are statistical devices to indicate disorder of those atoms in a given site when averaged over many unit cells. The atoms obviously cannot exist as hybrids in the structure, and especially near critical compositions such as $(Li_{0.5}Al_{0.5}) \times 2$ one might suspect that ordering within the M2 octahedra has taken place to lower the symmetry or to create a superlattice. Guggenheim and Bailey (1977) investigated this possibility for a zinnwaldite-1M crystal from the Erzgebirge. Despite lack of appreciable tetrahedral ordering they found that octahedral ordering had lowered the resultant symmetry from the ideal $C2/m$ symmetry to that of subgroup $C2$ with all three octahedra having different scattering powers. All of the octahedral Al is concentrated in one of the M2 sites and the remaining Fe, Li and other cations are distributed not quite equally over M1 and the second M2 site. The hybrid F,OH atom has moved off the mirror plane of the 1M structure in order to coordinate more closely with the small Al. Backhaus (1983) also found the three octahedral sites to be different in a lepidolite-1M from Wakefield, Canada. Refinement in subgroup symmetry $C2$ gave mean bond lengths and compositions for the three sites of M1 = 2.098 Å ($Li_{0.87}Al_{0.13}$), M2 = 2.058 Å ($Li_{0.61}Mn_{0.16}Fe_{0.15}^{2+}Fe_{0.03}^{3+}Ti_{0.01}\square_{0.04}$), and M2' or M3 = 1.914 Å ($Al_{1.0}$).

Etch pits on the basal surfaces have a symmetry that is said to contradict the higher symmetry $C2/m$ as a consequence of the octahedral ordering. The ideal space group of lepidolite-$3T$ permits all three octahedra to be of different composition, and the structure by Brown (1978) shows this to be the case for a crystal from Australia. Guggenheim (1981) found different amounts and patterns of octahedral ordering in different $1M$ lepidolite crystals, some in ideal symmetry and some in subgroup symmetry. This emphasizes the dangers inherent in making generalizations or extrapolations based on the structural refinement of a single specimen.

Several micas have been described with total octahedral occupancies halfway between dioctahedral and trioctrahedral. Levinson (1953) has shown in the muscovite-lepidolite series that bulk compositions in this intermediate range actually are intimate mixtures of separate dioctahedral and trioctahedral phases. But in other series the intermediate compositions appear to apply to a single phase, and it is of interest to know the nature of the structural adaptations.

Toraya et al. (1976, 1978a) refined the structures of two synthetic micas having octahedral occupancies close to $Mg_{2.5}\square_{0.5}$. In the silicate mica [$\sim K_{0.88}(Mg_{2.56}\square_{0.44})Si_4O_{10}F_2$] the vacancies were found to be distributed equally over all three octahedral positions, thus structurally simulating a disordered trioctahedral mica. In the germanate mica [$K(Mg_{2.5}\square_{0.5})Ge_4O_{10}F_2$] there is a greater concentration of vacancies in the larger M1 site, structurally simulating a true intermediate between dioctahedral and trioctahedral, and the authors postulated in this case that M1 needs to be expanded laterally by incorporating vacancies in order to fit better with the large Ge-rich tetrahedral sheet. Lin and Guggenheim (1983) reported a brittle mica intermediate between dioctahedral margarite-$2M_1$ and trioctahedral bityite-$2M_1$ that has an octahedral composition of ($Al_{2.044}Li_{0.547}Fe^{3+}_{0.007}\square_{0.402}$). Refinement of the structure in subgroup Cc indicates that two Al cations are concentrated in the M2 and M3 cis octahedra, and that the larger $trans$ M1 octahedron contains primarily Li and vacancies. This distribution is also that of a true intermediate on average, but the authors cite the evidence of split hydrogen protons on electron density difference maps to emphasize that the crystal actually is composed of both dioctahedral (Li-poor) and trioctahedral (Li-rich) unit cells in which the orientations of the

O...H vector are quite different (see Chapter 3). It is possible that cooperative forces would aggregate similar cells into two kinds of small domains that differ in their dioctahedral or trioctahedral nature, but the X-ray evidence is not definitive on the distribution of the unit cells. A similar splitting of hydrogen protons was found by Joswig et al. (1983) in a neutron and X-ray diffraction study of a sodian, lithian margarite that has 21% occupancy of M1 by Li+Mg. Neutron diffraction determined 79% of the unit cells to have an inclined $O-H^+$ orientation typical of dioctahedral micas and 21% to have a nearly vertical orientation typical of triochtahedral micas.

Another unusual octahedral ordering pattern involving an unequal distribution of cations and vacancies is that reported for a natural paragonite-$3T$ specimen in which the two octahedral Al cations are said to be distributed in differing amounts over all three independent octahedral positions. The normally vacant site M1 is reported to have a composition of $Al_{0.3}\square_{0.7}$ according to electron density maps and analysis of the mean bond lengths derived from a high voltage texture electron diffraction study by Sidorenko et al. (1977b), whereas M2 = $Al_{0.9}\square_{0.1}$ and M3 = $Al_{0.8}\square_{0.2}$. The high R value of 13.0% for this refinement makes the distinction between M2 and M3 questionable.

Ordering of interlayer cations

Any ordering of interlayer cations necessarily reduces the symmetry to a subgroup of the parent space group for the six standard mica polytypes of Smith and Yoder (1956). No ordering of these cations has been reported, but the possibility does not appear to have been seriously investigated.

Unmixing of different size or charge interlayer cations is well documented in the muscovite-paragonite-margarite ternary system. An especially interesting unmixing intergrowth also has been observed in the only known occurrence of wonesite. Veblen (1983) used transmission electron microscopy, electron diffraction, and X-ray analytical electron microscopy to show that wonesite having a bulk interlayer composition of $Na_{0.395}K_{0.073}Ca_{0.002}\square_{0.53}$ has exsolved into a very fine lamellar intergrowth of talc and a wonesite of a different interlayer composition of approximately $Na_{0.505}K_{0.093}Ca_{0.002}\square_{0.40}$. The exsolved wonesite also is enriched in Al, Ti, Cr, and Fe relative to the exsolved talc. An asymmetric solvus is depicted as lying between talc (with no interlayer cations) and a hypothetical mica "E" (with no interlayer vacancies) that lies on the join between Na-phlogopite and preiswerkite. Surprisingly, the intergrown lamellae are inclined to (001) by an average angle of 37°.

Local charge balance

The stability of a structure would be enhanced if, in addition to the presence of tetrahedral and octahedral cation ordering, the ordered constituents can be arranged in patterns that provide charge balance between the local sources of excess positive and negative charges created by ordering. Local charge balance between the ordered constituents of tetrahedral and octahedral sheets has been recognized in certain specimens of chlorite and vermiculite. In these cases the local balance occurs as a result of a particular arrangement of an octahedral interlayer relative to the tetrahedral sheets of 2:1 layers above and below. In micas there is the added complication of a positively charged interlayer cation that cannot contribute to local charge balance because its charge necessarily is distributed equally over all of its basal oxygen neighbors. But local charge balance might still be possible within a 2:1 layer in two kinds of micas: (1) those that have a high amount of tetrahedral substitution of R^{3+} for R^{4+}, and for which part of the excess negative charge thus created on the tetrahedra is reduced by a positive charge due to octahedral substitution of R^{3+} for R^{2+} or R^+ for $\square$, or (2) in trioctahedral micas of smaller tetrahedral substitution in which octahedral R^{3+} substitution in one site can be compensated in part by octahedral R^{1+} or vacancies in the other two sites. Logical candidates thus would include Al- or Fe^{3+}-rich biotites, lepidolite, zinnwaldite, masutomilite, wonesite, preiswerkite, and clintonite.

Soboleva and Mineeva (1981) have claimed the existence of a different type of charge balance in the structure of phengite-1M as determined by Sidorenko et al. (1975). Here the substitution of tetrahedral Al^{3+} for Si^{4+} and of octahedral Mg^{2+} for Al^{3+} creates two different sources of negative charge, and the ordering places the local sources of the two negative charges as far apart as possible. This effect might better be termed avoidance of local charge imbalance.

Bailey (1984) has analyzed mica structures in order to determine if the concept of local charge balance has any validity in this mineral group. He concludes that the available data do not prove any strong tendency for either true local charge balance or for the avoidance of local charge imbalance as a result of cation ordering in micas. The structures that violate these principles are nearly equal in number to those that follow them. Such analysis admittedly is handicapped by the small number of micas that exhibit both tetrahedral and octahedral cation ordering and by the small magnitudes of the ordering in some cases (as indicated by small differences in mean T-O or M-O bond lengths). More accurate structural refinements are needed for cases where it is important to be able to determine the reality of ordering involving small amounts of substitution. But even if the small differences are real, Baur (1970) has shown that variations in bond lengths compensate adequately for most observed variations in valence saturations. Thus, the driving force for local charge balance in micas must be minimal.

Standardized cation notation

More credence could be given to the significance of small bond length differences as a measure of ordering if they could be shown to be consistent for a given polytype, say with T1-O > T2-O in all cases of ordering within the $2M_1$ structure, or to be in accord with established crystal chemical factors that might favor localization of a given cation in a specific tetrahedral or octahedral site. A standard notation for the possible sites is a necessary first step in investigating these possibilities, because to date different authors have used T1, T2, M2, M3, etc. as labels for different tetrahedral and octahedral sites and one cannot analyze Tables 3 and 5 in terms of an absolute locus of the ordered substitutions.

Bailey (1984) has recommended a standard notation system for structural sites in micas and has applied his system to the known examples of ordered micas. It is necessary to evaluate each structural type separately, and this cannot be done with confidence because of the few ordered examples available and the small degree of ordering postulated in some cases. For tetrahedral cation ordering the only clear trend that emerges from Bailey's analysis is that tetrahedral Al prefers standard site T2 [located to the left of the symmetry plane of the first layer when viewed with the direction of intralayer stagger pointing away from the observer] in the $3T$ structures of muscovite, paragonite, lepidolite, and protolithionite and in the three ordered phengite structures. The reason for this preference is not known. For octahedral cation ordering, neglecting the usual preference of vacancies and larger cations for the *trans* site M1, there is a slight trend for the *cis* octahedron M(2) [located to the right of the symmetry plane] to be larger than the *cis* octahedron M3 [to the left] in cases where they are not equivalent. Verification of any trends for localization of cations into specific structural sites will require the results of many more structural refinements of high accuracy than are presently available.

It is not surprising that there is little or no preference for either the tetrahedral or octahedral cations to enter specific structural sites when the symmetry is reduced by ordering from that of the parent space group to that of a subgroup, such as in the change from $C2/m$ to $C2$ for the $1M$ structure. In the $1M$ structure tetrahedra T1 and T2 would be equivalent to one another in $C2/m$ symmetry, as would octahedra M2 and M3, and the selection of specific ordering sites in the higher symmetry where ordering must start should be a matter of random choice at first. It is to be expected that cooperative forces would tend to extend initial randomly scattered ordering "seeds" into larger local domains, and that the coalescence of adjacent domains would result in a true single crystal only if there were effective tendencies for ordering into the same specific structural sites in each domain. Favorable crystallization and cooling conditions could aid in this process, as suggested by Soboleva and Mineeva (1981) for phengite-$1M$. It is more likely that adjacent local domains would have different ordering patterns due to the random choice effect, and coalescence of these domains then can result in out-of-step relations or twinning that may or may not be discernible from the usual Bragg diffraction spectra. One possible explanation for the lack of observed long range tetrahedral ordering in the abundant $1M$ phlogopite-biotites and $2M_1$ muscovites is that the ordering for these compositions is entirely in local domains, and that the average of all of the domains is long range tetrahedral disorder as determined by the normal X-ray and electron diffraction techniques (see next section on short range order). This is equivalent to saying that the energies of all the ordering patterns present in the local domains are about the same, so that no one pattern is predominant. For reasons not yet understood the $3T$ structure is most conducive to tetrahedral ordering for a variety of compositions, and the phengite composition likewise for a variety of layer stacking arrangements.

Phengite ordering may be related to the restricted conditions of low temperature and high pressure metamorphism or of hydrothermal solutions under which the specimens studied are believed to have formed. The finding of Rule and Bailey (1984) that phengite from the amphibolite facies is disordered also suggests some environmental control.

If different ordering patterns, or similar patterns in different orientation, are adopted at random in different domains of the same crystal, they may be described as twinned or out-of-step relative to one another upon coalescence of the domains.   Layer silicates with tetrahedral Si:Al ratios near 1:1 have strong ordering tendencies and large resultant domains that are visible under crossed nicols of the petrographic microscope. Complex twinning of this sort that is believed to result from inversion to a lower symmetry is ubiquitous in crystals of margarite, bityite, and ephesite among the micas, and in amesite among the serpentines. Successful crystallographic refinement of these structures requires isolation of a single twin (domain) unit. For smaller domains that are not big enough to give their own X-ray patterns an average structure will result upon structural refinement.

SHORT RANGE ORDERING

In addition to ordering over long distances in the crystal, it has been mentioned that it is possible to have ordering in small domains that extend over only a few unit cells. These domains can be seen especially well by TEM techniques. Depending on the details of the ordering patterns and the distribution of domains, such local or short range ordering may or may not show up during structural refinement as a perturbation of any long range ordering that may be present. It is quite possible that 100% local order would show up as 0% long range order. Very often the domains tend to be antiphase in nature so that the normal diffraction evidence for their presence is cancelled. Local modulations of the average structure due to the domains will show up as non-Bragg scattering, i.e., diffuse diffracted intensity positioned between the normal Bragg reflections. Such non-Bragg scattering is present in diffraction records of many micas, but its interpretation is controversial. If the domains are spaced at regular intervals, however, extra diffraction satellite spots will appear, and some conclusions as to the size, distribution, and orientation of the domains can be drawn from the shapes, positions, and intensities of the satellites.

The lack of long range order of tetrahedral Si,Al in muscovite-$2M_1$ has already been mentioned. Gatineau (1964) interpreted the diffuse non-Bragg scattering in muscovite-$2M_1$ as due to short range ordering of Al in zigzag chains within the tetrahedral sheet. In a given small domain the direction of chain alignment may be along any one of the three pseudohexagonal X axes of the crystal. Chains of pure Al are said to alternate along the Y direction in partly ordered fashion with chains of pure Si. This interpretation must be viewed with some caution because of the later finding of Kodama et al. (1971) that in the local domains the tetrahedral network is distorted in linear waves characterized by alternating rows of tetrahedra of slightly differing dimensions even in the absence of any tetrahedral substitution, e.g., as in pyrophyllite and talc. Further study appears to be required to establish the contributions to the diffuse scattering that may be given by local ordering as well as by such phenomena as the oxygen network distortion, thermal motion, and the presence of dislocation arrays. Similar comments apply to the structures of phlogopite and biotite, for which Gatineau and Méring (1966) observed diffuse scattering very similar to that in muscovite.

Gatineau and Méring (1966) also have studied the diffuse X-ray scattering of a lepidolite of unstated structural type. The composition is not given, but by implication is that of polylithionite. Diffuse spots are observed at positions that are satellitic ($\pm a^*/3$) to the positions of certain of the $h + k$ = odd Bragg reflections that are forbidden by C centering. The spots are interpreted to indicate short range octahedral ordering of (1Al + 2Li) in local domains. The ordering takes the form of rows of pure Al and pure Li aligned along one or the other of the three pseudohexagonal Y axes [01], [31], or [3$\bar{1}$], such that each domain is characterized by only one ordering direction. The ordering is two-dimensional in that the Al rows are spaced regularly within a layer at intervals of $3a/2$ (every third row), but only irregularly between layers. The extent of each domain is small.

Emphasis in this chapter has been placed on the diffraction method of study of order and disorder. This method requires good quality crystals and considerable expenditure of time for refinement of the structural parameters. Spectroscopic methods, including Mössbauer, infrared, nuclear magnetic resonance, and Raman, do not have these disadvantages to the same degree and can be effective in determining the distribution of cations and vacancies over the available sites (see Chapter 5). These methods are sensitive to the local environments of the atoms and therefore can provide information on both long range and short range order or disorder. For example, NMR study of phlogopites by Sanz and Stone (1979) has shown that Fe^{2+} tends to be distributed randomly over M1 and M2 on a long range basis. But as the F-content increases local domains can be recognized in which Fe^{2+} concentrates in association with OH in one type of domain and Mg concentrates in association with F in another type of domain. In each domain type the cations occupy both M1 and M2 sites. Infrared patterns of micas can be interpreted to show the association of one, two, or three nearest-neighbor Fe^{2+} cations around a given OH group, as well as the association of OH with more highly charged cations. These local distribution often average out to long range disorder as seen by diffraction study.

Study of domain structures and of short-range order by a combination of investigative techniques is an area of growing research interest that has great future potential.

ACKNOWLEDGMENTS

This work has been supported in part by National Science Foundation grant EAR-8106124 and in part by grant 13157-AC2-C from the Petroleum Research Fund, as administered by the American Chemical Society. S. Guggenheim reviewed an early version of the manuscript, and M. Rieder supplied a copy of his paper with Weiss et al. prior to publication.

REFERENCES

Alexandrova, V.A., Drits, V.A. and Sokolova, G.V. (1972) Structural features of dioctahedral one-packet chlorite. Soviet Phys. Crystallogr. 17, 456-461 (English transl.).

Appelo, C.A.J. (1978) Layer deformation and crystal energy of micas and related minerals. I. Structural models for $1M$ and $2M_1$ polytypes. Am. Mineral. 63, 782-792.

_____ (1979) Layer deformation and crystal energy of micas and related minerals. II. Deformation of the coordination units. Am. Mineral. 64, 424-431.

Backhaus, K.O. (1983) Structure refinement of an $1M$-lepidolite. Crystal Research & Tech. 18, 1253-1260.

Bailey, S.W. (1975) Cation ordering and pseudosymmetry in later silicates. Am. Mineral. 60, 175-187.

_____ (1980) Structures of layer silicates. Ch. 1 in: Crystal Structures of Clay Minerals and their X-ray Identification, G.W. Brindley and G. Brown, eds. Mono. 5 Mineralogical Soc., London, 1-124.

_____ (1984) Review of cation ordering in micas. Clays & Clay Minerals 32, 81-92.

Bassett, W.A. (1960) Role of hydroxyl orientation in mica alteration. Bull. Geol. Soc. Am. 71, 449-456.

Baur, W.H. (1970) Bond length variation and distorted coordination polyhedra in inorganic crystals. Trans. Am. Crystallogr. Assoc. 6, 129-155.

Birle, J.D. and Tettenhorst, R. (1968) Refined muscovite structure. Mineral. Mag. 36, 881-886.

Bohlen, S.R., Peacor, D.R. and Essene, E.J. (1980) Crystal chemistry of a metamorphic biotite and its significance in water barometry. Am. Mineral. 65, 55-62.

Bookin, A.S., and Drits, V.A. (1982) Factors affecting orientation of OH-vectors in micas. Clays & Clay Minerals 30, 415-421.

_____, _____, Rozhdestvenskaya, I.V., Semenova, T.F. and Tsipursky, S.I. (1982) Comparison of orientations of OH-bonds in layer silicates by diffraction methods and electrostatic calculations. Clays & Clay Minerals 30, 409-414.

Bradley, W.F. (1959) Current progress in silicate structures. Clays & Clay Minerals 6, 18-25.

Brown, B.E. (1978) The crystal structure of a $3T$ lepidolite. Am. Mineral. 63, 332-336.

Brown, G. (1955) The effect of isomorphous substitutions on the intensities of (00ℓ) reflections of mica- and chlorite-type structures. Mineral. Mag. 30, 657-665.

Burnham, C.W. and Radoslovich, E.W. (1964) Crystal structures of coexisting muscovite and paragonite. Carnegie Inst. Wash., Yearbook 63, 232-236.

Burns, A.F. and White, J.L. (1963) The effect of potassium removal on the b-dimension of muscovite and dioctahedral soil micas. Proc. Int'l. Clay Conf. 1963, Stockholm I, 9-16.

Cruickshank, D.W.J. (1949) The accuracy of electron-density maps in X-ray analysis with special reference to dibenzyl. Acta Crystallogr. 2, 65-82.

Donnay, G., Donnay, J.D.H. and Takeda, H. (1964a) Trioctahedral one-layer micas. II. Prediction of the structure from composition and cell dimensions. Acta Crystallogr. 17, 1374-1381.

_____, Morimoto, N., Takeda, H. and Donnay, J.D.H. (1964b) Trioctahedral one-layer micas. I. Crystal structure of a synthetic iron mica. Acta Crystallogr. 17, 1369-1373.

Drits, V.A. (1969) Some general remarks on the structure of trioctahedral micas. Proc. Int'l. Clay Conf. 1969, Tokyo I, 51-59.

Filut, M.A., Rule, A.L. and Bailey, S.W. (1984) Refinement of the anandite-$2Or$ structure. (in preparation).

Gatineau, L. (1963) Localisation des remplacements isomorphiques dans la muscovite. Comptes rendus hebd. Séanc. Acad. Sci., Paris 256, 4648-4649.

_____ (1964) Structure réele de la muscovite. Répartition des substitutions isomorphes. Bull. Soc. franc. Minéral. Cristallogr. 87, 321-355.

_____ and Méring, J. (1966) Relations ordre-désordre dans les substitutions isomorphiques des micas. Bull. Groupe franc. Argiles 18, 67-74.

Giese, R.F., Jr. (1971) Hydroxyl orientation in muscovite as indicated by electrostatic energy calculations. Science 172, 263-264.

_____ (1979) Hydroxyl orientations in 2:1 phyllosilicates. Clays & Clay Minerals 27, 213-223.

Guggenheim, S. (1981) Cation ordering in lepidolite. Am. Mineral. 66, 1221-1232.

_____ and Bailey, S.W. (1975) Refinement of the margarite structure in subgroup symmetry: Am. Mineral. 60, 1023-1029.

_____ and _____ (1977) The refinement of zinnwaldite-$1M$ in subgroup symmetry: Am. Mineral. 62, 1158-1167.

_____ and _____ (1978) Refinement of the margarite structure in subgroup symmetry: correction, further refinement, and comments. Am. Mineral. 63, 186-187.

_____ and Kato, T. (1984) Kinoshitalite and Mn phlogopites: Trial refinements in subgroup symmetry and further refinement in ideal symmetry. Mineral. J. (Japan) 12, 1-5.

Güven, N. (1971a) The crystal structure of $2M_1$ phengite and $2M_1$ muscovite: Z. Kristallogr. 134, 196-212.

_____ (1971b) Structural factors controlling stacking sequences in dioctahedral micas: Clays & Clay Minerals 19, 159-165.

_____ and Burnham, C.W. (1967) The crystal structure of $3T$ muscovite. Z. Kristallogr. 125, 1-6.

Hazen, R.M. and Burnham, C.W. (1973) The crystal structure of one-layer phlogopite and annite. Am. Mineral. 58, 889-900.

_____, Finger, L.W. and Velde, D. (1981) Crystal structure of a silica- and alkali-rich trioctahedral mica. Am. Mineral. 66, 586-591.

Hendricks, S.B. and Jefferson, M.E. (1939) Polymorphism of the micas. Am. Mineral. 24, 729-771.

Hewitt, D.A. and Wones, D.R. (1972) The effect of cation substitutions on the physical properties of trioctahedral micas. Am. Mineral. 57, 103-129.

_____ and _____ (1975) Physical properties of some synthetic Fe-Mg-Al trioctahedral biotites. Am. Mineral. 60, 854-862.

Joswig, W. (1972) Neutronenbeugungsmessungen an einem $1M$-Phlogopit. Neues Jahrb. Mineral. Monatsh. 1-11.

_____, Takéuchi, Y. and Fuess, H. (1983) Neutron-diffraction study on the orientation of hydroxyl groups in margarite. Z. Kristallogr. 165, 295-303.

Kato, T., Miúra, Y., Yoshii, M. and Maeda, K. (1979) The crystal structures of $1M$-kinoshitalite, a new barium brittle mica and $1M$-manganese trioctahedral micas. Mineral. J. (Japan) 9, 392-408.

Knurr, R.A., Bailey, S.W. and Rossman, G.R. (1984) Refinement of Mn-substituted muscovite and phlogopite. (in preparation).

Kodama, H., Alcover, J.F., Gatineau, L. and Méring, J. (1971) Diffusions anormales (Rayons X et electrons) dans les phyllosilicates 2-1. Structure et Propriétes de Surface des Minéraux Argileux Symp., Louvain, 15-17.

Lee, J.H. and Guggenheim, S. (1981) Single crystal X-ray refinement of pyrophyllite-$1Tc$. Am. Mineral. 66, 350-357.

Levillain, C., Maurel, P. and Menil, F. (1981) Mössbauer studies of synthetic and natural micas on the polylithionite-siderophyllite join. Phys. Chem. Minerals 7, 71-76.

Levinson, A.A. (1953) Studies in the mica group: Relationship between polymorphism and composition in the muscovite-lepidolite series. Am. Mineral. 38, 88-107.

Lin, C.-yi and Bailey, S.W. (1984) The crystal structure of paragonite-$2M_1$. Am. Mineral. 69, 122-127.

Lin, J.-C. and Guggenheim, S. (1983) The crystal structure of a Li,Be-rich brittle mica: A dioctahedral-trioctahedral intermediate. Am. Mineral. 68, 130-142.

Mathieson, A.McL. and Walker, G.F. (1954) Crystal structure of magnesium-vermiculite. Am. Mineral. 39, 231-255.

McCauley, J.W. and Newnham, R.E. (1971) Origin and prediction of ditrigonal distortions in micas. Am. Mineral. 56, 1626-1638.

_____, _____ and Gibbs, G.V. (1973) Crystal structure analysis of synthetic fluorophlogopite. Am. Mineral. 58, 249-254.

Newnham, R.E. and Brindley, G.W. (1956) The crystal structure of dickite. Acta Crystallogr. 9, 759-764.

Norrish, K. (1973) Factors in the weathering of mica to vermiculite. Proc. Int'l. Clay Conf. 1972, Madrid, 417-432.

Ohta, T., Takeda. H. and Takéuchi, Y. (1982) Mica polytypism: Similarities in the crystal structures of coexisting $1M$ and $2M_1$ oxybiotite. Am. Mineral. 67, 298-310.

Peterson, R.C., Hill, R.J. and Gibbs, G.V. (1979) A molecular-orbital study of distortions in the layer structures brucite, gibbsite and serpentine. Canadian Mineral. 17, 703-711.

Profi, S., Sideris, C. and Filippakis, S.E. (1973) X-ray diffraction study of some biotites from volcanic rocks of W. Thrace, N. E. Greece. N. Jahrb. Mineral. Monatsh. 8-17.

Radoslovich, E.W. (1959) Structural control of polymorphism in micas. Nature 183, 253.

_____ (1960) The structure of muscovite, $KAl_2(Si_3Al)O_{10}(OH)_2$. Acta Crystallogr. 13, 919-932.

_____ (1961) Surface symmetry and cell dimensions of layer lattice silicates. Nature 191, 67-68.

_____ and Norrish, K. (1962) The cell dimensions and symmetry of layer-lattice silicates. I. Some structural considerations. Am. Mineral. 47, 599-616.

Rayner, J.H. (1974) The crystal structure of phlogopite by neutron diffraction. Mineral. Mag. 39, 850–856.

Richardson, S.M. and Richardson, J.W., Jr. (1982) Crystal structure of a pink muscovite from Archer's Post, Kenya: Implications for reverse pleochroism in dioctahedral micas. Am. Mineral. 67, 69–75.

Robert, J.-L. (1976) Titanium solubility in synthetic phlogopite solid solutions. Chem. Geol. 17, 213–227.

Rothbauer, R. (1971) Untersuchung eines $2M_1$-Muskovits mit Neutronenstrahlen. N. Jahrb. Mineral. Monatsh. 143–154.

Rozhdestvenskaya, I.V. and Frank-Kamenetskii, V.A. (1974) The structure of the dioctahedral mica chernykhite. Crystal Chemistry and Structure of Minerals, "Nauka", Leningrad, p. 18–23 (in Russian).

Rule, A.L. and Bailey, S.W. (1984) Crystal structure of a Cr-bearing phengite-$2M_1$. (in preparation).

Sanz, J. and Stone, W.E.E. (1979) NMR study of micas, II. Distribution of Fe^{2+}, F^-, and OH^- in the octahedral sheet of phlogopites. Am. Mineral. 64, 119–126.

Sartori, F. (1976) The crystal structure of a $1M$ lepidolite. Tschermaks Mineral. Petrogr. Mitt. 23, 65–75.

_____, Franzini, M. and Merlino, S. (1973) Crystal structure of a $2M_2$ lepidolite. Acta Crystallogr. B29, 573–578.

Semenova, T.F., Rozhdestvenskaya, I.V. and Frank-Kamenetskii, V.A. (1977) Refinement of the crystal structure of tetraferriphlogopite. Soviet Phys. Crystallogr. 22, 680–683 (English transl.).

_____, _____ and Pavlishin, V.I. (1983) Crystal structure of tetraferriphlogopite and tetraferribiotite. Mineral. Zhurnal 5, 41–49 (in Russian).

Shannon, R.D. (1976) Revised effective ionic radii and systematic studies of interatomic distances in halides and chalcogenides. Acta Crystallogr. 232, 751–767.

Sidorenko, O.V., Zvyagin, B.B. and Soboleva, S.V. (1975) Crystal structure refinement for $1M$ dioctahedral mica. Soviet Phys. Crystallogr. 20, 332–335 (English transl.).

_____, _____ and _____ (1977a) Refinement of the crystal structure of $2M_1$ paragonite by the method of high-voltage electron diffraction. Soviet Phys. Crystallogr. 22, 554–556 (English transl.).

_____, _____ and _____ (1977b) The crystal structure of $3T$ paragonite. Soviet Phys. Crystallogr. 22, 557–560 (English transl.).

Smith, J.V. and Yoder, H.S., Jr. (1956) Experimental and theoretical studies of the mica polymorphs. Mineral. Mag. 31, 209–235.

Soboleva, S.V. and Mineeva, R.M. (1981) Stabilité de différents polytypes de micas dioctaédriques en fonction des potentiels partiels sur les atomes. Bull. Minéral. 104, 223–228.

_____, Sidorenko, O.V. and Zvyagin, B.B. (1977) Crystal structure of paragonite $1M$. Soviet Phys. Crystallogr. 22, 291–293 (English transl.).

Sokolova, C.V., Aleksandrova, V.A. Drits, V.A. and Vairakov, V.V. (1979) Crystal structure of two lithian brittle micas. In Crystal Chemistry and Structures of Minerals, "Nauka", Leningrad, 55–66 (in Russian).

Steinfink, H. (1962) Crystal structure of a trioctahedral mica: phlogopite. Am. Mineral. 47, 886–896.

Swanson, T.H. and Bailey, S.W. (1981) Redetermination of the lepidolite-$2M_1$ structure. Clays & Clay Minerals 29, 81–90.

Takeda, H. and Burnham, C.W. (1969) Fluor-polylithionite: A lithium mica with nearly hexagonal $(Si_2O_5)^{2-}$ ring. Mineral. J. (Japan) 6, 102–109.

_____ and Donnay, J.D.H. (1966) Trioctahedral one-layer micas. III. Crystal structure of a synthetic lithium fluormica. Acta Crystallogr. 20, 638–646.

_____, Haga, N. and Sadanaga, R. (1971) Structural investigation of polymorphic transition between $2M_2$-, $1M$-lepidolite and $2M_1$ muscovite. Mineral. J. (Japan) 6, 203–215.

_____ and Morosin, B. (1975) Comparison of observed and predicted structural parameters of mica at high temperature. Acta Crystallogr. B31, 2444–2452.

_____ and Ross, M. (1975) Mica polytypism: Dissimilarities in the crystal structures of coexisting $1M$ and $2M_1$ biotite. Am. Mineral. 60, 1030–1040.

Takéuchi, Y. (1965) Structures of brittle micas. Clays & Clay Minerals 13, 1–25.

_____ and Sadanaga, R. (1966) Structural studies of brittle micas. (I) The structure of xanthophyllite refined. Mineral. J. (Japan) 4, 424–437.

Tateyama, H., Shimoda, S. and Sudo, T. (1974) The crystal structure of synthetic Mg^{IV} mica. Z. Kristallogr. 139, 196–206.

Tepikin, E.V., Drits, V.A. and Alexandrova, V.A. (1969) Crystal structure of iron biotite and construction of structural models for trioctahedral micas. Proc. Int'l. Clay Conf. 1969, Tokyo I. 43-49.

Toraya, H. (1981) Distortions of octahedra and octahedral sheets in $1M$ micas and the relation to their stability. Z. Kristallogr. 157, 173-190.

____, Iwai, S., Marumo, F., Daimon, M. and Kondo, R. (1976) The crystal structure of tetrasilicic potassium fluor mica. $KMg_{2.5}Si_4O_{10}F_2$. Z. Kristallogr. 144, 42-52.

____, ____, ____ and Hirao, M. (1977) The crystal structure of taeniolite, $KLiMg_2Si_4O_{10}F_2$. Z. Kristallogr. 146, 73-83.

____, ____, ____ and ____ (1978a) The crystal structure of germanate micas, $KMg_{2.5}Ge_4O_{10}F_2$ and $KLiMg_2Ge_4O_{10}F_2$. Z. Kristallogr. 148, 65-81.

____, ____, ____ and ____ (1978b) The crystal structure of a germanate mica, $KMg_3Ge_3AlO_{10}F_2$. Mineral. J. (Japan) 9, 221-230.

____, ____, ____, Nishikawa, T. and Hirao, M. (1978c) The crystal structure of synthetic mica, $KMg_{2.75}Si_{3.5}Al_{0.5}O_{10}F_2$. Mineral. J. (Japan) 9, 210-220.

____ and Marumo, F. (1981) Structure variation with octahedral cation substitution in the system of germanate micas, $KMg_{3-x}Mn_xGe_3AlO_{10}F_2$. Mineral. J. (Japan) 10, 396-407.

____ and ____ (1983) The crystal structure of a germanate mica $KMg_{2.5+x}Ge_{4-2x}Al_{2x}O_{10}F_2$ ($x \approx 0.14$) and distortion of $(Ge,Al)O_4$ tetrahedra. Mineral. J. (Japan) 11, 222-231.

____, ____ and Hirao, M. (1983) Synthesis and the crystal structure of a manganoan fluoromica, $K(Mg_{2.44}Mn_{0.24})(Si_{3.82}Mn_{0.18})O_{10}F_2$. Mineral. J. (Japan) 11, 240-247.

Tsipurskii, S.I. and Drits, V.A. (1977) Effectivity of the electronic method of intensity measurement in structural investigation by electron diffraction. Izvestiya Akad. Nauk SSSR, Ser. phys. 41, 2263-2271 (in Russian).

Veblen, D.R. (1983) Exsolution and crystal chemistry of the sodium mica wonesite. Am. Mineral. 68, 554-565.

Yoshii, M., Togashi, Y. and Maeda, K. (1973) On the intensity of basal reflections with relation to barium content in manganoan phlogopites and kinoshitalite. Bull. Geol. Surv. Japan 24, 543-550.

Zhoukhlistov, A.P., Zvyagin, B.B., Soboleva, S.V. and Fedotov, A.F. (1973) The crystal structure of the dioctahedral mica $2M_1$ determined by high voltage electron diffraction. Clays & Clay Minerals 21, 465-470.

____, ____, Lazarenko, E.K. and Pavlishin, V.I. (1977) Refinement of the crystal structure of ferrous celadonite. Soviet Phys. Cyrstallogr. 22, 284-288 (English transl.).

Weiss, Z., Rieder, M., Chmielová, M. and Krajiček, J. (1984) Geometry of the octahedral coordination in micas: A review of refined structures. (Submitted to Am. Mineral.).

Zvyagin, B.B. (1957) Determination of the structure of celadonite by electron diffraction. Soviet Phys. Crystallogr. 2, 388-394 (English transl.).

____ (1967) Electron Diffraction Analysis of Clay Mineral Structures. Plenum Press, New York (English transl.).

____ (1979) High Voltage Electronography in the Study of Layered Minerals. Science Publ. House, Moscow (in Russian).

____ and Mishchenko, K.S. (1962) Electronographic data on the structure of phlogopite-biotite. Soviet Phys. Crystallogr. 7, 502-505.

____ and Soboleva, S.V. (1979) Variabilité des longueurs de liaisons Si, Al_T-O dans les silicates lamellaires. Bull. Minéral. 102, 415-419.

____, ____, Vrublevskaya, Z.V., Zhoukhlistov, A.P. and Fedotov, A.F. (1972) Factors in the ditrigonal rotation of the tetrahedra in the structures of layer silicates. Soviet Phys. Crystallogr. 17, 466-469 (English transl.).

3. The BRITTLE MICAS
Stephen Guggenheim

INTRODUCTION

The brittle micas are distinguished from the true micas by a layer charge per formula unit of approximately -2.0. In consequence, their interlayer cation is Ca or Ba (Deer et al., 1963, p. 7). Brittle mica structures containing other divalent cations in the interlayer site are unknown, presumably because this site requires a large cation to prop adjacent 2:1 layers apart and because other such large cations (e.g., Sr) are rare in nature. Calcium appears to be the smallest interlayer cation acceptable, since micas with Ca at this site have adjacent oxygen layer surfaces nearly in contact.

The general structural aspects of the tetrahedral and octahedral sheets of true micas have been discussed in Chapters 1 and 2; they are applicable to the brittle micas as well. Special attention will be given in this chapter to the relatively small differences in the structures of the two mica subgroups.

NOMENCLATURE AND GENERAL CHEMICAL CLASSIFICATION

Table 1 gives details regarding brittle mica species, ideal compositions, polytypic forms and respective ideal space groups, and discredited variety names. In much of the older literature, *clintonite* was divided into two varieties (clintonite and xanthophyllite) based on the orientation of the optic

Table 1. Natural brittle micas

Name	Polytype	Ideal Space Group	Ideal Composition	Discredited or Previously Used Names
clintonite	$1\underline{M}$	$C2/m$	$Ca(Mg_2Al)(SiAl_3)O_{10}(OH)_2$	Xanthophyllite (yellow), valuevite or waluewite (green)
	$2\underline{M}_1$	$C2/c$		Brandisite or disterrite (green), seybertite (reddish), holmite or holmesite, chrysophane
	$3\underline{T}$	$P3_1 12$		None
margarite	$2\underline{M}_1$	$C2/c$	$CaAl_2(Si_2Al_2)O_{10}(OH)_2$	Lesleyite, diphanite corundellite, clingmanite, soda-margarite (=ephesite), emerylite
bityite	$2\underline{M}_1$	$C2/c$	$CaLiAl_2(Si_2AlBe)O_{10}(OH)_2$	Bowleyite
anandite	$2Or$	$Ccmm$	$BaFe_3(Si_3Fe^{3+})O_{10}(OH)S$	None
	$2\underline{M}_1$	$C2/c$		
	$1\underline{M}$	$C2/m$		
kinoshitalite	$1\underline{M}$	$C2/m$	$BaMg_3(Si_2Al_2)O_{10}(OH)_2$	None
	$2\underline{M}_1$	$C2/c$		

axes (Deer et al., 1962). However, the symmetry of a crystal, which determines partly its optical properties, is dependent on the polytypic modification. In addition, since there are no significant chemical differences between the two varieties, the name clintonite has been adopted for both (Forman et al., 1967a). Earlier variety names, such as valuevite, seybertite, etc., were based primarily on color, which is now thought to be caused by trace impurities. Spectroscopic studies (Manning, 1969) of a brown clintonite from Amity, New York, indicate the color of this sample is a result of 0.3 wt % Ti through $Ti^{3+}-Ti^{4+}$ electronic interactions by means of direct d-orbital overlap in the (001) plane. However, other similar studies (Forman et al., 1967b) could not relate the green and yellow colors of clintonite-1M to obvious chemical variations.

Natural clintonites do not vary greatly in composition: Tetrahedral Si:Al ranges from 1.11:2.89 to 1.36:2.64. The octahedral sites contain predominantly Mg, although Fe^{2+} and Fe^{3+} may substitute for up to 0.33 cations per formula unit (three octahedral sites). Substitution of Na and K in the interlayer site occurs in trace amounts only. Substantial amounts (up to 1.91 wt %) of F have been reported by Forman et al. (1967a). The narrow range in composition is probably caused by the restrictive bulk composition typical of the thermally metamorphosed Ca- and Al-rich rocks in which clintonites are found (see below for petrologic information). Olesch (1975) reported the synthesis of clintonite-1M with a greater tetrahedral Si:Al range (0.6:3.4 to 1.4:2.6) than natural samples. He reported also that maximum substitution toward dioctahedral compositions (margarite) was 24 mole %, somewhat higher than the maximum observed in natural samples of approximately 15 mole % as reported by Harada et al. (1965). Most natural samples have little of the dioctahedral component.

Margarite, the dioctahedral brittle mica, varies considerably in composition within specified limits: the Ca interlayer cation may be replaced in part by Na and, to compensate for the charge difference, Si replaces tetrahedral Al, thereby approaching in composition the dioctahedral Na-mica, paragonite. In a study of margarite occurrences in the Alps, Frey et al. (1977) studied the system synthetically and confirmed by infrared analysis the existence of a temperature-dependent solvus between 30 and 45 (at 600°C), or 20 and 55 (at 400°C) mole % margarite (the range of $P(H_2O)$ is 1-6 kbar). A similar solvus was observed in natural samples between 15 and 50 mole % by Ackermand and Morteani (1973), who suggested also a maximum value of 10 mole % muscovite entering the margarite structure.

An alternate method of charge compensation for Na substitution in margarite involves a coupled substitution of Li for a vacancy. Schaller et al. (1967) have shown that the "soda-margarite" as described in the older literature contains significant amounts of Li and has designated this material as ephesite, which ideally is $NaLiAl_2(Al_2Si_2)O_{10}(OH)_2$. Natural samples are limited in number and little is known about the extent of solid solution between ephesite and margarite.

Compositional variations in margarite do not always affect the interlayer site. A coupled substitution of Li for vacancy and Be for tetrahedral Al maintains charge balance also. The resulting trioctahedral end member, *bityite* (ideally $CaLiAl_2(Si_2AlBe)O_{10}(OH)_2$), was described originally by Lacroix (1908), and has been identified by Gallagher and Hawkes (1966) with various intermediate compositions along the bityite-margarite join. Wet chemical analyses (Gallagher and Hawkes, 1966) for Li and Be on crystals of bulk composition midway between the two end members cannot distinguish if a miscibility gap occurs. However, probe analyses by Lin and Guggenheim (1983) on a crystal near the center of the join suggest that Al content does not vary within this single crystal. Infrared studies by Farmer and Velde (1973) on crystals with tetrahedral Al:Be:Si ratios of 1:10:0.59:2.31 and 0.80:1.10:2.10 show no evidence of two phases, although Be and Li substitution in this series has only subtle effects on the infrared spectra and thereby reduces the possibility of finding a two-phase mixture. Although bityite and Be margarite localities are few, the apparent scarcity of these minerals is probably due to the difficulty in identifying them.

Deer et al. (1962) suggested that O^{2-} may be replaced partially by (OH) to charge balance Na substitution for Ca in hydrated sodian margarites. A similar mechanism has been suggested by Beus (1956) to charge compensate for the substitution of tetrahedral Be.

Margarite occurs naturally in the $2M_1$ form, but synthetic samples have been reported also as $1M_d$ and $1M$ (Velde, 1971; 1980). Velde (1971) suggested that the sequence of $1M_d \to 1M \to 2M_1$ occurs with increasing time (and with increasing crystallinity) in analogy with the muscovite sequence (Yoder and Eugster, 1955). However, additional work is required to prove the existence of both the $1M$ and $1M_d$ forms.

Anandite and *kinoshitalite* are Fe and Mg analogues, respectively, of the trioctahedral Ba brittle micas. Anandite is known from Wilagedera, Ceylon, where it was found both as the $2M_1$ and the rare $2Or$ polytypes in a drill core of a magnetite ore zone (Pattiaratchi et al., 1967; Giuseppetti and Tadini, 1972). Kato (pers. comm., 1984) has reported a $1M$ polytype from this locality also. Anandite has been tentatively identified at a second occurrence at Rush Creek and Big Creek, Fresno County, California (Appleman, pers. comm.). Lovering and Widdowson (1968) found substantial replacement of S (and lesser amounts of Cl) for (OH) and a deficiency of silica in the Ceylon material. Presumably, Fe^{3+} substitutes for Si, although an accurate Fe^{2+}/Fe^{3+} ratio has not been determined. Kinoshitalite is defined as having Ba:K > 1 and may be considered also the Ba analogue of phlogopite (Yoshii et al., 1973a,b). Kinoshitalite has been reported in the $1M$ and $2M_1$ forms from the Nodatamagawa Mine, Iwate Prefecture, Japan (Yoshii et al., 1973a), and in the $1M$ form at Hokkejino, Kyoto Prefecture, Japan (Matsubara et al., 1976). Kinoshitalite has not been found as a compositional end member; Mn^{2+} and Mn^{3+} may substitute octahedrally by more than 10 wt %, as well as lesser amounts of Al. Potassium readily enters the interlayer site and appears to form a complete solid solution series with manganoan phlogopite. For samples in which the ratio of Mn to total octahedral occupancy is nearly constant, there is a sharp break in optical and physical properties near the ratio of Ba:K = 1:2, thus delineating a manganoan phlogopite-kinoshitalite boundary (Yoshii and Maeda, 1975).

Mansker et al. (1979) have reported barian-titanian biotites (some more appropriately classified as ferroan-titanian kinoshitalites) from Hawaiian nephelinites. Ratios of Si to ^{IV}(Al,Ti) are near 2:2 with approximately 0.36 Ti substitution per 4 tetrahedral sites and 0.50 Ti substitution per 3 octahedral sites. The incorporation of Ti involves a complex series of substitutions of the type $Ti^{4+} + \square = 2(Mg,Fe)$ and $2Ti^{4+} + 2Al = 3(Mg,Fe) + 2Si$. The Ba content in these samples varies from 0.47 to 0.70 cations per site. No structural information is available.

Synthetic brittle micas with compositions not found in nature have been produced in quantity. In a study to determine the feasibility to produce brittle micas commercially for their dielectric properties, Hatch et al. (1957) successfully synthesized trioctahedral Ba fluormicas with Mg or (Mg,Li) octahedral cations, and with tetrahedral cations of (Si,Al) or combinations of Si and B or Be. In addition, Frondel and Ito (1968) reported the synthesis of a hydroxyl Ba brittle mica with Mg as the octahedral cation.

CRYSTAL CHEMISTRY OF THE BRITTLE MICAS

Introduction

In the previous chapter, Bailey has discussed a two-dimensional geometric model for the micas that involves tetrahedral and octahedral sheet configurations. Such a model, refined to include variations or distortions as discussed by Bailey, is appropriate for the brittle micas also and will be useful to predict the effect of changing environmental parameters on the brittle mica structure (see below). For purposes of continuity within this chapter, certain structural features described in the preceding chapter are discussed briefly here, although the emphasis is on how the brittle micas differ from the true micas.

Charge considerations. Because of the interlayer occupancy of a divalent cation and the requirement for overall charge neutrality, the resulting 2:1 layer charge of -2 per formula unit may originate in one of two ways. The 2:1 layer charge may be dependent on the tetrahedral sheet alone or may involve contributions from both tetrahedral and octahedral sheets. For cases in which the octahedral sheet contains cations of R^{3+} in a dioctahedral arrangement or cations of R^{2+} in a trioctahedral arrangement, the overall -2 layer charge is produced by a tetrahedral composition of Si_2Al_2, as in margarite or kinoshitalite. Variations in charges arising from substitutions in the octahedral sheet are sufficient to give a diversity of tetrahedral sheet compositions, ranging from $SiAl_3$ for an octahedral charge excess of +1 to Si_3Al for an octahedral charge deficiency of -1, as in clintonite and synthetic Ba,Li-fluormica ($BaLiMg_2AlSi_3O_{10}F_2$), respectively. Trivalent Fe, as found in anandite, satisfies the tetrahedral charge requirements in a way analogous to the Ba,Li fluormica. In contrast to clintonite, the addition of Li^+ in the octahedral site of bityite is compensated for by the coupled substitution of Be^{2+} for tetrahedral Al to produce a tetrahedral sheet composition of Si_2BeAl.

Sheet size considerations. In addition to the correlation of the distribution of charge between the tetrahedral and octahedral sheets, these sheets must also articulate into a layer with common planes of junction. Therefore, the lateral dimensions of both sheets must be equal. If not, one or both sheets must deform to allow congruence. Following the approach in the previous chapter, an approximation of the lateral dimensions of an unconstrained (i.e., free) octahedral sheet may be made by comparing the b cell parameters of various hydroxide minerals, which are chemically and structurally similar to dioctahedral and trioctahedral mica sheets (Table 2). Approximate lateral dimensions of an undeformed and unconstrained tetrahedral sheet may be calculated from:

$$b\ (Si_{1-x}Al_x) \cong 9.15\ A + 0.74x$$

Thus, it is apparent from Table 2 that most true micas, with tetrahedral sheet compositions of Si_3AlO_{10}, have tetrahedral sheet lateral dimensions nearly equal to the lateral width of a Mg-rich trioctahedral sheet. In contrast, because of charge considerations, tetrahedral sheets common to the brittle micas are typically very aluminous, and are much larger than their component octahedral sheet. Such differences are reconciled, in part, by tetrahedral

Table 2. Approximate lateral dimensions [b, in Å]
of unconstrained octahedral and tetrahedral sheets

Composition	Mineral	b(Å)	Mica example
Octahedral Sheet			
$Al(OH)_3$	Gibbsite, bayerite	8.64, 8.67	Margarite (dioctahedral)
$Mg(OH)_2$	Brucite	9.36	Kinoshitalite (trioctahedral)
			Phlogopite (trioctahedral true mica)
Tetrahedral Sheet			
Si_4O_{10}		9.15	Polylithionite (trioctahedral true mica)
$(Si_3Al)O_{10}$		9.335	Synthetic Ba,Li fluormica and most true micas
$(Si_2Al_2)O_{10}$		9.52	Margarite
$(SiAl_3)O_{10}$		9.705	Clintonite

rotation, the in-plane rotation of adjacent tetrahedra in opposite directions to reduce the lateral dimensions of the tetrahedral sheet (Fig. 1). The tetrahedral rotation angle α, as measured directly from structural refinements, is given for each of the brittle micas in Table 3. Comparison to those values given in Table 1 of Chapter 2 clearly shows the effect of Al (or Fe^{3+}) substitution in the tetrahedral sheet as compared to octahedral sheet size. In most cases, the true micas have tetrahedral rotations of less than half that of brittle micas. One notable exception in the true micas is given by paragonite ($\alpha \cong 17-19°$), which has an unusually large α because the interlayer cation (Na) is small. Other notable exceptions include the true micas ephesite, preiswerkite, and siderophyllite (Table 1, Chapter 1) because of high Al:Si ratios, although structure refinements for the latter two micas are not available. Large tetrahedral rotations probably have a considerable effect on the mechanism of thermal decomposition, which is explored in more detail below.

Cation ordering

The symmetries of mica structures are influenced by the geometry of stacking of successive layers (see Chapter 1), sheet and layer distortions caused by the misfit between tetrahedral and octahedral sheets, distortions from other sources such as cation-cation repulsions, and cation order. Appreciable ordering may occur with cations that commonly substitute in both tetrahedral and octahedral sites. (As with the true micas, interlayer cation ordering has not been observed.) In some cases the ordering pattern has no effect on the ideal symmetry, whereas in others the symmetry of the space group is reduced. However, in the process of a crystal structure refinement it is often necessary to assume the space group symmetry. Therefore, if an ordering pattern that

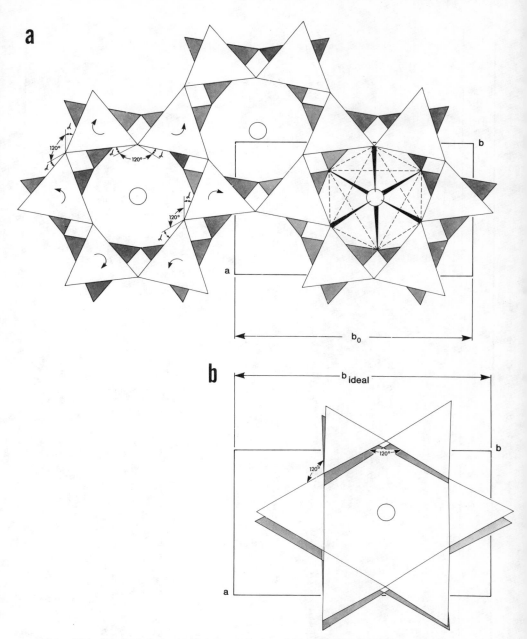

Figure 1(a). Portions of the margarite structure projected onto the (001) plane to illustrate the interlayer region and the effect of tetrahedral rotation. The apical oxygens are not shown. The ditrigonal ring on the left shows the direction the tetrahedra rotate and how the tetrahedral rotation angle ($\alpha = 21°$) may be measured directly from a pair of tetrahedra. The resulting coordination around the interlayer cation is an antiprism, or octahedron, as illustrated on the right.

(b) An "idealized" interlayer cavity without the effect of tetrahedral rotation. The coordination is 12-fold and the lateral dimensions, as shown by b_{ideal}, are considerably larger. A lateral expansion of similar magnitude occurs with the a axis. The 120° angles shown here are only approximate, because the Si tetrahedra (not labelled) are smaller than the Al tetrahedra, thereby preventing the formation of undistorted hexagons of tetrahedra around the interlayer cation.

Table 3. Structural details of the brittle micas

Reference	Species	Composition	# Refl.	Final R_1 %	α_{tet}
1. Takéuchi and Sadanaga (1966)	clintonite-1M	$Ca_{1.1}(Mg_{2.18}Al_{0.72})(Si_{1.05}Al_{2.95})O_{10}(OH)_2$	384	10.4	23.0
2. Joswig and Takéuchi (in prep.)	clintonite-1M	$Ca_{1.01}Na_{0.01}(Mg_{2.30}Al_{0.62}Fe_{0.17}Ti_{0.01})(Si_{1.22}Al_{2.78})O_{10}(OH)_2$	558	2.0	23.1
3. Akhundov, Mamedov, and Belov (1961)	clintonite-2M_1	$Ca(Mg_{1.7}Al_{1.1}Ca_{0.2})(Si_{1.2}Al_{2.8})O_{10}(OH)_2$	?	~19.5	8.6
4. Takéuchi (1965)	margarite-2M_1	$(Ca_{0.87}Na_{0.13})Al_2(Si_2Al_2)O_{10}(OH)_2$	632	16.5	20.4
5. Guggenheim and Bailey (1975, 1978)	margarite-2M_1	$(Ca_{0.81}Na_{0.19}K_{0.01})(Al_{1.99}Fe_{0.01}Mg_{0.03})(Si_{2.11}Al_{1.89})O_{10}(OH)_2$	1,071	4.0	20.9 / 20.6
6. Joswig, Takéuchi, and Fuess (1983)	margarite-2M_1	$(Ca_{0.73}Na_{0.21})(Al_{1.96}Fe_{0.03}Mg_{0.10}Li_{0.12})O_{10}(OH)_2 \cdot 2.12(Si_{1.92}Al_{2.08})$	1,003	1.7	20.9 / 20.9
7. Lin and Guggenheim (1983)	margarite/bityite-2M_1 intermediate	$(Ca_{0.95}Na_{0.02})(Li_{0.55}\square Al_{2.04}Fe_{0.01}^{3+})(Si_{2.02}Al_{1.34}Be_{0.64})O_{10}(OH)_2$	1,917	3.0	21.6 / 21.7
8. Sokolova, Aleksandrova, Drits, and Vairakov (1979)	margarite/bityite-2M_1 intermediate	$(Ca_{0.97}Na_{0.02}K_{0.02})(Li_{0.48}\square_{0.37}Al_{2.03}Mg_{0.10}Fe_{0.02})(Si_{2.00}Al_{1.29}Be_{0.71})O_{10}(OH_{1.97}O_{0.03})$	450	11.5	21.4 / 20.4
9. Giuseppetti and Tadini (1972)	anandite-2Or	$(Ba_{0.87}Mn_{0.04}K_{0.05}Na_{0.04})(Fe_{2.46}^{2+}Mg_{0.48}Mn_{0.06})(Si_{2.64}Fe_{0.7}^{3+}Fe_{0.58}^{2+}Al_{0.08})O_{10}(OH)(S_{0.85}Cl_{0.15})$	853	6.1	2.1 / 3.0
10. Filut, Rule and Bailey (in prep.)	anandite-2Or	$(Ba_{0.96}K_{0.03}Na_{0.01})(Fe_{2.02}^{2+}Fe_{0.31}^{3+}Mg_{0.45}Mn_{0.04}^{2+}Mn_{0.04}^{3+})(Si_{2.62}Fe_{1.38}^{3+})O_{10}S_{0.84}Cl_{0.16}F_{0.04}(OH)_{0.96}$	1,074	6.4	0.9 / 0.9
11. Kato, Miūra, Yoshii, and Maeda (1979) and Guggenheim and Kato (1984)	kinoshitalite-1M	$(Ba_{0.58}K_{0.35}Na_{0.11})(Mg_{2.07}Mn_{0.52}Mn_{0.21}Al_{0.22}Fe_{0.05}^{3+})(Si_{2.05}Al_{1.94}Ti_{0.01})O_{10}(OH)_2$	745	7.2	12.4
12. McCauley and Newnham (1973)	synthetic Ba-Li fluoromica-1M	$Ba_{0.97}(Mg_{2.23}Li_{0.77})(Si_{2.84}Al_{1.16})O_9.9F_{2.08}$	479	7.1	4.7
13. Kato, Miūra, Yoshii, and Maeda (1979) and Guggenheim and Kato (1984)	Ba-phlogopite-1M	$(Ba_{0.35}K_{0.58}Na_{0.09})(Mg_{2.10}Mn_{0.52}^{2+}Mn_{0.22}^{3+}Al_{0.35}Fe_{0.04})(Si_{2.33}Al_{1.65}Ti_{0.01})O_{10}(OH_{1.18}F_{0.07})$	627	3.8	11.0

Table 3, continued.

	ψ_{oct}	Sheet Thickness Tet.	Sheet Thickness Oct.	Interlayer Separation	Basal Oxygen Δz	Interlayer Mean Bond Distances Inner	Interlayer Mean Bond Distances Outer	Intralayer Shift	Layer Offset	Resultant Shift
1.	57.9(M1) 58.4(M2)	2.342	2.148	2.818	0.06	2.414	3.529(Ca)	$-0.321a_1$	$-0.010a_1$	$-0.331a_1$
2.	59.1(M1) 58.6(M2)	2.322	2.103	2.911	0.01	2.443	3.553(Ca)	$-0.336a_1$	$+0.000a_1$	$-0.336a_1$
3.	60.6(Ave)	2.391	1.958	2.674	0.23	2.730	3.137(Ca)	$+0.330a_{3,2}$	$-0.006a_{1,1}$	$-0.318a_1$
4.	58.9(M2)	2.330	2.074	2.832	0.19	2.458	3.427(Ca)	$+0.372a_{2,3}$	$+0.006a_{1,1}$	$-0.360a_1$
5.	56.9(M2) 57.1(M3)	2.270 2.240	2.080	2.876	0.19 0.18	2.454	3.432(Ca)	$+0.366a_{2,3}$	$+0.006a_{1,1}$	$-0.357a_1$
6.	57.0(M2) 57.1(M3)	2.288 2.293	2.079	2.874	0.19 0.20	2.454	3.438(Ca)	$+0.369a_{2,3}$	$+0.006a_{1,1}$	$-0.358a_1$
7.	61.4(M1) 57.4(M2) 57.4(M3)	2.275 2.277	2.051	2.910	0.16 0.16	2.432	3.437(Ca)	$+0.358a_{2,3}$	$+0.002a_{1,1}$	$-0.355a_1$
8.	62.0(M1) 57.3(M2) 57.2(M3)	2.282 2.236	2.054	2.935	0.13 0.30	2.464	3.442(Ca)	$+0.364a_{2,3}$	$+0.005a_{1,1}$	$-0.357a_1$
9.	56.2(M1) 59.0(M2) 58.7(M3) 56.2(M4)	2.301	2.338	3.041	0.19	3.026	3.241(Ba)	$\pm 0.331a_1$	0	0
10.	56.3(M1) 58.7(M2) 58.5(M3) 56.7(M4)	2.262	2.323	3.078	0.31	3.054	3.228(Ba)	$\pm 0.333a_1$	0	0
11.	58.6(M1) 58.4(M2)	2.294	2.184	3.328	0.02	2.866	3.438(Ba,K)	$-0.334a_1$	$-0.001a_1$	$-0.333a_1$
12.	58.7(M1) 58.5(M2)	2.277	2.155	3.172	0.01	2.975	3.192(Ba)	$-0.333a_1$	$-0.001a_1$	$-0.334a_1$
13.	58.8(M1) 58.7(M2)	2.273	2.161	3.380	0.002	2.910	3.413(K,Ba)	$-0.332a_1$	$-0.001a_1$	$-0.331a_1$

Table 4. Octahedral ordering for selected brittle micas.

Mica/Reference	Space Group	Site	$\overline{M\text{--}O}$ distance	[1]Occupancy
Clintonite-1$\underline{M}$ Takéuchi and Sadanaga (1966)	C2/$\underline{m}$	M(1) M(2)	2.019 2.050x2	$Al_{0.72}Mg_{0.18}\square_{0.10}$ $Mg_{1.0}$x2
Clintonite-1$\underline{M}$ Joswig and Takéuchi (in preparation)	C2/$\underline{m}$	M(1) M(2)	2.046 2.020	$Mg_{0.76}Al_{0.24}$ $Mg_{0.73}Al_{0.18}Fe_{0.08}Ti_{0.005}$
Margarite-2$\underline{M}_1$ Guggenheim and Bailey (1975, 1978)	Cc	M(1) M(2) M(3)	2.19 1.903 1.915	$\square_{0.96}Al_{0.04}$ $Al_{1.0}$ $Al_{0.96}Mg_{0.03}Fe^{2+}_{0.01}$
Margarite-2$\underline{M}_1$ Joswig, Takéuchi, and Fuess (1983)	Cc	M(1) M(2) M(3)	2.20 1.909 1.912	$\square_{0.79}Mg_{0.10}Li_{0.10}$ $Al_{0.99}Fe^{2+}_{0.01}$ $Al_{0.98}Fe^{2+}_{0.02}$
Margarite/bityite-2$\underline{M}_1$ intermediate Lin and Guggenheim (1983)	Cc	M(1) M(2) M(3)	2.14 1.902 1.903	$\sim Li_{0.55}\square_{0.45}$ $Al_{1.0}$ $Al_{1.0}$
Anandite-2$\underline{Or}$ Filut, Rule, and Bailey (in preparation)	Pnmn	M(1) M(2) M(3) M(4)	2.095 2.233 2.225 2.118	[2]$Mg_{0.59}Fe_{0.41}$ $Fe_{1.0}$ $Fe_{1.0}$ $Fe_{0.89}Mg_{0.11}$
Kinoshitalite-1$\underline{M}$ Kato, Miūra, Yoshii, and Maeda (1979) and Guggenheim and Kato (1984)	C2/$\underline{m}$	M(1) M(2)	2.095 2.087x2	[3]$Mg_{0.52}Mn^{2+}_{0.28}Al_{0.20}$ [3]$Mg_{0.75}Mn^{2+}_{0.11}Mg^{3+}_{0.10}Al_{0.015}Fe^{3+}_{0.02}$
Synthetic Ba,Li fluormica-1$\underline{M}$ McCauley and Newnham (1973)	C2/$\underline{m}$	M(1) M(2)	2.073 2.062x2	[4]$Li_{0.26}Mg_{0.74}$ $\sim Li_{0.26}Mg_{0.74}$
Ba-phlogopite-1$\underline{M}$ Kato, Miūra, Yoshii, and Maeda (1979) and Guggenheim and Kato (1984)	C2/$\underline{m}$	M(1) M(2)	2.086 2.078x2	[3]$\sim Mg_{0.86}Mn^{2+}_{0.09}Fe^{3+}_{0.04}$ $\sim Mg_{0.54}Mn^{2+}_{0.19}Al_{0.17}Mn^{3+}_{0.10}$

1. Site occupancies calculated from bond lengths (Shannon, 1976), scattering factors, and compositional constraints, when possible.
2. Assuming no vacancies.
3. An unambiguous site assignment cannot be made because of the large number of possible substitutions.
4. Calculated bond lengths are inconsistent with the determined chemistry.

causes a symmetry reduction is not anticipated, an incorrect space group may be used in the refinement procedure, leading to an ordering pattern that does not best describe the actual structure. So that the correct space group may be chosen, corroborating evidence (such as infrared spectra, etch figures, acentricity tests, etc.) may be useful to limit possible ordering patterns or symmetries. Bailey (1975; 1980; 1984) has summarized the results of structural refinements to date for both ideal and subgroup symmetries.

Octahedral ordering. Table 4 gives the bond lengths and octahedral occupancies determined for several brittle micas. Because of associated experimental errors with either probe or wet chemical data, bond length calculations, and site refinement procedures that are based on differences in scattering powers, small differences between site occupancies should not be judged

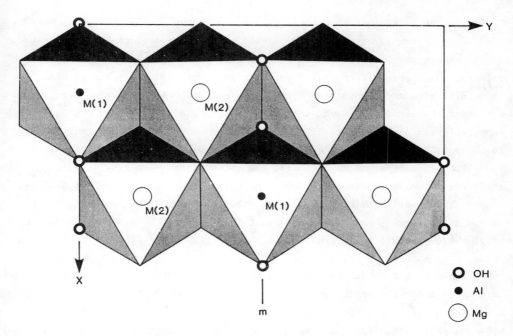

Figure 2. The projection of the octahedral sheet of clintonite-1*M* on the (001) plane as determined from the data of Takéuchi and Sadanaga (1966). Its unique ordering pattern of a smaller, high charge cation in M(1) and a larger, low charge cation in M(2) is illustrated. See the text for comments on the reliability of this refinement.

significant. The octahedral ordering arrangement for most of the true and brittle micas shows a general trend of a larger average size and lesser average charged cation in M(1) relative to M(2). For example, in dioctahedral micas (e.g., margarite), the vacancy is always located in the M(1) site, which is the *trans* octahedron (the OH,F groups are located at opposing octahedral corners). A vacancy, by analogy to a cation, has no charge and is considered large (approximately 0.8 Å in "radius"). In trioctahedral micas (e.g., kinoshitalite; Ba-phlogopite; Ba,Li-fluormica), the M(1) site is usually occupied by a hybrid (where several cations occupy a site when averaged over many units cells) cation, whose average charge is smaller than that found in M(2) and whose average size is larger.

As noted in Chapter 2, there are several true mica exceptions in which M(1) is the same size as either one or both of the other octahedra (e.g., zinn-waldite, lepidolite-3*T*, and a lepidolite-1*M* structure). However, octahedral ordering of a clintonite-1*M* from Japan appears to be a special case among micas (Takéuchi and Sadanaga, 1966). Mean octahedral bond lengths (see Table 4) indicate a smaller M(1) ($Al_{0.72}Mg_{0.18}$) and two larger and equivalent M(2) sites ($Mg_{1.0}$). Such an ordering scheme (Fig. 2) produces an interlayer shift (see

Table 3) that is smaller than the ideal of $a/3$, although the β angle is nearly ideal because the layer offset effectively cancels the deviation caused by the ordering pattern (Bailey, 1975). A clintonite-$1M$ (Joswig and Takéuchi, in preparation) specimen from Italy has the more normal ordering pattern of a large M(1) site and two smaller M(2) sites (see Table 4). These results suggest that the earlier Takéuchi and Sadanaga work should be viewed with caution, especially since the octahedral sizes differ by only 0.031 Å and small errors in the positional parameters could affect the ordering pattern significantly.

Although unusual in composition, the intermediate margarite/bityite-$2M_1$ structure is an important one because a coupled substitution involves a single compositional variable (Li ⇄ ☐) in the octahedral sheet and another (Al ⇄ Be) in the tetrahedral sheet. The Li substitution affects the M(1) site occupancy only, thereby providing information on how the topology of the octahedral sheet varies with changes in M(1). This Li ⇄ ☐ substitution contrasts sharply with other types of cation substitutions found in both natural and synthetic silicate micas intermediate in composition between dioctahedral and trioctahedral end members. In these cases, the vacancy or substituting cations are distributed over M(1) and M(2) to simulate structurally a disordered trioctahedral mica. Although the intermediate margarite/bityite-$2M_1$ mica has three independent octahedral sites instead of the more common configuration of an M(1) site and two equivalent M(2) sites, this configuration results from the reduction in symmetry due to appreciable tetrahedral ordering, which is discussed in detail below. M(2) and M(3), which are symmetry independent, are nearly identical in size and composition.

Temperature factors are a measure of the time-averaged displacement of electron density from the mean centric position. These parameters may reflect either thermal vibrations or positional disorder. Because unit cells are either Li-rich (and trioctahedral) or Li-poor (and dioctahedral) for the margarite/bityite intermediate, the shapes of the "thermal" ellipsoids of anions in the octahedral sheet are a reasonable indication of the positional differences of these anions between the two types of unit cells present in the crystal (see Fig. 3). Note in Figure 3 that the oxygen ellipsoids around M(1), the cation site affected by the substitution, may be described approximately in plan as radiating away from or toward the M(1) cation. (The OH group deviates from this pattern.) It is interpreted that these ellipsoids represent an average position for oxygen in Li-rich and Li-poor unit cells as could be expected with the size of M(1) varying in occupancy.

Also, the change in shape of M(1) has an effect on the M(2) site; a twisting occurs around M(2) [or M(3)] that is consistent with a rotation, in opposite

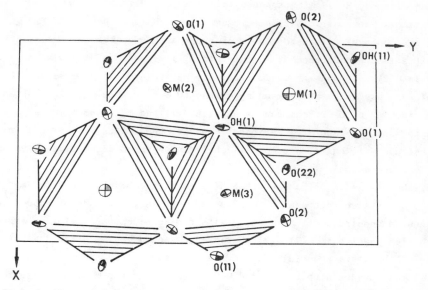

Figure 3. The octahedral sheet of the margarite/bityite-$2M_1$ intermediate projected down c^*. The lateral arrangement of anion ellipsoids is determined largely from a contribution to the effect of thermal vibrations by positional disorder, suggesting that the M(1) site isotropically expands (or contracts) as cations of differing size occupy the site. The representation of the M(1) cation has no physical significance (from Lin and Guggenheim, 1983).

directions, of the upper and lower anion triads (described in Chapter 2 as an octahedral counter-rotation). The amount of this octahedral rotation has been determined to be a linear function of the difference in size between two neighboring octahedra (Lin and Guggenheim, 1983). Not only does a change in shape of M(1) affect the neighboring octahedral sites, but Lin and Guggenheim (1983) have shown also that the field strength of *neighboring* octahedra may affect the octahedral flattening angle (ψ). The effect is discussed in Chapter 2 in more detail.

For the discussion below, two aspects regarding cation substitution are emphasized. First, for a simple substitution in M(1), one would expect M(1) to increase or decrease in size by anion movement in a direction that is radial to the cation. This configuration not only is observed in the average structure of the margarite/bityite intermediate, but also is implied by the observation that the M(1) site of clintonite (Takéuchi and Sadanaga, 1966), which contains a small cation, is regular and nearly undistorted. Second, it is important to recognize that substitutions in one site distort the neighboring sites because of the sheet-like arrangement of the coordinating polyhedra.

Toraya (1981) has discussed why the normal octahedral ordering scheme of a large cation of low charge in M(1) may be favored over other arrangements, such as that found in the clintonite from Japan. For normal octahedral ordering with the M(1) site expanded relative to M(2), shared oxygen to oxygen distances between two M(2) sites must lengthen (see Chapter 2, Fig. 1 for a similar topology). The shortening of the shared edge between two adjacent M(2) sites is a consequence of an enlarged M(1) site and of the high charge cation in M(2); high charge cations tend to maintain as large a separation as possible and try to have intervening anions to act as shields. In contrast, the lengthening of O-OH,F distances between M(1) and M(2) should be a poor shield for the M(1) and M(2) cations. However, the effect of charge is minimized because M(1) has either a small charge associated with it or is vacant (i.e., has no charge).

If the M(1) site is smaller than M(2) as in the clintonite from Japan, shared O-O distances between two M(2) sites must lengthen and shared O-OH,F distances between adjacent M(1) and M(2) sites must decrease. Toraya argues that the lengthening of M(2)-M(2) shared edges would cause charge instability because of the increased M(2)-M(2) cation repulsions. In addition, from his model, Toraya suggests that unshared O-O distances around M(1) decrease to unacceptably short values to get anion-anion repulsions.

Unfortunately, Toraya fails to take the redistribution of charge into account when assessing the model in which the M(1) site is smaller than M(2). The M(2) sites contain the lower charge cations in this model and therefore, the lengthening of M(2)-M(2) shared edges should not produce excessive cation-cation repulsions. In fact, two lower charge cations interacting across a lengthened shared edge as found in this model would be more stable than a low charge cation in M(1) and a high charge cation in M(2) interacting across a lengthened shared edge as found in the reverse model where M(1) is larger than M(2). In addition, the model used to change the size of M(1) and to distort neighboring octahedra is not consistent with the structural adjustments observed in either the margarite/bityite intermediate or the clintonite-$1M$ from Japan. Instead, the M(1) site is compressed anisotropically so that neither the (radial) movement of anions around M(1) is comparable, nor is the octahedral rotation similar. Although Toraya suggested his model for $1M$ polytypes, the adjustments around M(1) observed in the intermediate margarite/bityite-$2M_1$ structure are possible for $1M$ polytypes (and observed in the clintonite-$1M$). Based on these arguments, there is no reason to believe that anion-anion repulsions would increase substantially when unshared lateral edges of M(1) increase in length.

A third octahedral cation ordering pattern, much different than those represented by margarite [large, low-charge cation in M(1) and small, high-charge cation in M(2) or M(3)] or clintonite-$1M$, is found in anandite-$2Or$.

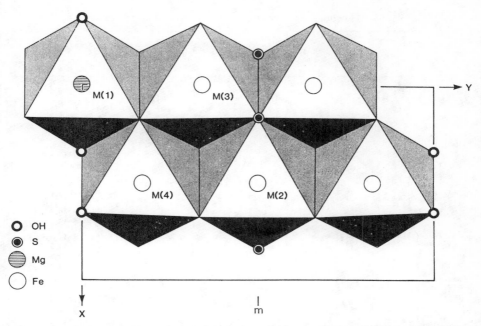

Figure 4. The anandite-$2Or$ octahedral sheet projected down the c axis. Note that the C-centering is violated by the positions of the S and OH (see also Fig. 6). Most of the Mg is contained in M(1) with lesser amounts in M(4).

The ideal space group of a two layer orthorhombic polytype is $Ccmm$, but tetrahedral cation ordering (see below) and the partially ordered substitution of ($S_{0.85}Cl_{0.15}$) for OH has reduced the symmetry to $Pnmn$ (Giuseppetti and Tadini, 1972), thereby producing four symmetry independent octahedral sites (see Fig. 4). Both M(1) and M(2) are in a *trans* configuration, with OH⁻ associated with the former and either S^{2-} or SH⁻ with the latter. A recently completed refinement (Filut, Rule, and Bailey, in preparation) has confirmed the general aspects of the structure as presented by Giuseppetti and Tadini (1972). The coordination polyhedra around M(2) and M(3), which involve S, are distorted (note ψ values in Table 3) and individual cation-anion bond lengths within these polyhedra vary by as much as 0.375 Å. Most of the Mg and vacancies are located in the M(1) site with lesser amounts in M(4), whereas the other octahedral sites, which are associated with S, contain iron (see Table 4). Because the Fe^{3+}:Fe^{2+} ratio for the crystal has not been ascertained to date, octahedral Fe occupancy remains uncertain.

Tetrahedral ordering. Since Si^{4+} and Al^{3+} scatter X-rays with similar efficiencies, the occupancy of tetrahedral sites is not usually determined from refinement procedures based on differences in scattering but by comparing mean T-O bond lengths. Following the procedure given in Chapter 2, the

Table 5. Tetrahedral ordering for selected brittle micas.

Mica/Reference	Space Group	Site	$\frac{1}{T-O}$ distance	Occupancy
Clintonite-1$\underline{M}$ Takéuchi and Sadanaga (1966)	C2/$\underline{m}$	T(1)	1.725x2	$Si_{0.28}Al_{0.72}$
Clintonite-1$\underline{M}$ Joswig and Takéuchi (in preparation)	C2/$\underline{m}$	T(1)	1.727x2	$Si_{0.27}Al_{0.73}$
Margarite-2$\underline{M_1}$ Guggenheim and Bailey (1975, 1978)	Cc	T(1) T(11) T(2) T(22)	1.747 1.623 1.633 1.736	$Si_{0.15}Al_{0.85}$ $Si_{0.91}Al_{0.09}$ $Si_{0.85}Al_{0.15}$ $Si_{0.21}Al_{0.79}$
Margarite-2$\underline{M_1}$ Joswig, Takéuchi, and Fuess (1983)	Cc	T(1) T(11) T(2) T(22)	1.748 1.626 1.628 1.748	$Si_{0.14}Al_{0.86}$ $Si_{0.89}Al_{0.11}$ $Si_{0.88}Al_{0.12}$ $Si_{0.14}Al_{0.86}$
Margarite/bityite-2$\underline{M_1}$ intermediate Lin and Guggenheim (1983)	Cc	T(1) T(11) T(2) T(22)	1.723 1.632 1.628 1.721	$^2Si_{0.05}Al_{0.71}Be_{0.24}$ $Si_{0.81}Al_{0.15}Be_{0.04}$ $Si_{0.88}Al_{0.12}$ $Al_{0.70}Be_{0.30}$
Anandite-2$\underline{Or}$ Filut, Rule, and Bailey (in preparation)	Pnmn	T(1) T(2)	1.618 1.797	$Si_{0.94}Al_{0.06}$ $^3Fe^{3+}_{0.7}Si_{0.3}$
Kinoshitalite-1$\underline{M}$ Kato, Miūra, Yoshii, and Maeda (1979) and Guggenheim and Kato (1984)	C2/$\underline{m}$	T(1)	1.684x2	$Si_{0.53}Al_{0.47}$
Synthetic Ba,Li fluormica-1$\underline{M}$ McCauley and Newnham (1973)	C2/$\underline{m}$	T(1)	1.645x2	$Si_{0.78}Al_{0.23}$
Ba-phlogopite-1$\underline{M}$ Kato, Miūra, Yoshii, and Maeda (1979) and Guggenheim and Kato (1984)	C2/$\underline{m}$	T(1)	1.672x2	$Si_{0.61}Al_{0.39}$

1. The tetrahedral cation occupancy is calculated from the regression equation of Hazen and Burnham (1973), which is not sufficiently accurate to precisely define true occupancy. For sites with low amounts of Si present, the equation overestimates the amount of silicon present by as much as 0.10. For tetrahedral compositions near pure Si, Al is overestimated. See the text for additional comments.

2. The presence of Be makes occupancy calculations impossible without simplifying assumptions. See original source for details.

3. Occupancy determination based on scattering factor refinement.

calculation of tetrahedral compositions for the brittle micas (Table 5) is based on the linear regression equation of Hazen and Burnham (1973). However, Baur (1981) recently showed through a survey of many silicate structures that the Si-O and Al-O bond lengths are closer to 1.623 Å and 1.752 Å respectively, rather than 1.608 Å and 1.771 Å as derived from the Hazen-Burnham equation. Although in need of updating, the Hazen-Burnham equation is of value to calculate the intermediate tetrahedral compositions. However, for tetrahedral compositions near pure Si, the Hazen-Burnham equation will tend to overestimate the amount of Al actually present and, for compositions near pure Al, it will tend to underestimate Al occupancy.

Two trioctahedral 1M brittle micas with Si:Al ∿ 3:1 have been studied in the ideal symmetry of $C2/m$ and all reasonable subgroups to detect cation order. Both the synthetic Ba,Li fluormica-1M (McCauley and Newnham, 1973) and the Ba phlogopite-1M (Kato et al., 1979; Guggenheim and Kato, 1984) structures have been shown to be tetrahedrally disordered, as is also the case for the more abundant varieties of 1M true micas. Clearly, for the mica group taken as a whole, the mechanism(s) which favors tetrahedral order does not have sufficient driving force at these Si to Al ratios.

Clintonite-1M has an unusually large amount of tetrahedral Al (tetrahedral composition of 1 Si + 3 Al). Such high Al compositions require a violation of the principle of avoidance of tetrahedral Al-Al neighbors given by Loewenstein (1954). Clintonite-1M has been refined by Takéuchi and Sadanaga (1959, 1966) and Joswig and Takéuchi (in preparation) in the ideal space group of $C2/m$. The $C2/m$ space group has only one symmetry independent tetrahedron and therefore, the detection of tetrahedral cation order from these results is not possible. However, infrared absorption evidence (Farmer and Velde, 1973) and optical second harmonic generation (SHG) acentricity tests (Bish et al., 1979; Guggenheim et al., 1983) suggest tetrahedral disorder, which would imply that the higher order space group of $C2/m$ is correct. In addition, the low R value (R = 0.020) obtained by Joswig and Takéuchi in $C2/m$ symmetry adds considerable support for the use of the ideal space group in the refinement. Although Akhundov et al. (1961) have studied a clintonite-$2M_1$ structure, they could not accurately determine atomic positions due to stacking disorder ($0kl$ reflections with $k \neq 3n$ were elongated parallel to z^*).

Margarite-$2M_1$ (Guggenheim and Bailey, 1975; 1978) and bityite-$2M_1$ are two-layer brittle micas with tetrahedral compositions of 2 Si + 2 Al or 2 Si + 2(Al,Be). Lin and Guggenheim (1983) and Sokolova et al. (1979) have structurally refined crystals that are chemically intermediate between bityite and margarite, although the latter refinement suffers from poor quality data. The two layer structure (e.g., margarite) has 16 tetrahedral sites per unit cell and may be described in the ideal space group of $C2/c$ by two non-equivalent tetrahedral sites with each site containing ½ Si and ½ Al hybrid cations. However, by eliminating the two-fold axis and the center of symmetry, the structure may be characterized as having four non-equivalent sites in subgroup Cc, with each site containing either Si or Al. The pattern of distribution found in each tetrahedral sheet is an ordered alternation of Si and Al within the ditrigonal tetrahedral ring. Adjacent tetrahedral sheets across the interlayer region have this pattern rotated by 60° so that two adjacent tetrahedra across the interlayer do not have the same tetrahedral occupancy (see Fig. 5).

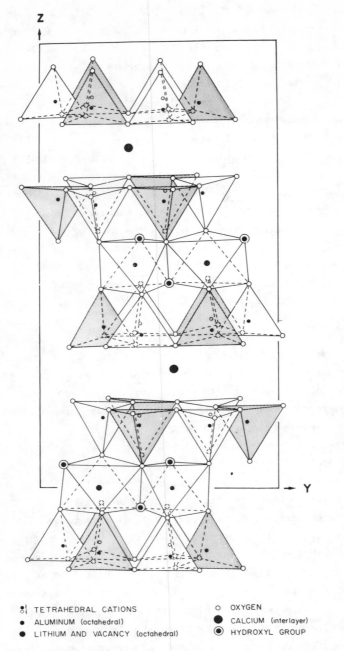

○		TETRAHEDRAL CATIONS	○ OXYGEN
●	ALUMINUM (octahedral)	● CALCIUM (interlayer)
●	LITHIUM AND VACANCY (octahedral)	⊙ HYDROXYL GROUP

Figure 5. An illustration of the margarite/bityite-$2M_1$ intermediate structure as projected down the a axis. Shaded portions represent Al-rich tetrahedra. An identical pattern exists in margarite-$2M_1$. Note that the alternation of Si and Al,Be tetrahedra of one sheet is offset by ±60° from the pattern of the adjacent tetrahedral sheet across the interlayer (from Lin and Guggenheim, 1983).

Alternatively, this pattern may be described by a pseudo-two-fold axis that passes laterally through the two octahedral Al cations. This axis relates compositionally similar tetrahedra, but does not properly describe the entire structure. This ordering distribution violates the center of symmetry normally present at the M(1) site in $2M_1$ structures of ideal symmetry.

Noncentric refinements using X-ray data of structures exhibiting pseudo-symmetry are often difficult and may lead to erroneous models, if the results are interpreted improperly. Particularly because of the similarities in scattering efficiencies of Si^{4+} and Al^{3+}, it is useful to have additional supporting evidence for subgroup cation ordering. Prior to the margarite X-ray work, Farmer and Velde (1973) interpreted the sharp infrared data as lacking the Al-O-Al vibrations that would occur for tetrahedral cation disorder, and Gatineau and Méring (1966) suggested Si,Al order from the lack of diffuse X-ray scattering. More recently, Bish et al. (1979) observed a positive response that indicates acentricity (and implies Si,Al order) from an optical second harmonic generation study. Neutron diffraction (Joswig et al., 1983), in which Si and Al do not have similar scattering efficiencies, has also corroborated the earlier X-ray work.

Slade and Radoslovich (pers. comm.) have found *very* weak X-ray reflections in margarite that violate the c-glide plane and reduce further the symmetry to $C1$. These reflections strengthen considerably as an ephesite component (NaLi) enters into solid solution with margarite. All specimens studied to date contain at least a small amount of Na and a small excess of octahedral cations that must enter the M(1) site. In margarite, such reflections indicate that even the completely Si,Al ordered model is an average of the true structure. Since additional tetrahedral cation ordering is not possible in subgroups of Cc and the amount of Na and excess octahedral cations in some samples studied is low, such reflections must be a result of distortions in the structure rather than a cation ordering phenomenon. There is no evidence for a violation of the C-centering.

Güven (1971), in discussing the differences between muscovite-$2M_1$ and muscovite-$3T$, has suggested a mechanism that might favor the pattern of ordering observed in margarite and margarite/bityite intermediates. For an apical oxygen that belongs to the coordination about two Al^{3+} octahedral cations and a tetrahedral cation, the charge on the oxygen may be balanced if the tetrahedral cation is Si^{4+} or undersaturated (i.e., insufficient positive charge) if the tetrahedral occupancy is Al^{3+}. Since shared octahedral edges tend to be short, two apical oxygens that are electrostatically unbalanced would tend to avoid these edges and the repulsion effects that would occur. Therefore, each shared edge about an octahedral site involving two apical oxygens is most stable when one apical oxygen belongs to a Si tetrahedron and the other to an Al tetrahedron (note Fig. 5).

Electrostatic repulsions are minimized also by the configuration of Si and Al tetrahedra across the interlayer region. The three large basal anions in a tetrahedron are not efficient shields for the high charge cation located directly across the interlayer in the opposing tetrahedron. The pattern of Si and Al alternation and the 60° rotation of the pattern across the interlayer reduces the repulsive forces that would occur if Si^{4+} cations oppose each other. This effect may be marginal for micas containing K and Ba which expand interlayer separations to greater than about 3 Å (electrostatic effects decrease in proportion to $1/r^2$, where r is the interatomic distance), but it becomes increasingly important when adjacent basal oxygen surfaces are nearly in contact, as in the Ca-micas. It is perhaps noteworthy that in pyrophyllite and talc, which have no interlayer cation and an interlayer separation of about 2.76 Å, the ditrigonal Si tetrahedral rings are offset across the interlayer so that tetrahedral rings do not directly superimpose.

Caution is required when considering electrostatic arguments and interlayer separation. At high temperatures where the ordering process would occur, the interlayer separation would be expected to expand at a greater rate for a mica with a univalent interlayer cation than for one with a divalent cation. (See the last section of this chapter for additional comments on the effect of temperature on the mica structure.) Therefore, the temperature of formation and resulting interlayer separation may determine whether a Na mica with Si:Al = 1:1 exhibits tetrahedral cation ordering. Also, repulsive effects can be reduced even if Si^{4+} cations oppose each other across the interlayer. In such cases, interlayer separation may increase further from the distance found in a similar structure having Si^{4+} and Al^{3+} opposing each other across the interlayer. Such a possibility is greater for micas with univalent interlayer cations because the bonding between layers is weaker than that found in the brittle micas. In contrast to the calcium micas, the true mica ephesite-1M has an apparent Al and Si alternation pattern with the Si^{4+} cations opposing each other across the interlayer (Sokolova et al., 1979). Such an ordering pattern may be relatively unstable, not only because of the opposing Si^{4+}, but also because two electrostatically unbalanced apical oxygens form shared octahedral edges. On the other hand, the ephesite-1M structure as refined by Sokolova et al. (1979) in $C2$ symmetry may be incorrect: The R factor improved by only approximately 0.01 when the $C2$ model is compared to the $C2/m$ model (Sokolova et al., 1979), although there are 39 varied parameters in a $C2$ refinement vs

22 in a $C2/m$ refinement. Furthermore, the R factor is high (0.115), five of the twelve atoms have negative isotropic thermal parameters, and variations of individual T-O distances within one tetrahedron vary by as much as 0.08 Å. Hence, the ephesite structure requires further refinement with high quality data. In addition, structural studies of micas with other interlayer cations and with Si and Al ratios of 1:1, such as preiswerkite and siderophyllite, are required to help determine the relationship between interlayer separation and tetrahedral cation order/disorder.

If the amount of interlayer separation is a critical factor for Si and Al ordering, then the lack of tetrahedral ordering in kinoshitalite-1M (Guggenheim and Kato, 1984) is not surprising, since the interlayer Ba cation is large. However, kinoshitalite may not be a very good test case for two reasons. First, environmental conditions have not been considered: High temperatures, rapid cooling and a relatively short period of crystallization tend to promote cation disorder. Bityite occurs as either a late stage magmatic or hydrothermal alteration product and such environments would tend to promote order. Margarite, which forms in low to medium grade metamorphic environments, is from higher temperature regimes. Infrared work (Langer et al., 1981) on margarite synthesized at 620°C and 11 kbar shows tetrahedral Si,Al order. Such evidence suggests that tetrahedral cation ordering is not environmentally induced and is a consequence of a more stable charge distribution (Lin and Guggenheim, 1983). The kinoshitalite crystal studied by single crystal X-ray methods is from a contact metamorphosed manganese deposit. This low pressure crystallization environment differs significantly from that of margarite as given above. It may be argued that the lack of tetrahedral cation ordering in kinoshitalite is a result of rapid crystallization at presumably high temperatures and *low* pressures followed by rapid cooling. However, such arguments appear tenuous without more supportive data. A second possibility which cannot be ruled out is that twinning may have affected the crystal studied, particularly since the residual value, R, for the refinement was relatively high (R_1 = 0.072, R_2 = 0.108). For a 1M polytype with an ordered Al and Si alternating pattern in the tetrahedral sheet similar to margarite, the most likely space groups are either $C2$ or $C\bar{1}$. However, if the twin law is a two-fold axis parallel to the *b* axis and/or a mirror plane perpendicular to *b* for the $C\bar{1}$ case, or a mirror plane perpendicular to *b* for the $C2$ case, then the refinement could proceed satisfactorily in space group $C2/m$ to produce an apparently disordered tetrahedral cation pattern. Such twinning is common in the micas and would be difficult to detect optically.

When tetrahedral cation ordering occurs in the brittle micas, the larger

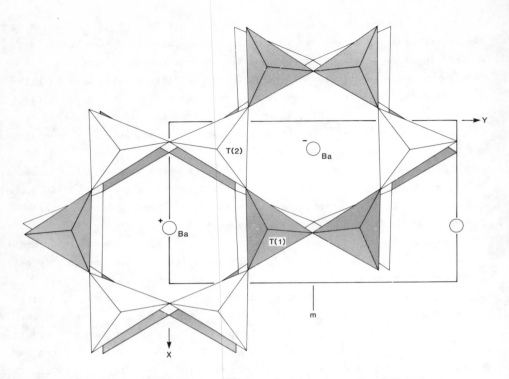

Figure 6. The interlayer region of anandite-$2Or$ projected down the c axis. Sulfur atoms either project on top (+) of the Ba cation or below (-). Two types of tetrahedral rings exist, one with four large tetrahedra [T(2)] and the other with four small tetrahedra [T(1)]. Such a pattern reduces the cell from C-centered to primitive. The large Ba cation is coordinated by 12 oxygens and 1 sulfur.

Al-rich tetrahedra alternate with the smaller Si-rich tetrahedra around the ditrigonal tetrahedral ring. In anandite (Giuseppetti and Tadini, 1972), each hexagonal ring about the interlayer cation has an unequal number of similar sized tetrahedra and a simple alternation pattern is not possible (see Fig. 6). This pattern is unique among micas. It not only lowers the space group symmetry from a C-centered cell to a primitive cell, but it also requires large distortions in individual tetrahedra (Filut, Rule, and Bailey, in preparation). The larger T(2) site contains substantial (see Table 5) amounts of iron.

Interlayer region

For larger cations such as K and Ba, tetrahedral rotation may be limited somewhat and the unit cell enlarged both laterally and vertically. Vertical adjustments are made by the propping apart of the basal oxygen surfaces to increase the interlayer separation. In addition, the tetrahedral sheet through tetrahedral rotation imposes partial restrictions on the interlayer site that influence its shape. For example, for a specific interlayer cation occupancy but different tetrahedral and octahedral occupancies, tetrahedral rotation may elongate (or compress) the shape of the interlayer cation cavity by propping apart the basal surfaces without significant lateral expansion of that site, as has been observed for Ca (Lin and Guggenheim, 1983) in $2M_1$ dioctahedral cases. However, such expansion primarily in the vertical direction along z^* probably leads to increasing instability as the charge distribution around the interlayer site becomes increasingly nonsymmetric and dipolar.

As noted in the previous chapter, the interlayer stacking angle may determine whether an anti-prism (octahedron) or trigonal prism of anions coordinates the interlayer cation. The former is a result for interlayer stacking angles of 0° or ±120°, whereas angles of ±60° or 180° produce the latter. The predilection of most micas for the $1M$ and $2M_1$ (and, to a lesser degree, the $3T$) polytypes is caused by the more stable anti-prism anion configuration about the interlayer cation (Takeda et al., 1971). The trigonal prism is normally less stable than an anti-prism because the three negatively charged basal oxygens of one layer closely approach those of the adjacent layer. The negative charge on the basal oxygens is reduced when tetrahedral substitutions for silicon are minimal, as in lepidolite-$2M_2$ and muscovite-$2M_2$. (See the last section of this chapter for additional comments.)

Before discussing the apparently enigmatic occurrence of a $2Or$ polytype in anandite where tetrahedral substitutions are prominent, it is also necessary to consider what attractive forces (in addition to the Ba-O bonds) may be available to counteract basal oxygen repulsions across the interlayer (when a coordinating hexagonal prism exists about the interlayer cation). In a study of synthetic fluor-polylithionite-$1M$ in which there is no tetrahedral substitution and the silicate rings are nearly hexagonal, Takeda and Burnham (1969) suggested that the major attractive force is between the K interlayer cation and F. In trioctahedral hydroxyl micas where the O--H vector tends to be parallel to c^*, cation repulsion of K by H^+ exists. Such repulsions are avoided when F substitutes for OH. These arguments are less appropriate for dioctahedral micas where H^+ is not directed toward the interlayer cation and interlayer cation to H^+ repulsions are attenuated. In such cases, the interlayer cation to OH distance may be quite short, approaching those found in F-rich trioctahedral micas. Likewise, very short interlayer cation to OH distances may occur also when there is an exceptionally large negative charge on the basal oxygens as in clintonite-$1M$ (Al_3Si tetrahedral composition), where very strong attractive forces are produced between the Ca and its coordinating oxygens.

The Ba interlayer cation in anandite-$2Or$ is coordinated by 12 basal oxygens forming an irregular hexagonal prism and a S, which is centered on one base (Ba-S = 3.194 Å). The hydroxyl ion is considered too distant (Ba-OH = 4.157 Å) to be a nearest neighbor. The OH group is located at the z level of the apical oxygens of the smaller (Si-rich) tetrahedral ring whereas the S is located only approximately in the corresponding position in the larger Fe-rich tetrahedral ring (see Fig. 6). The Ba cation is not located in the center of the interlayer cavity, but has moved closer to the S by about 0.1 Å. The S, in turn, has moved closer to the Ba by about 0.5 Å and is no longer coplanar with the apical oxygens.

In anandite, repulsive forces between basal oxygens of adjacent layers are minimized because the bridging oxygen between two large and Si-poor T(2) sites is most closely associated across the interlayer with the bridging oxygen of two smaller and Si-rich T(1) sites. In addition, the divalent charge of the Ba offsets the undersaturated nature of the basal oxygens; the shortest Ba-O distances involve those with greatest undersaturation (the T(2)-O-T(2) bridging oxygen has Ba-O distances (Filut, Rule, and Bailey, in preparation) of 2.967 Å and 3.024 Å, with mean Ba-O = 3.054 Å). There is no physical evidence to suggest that the S is present as SH^- or S^{2-}, although Giuseppetti and Tadini (1972) favor the latter because Ba and the three octahedral Fe^{2+} cations form a nearly regular tetrahedron around the S. It appears reasonable also based on the apparent attraction of Ba to the S. In summary, the instability caused by the relatively close approach of negatively charged basal oxygens of adjacent 2:1 layers is offset in anandite by the large and highly charged interlayer cation, by the substitution of S for OH, and by the association of the bridging oxygens.

Orientation of the O--H vector. Hydroxyl group protons are believed to be important in determining the strength of the attraction between the interlayer cation and the 2:1 layers, which in turn may influence mica alteration (Bassett, 1960; Norrish, 1973) and polytype formation (Takéuchi, 1966, and previous section). For the brittle micas, proton positions have been determined in clintonite-1M (Joswig and Takéuchi, in preparation) and margarite-2M_1 (Joswig et al., 1983) by neutron diffraction techniques and in the margarite-bityite-2M_1 intermediate (Lin and Guggenheim, 1983) by X-ray diffraction methods. In both margarite cases, the crystal studied had substantial octahedral cation substitutions for the normally vacant M(1) site and thus, has regions with dioctahedral and trioctahedral unit cells present. The significant differences in composition between the two crystals are that the margarite sample has approximately 0.2 substitution for the vacant site with about equal amounts of Mg and Li, and approximately 0.2 Na substitution per interlayer (Ca) cation, whereas the margarite/bityite intermediate has approximately 0.5 Li substitution for the vacancy and 0.6 Be substitution in the tetrahedral sites. Therefore, both crystals would be expected to have an O--H vector orientation characteristic of unit cells with a vacancy at M(1), similar to a pure end-member margarite, and a second O--H vector orientation corresponding to a trioctahedral end-member, either bityite in the case of the X-ray study and a $(Li,Mg)Al_2$ trioctahedral mica for the neutron diffraction study. At least qualitatively these cases follow the general trend of O--H orientations for dioctahedral and trioctahedral micas. The O--H orientation for the hydroxyl group associated with the vacant site is the same for both crystals, to within the limits of the data. The O--H vector is elevated by about 22° from the (001) plane, tilting away from the normal and closer to the vacant site, similar to that found in muscovite (Rothbauer, 1971). For unit cells with the bityite end-member characteristics, the O--H vector is substantially higher, at about 68°, although it should be noted that the X-ray data are at the limit of resolution, and these results are considered tentative. However, the observed elevated position is consistent with the charge of +1 in the M(1) site. In the margarite crystal described by Joswig et al., unit cells with Mg and Li substitution for the vacancy have the O--H vector at approximately 70° to the (001). Although a vector associated with Mg and one with Li might be anticipated, the limit of resolution is uncertain and it appears that the hydrogen positions associated with Li and Mg cannot be distinguished. It is uncertain also what the effect of Na substitution in the interlayer site would have on these results. The clintonite-1*M* structure has an O--H vector orientation perpendicular (89.7°) to the (001) plane, in accord with the expected position for a trioctahedral mica with Mg in M(1).

PHASE RELATIONS OF MARGARITE AND CLINTONITE

Introduction

In general, the phase relations of most brittle micas are not well known, primarily because the majority of brittle micas are rare; extensive field and experimental work are not available. However, margarite is now recognized as a rock-forming mineral of some importance (Sagon, 1967; 1970; 1978; Frey and Niggli, 1972; Höck, 1974a; Frey and Orville, 1974; Ackermand and Morteani, 1973; and many others) and experimental, theoretical, and field-related studies have provided information about its compatibility relations. Although less common, clintonite stability also has been studied in detail recently. Therefore, the emphasis here is on this recent work of the more common brittle micas, margarite, and clintonite. Additional comments on the metamorphic occurrences of the white micas may be found in Chapter 10.

Frey et al. (1982) have compiled the occurrences of margarite as a prograde metamorphic mineral and determined that margarite may be found in metamorphic pelites, marls, bauxites, basites, and anorthosites, and at metamorphic grade from lower greenschist to upper amphibolite facies. Margarite may be related also to polymetamorphic events to produce pseudomorphs commonly after andalusite, kyanite, sillimanite, or corundum and, less often, after chloritoid, staurolite, or muscovite (Guidotti and Cheney, 1976; Frey et al., 1982; Teale, 1979).

Clintonite occurrences are restricted to carbonate rocks that have undergone metasomatism (e.g., Knopf, 1953; Burnham, 1959; and others) except for one occurrence of a high Mg,Si-rich clintonite in a diorite-serpentinite contact zone (Stevenson and Beck, 1965). One metasomatic aureole, at Crestmore, California, where a quartz monzonite has intruded a magnesian limestone, shows a zonal arrangement with sharply contrasting amounts of Fe oxides, Si, Al, and Mg in three zones (Burnham, 1959). Burnham concluded that the contact metamorphic mineral assemblages at Crestmore are not strictly a result of gradations in P-T conditions, but are caused by a progressive metasomatism with consequent decarbonization at high temperatures (rather than a slow decarbonization as temperature increases).

Margarite

Hydrothermal studies involving margarite stability were made by Velde (1971), Chatterjee (1974), Storre and Nitsch (1974), Nitsch et al. (1981), and Jenkins (1983 and in preparation). For the reaction involving the breakdown of margarite, margarite = anorthite + corundum + H_2O, Figure 7 shows that the work by Chatterjee (1974) and Storre and Nitsch (1974) are similar, whereas that by Velde (1971) has a markedly different slope. Although the Chatterjee and Velde curves were each determined by starting with a mixture of synthetically prepared phases, a comparison of cell parameters of the resulting margarite shows a significant difference (c_v = 19.117 Å vs c_c = 19.166(3) Å). The cell parameters reported by Chatterjee are closer to natural end member margarites, suggesting that Velde's samples were contaminated. The natural margarite starting material used by Storre and Nitsch (1974) has paragonite and ephesite components which may account for the difference in slope when compared to that of Chatterjee. The Chatterjee results are based on reaction reversals.

For lower pressures, Figure 8 compares the breakdown of margarite with and without quartz and shows that the presence of quartz reduces margarite stability.

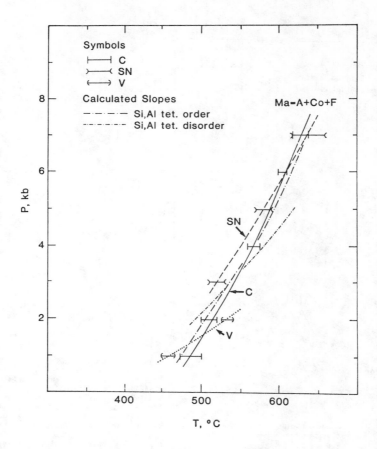

Figure 7. Experimental data of Chatterjee (1974), Storre and Nitsch (1974), and Velde (1971) showing margarite stability. The calculated slopes are from Perkins et al. (1980). Symbols: Ma = margarite, A = anorthite, Co = corundum, F = fluid, C = Chatterjee, SN = Storre and Nitsch, and V = Velde.

The curve labelled NST as given by Nitsch et al. (1981) is based on synthetic margarite and that labelled SN (data from Storre and Nitsch, 1974) is based on natural margarite. At higher pressures, the experimental data of Storre and Nitsch (1974) and Jenkins (in preparation) are shown. In addition, calculated Clapeyron slopes (Perkins et al., 1980) for the reactions in both Figures 7 and 8 are shown for tetrahedral Si,Al ordered margarite and for tetrahedral Si,Al disordered margarite. In Figure 7, there is excellent agreement between the experimental determinations and the calculated Si,Al ordered margarite slopes.

The experimentally determined curves for the reaction, margarite + quartz = anorthite + ky/and + H_2O, in Figure 8 fall between the two extremes in slope established by the calculated curves, although both are closer to the calculated ordered margarite slope. For the curve (SN) derived from experiments with

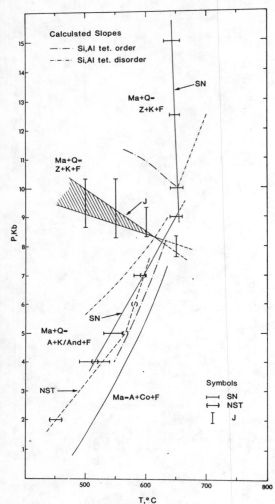

Figure 8. Experimental data from Storre and Nitsch (1974), Nitsch et al. (1981), and Jenkins (in preparation) and calculated data (Perkins et al., 1980) for the reaction involving margarite + quartz. The curve(s) from Nitsch et al. (1981) is reproduced directly from that source. The curve for margarite breakdown (without quartz) is from Chatterjee (1974). See the text for a discussion on the reliability of the experimental data. Note that one bracket for the curve derived by Jenkins is located in the metastable field of margarite + quartz (anorthite out). Symbols: Q = quartz, K = kyanite, And = andalusite, Z = zoisite, NST = Nitsch et al., J = Jenkins, other symbols as before.

natural margarite (from Chester, Massachusetts), it seems likely that any deviation from the calculated ordered Clapeyron slope may be attributed to other components in solid solution rather than a partially ordered sample. This conclusion is consistent with the infrared study of Langer et al. (1981) in which Chester margarite was determined to be Si,Al ordered.

The calculated Clapeyron slope for tetrahedral Si,Al ordered margarite for the reaction margarite + quartz = zoisite + kyanite + H_2O (Fig. 8) is in good agreement with the range of possible slopes (dP/dT = -7 to -15 bars/K) given by Jenkins (in preparation). Perkins et al. (1980) arbitrarily selected the experimental bracket (as shown in Fig. 8) given by Storre and Nitsch (1974) and then extrapolated the Clapeyron slopes to other pressures and temperatures. Therefore, the actual position in P-T space of the calculated Si,Al ordered margarite Clapeyron curve is of less importance than the slope. The large brackets associated with the curve derived by Jenkins are not errors associated with the pressure measurement, but are related to the slow reaction rate that prevented the direction of the reaction to be ascertained within these brackets. Runs were made for up to 27 days duration.

In contrast, the experimental work of Storre and Nitsch (1979) is in very poor agreement with the calculated slopes of Perkins et al. (1980). Although Perkins et al. suggested that the reactants might be ordered partially since the experimental curve is between two extremes, such a suggestion may now be discounted. Synthetic margarite quenched from 620°C and $P(H_2O)$ = 11 kbar has been determined to be Si,Al ordered by Langer et al. (1981). More likely, Storre and Nitsch may have had difficulties in judging the growth of the reactant assemblages by X-ray diffraction or optically, whereas Jenkins had little difficulty in discerning reaction reversals by use of an infrared absorption peak technique.

Experimental phase equilbria studies are hindered by very slow reaction rates and thus, considerable attention has been given to theoretical methods. The petrogenetic grid for the CASH ($CaO-Al_2O_3-SiO_2-H_2O$) system has been derived by Chatterjee (1976) on the basis of phase equilibria and calorimetric data and by Perkins et al. (1980) using data from adiabatic calorimetry and differential-scanning calorimetry. In the latter study (Fig. 9), $S°_{298}$ and $C°_p(T)$ data were determined for margarite, lawsonite, zoisite, and prehnite. Then, an experimentally well defined bracket on an equilibrium curve was selected. Using this bracket as a starting point, Clapeyron slopes for univariant equilibria were derived by extrapolations to other pressures and temperatures by using thermodynamic methods based on the calorimetrically obtained values of $S°_{298}$ and $V°_{298}$. Both the grids of Chatterjee (1976) and Perkins et al. (1980) show very similar results, especially since different sets of input data were used. However, the latter shows the stability fields of margarite and of margarite + quartz extended to higher pressures and temperatures (e.g., increasing maximum pressures from 8.6 to 10.4 kbar and temperatures from 560 to 620°C for the margarite + quartz fields). A similar topology has been calculated by Helgeson et al. (1978) for margarite stability fields, but lower grade assemblages differ and the grid is not as extensive.

Chatterjee (1976) has considered also the effect of the activity of water on the margarite + quartz stability fields (Fig. 10). He has shown that the reaction of margarite + quartz = zoisite + kyanite + H_2O is sensitive to the

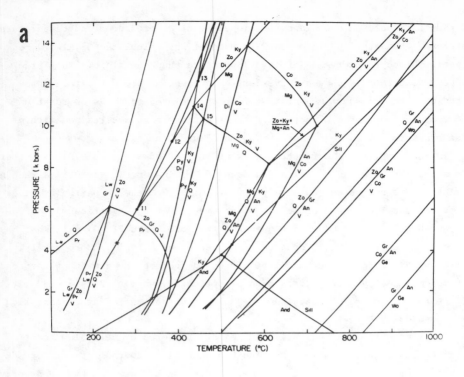
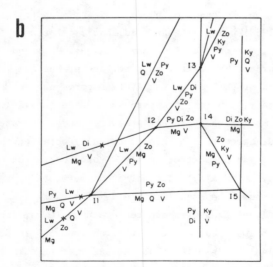

Figure 9(a). The petrogenetic grid (Perkins et al., 1980) for the CaO-Al_2O_3-SiO_2-H_2O system excluding zeolite- and kaolinite-bearing reactions. (b) A schematic enlargement of a portion of 9a. The unlabelled reaction emanating from invariant point I3 is $2Lw + Di = Ky + Zo + 4H_2O$. Note that the three reaction curves designated by X have been simplified in 9a. Symbols: Lw = lawsonite, Gr = grossular, Sill = sillimanite, Ky = kyanite, Pr = prehnite, Zo = zoisite, An = anorthite, v = vapor, Py = pyrophyllite, Mg = margarite, Di = diaspore, Wo = wollastonite, Ge = gehlenite, others as before.

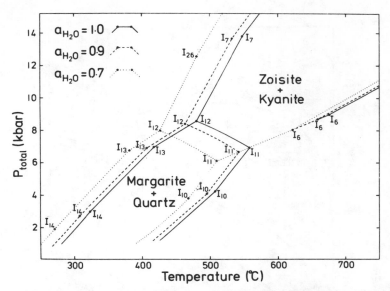

Figure 10. A simplified P-T diagram (Chatterjee, 1976) showing the effect of a reduced water activity (due to the presence of inert fluids) on the stability field of margarite + quartz. Most univariant reactions may be deduced by comparison to Figure 9, otherwise refer to the original paper (Chatterjee, 1976).

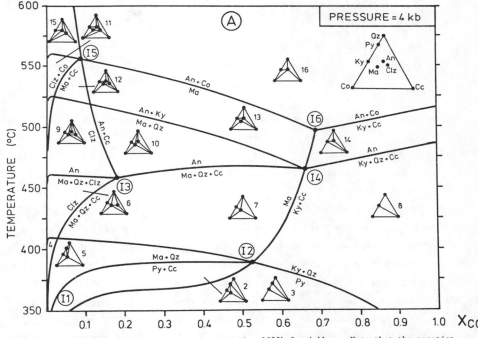

Figure 11. T-X(CO_2) diagram (Bucher-Nurminen et al., 1983) for 4 kbar. Note that the reaction anorthite = margarite + quartz + calcite is nearly isothermal. Symbols: Ma = margarite, Cc = calcite, Qz = quartz, Clz = clinzoisite, others as in Figure 9.

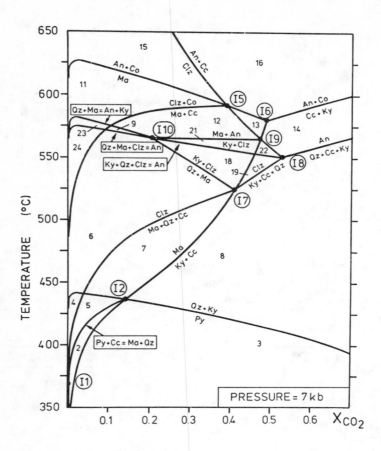

Figure 12. T-X(CO_2) diagram (Bucher-Nurminen et al., 1983) for 7 kbar. Symbols as in Figure 11.

activity of water with the upper limits of the stability field of margarite and quartz decreasing with decreased water activity (due to the presence of other inert gases).

Bucher-Nurminen et al. (1983) combined a field study in the Alps with a theoretical approach to derive the phase relations in the system $CaO-Al_2O_3-SiO_2-(C-O-H)$. They calculated isobaric T-X(CO_2) sections in the presence of a C-O-H fluid and graphite at 4 kbar (Fig. 11) and 7 kbar (Fig. 12) total pressure, which are consistent with recent barometric estimates for the area. Of particular interest (Fig. 13) are the two assemblages: margarite + quartz + calcite + plagioclase for the 4 kbar diagram and margarite + clinozoisite + kyanite + quartz + calcite for the 7 kbar topology. These are mutually exclusive based on the two topologies, but are abundant at the low and high-grade ends of the field area, respectively. Such occurrences are excellent

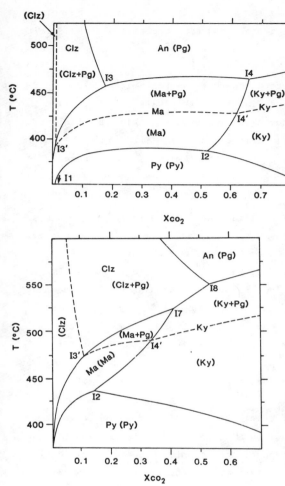

Figure 13(a). The 4 kbar topology (excess quartz + calcite) and the expansion of anorthite stability to lower temperature with increasing albite dilutions (compare to Fig. 11). For pure anorthite compositions, the solid curves define stability fields for Clz, An, Ma, Ky, and Py. Note that the kyanite and clinozoisite stability fields are separated by the margarite field and therefore kyanite and clinozoisite should not coexist at these pressures. Dashed curves represent the lower temperature limit for plagioclase at a composition of An_{33} and produces divariant assemblage fields of (Clz + Pg), (Ma + Pg), and (Ky + Pg). Assemblage fields labelled with parentheses are bounded by dashed and solid curves. The clinozoisite field is now at very low values of $X(CO_2)$. New invariant points are given at I3' and I4'.

Figure 13(b). Diagram analogous to 13a for the 7 kbar topology (excess quartz + calcite). For pure anorthite compositions, margarite and plagioclase cannot coexist. However, because the lower temperature limit (dashed curves) extends below the temperature for invariant point I7, plagioclase may coexist with margarite at high pressures if plagioclase is albitic (approximately An_{60}) and the assemblage (Ky + Clz + Pg) may exist when the plagioclase composition is greater than An_{60}. Both parts after Bucher-Nurminen et al. (1983). Symbols: Pg = plagioclase, others as in Figure 11.

geobarometers and indicate non-isobaric conditions in this field area. The occurrences of these assemblages, both in the $T-X(CO_2)$ sections and in the field, and the successful prediction of plagioclase composition variations (see also Frey and Orville, 1974) as observed in the field, support the validity of the derived diagrams. Comparisons of these diagrams to related equilibrium data show similar results. For example, phase relationships (Storre et al., 1982) involving (*ortho-*) zoisite (in the system $CaO-Al_2O_3-SiO_2-H_2O-CO_2$) locate the isobaric invariant point labelled I3 (Fig. 11) at 450°C and $X(CO_2) = 0.07$.

Bucher-Nurminen et al. (1983) were able to map the last occurrence of margarite + calcite + quartz as a well defined isograd. In Figure 11, the reaction 1 calcite + 2 quartz + 1 margarite = 2 anorthite + 1 CO_2 + H_2O is nearly isothermal as would be expected for such an isograd. It may be noted also that $f(O_2)$ is controlled in margarite schists by the silicate and calcite assemblages in the presence of graphite. For lower pressures, the presence of clinozoisite places limitations on the mole fraction of CO_2.

Bucher-Nurinen et al. (1983) determined that the stability field of margarite + calcite + quartz will decrease in area with albite solid solution in anorthite. (Fig. 13 shows the maximum expansion of the plagioclase field from the field study.) The isobarically invariant assemblage of margarite + quartz + calcite + clinozoisite + plagioclase (I3 in Fig. 11) varies considerably in temperature and $X(CO_2)$, and was frequently observed over a wide area in the field study.

Margarite/paragonite ± muscovite. Although the occurrence of margarite with (or in solid solution with) paragonite ± muscovite is well known (e.g., Ackermand and Morteani, 1973; Höck, 1974, Guidotti and Cheney, 1976; Frey, 1978; Guidotti et al., 1979; Teale, 1979; Baltatzis and Katagas, 1981; Frank, 1983; and many others), experimental and theoretical studies have not been extended to include $Na_2O \pm K_2O$ components to the $CaO-Al_2O_3-SiO_2-H_2O$ system. Although such systems are extremely complex, the petrologic implications of the presence of margarite (or margarite in solid solution) cannot be determined without additional data (see additional discussion in Chapter 10). Frank (1983), in accord with the graphical analysis of Frey and Orville (1974) for the system $CaO-Na_2O-Al_2O_3-SiO_2-H_2O-CO_2$, found that the assemblage of paragonite + calcite + quartz has a lower thermal stability than that of margarite + calcite + quartz in the absence of corundum. The margarite phase in this study has significant amounts of paragonite component.

The common occurrence of margarite + quartz (CASH system) in low to medium grade metamorphic rocks is readily explained by the calculated margarite phase relations showing that this assemblage is stable to approximately 600°C when $P(H_2O) = P_{total}$. The presence of a gas phase other than water will reduce water activity, thereby shifting the margarite or margarite + quartz assemblages to lower temperatures and pressures and increasing the area of the zoisite + kyanite stability field. Therefore, margarite occurrences are restricted to low to medium pressures. Chatterjee (1976) noted that the presence of paragonite in solid solution with margarite will offset the effect of a reduced water activity.

<u>Clintonite</u>

Although other workers (e.g., Velde, 1973; Ito and Arem, 1970; Christophe-Michel-Lévy, 1960; and others) have synthesized clintonite hydrothermally,

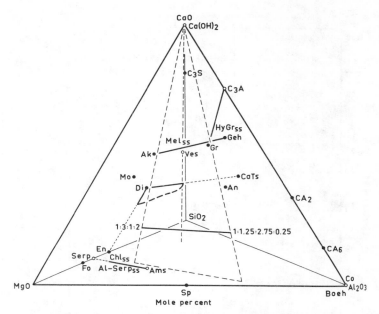

Figure 14. Diagrammatic representation of the phases in the system $CaO-MgO-Al_2O_3-SiO_2-H_2O$ as projected from H_2O (from Olesch and Seifert, 1976). Anhydrous phases are shown as solid circles. The join represented in the T-X diagram of Figure 15 is marked by molecular proportions in order of $CaO:MgO:Al_2O_3:SiO_2$. Symbols: Ak = akermanite, Al-Serp$_{ss}$ = aluminous serpentine solid solution, Ams = amesite, Boeh = boehmite, C_3A = tricalcium hexahydroxoaluminate, CA_2 = calcium dialuminate, CA_6 = calcium hexaluminate, C_3S = tricalcium silicate solid solution, CaTs = calcium Tschermak's molecule, Chl$_{ss}$ = chlorite solid solution, Di = diopside, En = enstatite, Fo = forsterite, Geh = gehlenite, HyGr = hydrogrossularite, Mel$_{ss}$ = melilite solid solution, Mo = monticellite, Serp = serpentinite, Sp = spinel, Ves = vesuvianite, others as in Figure 9a.

Olesch and Seifert (1976) studied clintonite phase relations systematically along the pseudo-binary join established by the substitution MgSi = 2Al ($CaO:3MgO:Al_2O_3:2SiO_2:xH_2O$ - $CaO:1.25MgO:2.75Al_2O_3:0.25SiO_2:xH_2O$) in the system $CaO-MgO-Al_2O_3-SiO_2-H_2O$. Figure 14 represents the phases of the system and Figure 15 shows the T-X diagram at 2 kbar. Note in Figure 14, that certain planes of stable assemblages (e.g., diopside-anorthite-forsterite) are located between clintonite-quartz, thereby preventing a clintonite-quartz equilibrium assemblage. Assemblages at temperatures below 600°C are not shown in Figure 15 because of apparent kinetic problems. None of the reactions have been reversed. Note from the shape of the clintonite solid solution field near the central part of the diagram that increasing Al content reduces the thermal stability of the mica, as would be expected. Because the basal oxygens augment their unsatisfied negative charge with additional tetrahedral Al substitutions, vertical structural adjustments are limited due to the increasing attraction of the Ca for the basal oxygens; Ca would also block supplemental tetrahedral rotations. Further instability occurs by the repulsion effects that would exist between the basal oxygens of adjacent layers. Although

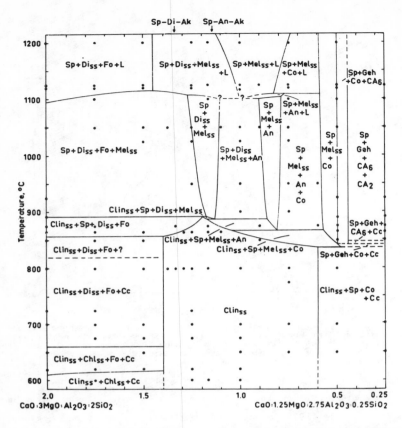

Figure 15. T-X diagram at 2 kbar under hydrous conditions. See Figure 14 for compositional information. Symbols: $Clin_{ss}$ = clintonite solid solutions, Cc = calcite, others as in Figure 14. From Olesch and Seifert (1976).

increasing tetrahedral Al (and increasing α) rotates basal oxygens from opposing tetrahedra to still closer, and perhaps more unstable, distances, this effect is minimized by increasing temperature (see the next section). The P-T diagrams (Fig. 16) illustrate the effect of Al content on stability also.

Olesch and Seifert (1976) considered high pressure regimes and found that clintonite solid solution persists to more than 20 kbar. They suggested, at least at the higher fluid pressures, that clintonites are restricted to low $X(CO_2)$ values of the coexisting gas phases. This extensive stability field emphasizes that the rarity of clintonite in nature is the result of restricted occurrences of alumina-rich and silica-poor environments (quartz nonexistent) and possibly a low $X(CO_2)$ coexisting gas phase. Such conditions preclude finding clintonites in most regionally metamorphosed impure limestones, although contact metamorphosed impure limestones associated with H_2O-rich solutions that dilute the CO_2 gas phase should be clintonite bearing.

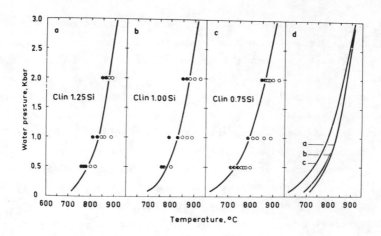

Figure 16. P-T diagrams from Olesch and Seifert (1976) show that increasing Al content of clintonite reduces the thermal stability of the mica. Symbols as in Figure 15.
(a) $Ca(Mg_{2.25}Al_{0.75})(Al_{2.75}Si_{1.25}O_{10})(OH)_2 = Sp + Di_{ss} + Fo + Clin_{ss} + H_2O$,
(b) $Ca(Mg_2Al)(Al_3SiO_{10})(OH)_2 = Sp + Mel_{ss} + An + Clin_{ss} + H_2O$,
(c) $Ca(Mg_{1.75}Al_{1.25})(Al_{3.25}Si_{0.75})O_{10}(OH)_2 = Sp + Mel_{ss} + Co + Clin_{ss} + H_2O$,
(d) summary.

THE RELATIONSHIP OF CRYSTAL STRUCTURE TO STABILITY

High temperature, single crystal studies of OH-rich layer silicates are difficult because hydroxyl loss appears to be diffusion controlled and heating experiments must be held at elevated temperatures for long periods. In order to avoid the complications involved with hydroxyl loss, Takeda and Morosin (1975) studied a fluorophlogopite-1M crystal at 700°C, well below its decomposition temperature of ~1200°C. The tetrahedral rotation angle decreased from 6.4° at room temperature to 2.7° in response to the thermal expansion of the octahedral sheet. These results are consistent with the numerous silicate structural analyses made in the mid-1970's which demonstrated that individual Si and Al tetrahedra neither expand nor compress with changing temperature and pressure. In the micas, the tetrahedral sheet has the freedom to change its lateral dimensions by tetrahedral rotation and therefore, may deform in this manner to adjust to the changes occurring in the octahedral sheet. Takeda and Morosin found also that the c axis length increases at a rate nearly twice that of the other axes, because of the relatively large mean thermal expansion coefficient of the K-O bond. In addition, Mg octahedra expand without changing shape to temperatures of approximately 400°C, above which they elongate more rapidly perpendicular to the layer. Several important conclusions may be drawn: (1) At high temperatures the silicate rings approach hexagonal symmetry, but the ideal hexagonal configuration is *not* necessarily the geometric limit of mica stability. Since the rotation angle decreases and the octahedral sheet distorts at higher temperatures, both must be considered important. Toraya (1981) has made a similar suggestion from more theoretical considerations. (2) For micas which form at high temperatures, arguments suggesting that polytypes with interlayer stacking angles of 0° or ±120° form a more stable octahedral coordination group around the interlayer cation (see Chapter 2 and this chapter, interlayer cation coordination sections) may have to be revised, since at elevated temperatures interlayer cation coordination for all polytypes approaches twelve. Thermal expansion along the c axis is primarily due to the K-O bond expansion. Such expansions would tend to minimize the repulsion effects across the interlayer region caused by basal oxygens with unsatisfied negative charges. (3) It is noteworthy that the M(2) site is larger in size than M(1) at 700°C (2.069 Å vs 2.095 Å), perhaps suggesting why Mössbauer results (Annersten, 1974; Bancroft and Brown, 1975) show a small preference of Fe^{2+} for M(2) and Fe^{3+} for M(1).

The crystal structure of muscovite was studied in detail at approximately 400°C (Chang, 1984; Chang and Guggenheim, in preparation) to determine the effect of the vacant site. Similar to phlogopite, the tetrahedral rotation angle decreases (from 11.3° at room temperature to 10.3°) and the c axis length increases substantially because of the relatively large mean thermal expansion coefficient of the K-O bond. However, in contrast to phlogopite, the (Al) octahedra do not expand as greatly; thermal expansion is much less than Al-containing octahedra of other structures. Instead, lateral expansion occurs mainly by the enlargement of the vacant M(1) site. Because the mean coefficient of thermal expansion is not necessarily a constant over a large temperature interval, higher temperature studies are in progress to determine if the trend continues.

The above results stress the need for caution in interpreting high temperature phenomena, such as cation ordering or stability relationships, with models derived from room temperature structures. On the other hand, room temperature data may be useful as a starting point for discussion, keeping in mind the inherent limitations of the data.

For the Ca-rich brittle micas, the tetrahedral sheet is deformed substantially ($\alpha \sim 20°$) because tetrahedral Al substitution greatly enlarges an unconstrained tetrahedral sheet's lateral dimensions. *If* octahedral sheet distortions and thermal expansion of the interlayer cation site are ignored, and *if* only the amount of tetrahedral rotation is considered as the limiting factor for stability (i.e., the tetrahedra rotate to $\alpha = 0°$ as the octahedral sheet enlarges at elevated temperatures), the structure will become unstable when the octahedral sheet expands beyond a critical point. Hazen (1977) has proposed calculations for determining (an upper limit for) biotite-phlogopite breakdown temperatures based on the limiting condition of tetrahedral rotation only, as given above. Using similar equations for clintonite, thermal breakdown based on this geometric limitation is predicted to be approximately 6972°C at 1 bar. This clearly is not the case and another mechanism must be responsible for thermal breakdown.

Particularly for micas with large tetrahedral rotations and small interlayer cations, instability may result when the tetrahedral rotation exceeds a critical lower value, but much greater than 0°. At this value the interlayer cation loses contact with its nearest basal oxygen neighbors. Although the break occurs at the interlayer junction, the topology is imposed by the tetrahedral/octahedral sheet geometry. In addition, for any mica, the Pauling bond strength is lowered as this interlayer cation coordination goes from approximately six to twelve anions with increasing temperature. This results in a large mean thermal expansion coefficient of the interlayer cation and the observed rapid linear thermal expansion along the c axis.

Generally, the octahedral sheet, by way of the tetrahedral rotation angle, imposes the cavity size on the interlayer site. For muscovite, the interlayer K is too large for its 6-coordinated site, thereby producing a much enlarged interlayer separation (compare 3.4 Å vs 2.76 Å observed in pyrophyllite, which does not have an interlayer cation). In contrast, the adjacent basal oxygen surfaces across the interlayer (2.87 Å) region in Ca micas are nearly in contact, suggesting that the cavity size in margarite cannot greatly expand as compared to muscovite, either laterally through tetrahedral rotation or vertically by increasing the interlayer separation. In part, the latter process is inhibited by the higher charge Ca^{2+}. Since both muscovite and margarite have Al dioctahedral sheets, margarite stability should be more restricted in temperature compared to muscovite. Following similar reasoning, paragonite thermal stability (interlayer separation of 3.05 Å, $\alpha = 16°$) should be between that of margarite and muscovite as has been noted by Chatterjee (1974) and Olesch and Seifert (1976). However, this sort of crystal chemical analysis has limitations; additional components to the systems can alter the relative thermal stability sequence. It should be noted also that without additional high temperature structural information, muscovite breakdown cannot be assessed as resulting from the misfit of the tetrahedral/octahedral junction or the weak K-O bond and the rapidly expanding interlayer separation.

Olesch and Seifert (1976), and many others as well, have noticed that trioctahedral micas generally have a higher thermal stability than those of the dioctahedral micas. In addition, Seifert and Schreyer (1971) observed that potassium micas with an octahedral occupancy intermediate between dioctahedral and trioctahedral micas also have a thermal stability limit between the corresponding dioctahedral and trioctahedral end member micas. Differences in thermal stability between dioctahedral and trioctahedral micas may be explained from a structural viewpoint, although it is first necessary to understand the dehydroxylation mechanisms involved and resulting reaction products. Wardle and Brindley (1972), in a study of pyrophyllite and its (metastable) dehydroxylate, describes the mechanism of reaction as homogeneous with adjacent pairs of $(OH)^-$ ions reacting to form O^{2-} ions and free water (which leaves the crystal). The reaction proceeds with aluminum in the dehydroxylate attaining five fold coordination with a shift of anions toward the vacant site. An additional oxygen must enter the structure at the same level along Z as the Al ions to form a shared corner between Al polyhedra. The vacant site plays an important role in this reaction since it allows the partial reorganization of the octahedral sheet anions (Wardle and Brindley, 1972). Such a mechanism is probably operating in muscovite (see Udagawa et al., 1974 for details on the muscovite dehydroxylate), paragonite, and margarite, although it is unclear as to the details of the effect of the interlayer cation on the reaction.

In contrast, the $(OH)^-$ group in trioctahedral micas is more strongly bonded and there is little room available for anion adjustment in the octahedral sheet at elevated temperatures. A dehydroxylate phase does not occur in trioctahedral structures and recrystallization occurs at higher temperatures than in dioctahedral micas. Vedder and Wilkins (1969) point out that for micas compositionally intermediate between dioctahedral and trioctahedral end members, $(OH)^-$ ions associated with vacant sites dissociate at lower temperatures than $(OH)^-$ ions elsewhere.

ACKNOWLEDGMENTS

I thank N.D. Chatterjee (Ruhr University, Bochum) and R.A. Eggleton (Australian National University, Canberra) for reviewing the manuscript. I also thank S.W. Bailey (University of Wisconsin-Madison), D.M. Jenkins (University of Chicago), W. Joswig (University of Frankfurt), and P.G. Slade and E.W. Radoslovich (Division of Soils, C.S.I.R.O., South Australia) for the use of unpublished data and J. Walshe (Australian National University, Canberra) for helpful discussions. This work has been supported by National Science Foundation grant EAR80-18222.

REFERENCES

Ackermand, D. and Morteani, G. (1973) Occurrence and breakdown of paragonite and margarite in the Greiner Schiefer Series (Zillerthal Alps, Tyrol). Contrib. Mineral. Petrol. 40, 293-304.

Akhundov, Y.A., Mamedov, K.S. and Belov, N.V. (1961) The crystal structure of brandisite. Dokl. Adad. Nauk SSSR, Earth Sci. 137, 438-440.

Annersten, H. (1974) Mössbauer studies of natural biotites. Am. Mineral. 56, 143-151.

Bailey, S.W. (1975) Cation ordering and pseudosymmetry in layer silicates. Am. Mineral. 60, 175-87.

_____ (1980) Structures of layer silicates. Ch. 1 in Crystal Structures of Clay Minerals and their X-ray Identification, G.W. Brindley and G. Brown, Eds. Monogr. 5, Mineral. Soc., London, 1-124.

_____ (1984) Review of cation ordering in micas. Clays and Clay Minerals 32, 81-92.

Baltatzis, Emm. and Katagas, C. (1981) Margarite pseudomorphs after kyanite in Glen Esk, Scotland. Am. Mineral. 66, 213-216.

Bancroft, G.M. and Brown, J.R. (1975) A Mössbauer study of coexisting hornblendes and biotites: Quantitative Fe^{3+}/Fe^{2+} ratios. Am. Mineral. 60, 265-272.

Bassett, W.A. (1960) Role of hydroxyl orientations in mica alteration. Bull. Geol. Soc. Am. 71, 445-456.

Baur, W.H. (1981) Interatomic distance predictions for computer simulation of crystal structures. In: Structure and Bonding in Crystals, M. O'Keefe and A. Navrotsky, Eds. Academic Press, New York, 31-52.

Beus, A.A. (1956) Characteristics of the isomorphous entry of berylleum into mineral structures. Geochemistry, Washington, 62-77.

Bish, D., Horsey, R.S. and Newnham, R.E. (1979) Acentricity in the micas: An optical second harmonic study. Am. Mineral. 64, 1052-1055.

Bookin, A.S. and Drits, V.A. (1982) Factors affecting orientation of OH-vector in micas. Clays and Clay Minerals 30, 415-421.

Bucher-Nurminen, K., Frank, E. and Frey, M. (1983) A model for the progressive regional metamorphism of margarite-bearing rocks in the central Alps. Am. J. Sci., Orville Volume, 370-395.

Burnham, C. Wayne (1959) Contact metamorphism of magnesian limestones at Crestmore, California. Bull. Geol. Soc. Am. 7C, 879-920.

Chang, Y-H. (1984) The crystal structure of muscovite at approximately 400°C and the variation of unit cell parameters with temperature to 1000°C. M.S. Thesis, University of Illinois at Chicago.

Chatterjee, N.D. (1974) Synthesis and upper thermal stability limit of 2M-Margarite, $CaAl_2[Al_2Si_2O_{10}/(OH)_2]$. Schweiz. Mineral. Petrogr. Mitt. 54, 753-767.

_____ (1976) Margarite stability and compatibility relations in the system $CaO-Al_2O_3-SiO_2-H_2O$ as a pressure-temperature indicator. Am. Mineral. 61, 699-709.

Chinner, G.A. (1974) Dalradian margarite: A preliminary note. Geol. Mag. 111, 75-78.

Christope-Michel-Lévy, M. (1960) Reproduction artificielle de l'idocrase. Bull. Soc. franc. Minéral. Cristallogr. 83, 23-25.

Deer, W.A., Howie, R.A. and Zussman, J. (1962) Rock-Forming Minerals, Vol. 3, Sheet Silicates. Wiley, New York.

Farmer, V.C. and Velde, B. (1973) Effects of structural order and disorder on the infrared spectra of brittle micas. Mineral. J. 39, 282-288.

Forman, S.A., Kodama, H. and Maxwell, J.A. (1967a) The trioctahedral brittle micas. Am. Mineral. 52, 1122-1128.

_____, _____ and Abbey, S. (1967b) A re-examination of xanthophyllite (clintonite) from the type locality. Can. Mineral. 9, 25-30.

Frank, E. (1983) Alpine metamorphism of calcareous rocks along a cross-section in the Central Alps: Occurrence and breakdown of muscovite, margarite and paragonite. Schweiz. Mineral. Petrogr. Mitt. 63, 37-93.

Franz, G., Hinrichsen, T. and Wannemacher, E. (1977) Determination of the miscibility gap on the solid solution series paragonite-margarite by means of the infrared spectroscopy. Contrib. Mineral. Petrol 59, 307-316.

Frey, M. (1978) Progressive low-grade metamorphism of a black shale formation, central Swiss Alps, with special reference to pyrophyllite and margarite bearing assemblages. J. Petrol. 19, 95-135.

_____, and Niggli, E. (1972) Margarite, an important rock-forming mineral in regionally metamorphosed low-grade rocks. Naturwiss. 59, 214-215.

_____ and Orville, P.M. (1974) Plagioclase in margarite-bearing rocks. Am. J. Sci. 274, 31-47.

_____, Bucher, K., Frank, E. and Schwander, H. (1982) Margarite in the central Alps. Schweiz. Mineral. Petrogr. Mitt. 62, 21-45.

Frondel, C. and Ito, J. (1968) Barium-rich phlotopite from Langban, Sweden. Arkiv. Mineral. Geol. 4, 445-447.

Gallagher, M.J. and Hawkes, J.R. (1966) Beryllium minerals from Rhodesia and Uganda. Bull. Geol. Surv. Gr. Brit. 25, 59-75.

Gatineau, L. and Méring, J. (1966) Relations ordre-désordre dans les substitution isomorphiques des micas. Bull. Grpe. fr. Argiles. 18, 67-74.

Giese, R.F., Jr. (1979) Hydroxyl orientations in 2:1 phyllosilicates. Clays and Clay Minerals 27, 213-223.

Giuseppetti, G. and Tadini, C. (1972) The crystal structure of 2O brittle mica: Anandite. Tschermaks Mineral. Petrogr. Mitt. 18, 169-184.

Guggenheim, S. and Bailey, S.W. (1975) Refinement of the margarite structure in subgroup symmetry. Am. Mineral. 60, 1023-1029.

_____ and _____ (1978) Refinement of the margarite structure in subgroup symmetry: Correction, further refinement and comments. Am. Mineral. 63, 186-187.

_____ and Kato, T. (1984) Kinoshitalite and Mn phlogopites: Trial refinements in subgroup symmetry and further refinement in ideal symmetry. Mineral. J. (Japan) 12, 1-5.

_____, Schulze, W.A., Harris, G.A. and Lin, J.-C. (1983) Noncentric layer silicates: An optical second harmonic generation, chemical and X-ray study. Clays and Clay Minerals 31, 251-260.

Guidotti, C.V. and Cheney, J.T. (1976) Margarite pseudomorphs after chiastolite in the Rangeley area, Maine. Am. Mineral. 61, 431-434.

_____, Post, J.L. and Cheney, J.T. (1979) Margarite pseudomorphs after chiastolite in the Georgetown area, California. Am. Mineral. 64, 728-732.

Güven, N. (1971) Structural factors controlling stacking sequences in dioctahedral micas. Clay and Clay Minerals 19, 159-165.

Harada, K., Kodama, H. and Sudo, T. (1965) New mineralogical data for xanthophyllite from Japan. Can. Mineral. 8, 255-262.

Hatch, R.A., Humphrey, R.A., Eitel, W. and Comeforo, J.E. (1957) Synthetic mica investigations IX: Review of progress from 1947 to 1955. U.S. Dept. of Interior, Bur. of Mines, Rep't of Invest. 5337.

Hazen, R.M. (1977) Temperature, pressure and composition: Structurally analogous variables. Phys. Chem. Minerals 1, 83-94.

_____ and Burnham, C.W. (1973) The crystal structure of one-layer phlogopite and annite. Am. Mineral. 58, 889-900.

Helgeson, H.C., Delaney, J.M., Nesbitt, H.W. and Bird, D.K. (1978) Summary and critique of the thermodynamic properties of rock-forming minerals. Am. J. Sci. 278A. 1-229.

Höck, V. (1974a) Coexisting phengite, paragonite and margarite in metasediments of the Mittlere Hohe Tauern, Austria. Contrib. Mineral. Petrol. 43, 261-473.

_____ (1974b) Zur metamorphose mesozoischer metasedimente in der mettleren Hohen Tauern. Schweiz. Mineral. Petrogr. Mitt. 54, 567-593.

Ito, J. and Arem, J.E. (1970) Idocrase: Synthesis, phase relations and crystal chemistry. Am. Mineral. 55, 880-912.

Jenkins, D.M. (1983) Upper stability of synthetic margarite plus quartz. Geol. Soc. Am. Progr. with Abst. 15, 604.

Jones, J.W. (1971) Zoned margarite from the Badshot Formation (Cambrian) near Kaslo, British Columbia. Can. J. Earth Sci. 8, 1145-1147.

Joswig, W., Takéuchi, Y. and Fuess, H. (1983) Neutron-diffraction study on the orientation of hydroxyl groups in margarite. Z. Kristallogr. 165, 295-303.

Kato, T., Miúra, Y., Yoshii, M. and Maeda, K. (1979) The crystal structures of 1M-Kinoshitalite, a new barium brittle mica, and 1M-manganese trioctahedral micas. Mineral. J. (Japan) 9, 392-408.

Knopf, A. (1953) Clintonite as a contact-metasomatic product of the Boulder batholith, Montana. Am. Mineral. 38, 1113-1117.

Lacroix, A. (1908) Les minéraux de filons de pegmatite a tourmaline lithique de Madagascar. Bull. Soc. fran. Mineral. 31, 218-247.

Langer, K., Chatterjee, N.D. and Abraham, K. (1981) Infrared studies of some synthetic and natural $2M_1$ dioctahedral micas. N. Jahrb. Mineral. Abh. 142, 91-110.

Lanphere, M.A. and Albee, A.L. (1974) $^{40}Ar/^{39}Ar$ age measurements in the Worcester Mountains: Evidence of Ordovician and Devonian metamorphic events in northern Vermont. Am. J. Sci. 274, 545-555.

Lin, J.-C. and Guggenheim, S. (1983) The crystal structure of a Li,Be-rich brittle mica: A dioctahedral-trioctahedral intermediate. Am. Mineral. 68, 130-142.

Loewenstein, W. (1954) The distribution of aluminum in the tetrahedra of silicates and aluminates. Am. Mineral. 39, 92-96.

Lovering, J.F. and Widdowson, J.R. (1969) Electron-microprobe analysis of anandite. Mineral. Mag. 36, 871-874.

Manning, P.G. (1969) On the origin of colour and pleochnoism of astrophyllite and brown clintonite. Can. Mineral. 9, 663-677.

Mansker, W.L., Ewing, R.C. and Keil, K. (1979) Barian-titanian biotites in nephelinites from Oahu, Hawaii. Am. Mineral. 64, 156-159.

Matsubara, S., Kato, A., Nagashima, K. and Matsuo, G. (1976) The occurrence of kinoshitalite from Hokkejina, Kyoto Prefecture, Japan. Bull. Nat'l Sci. Mus. Ser. C, 2, 71-78.

McCauley, J.W. and Newnham, R.E. (1971) Origin and prediction of ditrigonal distortions in micas. Am. Mineral. 56, 1626-1638.

_____ and _____. (1973) Structure refinement of a barium mica. Z. Kristallogr. 137, 360-367.

Nitsche, K-H., Storre, B. and Töpfer, U. (1981) Experimentelle Bestimmung der gleichgewichtsdaten der Reaktion 1 Margarit + 1 Quartz = 1 Anorthit + Andalusit/Disthen + $1H_2O$. Fortschr. Mineral. 59, 139-140.

Norrish, K. (1973) Factors in the weathering of mica to vermiculite. Proc. Int'l Clay Conf. 1972, Madrid, 417-432.

Olesch, M. (1975) Synthesis and solid solubility of trioctahedral brittle micas in the system $CaO-MgO-Al_2O_3-SiO_2-H_2O$. Am. Mineral. 60, 188-199.

_____ and Seifert, F. (1976) Stability and phase relations of trioctahedral calcium brittle micas (clintonite group). J. Petrol. 17, 291-314.

Pattiaratchi, D.B., Saari, E. and Sahama, Th.G. (1967) Anandite, a new barium iron silicate from Wilagedera, North Western Province, Ceylon. Mineral. Mag. 36, 1-4.

Perkins, D., III., Westrum, E.F., Jr. and Essene, E.J. (1980) The thermodynamic properties and phase relations of some minerals in the system $CaO-Al_2O_3-SiO_2-H_2O$. Geochim. Cosmochim. Acta 44, 61-84.

Rothbauer, R. (1971) Untersuchung eines $2M_1$-Muscovits mit Neutronenstrahlen. N. Jahrb. Mineral. Monatsh. 143-154.

Sagon, J.P. (1967) Le metamorphisme dans le Nord-Est du bassin de Châteaulin découverte de chloritoide et de margarite dans les schistes dévoniens. C. R. Soc. Géol. France 5, 206-207.

_____ (1970) Minéralogie des schistes paléozoiques du Bassin de Châteaulin (massif armoricain): Distribution de quelques minéraux phylliteux de metamorphisme. C. R. Acad. Sci. (Paris) 270, 1853-1856.

_____ (1978) Contribution á l'étude géologique de la partie orientale du bassin de Châteaulin (Massif armoricain): Stratigrapie, volcanisme, métamorphisme, tectonique. Thése de doctorat d'état, Univ. Paris, 671 p.

Schaller, W.T., Carron, M.K. and Fleischer, M. (1967) Ephesite, $Na(LiAl_2)(Al_2Si_2)O_{10}(OH)_2$, a trioctahedral member of the margarite group, and related brittle micas. Am. Mineral. 52, 1689-1699.

Seifert, F. and Schreyer, W. (1971) Synthesis and stability of micas in the system $K_2O-MgO-SiO_2-H_2O$ and their relations to phlogopite. Contrib. Mineral. Petrol. 30, 196-215.

Shannon, R.D. (1976) Revised effective ionic radii and systematic studies of interatomic distances in halides and chalcogenides. Acta Crystallogr. A32, 751-767.

Sokolova, G.V., Aleksandrova, V.A., Drits, V.A. and Vairakov, V.V. (1979) Crystal structures of two lithian brittle micas. Kristallokhimiya i Struktura Mineralov, Nauka, Moscow, 55-66.

Stevenson, R.C. and Beck, C.W. (1965) Xanthophyllite from the Tobacco Root Mountains, Montana. Am. Mineral. 50, 292-293 (abstr.).

Storre, B. and Nitsch, K-H. (1974) Zur Stabilität von Margarit in System $CaO-Al_2O_3-SiO_2-H_2O$. Contrib. Mineral. Petrol. 43, 1-24.

_____, Johannes, W. and Nitsch, K-H. (1982) The stability of zoisite in H_2O-CO_2 mixtures. N. Jahrb. Mineral. Monatsh. 9, 395-406.

Takeda, H. and Burnham, C.W. (1969) Fluor-polylithionite: A lithium mica with nearly hexagonal $(Si_2O_5)^{2-}$ ring. Mineral. J. (Japan) 6, 102-109.

_____ and Morosin, B. (1975) Comparison of observed and predicted structural parameters of mica at high temperature. Acta Crystallogr. B31, 2444-2452.

_____, Haga, N. and Sadanaga, R. (1971) Structural investigation of polymorphic transition between $2M_2$-, 1M-lepidolite and $2M_1$-muscovite. Mineral. J. (Japan) 6, 203-215.

Takéuchi, Y. (1965) Structures of brittle micas. Clays and Clay Minerals 13, 1-25.

_____ and Sadanaga, R. (1959) The crystal structure of xanthophyllite. Acta Crystallogr. 12, 945-946.

_____ and _____ (1966) Structural studies of brittle micas. (I). The structure of xanthophyllite refined. Mineral. J. (Japan) 4, 424-437.

Teale, G.S. (1979) Margarite from the Olary Province of South Australia. Mineral. Mag. 43, 433-435.

Toraya, H. (1981) Distortions of octahedra and octahedral sheets in 1M micas and the relation to their stability. Z. Kristallogr. 157, 173-190.

Udagawa, S., Urabe, K. and Hasu, H. (1974) The crystal structure of muscovite dehydroxylate. Japan. Assoc. Mineral., Petrol., Econ. Geol. 69, 381-389.

Vedder, W. and Wilkins, R.W.T. (1969) Dehydroxylation and rehydroxylation, oxidation and reduction of micas. Am. Mineral. 54, 482-509.

Velde, B. (1971) The stability and natural occurrence of margarite. Mineral. Mag. 38, 317-323.

_____ (1973) Détermination expérimentale de la limite de stabilite thermique superieure des clintonites magnésiennes. Réunion Annuelle Sci. de la Terre, Paris 409 (abstr.).

_____ (1980) Cell dimensions, polymorph type, and infrared spectra of synthetic white micas: The importance of ordering. Am. Mineral. 65, 1277-1282.

Wardle, R. and Brindley, G.W. (1972) The crystal structures of pyrophyllite-1Tc and of its dehydroxylate. Am. Mineral. 57, 732-750.

Yoder, H.S. and Eugster, H.P. (1955) Synthetic and natural muscovites. Geochim. Cosmochim. Acta 8, 225-280.

Yoshii, M., Maeda, K., Kato, T., Watanabe, T., Yui, S., Kato, A. and Nagashima, K. (1973a) Kinoshitalite, a new mineral from the Noda-Tamagawa mine, Iwate Prefecture, Chigaku Kenkyu 24, 181-190.

_____ and _____ (1975) Relations between barium content and the physical and optical properties in the manganoan phlogopite-kinoshitalite series. Mineral. J. (Japan) 8, 58-65.

_____, Togashi, Y. and Maeda, K. (1973b) On the intensity changes of basal reflections with relation to barium content in manganoan phlogopites and kinoshitalite. Bull. Geol. Soc. Japan 24, 543-550.

4. ELECTROSTATIC ENERGY MODELS of MICAS
R.F. Giese, Jr.

INTRODUCTION

The development of electrostatic potential energy calculations began shortly after the first crystal structure determinations of W.L. and W.H. Bragg (1913). Many of the compounds whose structures were determined in these early years were halides, and it was commonly assumed that they were ionic compounds. The picture of their structures at that time was of hard spheres with positive or negative charges arranged so that an ion of one type was surrounded by ions of the opposite charge. The entire crystal could be generated by translation, knowing the positions of the atoms in the unit cell (or asymmetric part of the unit cell). This type of structure lent itself readily to attempts at calculating the energy change involved in forming an ionic structure from the constituent ions. The theory, in its simplest form, said that the total potential energy was the sum of two terms; one described the attraction between the ions and the other was due to the repulsion between neighboring ions.

$$W_{total} = W_{attr} + W_{rep} \qquad (1)$$

The attraction energy was given by the Coulomb electrostatic energy but there was no comparably exact function for the repulsion term.

The fundamentals of the electrostatic energy calculation were described by Madelung (1918) and Born and Landé (1918a,b,c). The calculations of the electrostatic potential energy were known to be difficult because, as will be seen later, they involve the summation of an infinite series which converges slowly. Various attempts have been made to describe the nature of the repulsion energy, and these will be discussed briefly in a later section. For very accurate calculations of the potential energy, the van der Waals, as well as other terms, must be calculated. Here again, the form of these other relations is empirical and involves additional constants to be determined.

The energy, W_{total}, was originally called the "lattice energy" (*Gitterenergie*) since the calculations were for simple structures which could be treated as lattices of positive and negative charges. Most crystalline materials do not have such simple structures and the term "lattice energy" seems inappropriate. The term is, however, entrenched in the literature and will be used here. The term "electrostatic energy" will refer to W_{attr}, the Coulomb energy term.

Much of the work that followed the papers of Madelung and Born and Landé was concerned with calculations of the electrostatic energy of different crystal structure types (the Madelung constant), with improving the accuracy of the repulsion term, with determining the constants necessary to evaluate the repulsion term, and with using the electrostatic energies to calculate thermochemical data (Sherman, 1932). This latter effort is typified by the Born-Haber cycle (Born, 1919; Haber, 1919). As efforts to generalize the calculations to more complex structures were made, it became clearer that the approach was not going to be a universally applicable technique, at least not without great effort. The constants, for example, which appear in the repulsion term were found to vary from structure to structure in an unpredictable way and the computational simplifications possible in simple structures were not applicable to more complex structures in which the atoms were not completely fixed by the space group symmetry.

Since then, the purely computational problems have been solved; partly by the advent of the modern digital computer and partly by the development of algorithms which converge rapidly and are applicable to crystals of any symmetry. There are several computer programs available based on the method of Bertaut (1952) and Ewald (1921), and these will calculate the electrostatic energy of a structure such as phlogopite in 10 sec. or less on a modest-sized computer.

As a result of the intensive investigation of the alkali halides, a methodology emerged which could, in principle, be applied to any crystalline material. In spite of the fact that the electrostatic theory is only valid for purely ionic compounds, it has been applied to many other chemical systems, often with success. The fact that a correct result can be calculated using an ionic theory does not, in any sense, serve as a demonstration that the bonding in a specific system is ionic. Recently, electrostatic energy calculations have been used to investigate the energetics of defects in non-stoichiometric uranium oxide (Catlow, 1977) and corundum and rutile (Catlow et al., 1982). Baur (1965, 1972) has developed the electrostatic calculation as a tool for the location of the hydrogen atoms of water molecules in crystalline hydrates. The approach seems to handle hydrogen bonding very well. Jenkins and his colleagues (see for example: Jenkins and Waddington, 1971; Jenkins and Smith, 1975) have calculated electrostatic energies for a large group of inorganic compounds including cyanides, hexachloroplatinates, chromates, and perchlorates. The success of these efforts suggests that the calculations can be useful even in chemical systems far removed from the ionic ideal.

Of more interest are the numerous mineralogical problems which have been investigated. Catti (1981) has calculated the electrostatic energy of forsterite and the energy of the $[SiO_4]^{4-}$ ion. Ohashi and Burnham (1972) have calculated the electrostatic site energies for various cations in the M1 and M2 sites of pyroxenes. Ohashi (1976) examined the relation between the electrostatic energy and the degree of sharing of oxygens by silica tetrahedra. The relative stabilities of the pyroxenoid minerals have been correlated with the lattice energy by Catlow et al. (1982). A similar study has been done of the alkali feldspars by Brown and Fenn (1979). Among the clay minerals, the hydroxyl orientations in the kaolins (Giese, 1982) have been determined by methods similar to those used by Baur (1965).

This is certainly not a complete list of uses of the electrostatic energy calculations but it serves to illustrate the breadth of possible application. The older literature has been reviewed by Sherman (1932) and the more recent by Tosi (1964) and O'Keeffe (1981).

METHOD OF CALCULATION

Historically, the electrostatic energy calculations were developed for simple structures such as NaCl where all atoms are fixed on symmetry elements and have no variable parameters. In fact, most of the early studies were of structures where atoms were coincident with lattice points. These simple arrays of charged points in space were relatively simple to handle computationally. The general formulation of the electrostatic energy is given by the relation

$$\sum_{i=1}^{N} \sum_{j=1}^{\infty} Q_i Q_j / R_{ij} \qquad (2)$$

where the Q's are the ionic charges on the atoms and R is the distance between the i'th and j'th ions. The summation is over the N atoms in the unit cell and all other ions. Even with a large computer and generous amounts of computer time, the direct summation of the relation is not feasible. Still from a conceptual point of view the relation is useful in indicating the relative contribution of different ions to the electrostatic energy. As mentioned earlier, several algorithms have been devised which simplify the computational problem. These are based ultimately on the fact that the ions in a unit cell are repeated by the unit translations and thus the whole structure can be represented by a Fourier series. The direct sum in Equation 2 can be recast in a form which closely resembles the Patterson function of classical crystallography (Bertaut, 1952).

For simple structures with atoms only on lattice points, the structure can be separated into two parts, a geometrical one representing the lattice and the structural one representing the crystal structure. For these cases, the crystal structure is just a specification of the charges on the atoms. For example, NaCl has the face-centered Bravais lattice with the charges of +1 and -1. Similarly, MgO has the same lattice and charges of +2 and -2. Since the lattices are the same, the contribution to the electrostatic energy from the geometry of the atom positions will be identical. Mathematically, this is equivalent to factoring out the charges in the double summation given above. The resulting sum of alternating unit charges of opposite signs will yield a number which reflects the geometry alone. This is the Madelung number and is a constant for a specific structure type; e.g., for the NaCl structure the Madlung number is 1.74756. The electrostatic energy for this simple structure is

$$W_{attr} = A\, Q_1 Q_2 / R , \qquad (3)$$

where A is the Madelung number, the Q's are as before, and R represents a unit distance in the crystal structure, usually the distance between nearest neighbors.

It can be seen that the natural units for the electrostatic energy are e^2/Ångström. This is not a unit of energy in the systems of units commonly used, let alone in the SI units. The conversion to more common units can be made as follows:

from e^2/Ångström   to kcal/mol   multiply by   332.0701
"   to kJ/mol   multiply by 1389.382
"   to eV   multiply by   14.399

Some care must be used when comparing the numerical results generated by different computer programs because each may use slightly different values for these conversion factors (see Brown and Fenn, 1979, for a discussion of this and other problems). Some programs may actually calculate a "Madelung" number for a given structure by simply dividing the electrostatic energy by some arbitrary distance, for example, by the cube root of the unit cell volume. Such numbers are artificial and have little value.

Examination of the electrostatic energy relation will indicate that the energy will increase negatively (the structure becomes more stable) as the atoms approach each other. If this relation were the only factor, the structure would collapse to a point, but clearly atoms cannot easily approach each other beyond some minimum distance. The mathematical relation to represent this repulsion must increase positively from zero as the interionic distance decreases from the equilibrium value. The original formulation assumed that the energy varied as the inverse power of the interatomic distance

$$W_{rep} = B/R^n \tag{4}$$

where B and n are constants to be determined for specific ion pairs. There are trends for the values of n; in general the larger the ion, the larger will be n. Since the crystal structure is in equilibrium, the derivative of the energy with respect to interatomic distance must be zero, and this can be used to calculate B. For simple structures such as NaCl this leads to the relation

$$W_{total} = A\, Q_1 Q_2 \{1 - 1/n\} \tag{5}$$

(Born and Landé, 1918a).

Other formulations have been proposed for the repulsion term. Quantum mechanics suggests that an exponential is more realistic (Unsöld, 1927; Pauling, 1927), and since the 1930's, the repulsion generally has been used in the form

$$B \exp(-R/\rho) \, . \tag{6}$$

The constants B and ρ are evaluated as described above.

The difficulty with these functions is that they assume that the repulsion is a two-body interaction and that the hardness parameter (n or ρ) is the same for all ion pairs in the structure. More complex functions have been proposed and used. For example, Gilbert (1968) assigned a radius and hardness parameter to each species in the structure and used an exponential relation. This has worked very well for a group of alkaline earth chlorides with different crystal structures (Busing, 1970).

The minimum ionic description must have two terms, the electrostatic term and the repulsion term. In addition, one could add one or more dispersion or van der Waals terms of the type dipole-dipole, dipole-quadrupole and quadrupole-quadrupole. Each, of course, introduces more constants which are to be determined for each crystal.

The general form of the total electrostatic energy is shown in Figure 1. It is seen that for large distances between ions, the energy is almost exclusively due to the Coulomb interaction. For shorter distances, as the repulsion and van der Waals energies become important, the total energy passes through a minimum and increases (positively) very rapidly. The van der Waals energy is attractive and opposite in sign to the repulsion energy. Both are smaller in magnitude than the Coulomb energy at the minimum and their combined

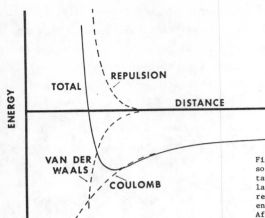

Figure 1. A plot of the lattice energy (TOTAL = solid line) as a function of the interionic distance. Also shown are the contributors to the lattice energy; the Coulomb, van der Waals, and repulsion energies. At equilibrium, the Coulomb energy is slightly larger than the lattice energy. After Busing (1970).

contribution is on the order of 10-15% of the Coulomb energy. For certain applications, one can ignore the repulsion and van der Waals energies, as will be seen later.

A novel approach to the calculation of electrostatic energies has been developed by Jenkins and Hartman (1979, 1980, 1981). Their procedure is akin to the Madelung calculation in that they take the structure of a mica and decompose the electrostatic energy into coefficients derived from the geometry of the structure (Madelung-like constants) and the magnitude of the charges on the ions of the structure. Having calculated the geometric coefficients, which is the time consuming part of the computation, they combine these with the charges to derive rapidly the electrostatic energies of a great variety of di- and trioctahedral micas. There are difficulties if, for example, the generic structure is trioctahedral and one uses it to determine the electrostatic energy of a dioctahedral mica. Their approach has not been widely used by others probably because the amount of computer time required by the direct algorithms is not a real barrier.

APPLICATIONS TO MICAS

Various aspects of the crystal structure and stability of the micas have been examined using lattice (or electrostatic) energy calculations. These studies have attempted to (1) explain the observed features of a given mica structure, (2) estimate the interlayer bond energy, (3) determine hydroxyl orientations, and (4) investigate the energetics of cation order/disorder. Some topics have been extensively studied (e.g., OH orientations) while others have been less popular (e.g., cation order/disorder). The primary difficulty in this type of research, as in most studies based on theoretical chemistry, lies in choosing a model for the calculations which is, at the same time, simple enough to do the calculations and realistic enough to make the results meaningful. In the following sections, I will devote a fair amount of space to a description of the problems involved in setting up each of the computational models. Only when these are understood will the results of the computations have meaning.

STRUCTURAL DISTORTIONS

In a topological sense, the anions in individual layers of phyllosilicates are equivalent to closest packed spheres. The deviations from the ideal arrangement of spheres and the origin of these distortions have interested mineralogists for quite some time. Introduction of cations into the polyhedral voids of the closest packed oxygen sheets creates two types of change; one affects the overall dimensions of the closest packed oxygen layer and the other affects the dimensions of individual polyhedra. The degree of change created by the cations seems to be a function of their size and charge. The commonly observed distortions are, for octahedral sites, shortening of shared edges, rotation of opposite faces, changes in the thickness of the octahedral sheet and various angular relations within the octahedron. The tetrahedral sheet, similarly, shows distortions such as a rotation of tetrahedra, tilting, and elongation or shortening of T-O. These and others are described in Chapter 2 by Bailey.

There exist at present a large number of more or less accurately refined crystal structures of micas of different composition and structure type. The discussion of the origin of the various distortions observed in these micas falls into two categories which one might term "rationalization" and "explanation." Both approaches attempt to predict (often after the fact) the observed distortions in terms of the arrangement of atoms in the individual layers and the manner in which they are stacked. With very few exceptions, the arguments advanced are essentially electrostatic in nature. For example, the mutual electrostatic repulsion of trivalent ions in the octahedral sites of a dioctahedral mica results in the shortening of the edge shared between two occupied octahedra. This statement is a rationalization unless there is an attempt to demonstrate that the distortion corresponds to a minimum in the lattice energy.

Apparently only two attempts have been made to demonstrate that the observed distortions correspond to minimum energy structures (Peterson et al., 1979; Appelo, 1979). The work of Peterson et al. is based on molecular orbital calculations and, hence, falls outside the scope of this paper.

Octahedral distortions

A complete description of a crystal structure based on some sort of bonding model should allow all variable parameters in the structure (unit cell dimensions as well as atomic positional parameters) to relax to the overall energy minimum. This is a formidable computational task and probably not feasible at present for a structure as complex as a mica, but isolated features of a structure can be studied. To demonstrate that a given configuration of atoms represents an energy minimum, as shown in Figure 1, requires that the repulsion term of the total energy be reasonably accurate. For his electrostatic calculations, Appelo (1979) used an exponential form with size and hardness parameters for each ionic species. Appelo examined the variation of the lattice energy as a function of the flattening of the octahedral sheet and the counter-rotation of the oxygen triads of the octahedral sites. He did this for both a $1M$ trioctahedral mica and a $2M_1$ dioctahedral mica. It is known that, in the trioctahedral case, flattening of the octahedral sheet will increase the lateral dimensions of the 2:1 layer but this does not necessarily occur for the dioctahedral case. For the trioctahedral structure, Appelo was unable to find a minimum in the lattice energy, nor was he able to detect any difference in the stability of the structure as the result of F exchange for OH, something which is observed in real mica structures (e.g., Munoz and Eugster, 1974).

Appelo had more success with the $2M_1$ structure for which he found a minimum in the lattice energy as a function of the octahedral rotation (Fig. 2). Separating the electrostatic and repulsion contributions to the total energy indicated that there was no minimum in the former; the minimum resulted from the rapid increase in repulsion as the octahedral rotation increased. This underlines the importance of the repulsion term in studies designed to allow the structure or parts of the structure to assume a minimum energy configuration. An injudicious choice of parameters for the repulsion or even an improper formulation of the repulsion will render the model unstable. It is curious that the calculations work for the $2M_1$ structure but not for the $1M$ structure. A similar disparity has been observed by the author in some calculations done with J.F. Alcover.

Tetrahedral distortion

Appelo also examined the relation between tetrahedral elongation, tetrahedral rotation, and the stability of the mica structure. Here we come upon a problem which is often glossed over but is, in the writer's opinion, of

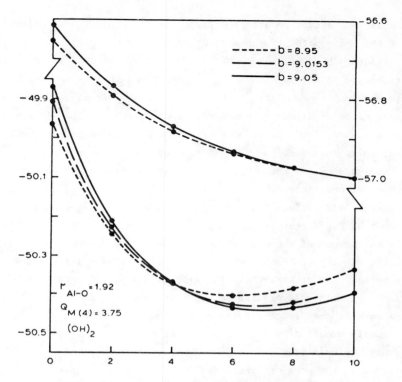

Figure 2. The lattice energy, electrostatic, and repulsion energies as a function of the octahedral rotation in a dioctahedral $2M_1$ mica. The distortion of the octahedron does not change the Al-O or the O-H distances so their contribution to the lattice energy does not contain the repulsion term. From Appelo (1979).

great importance. It has two apsects; what are the real charges on the atoms, and can correct energies be obtained by using formal ionic charges even though these are not chemically realistic. It is well known that the real charges are not the formal ionic charges, yet there is evidence that a reasonably correct total energy can be arrived at using formal charges (Giese et al., 1971; Harrison, 1980; O'Keeffe, 1981). If this is true, we still have a problem in the situation where a site is disordered and occupied by differently charged species. The classic example is Si and Al in tetrahedral sites. In the case of muscovite, the composite cation has a charge of 3.75. It is not at all clear that calculations based on this average charge will give the correct energy for the disordered structure (Giese, 1975a). On the contrary, as will be discussed later, there is good evidence to indicate that such a practice is to be avoided. Because Appelo has used composite charges, it is difficult for this writer to evaluate his discussion of the relative stabilities of the $1M$ and $2M_1$ polytypes.

An analogous situation arises in Appelo's calculation of the lattice energy of the phlogopite structure reported by Joswig (1972). The chemical analysis reported by Joswig showed ($K_{0.90}Na_{0.02}$) as the interlayer cation. This corresponds to a composite charge of +0.92 and this is the value used in Appelo's calculation. Similar deviations from the ideal occupancy of the other ionic sites resulted in "non-ideal" ionic charges. The difficulty, in this writer's view, is to describe correctly a site which is not, or apparently not, fully occupied. Is the discrepancy the result of errors in the chemical analysis or are there really sites in the crystal which are vacant? If the apparent deficiency of the interlayer cations is fictional, then the lattice energy calculations are not representative of the real structure. On the other hand, if there really are interlayer cation sites which are not filled, then the use of a composite interlayer cation cannot accurately represent either that part of the crystal which has a filled interlayer cation nor that part of the crystal where the site is empty. This problem is not restricted to the interlayer sites but exists commonly throughout the micas. A special case is the dioctahedral structure which contains a small occupancy of the normally vacant M1 site. This will be discussed later in connection with hydroxyl orientations.

In summary, the work of Appelo (1979) shows that, in spite of the difficulties in accurately describing the repulsion and other contributors to the lattice energy, some varieties of structural distortions in micas can be shown to be energetically favored.

HYDROXYL ORIENTATIONS

The octahedral and tetrahedral sheets of the micas share a plane of approximately closest packed oxygens. The oxygens are coordinated on the octahedral sheet side by two or three cations (dioctahedral or trioctahedral) and by a single cation on the tetrahedral sheet side. Since the tetrahedra are not closest packed, there are voids (the hexagonal or ditrigonal holes of the tetrahedral sheet) and the oxygen is normally bonded to a hydrogen to form a hydroxyl group. The manner in which the octahedral sheets are sandwiched between tetrahedral sheets results in hydroxyls forming edges shared by two adjacent octahedra. This architecture places the hydroxyls in a position where they can have a great influence on the physical and chemical properties of the mica; an influence which is remarkable because the hydrogen forms such a small part of the whole structure. This influence derives from the fact that the hydroxyl is a molecule which is attached to the structure by one end (the oxygen) and the other end (the hydrogen) is free to move. The hydroxyl lies at the bottom of a fairly large cavity, rather far from the atoms which form the cavity; the tetrahedral cations, the basal oxygens of the tetrahedra and the interlayer cation. In view of the distances between the hydrogen and the oxygens of the cavity, any hydrogen bonding is likely to be very weak. Since the hydroxyl oxygen is fixed, any motion of the hydrogen is equivalent to a change in orientation of the OH group. The hydrogen experiences a myriad of electrostatic attraction and repulsion forces. It seems reasonable to assume that the orientation chosen by the OH is such that these forces balance each other. This balance is equivalent to a minimum in the electrostatic energy of the structure.

Ionic model

Apparently Baur (1965) was the first to use electrostatic calculations to locate hydrogen positions in crystal structures. His particular interest was the orientation of water molecules in crystalline hydrates where one knew from a crystal structure refinement the location of the water oxygen but not the hydrogen positions themselves. The repulsion and van der Waals energies were ignored in these calculations because, as argued earlier, they are small compared to the electrostatic energy and, in the case of a water molecule, the interactions between the water hydrogens and neighboring atoms occur at relatively long distances where these energies should be smaller than for atoms in actual contact. There remained, however, the problem of the appropriateness of the ionic model for a water molecule hydrogen which participated in hydrogen bonding. It has been pointed out (Pauling, 1960; Coulson, 1961) that hydrogen has only one stable orbital which is used to form the principal bond to oxygen, in the case of water or OH. There is no other orbital with which to form a covalent bond with another atom. Since the electron of the hydrogen spends most of the time in the covalent bond, the hydrogen acts like a bare proton, i.e., a small, charged object. This is a prescription for electrostatic interactions. Coulson and Danielsson (1954) investigated the contributions to the total energy for three wave functions for the O–H$\cdots$O group; a non-bonded situation (I), an ionic hydrogen bond (II), and a covalent hydrogen bond (III). The contributions of these three cases to the total wave function was found to vary with the O$\cdots$O distance. For a long hydrogen bond (O$\cdots$O $\simeq$ 2.8 Å) wave function II was approximately six times more important than the covalent wave function. For shorter distances (short hydrogen bond) the ionic wave function was still important, being approximately three times the contribution of the covalent wave function.

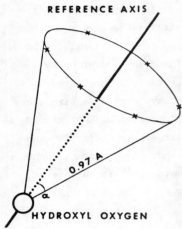

Figure 3. The coordinate system used to find the OH orientation which has the minimum electrostatic energy. The OH orientation is defined by two variables; α is the half angle of the cone which is the locus of all possible OH orientations, and an azimuthal angle which specifies the hydrogen position on the circle at fixed distance (here 0.97 Å) from the apex of the cone (the hydroxyl oxygen). From Giese and Datta (1973a).

In spite of the theoretical justification, the utility of the electrostatic calculation is determined by agreement between the predicted water molecule orientations and the actual water configuration. Baur (1965) investigated six hydrated materials with a total of 28 hydrogen atoms. These ranged from silicates (natrolite) and salts (lithium sulfate monohydrate) to an organic (violuric acid monohydrate). The average deviation between observed and predicted hydrogen positions was 0.1 Å. Considering the diversity of the crystals studied, this demonstrates that the method is useful.

For hydroxyls in phyllosilicates, the description of the orientation of an OH can be described in terms of two angles. If we consider an arbitrary axis through the hydroxyl oxygen, a cone can be drawn with the apex coincident with the oxygen (Fig. 3). The intersection of this cone with a sphere whose origin is the same oxygen and whose radius is the O-H distance (arbitrarily taken as 0.97 Å) is a circle. The circle describes the positions of possible hydroxyl orientations and, by varying the half angle of the cone, the surface of the sphere can be explored. In principle, the electrostatic energy can be calculated for a sampling of points on the sphere and the minimum obtained by interpolation.

Initially, several structures were examined to see if the electrostatic minimum correctly predicted the known OH orientations, the first of which was muscovite-$2M_1$ (Giese, 1971). At that time the best estimate of the OH orientation was that determined by Vedder and McDonald (1963) using infrared absorbance measurements from a single crystal. The orientations of hydroxyls in phyllosilicates are normally described in terms of the angle between the OH and the (001) plane (ρ) and the angle between the projection of the OH onto (001) and the b axis (ψ). In the case of a dioctahedral mica, the asymmetric arrangement of octahedral cations about the OH (two filled sites and one vacant) ensure that the OH will lie close to a plane which bisects the M2-O(H)-M2 and is perpendicular to (001). This means that the angle ψ is not going to vary very much from structure to structure unless the occupancy of the two M2 sites is such that they have very different charges. The electrostatic calculations of Giese indicated that the ρ angle for muscovite-$2M_1$ was 18° and the OH pointed away from the octahedral sheet into the ditrigonal cavity. This compared with the value of 16° from Vedder and McDonald. Muscovite, in common with many micas, has Al and Si disordered in the tetrahedral sites. For the purposes of these calculations, the charge on the T ions was taken as +3.75.

The reasonable agreement between the electrostatic orientation and that of Vedder and McDonald indicated that, at least in the case of muscovite, the calculational approach was valid. Shortly after the paper of Giese (1971) was published, Rothbauer (1971) published the results of a neutron diffraction crystal structure refinement of muscovite-$2M_1$. His orientation was given as ρ = 12°. A recalculation of the OH orientation using the non-hydrogen atom positions of Rothbauer gave a value of 15°. This is still reasonably good agreement.

Possible errors

In all this work, the underlying assumption is that the basic non-hydrogen atom structure is correct. Calculations by Giese and Datta (1973a) on kaolinite, dickite, and nacrite indicated that small errors in the positional parameters of the atoms would not contribute greatly to errors in the orientations of the hydroxyls. However, there is no guarantee that major errors in the crystal structure refinement will not result in incorrect orientations. A clear example of this is the structure of kaolinite which was refined by Zvyagin (1960) and used by Giese and Datta (1973a) for their calculations. A more accurate refinement of the kaolinite structure by Suitch and Young (1983) shows that the structure of Zvyagin is grossly inaccurate and the errors are such that they strongly influence the orientations of the inner surface hydroxyls. A recalculation of the OH orientations in kaolinite using the correct structure of Suitch and Young gives good agreement with the neutron orientations (Giese, 1982). Similarly, the hydroxyl orientation calculated for muscovite-$1M$ (Giese and Datta, 1973b) based on the electron diffraction structure of Soboleva and Zvyagin (1969) is incorrect because their refinement is very poor. Using a more realistic model of the structure gives a much more probable orientation (Giese, 1979; Bookin and Drits, 1982).

OH orientations in micas

The study of the OH orientations in micas began with an examination of muscovite-$2M_1$ (Giese, 1971) and was followed by a comparison of the orientations in the three polytypes of muscovite (Giese and Datta, 1973b) and, lastly, by a study of some twenty-seven 2:1 phyllosilicate structures (Giese, 1979). More recently, the electrostatic calculations of hydroxyl orientations have been taken up by Bookin et al. (1982) and Bookin and Drits (1982). Representative values from these papers are listed in Table 1.

Regarding the utility of the electrostatic calculations in predicting hydroxyl orientations with reasonable accuracy, Bookin et al. (1982) agree with Giese (1979) that the method works. Bookin et al. suggest that the method is not as reliable when applied to the 1:1 structures, particularly in the case of the inner hydroxyl, but this is not yet clear because there remains some ambiguity about the real orientation of the inner hydroxyl (compare, for example, the results of Adams, 1983, and Suitch and Young, 1983). If the people doing neutron diffraction refinements of the hydrogen positions in the 1:1 structures cannot agree among themselves on the correct orientations, it is hardly a definitive test of the electrostatic method.

Table 1 shows a clear difference in the OH orientations of the dioctahedral and trioctahedral micas. The environment of cations about the OH is dramatically different in these two groups; in the trioctahedral structures, three cations are disposed symmetrically about the OH as opposed to two higher charged cations and a vacant site in the dioctahedral case. One would expect that the trioctahedral structure would have the OH directed away from the three M sites toward the interlayer cation while the OH should be directed away from the two M sites and thus inclined to the (001) plane. The values of ρ corresponding to these two cases are $\sim 90°$ and $\sim 15°$, respectively. The real world is more complex, and the complexity largely results from ionic substitution, principally in the M sites but also in the T sites. Where the substitution is disordered, a structure refinement will yield only an average of all unit cells. The octahedral sites show some disorder, but there is a greater tendency for ordering, particularly in the dioctahedral micas, than is found for the tetrahedral sites. When Li is one of the octahedral ions, this ordering changes the electrostatic balance of the three M cations and

Table 1. Hydroxyl orientations determined by electrostatic energy calculations

Mica	a	T1/T1*	T2/T2*	M1	M2	M2*	IC	rho	
dioctahedral micas									
celadonite-1M	B	3.990	3.990	0.000	2.550	2.400	1.090	-14.0	Tsipursky (1979)
celadonite-1M	B	3.990	3.990	0.000	2.570	2.570	0.930	-14.0	Zvyagin (1979)
celadonite-1M	B	3.930	3.930	0.000	2.640	2.640	0.990	-5.0	ideal model (Bookin and Drits, 1982)
glauconite-1M	B	3.910	3.910	0.000	2.710	2.710	0.830	-6.0	ideal model (Bookin and Drits, 1982)
margarite-2M₁	G	3.500	3.500	0.000	3.000	3.000	2.000	4.6	Guggenheim and Bailey (1975)
margarite-2M₁	B	3.147	3.847	0.070	3.000	3.000	1.900	20.0	Guggenheim and Bailey (1975)
		3.908	3.215						
muscovite-1M	G	3.750	3.750	0.000	3.000	3.000	1.000	15.0	ideal model (Giese, 1979)
muscovite-1M	B	3.960	3.800	0.000	3.000	2.800	0.740	11.0	Zvyagin (1979)
muscovite-2M₁	G	3.750	3.750	0.000	3.000	3.000	1.000	17.2	Güven (1971)
muscovite-2M₁	G	3.750	3.750	0.000	3.000	3.000	1.000	15.0	Rothbauer (1971)
muscovite-2M₁	B	3.790	3.790	0.000	2.970	2.970	1.000	15.0	Tsipursky and Drits (1977)
muscovite-2M₁	B	3.875	3.875	0.000	2.865	2.865	0.770	9.5	Zhoukhlistov et al. (1973)
muscovite-3T₂	G	3.778	3.778	0.000	2.994	2.994	1.000	14.3	Güven and Burnham (1967)
paragonite-2M₁	G	3.750	3.750	0.000	3.000	3.000	1.000	20.6	Burnham and Radoslovich (1963)
phengite-1M	B	3.850	3.850	0.000	2.860	2.860	0.880	11.0	Tsipursky and Drits (1977)
phengite-2M₁	G	3.848	3.848	0.000	2.804	2.804	1.000	1.3	Güven (1971)
pyrophyllite-1Tc	G	4.000	4.000	0.000	3.000	3.000	0.000	23.1	Wardle and Brindley (1972)
trioctahedral micas									
Ba mica-1M	G	3.750	3.750	1.666	1.667	1.667	2.000	166.7	McCauley and Newnham (1973)
Li mica-1M	G	3.183	3.183	1.917	1.917	1.917	1.000	92.5	Takeda and Donnay (1966)
Mg mica-1M	G	3.830	3.830	1.893	1.893	1.893	1.000	91.5	Tateyama et al. (1974)
annite-1M	G	3.703	3.703	2.063	2.063	2.063	1.000	91.5	Hazen and Burnham (1973)
biotite-1M	G	3.706	3.706	2.058	2.058	2.058	1.000	88.6	Takeda and Ross (1975)
biotite-1M	B	3.710	3.710	2.340	2.340	2.340	0.980	90.0	Takeda and Ross (1975)
biotite-1M	G	3.706	3.706	2.058	2.058	2.058	1.000	84.3	Takeda and Ross (1975)
biotite-2M₁	B	3.710	3.710	2.340	2.340	2.340	0.980	85.0	Takeda and Ross (1975)
ferrianite-1M	G	3.750	3.750	2.000	2.000	2.000	1.000	87.4	Donnay et al. (1964)
lepidolite-2M₂	G	3.840	3.840	1.100	2.260	2.260	1.000	67.5	Sartori et al. (1973)
lepidolite-2M₂	G	3.848	3.848	0.690	2.460	2.460	1.000	39.9	Takeda et al. (1971)
phlogopite-1M₁	G	3.750	3.750	2.000	2.000	2.000	1.000	93.1	McCauley et al. (1973)
polylithionite-1M	G	4.000	4.000	1.200	1.900	1.900	1.000	73.5	Takeda and Burnham (1969)
protolithionite-3T	B	3.860	3.660	1.520	1.520	1.520	1.020	70.0	Pavlishin et al. (1981)
talc-1Tc	G	4.000	4.000	2.000	2.000	2.000	0.000	91.0	Raynor and Brown (1977)
zinnwaldite-1M	G	3.789	3.789	3.000	1.423	1.423	1.000	115.6	Guggenheim and Bailey (1977)

[a] B = Bookin and Drits (1982); G = Giese (1979)

Table 2. Hydroxyl orientations for trioctahedral micas determined for different M site occupancies.

Mica	M1	M2	M2*	rho	
Ba mica-1M	1.667	1.666	1.666	166.7°	(from Table 1)
	1.000	2.000	2.000	20.9	complete order in M sites[1]
	2.000	1.000	1.000	172.3	complete order in M sites[2]
polylithionite-1M	1.200	1.900	1.900	73.5	(from Table 1)
	3.000	1.000	1.000	183.3	complete order in M sites
zinnwaldite-1M	3.000	1.423	1.423	115.6	(from Table 1)
	1.949	1.949	1.949	87.4	complete disorder in M sites
	2.000	3.000	0.846	124.5	complete order in M sites

[1] the T sites are occupied by Si_3Al.
[2] the T sites are occupied by +4 charges only.
(from Giese, 1979)

will result in a shift of the OH away from the normal to (001) toward, in general, the lower charged M site. In the extreme case where Li in the octahedral sites is accompanied by a +2 interlayer cation, the OH orientation will be more like that of a dioctahedral mica (see, for example, the Ba mica-1*M* in Table 1).

For reasons given earlier, OH orientations based on average cation charges where a site is disordered are unlikely to correspond to real orientations in the crystal. Intuitively, the expectation is that the greater the difference in charge between the cations which are occupying a disordered site, the greater will be the discrepancy between the calculated "average" OH orientation and the real ones. Some indication of the magnitude of the deviations to be expected between ideal and real orientations can be gotten by comparing the "average" orientation with those for various ordered charge distributions. Since we do not actually know what these are in the structure, we can only try different combinations. This amounts to a certain degree of trickery since the resulting models do not correspond to an experimentally observed situation. The procedure can also be criticized because redistributing charges without also changing the sizes of the octahedra, for example, to reflect the new occupancy means that the model is incomplete. Still, it is a simple way to estimate the magnitude of the reorientations which are possible. Table 2 lists the hydroxyl orientations for several trioctahedral micas which contain some Li in the octahedral sites. The orientations are given for the observed disordered distribution and one or more ordered structures. The differences in orientation between the "average" and possible "real" structures are substantial, especially for a +2 interlayer cation. Thus, we should not be too surprised to find that the real hydroxyl orientations in micas are different and often very different than those derived either from a diffraction experiment or some sort of calculation based on the disordered charges.

Mixed tri- and dioctahedral micas

Use of composite ionic charges to represent disordered cations is risky because, as outlined earlier, these charges by definition cannot be accurate representations of any part of the structure. Nonetheless, the assumption has been made (Giese, 1979; Bookin and Drits, 1982) that use of composite charges will give "average" hydroxyl orientations. Since the average atom position in the unit cell is what comes out of a structure refinement, this seems reasonable. There are two cases where, to this author, the practice is bad; one has been alluded to earlier (i.e., use of a partial charge to represent vacancies in the interlayer cation site), and the second is the introduction of a partial charge to represent the small amount of scattering matter sometimes found in the normally vacant M1 site of a dioctahedral mica. These are risky practices because occupancy or vacancy of either type of site changes completely the character of the electrostatic field about the hydroxyl hydrogen. It has been suggested, for example, that if the interlayer cation were absent from a dioctahedral mica, the OH would reorient to $\rho = 60-70°$ (Radoslovich, 1963). Giese (1971, 1979) has verified this prediction for muscovite $2M_1$ although the amount of reorientation is a bit less than suggested by Radoslovich. A vacant interlayer site in a dioctahedral mica can cause a marked and difficult-to-predict deviation from the normal OH orientation. The difficulty in prediction stems from the high probability that the vacant interlayer cation site is accompanied by compensating changes in the charges on the T and perhaps M sites. The second factor, and in this writer's opinion the more important one, is the practice of Bookin and Drits (1982) to include the small charge on the M1 site resulting from some small amount of trioctahedral character in the mineral specimen. Since the occupancy or vacancy of the M1 site is so important for OH orientation, it seems particularly necessary to account correctly for the fact that a given crystal is strictly dioctahedral. Giese (1979) has taken the view that the best way to proceed is to ignore the small charge on the M1 site and determine the OH orientation for the purely dioctahedral part of the crystal. On the other hand, Bookin and Drits (1982) introduce the charge on the partial M1 cation into their calculations because they wish to determine the position of the hydrogen "mean-weighted over the whole crystal." The utility of a "mean-weighted" orientation when the contributors are so different is difficult to see. Margarite-$2M_1$ is a clear example. Joswig et al. (1983) have recently refined this structure using a combination of X-ray and neutron diffraction data. They found that the M1 site had a small amount of scattering matter corresponding to an occupancy of approximately 21% of the M1 sites. The refinement of the neutron data showed two peaks for the hydrogen; one corresponding to the normal trioctahedral orientation ($\rho = 78°$) and the other belonging to a dioctahedral hydroxyl ($\rho = 12°$). The occupancy of each of these two hydrogen sites matched very well the 21% trioctahedral character indicated by the chemical analysis and the scattering matter in the M1 site. A similar situation has been described by Lin and Guggenheim (1983) for the structure of a brittle mica. Here the M1 site is occupied by 0.55 Li so their crystal was close to being half trioctahedral and half dioctahedral. They observed two hydrogen positions for each hydroxyl, one having a ρ value of approximately 58° and the other a ρ of approximately 20°. These are the values to be expected for a trioctahedral brittle mica with Li in M1 (58°) and a dioctahedral brittle mica with vacant M1 (20°).

When one looks at the variety of chemistry and structure among the micas, it is natural to ask how the observed variation influences the hydroxyl orientation. Two attempts have been made to estimate the importance of various structural parameters on the hydroxyl orientation. Giese (1979) has used multiple regression based on 31 structures or modifications of structures to assess the factors which control the hydroxyl orientations. He chose to evaluate as predictors the charges on the M, T, and interlayer cation plus the tetrahedral rotation angle, α. The great dissimilarity in hydroxyl orientation between the dioctahedral and trioctahedral micas means that to obtain the most accurate prediction of the angle ρ, each should be estimated independently. For the dioctahedral micas, the angle ρ was determined by the charges on the three cation sites and not by the tetrahedral rotation. Trioctahedral micas were accurately predicted by the charge on the M1 site and the amount of tetrahedral rotation. Since then, more orientations have been published (Bookin and Drits, 1982), and it might be useful to redo the regressions using a larger data set. This is particularly important for the dioctahedral micas since the original data did not include micas with negative values of ρ (Giese, 1979). Giese (1979) found that the angle ρ could be predicted for the dioctahedral micas using the average T and M2 charges and the charge on the interlayer cation. The regressions reported here used the data in Table 1 and give the following predictor equation:

$$\rho = -624.93 + 84.46122(T) + 102.8414(M2) + 15.9195(IC) .$$

The equation has a correlation coefficient of 0.94, an average deviation between observed and predicted angle of 2.8° with a maximum deviation of 8° (for margarite-$2M_1$). That the largest error is found for margarite is not surprising because the T sites in margarite are completely ordered and this mica has the largest Al occupancy in the T sites of all the micas in Table 1. Thus, the model of the regression is least appropriate for this mica.

Giese (1979) found that the orientation of the OH in trioctahedral micas could be predicted by a regression equation based on the M1 charge and the amount of tetrahedral rotation. This equation is

$$\rho = 48.767 + 23.310 \text{ (M1)} - 0.851(\alpha)$$

and has a maximum deviation of 6.0°.

INTERLAYER BONDING

The chemical and physical properties of the phyllosilicates indicate that the bonding in the structure is highly anisotropic, and the anisotropy is related to the layer charge; low charge smectites spontaneously swell in the presence of water or organic liquids while the micas and brittle micas do not. Another manifestation of anisotropy is the Mohs hardness of the micas which varies with crystallographic direction. Switzer (1941) estimated the hardness of muscovite and biotite single crystals as 2.5 parallel to (001) and 4.0 perpendicular to (001). The traditional rationalization of these and similar observations has been that the ionic bonding between the layers is weaker than the bonding between atoms within the layers. This accounts qualitatively for the perfect (001) cleavage of the micas as well as the swelling properties of the 2:1 structures.

Ionic model

It has been assumed for some time that the bonding between layers of phyllosilicates (with the exception of talc and pyrophyllite) arises from the electrostatic attraction between the uncompensated ionic charges of the layer and the interlayer cation. In the case of muscovite, the uncompensated charge is due to Al for Si substitution in the tetrahedral sites. This net negative charge is physically very close to the interlayer K, leading to a strong electrostatic attraction between the layers. The role of the layer charge has been viewed as sufficiently important in determining the physical and chemical properties of the phyllosilicates that they have been classified partly on the magnitude of the layer charge (Mackenzie, 1965; Brindley, 1966; Pedro, 1967).

Substitution of cations can also occur in the octahedral sites and can lead to an uncompensated charge. In general, the net layer charge may have contributions from both the T and M sites, and some of these contributions may be, in fact, destabilizing (positive charge). Since the distances from the M sites to the interlayer cation are greater than for the T sites, it would be expected that an octahedral layer charge would result in a weaker interlayer bond than for the case of an equivalent charge on the T sites.

Calculation of interlayer bonding

As can be seen in Equation 2, the energy of the structure results from all possible two-body interactions. This includes much more than just the attraction between the interlayer cation and the net charge of the adjacent layers. It is not clear how the summation of Equation 2 for a specific mica would allow one to identify that part of the total energy which is due to the interlayer bonding. On the other hand, if one imagined the layers of the micas as being some sort of large ionic molecule, then the interlayer bonding would be the energy necessary to separate these molecules by an infinite distance. This is analogous to the definition of the electrostatic energy of an ionic crystal except that the ions are replaced by the mica layers. It would be a difficult task to derive an algorithm analogous to Equation 1 which would directly yield the interlayer bonding. An alternate approach has been used (Giese, 1974) based on the computation of the energies of a series of structures which contain a rigid layer but whose interlayer separation varies in a systematic manner. To take muscovite-$2M_1$ as an example, one would transform the normal structure into a larger unit cell whose a and b axes are fixed and the c axis and the interaxial angles are modified to yield a specific increase in the interlayer separation. Many of the positional parameters of the atoms have to be recalculated to maintain fixed interatomic distances and angles within the layer, but this is a straightforward procedure. The calculated energy for the modified structure will be less than for the original by an amount which estimates the energy necessary to separate the layers by the distance in question. By incrementally increasing the separation of the layers, a curve analogous to Figure 1 can be traced and, in many instances, the energies approach asymptotically a limiting energy. Extrapolation yields an estimate of the total energy involved. This is equivalent to estimating the (001) surface energy of the mica. Calculations of the surface energy of NaCl indicate excellent agreement with standard values (Giese, 1974).

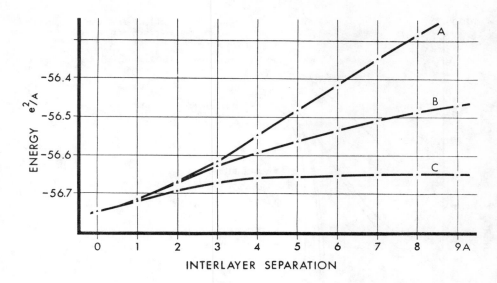

Figure 4. The electrostatic energy for muscovite-$2M_1$ as the rigid layers are separated in an orthogonal manner. The curve A has all the interlayer cations attached to one surface of the layer, curve B has all the interlayer cations fixed in a plane equidistant from the adjacent layers, and curve C has the interlayer cations distributed equally on both adjacent surfaces in an ordered manner. From Giese (1974).

This computational approach was first applied to muscovite-$2M_1$ (Giese, 1974). The calculation for the atoms belonging to the layers is simple, but the interlayer cations pose a problem since they do not, in a sense, belong more to one adjacent layer than to the other. Giese examined three possible distributions of the interlayer potassium for the separated layers; A with all the K on one surface, B with all the K restricted to a plane equidistant from the separated layers, and C an ordered distribution of half the K on one surface and the other half on the adjacent surface. Case A creates layers which have a net polar charge and hence is expected to be energetically unfavorable, B maintains electrostatic neutrality but is unfavorable because the repulsive K-K interactions remain unchanged while the K-O attractive interactions diminish in importance, and C reverses the trend of B and should be the most favorable. The energies plot as expected (Fig. 4). Case C is seen to reach a stable energy after expansion of the layers by about 7 Å giving a surface energy of 32 kcal/mol. Experimental measurements of the surface energy of muscovite in a hard vacuum (168 kcal/mol, Bryant, 1962) are larger than the calculated energy, indicating that the K ions are readily disordered during cleavage. Deryagin and Metsik (1959) have observed that the more rapidly the mica specimen is cleaved, the larger the energy needed to perform the cleavage. This corresponds to an experimental

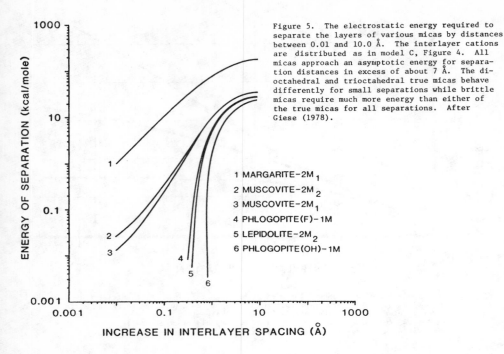

Figure 5. The electrostatic energy required to separate the layers of various micas by distances between 0.01 and 10.0 Å. The interlayer cations are distributed as in model C, Figure 4. All micas approach an asymptotic energy for separation distances in excess of about 7 Å. The dioctahedral and trioctahedral true micas behave differently for small separations while brittle micas require much more energy than either of the true micas for all separations. After Giese (1978).

curve lying between curves A and C in Figure 4 being displaced vertically as the abruptness of the cleavage increases.

These calculations assume, as outlined before, that the electrostatic part of the total ionic energy dominates over the repulsion and van der Waals energies. Estimates of the contributions of these three to the total indicate the essential correctness of the assumption, at least in the case of muscovite (Bailey and Daniels, 1973). In the case of talc and pyrophyllite where there is neither layer charge nor interlayer cation, the assumption has been that the layers are held together by van der Waals forces (see, for example, Ward and Phillips, 1971). A calculation similar to that for muscovite in the example above shows that, on the contrary, there is an ionic attraction between the layers of 6.5 kcal/mol for pyrophyllite and 4.1 kcal/mol for talc. These net attraction energies result from the summing of all the two-body interactions, even though there is no layer charge.

Interlayer bonding of micas

It would be a bit rash to assume that the energies calculated as described here were absolutely correct, but it seems reasonable to use the values in a comparative sense. This has been done by Giese (1978) for a group of representative phyllosilicates which included true micas, brittle micas, 1:1 minerals, and hydroxides. The results for the micas are shown in Figure 5 plotted as energy change versus the distance by which the layers have been separated. The curve for the brittle mica margarite lies at the highest energies reflecting the large layer charge and the divalent nature of the interlayer cation. The true micas form two groups; one being the dioctahedral micas and the other the trioctahedral micas. Both groups of micas require similar energies to separate the layers by 10 Å, but as the separations become smaller than 1 Å, the two groups diverge dramatically. There is sufficient variety of structure and chemistry to suggest strongly that the difference in behavior of the di- and trioctahedral micas is not related to obvious things such as F for OH substitution or polytype. The inference is that there is a fundamental difference between these two types of mica due to the fact that one has +3 charges in two-thirds of the M sites and the other has +2 charges in all the sites. Why this translates into a difference in the interlayer bonding is not at all clear. Earlier it was mentioned that the electrostatic model used by Appelo (1979) worked for dioctahedral but not trioctahedral micas. The interlayer bonding differences between these two groups may be another example of the dichotomy between the dioctahedral and trioctahedral micas.

Looking at the energies of the micas in Figure 5, it can be seen that the interlayer bonding of brittle micas is always greatest, and one would expect that the anisotropy of bonding in the layer compared to between the layers would be smallest resulting, probably, in the brittle nature of these minerals. Similarly, chemical exchange of the interlayer cations would be most difficult for the brittle micas followed by the dioctahedral micas and lastly the trioctahedral micas.

Jenkins and Hartman (198]) have concluded that a layer charge situated in the octahedral sites produces a stronger interlayer bond than if the charge were due to tetrahedral substitution. This seems contrary to one's expectations because the distance between the octahedral sheet and the interlayer cation is always greater than the tetrahedral sheet to interlayer cation distance. This question has not been examined elsewhere except indirectly by Giese (1978), who showed that the surface energy of a lepidolite-$2M_2$ was nearly identical to that of a phlogopite-$1M$, both in the form of the F-mica. The lepidolite-$2M_2$ had half the layer charge on the octahedral sites and half on the tetrahedral sites.

Substitution of F for OH

The OH in the trioctahedral structures, with the exception of the unusual charge distributions mentioned earlier, is essentially perpendicular to (001). This places the hydrogen much closer to the interlayer cation than is the case for the dioctahedral hydroxyl. The fact that the hydroxyl oxygen is displaced from the center of the ditrigonal cavity of dioctahedral structures is of secondary importance. It was pointed out earlier that the removal of the K from the muscovite-$2M_1$ structure (Giese, 1971) resulted in a substantial increase in ρ, indicating that the dioctahedral structure can accommodate the K-H repulsion by increasing the interionic distance as the OH orients away from the K. Such a mechanism is not available to trioctahedral structures except in the sense that a substitution of F for OH removes the K-H repulsion completely. A simplistic picture of the trioctahedral structure would envision the K-H repulsion as propping up the layers and destabilizing the interlayer cation. It is known that the weathering of biotite involves K exchange, and this is easiest if the octahedral iron remains in the +2 state. Oxidation of the iron to +3 results in an imbalance of charge in the octahedral sheet which is compensated for by loss of Fe and a change from trioctahedral to dioctahedral character adjacent to these newly vacant sites (Gilkes et al., 1972). This is accompanied by hydroxyl reorientation and a fixing of the interlayer potassium. These relations are particularly clear when F substitutes for OH; trioctahedral F-micas have a shorter c axis than the corresponding OH-micas (Noda and Roy, 1956; Noda and Ushio, 1966), but there is no equivalent change for the dioctahedral micas (Yoder and Eugster, 1955).

Giese (1975b) has modeled this dichotomy between trioctahedral and dioctahedral structures by calculating the energy required for a specific increase in the interlayer distance as described before. This was done for several structures, and the results are plotted in Figure 6. In this plot, A is a muscovite-$2M_1$, B and C are two different phlogopite-$1M$ structures, D is pyrophyllite-$1Tc$, and E is talc-$1Tc$. There are two curves for each structure and the vertical scale is in energy units with an arbitrary scale. One of the curves represents the case of the hydroxyl mica and the second is exactly the same structure but with F replacing OH. For the dioctahedral mica (A) and the structures with no interlayer cation (D, E), there is essentially no difference between the F and OH structures in terms of interlayer bond energy. There is a marked change for the two trioctahedral micas (B, C)

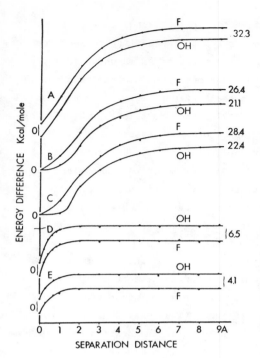

Figure 6. The electrostatic energy required to separate the layers of various F- and OH-micas. All interlayer cations are distributed as in Figure 4, model C. Curve A is muscovite-$2M_1$, curves B and C are phlogopite-$1M$, curve D is pyrophyllite-$1Tc$, and curve E is talc-$1Tc$. Replacement of OH by F does not change the interlayer bonding for the dioctahedral mica (A) or for the structures with no interlayer cation (D and E). The trioctahedral micas are strongly affected by the exchange.

of some 20-30%, the F-mica having the stronger interlayer bonding. The difference in energy reflects the strong H···K repulsion.

These energy differences are in agreement with the observations that the exchange of the interlayer cation is easier for trioctahedral micas and the c axis of a trioctahedra F-mica is shorter than the equivalent OH-mica. It can easily be seen that removal of some octahedral cations from a trioctahedral mica during weathering will increase the stability of the mineral to chemical weathering by stabilizing the interlayer cation.

Jenkins and Hartman (1981) agree that the F-micas have a greater interlayer bonding energy than the equivalent OH-micas, but they find the difference to be only a "few kJ mole^{-1}." This is for a model in which the interlayer cation (K in this case) remains halfway between the separating layers. This was shown (see above) to be a very energetically undesirable structure which, in all likelihood, dominates the total electrostatic energy, thus masking the difference between the F and OH forms of the micas.

CATION ORDERING

Much of our understanding of the properties of crystalline materials is based on our knowledge of crystal structures. The majority of crystal structure analyses are derived from diffraction experiments, most often using X-rays. The power of the technique rests on the ability to represent the crystal structure as a relatively small pattern of atoms which is reproduced in three dimensions. Such a representation is well suited to crystals which have a high degree of regularity in their structures. It is precisely the economy of representing an entire crystal structure this way which makes the technique much less useful in studying disordered crystals.

A crystal whose atomic arrangement does not vary from one region to another is said to be long-range ordered. Disorder over the dimensions of the crystal does not preclude the existence of regions in which a local or short-range ordering can exist. The distinction between long- and short-range is a bit arbitrary and is related to some extent to the technique used to examine the state of order.

There are different types of disorder, but the one which is most often encountered in silicates, and the one which is pertinent to this discussion, arises from the occupancy of an atomic site in a crystal structure by more than one type of atom. The examples of this substitutional disorder are numerous and well documented, Al and Si perhaps being the best known and most studied. A classical crystal structure refinement of a mineral in which substitutional disorder occurs yields a structure model which is averaged over the whole crystal. Thus, if a disordered site is occupied by atoms X and Y in equal proportion, the refinement will indicate that the site is occupied by a composite of $\frac{1}{2}$X and $\frac{1}{2}$Y. Without pressing the point too far, the disorder in a site could result from a variety of structures which have as the extremes the case where half the crystal is all X, the other half being all Y, and a completely random arrangement of A and B throughout the whole crystal. The segregated arrangement would probably be called a twinned crystal while the other extreme represents true disorder. Somewhere between the extremes lie the arrangements of A and B which we would refer to as short-range ordered.

Much attention has been paid to cation order/disorder in the micas. This is discussed in detail in Chapter 2 and in a paper by Bailey (1984). There are examples of disordered (muscovite-$2M_1$), partially ordered (muscovite-$3T$), and completely ordered (margarite-$2M_1$) Al and Si in tetrahedral sites, but disorder is more common. For the octahedral sites, dioctahedral micas generally show a large degree of order; i.e., the M1 site is vacant, while among the trioctahedral micas the M1 and M2 sites are more often partially disordered. We intuitively expect nature to be ordered so the origin and nature of the cation disorder in the micas, as among other minerals, is of interest. In addition, the disorder may tell us something about the thermal history of a mica sample.

Theoretical considerations

As has been seen earlier in this chapter, qualitative arguments which attempt to rationalize a given crystal structure are often cast in terms of the electrostatic energy of the system. The simplest form of the argument comes directly from Pauling's electrostatic rule which states that the sum of the strengths of the ionic bonds centered on an anion will equal the ionic charge of the anion (cf. Bloss, 1971). The charge on an oxygen shared by two tetrahedra will be completely satisfied if both tetrahedra contain silicon; it will be less well balanced if one of the cations is trivalent (Al for example) and the deviation from neutrality will become even greater for two trivalent cations. In principle, an oxygen bridging two Al tetrahedra is less stable than an equivalent oxygen between two Si tetrahedra. This is a very simple way of looking at stability because it examines only a small part of the structure and there may be other consequences of a particular ionic distribution which compensate for the local imbalance. The application of Pauling's electrostatic rule is, of course, valid for any oxygen bridging two or more polyhedra.

Loewenstein (1954) has argued that Al-O-Al linkages are avoided in crystal structures because Al can occur in octahedral coordination as well as tetrahedral and if two Al share an oxygen, at least one of them should be octahedrally coordinated to increase the Al···Al distance and decrease the electrostatic repulsion. Looked upon as an underbonded bridging oxygen or as excessive cation-cation repulsion, the result is the same; avoidance of adjacent Al tetrahedra.

The Al-avoidance rule indicates that, as the proportion of Al occupying a Si site increases, there will be a tendency to have Al and Si alternate in a locally ordered fashion. The limit would be half Al and half Si and would result in a completely long-range ordered structure. Examples in the micas are muscovite-$2M_1$ (Al-Si = 1:3; long-range disordered) and margarite-$2M_1$ (Al:Si = 1:1; long-range ordered). The Al-avoidance rule predicts the long-range ordering found in margarite-$2M_1$ but is of no help in determining whether short-range order exists in muscovite-$2M_1$.

A study of the electrostatic energy of various Al,Si distributions in zeolites indicated that, in addition to the avoidance of Al-Al tetrahedra, there is also an avoidance of Al-Si-Al triplets (Dempsey, 1968; Dempsey et al., 1969) in 4-rings and in 6-rings.

In order to predict short-range ordering patterns or to evaluate the energetic advantage of order over disorder, one must look at the whole crystal structure and the dependence of the energy on different ordering patterns. To the best of the author's knowledge, there are only two studies which attempt to evaluate the energetics of ordered versus disordered cation sites in micas. One deals with the octahedral sites of trioctahedral micas and the other concerns the tetrahedral sites of dioctahedral micas. Neither exhausts the possibilities for exploring this interesting topic.

Before describing the work which has been done in this area, one should discuss briefly the types of models which can and have been used for micas and other systems. As in all attempts to represent a real crystal structure by a mathematical model, various assumptions and simplifications must be made. Arbitrarily, we will define a zero-order model as one in which the cation disorder is represented by a single structure with an "average" charge on the disordered site; a first-order model will represent disorder by randomly distributing the correct charges over the available sites, and a second-order model will, in addition to the first-order model, take into account the varying sizes of the cations and the concomitant changes in the atomic parameters. Other factors can also be introduced but each increase in the complexity of the model increases enormously the length of the computation. Ultimately, the choice of model for a particular problem is determined by the amount of computer time available.

Order/disorder in trioctahedral M sites

In trioctahedral micas such as biotite, where there is extensive substitution of +3 and +2 cations as well as vacancies in the octahedral sites, the distribution of these species should to a large degree be influenced by the requirement of electrostatic neutrality. Krzanowski and Newman (1972) examined the patterns which resulted from imposing a more or less stringent rule concerning the allowable deviations from perfect charge balance (in the sense of Pauling's electrostatic rule). Their model is equivalent to the first-order type. Consider the hydroxyl oxygen in a trioctahedral mica. It is coordinated by a hydrogen and three octahedral cations and the ideal case, as in phlogopite, has three +2 cations in the M sites. It is unlikely that in a mineral with substantial octahedral substitution (+3 cations and vacant sites) there can be complete balance for every such oxygen. Still, it seems

Table 3. Octahedral cation combinations in a hypothetical biotite (after Krzanowski and Newman, 1972).

cation combinations	value	A	B	C	D	E
(+3), (+3), (+3)	9	0.003	0.000	0.000	0.000	0.000
(+2), (+3), (+3)	8	0.056	0.060	0.000	0.045	0.001
(+2), (+2), (+3)	7	0.304	0.313	0.424	0.291	0.346
(+2), (+2), (+2)	6	0.566	0.555	0.506	0.597	0.578
(+3), (+3), ( )	6	0.002	0.002	0.003	0.008	0.022
(+2), (+3), ( )	5	0.018	0.021	0.027	0.059	0.053
(+2), (+2), ( )	4	0.049	0.048	0.040	0.000	0.000
(+3), ( ), ( )	3	0.000	0.000	0.000	0.000	0.000
(+2), ( ), ( )	2	0.002	0.000	0.000	0.000	0.000
( ), ( ), ( )	0	0.000	0.000	0.000	0.000	0.000

A = random number generated
B = 4 < score < 8
C = 4 < score < 7
D = 5 < score < 8
E = 5 < score < 7
(underlined values are forbidden by the choice of score limits)

improbable that three +3 cations or three vacant sites would coordinate a single OH. Krzanowski and Newman formulated the problem in terms of the sum of the charges on the M sites plus the hydrogen. Ideally, this sum should be 7, and it will not change substantially for those oxygens which are co-ordinated by a tetrahedral cation rather than a hydrogen. By assuming a value of $\frac{1}{6}$ as the maximum deviation allowed for the sum of positive electrostatic bonds centered on an oxygen, Krzanowski and Newman determined that the sum of the valences on three M sites coordinating an oxygen must lie between 6 and 7 or 5 and 8 depending on the stringency required for the application of Pauling's rule.

They constructed their model by first specifying the proportion of +3, +2 and vacancies, such that the overall electrostatic neutrality of the chemical formula was maintained, and then randomly distributing cations in these proportions over an hexagonal lattice. This pattern is the same as the M sites in the octahedral sheet and between every three lattice points is an oxygen joined to either an OH or a tetrahedron. For a model with R^{3+} = 0.150, R^{2+} = 0.825, and R^0 = 0.025, the random distribution of cations yielded the triplet groups indicated in Table 2. It can be seen that there is a small but still finite probability of finding an oxygen coordinated by three +3 M cations and either a hydrogen or T cation. There is an even better chance of finding two +3 and a +2 cation in adjacent M sites. Thus the completely random model had regions which should not have been there. The second stage of their study involved rearranging the randomly distributed cations. They searched the hexagonal lattice representing the octahedral cations and each time they found an oxygen receiving ionic bonds which exceeded preset limits (e.g., less than 5 or greater than 8), they removed one of the three cations and exchanged it with another somewhere else in the lattice if the exchange removed the difficulty with the first triplet and did not create an imbalance of charge in the second site. Thus, exchange of a +3 cation from a triplet with too much charge with a vacancy from another triplet with too little charge will lead to a better balance in both sites. Krzanowski and Newman found that one complete pass through the lattice usually was sufficient. Their results for the model described above are also listed in Table 3. Comparison of the new distributions with the random one shows a clear improvement in the local charge balance.

Krzanowski and Newman found that the rearrangement of charges created local patterns of substituting cations. One of these patterns consisted of three +3 ions grouped about a vacant site.

```
       +3
       |
       0
    \ / \ /
    +3   +3
```

These coalesced into chains, the so-called "Y" chains. A second arrangement of +3 and vacant sites was observed to form the "X" chains.

```
   +3    +3    +3
  / \   / \   / \
     0     0
  \ /   \ /   \ /
   +3    +3    +3
```

A third, and less common, chain-forming element appeared in those configurations having substantial quantities of +3 and vacancies. This is the arrangement termed the "Z-configuration" by Krzanowski and Newman.

```
    +3           +3
   / \          / \
       0 - +3 - 0
   \ /          \ /
    +3           +3
```

Long chains of the Z-type are not often formed, probably because the ratio of +3 to 0 charges is too large and creates local imbalance in the structure.

Figure 7 shows the pattern of octahedral cations derived for a structure with the following probabilities: R^{2+} = 0.333, R^{3+} = 0.445, and R^0 = 0.277. The restriction was taken to be $5 \leq V \leq 7$. The plot shows vertical alignments

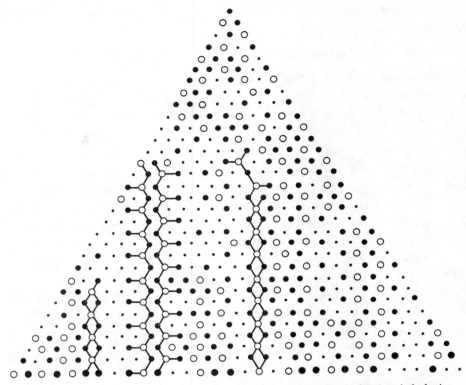

Figure 7. The pattern of octahedral cations generated for a hypothetical trioctahedral mica containing vacancies, divalent and trivalent species. The probabilities for each type of ion are; $R^{3+} = 0.445$, $R^{2+} = 0.333$, and $R^0 = 0.277$. The randomly distributed ions were rearranged according to the criterion that the total charge on the ions coordinating an oxygen should fall in the range $5 \leq V \leq 7$ as described in the text. Small dots are +2 ions, large dots are +3 ions and circles are vacant sites. Lines have been added to indicate the "X" chains (far left and far right), "Y" chains (center), and the "Z" configuration (top of right hand chain). After Krzanowski and Newman (1972).

of (from left to right) an "X" chain, two "Y" chains, and an "X" chain terminated above by a "Z" configuration. Not all of these patterns have been indicated on the figure in order to simplify the plot.

The work of Krzanowski and Newman (1972) is a very simple approach to a complex problem. There was no attempt in their work to calculate the lattice energy of the system and, in fact, it would be difficult to do such a calculation. Further, they assumed that by reducing the local charge imbalance, the energy of the system was reduced. This seems reasonable, but is open to question since they examined only first-neighbor interactions. In spite of the lack of sophistication in their computational model, the results are intriguing. They demonstrated that short-range order in the octahedral sites of am impure trioctahedral mica is to be expected, the short-range order should take the form of various types of linear elements made up of +3 and 0 sites, and the segregation of M site cations creates zones with a dioctahedral character.

Order/disorder in dioctahedral T sites

The second study of order/disorder in substituting cation species has been done by the author of this article and presented orally (Giese, 1975c) and, less completely, in a paper to appear soon (Lipsicas et al., 1984). This work was based on the equivalent of a second-order model and examined the Al, Si distribution in the structures of muscovite-$2M_1$ and margarite-$2M_1$. The latter has the T cations long-range ordered (Guggenheim and Bailey, 1975; Joswig et al., 1984) while the former is long-range disordered (Rothbaur, 1971). Short-range ordering of the Al and Si has been reported in muscovite (Gatineau, 1964) based on an analysis of non-Bragg X-ray scattering. The model segregates Al into strips, which is not a very likely arrangement, and it now seems that the non-Bragg scattering was more likely due to thermal diffuse effects rather than short-range ordering.

As discussed earlier, the mathematics of the lattice energy expression make it easy to calculate the energy for a crystal which is long-range ordered. Randomness or short-range ordering cannot be represented by a Fourier series, and the computation of the energy for these systems seems not to be possible. Rather than attempt to calculate accurately the lattice energy for a disordered structure, one could examine a series of ordered structures, each with a different distribution of the cations. The distribution of ions in each of these differently ordered structures ought to sample some portion of the real disordered structure. Thus, by examining the distribution of energies for the model structures, one might find, for example, that a large number of different cation distributions had roughly the same energy. This would suggest that there was little driving force for choosing one ordering scheme over all the others; i.e., long-range order would be unlikely.

It should be kept in mind that the space group of a particular mica seems to impose certain restrictions on the distribution of cations. Muscovite-$2M_1$ has space group $C2/c$ with two distinct tetrahedral sites each with a multiplicity of 8. According to the space group, ordering would be impossible for 12 Si and 4 Al, but it is not correct to conclude that the space group prevents the Si and Al from ordering; if muscovite wanted to have long-range ordered Si and Al it would do so and have a different space group. It is important to keep in mind that space groups are valid for long-range effects and can tell us nothing about the existence or nonexistence of short-range order.

The sizes of Al and Si tetrahedra are very different due to the differing T-O distances. Since the electrostatic energy depends strongly on interionic distance, a more sophisticated model should take these dimensional changes into account. The approach to the problem of disorder was to generate a random distribution of Al and Si in the proper ratio. Since this work dealt with differently ordered Al and Si sites, the normal space group symmetry was ignored and every atom in the unit cell became independent. For such a study, literally any volume could have been used but it was convenient to take the normal unit cell as the reference volume. This contained 16 tetrahedral sites and the Si and Al were distributed among them in a random manner. Since the aim of the study was to examine the variation in energy for different distributions of these two cations, some sampling of all possible combinations had to be done. There are 1820 ways of doing this for 12 Si and 4 Al (the muscovite ratio) and 12,870 ways for 8 Si and 8 Al (the margarite ratio). It is impossible to examine all combinations, so a sample of 100 for each structure was chosen. Each of these corresponds to a possible long-range ordered structure. Having assigned the T cations, the atomic and unit cell parameters appropriate to the known T-O distances were determined by the method of distance least-squares (DLS) developed by Meier and Villiger (1969). The DLS method is based on the fact that the number of adjustable parameters needed to specify a structure are less than the number of independent interatomic distances in the structure. Knowing the ideal interatomic distances (from refined structures or ionic radii), one can refine the positional and unit cell parameters to give the best fit between the ideal distances and those calculated from the refined structure. With some minor problems, the procedure works well and is not very costly (Giese, 1975d,e). Since the model structures, each with a different Al,Si distribution, reflect the real ionic sizes of the atoms, the repulsion energy should be roughly the same for all structures with the same bulk composition. This follows from the fact that repulsion is primarily a nearest-neighbor interaction.

The electrostatic energy was calculated for each of the 100 structures of the muscovite and margarite compositions. The hydroxyls presented a problem because they will reorient as the T sites are differently occupied and their reorientation can contribute strongly to the energy. It was too big a problem to refine the orientations of the OH's in the 200 structures, so the model micas were treated as F-micas. Figures 8a,b show the frequency distributions of the electrostatic energies for the different models while

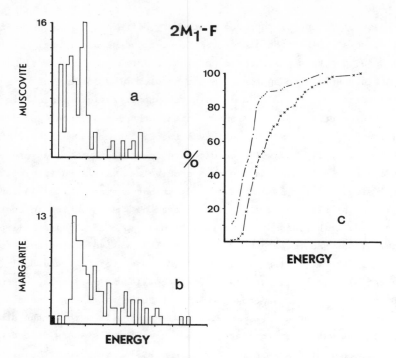

Figure 8. A histogram of the electrostatic energy for 100 different distributions of Al and Si in the tetrahedral sites of muscovite-$2M_1$ (a) and margarite-$2M_1$ (b). The cumulative frequency of the energies is shown in (c) with the muscovite-$2M_1$ lying above margarite-$2M_1$. The solid rectangle in (b) is the observed cation distribution in margarite and has the lowest energy of all the distributions examined.

Figure 8c shows the cumulative frequency of the energies of the two micas. The overall shape of the distributions for the muscovite and margarite compositions are similar but not identical. This is indicated clearly by the cumulative frequency curves (Fig. 8c) which shows that, at the low energy side (most stable), the muscovite composition has many different distributions with comparable energy but the margarite composition has few. If these two distributions are really representative of the total population of distributions, and 100 samples of 12,870 is not a very large sample, then one could reationalize the fact that muscovite is long-range disordered while margarite is not, by observing that there are many different distributions of Al and Si (at the lowest energy, i.e., most stable) in the muscovite composition which have very similar energies. Thus, there is little advantage in choosing one over all the others, and different parts of a crystal can choose whatever distribution pattern they prefer, perhaps based on local chemical variations. The overall result will be small domains with short-range order but collectively the crystal is long-range disordered. The margarite composition, on the other hand, has a tail on the low energy side of the distribution and, in fact, the real Al,Si distribution found by Guggenheim and Bailey (1975) has the lowest energy. It is shown by the solid bar in Figure 8b. Since this distribution was not chosen by the random selection process, it was added later and is the 101st structure in the figure. The existence of the tail indicates that the most stable distributions are few in number. If there were only one distribution at the lowest energy, then this would indicate an energetic preference for long-range order. This seems to be the case for the margarite composition. One could generalize, then, by saying that the tendency for a given structure and chemistry to long-range order can be determined from the distribution of energies of a sample of possible ordered distributions. A tail on the low energy side indicates a system which favors long-range order; lack of such a tail suggests that long-range disorder would be expected.

If these admittedly preliminary calculations are valid, then the muscovite composition in the $2M_1$ polytype will always be long-range disordered. Extensions of these calculations coupled with ^{29}Si magic angle spinning nuclear magnetic resonance spectroscopic experiments indicate that some sort of, as yet unspecified, short-range order exists in this system (Lipsicas et al., 1984).

SUMMARY

Lattice and electrostatic energy calculations have been applied, more or less successfully, to problems of crystal structure, hydroxyl orientation, interlayer bonding, and order/disorder in the micas. There are some very real difficulties in doing these calculations. One problem is the difficulty in specifying the non-electrostatic contributors to the lattice energy while the other problem is to determine what constitutes mineralogical reality and how it can be represented in a sufficiently simple computational model. The major problem of the latter type is caused by the substitutional disorder which is so often found in silicates.

Assuming that the non-electrostatic energies can be approximated reasonably correctly, or that the problem is such that they can be neglected, can these calculations give us any insight into the structure and properties of the micas, or any other mineral system? For the moment, it seems to the author that calculations designed to demonstrate that the observed distortions in the mica structure correspond to minima in the lattice energy can do no more than that. That is, they can confirm what we already know, but such a demonstration gives little insight into the underlying causes for the distortion. It is not possible to say, for such a summation, that a given interaction (e.g., Al-Al repulsion) is responsible for a given distortion (e.g., edge shortening between octahedral sites). Besides, as pointed out earlier, these calculations do not attempt to depict accurately the chemical bonding.

The predictions of hydroxyl orientations are potentially more useful. They predict, for example, that there should be two different OH orientations in structures which are mixed tri- and dioctahedral, and this has been confirmed by Joswig et al. (1983) and Lin and Guggenheim (1983). The calculations indicate that, in disordered structures, hydroxyls are very sensitive to the distribution of cations and one might be able to use them as probes. For example, by determining the specific OH orientations in a particular area, one might be able to indirectly observe the most common cation groupings in the structure.

The calculations also give one the ability to compare alternate structures to see if one is more stable than the other. This will work best if the structures being compared are very similar since any inaccuracy in the model will contribute equally to all. So far, the major application has been in comparing different cation distributions in the same crystal structure. The initial results indicate that the method may allow us to understand why there is or is not long-range order in micas. In addition, there is the hope that the actual short-range ordering might be identifiable if one could sample a sufficiently large number of possible cation distributions.

There may well be other applications of the lattice energy calculations which have not yet been attempted. These calculations have the great merit of being relatively inexpensive so that one could try new ideas without a major investment of either time or research funds.

ACKNOWLEDGMENTS

The author thanks Drs. S.W. Bailey and D. Bish for commenting on the manuscript and P.M. Costanzo and S. Stoops for invaluable help in preparing the text. Partial support from the National Science Foundation is gratefully acknowledged.

REFERENCES

Adams, J. (1983) Hydrogen atom positions in kaolinite by neutron profile refinement. Clays & Clay Minerals 31, 352-356.

Appelo, C.A.J. (1978) Layer deformation and crystal energy of micas and related minerals. I. Structural models for $1M$ and $2M_1$ polytypes. Am. Mineral. 63, 782-792.

Appelo, C.A.J. (1979) Layer deformation and crystal energy of micas and related minerals. II. Deformation of the coordination units. Am. Mineral. 64, 424-431.

Bailey, A.I. and Daniels, H. (1973) The determination of the intermolecular forces of attraction between macroscopic bodies for separations down to the contact point. J. Phys. Chem. 77, 501-515.

Bailey, S.W. (1984) Review of cation ordering in micas. Clays & Clay Minerals 32, 81-92.

Baur, W.H. (1965) On hydrogen bonds in crystalline hydrates. Acta Crystallogr. 19, 909-916.

Baur, W.H. (1972) Prediction of hydrogen bonds and hydrogen atom positions in crystalline solids. Acta Crystallogr. B28, 1456-1465.

Bertaut, F. (1952) L'energie electrostatic de reseaux ioniques. J. Phys. Radium 13, 499-505.

Bloss, F.D. (1971) Crystallography and Crystal Chemistry. Holt, Rinehart and Winston, New York.

Bookin, A.S. and Drits, V.A. (1982) Factors affecting orientation of OH-vectors in micas. Clays & Clay Minerals 30, 415-421.

Bookin, A.S., Drits, V.A., Rozdestvenskaya, I.V., Semenova, T.F., and Tsipursky, S.I. (1982) Comparison of orientations of OH-bonds in layer silicates by diffraction methods and electrostatic calculations. Clays & Clay Minerals 30, 409-414.

Born, M. (1919) Eine thermochemische Anwendung der Gittertheorie. Verhandl. Deut. Physik. Ges. 21, 13-24.

Born, M. and Lande, A. (1918a) Über die absolute Berechnung der Kristalleigenschaften mit Hilfe Bohrscher Atommodelle. Sitzb. Press. Akad. Wiss. Berlin 45, 1048-1068.

Born, M. and Lande, A. (1918b) Kristallgitter und Bohrsches Atommodell. Verhandl. Deut. Physik. Ges. 20, 202-209.

Born, M. and Lande, A. (1918c) Über die Berechnung der Kompressibilität regulärer Kristalle aus der Gittertheorie. Verhandl. Deut. Physik Ges. 20, 210-216.

Bragg, W.H. and Bragg, W.L. (1913) The reflection of X-rays by crystals. Proc. Roy. Soc. London 88A, 428-438.

Brindley, G.W. (1966) Discussions and recommendations concerning the nomenclature of clay minerals and related phyllosilicates. Clays & Clay Minerals 14, 27-34.

Brown, G.E. and Fenn, P.M. (1979) Structure energies of the alkali feldspars. Phys. Chem. Minerals 4, 83-100.

Bryant, P.J. (1962) Cohesion of clean surfaces and the effect of adsorbed gases. Trans. Ninth Vacuum Symp. 311-313.

Burnham, C.W. and Radoslovich, E.W. (1963) Crystal structures of coexisting muscovite and paragonite. Carnegie Inst. Wash. Yearbook 62, 232-236.

Busing, W.R. (1970) An interpretation of the structures of alkaline earth chlorides in terms of interionic forces. Trans. Am. Crystallogr. Assoc. 6, 57-72.

Catlow, C.R.A. (1977) Point defect and electronic properties of uranium dioxide. Proc. Roy. Soc. London A353, 533-561.

Catlow, C.R.A., James, R., Mackrodt, W.C., and Stewart, R.F. (1982) Defect energetics in a-Al_2O_3 and rutile TiO_2. Phys. Rev. B25, 1006-1026.

Catlow, C.R.A., Thomas, J.M., Parker, S.C., and Jefferson, D.A. (1982) Simulating silicate structures and the structural chemistry of pyroxenoids. Nature, Phys. Sci. 295, 658-662.

Catti, M. (1981) The lattice energy of forsterite. Charge distribution and formation enthalpy of the $[SiO_4]^{-4}$ ion. Phys. Chem. Minerals 7, 20-25.

Coulson, C.A. (1961) Valence. 2nd ed., Oxford Univ. Press, Oxford.

Coulson, C.A. and Danielsson, U. (1954) Ionic and covalent contributions to the hydrogen bond. Part I. Ark Fys. 8, 239-244.

Dempsey, E. (1968) In: Molecular Sieves. Soc. Chem. Ind.

Dempsey, E., Kuehl, G.H., and Olson, D.H. (1969) Variation of the lattice parameter with aluminum content in synthetic sodium faujasites. Evidence for ordering of the framework ions. J. Phys. Chem. 73, 387-390.

Deryagin, B.V. and Metsik, M.S. (1959) Role of electrical forces in the process of splitting of mica along cleavage planes. Soviet Phys.-Solid State 1, 1393-1399 (English transl.).

Donnay, G., Morimoto, N., Takeda, H., and Donnay, J.D.H. (1964) Trioctahedral one-layer micas. I. Crystal structure of a synthetic iron mica. Acta Crystallogr. 17, 1369-1373.

Ewald, P.P. (1921) Die Berechnung optischer und electrostatischer Gitterpotentiale. Ann. Physik 64, 253-287.

Gatineau, L. (1964) Structure réele de la muscovite. Repartition des substitutions isomorphes. Bull. Soc. franc. Minéral. 87, 321-355.

Giese, R.F. (1971) Hydroxyl orientation in muscovite as indicated by electrostatic energy calculations. Science 172, 263-264.

Giese, R.F. (1974) Surface energy calculations for muscovite. Nature, Phys. Sci. 248, 580-581.

Giese, R.F. (1975a) Electrostatic energy of columbite/ixiolite. Nature, Phys. Sci. 256, 31-32.

Giese, R.F. (1975b) The effect of F/OH substitution on some layer-silicate minerals. Z. Kristallogr. 141, 138-144.

Giese, R.F. (1975c) Crystal structure of phyllosilicates -- a geometrical approach. Proc. Int'l Clay Conf., Mexico City, Applied Publ., Wilmette, Illinois, p. 239.

Giese, R.F. (1975d) Crystal structure of ideal, ordered one-layer micas. ARCRL-TR-75-0438, Environmental Research Papers, No. 526.

Giese, R.F. (1975e) Crystal structure of ideal, ordered two-layer micas. ARCRL-TR-75-0471, Environmental Research Papers, No. 533.

Giese, R.F. (1978) The electrostatic interlayer forces of layer structure minerals. Clays & Clay Minerals 26, 51-57.

Giese, R.F. (1979) Hydroxyl orientations in 2:1 phyllosilicates. Clays & Clay Minerals 27, 213-223.

Giese, R.F. (1982) Theoretical studies of the kaolin minerals: Electrostatic calculations. Bull. Minéral. 105, 417-424.

Giese, R.F. and Datta, P. (1973a) Hydroxyl orientations in kaolinite, dickite and macrite. Am. Mineral. 58, 471-479.

Giese, R.F. and Datta, P. (1973b) Hydroxyl orientations in the muscovite polymorphs, $2M_1$, $3T$ and $1M$. Z. Kristallogr. 137, 436-438.

Giese, R.F., Weller, S., and Datta, P. (1971) Electrostatic energy calculations of diaspore, goethite and groutite. Z. Kristallogr. 134, 275-284.

Gilbert, T.L. (1968) Soft-sphere model for closed-shell atoms and ions. J. Chem. Phys. 49, 2640-2642.

Gilkes, R.J., Young, R.C., and Quirk, J.P. (1972) The oxidation of octahedral iron in biotite. Clays & Clay Minerals 20, 303-315.

Güven, N. (1971) The crystal structures of $2M_1$ phengite and $2M_1$ muscovite. Z. Kristallogr. 134, 196-212.

Güven, N. and Burnham, C.W. (1967) The crystal structure of $3T$ muscovite. Z. Kristallogr. 125, 163-183.

Guggenheim, S. and Bailey, S.W. (1975) Refinement of the margarite structure in subgroup symmetry. Am. Mineral. 60, 1023-1029.

Guggenheim, S. and Bailey, S.W. (1977) The refinement of zinnwaldite-$1M$ in subgroup symmetry. Am. Mineral. 62, 1158-1167.

Haber, F. (1919) Betrachtungen zur Theorie der Wärmetönungen. Verhandl. Deut. Physik. Ges. 21, 750-768.

Harrison, W.A. (1980) Electronic Structure and the Properties of Solids. Freeman, San Francisco, California.

Hazen, R.M. and Burnham, C.W. (1973) The crystal structure of one-layer phlogopite and annite. Am. Mineral. 58, 889-900.

Jenkins, H.D.B. and Hartman, P. (1979) A new approach to the calculation of electrostatic energy relations in minerals: the dioctahedral and trioctahedral micas. Phil. Trans. Roy. Soc. A293, 169-208.

Jenkins, H.D.B. and Hartman, P. (1980) Application of a new approach to the calculation of electrostatic energies of expanded di- and trioctahedral micas. Phys. Chem. Minerals 6, 313-325.

Jenkins, H.D.B. and Hartman, P. (1982) A new approach to electrostatic calculations for complex silicate structures and their application to vermiculites containing a single layer of water molecules. In: Developments in Sedimentology 35, 87-95.

Jenkins, H.D.B. and Smith, B.T. (1975) Extension and parameterisation of lattice potential energy equations for salts containing complex anions and cations. Chem. Phys. 11, 17-24.

Jenkins, H.D.B. and Waddington, T.C. (1971) Lattice energies, charge distributions and thermochemical data for salts containing complex ions. Nature, Phys. Sci. 232, 5-7.

Joswig, W. (1972) Neutronenbeugungsmessungen an einem 1*M*-phlogopit. N. Jahrb. Mineral., Monatsh. 1-11.

Joswig, W., Takéuchi, Y., and Fuess, H. (1983) Neutron-diffraction study on the orientation of hydroxyl groups in margarite. Z. Kristallogr. 165, 295-303.

Krzanowski, W.J. and Newman, A.C.D. (1972) Computer simulation of cation distribution in the octahedral layers of micas. Mineral. Mag. 38, 926-935.

Lin, J.-C. and Guggenheim, S. (1983) The crystal structure of a Li,Be-rich brittle mica: a dioctahedral-trioctahedral intermediate. Am. Mineral. 68, 130-142.

Lipsicas, M., Raythatha, R.H., Pinnavaia, T.J., Johnson, I.D., Giese, R.F., Costanzo, P.M., and Robert, J.-L. (1984) Silicon and aluminum site distribution in 2:1 layered silicate clays. Nature, Phys. Sci. (in press).

Loewenstein, W. (1954) The distribution of aluminum in the tetrahedra of silicates and aluminates. Am. Mineral. 39, 92-96.

Mackenzie, R.C. (1965) Nomenclature subcommittee of CIPEA. Clay Minerals Bull. 6, 123-126.

Madelung, E. (1918) Das elektrische Feld in Systemen von regelmässig angeordneten Punktladungen. Physik. Zeit. 19, 524-531.

McCauley, J.W. and Newnham, R.E. (1973) Structure refinement of a barium mica. Z. Kristallogr. 137, 360-367.

McCauley, J.W. Newnham, R.E., and Gibbs, R.V. (1973) Crystal structure of synthetic fluor-phlogopite. Am. Mineral. 58, 249-254.

Meier, W.M. and Villiger, H. (1969) Die Methode der Abstandsverfeinerung zur Bestimmung der Atomkoordinaten idealisierter Gerüetstrukturen. Z. Kristallogr. 129, 411-423.

Munoz, J.L. and Ludington, S.D. (1974) Fluoride-hydroxyl exchange in biotite. Am. J. Sci. 274, 396-413.

Noda, T. and Roy, R. (1956) OH-F exchange in fluorine phlogopite. Am. Mineral. 41, 929-932.

Noda, T. and Ushio, M. (1966) Hydrothermal synthesis of fluorine hydroxyl-phlogopite. Part II: Relationship between the fluorine content, lattice constants and the conditions of synthesis of fluorine-hydroxy-phlogopite. Coll. Papers, Synthetic Crystal Res. Lab., Faculty of Eng., Nagoya Univ. Japan 3, 96-104.

O'Keeffe, M. (1981) Some aspects of the ionic model of crystals. In: M. O'Keeffe and A. Navrotsky, eds., Structure and Bonding in Crystals, Vol. 1, Academic Press, New York.

Ohashi, Y. (1976) Lattice energy of some silicate minerals and the effect of oxygen bridging in relation to crystallization sequence. Carnegie Inst. Wash. Yearbook 75, 644-648.

Ohashi, Y. and Burnham, C.W. (1972) Electrostatic and repulsive energies of the M1 and M2 cation sites in pyroxenes. J. Geophys. Res. 29, 5761-5766.

Pauling, L. (1927) Sizes of ions and structure of ionic crystals. J. Am. Chem. Soc. 49, 765-792.

Pauling, L. (1960) The Nature of the Chemical Bond and the Structure of Molecules and Crystals. Cornell Univ. Press, Ithica, New York.

Pedro, G. (1967) Commentaires sur la classification et la nomenclature des minéruax argileux. Bull. Groupe Fr. Argiles 19, 69-86.

Peterson, R.C., Hill, R.J., and Gibbs, G.V. (1979) A molecular-orbital study of distortions in the layer structures brucite, gibbsite and serpentine. Canadian Mineral. 17, 703-711.

Radoslovich, E.W. (1963) The cell dimensions and symmetry of layer-lattice silicates IV. Interatomic forces. Am. Mineral. 48, 76-99.

Raynor, J. and Brown, G. (1973) The crystal structure of talc. Clays & Clay Minerals 21, 103-114.

Rothbauer, R. (1971) Untersuchung eines $2M_1$-Muskovits mit Neutronenstrahlen. N. Jahrb. Mineral., Monatsh. 4, 143-154.

Sartori, F., Franzini, M., and Merlino, S. (1973) Crystal structure of a $2M_2$ lepidolite. Acta Crystallogr. B29, 573-578.

Sherman, J.P. (1932) Crystal energies of ionic compounds and thermochemical applications. Chem. Rev. 11, 93-170.

Soboleva, S.V. and Zvyagin, B.B. (1969) Crystal structure of dioctahedral Al-mica. Sov. Phys. Crystallogr. 13, 516-519.

Suitch, P.R. and Young, R.A. (1983) Atom positions in highly ordered kaolinite. Clays & Clay Minerals 31, 357-366.

Switzer, C. (1941) Hardness of micaceous minerals. Am. J. Sci. 239, 316.

Takeda, H. and Burnham, C.W. (1969) Fluor-polylithionite; a lithium mica with nearly hexagonal $(Si_2O_5)^{2-}$ ring. Mineral. J. Japan 6, 102-109.

Takeda, H. and Donnay, J.D.H. (1966) Trioctahedral one-layer micas. III. Crystal structure of a synthetic lithium fluormica. Acta Crystallogr. 20, 638-646.

Takeda, H., Haga, N., and Sadanaga, R. (1971) Structural investigation of polymorphic transition between $2M_2$-, $1M$-lepidolite and $2M_1$ muscovite. Mineral. J. Japan 6, 203-215.

Takeda, H. and Ross, M. (1975) Mica polytypism: dissimilarities in the crystal structures of coexisting $1M$ and $2M_1$ biotite. Am. Mineral. 60, 1030-1040.

Tateyama, H., Shimoda, S., and Sudo, T. (1974) The crystal structure of synthetic Mg^{iv} mica. Z. Kristallogr. 139, 196-206.

Tosi, M.P. (1964) Cohesion of ionic solids in the Born model. Solid State Phys. 16, 1-120.

Tsipursky, S.I. (1979) The refinement of the celadonite structure by electron diffraction oblique texture method. In: Proc. 8th Conf. X-ray Study of Mineral Raw Materials, Moscow, 1979, IGEM Publ., Moscow (Abstract), 61.

Tsipursky, S.I. and Drits, V.A. (1977) The efficiency of electronometrical way of intensity measure in electron diffraction research. Izv. Akad. Nauk USSR Phys. Ser. 41, 2263-2271.

Unsöld, A. (1927) Quantentheorie des Wasserstoffmoleküions und der Born-Landeschen Abstanssungskräfte. Z. Physik 43, 563-574.

Vedder, W. and McDonald, R.S. (1963) Vibrations of the OH in muscovite. J. Chem. Phys. 38, 1583-1590.

Ward, W. and Phillips, J.M. (1971) Calculated lamellar binding I. van der Waals bonding in talc and pyrophyllite. Surface Sci. 25, 379-384.

Wardle, R. and Brindley, G.W. (1972) The crystal structures of pyrophyllite $1Tc$, and of its dehydroxylate. Am. Mineral. 57, 732-750.

Yoder, H.S. and Eugster, H.P. (1955) Synthetic and natural muscovites. Geochim. Cosmochim. Acta 8, 225-280.

Zhoukhlistov, A., Zvyagin, B.B., Soboleva, A.V., and Fedotov, A.F. (1973) The crystal structure of the dioctahedral mica $2M_2$ determined by high voltage electron diffraction. Clays & Clay Minerals 21, 465-470.

Zvyagin, B.B. (1960) Electron diffraction determination of the structure of kaolinite. Kristallografiya 5, 32-41.

Zvyagin, B.B. (1979) High Voltage Electronography in Layer Silicates Research. Nauka, Moscow.

5. SPECTROSCOPY of MICAS
George R. Rossman

INTRODUCTION

This chapter is concerned with spectroscopic studies of micas which address the origin of their color and pleochroism, the oxidation states and concentrations of cations, site occupancies and other structural details, thermodynamic properties, and broader aspects of the chemical-physics of spectroscopic interactions in general. The field of mica spectroscopy has not previously been reviewed as a whole, although chapters in Lazarev (1972) and Farmer (1974) discuss the infrared spectroscopy of layer silicates as a group.

The three primary spectroscopic methods used in this study are Mössbauer absorption, optical absorption, and infrared absorption. Mössbauer spectra measure the absorption of gamma rays by ^{57}Fe nuclei. The details of a spectrum are determined by the crystallographic site and oxidation state of the iron atoms (see review by Bancroft, 1973). Optical spectra in the ultraviolet, visible, and near-infrared spectral regions respond to oxidation state and coordination environment of a variety of metal ions, and can be strongly influenced by interactions with next-nearest neighbors, especially when adjacent ions are in different oxidation states (see review by Burns, 1970). Absorption in the infrared spectral region occurs when local units of a crystal are set in motion by the energy of incident radiation.

Electron spin resonance (ESR) spectra and especially nuclear magnetic resonance (NMR) spectra are finding greater use in the study of micas. ESR spectra respond to metal ions which have unpaired electrons and are especially applicable when they are at low concentrations. NMR spectra can monitor structural details of Al, Si, Na, H, and F.

OPTICAL SPECTRA AND COLOR

Studies of the optical spectra of micas were motivated by a desire to understand the reasons for their wide range of colors. Micas can be colorless, green to blue-green, greenish-brown, red-brown, brown, black, emerald green, pink, lavender, and red. Early observations correlated color with chemistry; for example, Hall (1974!) determined that the color of biotite depended upon the relative contents of Fe, Mg, and Ti. Titanium-free biotites were blue-green in the gamma polarization over a range of Fe concentrations. Greenish-brown biotites had either high or low Fe contents, but always had low, but non-zero, Ti contents, whereas brown biotites had high Fe and low Ti and Mg or low Ti and approximately equal Fe and Mg contents. The red-brown biotites inevitably had high Ti contents and either high or low Fe.

The initial studies of the optical spectra of micas were further motivated by the commercial classification of muscovite which proceeded upon the basis of color: "ruby" for those with a pinkish or cinnamon color, and "non-ruby" -- browns and greens -- or "green" for the others. Because at times the electrical properties correlated poorly with the color grade, two systematic quantitative examinations of the optical absorption spectra of muscovites were undertaken. They resulted in a more detailed classification and a realization that although several absorption features could be identified, many did not correlate in a straightforward fashion with the chemical composition (Finch, 1963; Ruthberg et al., 1963).

Attempts to make detailed assignment of the optical spectra of micas began with the studies of Faye (1968a,b) who interpreted the spectra of chemically analyzed micas on the basis of crystal field theory. The importance of Fe, Mg, Cr, and Ti were recognized, as well as the possibility of interactions between ions. While initial studies concentrated on identifying the spectroscopic features of the individual ions, later studies became increasingly concerned with interactions between ions in different oxidation states (e.g., Fe^{2+} and Fe^{3+}) and between dissimilar ions (e.g., Fe^{2+} and Ti^{4+}).

Because micas are generally monoclinic, they have three independent optical directions, and require three components to describe the absorption spectrum. They are commonly measured in the directions of the three principal refractive indices. The alpha spectrum is measured with polarized light vibrating in the direction in which the alpha index of refraction would be measured, the X vibration direction. The beta and gamma spectra are obtained in the corresponding two other directions. For common micas, the alpha spectrum is obtained with light vibrating normal to the cleavage plates. For convenience, micas have often been treated spectroscopically as uniaxial minerals so that just two spectra need be measured, one with the electric vector in the plane of the cleavage, and the other with the electric vector normal to the cleavage plane. This approximation greatly facilitates sample fabrication for the spectroscopic experiment, but as some of the examples in this chapter show, can also cause anisotropy in the cleavage plane to be missed.

Spectra of micas (and minerals in general) are commonly presented in either of two different units. Wavelength in nanometers, nm, (formerly in Ångström units, 1000 Å = 100 nm) is the unit in which most spectrometers present their data. Spectra are also converted to a unit linear in energy for graphical presentation because electronic absorption bands are symmetrical on an energy coordinate. This unit is normally the wavenumber, the reciprocal of wavelength

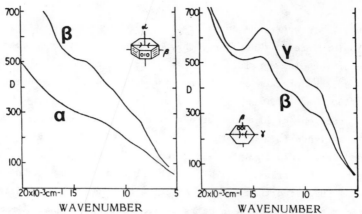

Figure 1 (left). Absorption spectra of a biotite with 7.93 wt % FeO and 3.14 wt % Fe_2O_3 showing the prominent bands at 9000, 11000, and 14000 cm^{-1} and the strong anisotropy in a section cut normal to the cleavage. Sample B10 from Figure 4 of Robbins and Strens (1972).

Figure 2 (right). Absorption spectra of a biotite with 14.41 wt % FeO and 4.12 wt % Fe_2O_3 showing the much weaker anisotropy in the plane of the cleavage. Sample B2 from Figure 5 of Robbins and Strens (1972).

(where wavelength is expressed in units of cm) such that 500 nm = 20000 cm^{-1}.

Individual ions

The spectra of individual ions are best observed in micas with low concentrations of the spectroscopically active metal ions to avoid complications from interactions between ions. When Fe^{3+} is present at high concentrations, or when Fe occurs in combination with Mn, Ti, or another oxidation state of Fe, the spectroscopic features become more complicated than would be predicted from the linear sum of the contributions of the individual components. Indeed, much of the current scientific interest in mica spectroscopy is directed at understanding the nature of these interactions and the underlying principles of chemical physics by which they are generated.

Fe^{2+} in phlogopite-biotite

The spectra of biotites containing iron are dominated by absorption bands near 9000 cm^{-1} (1150 nm), 11000 cm^{-1} (920 nm), and 14000 cm^{-1} (720 nm) which are superimposed upon a tail from a strong absorption centered in the ultraviolet region (Figs. 1 and 2). The two bands at 9000 and 11000 cm^{-1} have generally been assigned to Fe^{2+} (Faye, 1968; Robbins and Strens, 1972; Smith and Strens, 1976; Kliem and Lehmann, 1979). Smith (1978) and Smith et al. (1980), however, point out that the spectra taken in the plane of the cleavage show pronounced decreases in intensity after the sample is heat-treated under hydrogen (a procedure which should increase the amount of Fe^{2+} in the sample). Smith (1977) also observes that the intensity of these bands increases at low temperatures, a property he considers more typical of interacting pairs of ions than of isolated Fe^{2+}. Therefore, they conclude that much of the intensity in the plane of the cleavage is due to intensification of Fe^{2+} absorption features brought about by interaction with Fe^{3+} inevitably present in untreated samples.

The 14000 cm^{-1} (720 nm) band is important in establishing the intense dichroism of sections of biotite cut normal to the cleavage direction. This band is usually assigned to a $Fe^{2+}-Fe^{3+}$ interaction. Initial studies (Faye, 1968; Robbins and Strens, 1972; Smith and Strens, 1976) associated this band with intervalence charge transfer (IVCT), although Bakhtin and Vinokurov (1978) and Smith (1978) suggested that it may be caused by a (technically different) type of magnetic interaction between pairs of Fe^{2+} and Fe^{3+} ions. Kliem and Lehmann (1979) reject these assignments and propose instead that the 14000 cm^{-1} band along with the 9000 and 11000 cm^{-1} bands are due to Fe^{2+} alone in both the M1 and M2 sites. They base this conclusion in part on the correlation of the intensity of the 14000 cm^{-1} band with the total FeO concentration of a number of micas; however, they do not explain the strong anisotropy of the 14000 cm^{-1} band nor its complete absence in the direction perpendicular to the cleavage. For this reason, Smith (1980) again favored the charge transfer interpretation, but could not identify the potential contributions from the nonequivalent M1 and M2 sites.

Additional, low intensity absorptions from Fe^{2+}, the spin-forbidden bands, should occur at shorter wavelengths. They have been observed in the spectra of a variety of minerals such as olivines and pyroxenes in the 600 to 400 nm range. Faye (1968) suggests that the absorption in this region in the biotite spectrum has a similar origin. Kliem and Lehmann (1979) reject this interpretation on the basis that the intensity of these bands is orders of magnitude too high and propose instead that Fe^{3+} features are responsible. Faye (1968) pointed out, however, that there are mechanisms through which the normally weak spin-forbidden bands can gain intensity.

Fe^{2+} in muscovite

Surprisingly, there are few modern detailed studies of the optical spectra of Fe^{2+} in muscovites. Faye (1968a) examined two brown muscovites from Ontario, Canada, and located two shoulders in the spectral curves at about 11600 cm^{-1} (862 nm) and 9500 cm^{-1} (1053 nm) which he assigned to two components of an Fe^{2+} absorption. These are positions similar to where they occur in trioctahedral micas. All other studies concentrated on the visible spectral region and did not obtain full data on the Fe^{2+} region.

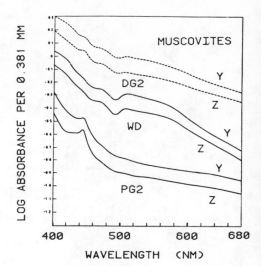

Figure 3. Visible spectra of three pegmatitic muscovite cleavages from the Harts Range, Australia. DG2, dark green-brown, Fe^{3+}/Fe^{2+} = 2.92; WD, red, Fe^{3+}/Fe^{2+} = 1.47; PG2, pure green, Fe^{3+}/Fe^{2+} = 6.87. From Figure 2 of Finch et al. (1982).

Fe^{3+} in muscovite

A characteristic feature in the spectra of Fe^{3+} in octahedral sites is a sharp absorption near 440 nm (22730 cm^{-1}). Faye (1968a) observed a comparatively sharp feature in the spectra of muscovites at about 26600 cm^{-1} (376 nm) and assigned it to Fe^{3+}. Finch et al. (1982) confirm this observation and assignment, but do not agree with Faye's assignment of weak features at 19650 cm^{-1} (508 nm), 17200 cm^{-1} (581 nm), and 14700 cm^{-1} (680 nm) to spin forbidden d-d transitions of Fe^{3+}.

Fe^{3+} in biotite

Chemical analyses have shown that major amounts of ferric iron occur in biotites. However, the characteristic features of isolated Fe^{3+} ions have never been recognized in the optical spectra of biotites.

The role of iron in the color of muscovite

Finch et al. (1982) examined the spectra of 14 pegmatitic muscovites in the visible (400-680 nm) region. They were chosen to represent a range of color types with a range of Fe^{2+}/Fe^{3+} ratios. Representative spectra are shown in Figure 3. The depth of color was determined by the background absorption which was in turn determined primarily by an Fe-O charge transfer absorption in the ultraviolet which correlated with the Fe^{2+} concentration in the normally unoccupied octahedral site. Samples with the highest Fe^{3+} content were pure green, and those with the highest Fe^{2+} content were red. The color was particularly dependent upon the intensity in the 500-600 nm region

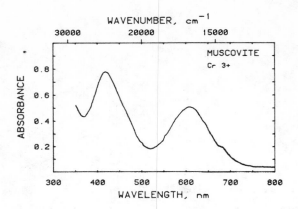

Figure 4. Optical spectrum of Cr^{3+} in a cleavage of muscovite from San Francisquito Canyon, California. Sample thickness: 0.18 mm.

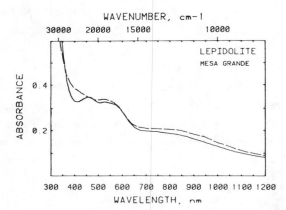

Figure 5. Optical spectrum of lavender lepidolite from the Himalaya mine, Mesa Grande, California. Sample thickness: 1.85 mm. Solid line E ∥ cleavage; dashed line ⊥ cleavage. Data of G.R. Rossman.

which Finch et al. assign to a metal-metal charge transfer system involving Fe^{2+}/Fe^{3+} possibly with contributions from Fe^{2+}/Ti^{4+}. Although Ti was present in all samples, none of the features correlated with the Ti content.

Cr^{3+} in micas

Chromium imparts green color to micas. Emerald-green chromian muscovite is so distinctive that it has been given the vertical names fuchsite and mariposite. The optical spectrum of Cr-containing micas (Fig. 4) shows a pattern consisting of two regions of absorption centered in the visible portion of the spectrum with maxima at about 600 nm (16700 cm^{-1}) and 400 nm (24000 cm^{-1}) (Faye, 1968). They are characteristic of Cr^{3+} in octahedral coordination. No indication exists that any other oxidation state of Cr occurs in micas.

Lepidolites: the role of Mn and Fe

Only two aspects of lepidolites have been published. Faye (1968a) identified features of Fe^{3+} at 22600 cm^{-1} (442 nm), and Fe^{2+} features at 11600 cm^{-1} (862 nm) and 9500 cm^{-1} (1053 nm) in the spectrum of a purple lepidolite with 0.1 wt % Fe and 1 wt % Mn from Lacorne Township, Quebec. Faye also assigned features at 24200 cm^{-1} (413 nm), 24000 cm^{-1} (417 nm), 23200 cm^{-1} (431 nm), and 18500 cm^{-1} (541 nm) to Mn^{2+}. A previously unpublished spectrum from our laboratory of a lepidolite from the Himalaya Mine in the Mesa Grande District, California, with 0.079 wt % MnO and 7.08:1 Mn:Fe shows a different spectrum (Fig. 5) with bands at 456 nm (21500 cm^{-1}), 590 nm (16900 cm^{-1}), and 860 nm (11600 cm^{-1}) nm.

Significant variation is evident in the color of lepidolites suggesting that additional spectroscopic study is needed to characterize the detailed origin of their color. Faye noted that lepidolites with a high Mn/Fe ratio are colored pink or purple, while those with a higher proportion of Fe are brown. It is specifically to be noted that, contrary to common belief, manganese, not lithium, is responsible for the pink color of lepidolites.

Ti in micas

Titanium has a strong influence on the color of micas, especially biotites. Micas with a percent or more of Ti absorb strongly in the violet and blue regions of the spectrum and are typically various shades of orange to red-brown. At high concentrations, the micas are crimson colored in the plane of the cleavage and produce a spectrum dominated by a tail extending across the visible from an intense band centered at high energy (Fig. 6). Faye (1968a) initially noticed a shoulder in the spectra of titaniferous micas in the 20000-24000 cm^{-1} region,

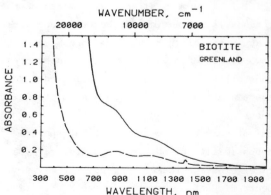

Figure 6. Absorption spectrum polarized in the plane of the cleavage of a dark red biotite from the Motzfeldt nepheline syenite body, South Greenland, containing about 4 wt % TiO$_2$ and 8 wt % FeO(total). The intense absorption at shorter wavelengths results from the presence of Ti. Data from G.R. Rossman and A.P. Jones.

and, like Chesnokov (1959), raised the possibility that it could occur as Ti^{3+}. His later studies (Faye, 1968b) indicated that an Fe-Ti interaction was more probable to cause this absorption. Interactions between Fe^{2+} and Ti^{4+} which now are recognized in a variety of minerals are the likely cause of this absorption; the details have yet to be worked out. Robbins and Strens (1972) observed that the short wavelength absorption in biotites is particularly sensitive to the Ti concentration. They argued that in the octahedral sheet of biotites the substitution of Ti^{4+}, probably accompanied by vacancies of zero charge, was particularly effective at broadening the $O-Fe^{2+}$ charge-transfer absorption band. Most likely, both Fe^{2+}/Ti^{4+} interactions and substitutional broadening of the Fe-O charge transfer are occurring simultaneously.

Interactions between cations

The striking pleochroism of biotite has long been held as a classic example of intervalance charge transfer between Fe^{2+} and Fe^{3+}. For incident light polarized in the plane of the cleavage sheet, intense absorption occurs. As has been summarized earlier in this chapter, this absorption has been attibuted to the exchange of electrons between adjacent edge-shared Fe^{2+} and Fe^{3+} octahedra. Because cations do not share valence orbitals in the direction normal to the cleavage direction, the incident light can not excite this interaction if it is polarized normal to the cleavage direction.

Detailed optical spectral measurements of the intense absorption indicate that it includes not only a band associated with the Fe^{2+}/Fe^{3+} intervalence charge transfer at 700 nm (14000 cm^{-1}), but also includes unusually intense absorption from Fe^{2+} as seen in Figure 1. Micas which are black in thin section for $E \perp c$ have these bands so intensified that essentially all visible light is absorbed (Fig. 7). From detailed measurements of the intensity of the 800 and 1200 nm bands as a function of the Fe^{2+} concentration, Smith et al. (1980) determined that in the black micas, the intensity of the Fe^{2+} absorption increases by up to two orders of magnitude when it interacts with Fe^{3+}. He heated samples in controlled atmospheres to oxidize and reduce Fe and was able to follow major, reversible changes in the Fe^{2+} absorption intensity which correlated with the amount of Fe^{3+} present.

Smith's results not only help to interpret the spectra of micas, but also indicate the near impossibility of obtaining quantitative Fe^{2+} concentrations from optical spectra of cleavage flakes of mica. It may still be possible to determine Fe^{2+} concentrations from spectra in the X direction (polarized perpendicular to the cleavage direction), but this is experimentally much more difficult and, as yet, essentially untested.

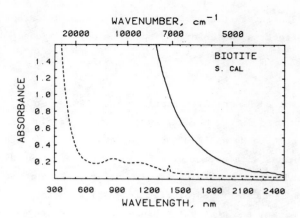

Figure 7. Optical spectrum of a thin section of a dark biotite from the Marble Mountains, Southern California, showing the pronounced anisotropy of the 700 nm (14000 cm^{-1}) IVCT band and the 920 and 450 nm Fe^{2+} bands which results from the interaction of Fe^{2+} with Fe^{3+}. The absorption intensity with light polarized in the cleavage plane is so intense that essentially all visible light is absorbed.

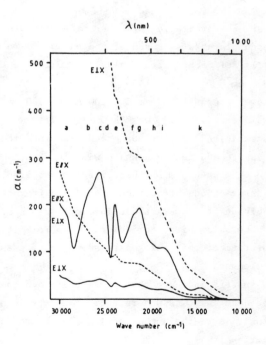

Figure 8. Optical spectra of manganian phlogopites with normal (dots) and reverse (solid line) pleochroism. Bands e, g, and i are assigned to transitions of Mn^{2+} which are intensified through interaction with Fe^{3+}. From Figure 2 of Smith et al. (1983).

The Fe^{2+}-Ti^{4+} interaction discussed in a preceding paragraph is also prominent for its effect on the color and spectroscopy of micas. Interactions involving Mn also occur.

Mn in phlogopite

In a study of eight Mn-bearing phlogopites, Smith et al. (1983) identified a variety of features in the optical spectra which originate from Mn. These samples contained both Mn and Fe, and both ions contributed to the spectra. The spectra of micas with normal pleochroism were interpreted in terms of octahedral Mn^{2+} and Fe^{3+}, whereas those with reverse pleochroism showed features of octahedral Mn^{2+} and tetrahedral Fe^{3+} (Fig. 8). In the case of the reversibly pleochroic micas, the absorption bands occur at energies close to those of the individual tetrahedral Fe^{3+} and octahedral Mn^{2+} ions, however, the intensity of the absorbing ion was intensified by interaction with other ions by factors up to 100. Absorption in the normal pleochroic Mn^{2+}-bearing phlogopites appeared to be due to the individual Fe^{3+} and Mn^{2+} ions.

Other micas

Few spectral data are available for other mica minerals. Sekino et al. (1975) presented an unpolarized spectrum of a cleavage flake of clintonite which showed primarily Fe^{2+} features at energies similar to those in phlogopite: 9800 and 11800 cm^{-1} for Fe^{2+}, 14500 cm^{-1} for Fe^{2+}/Fe^{3+} IVCT. Fe^{2+} bands occur at somewhat lower energy (8330 cm^{-1}, 1200 nm; 11360 cm^{-1}, 880 nm) in the spectrum of a zinnwaldite with 13.95 wt % FeO (Fig. 9). A significantly different spectroscopy is exhibited by a bright green mica intermediate in composition from ferri-muscovite and ferri-phengite with 16.79 wt % Fe from Mt. Ruker, Antarctica (Fig. 10). The spectrum of this site is undoubtedly dominated by bands arising from interactions between cations. A different spectroscopic behavior also probably due to interactions involving Fe is seen in the spectrum of a greenish-blue phlogopite from Edwards, New York (Fig. 11). The X-ray fluorescence spectrum of this phlogopite shows only Fe as a spectroscoptically active minor element.

These last two spectra indicate that there still are spectroscopic phenomena waiting to be described and understood. The details of these spectra may very well be sensitive to the number of ions which are interacting in a mixed valence cluster.

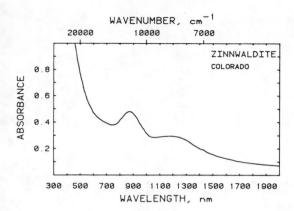

Figure 9. Optical spectrum of a light brown cleavage flake of zinnwaldite with 11.4 wt % FeO from Wigwam Creek, Colorado with prominent Fe^{2+} bands near 880 nm and 1200 nm but with almost no Fe^{2+}/Fe^{3+} IVCT at ~14300 cm^{-1} (~700 nm). Data from G.R. Rossman and E.E. Foord.

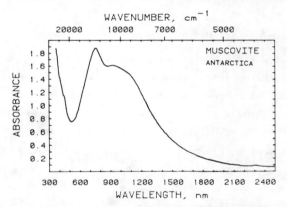

Figure 10. Optical spectrum of a bright green cleavage flake of ferri-muscovite with 16.8 wt % Fe from Mt. Ruker, Anarctica, showing bands tentatively assigned to Fe^{2+}/Fe^{3+} IVCT at 756 nm (13230 cm^{-1}) and intensified Fe^{2+} at ~1000 nm (10000 cm^{-1}). Sample thickness to 10 μm. Data from G.R. Rossman and W.A. Dollase.

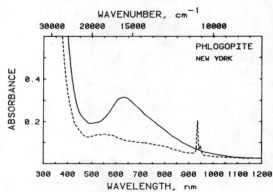

Figure 11. Optical spectrum of a 3.85 mm thick plate of blue phlogopite from Edwards, New York, which shows a band at 630 nm (15870 cm^{-1}) whose origin is not established, but which may be due to interactions between low concentrations of Fe^{2+} and Fe^{3+}.

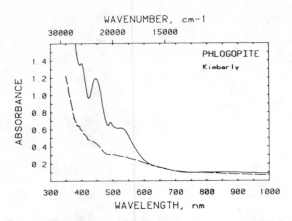

Figure 12. Optical spectrum of an orange-red phlogopite with reverse pleochroism from a South African kimberlite showing features from Fe^{3+} in tetrahedral coordination. Thin section of sample Kb5-2 from Farmer and Boettcher (1981).

Reverse pleochroism -- tetrahedral ferric iron

Some micas show reverse pleochroism which means they show strong absorption colors for light polarized normal to the cleavage plane ($E \| x \gg E \perp x$). Grum-Grzhimailo and Rimskaya-Korsakova (1964) showed that phlogopites with reverse pleochroism from ultrabasic rocks were low in tetrahedral Al and correspondingly high in tetrahedral Fe^{3+}. Faye and Hogarth (1969) studied the absorption spectrum of a phlogopite from the contact of a dolomitic marble and an aplite near Old Chelsea, Quebec. The basal sections were pale orange-brown and showed little pleochroism whereas intense red-brown color was associated with light polarized normal to the cleavage plane ($E \| c$). Similar samples were subsequently studied by Hogarth et al. (1970), and by Puustinen (1973) from a carbonatite and by Farmer and Boettcher (1981) from South African kimberlites and associated ultramafic xenoliths. Most commonly, phlogopites with tetrahedral Fe come from deep seated, Al- and Ti-poor, Fe-rich rocks.

The optical spectra of the reversibly pleochroic phlogopites are dominated in the alpha direction by three bands centered at 22700 cm^{-1} (441 nm), 20300 cm^{-1} (493 nm), and 19200 cm^{-1} (521 nm) which arise from tetrahedral Fe^{3+} (Fig. 12). They in turn are superimposed upon an intense tail of a band originating in the ultraviolet which is most likely due to Fe^{3+}-O charge transfer. The red color in the alpha direction is a direct result of these absorption features. Weak Fe^{2+} features are commonly also present at longer wavelengths.

Red to reddish-brown colors in the alpha polarization are commonly reported for micas which on the basis of stoichiometry are presumed to have Fe^{3+} in the tetrahedral sheet. Examples include: ferriannite (Wones, 1963; Miyano and Miyano, 1982), and synthetic phlogopites (Hazen and Wones, 1972).

Reverse pleochroism is also observed in some Mn-bearing phlogopites. Smith et al. (1983) studied phlogopites with both normal and reverse pleochroism from Harstigen and Långban, Sweden, and concluded that normal micas had octahedral Mn^{2+} and octahedral Fe^{3+} whereas reverse micas had octahedral Mn^{2+} and tetrahedral Fe^{3+}. The intense pleochroism of these micas came not only from the constituent ions alone, but was greatly enhanced by interactions between pairs of Mn and Fe ions. This emphasizes a recurring theme emerging from recent studies that interactions between ions can dominate the optical absorption properties of minerals. These micas were discussed in more detail earlier in this chapter under Mn in phlogopite.

Reverse pleochroism -- pink muscovite

Reverse pleochroism in dioctahedral micas is rarely mentioned. One example is a pink muscovite with reverse pleochroism from a hydrothermal Mn-bearing vein from Archer's Post, Kenya. It is pink-rose in the X direction and pale yellow to orange-pink in the plane of the cleavage (Richardson, 1975). The spectrum taken polarized perpendicular to the cleavage is dominated by a band at 22624 cm^{-1} (442 nm) and at 19608 cm^{-1} (509 nm) with a shoulder at 18248 cm^{-1} (548 nm). The spectrum is similar to an unpolarized spectrum of a pink muscovite from Sign, Norway (Askvik, 1972). Richardson proposed that iron, present as Fe^{3+} as indicated from the Mössbauer spectrum, was in tetrahedral coordination and, as such, was responsible for the reverse pleochroism. Annersten and Hålenius (1976), however, present additional spectra of pink muscovites and argue that the color and pleochroism are due to octahedral Mn^{3+}, and that the bands at $\sim$22600 cm^{-1} and $\sim$19600 cm^{-1} are due instead to Fe^{3+} in octahedral coordination. Tetrahedral iron, if present, is below the detection limit of the single-crystal X-ray structural determination by Richardson and Richardson (1982).

The ultraviolet region

Micas free of transition metal ions should be free of absorption in the visible spectral region but should show intense short wavelength absorption in the ultraviolet region. Concentrations of metal ions too low to cause appreciable absorption in the visible region should likewise cause oxygen-to-metal charge transfer bands in the ultraviolet. Although no systematic study of the UV region has been published, some exploratory data are available. Karickhoff and Bailey (1973) present data for thin sheets of muscovite, phlogopite, and biotite which show a variety of intense absorption bands in this region.

Future needs

There are many problems remaining regarding the optical spectroscopy of micas. The ultraviolet region just mentioned has barely been touched from a mineralogical perspective. The role of oxygen to metal charge transfer in this region, and its application for site occupancy determinations is an obvious candidate for study. There are many conflicting assignments of the visible features. This is due in part to the lack of recognition in the early studies of the powerful influence of cation interactions. The micas provide a fertile ground for basic study of these interactions. Before there is hope of developing satisfactory analytical methods based on optical spectroscopy for Fe^{2+} and Fe^{3+} concentrations in micas, we must first understand the nature of these cation-bearing interactions. The principles which are emerging from the study of micas are broadly applicable to the spectroscopic study of minerals in general. The role of the weak, spin-forbidden bands has not been satisfactorily delineated. They, too, are involved with intensification through cation-cation interaction. Complete anisotropy data are available for very few micas. As some of the examples in this section have illustrated, there are even phenomenologies which have yet to be described. The unusual samples may be most valuable for providing insight into the nature of the various interactions. And finally, elements other than Fe are still poorly characterized from the spectroscopic point of view. Manganese, in particular, is rich in its effects, but only recently have comprehensive studies of Mn-micas begun to appear.

MÖSSBAUER SPECTRA

The study of the Mössbauer spectra of micas has been limited to the study of the isotope ^{57}Fe. Ideally, it should be possible to distinguish not only between Fe^{2+} and Fe^{3+}, but also between octahedral and tetrahedral Fe, and even between the two fundamentally different octahedral sites, one with *trans* hydroxides and the other with *cis* hydroxides. The designations, M1 and M2 are routinely used for the two octahedral sites, however, some authors designate the *trans* site as M1 while other authors use this designation for the *cis* site. Needless to say, this double convention has led to confusion (Goodman, 1976b). The majority of Mössbauer studies of the mica group have been concerned with biotite.

Biotite

Initial studies of micas (Pollak et al., 1962; Bowen et al., 1969; Haggstrom et al., 1969a,b; Annersten et al., 1971; Hogg and Meads, 1970; and

Table 1. Summary of representative optical assignments
==

Muscovite
Iron bands

Wavenumber	Assignment	Reference
242 nm	41320 cm^{-1}	Fe^{3+} charge transfer	Karickhoff and Bailey (1973)
275	36360 cm^{-1}	Fe^{2+} charge transfer	"
365	27400	Fe^{3+}	Faye (1968a)
376	26600	Fe^{3+}	"
443	22570	Fe^{3+}	"
470	21280	Ti^{3+}	"
508	19680	unknown	Finch et al. (1982)
		Fe^{2+}/Fe^{3+}	Faye (1968a)
581	17200	Fe^{3+}/Fe^{3+}	Finch et al. (1982)
		Fe^{2+}/Fe^{3+}	Finch et al. (1982)
680	14700	Fe^{3+}	Faye (1968a)
862	11600	Fe^{2+}	Finch et al. (1982)
1050	9500	Fe^{2+}	Faye (1968a)

Manganese bands

442 nm	22620 cm^{-1}	Fe^{3+} Td	Richardson (1975)
458	21830	"	"
468	21370	"	"
510	19610	Fe^{3+} Td	"
548	18250	"	"
555	18020	Mn^{3+}	"
730	13700	Mn^{3+}	"

Chromium bands

407 nm	24600 cm^{-1}	Cr^{3+}	Faye (1968a)
415	24100	"	"
472	21200	"	"
599	16700	"	"
676	14800	"	"

Vanadium bands

| 431 nm | 23200 cm^{-1} | V^{3+} | Schmetzer (1982) |
| 595 | 16800 | " | " |

Biotite

270 nm	37040 cm^{-1}	Fe^{3+} charge transfer	Karickhoff and Bailey (1973)
385	25970	Fe^{2+}	"
456	21930	Fe^{2+}	"
414	24150	Fe^{3+} clusters	Kliem and Lehmann (1979)
488	20490	"	"
584	17120	"	"
720	14000	$Fe^{2+}-Fe^{3+}$	Robbins and Strens (1972)
920	11000	Fe^{2+}	"
1150	9000	Fe^{2+}	"

Phlogopite

212 nm	47170 cm^{-1}	Fe^{3+} charge transfer	Karickhoff and Bailey (1973)
280	39290	Fe^{2+} charge transfer	"
344	29100	$Fe^{3+}+Mn^{2+}$	Smith et al. (1983)
370	27000		"
392	25500	Fe^{3+} Td	"
403	24800	"	"
420	23800	Mn^{2+}	"
441	22700	Fe^{3+} Td	Faye (1969)
459	21800	Mn^{2+}	Smith et al. (1983)
469	21300	"	"
493	20300	Fe^{3+} Td	Faye (1969)
521	19200		
541	18500	Mn^{2+}	Smith et al. (1983)
690	14500	"	"
862	11600	Fe^{2+}	Faye (1969)
1100	9100	"	"

Lepidolite

413 nm	24200 cm^{-1}	Mn^{2+}	Faye (1968a)
417	24000	Mn^{2+}	"
431	23200	Mn^{2+}	"
439	22800	Fe^{3+}	"
472	21200	Fe^{2+}	"
541	18500	Mn^{2+}	"
658	15200	Fe^{3+}	"
862	11600	Fe^{3+}	"
1050	9500	Fe^{2+}	"

Hogarth et al., 1970) demonstrated that both Fe^{2+} and Fe^{3+} occur in biotites, and that considerable variation exists in the Mössbauer patterns. Annersten (1974) made a systematic attempt to correlate the Mössbauer parameters of a variety of biotites representing a range of compositions and conditions of formation, using 30-40 mg of sample. Special attention was paid to preparing a sample free of preferred orientation which would produce asymmetric intensities of the Fe doublets. The spectra could be satisfactorily fit only when two ferrous and two ferric doublets were modeled, indicating that Fe occupies both the *cis* and *trans* sites in each oxidation state. The assignments of the individual doublets in Figure 13 and Table 2 are based on detailed considerations of the distortions of the octahedral polyhedra from cubic symmetry and the electric field gradients due to neighboring cations and vacancies. Ferrous iron was found to be slightly enriched in the *cis* octahedral site (e.g., 61:44 *cis:trans*) in a number of the biotites, although in others, the ordering was reversed (e.g., 33:44 *cis:trans*). Ferric iron was either equally distributed between the two octahedral sites or enriched in the *trans* site (e.g., 9:6 *trans:cis*).

Sanz et al. (1978) examined a series of phlogopites and biotites and concluded that, on the average, the distribution of Fe^{2+} between the two octahedral sites was random. They and Annersten (1974) also found that the agreement between Fe^{3+} content determined through Mössbauer spectra and by wet chemical methods was poor. Generally, the wet chemical methods provided higher Fe^{3+} values, leading to the conclusion that because of unavoidable oxidation during dissolution, the wet chemical methods overestimated the Fe^{3+} content.

Few petrologic applications have made use of the site occupancy information available from Mössbauer spectroscopy of micas. Bancroft and Brown (1975) used Mössbauer spectra to relate Fe-ordering to crystallization temperatures in coexisting amphiboles and biotites from the Sierra Nevada batholith. Little, if any, *cis-trans* ordering was found. Mössbauer spectra remain a primary tool to study the physical properties of micas; for example, Ballet and Coey (1982) studied the magnetic ordering of biotite and glauconite at low temperatures with the Mössbauer effect. Micas are also used to explore the nature of the Mössbauer interaction and are especially convenient for work involving single crystal studies. For example, Bonnin and Muller (1981) studied the electric field gradient of a single crystal of muscovite.

Muscovite

The spectra of muscovites and biotites are similar. The initial studies of muscovite identified both Fe^{2+} and Fe^{3+} in the Mössbauer spectrum (Bowen et

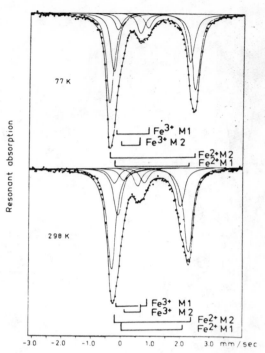

Figure 13. Mössbauer spectra of a biotite with 5.6 wt % Fe_2O_3 and 13.8 wt % FeO from amphibolite facies gneiss from SW Greenland. Figure 2 of Annersten (1974).

Table 2. Typical Mössbauer parameters of micas.

ref	Mica	Fe^{2+} cis		Fe^{2+} trans		Fe^{3+} cis		Fe^{3+} trans		Fe^{3+} Td	
		IS	QS	IS	QS	IS	QS	IS	QS	IS	QS
1	Glauconite	1.59	1.90	1.43	2.53	0.46	0.42	0.46	1.04	--	--
2	Muscovite	--	--	--	--	0.465	0.738	--	--	--	--
3	"	1.15	3.04	1.10	2.14	0.37	0.74	--	--	--	--
4	Biotite	1.06	2.56	1.03	2.10	0.43	0.50	0.41	1.02	--	--
5	"	1.20	2.61	1.16	2.20	0.58	0.60 *	--	--	--	--
5	Phlogopite	1.16	2.60	1.17	2.20	0.65	0.97 *	--	--	--	--
	"	1.17	2.65	--	--	--	--	--	--	0.30	0.45
6	Ferriphlogopite	--	--	--	--	--	--	--	--	0.17	0.50
7	Phlogopite	1.33	2.95	1.23	2.76	0.68	1.22 *	--	--	0.23	0.44
8	Celadonite	1.12	2.40	--	--	0.38	0.42	--	--	--	--
9	Clintonite	1.04	1.72	1.08	2.48	0.52	1.05 *	--	--	0.27	0.62

Reference: Sample temperature
1. McChonchie et al. (1979)   80 K
2. Goodman (1967a)   78 K
3. Hogg & Meads (1970)   RT
4. Annersten (1974)   RT
RT = room temperature

Reference: Sample temperature
5. Sanz et al. (1978)   RT
6. Annersten et al. (1971)   RT
7. Hogarth et al. (1970)   RT
8. Heller-Kallai & Rozenson (1980)   RT
9. Annersten & Olesch (1978)   RT

* cis and trans Fe^{3+} not distinguished

Isomer shifts quoted with respect to metallic iron.

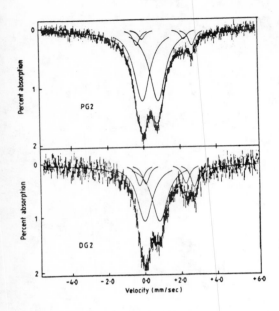

Figure 14. Mössbauer spectra of two muscovites showing intense Fe^{3+} absorption and weaker Fe^{2+} absorption. Two Fe^{2+} doublets and Fe^{3+} doublets were fit by computer. From Figure 1 of Finch et al. (1982). Iron contents of these samples are given in Figure 2.

et al., 1969; Hogg and Meads, 1970). The detailed, low temperature studies of Goodman (1976a) indicate that Fe^{3+} preferentially occupies the smaller *cis* octahedral site, that iron does enter the *trans* octahedral site, but it goes in with a strong preference as Fe^{2+}. Finch et al. (1982) correlated the Mössbauer spectra of a number of muscovites with their color (Fig. 14). They found similar Mössbauer parameters to those of Goodman (1976a).

Glauconite

In the Mössbauer spectra of glauconites taken at room temperature, major Fe^{2+} and minor Fe^{3+} peaks are evident (Hofmann et al., 1967; Weaver et al., 1967; Taylor et al., 1968; Raclavasky et al., 1975; Annersten, 1975; Rolf et al., 1977; Rozenson and Heller-Kallai, 1978; McConchie et al., 1979). The Fe^{2+} peaks are partially obscured by a broad shoulder which extends from 1.0 to 2.5 mm/sec, probably due to electron exchange between Fe^{2+} and Fe^{3+}. Rolf et al. (1977) determined that the resolution of the spectra is increased at low temperature. The spectra of glauconites from South Island, New Zealand, obtained at 80 K show four quadrupole doublets assigned to Fe^{2+} and Fe^{3+} in both the *cis* and *trans* sites (McConchie et al., 1979). Fe^{3+} shows a stronger preference for the *cis* site, whereas Fe^{2+} shows an even stronger preference for the *trans* site.

Clintonite

Annersten and Olesch (1978) examined clintonites from four localities and a synthetic sample and found that Fe^{2+} was in both the *cis* and *trans* sites but showed a strong preference for the *trans* site. The stronger preference of Fe^{3+} compared to Al for the tetrahedral site is in contrast to the behavior of Fe^{3+} in ferro-magnesium silicates. Annersten and Olesch attributed this to the unusually large tetrahedral sites in clintonites.

Other micas

Data exist for ferriannite (Weaver et al., 1967), celadonite (Heller-Kallai and Rozenson, 1980), phengite (Heller-Kallai and Rozenson, 1980), siderophyllite (Levillain et al., 1981, 1982; Heller-Kallai, 1982), and zinnwaldite (Herzenberg et al., 1968; Hogg and Meads, 1970; Levillain et al., 1981).

Commentary

Many Mössbauer spectra have now been obtained on micas, but they have been obtained by a variety of groups at different times, with different instruments, and fit with different computer programs and with different initial assumptions. These differences will be particularly apparent in the fit of the Fe^{3+} region where the doublets are poorly resolved experimentally. Much progress has been made in the interpretation of mica spectra, but there still remain inconsistencies, especially involving Fe^{3+}. There is need for a comprehensive analysis of the mica group and related layer silicates using a consistent modeling routine. Systematic data still need to be obtained over a wide range of compositions for the less common micas, and the possible effects of electron delocalization are probably not fully explored.

ELECTRON SPIN RESONANCE SPECTRA

There have been a few exploratory applications of electron spin resonance (ESR) spectroscopy to micas. Kemp (1969, 1972) observed Fe^{3+} signals from four environments in a Canadian phlogopite. Kemp's (1973) ESR spectra establish that the Fe^{3+} resonances arise from Fe^{3+} free of interactions with other Fe^{3+} ions from two sites in Australian muscovite. Hassib and Hedewy (1982) measured the ESR spectrum of muscovites from Sudan and Saudi-Arabia and observed the signals of Mn^{2+} and Fe^{3+} in two sites. They concluded that there was a higher concentration of Fe^{3+} in a greenish muscovite than in their brown sample. Other studies include those of Mank et al. (1974), that of Olivier et al. (1977), who found Fe^{3+} in four environments -- two in octahedral sites and two in tetrahedral sites in a Madagascar phlogopite, and those of Ivanitskey et al. (1974,

1975), who examined naturally irradiated biotites.

NUCLEAR MAGNETIC RESONANCE SPECTRA

Proton nuclear magnetic resonance (NMR) spectra which are routinely used to study the structure and bonding of H^+ in organic chemicals have also been applied to the study of hydrogen in micas. Kalinichenko et al. (1974, 1975) recognized that because proton NMR spectra are sensitive to the magnitude of local magnetic fields at the site of a paramagnetic ion such as Fe, the OH resonance of micas should shift in energy depending upon the number of Fe atoms bonded to it. Specifically, an OH bound to 3 Mg should be in a different magnetic environment than one bound to 2 Mg and 1 Fe, etc. After examining a series of biotites, they observed that there were noticeable differences in the distribution of cations in the octahedral sheets of samples from different geological environments. Their data suggested that cation order increased in biotites crystallized at higher temperatures. They also addressed the ordering of Al and V in chernykhite. In the spectra of large crystals (∼1 cm) of phlogopites and biotites, Sanz and Stone (1977) also observed multiple shifted lines in the H^+ spectra as a result of local environments representing various associations of OH groups with Mg and Fe^{2+} in the M1 and M2 sites. The complexity of the spectra increased as the Fe content increased. The experimental spectra vary greatly with respect to the orientation of the crystal in the external magnetic field (Fig. 15), but because these variations can be accurately modeled from the known crystal structure, it is possible to interpret the various components in the NMR spectrum (Fig. 16).

At low Fe concentrations, the association of OH with 3 Mg was most prominent, but as Fe content increased, first lines due to 2 Mg,Fe and then due to Mg, 2 Fe, and 3 Fe became more important. Sanz and Stone (1983) found that the trends deviated from a statistical distribution based on the amount of Fe and Mg indicated by chemical analysis. Iron was preferentially clustering around the OH groups. In a parallel study, Sanz and Stone (1979) observed the fluorine NMR in phlogopites with various degrees of (F,OH) substitution. They found that F occurred exclusively in association with 3 Mg. Even though Fe was also present in their samples, there was complete exclusion of F bound with Fe. Litovchenko et al. (1982) obtained the NMR spectrum of Li in zinnwaldites with about 10 or greater wt % Fe, and concluded that every Li ion has neighboring cation positions occupied by Fe, and that there was the possibility that there could be local clustering of Fe.

To narrow the normally broad solid state NMR spectra, a technique known as Magic Angle Spinning (MAS) can be utilized. This method spins the sample

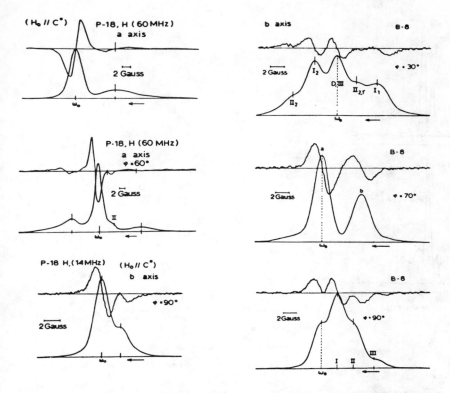

Figure 15. NMR spectra of phlogopite P-18 with 2.6 wt % Fe and biotite B-8 with 3.8 wt % Fe at various orientations around the a and b axes. The incident magnetic field is held parallel to c^* for the top left spectrum. The top curve of each pair corresponds to the experimentally determined derivative spectrum and the bottom curve of the pair is the calculated absorption spectrum. From Figures 2 and 6 of Sanz and Stone (1977). Line I corresponds to OH bonded to 2 Mg^{2+} and 1 Fe^{2+}; line II_2 to 1 Mg and 2 Fe^{2+}.

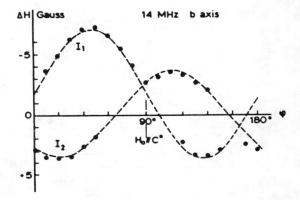

Figure 16. Calculated angular variation of shifts around the b axis for nearest neighbor H^+. ● Experimental points for a phlogopite with 2.6 wt % Fe. From Figure 4 of Sanz and Stone (1977).

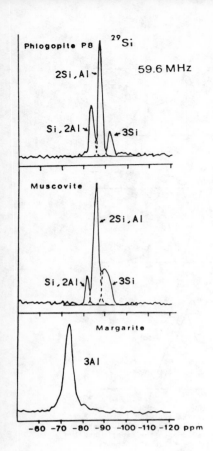

Figure 17 (left). ^{29}Si NMR spectra of three micas showing multiple lines arising from Si environments with different numbers of adjacent Al and Si. Modified from Figures 2 and 3 of Sanz and Serratosa (1984).

Figure 18 (below). Summary of ^{29}Si and ^{27}Al NMR chemical shifts indicating the high degree of discrimination among different environments. From Figure 5 of Sanz and Serratosa (1984).

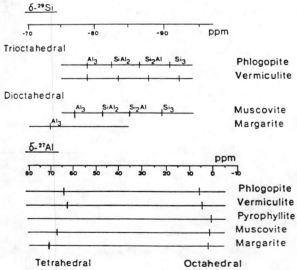

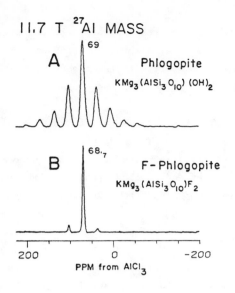

Figure 19. ^{27}Al "magic angle" NMR spectra of phlogopite and a synthetic fluorphlogopite The phlogopite shows multiple spinning-side bands on both sides of the central Al resonance at 69 ppm. They result from paramagnetic impurities and change position when the rate of sample spin is changed. From Figure 6 of Kinsey et al. (1984).

in the magnetic field of the NMR spectrometer at an angle (the "Magic" angle) which minimizes the interactions which broaden the signal. MAS-NMR has made it possible to obtain high quality spectra of Al, Si, and Na in mica. Sanz and Serratosa (1984) were able to analyze the Al,Si cation distribution, determine the tetrahedral Si/Al ratios, and confirm the Al-avoidance rule (avoidance of Al-O-Al linkages, Loewenstein, 1954). They were able to specifically identify and quantify signals from Si bonded to three O-Si groups, symbolized by Si(3Si), in addition to Si(2Si,Al), Si(Si,2Al), and Si(3Al) groups (Fig. 17). Tetrahedral and octahedral Al were distinguished by the large difference in their resonance positions measured as the chemical shift with respect to a standard of aqueous Al ion (Fig. 18). Tetrahedral values were 60 to 70 ppm whereas octahedral values were 0 to 10 ppm.

Kinsey et al. (1984) studied the Al and Si MAS-NMR of micas and other layer silicates and also found a large difference between octahedral and tetrahedral Al chemical shifts. Their ratios of tetrahedral to octahedral Al determined by NMR were in good agreement with those obtained by chemical analysis. Their NMR data suggest that more than one Si site is present in both phlogopite and muscovite, implying that the Al,Si ordering in the tetrahedral sheets is not perfect. Multiple sites were especially pronounced and well-resolved in their spectrum of a synthetic fluorophlogopite. Computer simulation of their fluorphlogopite spectrum indicated that only 64% of the silicon sites were Si(2Si,Al), whereas 100% would have been in a perfectly ordered phlogopite. The ^{27}Al spectra of the phlogopites indicate just tetrahedral Al. In addition to the central tetrahedral Al absorption at 69 ppm, spinning side bands are evident. Their positions vary with the frequency at which the sample is spun, and their intensity is related to the content of paramagnetic ions in the sample (Fig. 19).

The initial MAS-NMR studies are largely exploratory in nature and have served to demonstrate the power of the method for obtaining quantitative information on cation occupancy. Sanz and Stone (1983) and Sanz et al. (1978) have compared NMR with Mössbauer and infrared spectra and conclude that the NMR spectra provide a more accurate measure of site occupancy. The NMR spectra do not suffer from overlap to the extent that both the Mössbauer and IR spectra do and unlike the infrared spectra, appear to give site occupancies directly proportional to integrated area of the spectral band. These advantages are offset in part by the need for comparatively large crystals or large amounts of powder for the NMR method. This will be a problem when samples are available in limited quantity or when the samples are inhomogeneous. Paramagnetic ions such as iron if present in the sample at concentrations above a few percent can interfere with some types of NMR experiments because they degrade the homogeneity of the magnetic field applied to the sample in the NMR spectrometer and facilitate development of spinning side bands. Surveys of the NMR spectra of a number of silicates including micas continue to be conducted to find the applications of this versatile technique (e.g. Lippmaa et al., 1980; Magi et al., 1984).

X-RAY SPECTROSCOPY

Little has been reported regarding the application of sophisticated X-ray spectroscopic methods to mineralogical or structural aspects of micas. Rao et al. (1982) obtained the K-absorption edge spectra of muscovite from the Koderma mine, Bihar, India. They were able to identify Fe^{3+} and from the extended X-ray absorption fine structure on the edge, calculate that it was in a site with an average Fe-O bond distance of 2.04 Å. It should be only a matter of time before there appear systematic studies of micas with X-ray absorption edge spectroscopy and synchrotron based studies of absorption edge fine structure. These methods can address concentrations of minor elements, and their oxidation state and coordination number.

X-ray photoelectron spectroscopy

X-ray photoelectron spectroscopy involves exposing the sample to incident X-rays and measuring the energy of the ejected electrons. The observed energy for a particular element depends upon the core-electron binding energy which is a function of the oxidation state of the element. Studies of micas (Evans and Raftery, 1980; 1982) were concerned with the oxidation state of Ti in biotite and phlogopite and Mn in lepidolite. Although Ti was first reported to be present as Ti^{3+}, reinterpretation pointed to Ti^{4+}. Manganese was found to be present as Mn^{2+} in a pink-purple Norwegian lepidolite. The limits of sensitivity to minor amounts of other oxidation states was not stated.

INFRARED SPECTRA

Infrared spectroscopy can be used to identify individual micas, estimate chemical composition, determine ordering and orientation of small molecule-like units in a mica, follow the atomic details of decomposition and oxidation reactions, derive thermodynamic parameters, and examine the interactions of a wide variety of chemical substances with micas. Because excellent reviews of the infrared spectroscopy of micas already exist (Lazarev, 1972; Farmer, 1974), only highlights and more recent studies will be emphasized here.

OH groups and their orientation

Much attention has been given to the analysis of the OH-stretching bands of micas and other phyllosilicates because it appears that each of the various bands in the OH region (3800-3200 cm^{-1}) can be attributed to a specific atomic grouping around the OH site. The stretching frequency of the OH groups occurs in an isolated region of the spectrum facilitating identification and analysis. Chapter 15 in Farmer (1974) summarizes the spectra in this region. In the spectrum of trioctahedral micas, the primary OH stretching frequency is about 3710 cm^{-1}, but can vary with substitutions in the octahedral and tetrahedral sheets. Minor components which can occur as three broad bands in the 3450 to 3620 cm^{-1} region are a result of associations of OH with vacancies in the octahedral sheet, especially in high Fe-content biotites.

Velde (1983) examined the crystal chemical factors which influence the energy at which a particular OH stretch occurs. The average electronegativity of the group of octahedral ions bound to the OH gave the best correlation with the OH stretch frequency in micas. The frequencies shift 96 cm^{-1} per electronegativity unit with trioctahedral micas and 170 cm^{-1} for dioctahedral micas. Lesser effects were observed for substitution in the tetrahedral sheet where changes in unit cell size were more important than electronegativity effects.

The orientations of the OH groups in micas have been determined from the infrared spectra. Phlogopite provides a simple demonstration because in the ideal end-member composition each OH occurs in just one site associated with

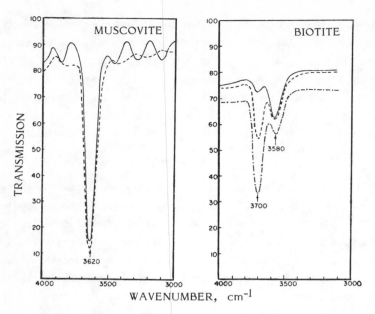

Figure 20. Infrared transmission spectra in the OH stretching region of cleavage plates of muscovite (30 μm thick) and biotite (100 μm thick) which have been rotated at various angles to show the strong anisotropy of the OH stretching mode in biotite (solid = 0°, dash = 30°, dot-dash = 60°) and the weak anisotropy of muscovite (solid = 0°, dash = 45°). From Figures 1 and 2 of Serratosa and Bradley (1958).

three Mg ions. Serratosa and Bradley (1958) showed that there is substantially no absorption when the electric vector of the incident light is normal to the cleavage plane, but that absorption increased as the mica flake is tilted so that a component of the electric vector is normal to the cleavage (Fig. 20). Because the OH group is excited into vibration only when the electric vector is aligned with the O-H bond, this result demonstrates that in phlogopite, the OH group is oriented essentially normal to the cleavage plane.

Comparable behavior was observed for other trioctahedral minerals. Vedder (1964) redetermined this orientation and concluded that there was a 10° angle between the OH unit and $c*$.

The spectrum of the dioctahedral micas indicates that the OH groups are inclined from the normal to the cleavage plane because of the vacancies in the octahedral sheet. In the case of muscovite, the OH groups are in a plane which includes b and is about 16° to 30° above a (Tsuboi, 1950; Serratosa and Bradley, 1958; Vedder and McDonald, 1963; Rouxhet, 1970). The dioctahedral OH frequencies are also typically lower due to stronger hydrogen bonding. In accord with the chemical analyses of lepidolite and many biotites which indicated cation occupancy intermediate between di- and tri-octahedral, the infrared spectra show two types of an OH absorption, one at higher energies indicating orientation

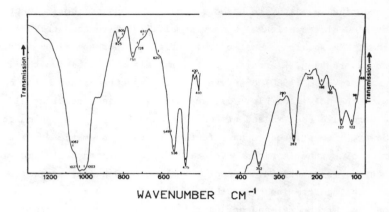

Figure 21. Infrared spectrum in the mid- and far-infrared regions of synthetic muscovite showing Si-O stretching modes near 1000 cm^{-1}, Si-O bending modes near 500 cm^{-1}, Al-O and Al-O-Si modes at 825 and 751 cm^{-1}, and K$^+$ modes below 150 cm^{-1}. Several of the minor features originate from the OH groups in combination with motions of the other ions. From Figure 1 of Velde (1978).

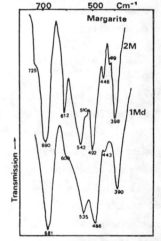

Figure 22. Infrared spectra of the $2M_1$ and $1Md$ polytypes of synthetic margarite showing bands at 612 and 446 cm^{-1} which are present in the $2M_1$ polytype but weak in the $1Md$. From Figure 6 of Velde (1980).

nearly normal to the cleavage, and one at lower energies indicating orientation inclined to the cleavage (Serratosa and Bradley, 1958).

Cation ordering

Because the frequency of the vibrations of the local groups of cations depend upon the mass of the atoms in motion, infrared spectra at lower frequencies than the OH region (Fig. 21) are also sensitive to cation ordering. For example, Farmer and Velde (1973) determined that there is tetrahedral Al,Si ordering in margarite based on the sharpness of the IR spectra and the lack of Al-O-Al vibrations. Kinsey et al. (1984) reached the same conclusion from their NMR spectra. Most of the current IR work on micas includes a discussion of ordering. Infrared spectra may also be sensitive to mica polytypes. For example, Velde (1980) observed that the spectra of synthetic margarites vary with the polytype (Fig. 22), and sensitive bands occur at 612 and 446 cm^{-1} in the margarite spectra and at 802, 737, and 643 cm^{-1} in the muscovite spectra.

Several recent studies have focused on the systematic study of IR spectra of series of synthetic micas. Velde (1980) correlated cell dimensions with the IR spectra of synthetic dioctahedral muscovite-phengite, margarite, and paragonite. The spectra indicated that changes in the b axis are due to ordering in both the tetrahedral and octahedral sheets. Muscovites order in the tetrahedral sites to yield a smaller cell size whereas phengites order in the octahedral sheet. Other studies include those of Velde (1983), Langer et al. (1981) -- muscovite, paragonite, and margarite, Levillian and Maurel (1980) -- biotite and eastonite, Velde (1979) -- celadonite, muscovite, margarite, clintonite, and derivative chemistries, Franz et al. (1977) -- paragonite-margarite, Tateyama et al. (1976) -- Al-free magnesian micas, and Pavlishin (1975) -- Li-Fe micas.

The far-infrared spectral region

Because the instrumentation and experimental methods used in the spectral region below about 300 cm^{-1} are different, the far infrared spectra (∼300 - 20 cm^{-1} which is ∼30 μm to 500 μm) is usually considered separately. Comparatively few mineralogical studies have been performed in this range which responds to vibrational motions of heavy ions and the motions of larger units of the structure. Larson et al. (1972) surveyed the far infrared (FIR) spectra of three micas and found that they were distinct, and Velde obtained spectra for members of the muscovite-magnesian celadonite series (Fig. 21) and examined the systematic variation of band positions with composition. Loh (1973) examined both far infrared and Raman spectra of biotite, phlogopite, muscovite, and lepidolite and identified MO_6 vibrations below 210 cm^{-1} and vibrations involving OH in micas containing octahedral Al^{3+} in the 190 cm^{-1} to 270 cm^{-1} range. The positions of low energy vibrations such as those found in these studies are needed for input parameters into thermodynamic model calculations such as those of Keiffer (1982).

RAMAN SPECTRA

Raman spectra and inelastic neutron scattering spectra, like infrared spectra, record the vibrational motions of samples being studied. Because the selection rules which govern whether a particular vibration is observed are different for infrared, Raman, and neutron scattering spectra, a complete experimental description of the vibrational motions of a crystal would require all three types of spectra. Very few minerals have been studied by inelastic neutron scattering. Because Raman spectra are more readily obtained, many minerals have been studied with this technique, although few studies of micas have been performed. Loh (1973) has surveyed the Raman spectra of micas and proposed assignments for many of the lower energy spectral bands, and Haley et al. (1982) obtained the Raman spectra of muscovite. These studies have demonstrated the feasibility of obtaining regular and micro-Raman spectra of micas, but systematic studies of micas as a function of composition and ordering have yet to be reported.

NEAR INFRARED SPECTRA

The phlogopite spectrum in Figure 23 illustrates a typical near infrared spectrum (NIR = 800 to 2500 nm). This spectral region is dominated by overtones and combination bands. The sharp bands at about 1380 ($\sim$7250 cm^{-1}) and 930 nm ($\sim$10750 cm^{-1}) are overtones of the OH stretching mode at about 2700 nm ($\sim$3700 cm^{-1}). A variety of combination modes occur between 1800 and 2700 nm. Almost no use has been made of this spectral region for mineralogical studies, although much potential exists. Because the absorption bands in this region are much less intense than in the mid-IR, it is easier to fabricate samples so that bands will remain "on scale" for quantitative measurements.

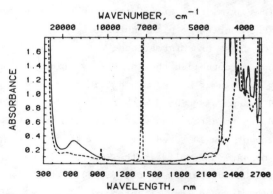

Figure 23. Visible and near-infrared spectrum of the phlogopite from Edwards, New York, which appeared in Figure 11. Overtones of the OH stretch occur at 920 and 1390 nm. Absorptions between 1800 nm and 2700 nm are overtones and combinations of fundamental modes at longer wavelengths. Solid line = polarized in the cleavage plane; dashed line = polarized normal to the cleavage. Sample thickness: 3.85 mm.

RADIATION HALOES AND OTHER RADIATION EFFECTS

A characteristic of micas in thin section is the appearance of haloes around radioactive inclusions. They frequently take the form of bleached areas around a dark or opaque inclusion. Vance (1978) and Seal et al. (1981) examined the optical and Mössbauer spectra of the giant haloes in Madagascar micas as part of a study into the possibility that they might represent the effects of extinct superheavy elements and found that they are formed by the diffusion of radiation-produced atomic hydrogen which reduces ferric iron. This would result in the elimination of the ferrous-ferric intervalance interaction and a corresponding loss of color.

Other radiation effects in micas have received little attention, at least from the spectroscopic point of view, even though radiation effects are often seen in other minerals and may be associated with pink pegmatitic micas which fade upon exposure to sunlight on the mine dumps. Ivanitskey et al. (1975) studied the effects of irradiation of iron in biotites with Mössbauer spectra and found that Fe^{2+} is converted to Fe^{3+} by natural radiation in zones of uranium mineralization. They proposed that the age of the orebody could be ascertained from a comparison of the Fe^{3+}/Fe^{2+} ratios determined by Mossbauer spectra on irradiated and adjacent non-irradiated biotites. Kotlicki et al. (1976) studied the laboratory irradiation of biotite with 3 MeV protons. They also observed oxidation of Fe^{2+} and ultimate formation of magnetite after heavy irradiation. They proposed that the effects were in response to the thermal energy associated with the irradiation.

OXIDATION AND DEHYDROXYLATION OF BIOTITE

The oxidation of Fe^{2+} in biotite is of interest from the point of view of fundamental mechanisms of mineral transformations and from the point of view of soil formation and the sources of plant nutrients. Spectroscopic methods have aided in the study of these transformations. Vedder and Wilkins (1969) used infrared spectroscopy to establish that dehydroxylation of biotite in air proceeds progressively by loss of hydrogen and oxidation of Fe^{2+} to Fe^{3+}, elimination of H_2O from OH groups associated with two cations and one vacancy, and finally, loss of OH groups associated with three cations. Gilkes et al. (1972) studied the OH region of the infrared spectra of laboratory oxidized biotites and concluded that oxidation results in the loss of octahedral sheet cations, a conclusion also reached by Farmer et al. (1971). These conclusions were based on the oxidation induced loss of intensity of the high-frequency OH band associated with 3 Mg(OH) units and the growth of the low-frequency OH band associated with octahedral vacancies and trivalent ions.

Mössbauer studies have figured prominently in the study of the weathering and dehydration of biotite and other micas (Goodman and Wilson, 1973; Heller-Kallai and Rozenson, 1980; Bagin et al. 1980). These studies show that Fe^{2+} is oxidized either before or during dehydroxylation, and that the local coordination environment around the iron can become very distorted during dehydroxylation. Large values of quadrupole splittings and line widths characterize dehydroxylated biotite. Chandra and Lokanathan (1982) used Mössbauer spectra to determine that over the range 400-1200°C, oxygen needed to be in contact with biotite for complete decomposition, a conclusion similar to that reached by Hogg and Meads (1975). Sanz et al. (1983) combined NMR, IR, and Mössbauer spectra to show that OH loss and Fe^{2+} oxidation are correlated, but that lack of O_2 delayed oxidation of Fe^{2+} and favored loss of OH^- associated with octahedral vacancies. In a related study, Rouxhet (1970) used IR spectra to follow quantitatively the rate of OH - OD exchange in micas. He found that the rates in muscovite and phlogopite were similar, and that the OH's in biotite all exchange at the same rate. Aines and Rossman (1984) followed the dehydration of muscovite with infrared spectroscopy with the sample at high temperatures and observed that there was a smooth loss of the OH group without evidence of multiple environments just prior to or during decomposition.

QUANTITATIVE INTENSITIES AND ANALYTICAL DETERMINATIONS

In addition to the use of IR spectra for phase identification such as the distinction of the brittle micas (Farmer and Velde, 1973), many attempts have been made to use IR spectra for quantitative determinations of mineral composition. Recent applications to micas have included the ammonium content of micas (Shigorova, 1982), composition of phengite micas (Kozryreva et al., 1984), distribution of octahedral cations and vacancies (Osherovich and Nikitina, 1975; Rausell-Colom et al., 1979).

Analytical use of infrared spectroscopic data, particularly the OH bands in the 3800-3200 cm^{-1} range, usually has proceeded under the assumption that the intensity of the OH absorption is proportional to the number of OH groups contributing to the absorption. This assumption has been shown to be justified when considering a single type of OH. For example, Gilkes et al. (1972) demonstrated that the intensity of the 3530 cm^{-1} band in biotite associated with Fe^{3+} was proportional to the Fe^{3+} content determined independently. However, there are indications that different cation-(OH) groupings have intrinsically different absorption intensities (Rouxhet, 1970; Gilkes et al., 1972; Sanz et al., 1983).

Quantitative use of optical spectra for Fe^{2+}/Fe^{3+} determinations should be possible in principle, but has yet to be reliably achieved in practice. The main problem arises from the interactions between ions which result in intensification of the Fe^{2+} bands. More effort needs to be directed to understand the nature of these interactions. It may well be possible to make quantitative optical spectroscopic determinations with light polarized normal to the cleavage (in the direction in which the interactions do not occur), but this remains to be demonstrated. Quantification of the near-infrared, particularly the OH overtones, for H determinations and cation ordering, also remains unexplored. Here too, the possibilities appear to be favorable.

CONCLUDING REMARKS

From the more mature application of IR spectroscopy to the rapidly emerging use of NMR spectroscopy, much information can be gleaned from spectroscopic study of micas. Greater collaboration among practitioners of the different methods would benefit the science. Greater use of carefully controlled synthetics would undoubtedly also prove advantageous, although it is clear that much remains to be done to cover the range of variability of natural samples. Just consider the emphasis on pegmatitic micas for optical studies as an example of need to examine a broader variety of natural materials. A variety of elaborate experiments have been performed on micas, but often they have been done more to demonstrate the physics of the situation than to address mineralogical problems. Again, future collaboration between the mineralogists and their colleagues in other disciplines should lead to additional advances in our ability to probe this group of minerals.

REFERENCES

Aines, R.D. and Rossman, G.R. (1984) The high temperature behavior of water and carbon dioxide in cordierite and beryl. Am. Mineral. 69, 319-327.

Annersten, H. (1973) A Mössbauer characteristic of ordered glauconite. N. Jahrb. Mineral. Monatsh. 8, 378-384.

_____ (1974) Mössbauer studies of natural biotites. Am. Mineral. 59, 143-151.

_____ and Hålenius, U. (1976) Ion distribution in pink muscovite, a discussion. Am. Mineral. 61, 1045-1050.

_____ and Olesch, M. (1978) Distribution of ferrous and ferric iron in clintonite and the Mössbauer characteristics of ferric iron in tetrahedral coordination. Canadian Mineral. 16, 199-203.

_____, Devanarganan, S., Haggstrom, L., and Wappling, R. (1971) Mössbauer study of synthetic ferri-phlogopite $KMg_3Fe^{3+}Si_3O_8(OH)_2$. Phys. Stat. Solidi B 48, K137-K138.

Askvik, H. (1972) Red muscovite from a metasedimentary gneiss, Sogn, Norway. Norges Geol. Undersøk. 273, 7-11.

Bagin, V.I., Gendler, T.S., Dainyak, L.G., and Kuz'min, R.N. (1980) Mössbauer, thermomagnetic and X-ray study of cation ordering and high-temperature decomposition in biotite. Clay & Clay Minerals 28, 188-196.

Bakhtin, A.I. and Vinokurov, V.M. (1978) Exchange-coupled pairs of transition metal ions and their effect on the optical absorption spectra of rock-forming silicates. Geokhimiya 1, 87-95. Trans. Geochem. Int'l. 1978, 53-60.

Ballet, O. and Coey, J.M.D. (1982) Magnetic properties of sheet silicates, 2:1 layered minerals. Phys. Chem. Minerals 8, 218-229.

Bancroft, G.M. and Brown, J.R. (1975) A Mössbauer study of coexisting hornblendes and biotites: quantitative Fe^{3+}/Fe^{2+} ratios. Am. Mineral. 60, 265-272.

Bonnin, D. and Muller, S. (1981) Étude du gradient de champ electrique dans la muscovite par spectrometrie Mössbauer du fer. Phys. Stat. Solidi B 105, 649-657.

Bowen, L.H., Weed, S.B., and Stevens, J.G. (1969) Mössbauer study of micas and their potassium-depleted products. Am. Mineral. 54, 72-84.

Burns, R.G. (1970) Mineralogical Applications of Crystal Field Theory. Cambridge University Press, Cambridge, England.

Chandra, U. and Lokanathan, S. (1982) A Mössbauer study of the effect of heat treatment on biotite micas. J. Physics D 15, 2331-2340.

Chesnokov, B.V. (1959) Spectral absorption curves of certain minerals colored by titanium. Translation: Akad. Nauk. SSSR, Doklady 129, 1162-1163.

Evans, S. and Raftery, E. (1980) X-ray photoelectron studies of titanium in biotite and phlogopite. Clay Minerals 15, 209-218.

_____ and _____ (1982) Determination of the oxidation state of manganese in lepidolite by x-ray photoelectron spectroscopy. Clay Minerals 17, 477-481.

Farmer, G.L. and Boettcher, A.L. (1981) Petrologic and crystal-chemical significance of some deep-seated phlogopites. Am. Mineral. 66, 1154-1163.

Farmer, V.C. (1974) Layer silicates. In: Infrared Spectra of Minerals, Farmer, V.C. (ed.) Mineral. Soc., London, p. 331-360.

_____ and Russell, J.D. (1964) The infra-red spectra of layer silicates. Spectrochimica Acta 20, 1149-1173.

_____ and Velde, B. (1973) Effect of structural order and disorder on the infrared spectra of brittle micas. Mineral. Mag. 39, 282-288.

_____, _____, McHardy, W.J., Newman, A.C.D., Ahlrichs, J.L., and Rimsaite, J.Y.H. (1971) Evidence of loss of protons and octahedral iron from oxidized biotites and vermiculites. Mineral. Mag. 38, 121-137.

Faye, G.H. (1968a) The optical absorption spectra of iron in six-coordinate sites in chlorite, biotite, phlogopite and vivianite. Some aspects of pleochroism in the sheet silicates. Canadian Mineral. 9, 403-425.

_____ (1968b) The optical absorption spectra of certain transition metal ions in muscovite, lepidolite, and fuchite. Canadian J. Earth Science 5, 31-38.

_____ and Hogarth, D.D. (1969) On the origin of 'reverse pleochroism' of a phlogopite. Canadian Mineral. 10, 25-34.

Finch, J. (1963) A colorimetric classification of Australian pegmatitic muscovite. Am. Mineral. 48, 525-554.

_____, Gainsford, A.R., and Tennant, W.C. (1982) Polarized optical absorption and ^{57}Fe Mössbauer study of pegmatitic muscovite. Am. Mineral. 67, 59-68.

Franz, G., Hinrichsen, T., and Wannemacher, E. (1977) Determination of the miscibility gap on the solid solution series paragonite-margarite by means of the infrared spectroscopy. Contrib. Mineral. Petrol. 59, 307-316.

Fripiat, J.J., Rouxhet, P., and Jacobs, H. (1965) Proton delocalization in micas. Am. Mineral. 50, 1937-1958.

Gilkes, R.J., Young, R.C., and Quirk, J.P. (1972) The oxidation of octahedral iron in biotite. Clays & Clay Minerals 20, 303-315.

Goodman, B.A. (1976a) The Mössbauer spectrum of a ferrian muscovite and its implications in the assignment of sites in dioctahedral micas. Mineral. Mag. 40, 513-517.

_____ (1976b) On the interpretation of Mössbauer spectra of biotites. Am. Mineral. 61, 169.

_____ and Wilson, M.J. (1973) A study of the weathering of biotite using the Mössbauer effect. Mineral. Mag. 39, 448-454.

Gresens, R.L. and Stensrud, H. L. (1977) More data on red muscovite. Am. Mineral. 62, 1245-1251.

Grum-Grzhimaylo, S.V., and Rimskaya-Korsakova, O.M. (1964) Absorption spectra of phlogopite holding trivalent iron in fourfold coordination. Transl. Doklady Akad. Nauk SSSR 156, 123-135. Transl. from Doklady Akad. Nauk SSSR 156, 847-850.

Haggstrom, L., Wappling, R., and Annersten, H. (1969a) Mössbauer study of iron-rich biotites. Chem. Phys. Letters 4, 107-108.

_____, _____, _____ (1969b) Mössbauer study of oxidized iron silicate minerals. Phys. Stat. Solidi 33, 741-748.

Hall, A.J. (1941) The relation between colour and chemical composition in the biotites. Am. Mineral. 26, 29-33.

Hassib, A. and Hedewy, S. (1982) EPR study of muscovite mica. Egypt. Jour. Phys. 13, 71-78.

Hazen, R.M. and Wones, D.R. (1972) The effect of cation substitutions on the physical properties of trioctahedral micas. Am. Mineral. 57, 103-129.

Heller-Kallai, L. (1982) Mossbauer studies of synthetic and natural micas on the polylithionite-siderophyllite join: Comment. Phys. Chem. Minerals 8, 98.

_____ and Rozenson, I. (1980) Dehydroxylation of dioctahedral phyllosilicates. Clays & Clay Minerals 28, 355-368.

Herzenberg, C.L., Riley, D.L., and Lamoureaux, R. (1968) Mössbauer absorption in zinnwaldite mica. Nature 219, 364-365; erratum, 219, 773.

Hofmann, U., Fluck, E., and Kuhn, P. (1967) Mössbauer spectrum of iron in glauconite. Angew. Chem. Int'l. Ed. Engl. 6, 561-562.

Hogarth, D.D., Brown, F.F., and Pritchard, A.M. (1970) Biabsorption, Mössbauer spectra, and chemical investigation of five phlogopite samples from Quebec. Canadian Mineral. 10, 710-722.

Hogg, C.S. and Meads, R.E. (1970) The Mössbauer spectra of several micas and related minerals. Mineral. Mag. 37, 606-614.

_____ and _____ (1975) A Mössbauer study of thermal decomposition of biotites. Mineral. Mag. 40, 79-88.

Ivanitskey, V.P., Matyash, I.V., and Rakovich, F.I. (1975) Effects of irradiation on the Mössbauer spectra of biotites. Geochem. Int'l. 12, no. 3, 151-157. Transl. from Geokhimiya 6, 850-857.

Kalinichenko, A.M., Matyash, I.V., Rozhdestvenskaya, I.V., and Frank-Kamenetskii, V.A. (1974) Refinement of the structural characteristics of chernykhite from proton magnetic resonance data. Soviet Physics - Crystallogr. 19, 70-71. Transl. from Kristallografiya 19, 123-125, 1974.

_____, _____, Khomyak, T.P., and Pavlishin, V.I. (1975) Distribution of octahedral cations in biotite according to the data of proton magnetic resonance. Geochemistry Int'l. 12, 18-24.

Karickhoff, S.W. and Bailey, G.W. (1973) Optical absorption spectra of clay minerals. Clays & Clay Minerals 21, 59-70.

Kemp, R.C. (1969) Orthorhombic iron centers in muscovite and phlogopite micas. J. Physics C4, L11-L13.

_____ (1972) Corrigendum. J. Physics C5, 792.

_____ (1973) Electron spin resonance of Fe^{3+} in muscovite. Phys. Stat. Solidi B57, K79-K81.

Kieffer, S.W. (1982) Thermodynamics and lattice vibrations of minerals. 5. Applications to phase equilibriums, isotopic fractionation, and high-pressure thermodynamic properties. Rev. Geophys. Space Phys. 20, 827-849.

Kinsey, R.A., Kirkpatrick, R.J., Hower, J., Smith, K.A., and Oldfield, E. (1984) High resolution aluminum-27 and silicon-29 nuclear magnetic resonance spectroscopic study of layer silicates, including clay minerals. Am. Mineral., in press.

Kleim, W. and Lehmann, G. (1979) A reassignment of the optical absorption bands in biotites. Phys. Chem. Minerals 4, 65-75.

Kotlicki, A., Olsen, N.B., and Olsen, J.S. (1976) Mössbauer and X-ray study of radiation effects in biotites. Rad. Effects 28, 1-4.

Kozyreva, I.V., Korenbaum, S.A., and Narnov, G.A. (1984) Feasibility of determining the composition of phengite micas with respect to their IR spectra. Zap. Vses. Mineral. Obshch. 113, 113-120.

Langer, K., Chatterjee, N.D., and Abraham, K. (1981) Infrared studies of some synthetic and natural $2M_1$ dioctahedral micas. N. Jahrb. Mineral. Abh. 142, 91-110.

Larson, S.J., Pardoe, G.W.F., Gebbie, H.A., and Larson, E.E. (1972) The use of far infrared interferometric spectroscopy for mineral identification. Am. Mineral. 57, 998-1002.

Lazarev, A.N. (1972) Vibrational Spectra and Structure of Silicates. English edition. Consultants Bureau, New York.

Levillain, C. and Maurel, P. (1980) Étude par spectrométrie infrarouge des frequences d'elongation du groupement hydroxyle dans des, micas synthetiquees de la série annite-phlogopite et annite-sidérophyllite. Comp. Rend. (Paris) D290, 1289-1292.

_____, _____, and Menil, F. (1981) Mössbauer studies of synthetic and natural micas on the polylithionite-siderophyllite join. Phys. Chem. Minerals 7, 71-76.

_____, _____, and _____ (1982) Mössbauer studies of synthetic and natural micas on the polylithionite-siderophyllite join: Reply. Phys. Chem. Minerals 8, 99-100.

Lippmaa, E. Mägi, M., Samoson, A., Engelhardt, G., and Grimmer, A.-R. (1980) Structural studies of silicates by solid-state high-resolution ^{29}Si NMR. J. Am. Chem. Soc. 102, 4889-4893.

Litovchenko, A.S., Brodovoi, A.V., and Mel'nikov, A.A. (1982) Study of the temperature dependences of magnetic susceptibilities and lithium-7 NMR spectra of ferriferrous micas. Phys. Stat. Solidi A73, K79-K82.

Loewenstein, W. (1954) The distribution of aluminum in the tetrahedra of silicates and aluminates. Am. Mineral. 39, 92-96.

Loh, E. (1973) Optical vibrations in sheet silicates. J. Phys. C6, 1091-1104.

Magi, M., Lippmaa, E., Samoson, A., Engelhardt, G., and Grimmer, A.R. (1984) Solid-state high-resolution silicon-29 chemical shifts in silicates. J. Phys. Chem. 88, 1518-1522.

Mank, V.V., Karushkina, A.Y., Ovcharencko, F.D., and Vasil'yev, N.G. (1974) EPR spectra of clay minerals. Dokl. Akad. Nauk. SSSR 218, 118-119. Transl. from Dokl. Akad. Nauk. SSSR 218, 921-923.

Marfunin, A.S., Mineyeva, R.M., Mkrtchyan, A.R., Nyussik, Y.M., and Fedorov, V.Y. (1969) Optical and Mössbauer spectroscopy of iron in rock-forming silicates. Int'l. Geology Review 11, 31-44.

McConchie, D.M., Ward, J.B., McCann, V.H., and Lewis, D.W. (1979) A Mössbauer investigation of glauconite and its geological significance. Clays & Clay Minerals 27, 339-348.

Miyano, T. and Miyano, S. (1982) Ferri-annite from the Dales Gorge Member iron-formations, Wittenoom area, Western Australia. Am. Mineral. 67, 1179-1194.

Olivier, D., Vedrine, J.C., and Pezerat, H. (1977) Application de la RPE à la localisation des substitutions isomorphiques dans les micas: Localisation des substitutions isomorphiques dan les micas: Localisation du Fe^{2+} dans le muscovites et les phlogopites. Jour. Solid State Chem. 20, 267-278.

Osherovich, E.Z. and Nikitina, L.P. (1975) Determination of content of elements in the octahedral positions in ferromagnesian micas by valence vibrations of OH^-. Gochem. Int'l. 5, 71-76. Transl. from Geokhimiya 5, 727-732.

Pavlishin, V.I. (1975) Infrared spectra of micas of the lithium-iron isomorphous series. Zap. Vses. Min. Obshch. 104, 70-74. [In Russian]

Puustinen, K. (1973) Tetraferriphlogopite from the Siilinjarvi carbonatite complex, Finland. Bull. Geol. Soc. Finland 45, 35-42.

Raclavasky, K., Sitek, J., and Lipka, J. (1975) Mössbauer spectroscopy of iron in clay minerals. 5th Int'l. Conf. Mössbauer Spec. Proc. Part II, 368-371.

Rao, V.J. and Chetal, A.R. (1982) An X-ray K-absorption study of muscovite mica. J. Physics D15, L195-L197.

Rausell-Colom, J.A., Sanz, J., Fernandez, M., and Serratosa, J.M. (1979) Distribution of octahedral ions in phlogopites and biotites. In: Mortland, M.M. and Farmer, V.C., eds., Developments in Sedimentology 27, 27-36. Elsevier, Amsterdam.

Richardson, S.M. (1973) A pink muscovite with reverse pleochroism from Archer's Post, Kenya. Am. Mineral. 60, 73-78.

_____ (1976) Ion distribution in pink muscovite: A reply. Am. Mineral. 61, 1051-1052.

_____ and Richardson, J.W., Jr. (1982) Crystal structure of a pink muscovite from Archer's Post, Kenya: Implications for reverse pleochroism in dioctahedral micas. Am. Mineral. 67, 69-75.

Robbins, D.W. and Strens, R.G.J. (1972) Charge-transfer in ferromagnesian silicates: The polarized electronic spectra of trioctahedral micas. Mineral. Mag. 38, 551-563.

Rolf, R.M., Kimball, C.W., and Odom, I.E. (1977) Mössbauer characteristics of Cambrian glauconite, central U.S.A. Clays & Clay Minerals 25, 131-137.

Rouxhet, P.G. (1970) Hydroxyl stretching bands in micas: A quantitative interpretation. Clay Minerals 8, 375-388.

_____ (1970) Kinetics of dehydroxylation of OH-OD exchange in macrocrystalline micas. Am. Mineral. 55, 841-853.

Rozenson, I. and Heller-Kallai, L. (1977) Mössbauer spectra of dioctahedral smectites. Clays & Clay Minerals 25, 94-101.

Russell, J.D., Farmer, V.C., and Velde, B. (1970) Replacement of OH by OD in layer silicates, and identification of the vibrations of these groups in infra-red spectra. Mineral. Mag. 37, 869-879.

Ruthberg, S., Barnes, M.W., and Noyes, R.H. (1963) Correlation of muscovite sheet mica and the basis of color, apparent optic angle, and absorption spectrum. J. Research Natl. Bur. Stand. A, 67A, 309-324.

Sanz, J., Meyers, J., Vielvoye, L., and Stone, W.E.E. (1978) The location and content of iron in natural biotites and phlogopites: A comparison of several methods. Clay Minerals 13, 45-52.

_____ and Serratosa, J.M. (1984) ^{29}Si and ^{27}Al high resolution MAS-NMR spectra of phyllosilicates. J. Am. Chem. Soc., in press.

_____ and Stone, W.E.E. (1977) NMR study of micas. I. Distribution of Fe^{2+} ions on the octahedral sites. J. Chem. Phys. 67, 3739-3743.

_____ and _____ (1979) NMR study of micas, II. Distribution of Fe^{2+}, F^-, and OH^- in the octahedral sheet of phlogopites. Am. Mineral. 64, 119-126.

_____ and _____ (1983) NMR study of minerals. III. The distribution of Mg^{2+} and Fe^{2+} around the OH groups in micas. J. Phys. C16, 1271-1281.

Schmetzer, K. (1982) Absorption spectroscopy and color of V^{3+}-bearing natural oxides and silicates - A contribution to the crystal chemistry of vanadium. N. Jahrb. Mineral., Abh. 144, 73-106.

Seal, M., Vance, E.R., and Demago, B. (1981) Optical spectra of giant radiohaloes in Madagascan biotite. Am. Mineral. 66, 358-361.

Sekino, H., Kanisawa, S., Harada, K., and Ishikawa, Y. (1975) Aluminian xanthophyllite and paragonite from Japan. Mineral. Mag. 40, 421-423.

Serratosa, J.M. and Bradley, W.F. (1958) Determination of the orientation of OH bond axes in layer silicates by infrared absorption. J. Phys. Chem. 62, 1164-1167.

Shigorova, T.A. (1982) Determination of ammonium content in micas by using IR spectroscopy. Geokhimiya 3, 458-562. Transl. from Geochem. Int'l. 19, 110-114.

Smith, G. (1977) Low-temperature optical studies of metal-metal charge-transfer transitions in various minerals. Canadian Mineral. 15, 500-507.

_____ (1978) Evidence for absorption by exchange-coupled Fe^{2+}-Fe^{3+} pairs in the near infra-red spectra of minerals. Phys. Chem. Minerals 3, 375-383.

_____ and Strens, R.G.J. (1976) Intervalence transfer absorption in some silicate, oxide and phosphate minerals. In: R.G.J. Strens, ed., The Physics and Chemistry of Minerals and Rocks, p. 583-612. Wiley, New York.

_____, Howes, B., and Hasan, Z. (1980) Mössbauer and optical spectra of biotite: A case for Fe^{2+}-Fe^{3+} interactions. Phys. Stat. Solidi A57, K187-K192.

_____, Hålenius, U., Annersten, H., and Ackermann, L. (1983) Optical and Mössbauer spectra of manganese-bearing phlogopites: $Fe^{3+}(IV)$-$Mn^{2+}(VI)$ pair absorption as the origin of reverse pleochroism. Am. Mineral. 68, 759-768.

Stubican, V. and Roy, R. (1961) Isomorphous substitution and infrared spectra of the layer lattice silicates. Am. Mineral. 46, 32-51.

Tateyama, H., Shimoda, S., and Sudo, T. (1976) Infrared absorption spectra of synthetic Al-free magnesium micas. N. Jahrb. Mineral., Monatsh. 128-140.

Taylor, G.L., Ruotsala, A.P., and Keeling, R.O. (1968) Analysis of iron in layer silicates by Mössbauer spectroscopy. Clays & Clay Minerals 16, 381-391.

Tsuboi, M. (1950) On the positions of the hydrogen atoms in the crystal structure of muscovite, as revealed by the infra-red absorption study. Bull. Chem. Soc. Japan 23, 83-90.

Vance, E.R. (1978) A possible mechanism for the formation of bleached, giant haloes in Madagascar mica. In: M.A.K. Lodhi, ed., Superheavy Elements. Pergamon Press, New York, p. 228-235.

Vedder, W. (1964) Correlations between IR spectrum and chemical composition of mica. Am. Mineral 49, 736-768.

_____ and McDonald, R.S. (1963) Vibrations of the OH ion in muscovite. J. Chem. Phys. 38, 1583-1590.

Velde, B. (1978) Infrared spectra of synthetic micas in the series muscovite-MgAl celadonite. Am. Mineral. 63, 343-349.

_____ (1979) Cation-apical oxygen vibrations in mica tetrahedra. Bull. Minéral. 102, 33-34.

_____ (1980) Cell dimensions, polymorph type, and infrared spectra of synthetic white micas: The importance of ordering. Am. Mineral. 65, 1277-1282.

_____ (1983) Infrared OH-stretch bands in potassic micas, talcs and saponites; influence of electronic configuration and site of charge compensation. Am. Mineral. 68, 1169-1173.

Weaver, C.E., Wampler, J.M., and Pecuil, T.E. (1967) Mössbauer analysis of iron in clay minerals. Science 156, 504-508.

Wones, D.R. (1963) Phase formation of "ferriannite", $KFe_3^{2+}Fe^{3+}Si_3O_{10}(OH)_2$. Am. J. Sci. 261, 581-596.

6. OPTICAL PROPERTIES of MICAS under the POLARIZING MICROSCOPE
Ray E. Wilcox

INTRODUCTION

This chapter deals with optical properties used in identifying micas under the polarizing microscope in the visible spectrum. Optical properties as related to crystal structure are treated in Chapter 5 of this volume, along with the rest of the electromagnetic spectrum and Mössbauer phenomena. It is assumed here that the reader is acquainted with the theory of optical crystallography and the operation of the polarizing microscope. Nomenclature in this chapter does not correspond in every respect to that of other chapters because much of the available optical data comes from literature published before detailed structural studies were introduced.

Micas, as phyllosilicates, have a perfect single cleavage, (001), resulting from the strict layered arrangement of strongly bonded $[Si_4O_{10}]$, $[Si_3AlO_{10}]$, and $[Si_2Al_2O_{10}]$ groups in the crystal and the weak interlayer bonding (Chapters 1 and 2). The cleavage stands out prominently under the microscope and at once aids in the recognition of the unknown as a probable phyllosilicate. The layered arrangement, however, also imposes a set of optical characteristics (negative optic sign, parallel extinction, positive elongation, and in many cases small optic angle) that are common to most phyllosilicates. Refractive indices of some phyllosilicates, especially the trioctahedral micas, the stilpnomelanes, and the chlorites, range widely and overlap. Thus, the strong similarities and extensive overlap of properties make it difficult or impossible in some cases to distinguish optically between certain micas and other phyllosilicates, as well as between certain species of micas themselves without additional information from structural or chemical studies.

OPTICAL PROPERTIES OF INDIVIDUAL MINERALS

With few exceptions phyllosilicate minerals are optically negative, have parallel or very nearly parallel extinction, and positive elongation in respect to the perfect cleavage, to which the acute bisectrix is nearly perpendicular. Typical examples of optical orientations in the mica group are shown in Figure 1 for muscovite and phlogopite as one might encounter them in thin section with the cleavage vertical. For both minerals the acute bisectrix is the vibration direction of α, within a very few degrees of the normal to the cleavage. Because the vibration direction of β and γ are in the plane of the cleavage or very close thereto, extinction for these positions or for intermediate

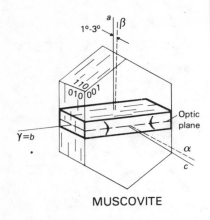 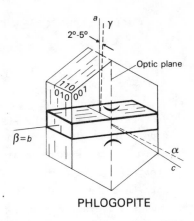

Figure 1. Sketches of crystal plates of muscovite and phlogopite with cleavage standing perpendicular to the plane of the thin section, showing relations of optical elements to crystallographic axes and the (001) cleavage.

positions will be parallel or very nearly parallel to the cleavage, and the sign of elongation will be positive. (Exceptions include the rare Ba-Fe mica, anandite, and several high-iron chlorites, for which the optic sign is positive and elongation is negative.) Tables 1 and 2 list the pertinent optical properties of micas and a number of other phyllosilicates, taking advantage of the properties that serve to distinguish between species, while at the same time drawing attention to the examples of overlap and similarity in other optical properties. The tables follow in most respects the format of Larsen and Berman (1934) but incorporate several helpful improvements from Winchell (1939) and Fleischer et al. (in press).

Explanation of tables

In Table 1 the phyllosilicates (with micas shown in upper case type) are listed in order of increasing intermediate principal index, β, a convenient starting property for identifying mineral fragments by the immersion method. Because refractive indices of a particular mineral species may vary between limits, due to partial or complete solid solution of constituent elements, several of the minerals are listed at intervals in the table through their respective refractive index ranges. Lines along the left side of the table connect these listings. Values of the extreme principal indices, α and γ, are given in separate columns. Other columns give maximum birefringence ($\gamma - \alpha$), optic axial angle ($2V_\alpha$ and dispersion), and pleochroism, with the absorption formula referred to vibration directions corresponding to the principal refractive indices (α, β, and γ). Finally the column for Remarks cites a literature source of the data and notes other useful properties or

Table 1. Optical properties of micas and selected other phyllosilicates arranged in order of increasing refractive index. For explanation of abbreviations, see the end of Table 2, page 191.

Refractive indices			Mineral name	Bire-frin-gence	2V$_\alpha$ disp	Pleochroism and absorption			Remarks
α	β	γ				α	β	γ	
1.475	1.499	1.500	montmorillonite	.025	5-25°	(colorless)			Trg #208
1.534-1.535	1.534-1.535	1.535-1.537	apophyllite	.002	180°	(colorless)			Opt (+), $\gamma \sim \|\|c$ brtl, DHZ 258
1.524	1.543	1.545	POLYLITHIONITE (lepidolite ser)	.021	0-30°	(colorless)			Trg #205
1.525	1.545	1.545	vermiculite	.020	0-8°	cols	< yl-bn	= yl-bn	Exfol, DHZ 246
1.522	1.548	1.549	PHLOGOPITE (biotite ser)	.027	14° r<v	cols	< bn	<? bn	Synth F-end member DHZ 46#1
1.546	1.551	1.552	antigorite	.006	60° r>v	(colorless)			Trg #217
1.522	1.553	1.553	TAENIOLITE	.031	0-5°	cols	< p pk	= yl	Vl 29
1.529	1.554	1.556	TRILITHIONITE (PAUCILITHIONITE) (lepidolite ser)	.027	25°	cols	< p pk	= p pk	WW 371
1.530	1.557	1.558	PHLOGOPITE (biotite ser)	.028	15° r<v	yl	< bn-r	= bn-r	DHZ 42
1.553	1.559	1.560	kaolinite	.007	25°	(colorless)			$\gamma\|\|b$, DHZ 194
1.537	1.563	1.566	POLYLITHIONITE (lepidolite ser)	.029	0°	(colorless)			Trg #205
1.562	1.565	1.565	penninite (chlorite grp)	.003	0-5° r>v	p yl-gn	(colorless) ~ p gn-yl	~ gn-yl	Abn int colors. Trg #212
1.565	1.567	1.570	kaolinite	.005	24-50°	(colorless)			$\gamma\|\|b$, DHZ 194
1.534-1.536	1.569-1.570	1.570-1.571	MASUTOMILITE	.036 .035	28-31° r>v wk	cols p pk	< pr < pr	> cols > p pk	AM 62, 594
1.54	1.57	1.57	beidellite	.03	v sm	(colorless)			DHZ 212
1.540	1.570	1.570	TAENIOLITE	.040	small	(?)			Vl 29
1.535	1.570	1.572	ZINNWALDITE	.037	10-15° r>v wk	?	?	?	DHZ 92
1.565	1.572	1.595	cookeite (chlorite grp)	.030	100-160° r>v	p gn-pk	~ p gn-pk	> cols	Opt (+), $\gamma \sim \|\|c$ Trg #216a
1.555	1.575	1.581	TOBELITE	.026	28°	(colorless)			AM 68, 850
1.543	1.576	1.576	stilpnomelane	.033	0°	yl	<< r-bn	~ r-bn	No brdsye DHZ 103
1.558	1.578	1.589	"hydromuscovite"	.031	10-20° (72±8° calc)	(colorless)			DHZ 218 #3
1.552	1.582	1.587	MUSCOVITE	.035	30° r>v	(colorless)			$\gamma\|\|b$, DHZ 11
1.563	1.583	1.583	vermiculite	.020	0-8° r>v wk	cols	< p gn	~ p gn-bn	Exfol DHZ 246
1.575	1.584	1.600	cookeite (chlorite grp)	.025	125° (106±10°)	(colorless, pk, p gn)			Opt (+), $\gamma \sim \|ic$ Vl 33
1.545	1.584	1.584	talc	.039	0-10°	(colorless)			AM 53, 758
1.547	1.584	1.587	PHENGITE	.040	24-36° r>v	(colorless)			u sericitic Trg #199a

Table 1, continued.

Refractive indices			Mineral name	Bire-fringence	$2V_\alpha$ disp	Pleochroism and absorption			Remarks
α	β	γ				α	β	γ	
1.559	1.585 (calc)	1.586	GLAUCONITE, var Skolite	.027	var		(colorless)		DHZ 38 #1
1.544-1.550	1.586-1.589	1.596-1.601	pyrophyllite	.062 .051	50-60° r>v wk	cols	(colorless) < p gy	~ p gy	γ∥b, DHZ 115
1.584	1.588	1.598	cookeite (chlorite grp)	.014	120°	p pk	~ p pk	> p yl	Opt(+), γ~∥c BeO 1.1% MA 19 308
1.588	1.588	1.611	amesite	.023	170° r>v		(colorless)		Opt(+), γ∥c, Trg #219a
1.558	1.589	1.590	ZINNWALDITE	.032	0-20° r>v wk	cols	< bn-gy	< bn-gy	DHZ 92 Trg #205a
1.564	1.594	1.600	PARAGONITE	.036	0-30° r>v		(colorless)		γ∥b, DHZ 31
1.551	1.594	1.594	stilpnomelane	.043	0°	yl	<< r-bn	~ r-bn	No brdsye DHZ 107 #1
1.597	1.597	1.612	amesite	.015	170° r>v		(colorless)		Opt(+), γ∥c Trg #219a
1.554	1.599	1.602	talc	.048	sm	cols	< p gn	~ p gn	γ∥b, CA 76 129852
1.595	1.603	1.604	antigorite	.009	30°	p gn-yl	~ p gn	~ p gn	Trg #217
1.600	1.603	1.610	corundophilite (chlorite grp)	.010	120°		(weak)		Opt(+) DHZ 139#10
1.565	1.605	1.605	BIOTITE (biotite ser)	.040	0-9° r<v	bn gn-bn	<< bn << gn	~ bn ~ gn	DHZ 55
1.580	1.605	1.605	MONTDORITE	.025	0-3°	?	?	?	AM 64, 1331
1.544	1.608	1.608	WONESITE	.064	0-5°	p bn	< bn	= bn	AM 66, 100
1.580	1.609	1.609	PARAGONITE	.029	0-30° r>v		(colorless)		γ∥b, DHZ 31
1.57	1.61	1.61	illite (hydromica)	.04	v sm		(colorless)		dioctahedral γ∥b, DHZ 213
1.574	1.610	1.616	MUSCOVITE	.042	45° r>v		(colorless)		γ∥b, DHZ 11
1.571	1.610	1.612	PHENGITE	.041	24-36° r>v		(colorless)		u sericitic γ∥b, Trg #199a
1.586	1.612	1.613	MARGARITE, Na-analogue	.027	50°		(colorless)		Brtl, no brdsye γ∥b, Na₂O 5.6% DHZ 97#6
1.592	1.614	1.614	CELADONITE (GLAUCONITE)	.022	0-25° r>v	bl-gn	< bn-yl	~ bn-yl	DHZ 37
1.560	1.614	1.615	PREISWERKITE	.055	5-7°		(colorless)		AM 65, 1135
1.576	1.615 (calc)	1.618	MUSCOVITE	.042	28-31° r>v		(pale green)		γ∥b, MA 80-3449
1.580	1.615	1.615	minnesotaite	.035	v sm	cols	< p gn	= p gn	γ∥b, fibrous Trg #197a
1.580	1.620	1.623	PHENGITE	.043	35°	cols	< p gn	~ p gn	γ∥b, AM 45,841
1.592	1.624	1.625	EPHESITE	.033	20°		(colorless)		Brtl, no brdsye γ∥b, PG 596
1.580	1.625 (calc)	1.625	nontronite	.045	v sm	yl	< yl-gn	< ol-gn	DHZ 233 #11 Trg #209
1.582	1.625	1.625	SIDEROPHYLLITE (biotite grp)	.043	sm	or-bn	<< r-bn	~ r-bn	Trg #202

Table 1, concluded.

Refractive indices			Mineral name	Bire-frin-gence	2V α disp	Pleochroism and absorption			Remarks
α	β	γ				α	β	γ	
1.59	1.63	1.64	ROSCOELITE	.05	med r<v		(?)		γ∥b USGS 150, 121
1.615-1.619	1.630-1.633	1.630-1.635	KINOSHITALITE	.015-.014	23°	p yl	< p yl	~ p yl	Brt1 AM 60, 486
1.592	1.631 (calc)	1.632	minnesotaite	.040	small	p gn	> p gn-yl	= p gn-yl	AM 50, 166
1.590	1.637	1.637	PHLOGOPITE (Biotite ser)	.047	0-20° r<v	yl yl	<< bn-r << gn	= bn-yl = gn	DHZ 42
1.632	1.638	1.638	Fe-chlorite (chlorite grp)	.006	0-3° r>v	p bn	< bn-gn	~ bn-gn	Trg #214
1.610	1.641	1.641	CELADONITE (GLAUCONITE)	.031	0-25° r>v	gn-yl	< gn	~ gn	DHZ 37
?	1.64	1.64	HENDRICKSITE	?	2-5°		(brown to red brown)		AM 51 1114
1.638	1.648	1.650	MARGARITE	.012	60° r<v		(colorless)		Brt1, γ∥b DHZ 95
1.643	1.652	1.654	BITYITE	.011	med		(colorless)		Brt1 LB 177
1.600	1.652	1.655 (calc)	willemseite (talc ser)	.055 (calc)	27°		(colorless)		AM 55, 33
1.643	1.655	1.655	CLINTONITE	.012	30°		(colorless)		Brt1 AM 52 1126
1.598	1.658	1.660	HENDRICKSITE	.062	8°		(dark colors)		CA 65 18345
1.651	1.659	1.661	BITYITE	.010	35-40°		(colorless)		Brt1 PG 610
1.584	1.661	1.661	stilpnomelane	.077	0°	yl	<< gn	~ gn	No brdaye DHZ 108#7
1.648	1.662	1.663	CLINTONITE	.015	30°	p or	< p gn	~ p gn	Brt1, γ∥b? Trg #206
1.644	1.662 (calc)	1.663	CELADONITE	.019	sm	yl-gn	< yl-gn	~ gn	CA 67 #10441
1.63	1.67 (calc)	1.67	illite	.04	sm		(colorless?)		Trioctahedral γ∥b, DHZ 222
1.616	1.670	1.670	SIDEROPHYLLITE (biotite ser)	.054	small	yl	< gn	< gn	WW 374
1.671	1.684	1.685	chamosite (chlorite grip)	.014	0-30°	p bn-gn	<gn	~ gn	Trg #215
1.610	1.685	1.704	ROSCOELITE	.094	25-40°	gn-bn	~ ol-gn	~ ol-gn	DHZ 21
1.629	1.686	1.686	HENDRICKSITE	.062	2-5°		(dark colors)		AM 51, 1114?
1.640	1.686	1.702	CHERNYKHITE	.062	12° (60±4°)		(ol-gn to dk-gn)		AM 58, 966
1.625	1.690	1.691	ANNITE (biotite ser)	.066	sm r>v	yl	<< bn	~ bn	PG 620
1.653-1.677	1.690-1.720 (calc)	1.691-1.721	FERRI-ANNITE (biotite ser)	.038-.044	0-10° sm	p r-br bn-r	~ p yl-gn ~ p gn-bn	~ p yl-gn ~ p gn-bn	AM 67, 1179
?	1.697	1.697	HENDRICKSITE	?	sm		(brown to black)		AM 51, 1109
1.610	1.722	1.730	OXYBIOTITE (biotite ser)	.120	20-30° r<v str	yl-bn rd-bn	<< bn-r << bn-r	~ bn-r ~ bn-r	Trg #203a
1.634	1.745	1.745	stilpnomelane	.111	0°	yl	<< dk bn	~ dk bn	DHZ 103
1.85	1.85	1.88	ANANDITE	.03	>>90°	?	> gn	~ gn	Opt (+), γ∥c AM 52, 1586

exceptions to the general rules.

Once the unknown mineral has been recognized as a phyllosilicate -- that is, it has been found to have but one cleavage (exceptionally perfect), parallel or nearly parallel extinction, and positive elongation -- an effective procedure for the use of the table is as follows: Determine by immersion methods the values of the principal refractive indices, α, β, and γ, from which is obtained the birefringence, $\gamma - \alpha$. Measure or estimate the optic angle, $2V_\alpha$, and note the pleochroism and absorption, if any. Then enter Table 1 at the determined value of β, noting the tie lines at the left that pass through this level. Trace each tie line in both directions to listings of its mineral name and record the name of each mineral whose birefringence, optic angle, and pleochroism are in reasonable agreement with the unknown. Then explore higher and lower in the table for nearby listed minerals not connected by tie lines, and record those of indices, birefringence, optic angle, and pleochroism similar to the unknown. Finally, for each of the minerals so recorded, examine the Remarks column for additional distinctive properties to look for in the unknown and for entry into the published literature for additional information.

In Table 2 the phyllosilicates are arranged in order of increasing birefringence as a convenience for working with micas in thin section, where refractive indices can be estimated only crudely but birefringence can be estimated fairly closely from the order of interference colors shown by those crystals with cleavage perpendicular to the section. For identification of an unknown, Table 2 may be used in the same manner as outlined above for Table 1, but entering the table with the determined birefringence of the unknown. As in Table 1, tie lines at the left connect the separate entries of each mineral series. In the first column are shown birefringences ($\gamma - \alpha$) and in the next column the corresponding maximum interference color to be observed in a grain in a thin section of standard thickness, here assumed to be 0.025 to 0.030 mm. For highly colored minerals, such as biotite, some difficulty may be expected in determining the higher orders of interference colors because of increasing mixing of interference colors and superposed pleochroic colors. Separate columns give data for refractive indices, optic angle, pleochroism, and the information of the Remarks column.

Some distinctions stand out in these tabulated data. For instance, micas are distinguished from the common chlorites by their higher birefringence, even though there is extensive overlap of refractive indices. In thin section biotites may be difficult to distinguish from stilpnomelane because of the wide overlap in birefringence, refractive indices, and pleochroism, leaving only

Table 2. Optical properties of micas and selected other phyllosilicates arranged in order of increasing birefringence. For explanation of abbreviations, see the end of this table (p. 191).

Birefringence	Interference color	Mineral name	Refractive Indices			$2V_\alpha$	Pleochroism and absorption			Remarks
			α	β	γ	disp	α	β	γ	
.001 .002	gray	apophyllite	1.534 1.535	1.534 1.535	1.535 1.537	180°	(colorless)			Brtl, opt(+) γ~||c, DHZ 258
.003		penninite (chlorite grp)	1.562	1.565	1.565	0-5°	p yl-gn	~ p gn-yl	~ p gn-yl	Abn int colors Trg #212
.005- .007	white	kaolinite	1.565- 1.553	1.567- 1.559	1.570- 1.560	24-50°	(colorless)			γ||b DHZ 194
.006- .007	yellow	antigorite	1.546- 1.595	1.551- 1.603	1.552- 1.604	60° 30°	p yl	(colorless) ~ p gn	~ p gn	Trg #217
.006	ORDER	Fe-chlorite (chlorite grp)	1.632	1.638	1.638	0-3°	p bn	< bn-gn	~ bn-gn	Trg #214
.010	FIRST	corundophilite (chlorite grp)	1.600	1.603	1.610	120°	(weak)			Opt (+) DHZ 139 #10
.010- .011	orange	BITYITE	1.651- 1.643	1.659- 1.652	1.661- 1.654	35-40° med	(colorless) (colorless)			Brtl, PG 619 LB 177
.012		MARGARITE	1.638	1.648	1.650	60°	(colorless)			Brtl DHZ 95
.012		CLINTONITE	1.643	1.655	1.655	30°	cols	< p yl	~ p yl	Brtl AM 52, 1126
.014		chamosite (chlorite grp)	1.671	1.684	1.685	0-30°	p bn-gn	< gn	~ gn	Trg #215
.014		cookeite (chlorite grp)	1.584	1.588	1.598	120°	p pk	~ p pk	>b yl	Opt(+), γ~||c BeO 1.1% MA 19, 308
.014- .015		KINOSHITALITE	1.619- 1.615	1.633- 1.630	1.635- 1.630	25°	p yl	< p yl	= p yl	Brtl AM 60, 486
.015	orange	CLINTONITE	1.648	1.662	1.663	30°	p or	< p gn	~ p gn-bn	Brtl, Trg #206
.015	ORDER	amesite	1.597	1.597	1.612	170°	(colorless)			Opt(+), Trg #219a
.019	FIRST red	CELADONITE	1.644	1.662 (calc)	1.663	sm	yl-gn	< yl-gn	~ gn	CA 67 #10441
.020 .020		vermiculite	1.525- 1.563	1.545- 1.583	1.545- 1.583	0-8°	cols cols	< yl-bn < p gn	= yl-bn ~ p gn-bn	Exfol DHZ 246
.021		POLYLITHIONITE (lepidolite ser)	1.524	1.543	1.545	0-30°	(colorless)			Trg #205
.022		CELADONITE (GLAUCONITE)	1.592	1.614	1.614	0-25°	bl-gn	< bn-yl	~ bn-yl	DHZ 37
.023		amesite	1.588	1.588	1.611	170°	(colorless)			Opt(+), Trg #219a
.025	blue	cookeite (chlorite grp)	1.575	1.584	1.600	125° (106±10° calc)	(colorless, pk, p gn)			Opt(+), γ~||c Vl 33
.025	ORDER	montmorillonite	1.475	1.499	1.500	5-25°	(colorless)			Trg #208
.025		MONTDORITE	1.580	1.605	1.605	0-3°	?	?	?	AM 64, 1331
.026	SECOND	TOBELITE	1.555	1.575	1.581	28°	(colorless)			AM 68, 850
.027	green	MARGARITE (Na-analogue)	1.586	1.612	1.613	50°	(colorless)			Brtl, no brdsye DHZ 96, #6
.027		GLAUCONITE var Skolite	1.559	1.585 (calc)	1.586	var	(colorless, p gn)			DHZ 38, #1
.027		TRILITHIONITE (PAUCILITHIONITE) (lepidolite ser)	1.529	1.554	1.556	25°	cols	< p pk	= p pk	WW 371

Table 2, continued.

Bire-fringence	Inter-ference color	Mineral name	Refractive Indices			$2V_\alpha$ disp	Pleochroism and absorption			Remarks
			α	β	γ		α	β	γ	
.027-.028	green	PHLOGOPITE (biotite ser)	1.522-1.530	1.548-1.557	1.559-1.558	14°	cols yl	< bn < bn-r	<? bn = bn-r	DHZ 46 #1 DHZ 42
.029		POLYLITHIONITE (lepidolite ser)	1.537	1.563	1.566	0°		(colorless)		Trg #205
.029		PARAGONITE	1.580	1.609	1.609	0-30°		(colorless)		DHZ 31
.030		cookeite (chlorite grp)	1.565	1.572	1.595	140°	p gn-pk	~ p gn-pk	> cols	Opt(+) Trg #216a
.03	yellow	ANANDITE	1.85	1.85	1.88	>>90°	?	> gn	~ gn	Opt(+), γ∥c AM 52, 1586
.03		beidellite	1.54	1.57	1.57	v sm		(colorless)		DHZ 212
.031		CELADONITE (GLAUCONITE)	1.610	1.641	1.641	0-25°		(colorless)		DHZ 37
.031		TAENIOLITE	1.522	1.553	1.553	0-5°	cols	< p pk	= yl	Vl 29
.031	SECOND ORDER	"hydromuscovite"	1.558	1.578	1.589	10-20° (72±8° calc)		(colorless)		DHZ 218 #3
.032		ZINNWALDITE	1.558	1.589	1.590	0-20°	cols	< bn-gy	< bn-gy	DHZ 92, Trg #205a
.033		EPHESITE	1.592	1.624	1.625	20°		(colorless)		Brtl, no brdsye γ∥b, PG 596
.033	orange	stilpnomelane	1.543	1.576	1.576	0°	yl	<< r-bn	~ r-bn	No brdsye DHZ 103
.035		minnesotaite	1.580	1.615	1.615	v sm	cols	< p gn	~ p gn	Fibrous, γ∥b Trg #197a
.035		MUSCOVITE	1.552	1.582	1.587	30°		(colorless)		γ∥b, DHZ 11
.035-.036		MASUTOMILITE	1.536-1.534	1.570-1.569	1.571-1.570	28-31°	p pk cols	< pr < pr	> p pk > cols	AM 62, 594
.036		PARAGONITE	1.564	1.594	1.600	0-30°		(colorless)		DHZ 31
.037	orange	ZINNWALDITE	1.535	1.570	1.572	10-15°	?	?	?	DHZ 92, Trg #205a
.038		FERRI-ANNITE (biotite ser)	1.653	1.690 (calc)	1.691	0-10°	p r-bn	~ p yl-gn	~ p yl-gn	AM 67, 1179
.039		talc	1.545	1.584	1.589	0-10°		(colorless)		AM 53, 758
.040	red	TAENIOLITE	1.540	1.570	1.570	sm	?	?	?	Vl 29
.040		BIOTITE	1.565	1.605	1.605	0-9°	p bn* gn-bn	<< bn << gn	= bn = bn	DHZ 55
.040	SECOND ORDER	minnesotaite	1.592	1.631 (calc)	1.632	sm	p gn p gn	> cols > p gn-yl	= cols ~ p gn-yl	Fibrous AM 50, 166
.04		illite (hydromica)	1.57-1.63	1.61-(1.67)	1.61-1.67	v sm		(colorless)		Dioctahedral Trioctahedral DHZ 213
.040-.041		PHENGITE	1.547-1.571	1.584-1.610	1.587-1.612	24-36°		(colorless)		u sericitic γ∥b, Trg #199a
.042		MUSCOVITE	1.574-1.576	1.610-1.615	1.616-1.618	45° 28-31°		(colorless) (pale green?)		γ∥b, DHZ 11 MA 80-3449
.043	blue	PHENGITE	1.580	1.620	1.623	35°	cols	< p gn	~ p gn	γ∥b, AM 45, 841
.043	THIRD ORDER	SIDEROPHYLLITE (biotite ser)	1.582	1.625	1.625	sm	or-bn<	< r-bn	~ r-bn	Trg #202
.043		stilpnomelane	1.551	1.594	1.594	0°	yl	<< r-bn	~ r-bn	Brtl, no brdsye DHZ 107 #1
.044		FERRI-ANNITE (biotite ser)	1.677	1.720 (calc)	1.721	sm	bn-r	~ p gn-bn	~ p gn-bn	AM 67 1179

Table 2, concluded.

Bire-fringence	Interference color	Mineral name	Refractive Indices			$2V_\alpha$ disp	Pleochroism and absorption			Remarks
			α	β	γ		α	β	γ	
.045	THIRD ORDER — red — yellow — green	nontronite	1.580	1.625 (calc)	1.625	sm	yl	< yl-gn	< ol-gn	DHZ 233 #11 Trg #209
.047		PHLOGOPITE (biotite ser)	1.590	1.637	1.637	0-20°	yl yl	<< bn-r << gn	= bn-yl = gn	DHZ 42
.048		talc	1.554	1.594	1.602	sm	cols	< p gn	~ p gn	γ‖b CA 76 #129852
.05		ROSCOELITE	1.59	1.63	1.64	med	?	?	?	γ‖b USGS 150, 121
.051		pyrophyllite	1.550	1.589	1.601	50-60°	(colorless?)			DHZ 115
.054		SIDEROPHYLLITE (biotite ser)	1.616	1.670	1.670	sm	yl	< gn	< gn	WW 374
.055		PREISWERKITE	1.560	1.614	1.615	5-7°	(colorless)			AM 65, 1135
.055 (calc)		willemseite (talc ser)	1.600	1.652	1.655 (calc)	27°	(colorless)			AM 55, 33
.062		CHERNYKHITE	1.640	1.686	1.702	12° (60±4°)	(ol-gn to dk-gn)			AM 58, 966
.062		pyrophyllite	1.544	1.586	1.596	50-60°	cols	< p gy?	= p gy?	DHZ 115
.062		HENDRICKSITE	1.598- 1.629	1.658- 1.686	1.660- 1.686	8° 2-5°	(dark colors)			CA 65, 18345 AM 51, 1114?
.064	HIGHER ORDERS	WONESITE	1.544	1.608	1.608	0-5°	p bn	< bn	= bn	AM 66, 100
.066		ANNITE (biotite ser)	1.625	1.690	1.691	sm	yl	<< bn	~ bn	PG 620
.077		stilpnomelane	1.584	1.661	1.661	0°	yl	<< gn	~ gn	Brtl, no brdsye DHZ 108 #7
.094		ROSCOELITE	1.610	1.685	1.704	25-40°	gn-bn	~ ol-gn	~ ol-gn	DHZ 21
.111		stilpnomelane	1.634	1.745	1.745	0°	yl	<< dk-bn	~ dk-bn	DHZ 103
.120		OXYBIOTITE (biotite ser)	1.610	1.722	1.730	20-30°	yl-bn r-bn	<< bn-r << bn-r	~ bn-r ~ bn-r	Trg #203a

LIST OF ABBREVIATIONS AND SYMBOLS FOR TABLES 1 AND 2

abn	- abnormal	ser	- series	α	- lowest principal refractive index, or corresponding vibration direction
bl	- blue	sm	- small		
bn	- brown	str	- strong		
brdsye	- "birdseye" mottling at extinction	synth	- synthetic	β	- intermediate principal refractive index, or corresponding vibration direction
		u	- usually		
brtl	- brittle	v	- violet, very		
calc	- calculated value	var	- variety, variable	γ	- highest principal refractive index, or corresponding vibration direction
cols	- colorless	wk	- weak		
disp	- dispersion	yl	- yellow		
dk	- dark			γ'	- refractive index in range of β to γ
exfol	- exfoliates on heating	AM	- American Mineralogist	<	- less than
gn	- green	CA	- Chemical Abstracts	>	- greater than
grp	- group	DHZ	- Deer, et al., vol. 3 (1962)	=	- equal to
gy	- gray	LB	- Larsen and Berman (1934)	~	- approximately
int	- interference	MA	- Mineralogical Abstracts	‖	- parallel to
med	- medium	PG	- Phillips and Griffen (1981)		
ol	- olive	Trg	- Troeger (1967, 1971)		
opt	- optically	USGS	- U.S. Geological Survey Professional Paper		
or	- orange				
p	- pale	Vl	- Vlasoff, vol. 2 (1966)		
pk	- pink	WW	- Winchell and Winchell (1951)		
pr	- purple				
r	- red				

the presence in biotite of the so-called "birdseye" mottled texture at extinction and its absence in stilpnomelane by which to tell them apart. Muscovite differs from talc in its larger optic angle and from pyrophyllite in its smaller optic angle and birefringence. Phengite, which otherwise appears indistinguishable from muscovite in respect to its optical properties alone, is usually finely grained ("sericitic").

Variation of optical properties with chemical composition

As seen in Tables 1 and 2, refractive indices and birefringence vary widely within some mica series, and this is due mainly to solid solution substitution of ions of high refractivity in place of those of low refractivity. Generally refractive indices are raised by substitution of Fe^{2+}, Fe^{3+}, Mn, Ti, V, Cr, Zn, Ca, and Ba in place of Mg, Al, K, Na, and Li, with valence compensated where necessary by substitution of Al for Si. (The influence on the resultant refractive indices are additive so that specimens of mica containing different combinations of these substituted elements may have similar refractive indices.) The greatest continuous variation is shown by the biotite series, mainly through substitution of Fe^{2+} for Mg and of Fe^{3+} for Al; both types of substitutions strongly increase the refractive indices and the birefringence. Substitution of F for (OH) is common in the mica series, but in itself produces only small changes in optical properties. The same applies to partial substitution of Na for K.

The optic angles of most mica solid solution series do not vary significantly with composition, but remain small throughout, thus may not be helpful in establishing the positions of members in the series. A complicating factor in the variation of apparent optic angle, not related to composition, has been encountered in synthetic fluorophlogopite by Bloss (1965, see also Bloss et al., 1963), who suggests that it might also exist in natural micas. Here the apparent optic angle, as determined by Mallard's method from the interference figure of a cleavage flake, may be smaller than the true optic angle, but as the cleavage flake is split into progressively thinner sheets, it increases, approaching the true angle. This is explained as caused by fine lamellar twinning on [310], more rarely on [110], with the composition plane as (001), whereby the superposed twin units act in the manner of compensator plates to reduce the distance between the imaged melatopes.

Among the variation diagrams that have been proposed to show the relations between optical properties and chemical composition is one for the biotite series (Fig. 2) showing the γ refractive index versus the sum of the weight percentages FeO + 2(Fe_2O_3 + TiO_2). Figure 3 employs a rectangular

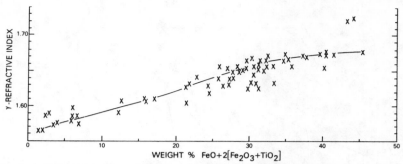

Figure 2. Variation diagram of γ-refractive index against chemical composition in the biotite series. Each cross represents a datum. From Heinrich (1946, Fig. 10).

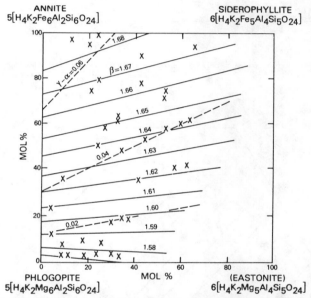

Figure 3. Variation of β-refractive index and birefringence in the biotite series shown by contours. Each cross represents the plotted position of a chemical analysis. From Winchell and Winchell (1951).

diagram on which the proportions of end members (phlogopite, annite, siderophyllite, and eastonite) are plotted, and shows by contours the associated β refractive index and birefringence. Figure 4 shows γ and birefringence by contours in the biotite series on the triangular grid of Foster (1960) with the sum of the R_3 oxides as one apex. Figure 5 shows the miscibilities and optical relations between siderophyllite, trilithionite, and muscovite and Figure 6 shows the known extent of miscibility in the lepidolite series between end members polylithionite, trilithionite, and muscovite. Figure 7, for the (partially miscible?) series margarite-ephesite, shows a total range of some 25° in the optic angle. The lines through the plotted points disregard

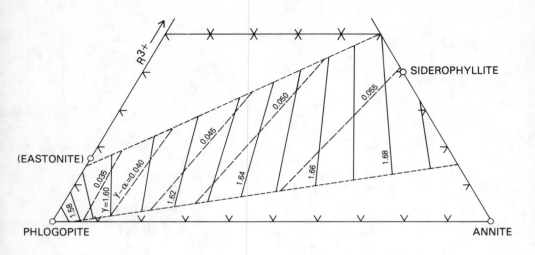

Figure 4. Variation of γ-refractive index and birefringence in the biotite series. From Tröger et al. (1971).

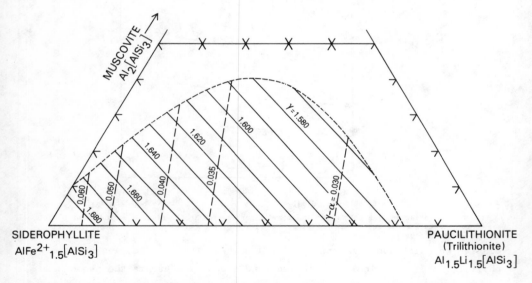

Figure 5. Variation of γ-refractive index and birefringence in the ternary system, siderophyllite-trilithionite-muscovite. From Tröger (1962).

the aberrant properties of one specimen of "sodium-margarite" from the Caucasus.

Caution is necessary when applying these variation diagrams in the opposite sense -- that is, to infer chemical composition of a member of a mica isomorphous series using the observed optical properties. The best fit line of Figure 2, for instance, has been reproduced (Heinrich, 1965, p. 288) without

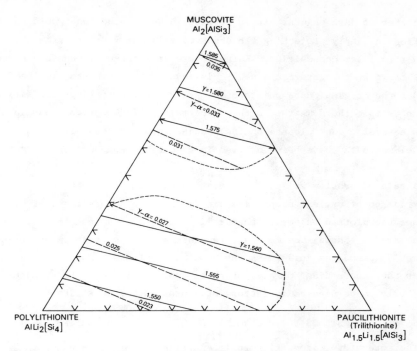

Figure 6. Variation of γ-refractive index and birefringence in the ternary system, polylithionite-trilithionite-muscovite. (From Tröger (1962).

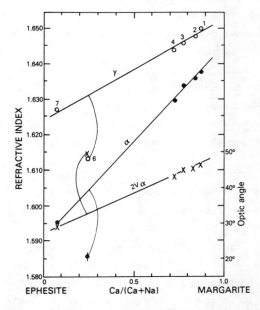

Figure 7. Variation of α- and γ-refractive indices and optic axial angle $2V_\alpha$ in the margarite-ephesite series. Numbers indicate analyses in Deer et al. (1962, p. 96-97).

the scatter of data points. It is obvious, however, that one would not be justified in simply projecting the observed value of γ directly to this line in order to infer composition. Rather there is a range of possible compositions, shown by the intersection of this index value with the wide band that includes the original data points of Figure 2. For inferring compositions in such an example there is the further uncertainty of the proportions in the sum of the several oxides plotted on the abscissa and the possibility that still other elements may have been involved in producing the observed refractive index.

A related problem exists for variation diagrams in which contours express the variation of optical properties with respect to composition. In those diagrams the plotted values of the original data used in construction of the diagram are not shown, and one must recognize that a certain amount of scatter has been accommodated in drawing the best-fit contours. The conclusion therefore is that these diagrams are not to be applied literally or precisely, rather only to be used as indicators of approximate composition.

METHODS FOR DETERMINATION OF OPTICAL PROPERTIES

The more useful properties for identification of micas under the microscope are refractive indices, birefringence, optic angle, and pleochroism in relation to cleavage and crystal form. The following briefly reviews the means for estimating or measuring these properties, first for the unknown as fragments (flakes) in immersion liquid mounts and second for the unknown in thin section. An advantage of immersion liquids is that more optical properties can be measured, generally with greater precision, because the flakes can be isolated and manipulated in selected immersion liquids. An advantage of thin sections, on the other hand, is that critical textures and relationships of the phyllosilicate crystals to the other minerals making up the rock can be observed (see also discussion in Chapters 8, 9 and 10.)

Determination in immersion liquids

Cleavage flakes of a phyllosilicate scattered loosely in immersion liquid unfortunately come to rest lying on the cleavage, so that the only optical properties accessible for measurement are two indices, β and γ (usually very close to each other) and the optic angle (often not diagnostic). Two techniques more suitable for phyllosilicates are available: (a) the coated slide technique, in which flakes are fixed in random orientation on the adhesive film, and (b) the spindle stage technique, in which a single flake mounted at the tip of a spindle is manipulated into desired orientations for measurements of diagnostic properties.

(a) *Coated slide technique*. To assure random orientation of cleavage flakes, one may coat the microscope slide with a tacky quick-setting adhesive, inert to the immersion liquids, to catch and hold flakes as they are sprinkled lightly over its surface. Among adhesives that have been suggested are waterglass (Lindberg, 1944), gelatin (Olcott, 1960), and epoxy (Langford, 1962). With such a coated slide carrying randomly oriented flakes, one may search out and locate flakes favorably oriented for each of the principal refractive indices, as shown by their interference colors and finally confirmed by their interference figures, following conventional random search techniques (Stoiber and Morse, 1972). The corresponding indices are then determined by changing liquids to obtain a match for each principal refractive index. Pleochroism can be estimated on these same flakes, and optic angle can be determined by Mallard's method on flat-lying flakes.

(b) *Spindle stage technique*. The spindle stage is admirably suited for determination of optical properties of substances whose dominating cleavage prevents random orientations in ordinary immersion mounts. This simple and inexpensive device (Rosenfeld, 1950; Wilcox, 1959; Hartshorne and Stuart, 1970; Bloss, 1981) consists of a spindle, at the tip of which the mineral fragment is mounted and inserted into an immersion cell on a base plate for attachment to the microscope stage. Proper orientation of the fragment for the determination of a principal refractive index is obtained by simply rotating the spindle and the microscope stage while observing the evolving interference figures.

For determination of the three principal refractive indices no special mounting of the flake on the tip of the spindle would be required. For phyllosilicates, however, one may determine the principal indices and also the optic angle and pleochroism on a cleavage flake mounted so that it extends out parallel to the spindle axis. (This is readily brought about for a randomly mounted flake by nudging it with a needle before the adhesive has set.) So situated, the flake can be rotated into parallelism with the microscope stage for direct measurement of the optic angle by Mallard's method. At or near this position, as judged by the interference figure, β can also be determined at one extinction position and γ at the other. Here also the body colors at the two positions give the pleochroism and the absorption formula for those two vibration directions.

To place a flake in position for matching the α index requires rotating on the spindle to bring the flake perpendicular to the microscope stage and then rotating the microscope stage to the extinction position in which the cleavage is transverse to the vibration direction of the polarizer. For the actual tests of match or mismatch of refractive index between crystal and liquid, there is a choice of several techniques. Of these the central focal masking technique (Wilcox, 1983) has several advantages over the conventional Becke line and oblique illumination techniques. All techniques encounter some difficulties however in the present case, in which the flake is viewed edge-on and has an exceptionally long steep interface at which the sought-for optical effect is formed. With central focal masking the situation can be improved by tilting the flake a few degrees from the edgewise position, whereupon the diagnostic dispersion colors are better displayed. The refractive index inferred at this amount of disorientation will be only slightly above the true value of α.

Here also the pleochroic color for the vibration direction of α can be compared with that of γ' (an intermediate vibration direction between β and α), taking into account the now much longer light path through the diameter of the flake. Under conditions of normal illumination it may be difficult to distinguish between the change due to faint pleochroic colors and the apparent color change due to differences in relief at the two extinction positions. This may be overcome by inserting the auxiliary substage condenser, which greatly reduces the effect of relief.

Determinations in thin section

In rock thin sections it is usually the prominent cleavage, parallel extinction, and sign of elongation that enable quick recognition of the unknown as a phyllosilicate. In crystals of mica and most phyllosilicates that stand with their excellent cleavage perpendicular to the plane of the section (Fig. 1) the vibration direction of α is essentially horizontal, perpendicular to the cleavage, whereas that of γ' lies essentially in the plane of the cleavage. Hence interference colors represent the difference between the refractive indices γ' and α. Inasmuch as the value of β of most micas is close to that of γ, the interference color will be similar whether the higher of the two indices involved is β, or γ, or an intermediate value, and the birefringence inferred will be equal or nearly equal to $(\gamma - \alpha)$.

Pleochroism, if present, is usually most conspicuous in phyllosilicate crystals when the cleavage is perpendicular to the thin section. The stronger absorption is generally seen when the cleavage is parallel to the vibration direction of the light, weaker when perpendicular thereto. Here again, it may be desirable to extract faint pleochroic colors from the effects of change in relief by inserting the auxiliary substage condenser. (Rare occurrences of rose colored muscovite, discussed in Chapter 5, have reverse pleochroism, that is, greater absorption in the vibration direction corresponding to α.) Mica grains lying such that their cleavage is parallel or at a small angle to the plane of the section do not usually show pronounced pleochroism, because here the comparison is between colors of light vibrating subparallel to β and to γ, both usually of about the same depth of color. Neither does the crystal in this position display its cleavage or a noticeable difference in relief, thus its appearance differs greatly from those crystals of the same species whose cleavage is inclined steeply to the section.

Refractive indices of a crystal in thin section usually can only be approximated. Here use must be made of the observed relief in respect to adjacent material of known index, either another mineral or the mounting medium itself. The procedure for locating properly oriented crystals is much the same as that outlined for the coated slide technique in immersion mounts. After locating a suitably oriented crystal the chosen principal refractive index is then compared with that of the known adjacent material, observing the strength of relief and the Becke line movement in order to arrive at an estimate of the refractive index.

Obviously a great deal of judgment is required here, especially for estimating crystal indices far removed from that of the mountant. A further uncertainty has arisen in the last couple of decades with the introduction of a variety of epoxy mounting media whose refractive indices differ widely from the former standard 1.54 possessed by Canada balsam and Lakeside 70. Although there is now a gratifying trend towards the use of epoxies of index near 1.54, many thin sections made during the early epoxy period are still in circulation with no indication of the index of their mountant.

SUMMARY

Several optical characteristics of most micas, such as parallel extinction and positive elongation in respect to the perfect (001) cleavage, negative optic sign, and in many cases small optic angle, are held in common with other phyllosilicates and are not available therefore for clear distinction between species under the microscope. Remaining properties to aid in identification are mainly the principal refractive indices, maximum birefringence, and pleochroism. Refractive indices and birefringence may vary appreciably due to solid solution, especially in the trioctahedral micas, and their measurements using the coated slide technique or the spindle stage may also serve to indicate the approximate position of a mica in its solid solution series. In rock thin sections, birefringence, as expressed by maximum interference colors, becomes an important means of identification, along with pleochroism.

REFERENCES

Bloss, F.D. (1965) Pitfall in determining 2V in micas. Am. Mineral. 50, 944-947.

Bloss, F.D. (1981) The Spindle Stage. Cambridge University Press, New York, 340 p.

Bloss, F.D., Gibbs, G.V., and Cummings, D. (1963) Polymorphism and twinning in synthetic fluorophlogopite. J. Geol. 71, 537-547.

Deer, W.A., Howie, R.A., and Zussman, J. (1962) Rock-Forming Minerals, Vol. 3. Sheet Silicates. John Wiley, New York, 270 p.

Fleischer, M., Wilcox, R.E., and Matzko, J. (in press) The Microscopic Determination of the Nonopaque Minerals, 3rd ed. U.S. Geol. Surv. Bull. 1627.

Foster, M.D. (1960) Interpretation of the Composition of Trioctahedral Micas. U.S. Geol. Surv. Prof. Paper 354-B, 11-49.

Hartshorne, N.H. and Stuart, A. (1970) Crystals and the Polarising Microscope, 4th ed. Edward Arnold, London, 614 p.

Heinrich, E.W. (1946) Studies in the mica group; the biotite-phlogopite series. Am. J. Sci. 244, 836-848.

Heinrich, E.W. (1965) Microscopic identification of minerals. McGraw-Hill, New York, 414 p.

Langford, F.F. (1962) Epoxy resin for oil immersion and heavy mineral studies. Am. Mineral. 47, 1478-1480.

Larsen, E.S. and Berman, H. (1934) The Microscopic Determination of the Nonopaque Minerals, 2nd edition. U.S. Geol. Surv. Bull. 848, 266 p.

Lindberg, M.L. (1944) A method for isolating grains mounted in index oils. Am. Mineral. 29, 323-324.

Olcott, G.W. (1960) Preparation and use of a gelatin mounting medium for repeated oil immersion of minerals. Am. Mineral. 45, 1099-1101.

Phillips, W.R. and Griffen, D.T. (1981) Optical Mineralogy. The Nonopaque Minerals. W.H. Freeman, San Francisco, 677 p.

Rosenfeld, J.L. (1950) Determination of all principal indices of refraction on difficultly oriented minerals by direct measurement. Am. Mineral. 35, 902-905.

Stoiber, R.E. and Morse, S.A. (1972) Microscopic Identification of Crystals. Ronald Press, New York, 278 p.

Tröger, W.E. (1962) Uber Protolithionit und Zinnwaldit. Beitr. Mineral. Petrogr. 8, 418-431.

Tröger, W.E. (1967) Optische Bestimmung der gesteinsbildenden Minerale. Teil 2, Textband. Schweitzerbart'sche Verlag, Stuttgart, 822 p.

Tröger, W.E., Bambauer, H.-U., Taborszky, F., and Trochim, H.D. (1971) Optische Bestimmung der gesteinsbildenden Minerale. Teil 1, Bestimmungstabellen, 4th ed. Schweizerbart'sche Verlag, Stuttgart, 188 p.

Vlasov, K.A., (editor) (1966) Geochemistry and Mineralogy of Rare Elements and Genetic Types of their Deposits. Vol. 2, Mineralogy of Rare Elements. Israel Program for Scientific Translations. Jerusalem.

Wilcox, R.E. (1959) Use of the spindle stage for determination of principal indices of refraction of crystal fragments. Am. Mineral. 44, 1272-1293.

Wilcox, R.E. (1983) Refractive index determination using the central focal masking technique with dispersion colors. Am. Mineral. 68, 1226-1236.

Winchell, A.N. (1939) Elements of Optical Mineralogy. Part III, Determinative Tables, 2nd ed. John Wiley, New York, 231 p.

Winchell, A.N. and Winchell, H. (1951) Elements of Optical Mineralogy. Part II, Description of Minerals, 2nd ed. John Wiley, New York, 551 p.

7. EXPERIMENTAL PHASE RELATIONS of the MICAS
David A. Hewitt & David R. Wones

INTRODUCTION

The micas have been the subject of many studies in the field of experimental petrology. Early work by Bowen and Tuttle (1949) on talc stability and by Yoder and Eugster (1954, 1955) and Eugster and Yoder (1954a,b) on the stability of paragonite, muscovite and phlogopite started the revolution towards calibrating dehydration reactions of common silicates for application to the petrogenesis of igneous and metamorphic rocks. Following the development of the solid buffer technique by Eugster (1957), the study of the stability of annite by Eugster and Wones (1962) was among the first dealing with the stability of iron-bearing minerals as functions of $T-P-f(O_2)$. This was followed by Wones and Eugster's (1965) study on the stability of mixed Fe-Mg biotites. In another direction Luth (1967) expanded our horizons with his investigations of the melting relations of hydrous phlogopite and their implications in the generation and crystallization of granitic melts.

The novelty of experimentation represented by these early studies on micas has faded somewhat. Much of the work carried out since the early sixties has been involved with the refinement of equilibria investigated earlier. More variables have been considered, and the reliability of the data for thermodynamic extrapolation and application to petrogenesis has improved markedly. Still, there are many new forefronts to be breached, and the micas remain a deceptively complex and difficult group of minerals to deal with. Only recently, except for an early study on the muscovite-paragonite solvus by Eugster et al. (1972), have microanalytical techniques been used commonly for synthetic mica run products. Micas analyzed in these studies are frequently close to, but rarely on, the ideally prescribed composition, although complete analyses including ferric iron and water are generally unavailable. Structurally little is known about synthetic micas. Only a small amount of structural data is available on the natural phases [see Chapter 2], and no studies to date have used single-crystal x-ray or TEM techniques on synthetic hydrous micas. We know from studies by Veblen and Buseck (1980), Iiyima and Buseck (1978) and Ross et al. (1966) of the great complexity in the stacking of natural micas but nothing of the inevitable occurrence or effects of such features in experimental run products.

In this chapter we will deal first with the subsolidus phase relations of the dioctahedral and trioctahedral micas and the available data on the stabilities of mica plus quartz. Second, we will look at more complex subsolidus equilibria involving micas and other phases, and finally, we will report on the melting relations of the micas. We will concentrate on the common micas but have included available references for many of the other micas in table form.

SUBSOLIDUS PHASE RELATIONS OF THE DIOCTAHEDRAL MICAS

Stability of muscovite and muscovite + quartz

Beginning with the study of Yoder and Eugster (1955) there has been a series of investigations on the decomposition of muscovite. Crowley and Roy (1964), Evans (1965), Velde (1966), Ivanov et al. (1973), and Chatterjee and Johannes (1974) have all revised the previous results, and several of the studies are in sharp contrast with each other. The equilibrium data for Reaction 1 from Chatterjee and Johannes (1974) are given in equation form below and are shown in Figure 1.

Muscovite → Sanidine + Corundum + Water

$$\log K(1) = \log f(H_2O) = 8.5738 - 5126/T + 0.0331(P-1)/T \ . \tag{1}$$

These data appear to be the most complete and self-consistent and pass through the high pressure muscovite-sanidine-corundum-liquid-vapor invariant point determined by Huang and Wyllie (1974). Data on Reaction 1 at values of $P(H_2O) < P(total)$ were determined by Ivanov et al. (1973). For values of $X(H_2O) = X(CO_2) = 0.5$ they report the following equilibrium values.

P(total) = 1 kbar	T = 540°C
= 2 kbar	= 580°C
= 3 kbar	= 610°C
= 4 kbar	= 635°C

The addition of quartz to this system and the determination of the muscovite + quartz stability is one of the cornerstones of experimental petrology. The petrologic significance of Reaction 2 has generated numerous investigations to refine the position of the equilibrium as precisely as possible.

Muscovite + Quartz → Sanidine + Al_2SiO_5 + Water (2)

Taking the 2 kbar equilibrium point for comparison, the studies of Evans (1965), Althaus et al. (1970), Kerrick (1972), Day (1973), and Chatterjee and Johannes (1974) yield equilibrium temperatures of 600°C (And), 622°C (And), 605°C (And), 640°C (Sill), and 600°C (And), respectively, where the aluminosilicate shown in parentheses was the phase used as seed material. Using either the Richardson et al. (1969) or Holdaway (1971) value for the aluminosilicate triple point, andalusite is the stable phase at 600°C and 2 kbar. However, the 640°C value

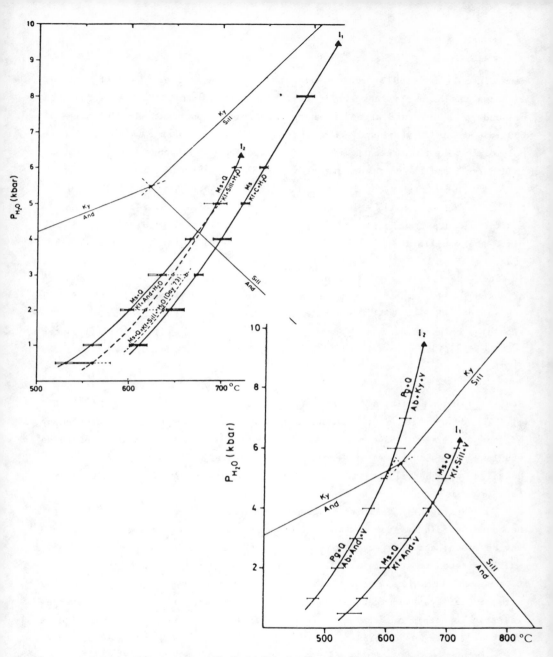

Figure 1 (upper). P-T plot of equilibria for Reactions 1 and 2 (Chatterjee and Johannes, 1974). Data on Reaction 2 from Day (1973) and the aluminosilicate data from Richardson et al. 1969 are also shown. Invariant point I_1 from Huang and Wyllie (1974) involves the phases Ms-San-Cor-L-V, and invariant point I_2 (Chatterjee and Johannes, 1974) involves the phases Ms-Q-San-Sill-L-V. From Chatterjee and Johannes (1974).

Figure 2 (lower). P-T plot of equilibria for Reactions 1 and 5 taken from the data of Chatterjee and Johannes (1974) and Chatterjee (1972). The aluminosilicate triple point and the invariant points I_1 and I_2 are the same as in Figure 1. From Chatterjee and Froese (1975).

determined by Day (1973) would fall within Holdaway's field for sillimanite. Although the higher values should not be dismissed without further study, the preponderance of data argue for a decomposition temperature in the range of 600°-610°C at 2 kbar and with andalusite as the stable aluminosilicate polymorph. The most recent curves from Chatterjee and Johannes (1974) are shown in Figures 1 and 2, and the equilibrium equations are given below.

$$\log K(2)(And) = \log f(H_2O) = 8.3367 - 4682/T + 0.0163(P-1)/T$$
$$\log K(2)(Sill) = \log f(H_2O) = 8.9197 - 5285/T + 0.0248(P-1)/T$$

Additional data from Kerrick (1972), under the conditions $X(H_2O) = X(CO_2) = 0.5$, place the equilibrium at P(fluid) = 2 kbar, T = 555 ± 10°C and P(fluid) = 3.5 kbar, T = 595 ± 10°C.

Considerable effort has also gone into the determination of the stability of muscovite and quartz in aqueous chloride solutions. The earlier data of Hemley (1959), Shade (1974), Ivanov et al. (1974) and Montoya and Hemley (1975) should be compared with the recent studies of Wintsch et al. (1980) and Gunter and Eugster (1980) for the reaction 3 Sanidine + 2 HCl = Muscovite + 6 Quartz + 2 KCl.

Stability of paragonite and paragonite + quartz

The decomposition of paragonite by reaction (3) has been investigated by Eugster and Yoder (1954a,b) and Sand et al. (1957), although neither of those studies represent reversed equilibrium. Chatterjee (1970) carried out a detailed hydrothermal investigation of the stability over the range of pressures from 1-7 kbar.

Paragonite → Albite + Corundum + Water (3)

He found, as did Eugster and Yoder (1954a,b) and other early workers with muscovite (Yoder and Eugster, 1955; Velde, 1966), that at low temperatures the 1M polytype tends to form metastably and gradually recrystallizes to the $2M_1$ polytype as the duration of crystallization increases. By using seeded runs and normally being at temperatures above where the 1M paragonite occurs, Chatterjee was able to determine Reaction 3 reversibly; see Figure 3. An additional data point at 1 kbar was determined by Ivanov and Gusynin (1970) and is identical to the point determined by Chatterjee.

The high pressure stability of paragonite has been investigated using a piston-cylinder apparatus by Holland (1979), who determined the equilibrium conditions for Reaction 4. This reaction has a gentle negative slope and is bracketed as given below over the temperature range 550°C to 700°C. The data

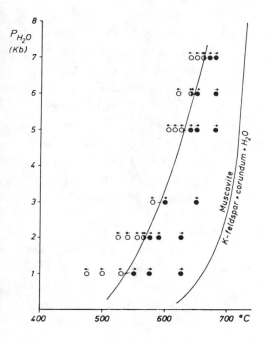

Figure 3. P-T plot of the equilibrium for Reaction 3. Data from Velde (1966) for Reaction 1 is shown for comparison. From Chatterjee (1970).

are consistent with the extrapolated data of Chatterjee (1970) and the stability of paragonite plus quartz.

$$\text{Paragonite} \rightarrow \text{Jadeite} + \text{Kyanite} + \text{Water} \tag{4}$$

$550°C \quad 24 \leq P \leq 26 \text{ kbar}$

$660°C \quad 24 \leq P \leq 25.5 \text{ kbar}$

$650°C \quad 24 \leq P \leq 25 \text{ kbar}$

$700°C \quad 23 \leq P \leq 24.5 \text{ kbar}$

As is the case with muscovite, the geologically important reaction for the stability of paragonite is the decomposition of paragonite plus quartz according to Reaction 5.

$$\text{Paragonite} + \text{Quartz} \rightarrow \text{Albite} + \text{Al}_2\text{SiO}_5 + \text{Water} \tag{5}$$

Ivanov and Gusynin (1970) and Chatterjee (1972) have studied this reaction; the former only determined a single bracket at 500 ± 5°C, and 1 kbar. Their datum is ~20°C higher than those of Chatterjee (1972) who determined the equilibrium for both the reaction to andalusite and to kyanite. Chatterjee's data are shown in Figure 2 and compared to the data for the muscovite + quartz reaction. It should be noted that the choice of aluminosilicate seeds for individual experiments was based on the aluminosilicate triple point of Richardson et al. (1969) and that some of the andalusite experiments are metastable with respect to sillimanite if the lower temperature and pressure triple point of Holdaway (1971) is correct.

The low temperature synthesis of paragonite plus quartz by Reaction 6 has been investigated by Chatterjee (1973) over the pressure range of 2-7 kbar.

$$3Na - Montmorillonite + 2\ Albite \rightarrow 3\ Paragonite + 8\ Quartz \qquad (6)$$

Although the reaction is slow and complicated by the occurrence of apparently metastable mixed layer sheet silicate phases, it appears that the transition occurs between 335°-315°C. Although the data are not reversed they are in good agreement with the hydrolysis measurements of Hemley et al. (1961). Additional aqueous chloride solution data for the assemblages paragonite-quartz-albite and paragonite-andalusite-quartz have been measured by Montoya and Hemley (1975).

The system muscovite-paragonite-quartz

The solvus along the join muscovite-paragonite received experimental consideration by Eugster and Yoder (1955), Eugster (1956), Iiyama (1964), and Blencoe and Luth (1973). The moderate to low temperatures where the solvus exists demand long run times. This problem, along with the metastable persistance of 1M phases at low temperature and the problems of analysis of the coexisting phases, delayed the comprehensive analysis of the system until the paper by Eugster et al. (1972). These workers were able to take the long term runs done by Eugster in 1956 and, by carrying out detailed x-ray and fine-particle electron microprobe analyses, arrive at a reasonable solvus from which they derived a third-order Margules formula to fit the data.

$$G_{Ex} = W_{G,Pa} X_{Pa} X^2_{Mu} + W_{G,Mu} X_{Mu} X^2_{Pa}$$

$W_{G,Pa}$	$W_{G,Mu}$
Eugster et al. (1972)	
$3082 + 0.170\ T(K) + 0.082\ P(bars)$	$4164 + 0.395\ T(K) + 0.126\ P(bars)$
Chatterjee & Froese (1975)	
$2923.1 + 0.1698\ T + 0.1590\ P$	$4650.1 + 0.3954\ T - 0.1090\ P$

Chatterjee and Froese (1975) modified the pressure dependence of the Margules parameters based on their review of the data in the literature on the molar volumes of muscovite-paragonite solid solutions. Blencoe (1977) presented a large amount of new molar volume data for 1M and $2M_1$ phases synthesized over a range of pressures. In general he found that the $2M_1$ polytype was slightly smaller than the 1M polytype and that the molar volume of the 1M phase decreased with the pressure of the synthesis. His plot of molar volume versus composition (Fig. 4) illustrates the significant differences among the investigations. Tabulated below are the values for Wv to be compared for the different data sets.

	$W_{V,Pa}$	$W_{V,Mu}$
Blencoe (1977) 1M, 2 kbar	.0867	.0753
1M, 4 kbar	.1076	.0142
1M, 8 kbar	.1241	-.0363
2M, 8 kbar	.0750	.0571
Eugster et al. (1972)	.0822	.1259
Blencoe and Luth (1973)	-.0340	-.0770
Chatterjee and Froese (1975)	.1590	-.1090

These different values lead to significantly different compositions of coexisting phases, particularly at high pressures (Blencoe, 1977). Additional data are needed to resolve the pressure effect as well as to refine the position of the solvus and its temperature dependence.

Based on their refinement of the Eugster et al. (1972) data set and using the Chatterjee and Johannes (1974) data on the muscovite quartz reaction, the Chatterjee (1972) data on paragonite + quartz and the data from Waldbaum and Thompson (1969) on the sanidine-high albite solution, Chatterjee and Froese (1975) have calculated the T-X phase diagram for the muscovite-paragonite join as a function of pressure under the conditions of quartz, aluminosilicate, and water saturation (Fig. 5). These diagrams form a useful basis for the analysis of mica-quartz-feldspar-aluminosilicate assemblages in low Ca-rocks, but are subject to the compounding of errors from the individual systems described previously.

Stability of margarite and margarite + quartz

Syntheses by Eugster and Yoder (1954b) and Tu (1956) represent the earliest experimental data on the Ca-dioctahedral mica, margarite. These investigations and those that followed were all hampered by the low temperatures and sluggish reaction rates that prevail in the system. It was not until the early seventies that equilibrium experiments were successfully carried out by Chatterjee (1971), Velde (1971), Storre and Nitsch (1974) and Chatterjee (1974) on Reaction 7. The equilibrium equation given below is from the self-consistent data of Chatterjee (1974), determined over the range from 1-7 kbar and is in agreement

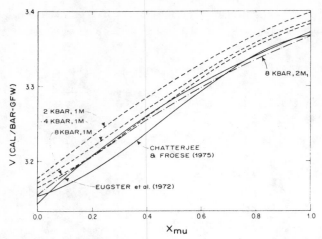

Figure 4. Molal volumes for Pa-Mu solid solutions calculated from the volume equations of Blencoe (1977) and compared to data from Eugster et al. (1972) and Chatterjee and Froese (1975). From Blencoe (1977).

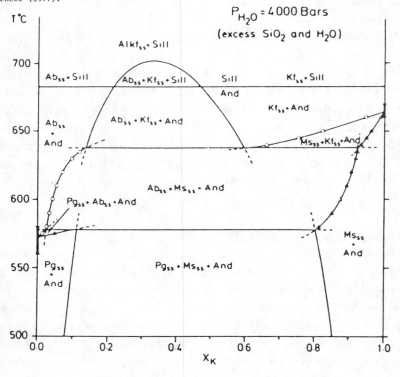

Figure 5. Calculated phase relations for the Mu-Pg join at 4 kbar under the conditions of SiO_2, Al_2SiO_5, and H_2O saturation. From Chatterjee and Froese (1975).

with the 2 kbar bracket of Velde (1971), but lies at ~25°C higher temperature than Velde's 1 kbar bracket.

$$\text{Margarite} \rightarrow \text{Anorthite} + \text{Corundum} + \text{Water} \tag{7}$$

$$\log K(7) = \log f(H_2O) = 8.8978 - 4758/T + 0.0171(P-1)/T$$

The data from Storre and Nitsch (1974) are generally at lower temperature but their starting material did not contain pure margarite.

The addition of quartz to the system has been investigated by Storre and Nitsch (1974) on Reactions 8 and 9 and by Jenkins (1983) on Reaction 8.

$$4 \text{ Margarite} + 3 \text{ Quartz} \rightarrow 2 \text{ Zoisite} + 5 \text{ Kyanite} + 3 \text{ Water} \tag{8}$$

$$\text{Margarite} + \text{Quartz} \rightarrow \text{Anorthite} + \text{Aluminosilicate} + \text{Water} \tag{9}$$

The two studies differ significantly, and this may be explained by the use of natural paragonite-bearing margarite by Storre and Nitsch (1974). Data from Jenkins (1983) on Reaction 8 place the equilibrium in the end-member system at T and P shown below.

Reaction 8			Reaction 9	
500°C		$P < 11.0$ kbar	4 kbar	$490 \leq T \leq 540°C$
550°C	$8.5 < P <$	10.0 kbar	5 kbar	$530 \leq T \leq 560°C$
600°C	$8.5? < P <$	9.5 kbar	7 kbar	$580 \leq T \leq 600°C$
			9 kbar	$640 \leq T \leq 660°C$

Data from Storre and Nitsch (1974) on Reaction 9 are given above, but it must again be noted that impure natural margarite was used as the starting material.

Chatterjee (1976) used the available experimental and thermochemical data to calculate a petrogenetic P-T grid for margarite reactions (Figs. 6a,b). Although the position of Reaction 8 (connecting invariant points I_{12} and I_{11} on Figs. 6a,b) is ~2 kbar lower pressure than the data of Jenkins (1983) and the positions of the other equilibria must be considered tentative, the diagram does establish the general relationships. Margarite and margarite + quartz are limited to pressures less than about 15 and 10 kbar, respectively, and have thermal stabilities ranging over bands of about 200°C and 100°C. The effect of decreasing water activity is shown in Figure 6b.

The major solid solution involving margarite is towards paragonite by the same substitution as occurs in the plagioclase feldspars between anorthite and albite. Data from numerous natural studies (e.g., Frey and Orville, 1974) indicate the presence of a solvus relationship in the margarite-paragonite binary. Experimental investigations by Hinrichsen and Schurmann (1971) and Franz et al. (1977) have determined a preliminary solvus (Fig. 7). The experiments consisted of synthesis runs along the join at pressures ranging from 1 to 6 kbar. Materi-

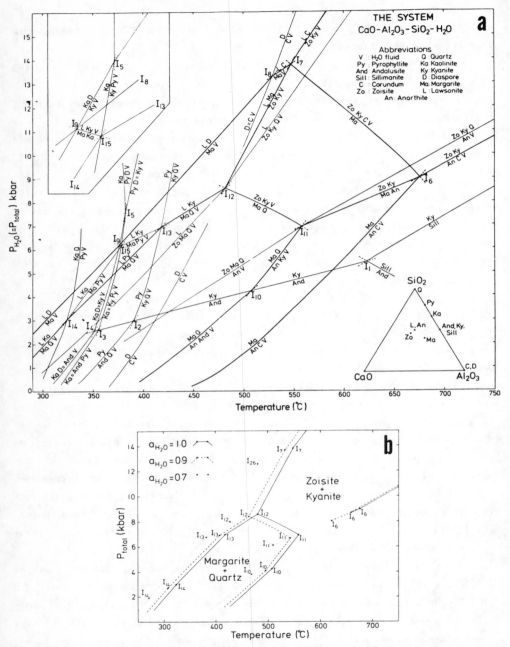

Figure 6. (a) Experimental and calculated P(H$_2$O)-T phase relations in the system CaO-Al$_2$O$_3$-SiO$_2$-H$_2$O. (b) Calculated P(total)-T phase relations in the system CaO-Al$_2$O$_3$-SiO$_2$-H$_2$O, outlining the location of the margarite + quartz stability field as a function of a(H$_2$O). From Chatterjee (1974).

als were analyzed by powder X-ray diffraction and IR-spectroscopy to establish the occurrence of the two-phase region.

Stability of pyrophyllite

The considerable effort of a number of workers including Reed and Hemley (1966), Matsushima et al. (1967), Althaus (1966a,b,c, 1969), Velde and Kornprobst (1969), Kerrick (1968), Hemley (1967), Thompson (1970), and Haas and Holdaway (1973) has gone into the investigation of the stability range of pyrophyllite. Experimental determinations of the major Reactions 10, 11, and 12 have yielded widely differing results.

Kaolinite + 2 Quartz → Pyrophyllite + Water (10)

Pyrophyllite + 6 Diaspore → 4 Aluminosilicate + 4 Water (11)

Pyrophyllite → Aluminosilicate + 3 Quartz + Water (12)

Many of the early studies, particularly those on Reaction 12, involved short run times and used amorphous or very fine-grained highly strained reactants. The most successful investigations used single-crystal weight change techniques (Kerrick, 1968; Haas and Holdaway, 1973). Figure 8 shows the data from Haas and Holdaway (1973) on Reactions 11 and 12, as well as data on Reaction 10 from Thompson (1970). All weight-change experiments on Reactions 11 and 12 were performed using andalusite seeds. The results within the kyanite stability field are recalculated from those metastable experiments (Fig. 8). The Haas and Holdaway (1973) data are ~25°C lower for Reaction 12 than the data from Kerrick (1968), with the differences being largely unexplained at this time. A thermodynamic analysis of the system by Perkins et al. (1979) is in good agreement with the Haas and Holdaway data for Reaction 11, but does not resolve these differences for Reaction 12. The position of Reaction 10 was also determined by

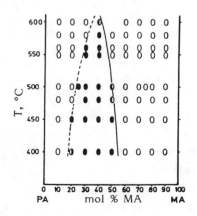

Figure 7. Preliminary outline of the Pa-Ma solvus based on IR spectra of synthesis experiments by Henrichsen and Schurmann (1971) and Franz et al. (1977). Pressure varied from 1 to 6 kbar. From Frantz et al. (1977).

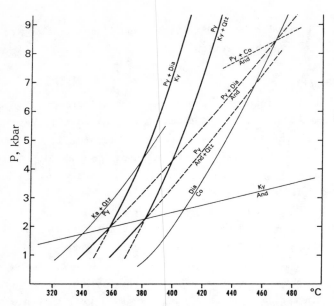

Figure 8. P-T plot of equilibria for Reaction 10 from Thompson (1970) and Reactions 11 and 12 from Haas and Holdaway (1973). The dashed lines represent metastable equilibria, and the Ky-And equilibrium is from Holdaway (1971). From Haas and Holdaway (1973).

weight-change techniques. Thompson (1970) reports a significantly lower temperature for the equilibrium (345°C at 2 kbar) than for all earlier studies except those of Velde and Kornprobst (1969) (310°C) and Reed and Hemley (1966) (300°C at 1 kbar). A thermochemical, topologic, and petrographic analysis of the system by Day (1976) suggests a similar picture to that of Figure 8 except that Day postulates that Reaction 11 is metastable at low pressures and that Reaction 10 occurs at 50°-75°C lower temperatures than those determined by Thompson (1970). Haas and Holdaway (1973) also comment that the reported position of Reaction 10 is "probably a little high." Perkins et al. (1979) review the discrepancies in the experimental data for Reaction 10 but offer no resolution to the problem.

Additional experimental data on pyrophyllite, which may have bearing on the disagreement of the stability experiments, have shown that minor solid solution of the type $Al^{+3} + H^+ \leftrightarrow Si^{+4}$ occurs in pyrophyllite (Rosenberg, 1974; Rosenberg and Cliff, 1980) and that monoclinic, triclinic and disordered pyrophyllite form in hydrothermal synthesis experiments (Eberl, 1979). In particular, Rosenberg (1974) suggests that this substitution in pyrophyllites synthesized from gels increases the stability of the phase and may account for the earlier higher temperature equilibria reported for Reaction 12.

SUBSOLIDUS PHASE RELATIONS FOR THE TRIOCTAHEDRAL MICAS

Stability of end-member biotites

Early investigations on the stability of phlogopite were carried out by Yoder and Eugster (1954), Crowley and Roy (1964), Luth (1967), and Wones (1967). A determination of the high temperature decomposition of phlogopite by Reaction 13 presented major problems in these studies.

$$2 \text{ Phlogopite} \leftrightarrow \text{Kalsilite} + \text{Leucite} + 3 \text{ Forsterite} + 2 \text{ Water} \qquad (13)$$

Except for those experiments in internally heated pressure vessels, the majority of runs were at or near maximum operating temperatures for the equipment. Quench phlogopite was a significant problem in the high temperature experiments (Wones, 1967), and only the data from Wones (1967) represent a reversed equilibrium. All other experiments were either synthesis runs with gels or oxides as starting materials or were decomposition experiments starting with phlogopite. Based on the two brackets obtained by Wones (1967), the equilibria can be expressed as follows:

$$\log K(13) = \log f^2(H_2O) = -16633/T + 18.10 + 0.105(P-1)/T$$

As shown by Wones (1967), these data are in agreement with many, but not all, of the other unreversed data.

In the quartz-bearing system Wones and Dodge (1977) place the invariant point phlogopite-quartz-enstatite-sanidine-liquid-vapor at $835 \pm 5°C$ and 450 ± 50 bars. Reaction (14) must radiate from this invariant point to lower pressures and temperatures.

$$\text{Phlogopite} + 3 \text{ Quartz} \rightarrow \text{Sanidine} + 3 \text{ Enstatite} + \text{Water} \qquad (14)$$

Recent data on Reaction 14 from Wood (1976) pass through the Wones and Dodge invariant point and yield the following equilibrium equation.

$$\log K(14) = \log f(H_2O) = -2356/T + 4.76 + 0.077(P-1)/T$$

At pressures above 450 bars and under water saturated conditions phlogopite + quartz decompose by incongruent melting. In the water undersaturated system, experiments by Bohlen et al. (1983) have determined another equilibrium point for Reaction 14 at $P(\text{total}) = 5$ kbar, $T = 790 \pm 10°C$ and $X^V(H_2O) = 0.35$.

Most subsolidus experimental work on biotite has been done in the iron-bearing system. The first major study on annite was reported by Eugster and Wones (1962). Using solid-oxygen buffers they were able to obtain equilibrium data directly or by calculation on Reactions 15 through 20.

$$2 \text{ Annite} + H_2O \rightarrow 2 \text{ Sanidine} + 3 \text{ Hematite} + 3H_2 \qquad (15)$$

$$\text{Annite} \rightarrow \text{Sanidine} + \text{Magnetite} + H_2 \qquad (16)$$

$$3 \text{ Annite} \rightarrow 3 \text{ Fayalite} + 3 \text{ Leucite} + \text{Magnetite} + 2H_2O + H_2 \qquad (17)$$

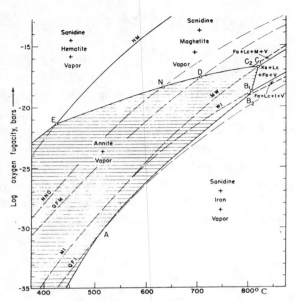

Figure 9. An $f(O_2)$-T plot of the phase relations for annite + vapor at 2070 bars. See Eugster and Wones (1962) for details on the experimental determinations of points E, N, D, C_2, C_1, B_1, B_2, and A. From Eugster and Wones (1962).

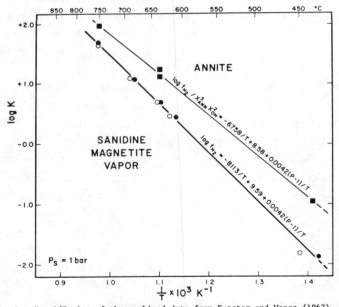

Figure 10. Log K - 1/T plot of the combined data from Eugster and Wones (1962), Rutherford (1969), and Hewitt and Wones (1981) for Reaction 16. The lower curve is based directly on the experimental data while the upper curve is corrected for the Fe^{3+} content of annite as measured by Partin et al. (1983). From Hewitt and Wones (1981).

$$2 \text{ Annite} \rightarrow 3 \text{ Fayalite} + \text{Leucite} + \text{Kalsilite} + 2H_2O \qquad (18)$$
$$\text{Annite} + H_2 \rightarrow \text{Fayalite} + \text{Leucite} + \text{Iron} + 2H_2O \qquad (19)$$
$$\text{Annite} + 3H_2 \rightarrow \text{Sanidine} + 3 \text{ Iron} + 4H_2O \qquad (20)$$

Figure 9 shows the data from Eugster and Wones (1962) at 2 kbar. Rutherford (1969) redetermined the equilibrium at the QFM buffer and presented an additional data point for Reaction 16 at the methane-graphite buffer at 2 kbar. Wones et al. (1971) presented a new determination for Reaction 16, using the Shaw (1967) hydrogen diffusion apparatus. Those data were in significant disagreement with the earlier data, and after Hewitt (1977) demonstrated that special procedures were necessary to obtain reliable data from the Shaw bombs, Hewitt and Wones (1981) proposed a revision of the equilibrium. They were able to demonstrate consistency among the NNO data from Eugster and Wones (1962), the QFM data of Rutherford, and new data from Hewitt and Wones (1981) at 50 bars $P(H_2)$ and at the HM buffer. A slightly modified version of those equilibrium data is shown in Figure 10 and given in equation form below.

$$\log K(16) = \log f(H_2) = -8113/T + 9.59 + 0.0042(P-1)/T$$

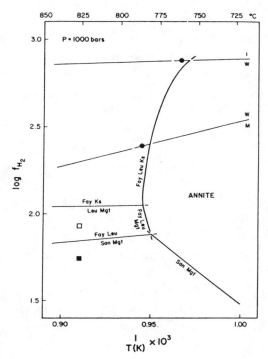

Figure 11. Experimental and calculated phase relations at 1 kbar for the upper stability of annite. Positions of Reaction 16 and the reaction sanidine + magnetite ↔ fayalite + leucite are from the experimental data of Hewitt and Wones (1981). Reactions 17 and 18 and the reaction leucite + magnetite → fayalite + kalsilite are calculated. Brackets for Reaction 18 on the WM and IW buffers are from the experiments of Eugster and Wones (1962). From Hewitt and Wones (1981).

Using these data along with an experimentally determined bracket on the reaction 3 Sanidine + 2 Magnetite ↔ 3 Fayalite + 3 Leucite + O_2, Hewitt and Wones (1981) were able to calculate the phase diagram for the maximum stability of annite (Fig. 11). Thermochemical data used in the calculations were taken from Robie et al. (1978) for fayalite, leucite, quartz, sanidine, kalsilite, and magnetite, from Burnham et al. (1969) for water, and from Shaw (1967) and Shaw and Wones (1964) for H_2. As shown in Figure 11 the calculations place the equilibrium for Reaction 18 within the original experimental error of the points determined by Eugster and Wones (1962) at the WM and WI buffers. The calculated equilibrium expressions for Reactions 17 and 18 are as follows:

$$\log K(17) = \log f^2(H_2O)f(H_2) = -26044/T + 32.41 + 0.075(P-1)/T$$
$$\log K(18) = \log f^2(H_2O) = -17705/T + 22.44 + 0.112(P-1)/T$$

Only very limited data are available on the stability of annite + quartz. Eugster and Wones (1962) present data at 2 kbar for the composition Annite + 3 Quartz (Fig. 12). Data at higher pressures involving reactions to sanidine + orthopyroxene are unavailable.

Eugster and Wones (1962), Wones (1963a), and Wones and Eugster (1965) emphasize the importance of ferric iron in biotite. They suggested on the basis of analyses of natural biotites and the results of their experiments that much of it could be accounted for by the ferriannite ($Al^{iv} = Fe^{iv+3}$) and oxyannite ($Fe^{2+} + OH^- = Fe^{3+} + O^=$) substitutions. Wones (1963a) presented phase equilibrium data for ferriannite (Fig. 13), indicating a slight increase in thermal stability over that of annite. However, the oxyannite substitution is probably the most important mechanism for Fe^{3+} substitution, although there are not sufficient data on hydrogen contents of natural or synthetic biotites to substantiate the hypothesis. Wones and Eugster (1965) suggest that the Fe^{3+} content (i.e., oxyannite content) of annite increases with $f(O_2)$. Refractive indices and unit cell parameters support that suggestion, and analytical data from Wones et al. (1971) on synthetic annites annealed at different $f(H_2)$ are also supportive.

Our present knowledge of the ferric iron contents of synthetic annite indicates that there are at least two factors other than the ferriannite substitution controlling the amount of Fe^{3+} in Fe-biotite. Hazen and Wones (1972) predicted on the basis of ionic radii and the tetrahedral rotation angle in biotite that pure ideal annite with only Fe^{2+} could not be stable because the octahedral sheet in the biotite would be too large to articulate with the tetrahedral sheets. Based on their calculations the most reduced annite would still

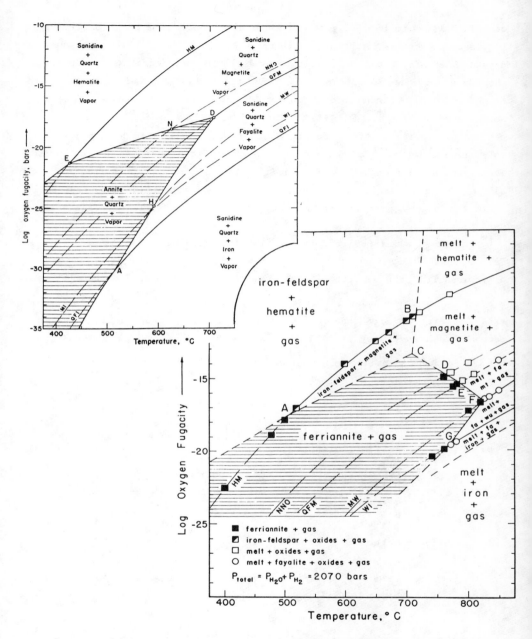

Figure 12. Log f(O$_2$) - T plot of the stability of annite + quartz at 2070 bars. For experimental details for points E, N, D, H and A see Eugster and Wones (1962) (their figure).

Figure 13. Log f(O$_2$)-T plot of the stability of ferriannite at 2070 bars pressure. From Wones (1963a).

contain ~12% octahedral Fe^{3+}, regardless of how reducing the conditions of crystallization were. Mössbauer data from Wones et al. (1971) on annites crystallized at 100 bars $P(H_2)$ showed at least 10% octahedral Fe^{3+}, supporting the Hazen and Wones (1972) arguments. Additional wet chemical measurements of Fe^{3+} in biotites synthesized at more oxidized conditions (Wones et al., 1971) showed increased Fe^{3+} and indicate that additional Fe^{3+} is dependent on the conditions of crystallization. Ferrow and Annersten (1984) used Mössbauer spectra to determine the Fe^{+3} contents of annealed synthetic annite and ferriannite. They found a relation between Fe^{+3} and $f(O_2)$ similar to the data from Wones et al. (1971) and Partin et al. (1983) and concluded that the substitution in annite occurred by the oxyannite mechanism and by the creation of interlayer vacancies.

Partin et al. (1983) and Partin (1984) carried out a systematic study of Fe^{3+} in annite and siderophyllite. They annealed crystalline synthetic biotites at various $f(H_2)$, using both solid buffers and the Shaw apparatus, and they used electron microprobe and wet chemical analyses to determine the $Fe^{2+}/Fe^{2+}+Fe^{3+}$ ratio. Their data show that annite contains a minimum amount of 11% Fe^{3+} in the octahedral sites at any $f(H_2) > 25$ bars and that at conditions more oxidizing the Fe^{3+} content increases proportionally.

T°C	P (kbar)	$f(H_2)$ (bars)	$Fe^{2+}/(Fe^{2+}+Fe^{3+})$
Annite			
791	1	101.6	.89
750	1	51.9	.89
700	1	26.1	.84-.90
615	1	6.0	.81-.82
600	2	4.0	.77
399	2	.004	.72
Siderophyllite			
750	1	51.9	1.0
700	1	26.1	.96-.98
615	1	6.0	.87-.93
600	2	4.0	.86
400	2	.004	.79

The data also clearly show that, in contrast to annite, the smaller octahedral and larger tetrahedral sheets in siderophyllite do not require Fe^{3+} to be present at very reducing conditions and that Fe^{3+} in siderophyllite is strictly dependent on the oxidizing conditions of the system. Based solely on size arguments, Partin (1984) suggests that biotites with >10% Tschermak's component or 24% phlogopite component do not require the presence of Fe^{3+} to satisfy the structural fit between the octahedral and tetrahedral sheets and will have an Fe^{+3} content dependent only on $f(O_2)$ and composition.

The Fe^{3+} contents of the annites of Partin (1984) have been used to revise the equilibrium expression for Reaction 16 as shown on Figure 10. The new log K expression is for ideal end-member annite and assumes ideal mixing of Fe^{3+} and Fe^{2+} and that the OH content is that prescribed by the oxyannite substitution.

Stability of biotite solid solutions

A wide range of cation substitutions have been suggested for the natural biotites, but only a few prominent substitutions have been studied experimentally. The major substitutions for which some data are available include the $FeMg_{-1}$ exchange, the Tschermak's exchange, the NaK_{-1} exchange, and some of the various possible Ti substitutions.

Wones and Eugster (1965) presented data on the stability of biotites on the join annite-phlogopite (Fig. 14). The data and particularly the equilibrium curve for the assemblage biotite-sanidine-magnetite-vapor are useful in defining $P-T-f(H_2O)-f(O_2)$ conditions in high grade metamorphic and granitic rocks. Discussions of the solution properties in the Fe-Mg biotites in Wones and Eugster (1965) were later modified by Wones (1972), based on experimental data, and by Mueller (1972), based on natural assemblages. The authors conclude that for biotites with low Fe^{3+} the join between phlogopite and annite behaves essentially as an ideal solution.

Data on Reaction 14 for the $annite_{50}$-$phlogopite_{50}$ + quartz composition suggest that the equilibrium lies at about 675°C at 2 kbar and that there is not much Fe-Mg fractionation between biotite and orthopyroxene (Hoffer and Grant, 1980).

The addition of Al to the system was studied by Rutherford (1973) along the join annite-siderophyllite (Fig. 15). Rutherford demonstrated that the addition of Al by Tschermak's substitution increases the stability of biotite. The primary decomposition reaction analogous to Reaction 16 for annite is similar to Reaction 16 but is divariant due to variable Al content of the biotite and is complicated by the production of magnetite-hercynite solid solutions and various mixtures of sanidine and leucite depending on the Fe, Al, and Si contents of

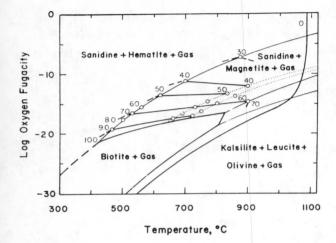

Figure 14 (left). Log $f(O_2)$-T plot of the stability of biotite as a function of Fe/(Fe+Mg) ratio at 2070 bars. Curve 0 represents the maximum stability of phlogopite. From Wones and Eugster (1965).

Figure 15 (below). Log $f(O_2)$-T plot of the stability of aluminous iron biotites in the silica undersaturated region at 2 kbar. Biotite and magnetite-hercynite compositions coexisting with sanidine and leucite are shown in medium weight contours. From Rutherford (1973).

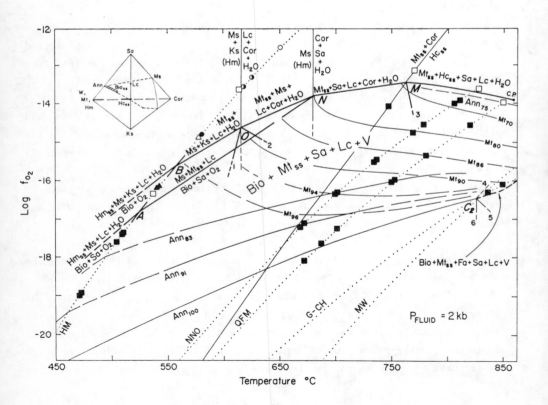

the coexisting biotite. In the silica saturated system, leucite would no longer be a problem for most geologically reasonable conditions.

The most aluminous biotite synthesized by Rutherford (1973) corresponded to the composition $K_2Fe_{4.5}Al_{1.5}Al_{3.5}Si_{4.5}O_{20}(OH)_4$ [Ann_{75}]. Unit cell parameters along the join Ann_{100}-Ann_{75} are presented by Rutherford at several different oxygen buffers. For all compositions there is a decrease in the unit cell volume and a general increase in refractive index with increasing $f(O_2)$, presumably reflecting the increase in Fe^{3+} content. Wones (1963b) reports physical properties for synthetic biotites along the join annite-phlogopite at several different oxygen buffers. Cell dimensions and refractive indices increase substantially from phlogopite to annite under the same buffering conditions. For the Fe-biotites, refractive indices increase and cell dimensions decrease as $f(O_2)$ is increased. Ferrow and Annersten (1984) reported on a detailed Mössbauer and x-ray investigation of synthetic annites over a similar range of oxidation conditions.

Hewitt and Wones (1975) redetermined the cell parameters and gamma index of refraction for the Fe-Mg-Al biotites synthesized at 1000 bars, 700°-850°C, and at hydrogen fugacities of ~100 bars for the Fe-bearing biotites. They found that the maximum Al substitution along the join phlogopite-aluminous eastonite ($KMg_2AlAl_2Si_2O_{10}(OH)_2$) occurred at the composition phlogopite$_{38}$. This is considerably less aluminous than the pure Al-eastonite reported by Crowley and Roy (1964). For the iron-bearing biotites the limit occurred at phlogopite-annite$_{25}$, the same limit that was reported by Rutherford (1973). Hewitt and Wones (1975) argued that the limit on the Tschermak's substitution in biotite was controlled structurally by the α-rotation of the tetrahedral sheet and the size of the interlayer cation. The contraction of the octahedral sheet and increase of the mean T-O bond length with Tschermak's substitution is accommodated in the biotites by tetrahedral rotation. This decreases the size of the interlayer site and thereby limits the amount of Tschermak's substitution in potassic biotites at a mean interlayer bond length of ~2.8 Å. Their hypothesis was supported by the synthesis of a pure sodium aluminous eastonite.

Data from Hewitt and Wones (1975) for $\underline{a}$, V, and n_γ are given below. Plots of unit cell volumes indicate that the solution properties are consistent with ideal mixing of Fe and Mg, and Mg and Al, but the non-linearity of volumes, seen when varying the Fe/Al ratio, suggests that the annite-siderophyllite solid solution may show some significant non-ideal mixing.

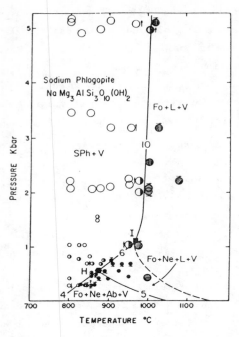

Figure 16. P-T plot of the stability of Na-phlogopite. From Carman (1974).

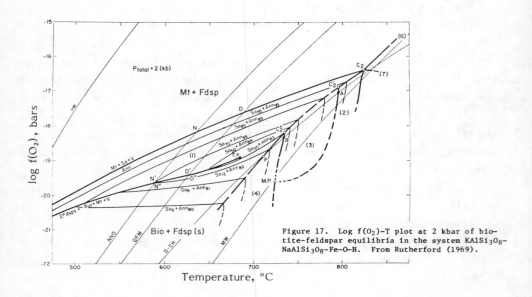

Figure 17. Log $f(O_2)$-T plot at 2 kbar of biotite-feldspar equilibria in the system $KAlSi_3O_8$-$NaAlSi_3O_8$-Fe-O-H. From Rutherford (1969).

$$\underline{a}(\text{Å})(0.002) = 5.406 - 0.089 \, x/x+y - 0.056z + 0.022 \, zx/x+y$$
$$V(\text{Å}^3)(0.3) = 514.2 - 17.3 \, x/x+y - 14.3z + 7.3 \, zx/x+y$$
$$n_\gamma(.001) = 1.686 - 0.103 \, x/x+y - 0.029z + 0.037 \, zx/x+y$$

x, y and z refer to the formula $K(Mg_xFe_yAl_z)Al_{1+z}Si_{3-z}O_{10}(OH)_2$; where $z = (3 - x - y)$ and $0 \leq z \leq 0.75$.

Several other solid solutions have been investigated. Robert (1976a) looked at phlogopite solutions in the system $K_2O\text{-}MgO\text{-}Al_2O_3\text{-}SiO_2\text{-}H_2O$ with compositions according to the formula

$$K_2(Mg_{6-x-y}Al_x\square_y)(Si_{6-x+2y}Al_{2+x-2y})O_{20}(OH)_4.$$

Robert concluded that the Tschermak's substitution and the $2Al^{iv} + Mg = 2Si^{iv} + \square^{vi}$ substitution were both operational. He found a large amount of substitution at 600°C and 1 kbar where the most aluminous phase had $x = 1.625$ and $y = 0.35$. As temperature increases the compositional range decreases. At 800°C and 1000 bars the most aluminous phlogopite contained $x = 0.75$ and $y = 0.25$.

The substitution of Na for K in phlogopite was studied by Carman (1974) who determined the maximum stability for Na-phlogopite (wonesite) (Figure 16). Carman also investigated the stability of two Na-phlogopite hydrates (I and II) that form at temperatures below ~350-400°C and 90-110°C, respectively. Franz and Althaus (1976) report the synthesis of Na-analogs of eastonite and Al-eastonite. Rutherford (1969) investigated the phase relations involving Fe-biotites, alkali feldspars (and/or feldspathoids) and magnetite (or fayalite) in the system $KAlSiO_4\text{-}NaAlSiO_4\text{-}SiO_2\text{-}Fe\text{-}O\text{-}H$. He found that the previously described annite and annite + quartz equilibria occurred at significantly lower temperatures in the Na-bearing system and that the biotite contained up to 20% Na-annite. The phase relations in the quartz-free and excess quartz systems are shown in Figures 17 and 18.

The Ti-substitution in biotite has stirred great interest and controversy. Numerous analytical studies including recent investigations by Dymek (1983) and Labotka (1983) have argued for various substitutions to account for the compositions of natural Ti-bearing biotites. Typically, plots of two or more component variables or the expression of a phase composition in terms of an additive component and a set of exchange vectors (Thompson, 1982) have been used to support one substitution or another. This approach can be misleading because the chemical analyses used are often incomplete (typically no Fe_2O_3 or H_2O data), making the calculation of the mineral formula suspect. In addition, Hewitt and Abrecht (submitted to Contributions to Mineralogy and Petrology) have shown that even with a complete analysis there are normally multiple sets of exchange vec-

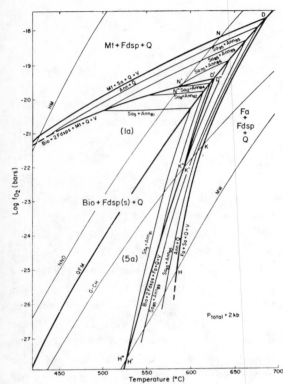

Figure 18. Log f(O_2)-T plot at 2 kbar of biotite-feldspar equilibria in the system $KAlSi_3O_8$-$NaAlSi_3O_8$-SiO_2-Fe-O-H. From Rutherford (1969).

tors (or proposed substitutions) that can be used to express the composition of the phase starting from an initial additive component. They argue that exchange vectors are not necessarily valid substitution mechanisms and that the establishment of a relationship between the two requires special compositions, specific knowledge of how an individual phase has changed composition or detailed experimental data on the system in question.

There have been three experimental studies involving the synthesis and stability of Ti-biotites. Forbes and Flower (1974) synthesized a Ti-phlogopite corresponding to the substitution $Ti + \square^{vi} = 2Mg$ (Ti-vacancy type). They report 100% synthesis of a biotite of composition $K_2Mg_4TiAl_2Si_6O_{20}(OH)_4$ in a band ~100°C wide just below the solidus at high pressure and at temperatures in excess of 1100°C. At higher temperatures the Ti-phlogopite decomposed to forsterite + rutile + liquid + vapor and at lower temperatures they report a stable assemblage of a phlogopite solid solution + rutile. No analytical data are available for the phlogopite solid solutions and the synthesis experiments in this field typically produced mixtures of mica, sanidine, and rutile. Therefore, no information on the amount and nature of Ti substitution can be derived from the lower temperature syntheses. It does appear, however, that at high temperatures the Ti-vacancy substitution is valid and that it extends the stability of Ti-phlogopite ~150°C above that of phlogopite at 30 kbar (Yoder and Kushiro, 1969).

Robert (1976b) synthesized a number of Ti-phlogopites based on the substitution $Ti^{vi} + 2Al^{iv} = 2Si^{iv} + Mg^{vi}$ (Ti-Tschermak's type). Run products were not analyzed, but several runs with >95% Ti-mica were reported. Normally these yields would be sufficient evidence for establishing the composition of the phase, but with TiO_2 contents of the starting materials generally less than 2% the composition of the Ti-phlogopite present is not well defined. In addition to phlogopite as the starting component, a number of experiments were carried out by adding small amounts of the Ti-Tschermak's type substitution to a starting material made of mixtures of phlogopite plus Tschermak's substitution and muscovite substitution ($2Al^{vi} + \square^{vi} = 3Mg$). Again, even with high yields, the low Ti content of the mix makes it difficult to establish the Ti-composition of the final material. Robert's (1976b) data at high temperature (1000°C) show high yields at TiO_2 contents up to 5.26 wt %. These provide good evidence for the Ti-Tschermak's substitution and along with the Forbes and Flower (1974) data establish that there are at least two important Ti-substitution mechanisms in biotite at high temperature and that the Ti content of biotite increases significantly with temperature in the Fe-free system.

Abrecht and Hewitt (1980, 1981, and manuscript submitted) synthesized micas based on several substitutions in the Fe-Al-bearing and the mixed Fe-Mg-Al biotite systems at temperatures up to 800°C. By obtaining electron microprobe analyses of the biotites synthesized they were able to make interpretations based on experiments where yields of less than 100% biotite were common. They found in the Fe-rich system that only the Ti-Tschermak's substitution produced high yields of biotite with the composition of the bulk mix. Ti-biotites with as much as 3.5 wt % TiO_2 could be synthesized on composition for mixtures of annite + Ti-Tschermak's exchange component. In the Fe-rich system the Ti-vacancy substitution and the Ti-oxy substitution ($Ti^{vi} + 2\ O^= = Fe^{2+} + 2(OH)^-$) typically produced low yields of biotite that were significantly off the composition of the mix. Often the Ti-contents of the biotites were high (up to 6 wt %), but the compositions of the phases involved various mixtures of possible substitutions. In the Mg system the Ti-Tschermak's substitution again produced high yields of biotite on composition as did the Ti-vacancy substitution. The authors conclude that the Ti-Tschermak's substitution is a valid mechanism in both the Fe and Mg biotites and that the Ti-vacancy substitution is valid in the Mg-rich biotites at magmatic or high-grade metamorphic conditions. Additional Ti-substitutions occur, but seemingly only in combinations with other substitutions that allow some crystal chemical constraints to be maintained.

Stability of talc

The decomposition of talc by Reaction 21 was one of the earliest experimentally determined hydrothermal decomposition reactions.

$$\text{Talc} \rightarrow \text{Enstatite} + \text{Quartz} + \text{Water} \tag{21}$$

Bowen and Tuttle (1949) in their classic study of the system $MgO\text{-}SiO_2\text{-}H_2O$ reported an equilibrium curve for Reaction 21 as well as several other equilibria in the system. The position of that equilibrium was redetermined using classical techniques by Greenwood (1963), Skippen (1971) and Chernosky (1976). Figure 19 shows the original curve as well as the reversals determined in the more recent studies. Additional high pressure data are available from Yamamoto and Akimoto (1977). A bracket on Reaction 22 was determined by Greenwood (1963), and the controversial relationships involving the [Fo] and [Qtz] invariant points in the talc-enstatite-anthophyllite-forsterite-quartz-H_2O system are discussed by Chernosky (1976).

$$7 \text{ Talc} \rightarrow 3 \text{ Anthophyllite} + 4 \text{ Quartz} + 4 \text{ Water} \tag{22}$$

Hemley et al. (1977b) have investigated the system using aqueous solution silica concentration methods. Their determination of the 1 kbar equilibrium point for Reaction 21 is in excellent agreement with the data of Chernosky (1976). Using the data in Figure 20 and additional thermochemical data from the literature, they have calculated the P-T diagram shown in Figure 21. There are significant differences between this grid and the ones proposed by Greenwood (1963), Chernosky (1976) and Zen and Chernosky (1976). Resolution of the differences and the correspondences with natural assemblages are still uncertain.

Low temperature phase relations involving the stability of talc were discussed recently by Hemley et al. (1977) and Bricker et al. (1973). Quench pH data at 500°C and 1 kbar are reported by Poty et al. (1972) for talc-quartz and talc-forsterite assemblages. Limited data on the stability of iron-bearing talc were given by Forbes (1971).

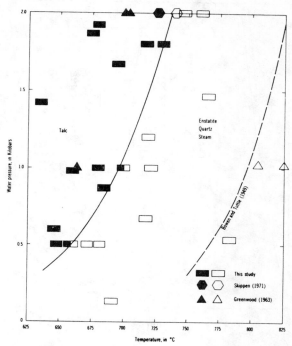

Figure 19. P-T plot of the equilibrium for Reaction 21 from Chernosky (1976) compared with the data from Bowen and Tuttle (1949), Skippen (1971) and Greenwood (1963). From Chernosky (1976).

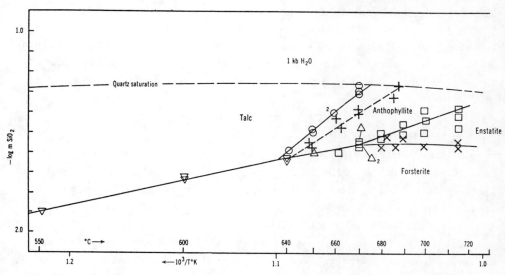

Figure 20. Silica concentration data at 1 kbar for the assemblages talc-anthophyllite, talc-forsterite, talc-enstatite, anthophyllite-enstatite, and enstatite-forsterite. From Hemley et al. (1977b).

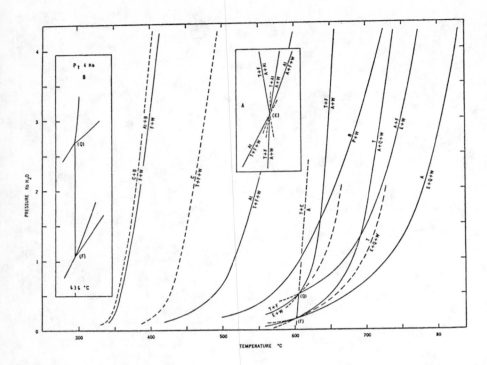

Figure 21. Calculated P-T plot for the systems $MgO-SiO_2-H_2O$ at 1 kbar. From Hemley et al. (1977b).

Stability of other micas and mica reactions

A wide variety of experimental data on the less common micas and on reactions involving micas plus other phases are available in the literature. Table 1 is a summary of much of that data. Only the more recent references are included and no assessment on the quality of the data is implied by the listing.

MELTING EQUILIBRIA OF MICAS

Melting of phlogopite

Melting relationships of the micas have been investigated by studies on the individual minerals, studies of systems, studies of single bulk compositions (rock-melting), and by theoretical deduction. Phlogopite is the most thoroughly studied of the micas, and a general overview of its melting reactions provides a basis for a discussion of the other micas.

The initial pioneering work of Yoder and Eugster (1954) was followed by Crowley and Roy (1964), Luth (1967), Wones (1967), Yoder and Kushiro (1969), Modreski and Boettcher (1972, 1973), Wood (1976), Wones and Dodge (1977), Wend-

Table 1. (Continued on the next page)

Synthesis and Stability of Other Micas

Phase	Reference
Ephesite ($Na(LiAl_2)Al_2Si_2O_{10}(OH)_2$)	Warhus & Chatterjee (1984); Chatterjee & Warhus (1984)
Na-Al mica ($NaAl_{7/3}Al_2Si_2O_{10}(OH)_2$)	Franke et al. (1982)
2.5 octahedral micas (e.g., $KMg_{2.5}Si_4O_{10}(OH)_2$)	Green (1981); Kwak (1971); Franz & Althaus (1974); Seifert & Schreyer (1971)
Polylithionite ($K_2Li_4Al_2Si_8O_{20}(F,OH)_4$)	Rieder (1971); Munoz (1971)
Trilithionite ($K_2Li_3Al_3Al_2Si_6O_{20}(OH)_4$)	Munoz (1971)
Clintonite ($Ca(Mg,Al)_3(Al,Si)_4O_{10}(OH)_2$)	Olesch (1975);
Celadonite	Olesch & Seifert (1976) Velde (1972); Wise & Eugster (1962)
Phengite	Velde (1965)
Ammonium muscovite ($NH_4Al_3Si_3O_{10}(OH)_2$)	Hallam & Eugster (1976)
Ga-muscovite ($KGa_2(Ga,Si)_3O_{10}(OH)_2$)	Kimata (1980)
Ammonium phlogopite ($NH_4Mg_3AlSi_3O_{10}(OH)_2$)	Eugster & Munoz (1966)
B-phlogopite ($KMg_3BSi_3O_{10}(OH)_2$)	Hazen & Wones (1972)
Cs-phlogopite ($CsMg_3AlSi_3O_{10}(OH)_2$)	Hazen & Wones (1972)
Co-phlogopite ($KCo_3AlSi_3O_{10}(OH)_2$)	Hazen & Wones (1978); Lindqvist (1966)
Cobaltous ferric-phlogopite ($KCo_3FeSi_3O_{10}(OH)_2$)	Hazen & Wones (1972); Ferrow (1984)
Cu-phlogopite ($KCu_3AlSi_3O_{10}(OH)_2$)	Hazen & Wones (1972)
Ferriphlogopite ($KMg_3FeSi_3O_{10}(OH)_2$)	Lindqvist (1966); Ferrow (1984)
Ga^{+3}-phlogopite ($KMg_3GaSi_3O_{10}(OH)_2$)	Hazen & Wones (1972)
Ge-phlogopite ($KMg_3AlGe_3O_{10}(OH)_2$)	Hazen & Wones (1972)
Mn-ferriphlogopite ($KMn_3FeSi_3O_{10}(OH)_2$)	Lindqvist (1966)
Ni-ferriphlogopite ($KNi_3FeSi_3O_{10}(OH)_2$)	Lindqvist (1966); Ferrow (1984)
Ni-phlogopite ($KNi_3AlSi_3O_{10}(OH)_2$)	Hazen & Wones (1972); Lindqvist (1966)
Rb-phlogopite ($RbMg_3AlSi_3O_{10}(OH)_2$)	Hazen & Wones (1972)
Zn-phlogopite ($KZn_3AlSi_3O_{10}(OH)_2$)	Perrotta & Garland (1975); Hazen & Wones (1972)
Zn-ferriphlogopite ($KZn_3FeSi_3O_{10}(OH)_2$)	Ferrow (1984)
Ge-biotites ($KFeMg_2AlGe_{3-x}Si_xO_{10}(OH)_2$)	Ferrow (1984)
Nickelous Ge-biotites ($KNi_3AlGe_{3-x}Si_xO_{10}(OH)_2$)	Ferrow (1984)

Mica-Carbonate Reactions

Reaction	Reference
Muscovite + cc + qtz → K-feldspar + anorthite	Hewitt (1973); Johannes & Orville (1972); Jacobs & Kerrick (1981)
Phlogopite + cc + qtz → tremolite + K-feldspar	Hoschek (1973); Hewitt (1975)
Margarite + cc + qtz → zoisite	Storre et al. (1982)
Talc + cc + qtz → tremolite	Slaughter et al. (1975)
Talc + cc + qtz → diopside	Skippen (1971)
Dolomite + talc + qtz → tremolite	Eggert & Kerrick (1981)
Dolomite + qtz → talc + cc	Gordon & Greenwood (1970); Metz & Puhan (1970, 1971); Jacobs & Kerrick (1981); Skippen (1971); Skippen & Hutcheon (1974); Eggert & Kerrick (1981)
Talc + dolomite → diopside + forsterite	Skippen (1971)
Dolomite + K-feldspar → phlogopite + cc	Puhan & Johannes (1974); Puhan (1978)

Element Partitioning

Mineral Pairs	Reference
Biotite-vapor (F-OH)	Munoz & Ludington (1974)
Muscovite-vapor (F-OH)	Munoz & Ludington (1977)
Biotite-amphibole-granitic melt (F-OH)	Anfilogov et al. (1977)
Biotite-apatite (F-OH)	Stormer & Carmichael (1971); Ludington (1978)
Biotite-vapor (K-Ba)	Krausz (1974)
Phlogopite-sanidine (Na-Rb-Tl)	Fung & Shaw (1978); Beswick (1973)
Biotite-pyrrhotite-K-feldspar-magnetite (Fe-Mg-f_{O_2}-f_{S_2})	Tso et al. (1979); Hammarback & Lindqvist (1972)
Garnet-biotite (Fe-Mg)	Ferry & Spear (1978); Goldman & Albee (1977); Schneider (1975)
Biotite-cordierite-sillimanite-quartz-K-feldspar (Fe-Mg)	Holdaway & Lee (1977)
Biotite-garnet-sillimanite-quartz (Fe-Mg)	Holdaway & Lee (1977)
Biotite-clinopyroxene (Fe-Mg)	Gunter (1974)

Mica-Silicate Reactions

Reaction	Reference
Muscovite + chlorite → kyanite + phlogopite + qtz	Bird & Fawcett (1973)
Muscovite + chlorite + qtz → phlogopite + cordierite	Bird & Fawcett (1973); Seifert (1970)
Muscovite + zoisite + qtz → anorthite + K-feldspar	Johannes & Orville (1972)
Muscovite + cordierite → phlogopite + aluminosilicate + qtz	Bird & Fawcett (1973); Seifert (1970)
Muscovite + talc → phlogopite + kyanite + qtz	Schreyer & Baller (1977)
Paragonite + zoisite + qtz → plagioclase	Franz & Althaus (1977)
Zoisite + margarite + qtz → anorthite	Storre et al. (1982)
Phlogopite + diopside + qtz → tremolite + K-feldspar	Wones & Dodge (1977)
Biotite + sillimanite + albite + qtz → cordierite + K-feldspar ± melt	Hoffer (1976,1978)
Biotite + plagioclase + cordierite + quartz → orthopyroxene + K-feldspar	Hoffer & Grant (1980)
Cordierite + K-feldspar + phlogopite + qtz → osumilite	Olesch & Seifert (1976)
Na-phlogopite + enstatite + albite + forsterite	Carman & Gilbert (1983)
Glaucophane → albite + Na-phlogopite - talc	Carman & Gilbert (1983)
Talc + forsterite → enstatite	Chernosky (1976)
Clinochrysotile → forsterite + talc	Chernosky (1982); Hemley et al. (1977); Scarfe & Wyllie (1967)

landt and Eggler (1980a,b), and Bohlen et al. (1983). Luth's summary of the system $KAlSiO_4$-Mg_2SiO_4-SiO_2-H_2O provides a useful overview of phlogopite stability at lower pressures. Figure 22 is taken largely from Luth's work with modifications based on other workers.

Phlogopite stability is bounded by a series of subsolidus dehydration reactions that terminate in a series of invariant points that lie at total pressures less than 1.2 kbar. At higher total pressure, the phlogopite stability curves are a series of melting reactions. In this series of invariant points the $a(SiO_2)$ decreases with increasing temperature, and $f(H_2O)$ increases with increasing temperature. Wones and Gilbert (1982) compared this effect with the amphiboles, and these relations are shown in Figure 23. Figure 24 shows the arrangement of univariant assemblages around each of the invariant points.

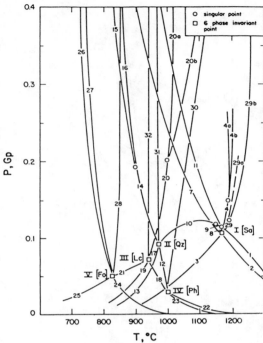

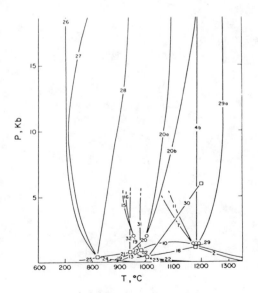

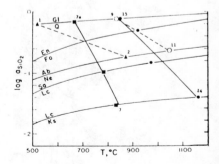

Figure 22 (right). The system $KAlSiO_4$-Mg_2SiO_4-SiO_2-H_2O in a P-T projection. (a) Pressures below 4 kbar. (b) Pressures below 25 kbar. The reactions represented by the univariant curves are given in Table 2. Invariant points are indicated by open squares, and singular points are shown by circles. The univariant curves are numbered, insofar as possible, after the scheme of Luth (1967). The figure is largely based on Luth's work (1967). Sources for other equilibria are given in Table 2.

Figure 23 (below). Stability relations of selected amphiboles and biotites as a function of temperature and SiO_2 activity at 400 bars.
Δ -- Ferrotremolite and Ferropargasite.
o -- Tremolite and Pargasite. □ -- Annite.
● -- Phlogopite. From Wones and Gilbert (1982).

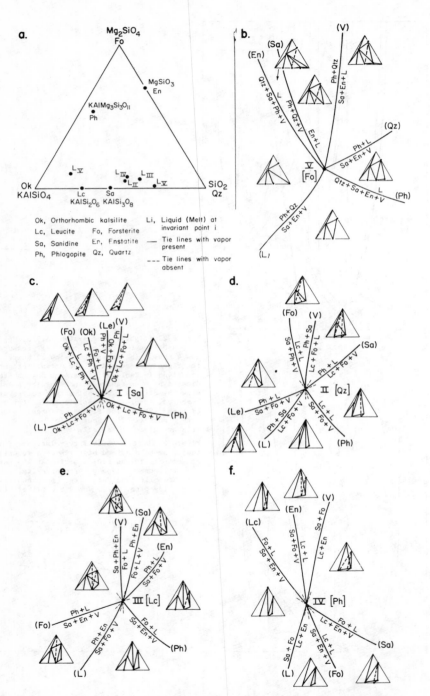

Figure 24. Details of the univariant equilibria around each invariant point of Fig. 22. After Luth (1967) and Yoder and Kushiro (1969).

Table 2 lists the reactions represented by the curves and the source of data for the particular reaction.

The subsolidus reactions that involve phlogopite are, in order of ascending thermal stability:

   Phlogopite + Quartz → K-feldspar + 3 Enstatite + Steam
   Phlogopite + 3/2 Quartz → K-feldspar + 3/2 Forsterite + Steam
   Phlogopite + 1/2 K-feldspar → 3/2 Leucite + 3/2 Forsterite + Steam
   Phlogopite → 1/2 Kalsilite + 1/2 Leucite + 3/2 Forsterite + Steam

Figure 23 shows these reactions at 400 bars as a function of SiO_2 activity and temperature. This dependence of phlogopite stability on silica activity explains the wide variations in the occurrence of the mineral and why it is a common interstitial mineral in gabbroic rocks, yet uncommon as a phenocryst in andesites.

Figure 24 (after Luth, 1967) is a plot of the liquid saturation surfaces projected onto the $KAlSiO_4$-Mg_2SiO_4-SiO_2 plane from the H_2O apex. The solubility of MgO in a feldspathic-siliceous melt is not enhanced by H_2O, whether the saturation phases is enstatite or phlogopite. This observation is in contrast to the analogous FeO- and Al_2O_3-bearing systems (Clemens and Wall, 1981). Any melt that contains several percent H_2O, a high K_2O activity, and greater than 0.2% MgO should crystallize biotite early in the crystallization sequence.

At each invariant point in Figure 22, there are pairs of melting curves that intersect the subsolidus dehydration curves. These are the vapor saturated melting curves, one of which lies below the invariant point where phlogopite is unstable, and the other above the invariant point where phlogopite is stable. The anhydrous composition of the liquids are given in Figure 25 for the appropriate invariant points. The reactions in order of increasing temperature are:

   Quartz + Sanidine + Enstatite + Steam = Melt
   Quartz + Sanidine + Phlogopite + Steam = Melt
   Sanidine + Forsterite + Steam + Leucite = Melt
   Sanidine + Phlogopite + Steam + Leucite = Melt
   Kalsilite + Leucite + Forsterite + Steam = Melt
   Kalsilite + Leucite + Phlogopite + Steam = Melt

All of these assume saturation with H_2O, a condition that is unlikely in most of the earth's crust and upper mantle. However, these reactions are indicative of the pressure at which vapor release of a magma occurs.

Emanating from the four invariant points that contain phlogopite as a phase are the following vapor-absent reactions:

Table 2.

Univariant equilibria intersecting in six-phase invariant points in the system $KAlSiO_4$-Mg_2SiO_4-SiO_2-H_2O.
Numbers refer to Figure 22. (After Luth, 1967, with modifications by other investigators.)

I [Sa] Ph-Ok-Lc-Fo-L-V

1	Lc + Fo + V = L	Luth (1967)
2*	Ok + Lc + Fo + V = L	Luth (1967)
3*	Ph = Ok + Lc + Fo + V	Wones (1967)
4*	Ph + V = Ok + Fo + L	Yoder & Kushiro (1969)
4a	Ph + Ok + V = Fo + L	Yoder & Kushiro (1969)
4b	Ph + V = Fo + L	Yoder & Kushiro (1969)
7*	Ph + Ok + Lc + V = L	Luth (1967)
8*	Ph + Lc + V = Fo + L	Luth (1967)
9	Ph + Lc + Fo + V = L	Luth (1967)
10	Ph + L = Fo + Lc + V	Luth (1967)
11	Ph + Lc + V = L	Luth (1967)
29	Ph = Ok + Lc + Fo + L	Yoder & Kushiro (1969)
29a	Ph = Fo + L	Yoder & Kushiro (1969)

II [Qz] Ph - Lc - Fo - En - L - V

10	Ph + L = Fo + Lc + V	Luth (1967)
12	Or + Fo + V = Lc + L	Luth (1967)
13	Ph + Sa = Fo + Lc + V	Luth (1967)
14	Ph + Sa + V = Lc + L	Luth (1967)
15	Ph + Sa + Lc + V = L	Luth (1967)
16	Ph + Sa + V = L	Luth (1967)
17	Ph + L = Sa + Fo + V	Luth (1967)
31	Ph + Sa = Lc + Fo + L	Luth (1967)

III [Lc] Ph - Sa - En - Fo - L - V

17	Ph + L = Sa + Fo + V	Luth (1967)
18	Sa + En + V = Fo + L	Luth (1967)
19	Ph + En = Sa + Fo + V	Luth (1967)
20	Ph + En = Fo + L + V	Modreski & Boettcher (1972)
20a	Ph + En + V = Fo + L	Modreski & Boettcher (1972)
20b	Ph + En = Fo + L	Modreski & Boettcher (1972)
21	Ph + L = Sa + En + V	Luth (1967)
32	Ph + Sa + En = Fo + L	Luth (1967)

IV [Ph] Sa - Lc - En - Fo - L - V

12	Sa + Fo + V = Lc + L	Luth (1967)
18	Sa + En + V = Fo + L	Luth (1967)
22	Lc + En + V = Fo + L	Luth (1967)
23	Sa + En + V = Lc + L	Luth (1967)
30	Sa + Fo = Lc + En	Luth (1967); Wendlandt & Eggler (1980)

V [Fo] Ph - Qz - Sa - En - L - V

21	Ph + L = Sa + En + V	Luth (1967)
24	Qz + Sa + En + V = L	Wones & Dodge (1977); Shaw (1965)
25	Ph + Qz = Sa + En + V	Wood (1976)
26	Qz + Sa + Ph + V = L	Wones & Dodge (1977); Bohlen et al. (1983)
27	Qz + Ph + V = En + L	Wones & Dodge (1977); Bohlen et al. (1983)
28	Qz + Ph = Sa + En + L	Bohlen et al. (1983)

Qz, quartz; Sa, sanidine; Lc, leucite; Ok, orthorhombic kalsilite; Fo, forsterite; En, enstatite; Ph, phlogopite; L, liquid; V, H_2O-rich vapor.

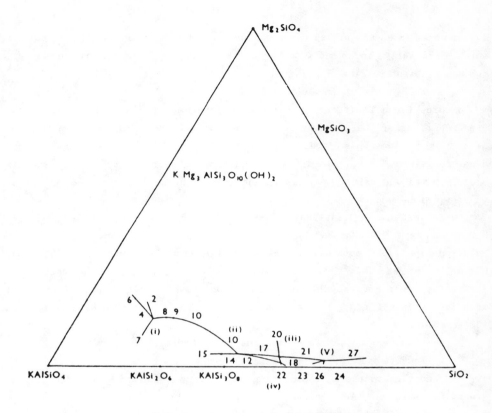

Figure 25. Polythermal, polybaric, projection of H_2O saturated liquidus surfaces in the system $KAlSiO_4$-Mg_2SiO_4-SiO_2. From Luth (1967).

       Phlogopite + Quartz   = Sanidine + Enstatite + Melt
       Phlogopite + Enstatite + Sanidine = Forsterite + Melt
       Phlogopite + Sanidine   = Leucite + Forsterite + Melt
       Phlogopite = Kalsilite + Leucite + Forsterite + Melt

These reactions are useful models for the anatectic reactions that involve micas in different regions of the earth's crust and mantle. Modreski and Boettcher (1972) demonstrated that at 2 kbar the reaction,

       Phlogopite + Enstatite = Forsterite + Melt + Steam

enters a singular point so that, at higher pressures, there are two discrete equilibria,

       Phlogopite + Enstatite + Steam = Forsterite + Melt,   and
       Phlogopite + Enstatite = Forsterite + Melt,

one of which is an additional vapor-absent equilibrium occurring at higher pressures. Similarly, Yoder and Kushiro (1969) demonstrated that at high pressures, the vapor-absent melting of phlogopite is

Phlogopite = Forsterite + Melt.

Modreski and Boettcher (1973) examined the melting of phlogopite in the presence of other phases such as diopside, pyrope, and corundum. At pressures below 35 kbar, two types of reactions exist: those that react with steam to produce forsterite or spinel with melt and those that react to make forsterite and/or spinel and melt. As pressure increases, the silica content of the reactant liquid decreases.

Wendlandt and Eggler (1980a,b) investigated the system $KAlSiO_4$ - $MgO-SiO_2$ - H_2O - CO_2 up to 50 kbar and explored the invariant, univariant, and divariant relationships among potassic phases at mantle-equivalent pressures (Fig. 26). Solid state reactions involving sanidine, leucite, kalsilite, enstatite and forsterite intersect the phlogopite reaction curves and generate a series of invariant points (Fig. 26). Leucite becomes unstable at high pressures through the reactions

Leucite + Enstatite = Sanidine + Forsterite, and

Leucite = Kalsilite + Sanidine.

In addition to these, the melting reaction of sanidine changes from incongruent to congruent. At higher pressures, the reaction

Sanidine + Forsterite = Kalsilite + Enstatite

further shifts the nature of the liquids coexisting with phlogopite to be less and less siliceous as pressure increases. Garnet becomes a reaction product above 40 kbar and phlogopite itself is no longer stable above 50 kbar.

An important result of the several studies in the model system is the change in the composition of the liquid generated by the vapor absent melting of the different phlogopite-bearing assemblages. Not only is the liquid a function of the phase assemblage, but it varies significantly with pressure. The higher the pressure, the lower the SiO_2 content of the liquid. Under high CO_2 pressures, the final liquid generated in the studies by Wendlandt and Eggler (1980b) is carbonate-bearing.

Studies have been carried out on the melting of ultrapotassic rocks (Arima and Edgar, 1983a,b; Barton and Hamilton, 1978, 1979, 1982; Edgar et al., 1976, 1980; Ryabchikov and Green, 1978). These data match the data for the

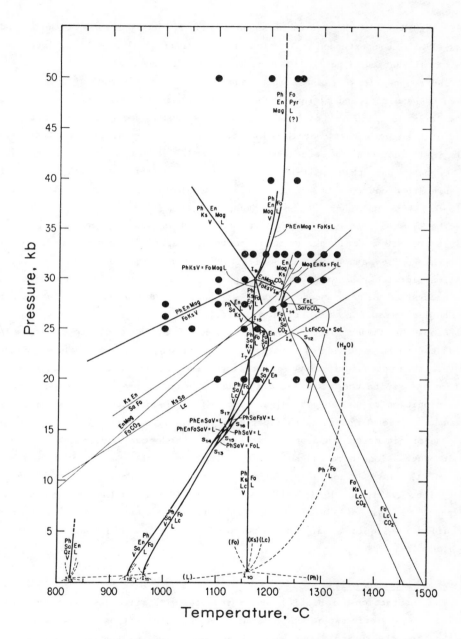

Figure 26. P-T projection of phase relations involving phlogopite, enstatite, forsterite, magnesite, sanidine, kalsilite, leucite, liquid and vapor in the system KAlSiO$_4$-MgO-SiO$_2$-H$_2$O-CO$_2$. Abbreviations: Fo, forsterite; En, enstatite; Ph, phlogopite; Ks, kalsilite; Lc, leucite; Sa, sanidine; Qz, quartz; L, liquid; V, vapor. From Wendlandt and Eggler (1980b).

$KAlSiO_4$-MgO-SiO_2-H_2O-CO_2 system reasonably well, and have been used by Edgar and Arima (1983) in an extended discussion of the origins of ultrapotassic lavas and the crystallization of phlogopite from ultrapotassic magmas.

The compositions of these rocks mimic those of the invariant points discussed by Wendlandt and Eggler (1980a,b) and by Modreski and Boettcher (1973). The major conclusion is that the ultrapotassic magmas are generated from portions of the mantle that contain phlogopite, but at different depths. There are no logical fractionation sequences that relate the several kinds of ultrapotassic magmas, but phlogopite-bearing peridotites will react at different pressures to produce melts of higher silica contents at low pressures, and lower silica contents at higher pressures (Kuehner et al., 1981).

The crystallization of phlogopite from such magmas is dependent upon the activity of H_2O as well as the magma containing significant concentrations of K_2O. Phlogopites can crystallize directly from the melt, or be in a reaction relationship with forsterite and/or diopside. These types of relationships are best understood in reference to Figure 22, where the various reaction curves are crossed during cooling at crustal pressures.

Drops in temperature while maintaining H_2O pressure and total pressure would lead to the crossing of several reaction curves in Figure 22 where olivine reacts with a potassic liquid to form phlogopite. Orthopyroxene will also react to form phlogopite under similar conditions. Drops in pressure of magmas that contain phlogopite phenocrysts might also lead to resorption of phlogopite, or its decomposition to glass or feldspar and pyroxene or olivine.

The origin of the phlogopite rich regions of the mantle is not well understood. Suggestions have ranged from metasomatism of the upper mantle (Bailey, 1982) or the interaction of siliceous liquids with peridotitic mantle (Sekine and Wyllie, 1982a,b,c, 1983; Wyllie and Sekine, 1982). In the latter studies, the phase diagrams of Figures 25 and 26 figured prominently in the arguments that siliceous liquids rising into a periodotitic mantle would precipitate zones of phlogopite.

Micas and siliceous melts

The interactions of biotites and muscovites with siliceous melts have been studied in systematic experimental studies, single composition ("rock-melting") studies, and theoretical studies. All of these studies have converged on the description of specific invariant points of which point V of Figure 22 may serve as a model, as well as being the limiting case for biotites and quartz. Figure

27 sketches the relationships for annite, which are based on extrapolation of subsolidus data of Hewitt and Wones (1981). Figure 28 gives the arrangement of

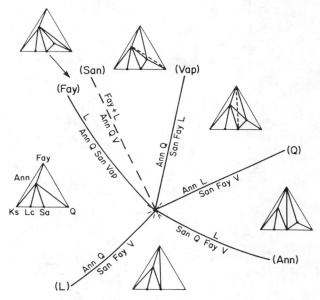

Figure 27. Proposed sequence of univariant equilibria around the invariant point annite-fayalite-sanidine-quartz-liquid-vapor.

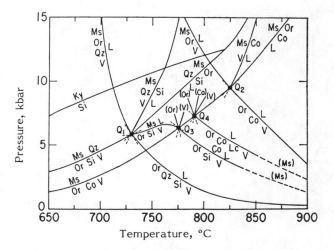

Figure 28. P-T projection for the system $K_2O-Al_2O_3-SiO_2-H_2O$. Invariant points are Q_1, Ms-Sill-Or-Qz-L-V; Q_2, Ms-Or-Co-Sill. From Huang and Wyllie (1974).

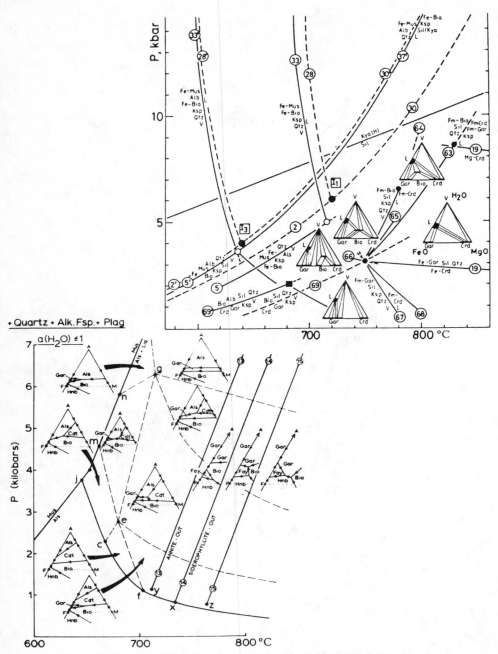

Figure 29 (above). P-T projections of the stability of selected micas interacting with siliceous melts. From A. B. Thompson (1982).

Figure 30 (below). P-T projection of the stability of biotites with varying Fe-Al composition and the "minimum melting" point of granite. From Abbott (1981).

curves after Huang and Wyllie (1974) for muscovite. The interaction of both biotite and muscovite with the aluminosilicates, garnet and cordierite generates an entire collection of invariant points and has been discussed by Abbott (1981) and Thompson (1982).

Each mica-bearing assemblage requires a minimum H_2O pressure for a particular mica to coexist with a hydrated siliceous melt and a particular phase assemblage. This H_2O pressure also fixes the H_2O content for that melt. Because of the presence of other volatiles in the melt (Wyllie and Tuttle, 1964) and suspended phenocrysts, a magma usually will contain less H_2O than that dissolved in the melt portion.

Emanating from each of the appropriate invariant points are reactions that are vapor absent. In the deep crust and mantle, where vapor is usually not present as a phase, these curves mark the fusion of a particular mineral assemblage to yield a hydrated, but not H_2O-saturated melt. These curves are presented in Figure 29. Because minerals exist as solutions, these reactions, although portrayed here as univariant equilibria, are often divariant or trivariant. Thompson (1982) and Abbott (1981) have dealt with these complexities in two different ways.

Thompson set up a network of invariant points based on the intersections of the dehydration reactions in the subsolidus, with the vapor saturated melting curves for the anhydrous phase assemblage. Thompson shows the "granite minimum melting curve" of Tuttle and Bowen (1958) about 20°C lower because of the extra dissolved components FeO, MgO, and Al_2O_3. MgO is probably not a contributor to this effect (Wones and Dodge, 1977; Bohlen et al., 1983). Abbott also lowered the "minimum melting curve," and sketched the boundary curves of the minerals within the AFM triangles (Fig. 30).

A useful series of experiments has been made by examining the reactions of the minerals within a given rock composition as a function of temperature and H_2O content under isobaric conditions. Maaloe and Wyllie (1978) showed that the crystallization sequence for biotite changed as a function of H_2O content, with biotite crystallizing out earlier at higher H_2O contents (Fig. 31). The increased water content lowers the melting points of the anhydrous minerals and enhances the stability of biotite. The combined effect is to give different sequences of crystallization, and this becomes a valuable indicator of the nature of the fugitive components within a crystallizing magma.

Whitney (1975) systematized this approach for quartz and the feldspars in a series of experiments on synthetic rock compositions. Naney (1983) has followed

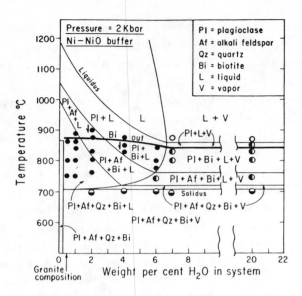

Figure 31. T-H$_2$O projection of the Norwegian monzogranite (Mg-1) at 2 kbar, showing the co-existing phases (Maaloe and Wyllie, 1975).

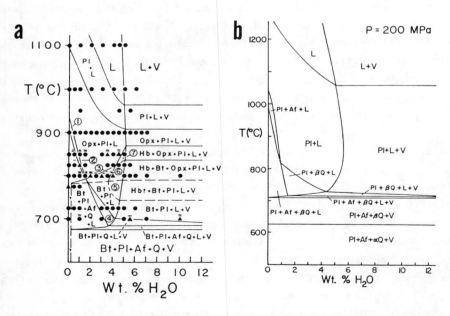

Figure 32. T-H$_2$O projections of a synthetic granodiorite at 2 kbar. (a) With FeO and MgO. (b) Without FeO and MgO. From Naney (1983).

up on this work, and added FeO and MgO to Whitney's compositions. Naney's results are given in Figure 32, which is interesting in that hornblende occurs as well as biotite as a hydrous mafic phase. Hornblende only occurs at high H_2O contents, and is present with increased anorthite content in the composition at a given pressure. This result, similar to that found by Cawthorn (1976), is strong evidence for enhanced Al_2O_3 contents of the melt with higher $f(H_2O)$. This suggests that Thompson and Abbott's assumptions about enhanced Al_2O_3 contents in the "minimum" melts is sound.

Clemens and Wall (1981) investigated the melting relations of the minerals in a peraluminous granitic rock composition, and also discovered that TiO_2 and MgO greatly enhance the stability of biotite in such melts. They also demonstrated that the melts formed in their experiments contained significant Al_2O_3 and FeO (Fig. 33). Piwinskii (1968), working under H_2O-saturated conditions, showed that biotite usually crystallizes out at lower temperatures than amphibole at pressure greater than 2 kbar (Fig. 34) In his experiments, the rocks were diopside normative compositions.

The relationships between biotites and amphiboles are quite complex. Wones and Gilbert explored these in a brief review in 1982. The relative stabilities depend on the activity of K_2O, CaO, H_2O, and SiO_2. Biotites by themselves are equal in thermal stability to amphiboles, so that the general reason that Bowen's reaction series is as successful as it is, is that K_2O activities are usually lowered by Na_2O contents of melts and feldspars. CaO is relatively insoluble in silicic melts, so that clinopyroxene or hornblende crystallizes earlier than biotites in most magmas. However, in ultrapotassic rocks, the sequence can be reversed, and rocks exist in which amphibole is only present as a late stage alteration product (Eriksson and Wones, 1983). Wones and Dodge (1977) examined the relations between phlogopite, quartz and diopside in an idealized system which is pictured in Figure 35. These results are very similar to those of Naney (1983), so that the occurrence of biotite euhedra as inclusions in hornblende, and the presence of biotite and diopsidic pyroxenes in ashflow tuffs indicate that the early part of the magmatic crystallization in these instances took place under conditions of relatively low H_2O content of the melt. Clemens (in press) reviewed the estimates of H_2O contents of magmas and concluded that most begin to crystallize at low water contents, gradually building up to saturation with the crystallization of anhydrous minerals. Burnham (1979) has written an extensive review of this general process. The point to be made

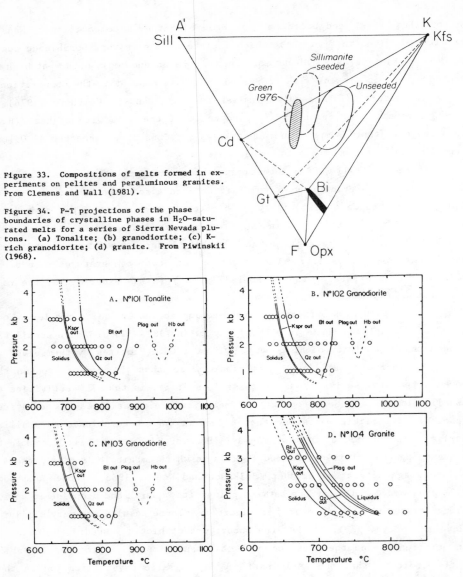

Figure 33. Compositions of melts formed in experiments on pelites and peraluminous granites. From Clemens and Wall (1981).

Figure 34. P-T projections of the phase boundaries of crystalline phases in H_2O-saturated melts for a series of Sierra Nevada plutons. (a) Tonalite; (b) granodiorite; (c) K-rich granodiorite; (d) granite. From Piwinskii (1968).

here is that micas are excellent monitors of the processes of H_2O interactions with magmas.

Muscovites in granitic plutons -- an attempt at geobarometry

The intersection of the muscovite + quartz subsolidus reaction with the Ab-Or-Q-H_2O minimum solidi has been used to estimate H_2O pressure (and total

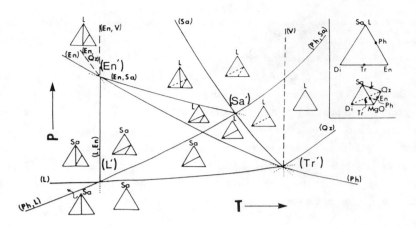

Figure 35. P-T projection of the system sanidine-enstatite-diopside-quartz at H₂O saturation. From Wones and Dodge (1977).

pressure) of crystallizing magmas that produce muscovite-bearing granites. Recent work (Miller et al., 1981) has shown that such granites are intruded at shallower depths than the 10-12 km estimated from the muscovite invariant point. The explanations for this apparent contradiction are (1) components in the muscovite that enhance its stability field, (2) components in the melt that lower the solidus temperature, and (3) the metastable persistence of the mica as the plutonic mass moves upward in the crust during intrusion.

Components present in muscovite are the phengite substitutions, $MgSiAl_{-1}Al_{-1}$, $FeSiAl_{-1}Al_{-1}$, $Fe^{3+}Al_{-1}$ and the substitution of Ti coupled with Fe or Mg, $FmTiAl_{-1}Al_{-1}$. Velde (1965) demonstrated that the phengite components have significantly lower thermal stabilities than pure muscovite. Guidotti (1978) demonstrated that with increasing grade of metamorphism the phengite component decreases, while the Ti component increases. Munoz (1971) showed the enhanced stability of Li-bearing micas over muscovite, although whether or not Li is a significant factor is unknown because with the advent of the microprobe, all too few investigators have determined the Li contents of micas and amphiboles.

The presence of FeO and Al_2O_3 in the granitic magma will lower the solidus temperature from that of ideal $Ab-Or-Q-H_2O$.. Burnham and Jahns (1962), Wyllie and Tuttle (1964), and Chorlton and Martin (1978) have demonstrated that the elements Li_2O, BeO, B_2O_3 and the halogens will all significantly lower the solidus of granitic magma. Many muscovite-bearing granites contain tourmaline, so that the presence of muscovites at anomalously low pressures in some granites may be ascribed to B_2O_3.

Hogan (1984) has suggested that muscovite may persist at shallower pressures during ascension of the magma to lower pressures. Muscovite, originally found at depth, will react to form K-feldspar and/or biotites, unless armored by early crystallized phases, but if armored, will persist metastably at higher levels in the crust.

Muscovite is not a panacea for the geobarometry of intrusive rocks, but careful petrographic observations coupled with interpretations based on the pertinent phase equilibria, can establish many of the parameters involving the formation, evolution and crystallization of siliceous melts.

REFERENCES

Abbott, R.N., Jr. (1981) AFM liquidus projections for granitic magmas, with special reference to hornblende, biotite and garnet. Canadian Mineral. 19, 103-110.

Abrecht, J. and Hewitt, D.A. (1980) Ti-Substitution in synthetic Fe-biotites. Geol. Soc. Am. Abstr. with Programs 12, 377.

Abrecht, J. and Hewitt, D.A. (1981) Substitutions in synthetic Ti-biotites. Geol. Soc. Am. Abstr. with Programs 13, 393.

Althaus, E. (1966a) Die Bildung von Pyrophyllit und Andalusit zwischen 2000 und 7000 bar H_2O druck. Naturwissen. 53, 105-106.

Althaus, E. (1966b) Der Stabilitätsbereich des Pyrophyllits unter dem Einfluss von Sauren; I: Beitr. Mineral. Petrol. 13, 31-50.

Althaus, E. (1966c) Der Stabilitätbereich des Pyrophyllits unter dem Einfluss von Sauren; II: Beitr. Mineral. Petrol. 13, 97-107.

Althaus, E. (1969) Das System Al_2O_3 - SiO_2 - H_2O. Experimentelle Untersuchungen und Folgerungen fur die Petrogenese der metamorphen Gesteine, Teil I und II. N. Jahrb. Mineral. Abh. 111, 74-143.

Althaus, E., Karotke, E., Nitsch, K.H., Winkler, H.G.F. (1970) An experimental re-examination of the upper stability limit of muscovite plus quartz. N. Jahrb. Mineral. Monatsh., 325-336.

Anfilogov, V.N., Bushlyakov, I.N., Vilisov, V.A., and Bragina, G.I. (1977) Distribution of fluorine between coexisting biotite and amphibole and granitic melt at 780°C and 1000 atm pressure. Transl. from Geokhimiya 3, 471-475.

Arima, M. and Edgar, A.D. (1983a) A high pressure experimental study on a magnesium-rich leucite-lamproite from the West Kimberley area, Australia: petrogenetic implications. Contrib. Mineral. Petrol. 84, 228-234.

Arima, M. and Edgar, A.D. (1983b) High pressure experimental studies on a katungite and their bearing on the genesis of some potassium-rich magmas of the west branch of the African rift. J. Petrol. 24, 166-187.

Bailey, B.K. (1982) Mantle metasomatism -- continuing chemical change within the Earth. Nature 296, 525-530.

Barton, M. and Hamilton, D.L. (1978) Water-saturated melting relations to 5 kb of three Leucite Hills lavas. Contrib. Mineral. Petrol. 66, 41-49.

Barton, M. and Hamilton, D.L. (1979) The melting relationships of a madupite from the Leucite Hills, Wyoming, to 30 kb. Contrib. Mineral. Petrol. 69, 133-142.

Barton, M. and Hamilton, D.L. (1982) Water-undersaturated melting experiments bearing upon the origin of potassium-rich magma. Mineral. Mag. 45, 267-278.

Beswick, A.E. (1973) An experimental study of alkali metal distributions in feldspars and micas. Geochim. Cosmochim. Acta 37, 183-208.

Bird, G.W. and Fawcett, J.J. (1973) Stability relations of Mg - chlorite - muscovite and quartz between 5 and 10 kb water pressure. J. Petrol. 14, 415-428.

Blencoe, J.G. (1977) Molal volumes of synthetic paragonite-muscovite micas. Am. Mineral. 62, 1200-1215.

Blencoe, J.G. and Luth, W.C. (1973) Muscovite-paragonite solvi at 2.4 and 8 kb pressure. Geol. Soc. Am. Abstr. with Programs 5, 553-554.

Bohlen, S.R., Boettcher, A.L., Wall, V.J., and Clemens, J.D. (1983) Stability of phlogopite - quartz and sanidine - quartz: a model for melting in the lower crust. Contrib. Mineral. Petrol. 83, 270-277.

Bowen, N.L. and Tuttle, O.F. (1949) The system $MgO-SiO_2-H_2O$. Bull. Geol. Soc. Am. 60, 439-460.

Bricker, O.P., Nesbitt, H.W., and Gunter, W.D. (1973) The stability of talc. Am. Mineral. 58, 64-72.

Burnham, C.W. (1979) Magmas and hydrothermal fluids. In: H.L. Barnes, ed., Geochemistry of Hydrothermal Ore Deposits, 2nd ed. New York: Wiley Interscience.

Burnham, C.W. and Jahns, R.H. (1962) A method for determining the solubility of water in silicate melts. Am. J. Sci. 260, 721-745.

Burnham, C.W., Holloway, J.R., and Davis, N.F. (1969) Thermodynamic properties of water to 1,000°C and 10,000 bars. Geol. Soc. Am. Spec. Paper 132, 96 p.

Carman, J.H. (1974) Synthetic sodium and its two hydrates: stabilities, properties, and mineralogic implications. Am. Mineral. 59, 261-273.

Carman, J.H. and Gilbert, M.C. (1983) Experimental studies on glaucophane stability. Am. J. Sci. 283-A, 414-437.

Cawthorn, R.G. (1976) Melting relations in part of the system $CaO - MgO - Al_2O_3 - SiO_2 - Na_2O - H_2O$ under 5 kb pressure. J. Petrol. 17, 44-72.

Chatterjee, N.D. (1970) Synthesis and upper stability of paragonite. Contrib. Mineral. Petrol. 27, 244-257.

Chatterjee, N.D. (1971) Preliminary results of the synthesis and upper stability limit of margarite. Naturwissen. 58, 147.

Chatterjee, N.D. (1972) The upper stability limit of the assemblage paragonite + quartz and its natural occurrences. Contrib. Mineral. Petrol. 34, 288-303.

Chatterjee, N.D. (1973) Low-temperature compatibility relations of the assemblage quartz-paragonite and the thermodynamic status of the phase rectorite. Contrib. Mineral. Petrol. 42, 259-271.

Chatterjee, N.D. (1974) Synthesis and upper thermal stability limit of 2M-margarite, $CaAl_2[Al_2Si_2O_{10}(OH)_2]$. Schweiz. Mineral. Petrogr. Mitt. 54, 753-767.

Chatterjee, N.D. (1976) Margarite stability and compatibility relations in the system $CaO - Al_2O_3 - SiO_2 - H_2O$ as a pressure-temperature indicator. Am. Mineral. 61, 699-709.

Chatterjee, N.D. and Froese, E. (1975) A thermodynamic study of the pseudobinary join muscovite-paragonite in the system $KAlSi_3O_8 - NaAlSi_3O_8 - Al_2O_3 - SiO_2 - H_2O$. Am. Mineral. 60, 985-993.

Chatterjee, N.D. and Johannes, W. (1974) Thermal stability and standard thermodynamic properties of synthetic $2M_1$ muscovite, $KAl_2[AlSi_3O_{10}(OH)_2]$. Contrib. Mineral. Petrol. 48, 89-114.

Chatterjee, N.D. and Warhus, U. (1984) Ephesite, $NaLiAl_2[Al_2Si_2O_{10}](OH)_2$: II. Thermodynamic analysis of its stability and compatibility relations, and its geological occurrences. Contrib. Mineral. Petrol. 85, 80-84.

Chernosky, J.V., Jr. (1976) The stability of anthophyllite - A reevaluation based on new experimental data. Am. Mineral. 61, 1145-1155.

Chernosky, J.V., Jr. (1982) The stability of clinochrysotile. Canadian Mineral. 20, 19-27.

Chorlton, L.B. and Martin, R.F. (1978) The effect of boron on the granite solidus. Canadian Mineral. 16, 239-244.

Clemens, J.D. and Wall, V.J. (1981) Origin and crystallization of some peraluminous (S-type) granitic magmas. Canadian Mineral. 19, 111-131.

Crowley, M.S. and Roy, R. (1964) Crystalline solubility in the muscovite and phlogopite groups. Am. Mineral. 49, 348-363.

Day, H.W. (1973) The high temperature stability of muscovite plus quartz. Am. Mineral. 58, 255-262.

Day, H.W. (1976) A working model of some equilibria in the system alumina-silica-water. Am. J. Sci. 276, 1254-1284.

Dymek, R.F. (1983) Titanium, aluminum and interlayer cation substitutions in biotite from high-grade gneisses, West Greenland. Am. Mineral. 68, 880-899.

Eberl, D. (1979) Synthesis of pyrophyllite polytypes and mixed layers. Am. Mineral. 64, 1091-1096.

Edgar, A.D. and Arima, M. (1983) Conditions of phlogopite crystallization in ultrapotassic volcanic rocks. Mineral. Mag. 47, 11-19.

Edgar, A.D., Green, D.H., and Hibberson, W.O. (1976) Experimental petrology of a highly potassic magma. J. Petrol. 17, 339-356.

Edgar, A.D., Condliffe, E., Barnett, R.L., and Shirran, R.J. (1980) An experimental study of an olivine ugandite magma and mechanisms for the formation of its K-enriched derivatives. J. Petrol. 21, 475-497.

Eggert, R.G. and Kerrick, D.M. (1981) Metamorphic equilibria in the siliceous dolomite system: 6 kbar experimental data and geologic implications. Geochim. Cosmochim. Acta 45, 1039-1049.

Eriksson, S.C. and Wones, D.R. (1983) The Parks Pond pluton, eastern Maine: evidence for cratonization or orogenic activity? Geol. Soc. Am. Abstr. with Programs 15, 118.

Eugster, H.P. (1956) Muscovite-paragonite join and its use as a geological thermometer. Bull. Geol. Soc. Am. 67, 1693.

Eugster, H.P. (1957) Heterogeneous reactions involving oxidation and reduction at high pressures and temperatures. J. Chem. Physics 26, 1160.

Eugster, H.P. and Munoz, J. (1966) Ammonium micas: possible sources of atmospheric ammonia and nitrogen. Science 151, 683-686.

Eugster, H.P. and Wones, D.R. (1962) Stability relations of the ferruginous biotite, annite. J. Petrol. 3, 82-125.

Eugster, H.P. and Yoder, H.S. (1954a) Stability and occurrence of paragonite. Bull. Geol. Soc. Am. 65, 1248-1249.

Eugster, H.P. and Yoder, H.S. (1954b) Annual Report of the Director of the Geophysical Laboratory. Carnegie Inst. Wash. Yearbook 53, 95-145.

Eugster, H.P., Albee, A.L., Bence, A.E., Thompson, J.B., Jr., and Waldbaum, D.R. (1972) The two-phase region and excess mixing properties of paragonite-muscovite crystalline solutions. J. Petrol. 13, 147-179.

Evans, B.W. (1965) Application of a reaction-rate method to the breakdown equilibria of muscovite and muscovite plus quartz. Am. J. Sci. 263, 647-667.

Ferrow, E. (1984) The Use of Mossbauer Quadropole Splitting to the Study of the Crystal Chemistry of Trioctahedral Micas. An Experimental and Theoretical Approach. Univ. Uppsula Dept. Mineralogy & Petrology Res. Rep. 37.

Ferrow, E. and Annersten, H. (1984) Ferric Iron in Tri-Octahedral Micas. Univ. Uppsula Dept. Mineralogy & Petrology Res. Rep. 39.

Ferry, J.M. and Spear, F.S. (1978) Experimental calibration of the partitioning of Fe and Mg between biotite and garnet. Contrib. Mineral. Petrol. 66, 113-117.

Forbes, W.C. (1971) Iron content of talc in the system $Mg_3Si_4O_{10}(OH)_2$ - $Fe_3Si_4O_{10}(OH)_2$. J. Geol. 79, 63-74.

Forbes, W.C. and Flower, M.F.J. (1974) Phase relations of titan-phlogopite, $K_2Mg_4TiAl_2Si_6O_{20}(OH)_4$: a refractory phase in the upper mantle? Earth Planet. Sci. Lett. 22, 60-66.

Franke, W., Jelinski, B. and Zarei, M. (1982) Hydrothermal synthesis of an ephesite-like sodium mica. N. Jahrb. Mineral. Monatsh. 1982, 337-340.

Franz, G. and Althaus, E. (1974) Synthesis and thermal stability of 2.5-octahedral sodium mica, $NaMg_{2.5}(OH)_2|Si_4O_{10}$. Contrib. Mineral. Petrol. 46, 227-232.

Franz, G. and Althaus, E. (1976) Experimental investigation on the formation of solid solutions in sodium-aluminium-magnesium micas. N. Jahrb. Mineral. Abh. 126, 253-253.

Franz, G. and Althaus, E. (1977) The stability relations of the paragenesis paragonite-zoisite-quartz. N. Jahrb. Mineral. Abh. 130, 159-167.

Franz, G., Hinrichsen, T. and Wannemacher, E. (1977) Determination of the miscibility gap on the solid solution series paragonite - margarite by means of the infrared spectroscopy. Contrib. Mineral. Petrol. 59, 207-336.

Hinrichsen, Th. and Schurmann, K. (1971) Synthese und Stabilitat von Glimmern im System CaO - Na_2O - Al_2O_3 - SiO_2 - H_2O. Fortschr. Mineral. 49, 21.

Hoffer, E. (1976) The reaction sillimanite + biotite + quartz → cordierite + K-feldspar + H_2O and partial melting in the system K_2O-FeO-MgO-Al_2O_3-SiO_2-H_2O. Contrib. Mineral. Petrol. 55, 127-130.

Hoffer, E. (1978) Melting reaction in aluminous metapelites, stability limits of biotite + sillimanite + quartz in the presence of albite. N. Jahrb. Mineral. Monatsh. 9, 396-407.

Hoffer, E. and Grant, J.A. (1980) Experimental investigation of the formation of cordierite-orthopyroxene parageneses in pelitic rocks. Contrib. Mineral. Petrol. 73, 15-22.

Hogan, J.P. (1984) Petrology of the Northport Pluton, Maine. M.S. thesis, Virginia Polytechnic Inst. & State Univ.

Holdaway, M.J. (1971) Stability of andalusite and the aluminum silicate phase diagram. Am. J. Sci. 271, 97-131.

Holdaway, M.J. and Lee, S.M. (1977) Fe-Mg cordierite stability in high-grade pelitic rocks based on experimental, theoretical, and natural observations. Contrib. Mineral. Petrol. 63, 175-198.

Holland, T.J.B. (1979) Experimental determination of the reaction paragonite = jadeite + kyanite + H_2O, and internally consistent thermodynamic data for part of the system Na_2O - Al_2O_3 - SiO_2 - H_2O, with applications to eclogites and blueschists. Contrib. Mineral. Petrol. 68, 293-301.

Hoschek, G. (1973) Die Reaktion Phlogopite + Calcit + Quarz = Tremolit + Kalifeldspat + H_2O + CO_2. Contrib. Mineral. Petrol. 39, 321-237.

Huang, W.L. and Wyllie, P.J. (1974) Melting relations of muscovite with quartz and sanidine in the K_2O - Al_2O_3 - SiO_2 - H_2O system to 30 kilobars and an outline of paragonite melting relations. Am. J. Sci. 274, 378-395.

Iiyama, J.T. (1964) Etude des reactions d'echange d'ions Na-K dans la serie muscovite-paragonite. Bull. Soc. franc. Mineral. Cristallogr. 87, 532-541.

Iiyama, S. and Buseck, P.R. (1978) Experimental study of disordered mica structures by high-resolution electron microscopy. Acta Crystallogr. A34, 709-719.

Ivanov, I.P., Belyaevska, O.N., and Potekhin, V.Yu. (1974) Improved diagram of hydrolysis and hydration equilibria in the open multi-system KCl - HCl - Al_2O_3 - SiO_2 - H_2O at P = 1,000 kg/cm². Dokl. Akad. Nauk. SSSR 219, 164-166.

Ivanov, I.P. and Gusynin, V.F. (1970) Stability of paragonite in the system SiO_2-NaAlSi$_3O_8$-Al_2O_3-H_2O. Trans. from Geokhimiya 7, 801-811.

Ivanov, I.P., Potekhin, V.Y., Dmitriyenko, L.T., and Beloborodov, S.M. (1973) An experimental study of T and P conditions of equilibrium of reaction: muscovite - K-feldspar + corundum + H_2O at P(H_2O) < P(total). Trans. from Geokhimiya 9, 1300-1310.

Jacobs, G.K. and Kerrick, D.M. (1981) Devolatilization equilibria in H_2O-CO_2 and H_2O-CO_2-NaCl fluids: an experimental and thermodynamic evaluation at elevated pressures and temperatures. Am. Mineral. 66, 1135-1153.

Jenkins, D.M. (1983) Upper stability of synthetic margarite plus quartz. Geol. Soc. Am. Abstr. with Programs 15, 604.

Johannes, W. and Orville, P.M. (1972) Zur Stabilitat der Mineral-paragenesen Muskovit + Calcit + Quarz, Zoisit + Muskovit + Quartz, Anorthit + K-feldspar und Anorthit + Calcit. Fortschr. Mineral. 50, 46-47.

Kerrick, D.M. (1968) Experiments on the upper stability limit of pyrophyllite at 1.8 kilobars and 3.9 kilobars water pressure. Am. J. Sci. 266, 204-214.

Kerrick, D.M. (1972) Experimental determination of muscovite + quartz stability with P(H_2O) < P(total). Am. J. Sci. 272, 946-958.

Frey, M. and Orville, P.M. (1974) Plagioclase in margarite-bearing rocks. Am. J. Sci. 274, 31-47.

Fung, P.C. and Shaw D.M. (1978) Na, Rb and Tl distributions between phlogopite and sanidine by direct synthesis in a common vapour phase. Geochim. Cosmochim. Acta 42, 703-708.

Goldman, D.S. and Albee, A.L. (1977) Correlation of Mg/Fe partitioning between garnet and biotite with $^{18}O/^{16}O$ partitioning between quartz and magnetite. Am. J. Sci. 277, 750-767.

Gordon, T.M. and Greenwood, H.J. (1970) The reaction: dolomite + quartz + water = talc + calcite + carbon dioxide. Am. J. Sci. 268, 225-242.

Green, T.H. (1981) Synthetic high-pressure micas compositionally intermediate between the dioctahedral and trioctahedral mica series. Contrib. Mineral. Petrol. 78, 452-458.

Greenwood, H.J. (1963) The synthesis and stability of anthophyllite. J. Petrol. 4, 317-351.

Guidotti, C.V. (1978) Compositional variation of muscovite in medium- to high-grade metapelites of northwestern Maine. Am. Mineral. 63, 878-884.

Gunter, A.E. (1974) An experimental study of iron-magnesium exchange between biotite and clinopyroxene. Canadian Mineral. 12, 258-261.

Gunter, W.D. and Eugster, H.P. (1980) Mica-feldspar equilibria in supercritical alkali chloride solutions. Contrib. Mineral. Petrol. 75, 235-250.

Haas, H. and Holdaway, M.J. (1973) Equilibria in the system Al_2O_3 - SiO_2 - H_2O involving the stability limits of pyrophyllite, and thermodynamic data of pyrophyllite. Am. J. Sci. 273, 449-461.

Hallam. M. and Eugster, H.P. (1976) Ammonium silicate stability relations. Contrib. Mineral. Petrol. 57, 227-244.

Hammarback, S. and Lindqvist, B. (1972) The hydrothermal stability of annite in the presence of sulfur. Geol. Fören. Förh. 94, 549-564.

Hazen, R.M. and Wones, D.R. (1972) The effect of cation substitutions on the physical properties of trioctahedral micas. Am. Mineral. 57, 103-129.

Hazen, R.M. and Wones, D.R. (1978) Predicted and observed compositional limits of trioctahedral micas. Am. Mineral. 63, 885-892.

Hemley, J.J. (1959) Some mineralogical equilibria in the system K_2O-Al_2O_3-SiO_2-H_2O. Am. J. Sci. 257, 241-270.

Hemley, J.J. (1967) Stability relations of pyrophyllite, andalusite and quartz at elevated pressures and temperatures. Am. Geophys. Union Trans., p. 224.

Hemley, J.J., Meyer, C., and Richter, D.H. (1961) Some alteration reactions in the system Na_2O - Al_2O_3 - SiO_2-H_2O. U.S. Geol. Survey Prof. Paper 424-D, 338-340.

Hemley, J.J., Montoya, J.W., Christ, C.L., and Hostetler, P.B. (1977) Mineral equilibria in the MgO-SiO_2-H_2O system: I Talc chrysotile - forsterite - brucite stability relations. Am. J. Sci. 277, 322-351.

Hemley, J.J., Montoya, J.W., Shaw, D.R., and Luce, R.W. (1977) Mineral equilibria in the MgO - SiO_2 - H_2O system: II Talc - antigorite - forsterite - anthophyllite - enstatite stability relations and some geologic implications in the system. Am. J. Sci. 277, 353-383.

Hewitt, D.A. (1973) Stability of the assemblage muscovite-calcite-quartz. Am. Mineral. 58, 785-791.

Hewitt, D.A. (1975) Stability of the assemblage phlogopite-calcite-quartz. Am. Mineral. 60, 391-397.

Hewitt, D.A. (1977) Hydrogen fugacities in Shaw bomb experiments. Contrib. Mineral. Petrol. 65, 165-169.

Hewitt, D.A. and Wones, D.R. (1975) Physical properties of some synthetic Fe-Mg-Al trioctahedral biotites. Am. Mineral. 60, 854-862.

Hewitt, D.A. and Wones, D.R. (1981) The annite-sanidine-magnetite equilibrium. GAC-MAC Joint Annual Meeting, Calgary, Abstracts, Vol. 6, p. A-66.

Kimata, M. (1980) Synthesis and properties of monoclinic $KGaSi_3O_8$ feldspars coexisting with Ga-muscovite, $KGa_2(Ga,Si_3)O_{10}(OH)_2$. N. Jahrb. Mineral. Monatsh. 1980, 37-48.

Kuehner, S.M., Edgar, A.D. and Arima, M. (1981) Petrogenesis of the ultrapotassic rocks from the Leucite Hills, Wyoming. Am. Mineral. 66, 663-677.

Krausz, K. (1974) Potassium-barium exchange in phlogopite. Canadian Mineral. 12, 394-398.

Kwak, T.A.P. (1971) An experimental study on Fe-Mg micas transitional between dioctahedral and trioctahedral compositions. N. Jahrb. Mineral. Monatsh. 1971, 326-335.

Labotka, T.C. (1983) Analysis of the compositional variations of biotite in pelitic hornfelses from northeastern Minnesota. Am. Mineral. 68, 900-914.

Lindqvist, B. (196) Hydrothermal synthesis studies of potash-bearing sesquioxide-silica systems. Geo. Fören. 1, Stockholm Förh. 88, 133-178.

Ludington, S. (1978) The biotite-apatite geothermometer revisited. Am. Mineral. 63, 551-553.

Luth, W.C. (1967) Studies in the system $KAlSiO_4$ - Mg_2SiO_4 - SiO_2 - H_2O: I, Inferred phase relations and petrologic applications. J. Petrol. 8, 372-416.

Maaløe, S. and Wyllie, P.J. (1975) Water content of a granite magma deduced from the sequence of crystallization determined experimentally with water-undersaturated conditions. Contrib. Mineral. Petrol. 52, 175-191.

Matsushima, S., Kennedy, G.C., Akella, J., and Haygarth, J. (1967) A study of equilibrium relations in the systems Al_2O_3 - SiO_2 - H_2O and Al_2O_3-H_2O. Am. J. Sci. 265, 28-44.

Metz, P. and Puhan,D. (1970) Experimentelle Untersuchung der Metamorphose von kieselig dolomitischen Sedimenten. I. Die Gleichgewichtsdaten der Reaktion 3 Dolomit + 4 Quarz + 1 H_2O ↔ 1 Talk + 3 Calcit + 3 CO_2 fur die Gesamtgasdrucke von 1000, 3000 und 5000 Bar. Contrib. Mineral. Petrol. 26, 302-314.

Metz, P. and Puhan, D. (1971) Korrektur zur Arbeit "Experimentelle Untersuchung der Metamorphose von kieselig dolomitischen Sedimenten I." Contrib. Mineral. Petrol. 31, 169-170.

Miller, C.F., Stoddard, E.F., Bradfish, L.J., and Dollase, W.A. (1981) Composition of plutonic muscovite: genetic implications. Canadian Mineral. 19, 25-34.

Modreski, P.J. and Boettcher, A.L. (1972) The stability of phlogopite and enstatite at high pressures: a model for micas in the interior of the Earth. Am. J. Sci. 272, 852-869.

Modreski, P.J. and Boettcher, A.L. (1973) Phase relationships of phlogopite in the system K_2O - MgO - CaO - Al_2O_3 - SiO_2 - H_2O to 35 kilobars: a better model for micas in the interior of the Earth. Am. J. Sci. 273, 385-414.

Montoya, J.W. and Hemley, J.J. (1975) Activity relations and stabilities in alkali feldspar and mica alteration reactions. Econ. Geol. 70, 577-583.

Mueller, R.F. (1972) Stability of biotite: a discussion. Am. Mineral. 57, 300-316.

Munoz, J.L. (1971) Hydrothermal stability relations of synthetic lepidolite. Am. Mineral. 56, 2069-2087.

Munoz, J.L. and Ludington, S.D. (1974) Fluoride-hydroxyl exchange in biotite. Am. J. Sci. 274, 396-413.

Munoz, J.L. and Ludington, S. (1977) Fluorine-hydroxyl exchange in synthetic muscovite and its application to muscovite-biotite assemblages. Am. Mineral. 62, 304-308.

Naney, M.T. (1983) Phase equilibria of rock-forming ferromagnesian silicates in granitic systems. Am. J. Sci. 283, 993-1033.

Olesch, M. (1975) Synthesis and solid solubility of trioctahedral brittle micas in the system CaO - MgO - Al_2O_3 - SiO_2 - H_2O. Am. Mineral. 60, 188-199.

Olesch, M. and Seifert, F. (1976) Stability and phase relations of trioctahedral calcium brittle micas (clintonite group). J. Petrol. 17, 291-314.

Olesch, M. and Seifert, F. (1981) The restricted stability of osumilite under hydrous conditions in the system K_2O - MgO - Al_2O_3 - SiO_2 - H_2O. Contrib. Mineral. Petrol. 76, 362-367.

Partin, E. (1984) Ferric/Ferrous Determinations in Synthetic Biotite. M.S. thesis, Virginia Polytechnic Institute and State University, Blacksburg.

Partin, E., Hewitt, D.A., and Wones, D.R. (1983) Quantification of ferric iron in biotite. Geol. Soc. Am. Abstr. with Programs 15, 659.

Perkins, D., III, Westrum, E.F., Jr., and Wall, V.J. (1979) New thermodynamic data for diaspore and their application to the system Al_2O_3-SiO_2-H_2O. Am. Mineral. 64, 1080-1090.

Perrotta, A.J. and Garland, T.J. (1975) Low temperature synthesis of zinc-phlogopite. Am. Mineral. 60, 152-154.

Piwinskii, A.J. (1968) Experimental studies of igneous rock series: Central Sierra Nevada batholith, California. J. Geol. 76, 548-570.

Poty, B., Holland, H.D., and Borcsik, M. (1972) Solution-mineral equilibria in the system MgO - SiO_2 - H_2O - $MgCl_2$ at 500°C and 1 kbar. Geochim. Cosmochim. Acta 36, 1101-1113.

Puhan, D. (1978) Experimental study of the reaction: dolomite + K-feldspar + H_2O ↔ phlogopite + calcite + CO_2 at the total gas pressures of 4000 and 6000 bars. N. Jahrb. Mineral. Monatsh. 1978, 110-127.

Puhan, D. and Johannes, W. (1974) Experimentelle Untersuchung der Reaktion Dolomit + Kalifeldspat + H_2O ↔ Phlogopit + Calcit + CO_2. Contrib. Mineral. Petrol. 48, 23-31.

Reed, B.L. and Hemley, J.J. (1966) Occurrence of pyrophyllite in the Kekiktuk Conglomerate, Brooks Range, northeastern Alaska. U.S. Geol. Survey Prof. Paper 550-C, C-162-C166.

Richardson, S.W., Gilbert, M.C., and Bell, P.M. (1969) Experimental determination of kyanite-andalusite and andalusite-sillimanite equilibria: the aluminum silicate triple point. Am. J. Sci. 267, 259-272.

Rieder, M. (1971) Stability and physical properties of synthetic lithium-iron micas. Am. Mineral. 56, 256-280.

Robert, J.L. (1976a) Phlogopite solid solutions in the system K_2O - MgO - Al_2O_3 - SiO_2 - H_2O. Chem. Geol. 17, 195-212.

Robert, J.L. (1976b) Titanium solubility in synthetic phlogopite solid solutions. Chem. Geol. 17, 213-227.

Robie, R.A., Hemingway, B.S., and Fisher, J.R. (1978) Thermodynamic properties of minerals and related substances at 298.15 K and 1 bar (10^5 pascals) pressure and at higher temperatures. U.S. Geol. Survey Bull. 1452, 456 p.

Rosenberg, P.E. (1974) Pyrophyllite solutions in the system Al_2O_3 - SiO_2 - H_2O. Am. Mineral. 59, 254-260.

Rosenberg, P.E. and Cliff, G. (1980) The formation of pyrophyllite solid solutions. Am. Mineral. 65, 1217-1219.

Ross, M., Takeda, H., and Wones, D.R. (1966) Mica polytypes: systematic description and identification. Science 151, 191-193.

Ryabchikov, I.D. and Green, D.H. (1978) The role of carbon dioxide in the petrogenesis of highly potassic magmas. In: Problems of Petrology of Earth's Crust and Upper Mantle, Trudy Inst. Geol. Geofiz., So An SSR, 403, Nauka, Novosibirsk.

Rutherford, M.J. (1969) An experimental determination of iron biotite - alkali feldspar equilibria. J. Petrol. 10, 381-408.

Rutherford, M.J. (1973) The phase relations of aluminous iron biotites in the system $AlSi_3O_8$ - $KAlSiO_4$ - Al_2O_3 - Fe - O - H. J. Petrol. 14, 159-180.

Sand, L.B., Roy, R., Osborn, E.F. (1957) Stability relations of some minerals in the Na_2O - Al_2O_3 - SiO_2 - H_2O system. Econ. Geol. 52, 169-179.

Scarfe, C.M. and Wyllie, P.J. (1967) Experimental redetermination of the upper stability limit of serpentine to 3 kb. Am. Geophys. Union Trans. 48, 225.

Schneider, G. (1975) Experimental replacement of garnet by biotite. N. Jahrb. Mineral. Monatsh. 1975, 1-10.

Schreyer, W. and Baller, Th. (1977) Talc-muscovite: synthesis of a new high-pressure phyllosilicate assemblage. N. Jahrb. Mineral. Monatsh. 1977, 421-425.

Seifert, F. (1970) Low-temperature compatibility relations of cordierite in haplopelites of the system $K_2O - MgO - Al_2O_3 - SiO_2 - H_2O$. J. Petrol. 11, 73-99.

Seifert, F. and Schreyer, W. (1971) Synthesis and stability of micas in the system $K_2O - MgO - SiO_2 - H_2O$ and their relations to phlogopite. Contrib. Mineral. Petrol. 30, 196-215.

Sekine, T. and Wyllie, P.J. (1982a) Phase relationships in the system $KAlSiO_4 - Mg_2SiO_4 - SiO_2 - H_2O$ as a model for hybridization between hydrous siliceous melts and peridotite. Contrib. Mineral. Petrol. 79, 368-374.

Sekine, T. and Wyllie, P.J. (1982b) Synthetic systems for modeling hybridization between hydrous siliceous magmas and peridotite in subduction zones. J. Geol. 90, 734-741.

Sekine, T. and Wyllie, P.J. (1982c) Phase relationships for mixtures of granite and peridotite, with applications to subduction zone hybridization. Contrib. Mineral. Petrol. 81, 190-202.

Sekine, T. and Wyllie, P.J. (1983) Experimental simulation of mantle hybridization in subduction zones. J. Geol. 91, 511-528.

Shade, J.W. (1974) Hydrolysis reactions in the SiO_2 excess portion of the system $K_2O-Al_2O_3-SiO_2-H_2O$ in chloride fluids at magmatic conditions. Econ. Geol. 69, 218-228.

Shaw, H.R. (1967) Hydrogen osmosis in hydrothermal experiments. In: P.H. Abelson, ed., Researches in Geochemistry, Vol. 2. New York: John Wiley & Sons, pp. 521-541.

Shaw, H.R. and Wones, D.R. (1964) Fugacity coefficients for hydrogen gas between 0°C and 1000°C, for pressures to 3000 atm. Am. J. Sci. 262, 918-929.

Skippen, G.B. (1971) Experimental data for reactions in siliceous marbles. J. Geol. 79, 457-481.

Skippen, G. and Hutcheon, I. (1974) The experimental calibration of continuous reactions in siliceous carbonate rocks. Canadian Mineral. 12, 327-333.

Slaughter, J., Kerrick, D.M. and Wall, V.J. (1975) Experimental and thermodynamic study of equilibria in the system $CaO - MgO - SiO_2 - H_2O - CO_2$. Am. J. Sci. 275, 143-162.

Stormer, J.C. and Carmichael, I.S.E. (1971) Fluorine-hydroxyl exchange in apatite and biotite: a potential igneous geothermometer. Contrib. Mineral. Petrol. 31, 121-131.

Storre, B. and Nitsch, K.-H. (1974) Zur Stabilitat von Margarit im System $CaO-Al_2O_3-SiO_2-H_2O$. Contrib. Mineral. Petrol. 43, 1-24.

Storre, B., Johannes, W., and Nitsch, K.-H. (1982) The stability of zoisite in H_2O-CO_2 mixtures. N. Jahrb. Mineral. Monatsh. 1982, 395-406.

Thompson, A.B. (1970) A note on the kaolinite-pyrophyllite equilibrium. Am. J. Sci. 268, 454-458.

Thompson, A.B. (1982) Dehydration melting of pelitic rocks and the generation of H_2O-undersaturated granitic liquids. Am. J. Sci. 282, 1567-1595.

Thompson, J.B., Jr. (1982) Composition space: an algebraic and geometric approach. In: J.M. Ferry, ed., Reviews in Mineralogy 10, Mineral. Soc. Am., pp. 1-32.

Tso, J.L., Gilbert, M.C., and Craig, J.R. (1979) Sulfidation of synthetic biotites. Am. Mineral. 64, 304-316.

Tu, K. (1956) Preliminary results on hydrothermal synthesis of the brittle micas (in Chinese). Acta Geol. Sinica 36, 229-237.

Tuttle, O.F. and Bowen, N.L. (1958) Origin of granite in light of experimental studies in the system $NaAlSi_3O_8 - KAlSi_3O_8 - SiO_2 - H_2O$. Geol. Soc. Am. Memoir 74, 153.

Veblen, D.R. and Buseck, P.R. (1980) Microstructures and reaction mechanisms in biopyriboles. Am. Mineral. 65, 599-623.

Velde, B. (1965) Phengite micas: synthesis, stability, and natural occurrence. Am. J. Sci. 263, 886-913.

Velde, B. (1966) Upper stability of muscovite. Am. Mineral. 51, 924-929.

Velde, B. (1971) The stability and natural occurrence of margarite. Mineral. Mag. 38, 317-333.

Velde, B. (1972) Celadonite mica: solid solution and stability. Contrib. Mineral. Petrol. 37, 235-247.

Velde, B. and Kornprobst, J. (1969) Stabilite des silicates d'alumine hydrates. Contrib. Mineral. Petrol. 21, 63-74.

Waldbaum, D.R. and Thompson J.B. (1969) Mixing properties of sanidine crystalline solutions: IV. Phase diagrams from equations of state. Am. Mineral. 54, 1274-1298.

Warhus, U. and Chatterjee, N.D. (1984) Ephesite, $Na(LiAl_2)[Al_2Si_2O_{10}](OH)_2$: I. Thermal stability and standard state thermodynamic properties. Contrib. Mineral. Petrol. 85, 74-79.

Wendlandt, R.F. and Eggler, D.H. (1980a) The origins of potassic magmas: 1. Melting relations in the systems $KAlSiO_4$ - Mg_2SiO_4 - SiO_2 and $KAlSiO_4$ - MgO - SiO_2 - CO_2 to 30 kilobars. Am. J. Sci. 280, 385-420.

Wendlandt, R.F. and Eggler, D.H. (1980b) The origin of potassic magmas: 2. Stability of phlogopite in natural spinel lherzolite and in the system $KAlSiO_4$ - MgO - SiO_2 - H_2O - CO_2 at high pressures and high temperatures. Am. J. Sci. 280, 421-458.

Whitney, J.A. (1975) The effect of pressure, temperature and $X(H_2O)$ on phase assemblages in four synthetic rock compositions. J. Geol. 83, 1-31.

Wintsch, R.P., Merino, E., and Blakely, R.F. (1980) Rapid-quench hydrothermal experiments in dilute chloride solutions applied to the muscovite-quartz-sanidine equilibrium. Am. Mineral. 65, 1002-1011.

Wise, W.S. and Eugster, H.P. (1964) Celadonite: synthesis, thermal stability and occurrence. Am. Mineral. 49, 1031-1083.

Wones, D.R. (1963a) Phase equilibria of "ferriannite," $KFe_3 KFe_3{}^2Fe^3Si_3O_{10}(OH)_2$. Am. J. Sci. 261, 581-596.

Wones, D.R. (1963b) Physical properties of synthetic biotites on the join phlogopite-annite. Am. Mineral. 48, 1300-1321.

Wones, D.R. (1967) A low pressure investigation of the stability of phlogopite. Geochim. Cosmochim. Acta 31, 2248-2253.

Wones, D.R. (1972) Stability of biotite: a reply. Am. Mineral. 57, 316-317.

Wones, D.R. and Dodge, F.C.W. (1977) The stability of phlogopite in the presence of quartz and diopside. In: D.G. Fraser, ed., Thermodynamics in Geology. Dordrecht-Holland: D. Reidel Publishing Co., pp. 229-247.

Wones, D.R. and Eugster, H.P. (1965) Stability of biotite: experiment, theory, and application. Am. Mineral. 50, 1228-1272.

Wones, D.R. and Gilbert, M.C. (1982) Amphiboles in the igneous environment. Reviews in Mineralogy 9B, Mineral. Soc. Am., 355-390.

Wones, D.R., Burns, R.G., and Carroll, B.M. (1971) Stability and properties of synthetic annite. Am. Geophys. Union Trans. 52, 369.

Wood, B.J. (1976) The reaction phlogopite + quartz = enstatite + sanidine = H_2O. In: G.M. Biggar, ed., Progress in Experimental Petrology. Natural Environment Res. Council, Publ. Series D.

Wyllie, P.J. and Sekine, T. (1982) The formation of mantle phlogopite in subduction zone hybridization. Contrib. Mineral. Petrol. 79, 375-380.

Wyllie, P.J. and Tuttle, O.F. (1964) Experimental investigation of silicate systems containing two volatile components. Part III. The effects of SO_3, P_2O_5, HCl, and Li_2O, in addition to H_2O, on the melting temperatures of albite and granite. Am. J. Sci. 262, 930-939.

Yamamoto, K. and Akimoto, S.-I. (1977) The system MgO - SiO_2 - H_2O at high pressures and temperatures—Stability field for hydroxyl-chondrodite, hydroxyl-clinohumite and 10 A phase. Am. J. Sci. 277, 288-312.

Yoder, H.S. and Eugster, H.P. (1954) Phlogopite synthesis and stability range. Geochim. Cosmochim. Acta 6, 157-185.

Yoder, H.S. and Eugster, H.P. (1955) Synthetic and natural muscovites. Geochim. Cosmochim. Acta 8, 225-280.

Yoder, H.S. and Kushiro, I. (1969) Melting of a hydrous phase: phlogopite. Am. J. Sci. 267-A, 558-?.

Zen, E-an and Chernosky, J.V., Jr. (1976) Correlated free energy values of anthophyllite, brucite, clinochrysotile, enstatite, forsterite, quartz, and talc. Am. Mineral. 61, 1156-1166.

8. PARAGENESIS, CRYSTALLOCHEMICAL CHARACTERISTICS, and GEOCHEMICAL EVOLUTION of MICAS in GRANITE PEGMATITES
Petr Černý & Donald M. Burt

INTRODUCTION

The mica minerals are the third most abundant constituent of many granitic pegmatites, after feldspars and quartz. Ferromagnesian micas are commonly the only mafic minerals in the most primitive pegmatite types, while the "white" micas are usually the main component responsible for the peraluminous nature of pegmatite bulk compositions. Pegmatitic micas are diverse in their chemical composition, typomorphic properties, and mineral associations, and they contribute several unique species and varieties to the structural and crystallochemical spectrum of the mica group. Individual mica types are characteristic of different stages of pegmatite evolution, and their precipitation, equilibria, and/or reaction with other phases aid in delineating pegmatite crystallization paths through PTX space. Trace elements in micas serve as indicators of the economic potential of their parent pegmatites or individual internal pegmatite units. Last but not least, the mica minerals themselves have a variety of industrial uses, ranging from optical, dielectric, and construction applications to the extraction of Li, Rb, Cs and Ga.

The aim of this chapter is to provide a general review of the mica species and their paragenetic position in granitic pegmatites, their principal crystallochemical features and polytypic relationships, and their geochemical evolution in different classes and types of pegmatites. Rather than to fragment the above information by quoting numerous specific examples, we attempted to summarize the available data in a generalizing manner. This goal is made difficult by the fact that pegmatitic micas have not been examined in as detailed and systematic a fashion as the micas of plutonic and metamorphic rocks. We have tried to avoid overgeneralizations, and to stress gaps in our knowledge wherever appropriate.

This chapter is not an exhaustive review of *all* references on pegmatite micas; however, it is based on a selection of the more important papers which represent our current understanding of the subject. The extreme variety of topics to be covered would render further accumulation of references unmanageable. The chapter is expected to be an improvement on the recent reviews by Hawthorne and Černý (1982), mainly because of updated coverage of the literature and incorporation of some topics that were intentionally omitted from the 1982 publication.

SOME DEFINITIONS AND OBJECTIVES

A brief review of specialized subjects, terms, and approaches is presented here to facilitate non-specialist's understanding of "affairs pegmatologic".

Classification of granitic pegmatites

The diversity of pegmatite varieties within the limits of generally granitic bulk compositions has been subject to numerous (and occasionally confusing) attempts at classification (as reviewed by Jahns, 1955, and Černý, 1982a). The most recent proposals of Rudenko et al. (1975) and Ginsburg et al. (1979) seem to be the most universal in their applicability. They have the advantage of separating pegmatite classes by their gross geological environments first, followed by the paragenetic and geochemical pigeon-holing of individual types afterwards. The four principal categories of granitic pegmatites, as modified in Černý (1982a, 1984) are the

(1) abyssal class: typical of anatectic zones in kyanite- and sillimanite-bearing upper amphibolite to granulite facies of metamorphism; near-autochthonous; derived by partial melting of enclosing high-grade metamorphic rocks at 5-8 kbar; locally enriched in U, Th, REE, Nb, Ti, and Zr;

(2) muscovite class: characteristic of Barrovian high-pressure metamorphic facies series, and hosted by schists of the kyanite + almandine subfacies of Winkler's (1967) almandine-amphibolite facies (mainly above the staurolite-out isograd and below the first sillimanite isograd); originate either by anatexis or by restricted fractionation of primitive, more or less autochthonous granites at 4-6 kbar, and consolidate close to the loci of magna generation; muscovite and feldspar deposits occasionally carry subordinate Be, Nb, REE, U, and Th;

(3) rare-element class: occur in Abukuma-type low-pressure metamorphic facies series; generated by fractionation of allochthonous differentiated granites, and consolidated at 2-4 kbar, mainly in rocks of the andalusite + cordierite + muscovite subfacies of Winkler's (1967) cordierite-amphibolite facies (mostly above the cordierite and staurolite isograds and below the first sillimanite isograd); one or more of Li, Rb, Cs, Be, Sn, Nb, Ta, Zr, Hf, Ga, Bi, and Mo, with or without significant quantities of B, P, and/or F, are typically enriched in this class which can be subdivided into several types:

(a) gadolinite type (Be, Y, REE, Nb $\gtrless$ Ta, Ti, U, Th),

(b) beryl-columbite type (Be, Nb $\gtrless$ Ta; ± Sn, REE, P, B),

(c) zoned to layered complex type (Li, Rb, Cs, Be, Ta>Nb, Sn; ± P, B),

(d) quasi-homogeneous spodumene type (Li, Be, Nb⩾Ta; ± Sn), and

(e) lepidolite type (F, Li, Rb, Be, Ta>Nb; ± Cs, Sn).

More detailed information on the characteristic features of the above pegmatite classes is available in Černý (1982c; 1984).

(4) miarolitic class: confined to cupolas of allochthonous, epizonal to subvolcanic, occasionally hypersolvus granites and their close vicinity, and consolidated at 1-2 kbar; fracture-filling veins or pods with the same pattern of trace-element enrichment as the rare-element class.

The four geological classes as characterized above refer to the pegmatites associated with late- to post-tectonic granitoid magmatism or orogenic environments. However, the third and fourth class also have subalkalic counterparts generated by anorogenic, mostly rift-related granites of the bimodal gabbro + granite suites. These pegmatites, while similar to the orogenic types in terms of depths of emplacement, are frequently amazonite-bearing and have a strikingly different geochemical signature (Y, REE, Nb, Ta, Fe, F, and much lower levels of Li, Rb, Cs, B, P, Be, Sn, and Ga). In the following text these two major families of pegmatites, differing in geotectonic affiliation and petrogenesis plus geochemistry of their granitic parents, will be referred to as "orogenic" and "anorogenic".

Internal structure of granitic pegmatites

In the following text, the paragenetic types of micas are related to the internal units of granitic pegmatites. The nomenclature used here is as standardized in North-American literature (Cameron et al., 1949; Jahns, 1955; Černý, 1982): zones of primary crystallization (border, wall, intermediate, core-margin, core), replacement bodies (albitization, greisen, lepidolite), and fracture-filling units. The cited references provide further information on this classification.

Mica species in granitic pegmatites

The following end-member compositions apply to the terms used in this chapter, as given in Hawthorne and Černý (1982) and Bailey (1984, this volume):

Dioctahedral micas:

muscovite $KAl_2(Si_3Al)O_{10}(OH,F)_2$
paragonite $NaAl_2(Si_3Al)O_{10}(OH,F)_2$
phengite $K[Al_{1.5}(Mg,Fe^{2+})_{0.5}](Si_{3.5}Al_{0.5})O_{10}(OH,F)_2$
celadonite $K(Mg,Fe^{2+})(Fe^{3+},Al)Si_4O_{10}(OH)_2$

Trioctahedral micas:

phlogopite $KMg_3(Si_3Al)O_{10}(OH,F)_2$
biotite $K(Mg_{0.6-1.8}Fe_{2.4-1.2})(Si_3Al)O_{10}(OH,F)_2$
annite $KFe_3^{2+}(Si_3Al)O_{10}(OH,F)_2$
siderophyllite $K(Fe_2^{2+}Al)(Si_2Al_2)O_{10}(OH,F)_2$
"lepidomelane" $K(Fe_2^{2+}Fe^{3+})(Si_2Al_2)O_{10}(OH,F)_2$
polylithionite $K(Li_2Al)Si_4O_{10}(F,OH)_2$
trilithionite $K(Li_{1.5}Al_{1.5})(Si_3Al)O_{10}(F,OH)_2$
zinnwaldite $K[Fe_{1.5-0.5}^{2+}Li_{0.5-1.5}(Al,Fe^{3+})](Si_{3.5-2.5}Al_{0.5-1.5})O_{10}(OH,F)_2$
masutomilite $K(Mn_{1.0-0.5}Li_{1.0-1.5}Al)(Si_{3.5-3.0}Al_{0.5-1})O_{10}(OH,F)_2$

The term lithian muscovite is used for Li-enriched muscovite with the $2M_1$ structure (as in Foster, 1960a, although its distinction from $2M_1$ lepidolite is not as clear as it used to be). Magnesian biotite and ferroan phlogopite are applied as in Foster (1960b). Taeniolite, $K(Mg_2Li)Si_4O_{10}(F,OH)_2$, and end-member polylithionite compositions are not considered in this chapter because both occur only in alkalic pegmatites of nepheline syenite (and related) parentage (Miser and Stevens, 1938; Semenov, 1959, and Gerasimovskyi, 1965, for taeniolite; Stevens, 1938; Shilin, 1953; Perrault, 1966, and Raade and Larsen, 1980, for polylithionite).

Graphic representation

In order to facilitate comparison with the existing literature, most of the graphic representations of bulk compositions and related features are presented in the classic (but far from perfect) diagrams of Foster (1960a,b). However, one of the central sections presents a new graphical representation designed for lithium-bearing micas, including their lithium-free end-member components (Burt and Burton, 1984). This system has the capacity to encompass all important compositional variables in the 3-dimensional space of a single polyhedron, and it is expected to find extensive use in the future.

Limitations of scope

The abyssal pegmatites are relatively poorly characterized, the interpretation of some of their occurrences is controversial, and nothing specific is known about their micas (which should be close to high-grade metamorphic and migmatite micas in any case). Thus the micas of the abyssal class are not considered here.

The discussion is focussed at the micas of the muscovite and rare-element classes of the orogenic series. The micas of the miarolitic pegmatites are treated jointly with those of the rare-element category, because of the general paragenetic and geochemical similarity of these classes. So far, no indication of systematic differences has been noticed between their mica assemblages and varieties.

The understanding of micas of the anorogenic pegmatites is rather limited; nevertheless, the dramatic differences between them and the micas of orogenic parentage can be adequately illustrated.

MICA ASSEMBLAGES IN OROGENIC PEGMATITES

Muscovite pegmatites

Biotite and muscovite are the only mica species encountered in this class. However, they occur in most zones and replacement units in several morphological varieties and generations. In texturally well-differentiated pegmatites of the southeastern Piedmont, Jahns et al. (1952) distinguish (1) border zone biotite (+ muscovite) in quartz ± plagioclase; (2) wall zone muscovite (+ biotite) in quartz + K-feldspar of granitoid to blocky texture; (3) core-margin muscovite + biotite, typically along the boundary of blocky K-feldspar and quartz core, or in coarse quartz + K-feldspar + mica aggregates; and (4) fracture-filling and replacement muscovite associated with quartz and albite.

Gordiyenko and Leonova (1976) classified the micas of the muscovite-class pegmatites of Northern Karelia into (1) lath-shaped biotite of outer granitoid zones; (2) book-type biotite and pseudographic quartz + muscovite intergrowths of the intermediate and blocky zones; (3) platy biotite and euhedral "ruby" muscovite along core-margins and within quartz cores; (4) muscovite associated with metasomatic albite, cutting across and replacing mineral assemblages of all zones (including quartz core); and (5) "gilbertite", a fine-grained, late muscovite, dispersed in all zones but replacing mainly K-feldspar.

Exomorphic muscovite is sparse to abundant in the wallrock adjacent to pegmatite contacts (Jahns et al., 1952; Shmakin, 1975). Fracture-controlled growth of micas in the pegmatites locally extends into the country rocks (Sokolov, 1959; Milovskyi, 1962), and the distribution and attitude of micas inside the pegmatites is commonly controlled by prominent tectonic features of the adjacent country rock (Jahns et al., 1952).

Rare-element (and miarolitic) pegmatites

Different types of rare-element pegmatites exhibit contrasting assemblages of mica species and distributions. Within most of these pegmatites, the kind and number of mica generations are too variable to systematize, even for single pegmatite fields or groups. Thus the micas are reviewed only in the most general manner, and specific cases can be found in the appended references.

Gadolinite type. Biotite is predominant, occurring with increasing crystal size from the border zone into the intermediate zones, and commonly also into the core. Muscovite is confined to the intermediate and core-margin zones (Nordenskjöld, 1906; Brotzen, 1959; Beus, 1960).

Beryl-columbite type. Biotite is a common but not abundant constituent of border zones; it only rarely occurs farther inside the pegmatites of this type. Muscovite may be abundant in the wall, intermediate, and particularly core-margin zones, and late generations accompany albitization and greisenization (Beus, 1960; Cameron and Shainin, 1965; Černý et al., 1981).

Complex type. Biotite, if present at all, is restricted to the border zone. Muscovite ± lithian muscovite or lepidolite are typical of the intermediate zones, and they are also associated with albitization and greisen-like replacements. Lithian muscovite and lepidolite are also found as late masses in near-central parts of this pegmatite type (Quensel, 1956; Beus, 1960; Norton et al., 1962; Cooper, 1964; Rinaldi et al., 1972; Heinrich, 1978; Wang et al., 1981).

Replacements are commonly observed among different mica generations and compositions, such as muscovite metasomatic after biotite, and lepidolite replacing muscovite (e.g., Ginsburg and Berkhin, 1953). However, these replacements may be locally difficult to distinguish from products of epitaxic coprecipitation. Zoned overgrowths of lithian micas on muscovite are widespread, particularly in the miarolitic cavities (Brock, 1974; Foord, 1976; Lahti, 1981).

Zinnwaldite (Babu, 1969) and masutomilite (Harada et al., 1976) are rare, and mostly restricted to specific Fe, Mn-enriched assemblages (associated with, e.g., triphylite, spessartine: Němec, 1983a,b; Ferreira, 1984). However, a late pink (to brownish) variety of muscovite is typical of complex pegmatites, known as rose muscovite (Heinrich and Levinson, 1953; Lahti, 1981).

Spodumene type. Sparse biotite is confined to the border zone in contact with country rocks. Muscovite is the only species of any significance here, commonly occurring in rather low quantities (Sundelius, 1963; Pye, 1965; Rijks and v.d. Veen, 1972; Kunasz, 1982). Gordiyenko (1970) distinguished three paragenetic types of muscovite in the spodumene pegmatites of "one of the pegmatite fields in the European part of the USSR": (1) platy muscovite dispersed through the medium-grained quartz + spodumene + albite assemblages; (2) coarse platy muscovite in local pods of blocky quartz + spodumene + K-feldspar and segregations of cleavelandite; and (3) fine-grained to sericitic muscovite associated with saccharoidal albite, locally recrystallized.

Lepidolite type. Biotite is absent, except for sparse flakes along the contacts with mafic wallrocks. Muscovite is subordinate, whereas lithian muscovite and lepidolite are predominant. The distribution and sequence of mica crystallization may be confusing even in a single pegmatite (Heinrich, 1967), and the number of paragenetic, morphological and color varieties of micas in different pegmatites of a field is highly variable (Vladykin et al., 1974).

Late muscovite alteration. "Gilbertite" and sericite commonly replace aluminum-bearing primary pegmatite minerals that have remained in contact with late, low temperature fluids. These minerals include feldspars, the lithium aluminosilicates (cf. London and Burt, 1982a,b), lithium aluminophosphates, aluminosilicates, topaz, garnet, cordierite, gahnite, tourmaline, beryl, and others.

Muscovite is associated with other alteration products as dictated by the composition of the primary phase. For example, chlorite and/or biotite accompany muscovite in the breakdown of cordierite (cf. Layman, 1963 and Povondra et al., 1984, for more exotic assemblages). A not uncommon byproduct of muscovite replacement of andalusite is corundum (Rose, 1957; Burt and Stump, 1984).

Exomorphic micas. These are commonly developed around the rare-element pegmatites as metasomatic aureoles. They are represented by muscovite and/or biotite *sensu lato* in gneissic or metapelitic wallrocks (e.g., Kretz, 1968; Neiva, 1980; Černý et al., 1981). However, this mica type exhibits a broader compositional range when replacing wallrocks of contrasting compositions. Phlogopite is common in contacts with peridotites, pyroxenites, and serpentinites (Hadley, 1949; Borodayevskaya, 1951; Kulp and Brobst, 1954; Morel, 1955; Omori, 1958; Bassett, 1959; Bogomolova, 1962). It also forms among pegma-- tites emplaced into marbles (Lacroix, 1898; v. Eckermann, 1922, 1923; Gevers,

1948; Rossovskyi, 1963; Černý and Miškovský, 1966; Černý, 1972). In both ultrabasic and carbonate environments, the phlogopite is occasionally associated with endocontact margarite, replacing plagioclase (Vlasov and Kutukova, 1959; Beus, 1960).

Special types of exomorphic micas with substantial contents of Li, Rb, and Cs occur mainly in metabasaltic wallrocks of lithium-bearing pegmatites (Hess and Stevens, 1937; Ginsburg et al., 1972; Zagorskyi et al., 1974; Černý et al., 1981). They are predominantly ferro-magnesian, variable in composition between magnesian biotite, ferroan phlogopite, and zinnwaldite, and commonly associated with tourmaline, holmquistite, phosphates, and sulphides. However, muscovite enriched in Fe, Mg, Mn, Li, Rb, and Ba is also encountered in this assemblage (Lahti, 1981).

Similar micas are also known to form extensive metasomatic zones in basic country rocks, only remotely related to spodumene-bearing pegmatites (Gordiyenko et al., 1975). These zones, related to fault and shear lineaments possibly intercepting the nearby pegmatites at depth, exhibit symmetric zoning of several assemblages involing chlorite, holmquistite, plagioclase, hornblende, and epidote.

MICA ASSEMBLAGES IN ANOROGENIC PEGMATITES

This pegmatite lineage is much less widespread than the orogenic sequences, and it has also been much less examined. Consequently, the mica-bearing assemblages remain poorly understood.

Rare-element pegmatites

Biotite is a subordinate component of the South Platte pegmatites, Colorado, restricted to lath-shaped crystals in the graphic wall zones and gigantic books in the blocky outer intermediate zones. In contrast, primary muscovite is absent; minor sericitic micas were generated along with rare-element minerals during albitization and replacement of topaz (Simmons and Heinrich, 1980).

Miarolitic pegmatites

Biotite and zinnwaldite are the typical micas of the Crystal Peak pegmatites, Colorado, with only minor muscovite in the central parts and vugs (Foord and Martin, 1979). Except for epitaxic overgrowths of ferroan muscovite on zinnwaldite from open vugs, primary muscovite may be totally absent from this pegmatite type, as shown by the miarolitic pegmatites of the Korosten pluton, Volynia (Lazarenko et al., 1973), biotite is the typical early mica here, grading into zinnwaldite and ultimately lepidolite in the

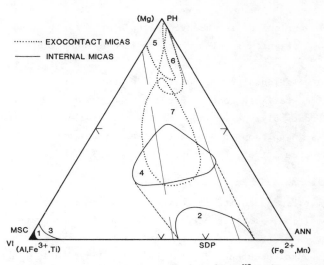

Figure 1. Composition of some pegmatite micas in the $(Al+Fe^{3+}+Ti)^{VI}-(Fe^{2+},Mn)-Mg$ triangle of Foster (1960b). The thin lines bordering a compositional field between phlogopite (PH) and siderophyllite (SDP) indicate the ranges of phlogopite, ferroan phlogopite to magnesian biotite, and biotite after Foster (1960b). 1 - muscovite and 2 - biotite from rare-element pegmatites; 3 - muscovite and 4 - biotite from the muscovite-class pegmatites; 5 - exocontact phlogopite from marbles; and 6 - ultramafic wallrocks; 7 - rare-alkali ferromagnesian micas from pegmatite exocontacts in amphibolites. Based on data by Hess and Stevens (1937); Foster (1960b); Bogomolova (1962); Shurkin et al. (1962); Černý and Miskovsky (1966); Ginsburg et al. (1972); Černý (1972); Zagorskyi et al. (1973); Gordiyenko and Leonova (1976); and unpublished data by P. Černý.

central parts of the pegmatite bodies. Sericitic micas form veinlets in quartz and replace K-feldspar, topaz, and beryl.

CRYSTALLOCHEMICAL CHARACTERISTICS AND POLYTYPISM

With few exceptions, the crystal chemistry and structural relationships of pegmatitic micas are best discussed in terms of species or substitutional series, rather than by the paragenetic assemblages. However, specific references will be made to the provenance of mica types where necessary.

Phlogopite - biotite

In terms of octahedral cation population, internal ferromagnesian micas fall into the general field established by Foster (1960b), straddling the phlogopite - siderophyllite join (Fig.1). The only exceptions are the ferroan phlogopites and magnesian biotites from the muscovite class of pegmatites that are, in part, distinctly more aluminous and less than trioctahedral (Shurkin et al., 1962). This mica type also is the most magnesian (Fig. 1); biotites from pegmatites of the rare-element class are considerably to extremely Mg-depleted (Heinrich, 1946; Heinrich et al., 1953; Foster, 1960b). Border-zone biotite is, however, commonly contaminated by the reaction with wallrocks and it may be considerably enriched in Mg (e.g. Gordiyenko, 1970).

Extreme Fe-enrichment is observed particularly in biotite of the anorogenic pegmatites, some of which is also characterized by substantial Fe^{3+} (Lazarenko et al., 1973).

Polytypic relationships are poorly understood. The available data are so scarce that no reliable conclusions can be drawn about the relative abundance of different polytypes. It seems likely, however, that the 1M polytype is dominant, as in phlogopite - biotite micas in general (Heinrich et al., 1953; Ross and Wones, 1965; Lazarenko et al., 1973).

Muscovite - lithian muscovite - lepidolite

Most pegmatitic muscovite is very close to the ideal dioctahedral and R^{2+}-free composition. This is particularly true for muscovite from the rare-element pegmatite class. However, muscovite-class micas show a slight increase in octahedral occupancy due to (Fe^{2+},Mg) substitution, and limited "phengite"-type substitution is common in muscovite of both classes (Fig. 1, 2; Salye, 1975; Lopes Nunes, 1973).

Paragonite component is generally low, reaching only exceptionally 30 mol% (and mostly only 8-13 mol%) in muscovite of the muscovite pegmatite class (Shurkin et al., 1962; Gordiyenko and Leonova, 1976). Muscovites of the rare-element pegmatites have perceptibly lower paragonite contents (3-20 but predominantly only 6-10 mol%), and the sodic component is further reduced in lithian muscovite and lepidolite (typically 3-7 mol%; Quensel, 1956; Gordiyenko, 1970; Rinaldi et al., 1972; Chaudhry and Howie, 1973; Lopes Nunes, 1973; Vladykin et al., 1974).

In the octahedral population triangle (Al, Fe^{3+}, Ti) - (Fe^{2+}, Mg, Mn) - Li of Foster (1960a), the discussed micas plot continuously from the muscovite corner to about halfway between trilithionite and polylithionite (Fig. 3a). A more complex but also more revealing distribution is seen in the muscovite - trilithionite - polylithionite diagram of Rieder (Fig. 4a; pers. comm., 1969), similar to that used by Munoz (1968, 1971, who did not provide information on the method of calculating the end-member precentages). A sizeable compositional gap appears along the muscovite - trilithionite sideline, and a vacant region adjoins the polylithoinite corner. Whereas the muscovite - trilithionite gap is of general significance for the mica group at large, the absence of poly- lithionite compositions holds only for granitic pegmatites. The polylithionite corner is otherwise occupied by lithium micas from peralkaline parageneses.

It is interesting to note that the extent of R^{2+} substitution generally increases with Li (Fig. 3a). Ferroan lepidolites seem to be realtively common, grading into marginal zinnwaldites (Fig. 3b).

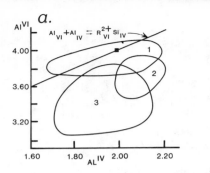
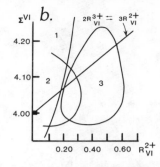

Figure 2. Octahedral-tetrahedral substitutions in muscovite from rare-element (1), rare-element-to-muscovite (2), and muscovite (3) pegmatite classes. Phengite substitution (a) and the di-trioctahedral trend (b). Modified after Salye (1975).

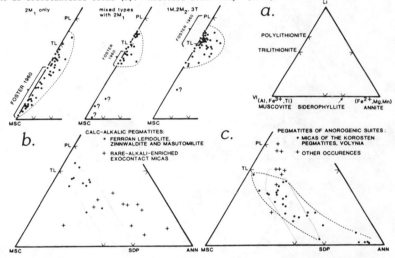

Figure 3. Compositions of muscovite-lepidolite micas (a) and zinnwaldite - biotite micas from orogenic (b) and anorogenic (c) environments in the $(Al+Fe^{3+}+Ti)^{VI}-(Fe^{2+},Mg,Mn)$-Li triangle of Foster (1960a). Figure 3a shows the compositional fields of $2M_1$, mixed types with $2M_1$, and $1M$, $2M_2$, $3T$ polytypes, compared to the muscovite - polylithionite ranges of the same from Foster (1960a). Dotted fields in (b) and (c) show the extent of zinnwaldite field after Foster (1960a). Based on data by Glass (1935); Hess and Stevens (1937); Berggren (1941); Quensel (1956); Foster (1960a); Foster and Evans (1962); Franzini and Sartori (1969); Babu (1969); Černý et al. (1970); Rinaldi et al. (1972); Lopes Nunes (1973); Chaudhry and Howie (1973a); Sartori et al. (1973); Lazarenko et al. (1973); Vladykin et al. (1974); Zagorskyi et al. (1974); Harada et al. (1976); Sartori (1976); Levillain et al. (1977); Foord (pers. comm., 1981); Guggenheim (1981); Nemec (1983a); and Ferreira (1984).

Lithian muscovites and lepidolites show a very good 1:1 correlation of Li and F (at. cont.) documented by Foster (1960a) and confirmed by later analyses (Levillain, 1980).

With accumulation of new data during the past 15 years, polytypic relationships in the discussed mica phases have become much more complex than the relatively simple (but admittedly incomplete) patterns presented by Levinson (1953), Foster (1960a), Tröger (1962), and Munoz (1968, 1971). First of all, the earlier discoveries of the natural $2M_1$ (Ukai et al., 1956) and $3M_2$ (Heinrich et al., 1953; Bailey and Christie, 1978) polytypes were acknowledged.

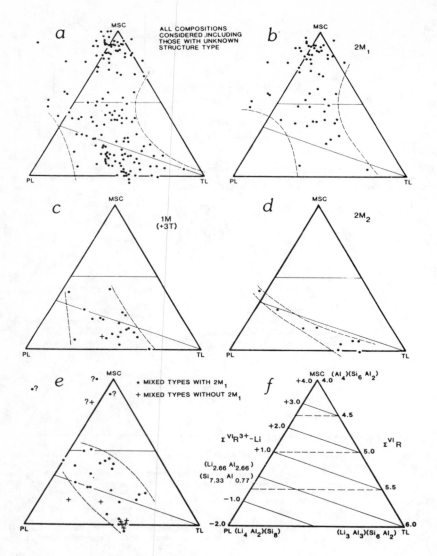

Figure 4. Compositions and polytypism of muscovite-lithian muscovite-lepidolite in Rieder's (1969, pers. comm.) muscovite-trilithionite-polylithionite diagram (see F for tetrahedral-octahedral compositions and method of plotting). The polylithionite corner is populated by 1M micas from peralkaline pegmatites that are not considered in this chapter. Note the compositional gap along the muscovite-trilithionite side of the diagram in (a) and the largely overlapping fields of individual polytypes. Based on data by Berggren (1941); Quensel (1956); Foster (1960a); Franzini and Sartori (1969); Černý et al. (1970); Rinaldi et al. (1972); Chaudhry and Howie (1973a); Lopes Nunes (1973); Sartori et al. (1973); Vladykin et al. (1974); Levillain et al. (1977); and Guggenheim (1981).

Secondly, the compositional ranges of individual polytypes and mixed types expanded, rather dramatically in some cases. This is shown in Figure 3a in direct comparison with Foster's (1960a) ranges, and in Figure 4b to 4e on Rieder's muscovite – trilithionite – pollithionite diagram.

The compositional ranges of the $3T$ and $3M_2$ polytypes are still obscure because of their general rarity (Stevens, 1938; Yakovleva et al., 1965; Vladykin et al., 1974; Brown, 1978 for $3T$, and Heinrich et al., 1953; Neumann et al., 1957; Bailey and Christie, 1978 for $3M_2$), and an even greater lack of chemical analyses. However, the $1M$, $2M_1$, and $2M_2$ polytypes are well represented.

It is mainly the compositional expansion of the "muscovite" $2M_1$ structure that changed the relationships so profoundly. $2M_1$ lithian muscovites are evidently transitional into $2M_1$ compositions intermediate between trilithionite and polylithionite. Lepidolites-$2M_1$ are being identified in increasing numbers (Černý et al., 1970; Lopes Nunes, 1973; Vladykin et al., 1974; Bailey and Christie, 1978; Levillain, 1981; Swanson and Bailey, 1981). In the future, Figure 4b will probably become even more densely populated in the polylithionite - trilithionite region, because many of the Li-rich mixed types with $2M_1$ component shown in Figure 4e are reported as $2M_1$-dominated (Lopes Nunes, 1973; Valdykin et al., 1974).

The field of the $1M$ polytype has also expanded, and that of $2M_2$ closely follows the join trilithionite (Li_3Al_3)-"octahedrally deficient trilithionite" ($Li_{2.66}Al_{2.66}$) (Fig. 4d; cf. Rieder et al., 1970). In general, however, the compositional ranges for all individual polytypes show extensive overlaps, and compositional control on polytypism seems to be much less distinctive than considered by earlier investigators (cf. the concluding section on genetic relationships).

Zinnwaldite - masutomilite

With only a couple of exceptions (Babu, 1969; Vladykin et al., 1974), zinnwaldites from orogenic rare-element pegmatites are Li-rich and Fe-poor members of the zinnwaldite series (Fig. 3b). The same holds for the few occurrences of masutomilite (Harada et al., 1976; Němec, 1983b). All of these micas have the $1M$ structure, as do some extreme ferroan lepidolites (Chaudhry and Howie, 1973a; Ferreira, 1984).

In contrast, zinnwaldites of the anorogenic, subalkalic pegmatites are relatively Fe-rich and Li-poor, populating mainly the central parts of the zinnwaldite field in Foster's (1960a) diagram (Fig. 3c). However, Li-enriched compositions also occur (Foster and Evans, 1962; Lopes Nunes, 1973). The $1M$ polytype seems to be characteristic, although the Li-poor "protolithionites" of the Korosten pegmatites favor $3T$ (Lazarenko et al, 1973; Pavlishin et al., 1981).

In zinnwaldites, Li and F are not as well correlated as in the muscovite-lepidolite group, but F is mostly in excess of Li. In the anorogenic pegmatites of the Korosten pluton, Fe-rich micas have F > Li but with increasing Li content the ratio becomes reversed (Lazarenko et al., 1973). Levillain (1980) has shown that zinnwaldite-siderophyllite micas have (OH,F) significantly in excess of the ideal amount, and that the excess recalculated as H_3O^+ equals the deficiency in the interlayer cations.

Exomorphic micas

Micas found along the contacts of orogenic pegmatites with marbles and ultrabasic silicate rocks are invariably Mg-rich phlogopites *sensu stricto* (Fig. 1). Their fields partly overlap, but those from contacts with ultramafic lithologies are distinctly Fe-enriched.

In contrast to this simple and straightforward composition, the Li, Rb, Cs-enriched ferromagnesian micas from contacts of rare-element pegmatites with rocks of gabbroic (metabasaltic, amphibolitic) composition are much more diversified. The Fe/Mg ratio marks them as magnesian biotites to ferroan phlogopites in the (Al, Fe^{3+}, Ti) - (Fe^{2+}, Mn) - Mg diagram (Fig. 1). Most of them are, however, Li-rich enough to warrant plotting in Foster's (1960a) (Al, Fe^{3+}, Ti) - (Fe^{2+}, Mg, Mn) - Li triangle. Here these micas show a wide scatter over the center of the zinnwaldite field and into the region of Li-bearing siderophyllite - annite (Fig. 3b; Hess and Stevens, 1938; Ginsberg et al., 1972; Zagorskyi et al., 1974).

More data are required on this mica type because much information available today is based only on partial analyses (e.g., Černý et al., 1981), bulk compositions suggest new end-members, and polytypic relationships are virtually unknown.

VECTOR REPRESENTATION OF LITHIUM MICA COMPOSITIONS

As evident from the preceding discussion and from Figures 1 to 4, the classic composition diagrams for micas provide the desired information only in a fragmented, and partly incorrect, manner. The crystallochemical accuracy is improved by Rieder's (1970) tetrahedron but the coverage of end-member compositions is limited. This section describes a vector system introduced by Burt and Burton (1984) which offers a unified representation of all the major features of Li-bearing micas, and can be used for plotting distributions of some of their composition-related properties (such as polytypism) as well. Current research by the above authors is aimed at plotting mica compositions from the literature and at looking for correlations with geologic environments.

The concept

Isomorphic substitutions in mineral groups such as the micas have both a direction (or sense) and a magnitude, and thus can be thought of as vector quantities. The analytical and graphical representation of such substitutions is facilitated by the use of exchange operators, or components containing negative quantities of certain elements (that perform the operation of exchange if added to a mineral formula: Burt, 1974, 1976; these are sometimes called "exchange components"). Simple examples of exchange operators (isomorphic substitutions) that are important in micas include KNa_{-1}, $FeMg_{-1}$, and $F(OH)_{-1}$.

If we considered all of such possible exchanges in the mica group, a graphical representation would become impossible. Inasmuch as we are mainly interested in coupled substitutions involving lithium, we can simplify our task considerably by "condensing" down vectors that involve simple substitution, such as those listed above. Once we understand the numerous coupled substitutions in this "condensed" system, it should be relatively simple to separate out Na from K, Mg and Mn from Fe^{2+}, Fe^{3+} from Al, or F from (OH) in natural micas. A similar approach has been demonstrated for the amphiboles by Smith (1959) and Thompson (1979, 1981); the general method is described by Thompson (1982).

Capacity and limitations

Using the exchange vector approach, all of the more than ten end-members of the lithium micas in "condensed" mica composition space can be represented in terms of only three vectors (or exchange operators, or coupled substitutions), starting from a single composition. The choice of vectors and initial composition is somewhat arbitrary (although many of what appear to be independent coupled substitutions turn out to be coplanar vectors in three dimensions). Burt and Burton (1984) chose to use annite as the starting composition and $\square Al_2Fe_{-3}$, $FeSiAl_{-2}$, and $LiAlFe_{-2}$ as the linearly independent exchange vectors. The justification for using Fe^{2+} in place of Mg is twofold - first, the micas in Li-rich pegmatites, granites, and greisens are normally quite Mg-depleted (and Mn-poor as well) and, second, they seem to grow under rather reducing conditions so that Fe^{3+} contents are minor. Thus the major components "lost through condensation" are Na, Mg, Mn, Fe^{3+} and F.

An interesting result is that the accessible lithium mica composition space becomes an irregular polyhedron, rather than the triangle proposed by Foster (1960a). Individual element (and octahedral vacancy) values per

24 (O plus OH) become planes through the resulting polyhedron and are intersecting straight lines on planar sections through it. Inasmuch as exchange operators or vectors intrinsically involve only elements, rather than ions (they normally are chosen to be electrically neutral, however), lines of equal Al, for example, do not depend for their position on whether the Al is octahedral or tetrahedral and lines of equal Fe do not depend on its valence. This feature should facilitate the plotting of mica compositions for which only partial analytical data are available; if complete data are available it should facilitate checking for internal consistency.

Planar subsystems

The basal or *Li-free portion of the condensed mica composition space* (with names for $K-Fe^{2+}-Al-Si$ micas) is shown on Figure 5, with formulas expressed per 24 (O plus OH). A somewhat similar but less complete diagram, without the vector notation, has been presented by Green (1981). The accessible compositions define an irregular polygon, all points of which can be derived from a single point such as the annite composition by application of the exchange vectors drawn to scale in the upper left.

The vertical axis is the exchange vector $FeSiAl_{-2}$ that relates siderophyllite (lower left corner) to annite above it and muscovite to phengite and Fe-Al-celadonite above it. The horizontal axis is the exchange vector $\Box Al_2 Fe_{-3}$ that relates annite to muscovite and montdorite to Fe-Al-celadonite. These two vectors respectively coincide with vertical lines of constant numbers of octahedral vacancies ("$\Box$") and with horizontal lines of constant Si. Lines of constant Fe have a slope of +3 and coincide with the vector $\Box Si_3 Al_{-4}$, whereas lines of constant total Al have a slope of +1 and coincide with the vector $\Box SiFe_{-2}$, as shown on the vector diagram.

Normally, one might assume that a line between annite and montdorite would mark the outer limit of accessible compositions on the polygon, but Seifert and Schreyer (1971) sythesized an Mg mica with 0.5 mole of tetrahedral Mg per 24 (O plus OH); the plotted composition is its Fe analog. (Recall that this is a condensed diagram, so that we are not differentiating between Fe^{2+}, Mn and Mg, Al and Fe^{3+}, K and Na, or F and OH.) Note that the fact that some of the Fe (or Mg) is tetrahedral does not detract from our ability to plot the resulting composition in the same plane. Similarly, Francke et al. (1983) synthesized a Na-mica of the composition shown at the lower right, derived from the siderophyllite composition by the substitution of $\Box Al_2 Fe_{-3}$.

In general, one might define the Li-free composition plane as that resulting from the simultaneous restrictions that, on the basis of 24 (O plus OH),

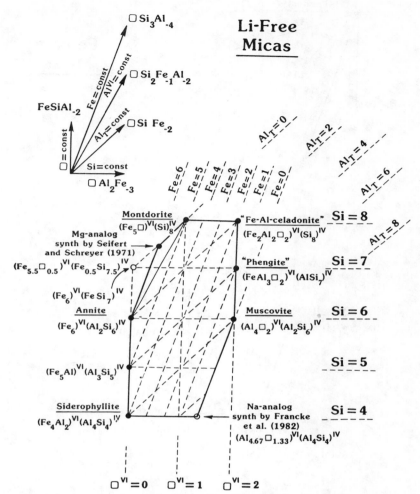

Figure 5. Schematic depiction of the K-Fe-Al-Si mica composition plane (Li-free micas), obtained by "condensation" down the compositional exchange vectors KNa_{-1}, $FeMg_{-1}$, $AlFe_{-1}$, and $F(OH)_{-1}$. The basis vectors, drawn to scale in the upper left, are $FeSiAl_{-2}$ ($\square FeMn_{-1}$ = const.) and $\square Al_2Fe_{-3}$ (Si = const. Other vectors, of constant Fe, total Al, and Al^{IV} are linear combinations of the two basis vectors. All mica compositions shown can be derived from a single composition such as annite by the operation of the exchange vectors. Lines of equal Fe, Al, Si, and $\square$=octahedral vacancies, can be drawn on the composition plane because they are parallel to their respective vectors. See text for further discussion.

Fe lies between 0 and 6, Al between 0 and 8.33, Si between 4 and 8, and vacancies between 0 and 2. These restrictions turn out to be too broad when applied to the K-Fe-Al-Si micas themselves. For example, Fe-Al celadonite is unstable (Velde, 1972), as are, due to geometric restrictions (Hazen and Wones, 1978), end-member annite, and siderophyllite or eastonite (Hewitt and Wones, 1975). In nature such micas are stabilized by numerous additional substitutions, including those involved in the operation of condensation.

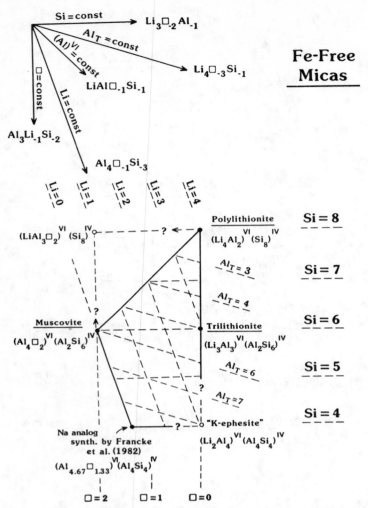

Figure 6. Schematic depiction similar to Figure 5 of the "condensed" K-Li-Al-Si mica composition plane (Fe-free micas). The basis exchange vectors, drawn to scale in the upper left, are $Al_3Li_{-1}Si_2$ ($\square$ = constant) and $Li_3\square_{-2}Al_{-1}$ (Si = constant). The other mica compositions can be derived from any single composition, such as that of muscovite, by the operation of these vectors. The silica-deficient micas along the base, ephesite and the synthetic product produced by Francke et al. (1982), are sodic. The silica-rich tetrasilicic composition to the upper left is only a hypothetical component of natural micas. See text for further discussion.

The "condensed" composition plane for the *Fe-free lithium micas* (Li-Al-Si micas) is shown on Figure 6. Inasmuch as this plane does not include annite, muscovite has been used as a starting composition. The exchange vectors chosen are again those of constant octahedral vacancy level ($Al_3Li_{-1}Si_{-2}$, shown increasing downwards, with vacancies decreasing to the right) and of constant Si ($Li_3\square_{-2}Al_{-1}$, increasing to the right, with Si increasing upwards). Linear combinations of these produce vectors of constant total Al and of constant Li, as shown to scale at the upper left of the diagram.

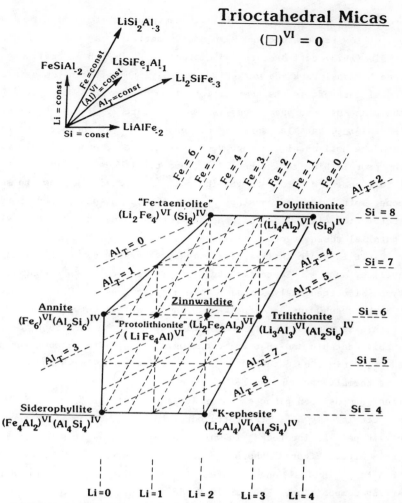

Figure 7. Schematic depiction similar to Figure 5 of the condensed" trioctahedral K-Li-Fe-Al-Si mica composition plane (micas with zero octahedral vacancies). The basis exchange vectors, drawn to scale at the upper left, are $FeSiAl_{-2}$ (Li = constant) and $LiAlFe_{-2}$ (Si = constant). The other mica compositions can be derived from any single composition, such as that of annite, by the operation of these vectors. Ephesite, to the lower right, is a sodium mica, and taeniolite, to the upper left, is a magnesium mica. See text for further discussion.

The result is again an irregular polygon, rather than the triangle muscovite-trilithionite-polylithionite shown by Foster (1960a). Additional points are again the mica synthesized by Francke et al. (1982), "K-ephesite" (remember, this is a condensed diagram that doesn't differentiate Na from K), and possibly a mica of composition $K_2LiAl_3\square_2Si_8(OH,F)_2$ derived from muscovite or polylithionite by the reverse vertical or horizontal exchange vectors, respectively. Such an end-member is probably unstable (it is compositionally equivalent to a mixture of eucryptite, feldspar, quartz and

H$_2$O), but it might be a component of low temperature white mica formed under conditions of low temperature and high silica activity, such as in a hot spring. Bargar et al. (1973) estimated that a lepidolite from such an environment in Yellowstone National Park was very close to polylithionite in composition.

Of equal interest is the composition plane of the *trioctahedral lithium micas* (the plane of zero vacancies), shown on Figure 7. The vertical exchange vector is FeSiAl$_{-2}$, parallel to lines of constant Li (Li increases to the right), and the horizontal exchange vector is LiAlFe$_{-2}$, progressively relating annite to "protolithionite", zinnwaldite, and trilithionite. This vector is parallel to lines of constant Si (Si increases upwards). As shown to scale at the upper left, lines of constant Fe have a slope of plus 2 and are parallel to the vector LiSi$_2$Al$_{-3}$; lines of constant total Al have a slope of plus 1/2 and are parallel to the vector Li$_2$SiFe$_{-3}$.

The result is again an irregular polygon (a pentagon), with vertices at siderophyllite, annite, "Fe-taeniolite", polylithionite, and "K-ephesite". This polygon simulataneously satisfies the restrictions that Li lie between 0 and 4, Fe between 0 and 6, Al between 0 and 8, and Si between 4 and 8. The fact that this is a condensed representation excuses the fact that "K-ephesite" is presumably unstable and that an Fe analog of taeniolite has not yet been described. (It might be expected in a peralkaline rock, in which most Fe, however, is normally present as Fe^{3+}.)

A final composition plane of interest is that of the *"tetrasilicic" micas*, in which there is only Si in the tetrahedral position. Such micas tend to occur only in peralkaline rocks, in which there is very little unattached Al to fill this site. Figure 8 shows the results of letting the two exchange vectors LiAlFe$_{-2}$ (vertical) and □Al$_2$Fe$_{-3}$ (horizontal) operate on montdorite as a starting composition. The result is an irregular polygon (a quadrilateral) for which the vertices are montdorite, Fe-Al celadonite, polylithionite, and Fe-taeniolite. As mentioned in the discussion related to Figure 6, there may be some extension towards a hypothetical mica of octahedral occupany (LiAl$_3$□$_2$). This plane satisfies the restrictions that Li be between 0 and 4, Fe between 0 and 5, Al between 0 and 2 (or possibly 3), vacancies between 0 and 2, and Si remain constant at 8.

The 3-dimensional polyhedron

Other composition planes could be drawn, but those depicted in Figures 5 through 8 are probably sufficient to assist in understanding the condensed mica composition space, depicted in Figure 9. The three exchange vectors are as depicted to the upper left. The result is an irregular polygon. The

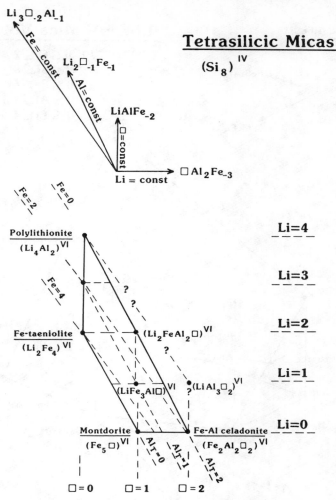

Figure 8. Schematic depiction similar to Figure 5 of the "condensed" tetrasilicic K-Li-Fe-Al-Si mica composition plane (micas with 8 Si atoms per 24 O + OH). The basis exchange vectors, drawn to scale at the upper left, are LiAlFe$_{-2}$ ($\square$ = constant) and $\square$Al$_2$Fe$_{-3}$ (Li = constant). The other mica compositions can be derived from any single composition, such as that of montdorite, by the operation of these vectors. Taeniolite, to the left, is a magnesium-mica and celadonite, to the lower right, is a magnesium-ferric iron-mica. The composition immediately above celadonite is only a hypothetical component of natural micas. See text for further discussion.

Li-free plane of Figure 5 is the base, the Fe-free plane of Figure 6 is the oblique front face, the vacancy-free (trioctahedral) plane of Figure 7 is the back face on the left, and the tetrasilicic plane of Figure 8 is the back face on the right. Values of Li increase upwards, octahedral vacancies increase from the rear to the front, and Si increases from the left front to the right rear, as shown.

The model presented assumes that all of the substitutions are occurring in the tetrahedral and octahedral sheets of the mica structure - that is,

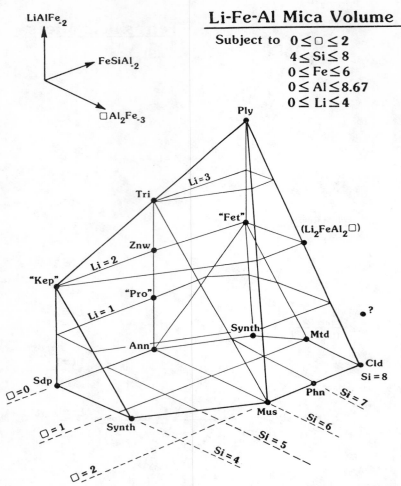

Figure 9. Schematic depiction of the "condensed" K-Li-Fe-Al-Si mica composition polyhedron, derived from any single mica composition, such as that of annite, by the operation of the three exchange vectors LiAlFe$_{-2}$, FeSiAl$_{-2}$, and □Al$_2$Fe$_{-3}$. These vectors, drawn to scale in the upper left, determine planes of constant Li (horizontal), Si (vertical, trending to the right), and octahedral vacancies (vertical, trending to the left). Planes of constant Al and Fe (not shown, except in the sections in Fig. 5-8) are likewise determined by linear combinations of these vectors. On this diagram Figure 5 is the basal plane, Figure 6 the large front face, Figure 7 the large back face, and Figure 8 the right face. See text for further discussion. Abbreviations: Sdp = siderophyllite, Synth = synthetic mica, Kep = ephesite, Ann = annite, Pro = "protolithionite", Znw = zinnwaldite, Tri = trilithionite, Mus = muscovite, Phn = phengite, Cld = celadonite, Mtd = montdorite, Fet = taeniolite, and Ply = polylithionite.

that the interlayer or K sites are completely filled, as are the (OH) sites. Neither of these assumptions is completely justified in natural micas, and in fact numerous other minor coupled substitutions occur (Bailey, 1984, this volume); strictly speaking, each of these substitutions would add a dimension to our composition space. Nevertheless, the major features can be seen in only three dimensions.

GEOCHEMICAL EVOLUTION OF PEGMATITE MICAS

Trace-element contents of pegmatite micas have been extensively studied by a variety of methods, for different element assemblages, and in different contexts. Compared to the abundance of information on the Li, Al-micas, the data for ferromagnesian micas (including zinnwaldite and exocontact micas) are scarce. Consequently, it is only the micas that are graphically illustrated in this section, on the basis of the classic fractionation ration K/Rb.

Micas of orogenic pegmatites

Muscovite pegmatites. "Primitive" compositions are typical of the micas in accord with the geochemistry of other rock-forming and accessory minerals in this pegmatite class.

Biotite is typically rich in Ba (370-5000 ppm), Sc (10-30 ppm), V (20-650 ppm), Ni (60-330 ppm), Co (20-130 ppm), and Cr (150-900 ppm). In contrast, very low concentrations are typical of Li (200-500 ppm), Rb (300-1200 ppm), Cs (10-220 ppm), Be (< 1-5 ppm), and Sn (< 1-16 ppm) (according to Gordiyenko and Leonova, 1976).

Muscovite shows a similar distribution of trace elements modified, of course, by the difference in the main octahedral cations: Ba, Ti (600-9000 ppm), Sc (5-90 ppm), and V (< 5-220 ppm) are enriched, Ni (10-30 ppm) is moderate, and Li, Rb, Cs, Be, Zn, Ga, and Sn are very low (Fig. 10, 11; Manuylova et al., 1966; Shmakin, 1973, 1975; Gordiyenko, 1975; Gordiyenko and Leonova, 1976). Not only the micas of the muscovite-bearing pegmatites proper but also those of the associated barren (ceramic) and rare-element-bearing pegmatites types within the Barrovian terranes share this primitive geochemical signature (Fig. 10, 11; Manuylova et al. 1966).

Within the 3-dimensional configuration of pegmatite groups and fields, Ba, Mg, Fe, Ni, Co, Sc, V, Cr, Ti, Ba/Rb, and K/Rb decrease in both biotite and muscovite outwards and upwards, whereas Li, Rb, Cs, Tl, Be, Sn, and Fe/Mg increase in the same directions, and Sr does not show any specific trend (Gordiyenko, 1975; Gordiyenko and Leonova, 1976). The same applies to the sequence of primary and metasomatic mica crystallization within individual pegmatites (Dyadkina, 1969; Gordiyenko and Leonova, 1976).

Rare-element pegmatites. The low fractionation levels of micas from the muscovite-pegmatite class become particularly conspicuous when compared to the mica minerals of the rare-element class. It must be kept in mind, however, that the apparent continuation of fractionation trends shown in Figures 10 and 11 is, to a degree, formal (i.e., an artifact). The illustrated mica compositions

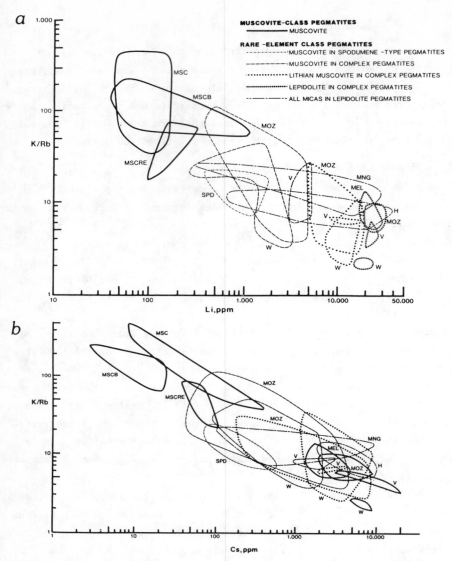

Figure 10. Li and Cs in muscovite - lithian muscovite - lepidolite versus the K/Rb ratio in pegmatites of orogenic affiliation: MSCB-barren, MSC-muscovite-bearing and MSCRE-rare-element-enriched pegmatites of the muscovite class (Manuylova et al., 1966). Rare-element pegmatite class: MOZ-Mozambique (Lopes Nunes, 1973); V-Varutrask (Quensel, 1956); MNG-Mongolia (Vladykin et al., 1974); MEL-Meldon (Chaudhry and Howie, 1973a); H-Harding (Rimal, 1962); SPD-spodumene pegmatite type (Gordiyenko, 1970); W-Winnipeg River district (Černý et al., 1981).

belong to pegmatites confined to Barrovian and Abukuma-type terranes, respectively, with no mutual physical (material) link. The only connection is by the pegmatite-generating process which is, of course, also active along geothermal gradients intermediate between the two typical cases above, and which promotes fractionation levels increasing with increasing geothermal gradient.

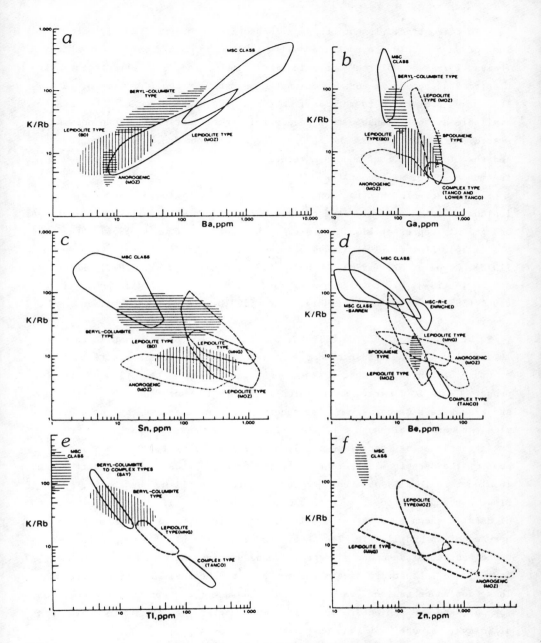

Figure 11. Ba, Ga, Sn, Be, Tl and Zn contents of muscovite-lithian muscovite-lepidolite versus the K/Rb ratio. Ruled fields are approximations based on limited numbers of analyses. Based on data by Quensel (1956); Slepnev (1961, 1964); Manuylova et al. (1966); Heinrich (1967); Gordiyenko (1970, 1975); Solodov (1971); Rinaldi et al. (1972); Lopes Nunes (1973); Shmakin (1973, 1975); Vladykin et al. (1974); Gordiyenko and Leonova (1976); Černý et al. (1981); Ferreira (1984), and unpublished data of P. Černý. SAY-Sayan Mts. (Slepnev, 1961, 1964); other locality symbols as in Figure 10.

To date, trace element behavior in biotite has been virtually unexplored. The few data on biotite that are available concern mainly barren pegmatites (Neiva, 1975) and wallrock contamination of the border zone affects the only biotite that occurs in more fractionated pegmatite types (Gordiyenko, 1970). It is thus preferable to refrain from discussing biotite because the data available are not amenable to a meaningful generalization. It can only be stated that in most cases the Ba, Sc, V, Ni, Co, and Cr contents are lower, and the Li, Rb, Cs, Be, and Sn contents are higher than in the biotites of the muscovite-class pegmatites.

The trace-element fractionation in muscovite - lithian muscovite - lepidolite is extensively covered in the literature; the following discussion and graphic representation are based on the work of Quensel (1956), Ukai et al. (1956), Slepnev (1961, 1964), Rimal (1962), Heinrich (1967), Gordiyenko (1970), Černý et al. (1970), Solodov (1971), Rinaldi et al. (1972), Chaudhry and Howie (1973a,b), Lopes Nunes (1973), Vladykin et al (1974), Foord (1976), Lahti (1981), Černý et al. (1981), Černý (1982b), Ferreira (1984), and Anderson (1984).

The behavior of Li, Cs, Tl, Ba, Be, Zn, Ga, and Sn is shown in Figures 10 and 11. All the illustrated trends exhibit continuity in fractionation of trace elements from muscovite through lithian muscovite to lepidolite proper. This is clearly demonstrated in Figure 10 where the K/Rb versus Li and K/Rb versus Cs data from some occurrences are separated for the above mica categories (although the boundaries are somewhat arbitrary).

Fractionation trends are mostly parallel and largely overlapping for most of the data from different pegmatite types, although notable differences in slope are locally observed (e.g., among the Mongolian, Mozambique, and Brown Derby pegmatites for Be, Zn, Ga, and Sn; Fig. 11). The best alignment of data is shown by K/Rb versus Tl, due to extremely close geochemical coherence of Rb and Tl (Fig. 11e; see also Černý, 1982b). On the other hand, broad scatter of data and divergent trends are typical of elements such as Sn, Ga, and Zn that are dissimilar to Rb in geochemical migration patterns. These depend on regional variations in abundance (cf. Černý and London, 1983, for Ga), and are subject to multiple complexing styles (Sn) or to dual anion affinity (chalcophile versus oxyphile tendencies of Zn).

The variation in Sr content is rather erratic. Its abundances are generally very low (< 5-100 ppm), and they seem to increase slightly with decreasing K/Rb. This may be partly due to an age-dependent contribution of ^{87}Sr to the total analyzed Sr, which effect should be more pronounced in geologically old and Rb-rich (lepidolite) micas.

Many data sets are available on the Be (Odikadze, 1958; Beus, 1956; Heinrich, 1967; Černý et al., 1981), Nb and Ta (Odikadze, 1958; Heinrich, 1962; Beus, 1966; Gordiyenko, 1970), and Sn (Ginsberg, 1954) contents of muscovite. However, the data are of variable quality, probably because of different analytical and calibration methods. Useful as they are for exploration purposes (cf. Trueman and Černý, 1982), a careful selection from existing data and much additional work are required to establish reliable ranges of absolute contents of these elements.

The data on the Sc, V, Ni, Co, and Cr contents of the Li, Al-micas are very scarce, but the available information indicates XO ppm or less; in Li-rich micas their abundances are commonly below the detection limits of emission spectrography and atomic absorption spectroscopy.

A conspicuous reversal of the general trends shown in the bulk composition and trace element evolution of muscovite - lithian muscovite - lepidolite is typical of the late muscovite varieties: rose muscovite, "gilbertite", and serictic alteration. These micas are close to ideal dioctahedral (OH)-muscovite with very low to negligible contents of Li, Rb, Cs, and F (Quensel, 1956; Chaudhry and Howie, 1973b; Lahti, 1981).

Finally, it should be pointed out that generalized summaries such as most of those presented in Figures 10 and 11 tend to obscure smaller-scale differences that are commonly encountered among individual pegmatites and pegmatite groups, even on a local scale of genetically related bodies. For example, the increase in Cs with decreasing K/Rb tends to proceed at about the same rate in all groups within a given pegmetite field (i.e., all groups exhibit the same slope in Fig. 10b), but the K/Rb:Cs ratio in individual pegmatite groups may be consistently different, breaking the field trend down into a series of parallel, laterally shifted patterns (Fig. 12). Examples of such parallel trends among K, Li, Rb, Cs, and Be were documented by, e.g., Lopes Nunes (1973), Černý et al. (1981), and Anderson (1984).

Micas of anorogenic subalkalic pegmatites

A scarcity of systematic data forbids any extensive discussion of mica species in both rare-element and miarolitic pegmatite classes. Amazonite pegmatites of Mozambique (Lopes Nunes, 1973) are used in Figure 11 to demonstrate characteristic differences between their ferroan lepidolites and the lithian micas of orogenic-pegmatite parentage: low Ga and Sn but high Be, Zn, and Pb (not shown). The Cape Ann zinnwaldites of Foster and Evans (1962) show the same trace-element abundances, and high Nb (850-1200 ppm) and Cu

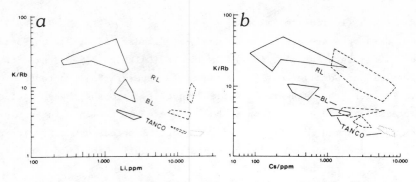

Figure 12. K/Rb versus Li, K/Rb versus Cs trends in micas of the Rush Lake (RL) and Bernic Lake (BL) groups, and of the Tanco pegmatite, southeastern Manitoba (Černý et al., 1981). Note the parallel, nonsequential fractionation patterns from muscovite (solid outline), through lithian muscovite (dashed outline) to lepidolite (dotted outline) in the individual groups and Tanco.

(120-480 ppm) as well. A similar pattern of trace-element contents is shown by the micas of the miarolitic pegmatites of the Korosten pluton (Lazarenko et al., 1973) but the data obtained in this study by emmission spectrography are too inexact for a meaningful interpretation. Much additional work is required to establish a general pattern of the geochemical evolution in pegmatite micas of anorogenic descent.

GENETIC ASPECTS OF MICA CRYSTALLIZATION IN GRANITIC PEGMATITES

In general, early generations of biotite and muscovite in granitic pegmatites represent a configuration of mica crystallization in plutonic granites. Biotite is commonly the only ferromagnesian silicate; muscovite indicates a peraluminous composition of the parent melt, increased activity of hydrous species, and a relatively high a_{H^+}/a_{K^+}. However, micas of the highly fractionated pegmatites, and especially the late mica generations reflect the more complex relationships in the consolidating pegmatite; of particular importance is the separation of a supercritical fluid, fluctuation of the activities of H, K, Na, Cl, and F bearing species, and the presence of other components such as P and B (Burt, 1976, 1981; London, 1982).

The following paragraphs comment on mica formation in specific classes and types of pegmatites.

Muscovite pegmatite class

Both biotite and muscovite reflect the primitive geochemical features of their parent pegmatites. Micas of the pegmatites, associated ("pegmatoid") granites, and surrounding gneissic rocks are all of similar composition: biotite is enriched in Mg and Al, muscovite in Fe and Mg, and both micas show a poorly fractionated spectrum of trace elements. These and other compositional similarities (e.g., in garnet and plagioclase: Leonova, 1966; Stern, 1966; Gresens, 1967b) have led some authors to interpret muscovite-class pegmatites as direct products of local recrystallization or anatexis (Sokolov, 1959; Sutchkov, 1961; Sokolov et al., 1975; see also Gresens, 1967a,b). However, mild fractionation from "pegmatoid" granites is advocated by other investigators, as reviewed by Ginsburg et al. (1979)

Textural relationships indicate that muscovite crystallizes late relative to its associated minerals, and in many cases convincing evidence is available that the late muscovite "mineralization" is contemporaneous with that developed in adjacent host rocks (Sokolov, 1959; Milovskyi, 1962). Two interpretations are again possible: pegmatite autometasomatism plus fracture-controlled exomorphism (e.g., Gresens, 1967a,b) versus regional retrogressive muscovitization subsequent to the consolidation of pegmatites (Sokolov, 1959).

Despite the differences in genetic interpretation of both pegmatites and the muscovites, structural, paragenetic, and geochemical features indicate that they consolidated under conditions close to those of the peak of regional metamorphism. The pressure regime (4-6 kbar) was sufficiently high to stablize primary muscovite in the presence of a granitic melt.

Rare-element (and miarolitic) class of orogenic suites

Biotite. In the least fractionated pegmatite types, biotite represents the final product of igneous Mg/Fe fractionation, approaching ideal annite in some of its occurrences (Nordenskjöld, 1906; Foster, 1960b).

Muscovite. In view of the available experimental data and chemographic analyses in the simple $K_2O - Na_2O - Al_2O_3 - SiO_2 - H_2O$ system, primary muscovite precipitated from a granitic melt should not occur in this relatively low-pressure pegmatite class (4-2 kbar), and particularly not in the miarolitic-class pegmatites (2-1 kbar). However, Anderson and Rowley (1981) argue that the Fe,Mg substitution commonly encountered in the early micas of plutonic leucogranites should stabilize muscovite in the presence of a peraluminous, volatile-rich granitic melt to pressures as low as 2 kbar (cf. Miller et al., 1981). This should also apply to the early, (Fe,Mg)-bearing micas of the discussed pegmatite classes (difficult as it may be to recognize some primary muscovites on textural grounds). Muscovite associated with cleavelandite and saccharoidal albite poses no stability problem as these feldspars and associated minerals are of subsolidus origin.

Lepidolite. Several aspects of lepidolite (including lithian muscovite) invite comment: (1) the occurrence of lepidolite seems to be controlled by several factors additional to the mere availability of Li and F; (2) the primary versus metasomatic nature of massive lepidolite units in central parts of complex pegmatites is disputable; and (3) the relative importance of composition and conditions of crystallization as controls on polytypism is not clear.

The principal role of fluorine in generating Li-rich micas is best illustrated by the contrasting mineralogy of Li in the quasi-homogeneous, F-depleted spodumene type and in the lepidolite type of rare-element pegmatites. In the first case, exemplified by the Kings Mountain, North Carolina, spodumene deposits (Kunasz, 1982), the activity of F is negligible throughout pegmatite crystallization. Spodumene (and minor triphylite) are the only primary Li-minerals, and lepidolite is absent. In contrast, lepidolite and minor elbaite are the only Li-bearing minerals in lepidolite pegmatites such as the Brown Derby swarm, Colorado (Heinrich, 1967). High activity of F and H throughout the pegmatite consolidation prevents crystallization of spodumene or petalite, and leads to precipitation of not only lepidolite but also other F-rich minerals such as topaz, fluorite, and microlite (cf. London, 1982).

High activity of P-bearing species during the primary pegmatite crystallization may prevent the formation of lepidolite by extracting fluorine in early phosphates. London and Burt (1982b) and Burt and London (1982) have shown that in environments with high activities of P and F, all lithium aluminosilicates (spodumene, petalite, eucryptite) become unstable relative to amblygonite (- montebrasite) + quartz. If amblygonite - montebrasite effectively exhausts F from the pegmatite melt and fluids, lepidolite may not form even at late stages, or only form locally in negligible amounts (e.g., in the Viitaniemi pegmatite; Lahti, 1981). If any sizable mica unit forms in the late consolidation stages of such an amblygonite-rich (and spodumene- or petalite-free) pegmatite, it has the composition of a very Li,F-poor lithian muscovite, rather than lepidolite proper (e.g., in the Peerless pegmatite, Sheridan et al., 1957; see also London and Burt, 1982b).

Large, more or less monomineralic units of lepidolite (and/or lithian muscovite) are common in the central parts of complex rare-element pegmatites, such as Bikita, Pidlite, Tanco, Hugo, Varuträsk, Mongolian Altai No. 3, and others (Jahns, 1953; Hutchinson, 1959; Quensel, 1956; Beus, 1960; Norton et al. 1962; Cooper, 1964; Wang et al., 1981; Černý, 1982b). These units are traditionally regarded as subsolidus metasomatic phenomena, due to abundant evidence of replacement along their margins, and enclosed relics of primary mineral assemblages of pre-existing zones (e.g., Ginsburg, 1960; Hutchinson, 1959;

Beus, 1960; Rinaldi et al., 1972; Crouse and Černý, 1972; London and Burt, 1982b). Stewart (1978) regards most lepidolite as a subsolidus product of K-feldspar and Li,Al-silicate reaction in the presence of F-rich fluids. This is supported by most of the experimental work of Munoz (1971), but some of his data and those of Glyuk et al. (1980) indicate that lepidolite is stable in the presence of a Li,F-bearing granitic melt, and may precipitate from it. Thus Norton (1983) argues in favor of lepidolite cores as products of residual magma crystallization (magma-like phase, final aluminous and alkali-rich product of magma), which also produces adjacent quartz units via silica extraction by a separating aqueous fluid. On the other hand, Melentyev and Delitsyn (1969) consider separation of an immiscible F-rich liquid as the primary cause of lepidolite core formation. The question will evidently remain unsettled until additional experimental evidence becomes available.

Early attempts to correlate the Li content of lithian muscovite and lepidolite with polytypism suggested a simple correlation (Levinson, 1953; Foster, 1960a). This was evidently due to a small data base. Detailed studies of lithian micas from individual pegmatites or pegmatite fields have shown that the correlation is highly variable and locality-dependent (Heinrich, 1967; Rinaldi et al, 1972; Lopes Nunes, 1973; Chaudhry and Howie, 1973a). Comparison with synthetic micas also gave uncertain and partly contradictory results (Munoz, 1968). The updated compilation presented in Figure 4 strengthens this impression. The restriction of $1M$ structures to the polylithionite segment of the compositional field (including polylithionites *sensu stricto* of peralkaline-pegmatitic origin), and the close adherence of $2M_2$ polytype to the trilithionite - "octahedrally depleted trilithionite" join (Rieder et al., 1970) appear to be the only cases with explicit components of compositional control. Otherwise it seems likely that factors other than composition exert primary control on polytype formation. Chaudhry and Howie (1973a) and Swanson and Bailey (1978) noticed that different polytypes of lepidolite occurring in single pegmatite bodies have strikingly different crystal habits, associated minerals, and general character of enclosing assemblages. Both these authors and Sartori (1976) concluded that P,T conditions, volatile phases, degree of saturation, and rate of cooling are more important than mica composition in determining the stability and occurrence of lepidolite polytypes. Changes in crystal growth mechanisms in micas synthesized under different conditions provide experimental support for this conclusion (Baronnet, 1975).

Zinnwaldite. This species rarely forms in orogenic pegmatites, representing an extreme case of ferroan lepidolite crystallization (Fig. 3a,b).

However, in many cases it forms not in consequence of a gradual increase in activities of Fe, Mn, and Mg during the evolution of pegmatite fluids but in response to local mobilization of these elements. For example, zinnwaldites-masutomilites from Czechoslovakian pegmatites inherit Fe and Mn from the almandine - spessartine garnets they replace (Němec, 1983a,b), and the borderline ferroan lepidolite - zinnwaldite from the Lower Tanco pegmatite is found only along contacts of K-feldspar and triphylite (Ferreira, 1984). Type masutomilite and associated Mn-rich phases from the Tanakamiyama pegmatite, Japan, are considered to result from assimilation of Mn by the parent granite which cuts stratiform manganese deposits in the enclosing sedimentary sequence (Harada et al., 1976).

Late muscovite alteration. Sericitic and "gilbertite"-type micas form either alone or as members of more complex and sequential assemblages (such as the replacement of spodumene by eucryptite + albite followed by muscovitization of eucryptite; London and Burt, 1982a,b). Late fluids may produce a diverse array of alteration effects: hydrolytic H^+ metasomatism (sericitization of K-feldspar), alkali exchange (sericitization of spodumene), K^+ and H^+ metasomatism (muscovite after gahnite, cordierite, andalusite, sillimanite, or topaz). However, all these processes can be triggered by low-T fluids of essentially the same K^+/H^+ activity ratio (cf. Burt and London, 1982b; Burt, 1981).

Of the generally Al-rich minerals prone to replacement by muscovite, it is particularly those with muscovite-like Si/Al_T ratios of about 1.0 that are most susceptible to this alteration (e.g., eucryptite). Minerals with different Si/Al ratios also yield quartz or highly peraluminous minerals as byproducts. For example, muscovite replacement of andalusite produces corundum - a suprising occurrence for such silica-rich rocks as granitic pegmatites. The explanation is that subsolidus K-metasomatism of the andalusite leads to a *local* deficiency of silica within the host crystal, due to the low Si/Al ratio in andalusite relative to that of muscovite (Rose, 1957). The overall reaction can be simplified to the following:

$$6 \text{ andalusite} + 2 K^+ + 3H_2O \rightarrow 2 \text{ muscovite} + 2 \text{ corundum} + 2H^+$$

(Burt and Stump, 1984). This is completely analogous to the formation of the silica-deficient eucryptite during the late albitization of spodumene (London and Burt, 1982b,c); the eucryptite commonly altered still later to muscovite.

Exomorphic micas. Metasomatic aureoles formed along the contacts of rare-element pegmatites with wallrocks of contrasting composition (serpentinites, marbles, amphibolites) are considered either as products of direct reaction of the pegmatite melt with the host rock, or as a result of late equilibration between solid phases facilitated by aqueous fluids. These fluids may be generated from within the pegmatite, or they may be mobilized from the host rocks after pegmatite consolidation. It is probable that both types of reaction occur, but their relative importance is variable in different environments.

For example, the contact reaction between pegmatites and ultrabasic wallrocks that generates phlogopite and associated tremolite, anthophyllite, and talc layers is facilitated by volatile agents such as water and fluorine (Korzhinskyi, 1953; Brownlow, 1961; Brady, 1977). Development of this assemblage along pegmatite contacts with anhydrous, non-serpentinized peridotite (P.M. Larrabee, pers. comm., 1970) indicates that the pegmatite melt alone can provide the fluids promoting the metasomatism. This reaction evidently proceeds above the temperature range of serpentinization. Serpentinization occasionally postdates pegmatite intrusion and contact reaction, and it results in retrogression of the contact assemblage to vermiculite, talc, and chlorite, and even in chloritization of the pegmatite margins (Černý and Povondra, 1965; Černý, 1968). Thus the metasomatic exchange could possibly accompany the initial pegmatite crystallization, and it does not need to be a postmagmatic event driven by genetically unrelated fluids (as argued by Phillips and Hess, 1936; Korzhinskyi, 1953; Sherstyuk, 1965). A pegmatite-generated late fluid phase also seems to be excluded, because massive reaction zones commonly envelop small pegmatite veins of simple granitic composition and texture that lack internal metasomatic features.

Phlogopite rimming pegmatite bodies intruded into calcareous and dolomitic marbles may be primary in some cases, but formation by fluids penetrating along pegmatite/marble contacts is documented in many localities. The Kugi-Lyal pegmatites of southwest Pamir are bordered by phlogopite associated with spinel, forsterite, enstatite, tremolite, anthophyllite, and talc (Rossovskyi, 1963). At least two stages of mineralization, and probably two generations of phlogopite are probable in the Loolekop and Grenvillian localities (Gevers, 1949; Currie, 1951; Hoadley, 1960). The diversified origin of phlogopite in the Slyudyanka district due to pegmatite intrusion, scapolite-bearing silicic veining, and metasomatic reaction along marble contacts with gneissic lithologies was recognized by Fersman (1940) and interpreted by Korzhinskyi (1947). Post-tectonic crystallization of phlogopite along contacts of aplite boudins in dolomitic marbles leaves no doubt about its origin by metasomatic exchange, unrelated to the aplite intrusion (Sekanina, 1965).

The ferromagnesian Li,Rb,Cs-enriched micas that occur along the contacts

of rare-element pegmatites with amphibolite-like host rocks were generated
largely by late fluids migrating from the pegmatites during the metasomatic
stages of their internal evolution. The mica aggregates are reduced to thin
seams, and contain low concentration of rare alkalis, when associated with
simple pegmatites of low fractionation levels, or with contacts remote from
extensively replaced pegmatite segments. In contrast, high concentrations of
Li, Rb, and Cs in extensive metasomatic aureoles are found along margins of
highly fractionated complex pegmatites, in which the primary rare-alkali-
bearing species such as spodumene, K-feldspar and pollucite are in part decomposed by albitization and micaceous replacement (Ginsburg et al., 1972).
In some cases, two or more stages of metasomatism may have contributed to
the formation of the exocontact assemblages, related to different internal
pegmatite processes (Zagorskyi et al., 1974; Shearer et al., 1984).

Rare-element (and miarolitic) class of anorogenic suites

Biotite and lepidolite reflect the general compositional characteristics of anorogenic granites and their pegmatites, so dramatically different
from the orogenic-related granitoids: extreme Fe-enrichment, relatively low
Fe^{2+}/Fe^{3+} ratio, high activity of F and relatively subordinate role of H_2O.
The dry nature of the anorogenic granitoid melts is indicated, i.e., by the
restriction of micas to the most fractionated granitic differentiates and
their pegmatites. The alkali-saturated chemistry of the melts is reflected
in the Al-poor composition of biotites (annite - lepidomelane) and in the
absence of primary muscovite.

Zinnwaldite (and extremely rare lepidolite) is the only Li-bearing species
in the anorogenic pegmatites that are generally Li-, Cs-, Be-, and Ga-poor.
On the other hand, increased contents of Nb, Y, and Zr are to be expected in
minerals of anorogenic pegmatite suites, as are those of Cu, Pb, and Zn.
Anorogenic granitoids are typically sulphur-depleted, with consquent incorporation of chalcophile elements into silicate phases (Foord and Martin, 1979).
The trace-element signature of pegmatitic biotites and zinnwaldites is in good
accord with the above geochemistry of anorogenic granites (Foster and Evans,
1962; Lazarenko et al., 1973).

CONCLUDING NOTE

Much information on micas from granitic pegmatites is scattered in the
literature. The fundamental crystallochemical features of all species are
well understood, and the principal geochemical characteristics are also

reasonably established. However, petrologic understanding of pegmatite micas is rather poor. Systematic studies of the role of mica minerals in pegmatite crystallization, and information on the nature of mica-generating melts and fluids, are extremely scarce. A detailed analysis of paragenetic position, co-existing mineral associations, bulk composition, trace elements, and isotopic ratios of all micas in a given pegmatite is required to utilize fully the potential of mica minerals for petrologic interpretation. In view of the diversity of pegmatite classes and types, and of the paragenetic and compositional complexity of the micas themselves, the whole task is rather formidable. However, this complexity should provide an abundance of petrogenetic information when fully understood.

REFERENCES

Anderson, A.J. (1984) The Cross Lake Pegmatite Field, Manitoba. M.Sc. thesis, Univ. Manitoba, Winnipeg, Canada (in preparation).

Anderson, J.L. and Rowley, M.C. (1981) Synkinematic intrusion of peraluminous and associated metaluminous granitic magmas, Whipple Mountains, California. Canadian Mineral. 19, 83-101.

Babu, S.K. (1969) Zinnwaldite from a pegmatite near Rajagarh Village, Ajmer District. Indian Mineral. 10, 205-209.

Bailey, S.W. (1984) Structures, classification, and crystal chemistry of micas. This volume.

_____ and Christie, O.H.J. (1978) Three-layer monoclinic lepidolite from Tørdal, Norway. Am. Mineral. 63, 203-204.

Bargar, K.E., Beeson, M.H., Fournier, R.O. and Muffler, L.J.P. (1973) Present-day deposition of lepidolite from thermal waters in Yellowstone National Park. Am. Mineral. 58, 901-904.

Baronnet, A. (1975) L'aspect croissance du polymorphisme et du polytypisme dans les micas synthétiques d'intérêt pétrologique. Fortschr. Mineral. 52, 203-216.

Bassett, W.A. (1959) The origin of the vermiculite deposit at Libby, Montana. Am. Mineral. 44, 282-293.

Berggren, T. (1941) Minerals of the Varuträsk pegmatite. XXV. Some new analyses of lithium-bearing mica minerals. Geol. Fören. Förh. 63, 262-278.

Beus, A.A. (1956) Beryllium: Evaluation of Deposits during Prospecting and Exploratory Work. Gosgeoltekhizdat, Moscow (in Russian), Freeman & Co. translation 1962, 161 p.

_____ (1960) Geochemistry of Beryllium and the Genetic Types of Beryllium Deposits. Acad. Sci. USSR Moscow (in Russian), Freeman & Co. translation 1966, 329 p.

_____ (1966) Distribution of tantalum and niobium in muscovites of granitic pegmatites. Geokhimiya 1966, 1216-1220 (in Russian).

Bogomolova, L.K. (1962) Micas from reaction rims of the pegmatites in vein fields of Central Ural Mts. Mineral. Sbornik 5; Trudy Gorno-Geol. Inst.; Ural Section of the Vses. Mineral. Obshtch. Sverdlovsk, 3-18 (in Russian).

Borodayevskaya, M.E. (1951) Contact phenomena connected with the vein granitoids of the Berezovsk District, Central Urals. Izvestiya Acad. Sci. USSR, Ser. Geol., No. 1, 38-46 (in Russian).

Brady, B.J. (1977) Metasomatic zoning in metamorphic rocks. Geochim. Cosmochim. Acta 41, 113-125.

Brock, K.J. (1974) Zoned lithium-aluminum mica crystals from the Pala pegmatite district. Am. Mineral. 59, 1242-1248.

Brotzen, O. (1959) Mineral association in granitic pegmatites: A statistical study. Geol. Fören. Förh. 81, 231-296.

Brown, B.E. (1978) The crystal structure of a $3T$ lepidolite. Am. Mineral. 63, 332-336.

Brownlow, A.H. (1961) Variation in composition of biotite and actinolite from monomineralic contact bands near Westfield, Massachusetts. Am. J. Sci. 259, 353-370.

Burt, D.M. (1974) Concepts of acidity and basicity in petrology — the exchange operator approach. Geol. Soc. Am. Abstr. with Progr. 6, 674-676.

_____ (1976) Hydrolysis equilibria in the system $K_2O-Al_2O_3-SiO_2-H_2O-Cl_2O_{-1}$: Comments on topology. Econ. Geol. 71, 665-671.

_____ (1981) Acidity-salinity diagrams — application to greisen and porphyry deposits. Econ. Geol. 76, 832-843.

_____ and Burton, J.H. (1984) Vector representation of lithium and other mica compositions using exchange operators. Geol. Soc. Am. Abstr. with Progr. 16 (submitted).

_____ and London, D. (1982) Subsolidus equilibria. Mineral. Assoc. Canada Short Course Handbook 8, 329-346.

_____ and Stump, E. (1984) Mineralogical investigation of andalusite-rich pegmatites from Szabo Bluff, Scott Glacier area, Antarctica. Antarctica J. of the U.S., 1983 Ann. Rev. issue (in press).

Cameron, E.N., Jahns, R.H., McNair, A. and Page, L.R. (1949) Internal Structure of Granitic Pegmatites. Econ. Geol. Monogr. 2, 115 p.

_____ and Shainin, V.E. (1947) The beryl resources of Connecticut. Econ. Geol. 42, 353-367.

Černý, P. (1968) Comments on serpentinization and related metasomatism. Am. Mineral. 53, 1377-1385.

_____ (1972) Phlogopite, hydrophlogopite and vermiculite from Heřmanov, Czechoslovakia. N. Jahrb. Mineral. Monatsh. 1972, 203-209.

_____ (1982a) Anatomy and classification of granitic pegmatites. Mineral. Assoc. Canada Short Course Handbook 8, 1-39.

_____ (1982b) The Tanco pegmatite at Bernic Lake, southeastern Manitoba. Mineral. Assoc. Canada Short Course Handbook 8, 527-543.

_____ (1982c) Petrogenesis of granitic pegmatites. Mineral. Assoc. Canada Short Course Handbook 8, 405-461.

_____ (1984) Granitic Pegmatites. Monograph Series on Geology of Ore Deposits, A.C. Brown, ed., in press.

_____ and London, D. (1983) Crystal chemistry and stability of petalite. Tschermak's Mineral. Petrogr. Mitt. 31, 81-96.

_____ and Miškovský, J. (1966) Ferroan phlogopite and magnesium vermiculite from Věžná, Western Moravia. Acta Univ. Carolinae-Geol., No. 1, 17-32.

_____ and Povondra, P. (1965) Harmotome from desilicated pegmatites at Hrubšice, Western Moravia. Acta Univ. Carolinae-Geol., No. 1, 31-43.

_____, Rieder, M. and Povondra, P. (1970) Three polytypes of lepidolite from Czechoslovakia. Lithos 3, 319-325.

_____, Trueman, D.L., Ziehlke, D.V., Goad, B.E. and Paul, B.J. (1981) The Cat Lake-Winnipeg River and the Wekusko Lake Pegmatite Fields, Manitoba. Man. Mineral Res. Div. Econ. Geol. Rept. ER80-1, 240 p.

Chaudhry, M.N. and Howie, R.A. (1973a) Lithium-aluminum micas from the Meldon aplite, Devonshire, England. Mineral. Mag. 39, 289-296.

_____ and _____ (1973b) Muscovite ("gilbertite") from the Meldon aplite. Proc. Ussher Soc. 2, 480-481.

Cooper, D.G. (1964) The geology of the Bikita pegmatites. The Geology of Some Ore Deposits in Southern Africa II, 441-462.

Crouse, R.A. and Černý, P. (1972) The Tanco pegmatite at Bernic Lake, Manitoba. I. Geology and paragenesis. Canadian Mineral. 11, 591-608.

Currie, J.B. (1951) The occurrence and relationship of some mica and apatite deposits in Southeastern Ontario. Econ. Geol. 46, 765-778.

Dyadkina, I.YA. (1969) On the trend of mica formation in pegmatites. Zapiski Vses. Mineral. Obshtch. 98, 280-287 (in Russian).

Eckermann, H.v. (1922) The rocks and contact minerals of the Mansjö-Mountain. Geol. Fören. Förh. 44, 205-241.

_____ (1923) Rocks and contact minerals of Tenneberg. Geol. Fören. Förh. 44, 465-538.

Ferreira, K.J. (1984) The Mineralogy and Geochemistry of the Lower Tanco Pegmatite, Bernic Lake, Manitoba, Canada. M.Sc. thesis, Univ. Manitoba, Winnipeg, Canada.

Fersman, A.E. (1940) Pegmatites. Reprinted 1960: Selected Works VI., Acad. Sci. USSR Moscow, 742 p. (in Russian).

Foord, E.E. (1976) Mineralogy and Petrogenesis of Layered Pegmatite-Aplite Dikes in the Mesa Grande District, San Diego County, California. Ph.D. dissertation, Stanford Univ., Palo Alto, California.

_____ and Martin, R.F. (1979) Amazonite from the Pikes Peak batholith. Mineral. Record 10, 373-384.

Foster, M.D. (1960a) Interpretation of the composition of lithium-micas. U.S. Geol. Surv. Prof. Paper 354-E, 115-147.

_____ (1960b) Interpretation of the composition of tioctahedral micas. U.S. Geol. Surv. Prof. Paper 354-B, 11-49.

_____ and Evans, H.T. (1962) New study of cryophyllite. Am. Mineral. 47, 344-352.

Francke, W., Jelinski, B. and Zarei, M. (1982) Hydrothermal synthesis of an ephesite-like sodium mica. N. Jahrb. Mineral. Monatsh. 337-340.

Franzini, M. and Sartori, F. (1969) Crystal data on $1M$ and $2M_2$ lepidolites. Contrib. Mineral. Petrol. 23, 257-270.

Gerasimovskyi, V.V. (1965) On taeniolite from carbonate formations and albitites. Trudy Mineral. Museum Acad. Sci. USSR 16, 215-218 (in Russian).

Gevers, T.W. (1949) Vermiculite at Loolekop, Northeast Trausvaal. Trans. Geol. Soc. South Africa 51, 133-173.

Ginsburg, A.I. (1954) On minerals - geochemical indicators and their significance in exploration for rare-element ores in pegmatites. Dokl. Acad. Sci. USSR 98 (in Russian).

_____ (1960) Specific geochemical features of the pegmatitic process. Int'l. Geol. Congress, 21st Sess. Norden, Rept. Pt. 17, 111-121.

_____ and Berkhin, S.I. (1953) On composition and chemical constitution of lithian micas. Trudy Mineral. Museum Acad. Sci. USSR 5, 90-131 (in Russian).

_____, Lugovskoi, G.P. and Ryabenko, V.E. (1972) Cesian slyudites - a new type of mineralization. Razvedka i Okhrana Nedr. 8, 3-7 (in Russian).

_____, Timofeyev, I.N. and Feldman, L.G. (1979) Principles of Geology of the Granitic Pegmatites. Nedra, Moscow, 296 p. (in Russian).

Glass, J.J. (1935) The pegmatite minerals from near Amelia, Virginia. Am. Mineral 20, 741-768.

Glyuk, D.S. Trufanova, L.G. and Bazarova, S.B. (1980) Phase relations in the granite-H_2O-LiF system at 1,000 kg/cm^2. Geochem. Int'l. 17, 35-48.

Gordiyenko, V.V. (1970) Mineralogy, Geochemistry and Genesis of the Spodumene Pegmatites. Nedra, Leningrad, 237 p. (in Russian).

_____ (1975) Geochemistry of the processes of pegmatite formation and muscovite generation. In: Muscovite Pegmatites of the U.S.S.R., Nauka, Leningrad, 107-117 (in Russian).

_____ and Leonova, V.A., ed. (1976) Mica-bearing Pegmatites of Northern Karelia. Nedra, Leningrad, 367 p. (in Russian).

_____, Syritso, L.F. and Krivovichev, V.G. (1975) A new type of rare-metal apobasite metasomatites and the regular distribution of Cs, Li and Rb in them. Dokl. Acad. Sci. USSR 224, 198-200 (in Russian).

Green, T.H. (1981) Synthetic high-pressure micas compositionally intermediate between the dioctahedral and trioctahedral mica series. Contrib. Mineral. Petrol. 78, 452-458.

Gresens, R.L. (1967a) Tectonic-hydrothermal pegmatites. I. The model. Contrib. Mineral. Petrol. 15, 345-355.

_____ (1967b) Tectonic-hydrothermal pegmatites. II. An example. Contrib. Mineral. Petrol. 16, 1-28.

Guggenheim, S. (1981) Cation ordering in lepidolite. Am. Mineral. 66, 1221-1232.

Hadley, J.B. (1949) Preliminary report on corundum deposits in the Buck Creek peridotite, Clay County, North Carolina. U.S. Geol. Surv. Bull. 948-E, 103-128.

Harada, K., Honda, M., Nagashima, K. and Kanisawa, S. (1976) Masutomilite, manganese analogue of zinnwaldite, with special reference to masutomilite-lepidolite-zinnwaldite series. Mineral. J. (Japan) 8, 95-109.

Hazen, R.M. and Wones, D.R. (1978) Predicted and observed compositional limits of trioctahedral micas. Am. Mineral. 63, 885-892.

Heinrich, E.W. (1946) Studies in the mica group; the biotite-phlogopite series. Am. J. Sci. 244, 836-848.

_____ (1962) Geochemical prospecting of the beryl and columbite. Econ. Geol. 57, 616-619.

_____ (1967) Micas of the Brown Derby pegmatites, Gunnison County, Colorado. Am. Mineral. 52, 1110-1121.

_____ (1978) Mineralogy and structure of lithium pegmatites. Jornal de Mineralogia, Recife, Vol. Djalma Guimaraes, 7, 59-65.

_____ and Levinson, A.A. (1953) Mineralogy of rose muscovites. Am. Mineral 38, 25-34.

_____, _____, Lewandowski, D.W. and Hewitt, C.H. (1953) Studies in the Natural History of Micas. Univ. Michigan Eng. Res. Inst. Proj. M978, Final Rept., 241 p.

Hess, F.L. and Stevens, R.E. (1937) A rare-alkali biotite from Kings Mountain, North Carolina. Am. Mineral. 22, 1040-1044.

Hewitt, D.A. and Wones, D.R. (1975) Physical properties of some synthetic Fe-Mg-Al trioctahedral biotites. Am. Mineral. 60, 854-862.

Hoadley, J.W. (1960) Mica Deposits of Canada. Geol. Surv. Canada, Econ. Geol. Series 19, 141 p.

Hutchison, R.W. (1959) Geology of the Motgary pegmatite. Econ. Geol. 54, 1525-1542.

Jahns, R.H. (1953) The genesis of pegmatites. II. Quantitative analysis of lithium-bearing pegmatite, Mora County, New Mexico. Am. Mineral 38, 1078-1112.

_____ (1955) The study of pegmatites. Econ. Geol., 50th Anniv. Vol., 1025-1130.

_____, Griffiths, W.R. and Heinrich, E.W. (1952) Mica Deposits of the Southeastern Piedmont. Part I. General Features. U.S. Geol. Surv. Prof. Paper 248-A, 99 p.

Korzhinskyi, D.S. (1947) Bimetasomatic phlogopite and lazurite deposits of the Baikal Archean. Trudy Inst. Geol. Sci. Acad. USSR, 29 (in Russian).

_____ (1953) An outline of metasomatic processes. In: Fundamental Problems in Research of the Magmatogenic Ore Deposits. Acad. Sci. USSR, 332-452 (in Russian).

Kretz, R. (1968) Study of Pegmatite Bodies and Enclosing Rocks, Yellowknife-Beaulieu Region, District of Mackenzie. Geol. Surv. Canada Bull. 159, 109 p.

Kulp, J.L. and Brobst, J.A. (1954) Notes on the dunite and the geochemistry of vermiculite at the Day Book dunite deposit, Yancey County, North Carolina. Econ. Geol. 49, 438-461.

Kunasz, I. (1982) Foote Mineral Company, Kings Mountain operation. Mineral. Assoc. Canada Short Course Handbook 8, 505-511.

Lacroix, A. (1898) Les filons granitiques et pegmatitiques des contacts granitiques de l'Ariège; leur importance théorétique. C. R. Acad. Sci. Paris 127, 570-572.

Lahti, S. (1981) On the granitic pegmatites of the Eräjärvi area in Orivesi, southern Finland. Bull. Geol. Surv. Finland 314, 82 p.

Layman, F.G. (1963) Alteration of cordierite. Geol. Soc. Am. Ann. Mtg. Progr. Abstr., New York, 100A.

Lazarenko, E.K., Pavlishin, V.I., Latysh, V.T. and Sorokin, YU.G. (1973) Mineralogy and genesis of the chamber pegmatites of Volynia. Lvov State Univ. Publ. House, Lvov, 359 p. (in Russian).

Leonova, V.A. (1965) Some problems of the geochemistry and genesis of the Tchupa pegmatite veins, Northern Karelia. Zapiski Vses. Mineral. Obschtch. 94, 272-287 (in Russian).

Levillain, C. (1980) Étude statistique des variations de la teneur en OH et F dans les micas. Tscherm. Mineral. Petrogr. Mitt. 27, 209-223.

_____ (1981) Présence d'un polytype $2M_1$ dans les lépidolites du granite de Beauvoir, Massif Central, France. Bull. Minéral. 104, 690-693.

_____, Maurel, P. and Menil, F. (1977) Localisation de fer, par voie chimique et par spectrométrie Mössbauer, dans la lépidolite du granite de Beauvoir, Massif Central, France. Bull. Soc. franç.Minéral. Cristallogr. 100, 137-142.

Levinson, A.A. (1953) Studies in the mica group; relationship between polymorphism and composition in the muscovite-lepidolite system. Am. Mineral. 38, 88-107.

London, D. (1982) Stability of spodumene in acidic and saline fluorine-rich environments. Carnegie Inst. Wash. Yearbook 81, 331-334.

_____ and Burt, D.M. (1982a) Alteration of spodumene, montebrasite, and lithiophilite in pegmatites of the White Picaho District, Arizona. Am. Mineral. 67, 97-113.

_____ and _____ (1982b) Lithium minerals in pegmatites. Mineral. Assoc. Canada Short Course Handbook 8, 99-133.

_____ and _____ (1982c) Chemical models for lithium aluminosilicate stabilities in pegmatites and granites. Am. Mineral. 67, 494-509.

Lopes Nunes, J.E. (1973) Contribution à l'étude minéralogique et géochimique des pegmatites du Mozambique. Sci. de la Terre, Nancy, Mém. 26, 261 p.

Manuylova, M.M., Petrov, L.L., Rybakova, M.M., Sokolov, Yu.M. and Shmakin, B.M. (1966) Distribution of alkali metals and beryllium in pegmatite minerals from the North Baikalian pegmatite belt. Geokhimiya 410-432 (in Russian).

Melentyev, B.N. and Delitsyn, L.M. (1969) Problem of liquation in magma. Dokl. Acad. Sci. USSR, Earth Sci. Ser. 186, 215-217, AGI New York.

Miller, C.F., Stoddard, E.F., Bradfish, L.J. and Dolase, W.A. (1981) Composition of plutonic muscovite: Genetic implications. Canadian Mineral.19, 25-34.

Milovskyi, A.V. (1962) On the evolution of the quartz-muscovite complex in mica-bearing pegmatites and their enclosing rocks. Zapiski Vses. Mineral. Obshtch. 91, 360-362 (in Russian).

Miser, H. and Stevens, R.E. (1938) Taeniolite from Magnet Cove, Arkansas. Am. Mineral 23, 104-110.

Morel, S.W. (1955) Biotite in the Basement complex of Southern Nyasaland. Geol. Mag. 92, 241-248.

Munoz, J.L. (1968) Physical properties of synthetic lepidolite. Am. Mineral. 53, 1490-1512.

_____ (1971) Hydrothermal stability relations of synthetic lepidolite. Am. Mineral. 56, 2069-2087.

Neiva, A.M.R. (1975) Geochemistry of coexisting aplites and pegmatites and their minerals from central northern Portugal. Chem. Geology 16, 153-177.

_____ (1980) Chlorite and biotite from contact metamorphism of phyllite and metagraywacke by granite, aplite-pegmatite and quartz veins. Chem. Geol. 29, 49-71.

Němec, D. (1983a) Zinnwaldit in moldanubischen Lithium-Pegmatiten. Chem. Erde 42, 197-204.

_____ (1983b) Masutomilite in lithium pegmatites of West-Moravia, Czechoslovakia. N. Jahrb. Mineral. Monatsh. 537-540.

Neumann, H., Sverdrup, T. and Saebø, P.C. (1957) X-ray powder patterns for mineral identification. Part III. Silicates. Avh. Norske Vid. Akad. Oslo I., Mat. Nat. Kl. no. 6.

Nordenskjöld, J. (1906) Der Pegmatit von Ytterby. Bull. Soc. Geol. Inst. Uppsala 9, 183-228.

Norton, J.J., Page, R.L. and Brobst, D.A. (1962) Geology of the Hugo pegmatite, Keystone, South Dakota. U.S. Geol. Surv. Prof. Paper 297-B, 49-128.

Odikadze, G.L. (1958) On the presence of niobium and tantalum in muscovites from pegmatites of the Dzirulsk crystalline massif. Geokhimiya 479-485 (in Russian).

Omori, K. (1958) Mode of occurrence and chemical composition of Mg-vermiculite from Odaka and Uzumine, Fukushima Prefecture. J. Mineral. Soc. Japan 3, 478-485 (in Japanese; English abstr.).

Pavlishin, V.I., Semenova, T.F. and Rozhdestvenskaya, I.V. (1981) Protolithionite-3T: Structure, typomorphism, and practical significance. Mineral. Zhurnal 3, 67-70 (in Russian).

Perrault, G. (1966) Polylithionite from St. Hilaire, P.Q. (abstr.). Canadian Mineral. 8, 671.

Phillips, A.H. and Hess, H.H. (1936) Metamorphic differentiation between serpentine and siliceous country rocks. Am. Mineral. 21, 333-362.

Povondra, P., Čech, F. and Burke, E.A.J. (1984) Sodian-beryllian cordierite from Gammelmorskärr, Kemiö Island, Finland, and its decomposition products. N. Jahrb. Mineral. Monatsh. 125-136.

Pye, E.G. (1965) Georgia Lake Area. Ontario Dept. Mines. Geol. Rpt. 31, 113 p.

Quensel, P. (1956) The paragenesis of the Varuträsk pegmatite, including a review of its mineral assemblage. Ark. Mineral. Geol. 2, 9-126.

Raade, G. and Larsen, A.O. (1980) Polylithionite from syenite pegmatite at Vøra, Sandefjord, Oslo Region, Norway. Norsk Geol. Tiddskr. 60, 117-124.

Rieder, M. (1970) Chemical composition and physical properties of lithium-iron micas from the Krušné Hory Mts., Erzgebirge. Part A: Chemical composition. Contrib. Mineral. Petrol. 27, 131-158.

Rijiks, H.R.P. and v.d. Veen, A.H. (1972) The geology of the tin-bearing pegmatites in the eastern part of the Kamativi District, Rhodesia. Mineral. Deposita 7, 383-395.

Rimal, D.N. (1962) Mineralogy of Rose Muscovite and Lepidolite from the Harding Pegmatite, Taos County, New Mexico. Ph.D. dissertation, Univ. New Mexico, Albuquerque, New Mexico.

Rinaldi, R., Černý, P. and Ferguson, R.B. (1972) The Tanco pegmatite at Bernic Lake, Manitoba. VI. Lithium-rubidium-cesium micas. Canadian Mineral. 11, 690-707.

Rose, R.L. (1957) Andalusite- and corundum-bearing pegmatites in Yosemite National Park, California. Am. Mineral. 42, 635-637.

Ross, M., Takeda, H. and Wones, D.R. (1966) Mica polytypes: Systematic description and identification. Science 151, 191-193.

Rossovskyi, L.N. (1963) Pegmatites in magnesian marbles from Kugi-Lyal in the southwest Pamir. Trudy Mineral. Museum Acad. Sci. USSR 14, 166-181 (in Russian).

Rudenko, S.A., Romanov, V.A., Morakhovskyi, V.N., Tarasov, E.B., Galkin, G.A. and Dorokhin, V.K. (1975) Conditions of formation and controls of distribution of muscovite objects of the North-Baikal muscovite province, and some general problems of pegmatite consolidation. In: Muscovite Pegmatites of the U.S.S.R., Yu.M.Sokolov, ed., Nauka, Leningrad, 174-182 (in Russian).

Salye, M.E. (1975) Metallogenic formations of pegmatites in the eastern part of the Baltic Shield. In: Muscovite Pegmatites of the U.S.S.R., Yu.M. Sokolov, ed., Nauka, Leningrad, 15-35 (in Russian).

Sartori, F. (1976) The crystal structure of 1M lepidolite. Tschermaks Mineral. Petrol. Mitt. 23, 65-75.

_____, Franzini, M. and Merlino, S. (1973) Crystal structure of a 2M_2 lepidolite. Acta Crystallogr. B29, 573-578.

Seifert, F. and Schreyer, W. (1971) Synthesis and stability of micas in the system K_2O-MgO-SiO_2-H_2O and their relations to phlogopite. Contrib. Mineral. Petrol. 30, 196-215.

Sekanina, J. (1965) Vermiculit am Kontakt zwischen Aplit und Dolomit bei Prosetín, Nordteil der Schwarzawakuppel. Acta Univ. Carolinae-Geol., Suppl. 2, 17-30.

Semenov, E.I. (1959) Lithian and other micas and hyromicas in the alkalic pegmatites of Kola peninsula. Trudy Mineral. Museum Acad. Sci., USSR 9, 68-83 (in Russian).

Shearer, C.K., Papike, J.J., Simon, S.B., Laul, J.C. and Christian, R. (1984) Pegmatite/wall-rock interactions, Black Hills, South Dakota: Progressive boron metasomatism adjacent to the Tip Top pegmatite. Geochim. Cosmochim. Acta, in press.

Sheridan, D.M., Stephens, H.G., Staatz, M.H. and Norton, J.J. (1957) Geology and Beryl Deposits of the Peerless Pegmatite, Pennington County, South Dakota. U.S. Geol. Surv. Prof. Paper 297-A, 47 p.

Sherstyuk, A.I. (1965) Slyudite complexes and their classification. Zapiski Vses. Min. Obshtch. 94, 62-70 (in Russian).

Shilin, L.L. (1953) On lithium micas from pegmatites of alkaline magmas. Trudy Mineral. Museum Acad. Sci. USSR 5, 153-163 (in Russian).

Shmakin, B.M. (1973) Geochemical specialization of Indian Precambrian pegmatites in relation to alkali and ore-element contents of the minerals. Geochemistry, 890-899.

_____ (1975) Geochemical model of the formation of muscovite pegmatites. In: Muscovite Pegmatites of the U.S.S.R., Yu.M. Sokolov, ed., Nauka, Leningrad, 98-106 (in Russian).

Shurkin, K.A., Gorlov, N.V., Salye, M.E., Dyk, V.L. and Nikitin, Yu.V. (1962) Belomorskyi complex of northern Karelia and southwestern Kola Peninsula - Geology and pegmatite potential. Acad. Sci. USSR Moscow-Leningrad, 306 p. (in Russian).

Simmons, W.B. and Heinrich, E.W. (1980) Rare-earth pegmatites of the South Platte District, Colorado. Colorado Geol. Surv., Resource Ser. 11, 131 p.

Slepvev, Yu.S. (1961) The ration of thallium to rubidium, cesium and potassium in the schists, granites and rare-metal pegmatites of the Sayan Mountains. Geochemistry 382-385.

_____ (1964) Geochemical characteristics of the rare-metal granitic pegmatites of the Sayan Mountains. Geochemistry 221-228.

Smith, J.V. (1959) Graphical representation of amphibole compositions. Am. Mineral. 44, 437-440.

Sokolov, Yu.M. (1959) Some problems of the distribution of micas and other minerals in the Tchuya muscovite deposit. Zapiski Vses. Mineral. Obshtch. 88, 191-204 (in Russian).

_____, Kratz, K.O. and Glebovitskyi, V.A. (1975) Regularities in the formation and distribution of the muscovite and muscovite-rare metal pegmatite formations in metamorphic belts. In: Muscovite Pegmatites of the U.S.S.R., Yu.M. Sokolov, ed., Nauka, Leningrad, 5-15 (in Russian).

Solodov, N.A. (1971) Scientific Principles of Perspective Evaluation of Rare-element Pegmatites. Nauka, Moscow, 591 p. (in Russian).

Stern, W.B. (1966) Zur Mineralchemie von Glimmern aus Tessiner Pegmatiten. Schweiz. Mineral. Petrogr. Mitt. 46, 137-188.

Stevens, R.E. (1938) New analyses of lepidolites and their interpretation. Am. Mineral. 23, 607-628.

Stewart, D.B. (1978) Petrogenesis of lithium-rich pegmatites. Am. Mineral. 63, 970-980.

Sundelius, H.W. (1963) The Peg claims spodumene pegmatites, Maine. Econ. Geol. 58, 84-106.

Sutchkov, P.N. (1961) To the question of the mica content of the Mama granite-pegmatite fields. Mineral. Syrye 47-67 (in Russian).

Swanson, T.H. and Bailey, S.W. (1981) Redetermination of the lepidolite-$2M_1$ structure. Clays & Clay Minerals 29, 81-90.

Thompson, J.B., Jr. (1979) Tschermak component and reactions in pelitic schists. In: Problemy Fiziko-khimicheskoi Petrologii, V.A. Zharikov, V.I. Fonarev and S.P. Koikovskii, eds., Nauka Press, Moscow, 146-159.

_____ (1981) An introduction to the mineralogy and petrology of the biopyriboles. In: Amphiboles and Other Hydrous Pyriboles - Mineralogy, D.R. Veblen, ed., Reviews in Mineralogy 9A, 141-188.

_____ (1982) Composition space: An algebraic and geometric approach. In: Characterization of Metamorphism through Mineral Equilibria, J.M. Ferry, ed., Reviews in Mineralogy 10, 1-31.

Tröger, W.E. (1962) Über Protolithionit und Zinnwaldit. Beitr. Mineral. Petrogr. 8, 418-431.

Trueman, D.L. and Černý, P. (1982) Exploration for rare-element granitic pegmatites. Mineral. Assoc. Canada Short Course Handbook 8, 463-494.

Ukai, Y., Nishimura, S. and Hashimoto, Y. (1956) Chemical studies of lithium micas from the pegmatite of Minagi, Okayama Prefecture. Mineral. J. (Japan) 2, 17-28.

Velde, B. (1972) Celadonite mica: Solid solution and stability. Contrib. Mineral. Petrol. 37, 235-247.

Vladykin, N.V., Dorfman, M.D. and Kovalenko, V.I. (1974) Mineralogy, geochemistry and genesis of rare-element topaz-lepidolite-albite pegmatites of the Mongolian People's Republic. Trudy Mineral. Museum Acad. Sci. USSR, 23, 6-49 (in Russian).

Vlasov, K.A. and Kutukova, E.I. (1959) Izumrudnye Kopi. Acad. Sci. USSR Moscow, 246 p. (in Russian).

Wang, X.-J., Zou, T.-R., Xu, J.-G., Yu, X.-Y. and Qiu, Y.-Z. (1981) Study of Pegmatite Minerals from the Altai region. Scientific Publ. House, Beijing, 140 p. (in Chinese).

Winkler, H.G.F. (1967) Petrogenesis of Metamorphic Rocks; 2nd Ed., Springer-Verlag, New York, 334 p.

Yakovleva, M.E., Razmanova, Z.P. and Smirnova, M.A. (1965) Lepidolite with small optic axial angle. Trudy Mineral. Museum Acad. Sci. USSR, 16, 287-292 (in Russian).

Zagorskyi, V.E., Makrygin, A.I. and Matveeva, L.N. (1974) On the rubidium-rich Li-Fe-Mg-micas from exocontacts of rare-element pegmatites. 1973 Ann. Rpt., Geochem. Inst., Sibir. Branch, Acad. Sci. USSR, 143-147 (in Russian).

9. MICAS in IGNEOUS ROCKS
J. Alexander Speer

INTRODUCTION

The minerals covered in this chapter include only those micas believed to have crystallized from a melt. These include the trioctahedral mica, biotite, and the dioctahedral mica, muscovite. Igneous muscovite is restricted in occurrence to peraluminous plutonic granitoids and, more rarely, extrusive rhyolites. Biotite can occur over nearly the entire spectrum of igneous rocks, from peridotites to granitoids to peralkaline rocks. While more frequently a mineral of plutonic or hypabyssal rocks, it is commonly observed in extrusive rocks. The literature on igneous micas is enormous, and many topics are better included in other chapters in this volume where they can be combined with information from metamorphic rocks, crystal chemical studies, and experimental work. This chapter is confined to studies of naturally occurring micas, often in conjunction with experimental phase equilibria, which have contributed to an understanding about igneous rocks; their origin, emplacement, crystallization history and conditions, timing, subsolidus events, etc.

On the basis of the several topics covered in this chapter, it can be concluded that igneous micas can experience extensive changes in their major element, trace element, and isotopic compositions by postmagmatic processes. This makes it difficult to use biotites in studying magmatic events but does open up the possibility of studying subsequent events which are not traditionally considered a part of igneous petrology. Plutonic biotites would be expected to be more seriously affected than volcanic biotites because of their longer cooling histories. To a certain extent this is true, but an examination of the smaller number of available volcanic biotite analyses suggest they, too, are altered.

OXYGEN AND WATER FUGACITIES

The biotite composition in an igneous rock can be used to estimate either the maximum oxygen fugacity or temperature or the minimum water or hydrogen fugacity at which the biotite formed. If the biotite is part of an equilibrium mineral assemblage such as biotite + magnetite + alkali feldspar, biotite + muscovite + magnetite + aluminum silicate + quartz, or biotite + magnetite + alkali feldspar + leucite, the univariant $f(H_2)$-T or divariant $f(O_2)$-$f(H_2O)$-T conditions can be determined on the basis of the experimental work of Wones and Eugster (1965) and Rutherford (1973). If one or two of the parameters can be determined by an independent technique, the remaining value can be obtained. This experimental work is discussed in more detail by Hewitt and Wones (Chapter 7, this volume). For igneous rocks, most authors use biotite compositional data to determine the relative oxygen fugacity buffer at which the biotite crystallized or use the divariant equilibria involving $f(O_2)-f(H_2O)-T$ to determine the value of water fugacity after obtaining oxygen fugacity and temperature estimates from the coexisting ilmenite + magnetite. In order to obtain oxygen and water fugacities, the biotite must be part of the assemblage biotite + magnetite + alkali feldspar, a condition easy to meet in most igneous rocks containing biotite. The experimental work focuses on the annite end member; the mole percent annite is determined by measuring the maximum concentrations of K, Fe^{2+}, (Al,Si_3), and OH in their respective sites and making the assignment of the least of them the concentration of $KFe_3AlSi_3O_{10}(OH)_2$. This usually is the ferrous iron.

Oxygen fugacity

An approximation of a rock's oxygen fugacity can be made from Wones and Eugster's (1965) estimates of the composition of biotite solid solutions in the ternary system $KFe_3^{2+}AlSi_3O_{10}(OH)_2-KMg_3AlSi_3O_{10}(OH)_2-KFe_3^{3+}AlSi_3O_{12}(H_{-1})$ that are stable at a variety of oxygen buffers. The compositions of the "buffered" biotites in the ternary system are shown in Figure 1 for the magnetite - hematite, Ni - NiO, and $FeSiO_4 - SiO_2 - Fe_3O_4$ buffers (Fig. 1a). A number of investigators have plotted their biotite analyses on this diagram in order to obtain an idea of the relative oxygen fugacity conditions during crystallization. de Albuquerque (1973) found that the biotites of the Aregos granitoids, Portugal, plot close to the fayalite - quartz - magnetite buffer curve. The oxide mineral is ilmenite. In the granitoids of the Sierra Nevada batholith, Dodge et al. (1969) found that biotites coexisting with magnetite and ilmenite define buffering conditions between the Ni - NiO and $Fe_3O_4 - Fe_2O_3$ buffers (Fig. 1b). A few reported igneous biotites have compositions that plot near the $Fe_3O_4 - Fe_2O_3$ line. The examples in Figure 1c are from the Sithonia igneous complex, Greece (Sapountzis, 1976); others include biotites from granitoids of Yemen (Kabesh and Aly, 1980) and the El Atawi granitoids, Egypt (Kabesh and Ragab, 1974). The oxide mineral assemblages are not given in these papers, but the biotite compositions suggest fairly oxidizing conditions.

The distinctiveness of the ferric-ferrous ratio of biotites and the relation between the ratio and the oxide assemblage have been used by Ishihara

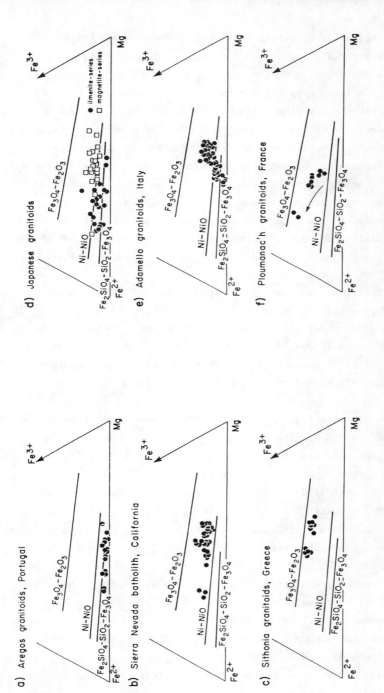

Figure 1. Compositions of biotites from granitoid rocks projected onto the annite-phlogopite-$KFe^{3+}AlSi_3O_{12}(H_{-1})$ ternary system. Lines labeled with the solid oxygen buffers are compositions of the "buffered" biotites from Wones and Eugster (1965). Data are from (a) granitoids of the Aregos region, northern Portugal (de Albuquerque, 1973), (b) granitoids of the central Sierra Nevada batholith, California (Dodge et al., 1969), (c) granitoids of the Sithonia igneous complex, Greece (Sapountzis, 1976), (d) magnetite- and ilmenite-series granitoids of Japan (Ishihara, 1977), (e) granitoids of the Adamello massif, Italy (de Pieri and Jobstraibizer, 1977, 1983), and (f) granitoids of the Ploumanac'h complex, France (Barrière and Cotten, 1979).

301

(1977) to characterize his magnetite (+ ilmenite)- series and ilmenite-series of granitoid rocks. The low Fe^{3+} content of the biotites and the presence of trace amounts of ilmenite, pyrrhotite, graphite, and muscovite in the rocks imply a lower oxygen fugacity in the ilmenite-series granitoids than in the magnetite-series granitoids, which contain biotites with a higher ferric iron content, and abundant magnetite + ilmenite ± hematite, pyrite, titanite, and epidote. Biotite compositions from Japanese magnetite- and ilmenite-series granitoids suggest that the oxygen fugacity boundary between the two types of granites is the Ni - NiO buffer (Fig. 1d). Ishihara (1977) attributed the oxygen fugacity difference to differing source regions. He suggested that magmas in the magnetite-series are generated in the lower crust and upper mantle, whereas the ilmenite-series magmas are generated in or mixed with carbon-bearing rocks of the crust. The hypothesis that granitoid "series" reflects the source area is an approach similar to that of Chappell and White (1974), who have shown that granitoids of southeastern Australia can be classified as being derived either from mostly igneous material (I-type granitoids) or from sedimentary material (S-type granitoids). Whelan (1980) subsequently showed that I-type granites have higher oxygen fugacities; mineralogic studies of I- and S-type granitoids, however, have not been sufficiently detailed to include ferric-ferrous ratios of the biotites.

The examples given in Figures 1a to 1c define trends that parallel the estimated compositions of biotite solid solutions for individual buffers. These trends suggest that consanguinous granitoids were buffered during crystallization; oxygen fugacity increases with decreasing temperature. Figure 1e shows the work of de Pieri and Jobstraibizer (1977, 1983) for biotites of the Ademello massif, Italy. The trend cuts across the "buffered" biotite compositions, suggesting that the biotites crystallized at differing oxygen fugacities. The massif comprises four masses, each with several lithologic varieties which, among other criteria, can be characterized by the ferrous iron content of their biotites. The modal amount of coexisting oxides decreases with decreasing Fe^{3+} content of the biotites. However, rock types linked by possible differentiation do not show trends of oxidation in their biotites and the cause of the various oxygen fugacities is unknown. Barriere and Cotton (1979) found that ferric iron contents of biotites in the Ploumanac'h granitoids increased during magmatic differentiation, reflecting the early fractionation of ilmenite. This effect, combined with buffering by fluids, produced a boomerang-shaped Fe^{2+}/Fe^{3+} trend (Fig. 1f). The trend of increasing Fe^{3+} content would have been more curved, but late-stage percolating fluids caused postmagmatic oxidation of some of the original Fe^{2+}.

The applicability of the ternary biotite compositional projection described above depends on the Fe^{3+} content of the biotites; the projection gives only a qualitative idea of the oxygen fugacity. A quantitative estimate of the oxygen fugacity or temperature can be obtained using the experimental work of Wones and Eugster (1965) if the biotite coexists with magnetite + alkali feldspar and an independent estimate can be made of either temperature or oxygen fugacity. This experimental work is summarized in Figure 2, projected onto the $f(O_2)$-T plane at 2070 bars. The biotite stability field is contoured with values of constant $Fe^{2+}/(Fe^{2+}+Mg)$ values. A biotite with a given $Fe^{2+}/(Fe^{2+}+Mg)$ content is stable at oxygen fugacity values below the corresponding contour. Above that contour, the biotite will decompose by the reaction: biotite = more magnesian biotite + alkali feldspar + magnetite. Included in Figure 2 are the biotite compositional trends of several plutons. Biotites from the Aregos and Sierra Nevada rocks parallel the solid oxygen-buffer curves which suggests that they crystallized under buffered conditions, with oxygen fugacity decreasing with temperature. The trend of the Ploumanac'h biotites follows the same general trend of decreasing oxygen fugacity with decreasing temperature, but the crystallization was not oxygen-buffered because the solid buffers are crossed. In each of these cases, the biotites increase their iron content with progressive differentiation following trend II, described by Wones and Eugster (1965). If biotites become more magnesian with progressive crystallization, they are following trend I. The occurrence of both these trends in calc-alkaline rock series was elegantly documented by Murakami (1969) in Japanese granitoid plutons. Using the solidification index as a measure of progressive crystallization, he compared the Fe/(Fe+Mg) trend of the rocks and their contained biotites (Fig. 3). In most instances both trends coincided (Fig. 3a), but in the Tamagawa pluton the trends diverge: the rocks increase in iron-enrichment whereas the contained biotites become more magnesian (Fig. 3b). There is a concurrent increase in the modal amount of magnetite as well. In situations with trend II biotites, the oxygen fugacity is interpreted as increasing with a drop in temperature. Similar trend II biotites have been reported for the Kolyvan complex, USSR (Potop'yev, 1964); the Ben Nevis complex, Scotland (Haslam, 1968); the Finnmarka complex, Norway (Czamanske and Wones, 1973); the Klokken complex, Greenland (Parsons, 1979, 1981), the Koloula Igneous Complex, Solomon Islands (Chivas, 1981); and the Baie-des-Moutons syenitic

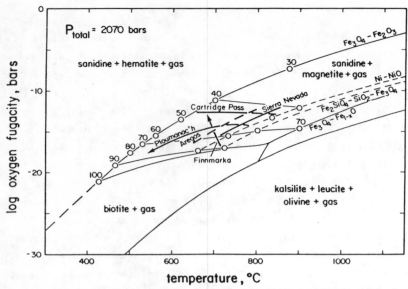

Figure 2. Stability of biotites of differing Fe/(Fe+Mg) compositions as a function of temperature and oxygen fugacity at 2070 bars total pressure, after Wones and Eugster (1965). Fe is total iron. The numbers are contours of constant 100 Fe/(Fe+Mg) values for biotites in the assemblage biotite + sanidine + magnetite. Heavy lines are trends for natural biotites from the Aregos granitoids, Portugal (de Albuquerque, 1973), Sierra Nevada batholith, California (Dodge et al., 1969), Ploumanac'h igneous complex, France (Barriere and Cotten, 1979), Finnmarka complex, Norway (Czmanske and Wones, 1973), and the Cartridge Pass pluton, California (Dodge and Moore, 1968). The temperature and oxygen fugacity conditions for various solid buffers are labeled.

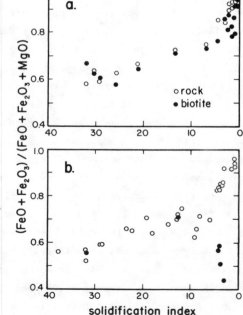

Figure 3. Relation between the $(FeO + Fe_2O_3)/(FeO + Fe_2O_3 + MgO)$ ratio of biotites and their host rocks and the solidification of the rock. The solidification index is

$$100 \, MgO/(MgO + FeO + Fe_2O_3 + Na_2O + K_2O),$$

where the oxides are in weight percent; this is a good way to separate rocks in a series. The data are from Murakama (1969) and include (a) Late Cretaceous to Tertiary granites and the Ichinohe alkali plutonic rocks of Japan and (b) the Tamagawa plutonic rocks of Japan.

complex, Quebec (LaLonde and Martin, 1983). For the Finnmarka complex, Czamanske and Wones (1973) found that the biotites from increasingly silicic rock types decrease in Fe/(Fe+Mg) content from 0.64 to 0.28 (Fig. 2). In the Finnmarka complex the extreme oxidation during differentiation is believed to be a result of vesiculation, where the separated water could act as an oxidizing medium through dissociation and loss of hydrogen. In the Inamumu zoned pluton of the Koloula Igneous Complex, Chivas (1981) found that from the oldest to youngest magmatic units, the Fe/(Fe+Mg) content of the biotites decrease from 0.45 to 0.33. This is interpreted as increasing oxidation resulting from the increased partial pressure of water caused by increasing amounts of fluid separating from the magma with crystallization. Haslam (1968) suggested that increased oxidation in the Ben Nevis was a result of atmospheric oxygen introduced by way of fissures into the hypabyssal pluton. Murakami (1969) also presumed that the increased oxygen fugacity of the Tamagawa pluton was atmospherically derived; this may be common in cases of cauldron subsidence such as the Tamagawa, Ben Nevis, and Finnmarka plutons.

Igneous plutons with uniform $Fe^{2+}/(Fe^{2+}+Mg)$ biotite compositions throughout the body could be interpreted several ways. Either all the biotites crystallized at one oxygen fugacity and temperature; the oxygen fugacity remained constant during cooling, perhaps by external buffering; or all the biotite compositions reequilibrated under the same conditions during a uniform, post-crystallization event. Dodge and Moore (1968) found that the biotites of the Cartridge Pass pluton, California, have uniform major element compositions with $Fe^{2+}/(Fe^{2+}+Mg)$ of about 0.45. The pluton is zoned from quartz diorite at its margins to quartz monzonite at its core. Because of the variation in the rock type and the greater Cr, Ni, and V and lesser Sc, Sr, and Zn contents in the biotites of the melanocratic rocks, Dodge and Moore concluded that the biotites crystallized over a range of temperatures. If so, the oxygen fugacity would have been constant at about 10^{-13} bars (Fig. 2). Dodge and Moore conclude that the Cartridge Pass pluton biotites acquired their uniform major-element compositions late in the crystallization history of the body as a result of a permeating late-stage fluid.

Water fugacity

If the oxygen fugacity and temperature of equilibration for an assemblage biotite + alkali feldspar + magnetite can be determined, water fugacity can be calculated (Wones and Eugster, 1965). Wones (1972) redetermined the free energy change of the reaction annite + O_2 = sanidine + magnetite + water after reexamining the stability of annite (Wones et al., 1971) and presented the relation

$$\log f_{H_2O} = 7409/T + 4.25 + 0.5 \log f_{O_2} + 3 \log X_{annite} - \log a_{sanidine} + \log a_{Fe_3O_4}, \quad [T \text{ in Kelvins}], \quad (1)$$

where X_{annite} is the fraction of Fe^{2+} in the octahedral site. On the basis of additional work on annite stability, Wones (1981) suggested the relation

$$\log f_{H_2O} = 4819/T + 6.69 + 3 \log X_{annite} + 0.5 \log f_{O_2} - 0.011(P-1)/T(°K) + \log a_{sanidine} + \log a_{Fe_3O_4} \quad [T \text{ in Kelvins}], \quad (2)$$

which is valid at oxygen fugacities between Ni - NiO and Fe_3O_4 - Fe_2O_3. Equation 1 has been used by several authors to obtain water fugacity estimates.

Czamanske and Wones (1973) applied Equation 1 to biotites of the Finnmarka complex. Because of the F-rich compositions of the biotites with between 0.8 and 5.6 wt % F, the activity of the annite was assumed to be $a_{annite} = X_{Fe^{2+}}^3 X_{OH}^2$, and the $3 \log X_{annite}$ term in Equation 1 is replaced with $3 \log X_{Fe^{2+}} + 2 \log X_{OH}$. The activity of Fe_3O_4 in magnetite was taken as 1 and the activity of $KAlSi_3O_8$ at 0.6 based on the work of Waldbaum and Thompson (1969) where the activity of sanidine is about 0.6 in the 600-800°C temperature range for feldspars with compositions between Or_{25} and Or_{75}. Discrete estimates of temperature and oxygen fugacity were not possible, so Czamanske and Wones calculated stability curves for the Finnmarka biotites in $f(H_2O)$-T space using oxygen fugacity estimates between Ni - NiO and Fe_3O_4 - Fe_2O_3. Approximate values of water fugacity and temperature of 1150 bars $f(H_2O)$ and 720°C for the monzonites and 2200 bars $f(H_2O)$ and 670°C for the granites were obtained from the intersection of the biotite stability curves and the minimum melting of granite (Or-Ab-Qz). These estimates are extreme values and depend on choice of oxygen fugacity, partial pressure of the fluid phase, uncertainty on subsolidus loss of F, and location of the Ab-Or-Qz minimum melting. For the Ploumanac'h biotites, Barriere and Cotten (1979) used ilmenite - magnetite pairs to determine temperature and oxygen fugacity values of 550°C and $\log f(O_2) = -19.5$ for the residual granite facies. Using Equation 1, ignoring the F content of the biotite, and assuming ideal mixing for the alkali feldspar ($a_{sanidine} = 0.85$) and magnetite ($a_{magnetite} = 0.95$), they obtained a value for the water fugacity of 490-550 bars which corresponds to a partial water pressure of 800-1000 bars. This value is much less than the fluid pressure derived from the contact aureole, and if not attributable to the uncertainty of the calculations, suggests variable partial pressures of the fluid during the crystallization or a subsolidus drop in the fluid pressure when the minerals may have reequilibrated.

BIOTITES COEXISTING WITH OTHER MINERALS

Amphibole

The distribution of elements between coexisting biotite and amphibole is a popular topic of study because these minerals are the most common ferromagnesian silicates in granitoid rocks. Of most interest has been the Fe-Mg distribution, but Mn, Al, F + Cl, Fe^{2+}/Fe^{3+}, and trace element distributions have also been investigated (Table 1). It was hoped that these elemental distributions would be reliable indicators of pressure, temperature, oxygen fugacity, fluid composition and other petrologic parameters. Although the distribution of elements between the two minerals has been found to be regular, interpretation has been hindered by the structural and chemical complexity of the two minerals. However, sufficient work has been done to show that it is the structure of the amphibole, rather than any physical condition, that is the most important factor in determining the distribution coefficients.

The distribution of $Fe^{2+}/(Fe^{2+}+Mg)$ between amphibole and biotite from a number of igneous rocks is shown in Figure 4. Most igneous amphibole + biotite pairs have intermediate compositions; the iron-rich pairs are from alkali granites. The data define a line with a slope of one. The distribution coefficient defined as $(Mg/Fe)_{biotite}/(Mg/Fe)_{amphibole}$ ranges between 1.1 and 0.66. The value of the distribution coefficient is thought to depend on a variety of parameters: rock chemistry, mineral assemblage, mineral chemistry, and physical conditions. Gorbatschev (1969) demonstrated the regularity of the Fe-Mg distribution between amphibole and biotite and found a statistical relationship with the composition of the tetrahedral site in both minerals. Kanisawa (1972) concluded that the variation in Fe-Mg distribution appeared dependent on the Al content of the biotite and amphibole. More specifically, Tanaka (1975) found that the distribution is directly proportional to the tetrahedral Al content of the amphibole (Fig. 5). Because of the systematic relation among the several chemical substitutions in amphiboles (Gorbatschev, 1977), the Fe-Mg distribution coefficient has a linear relationship with the amphibole's Mg/Fe, Na+K, $Al^{iv} + Fe^{3+}$, and Ti contents as well (Kato et al., 1977). Ramberg (1952) first suggested that the increasing Al^{iv} content in amphiboles causes a decrease in the electronegativity of the oxygen, and favors an increase in Fe/Mg.

Table 1. Studies on the distribution of elements between coexisting amphibole and biotite.

	Fe-Mg	Mn	Al	F+Cl	Fe^{+2}/Fe^{+3}	Cu	Zn	Ti	Na
Czamanske et al. (1977)		X		X				X	
Banks (1976)				X					
Gilberg (1964)				X					
Dodge & Ross (1971)	X	X	X	X	X				
Kanisawa (1972)	X	X	X		X				
Graybeal (1973)		X				X	X		
Greenland et al. (1968)		X							
Gorbatschev (1970,1977)	X		X						
Hietanen (1971)*	X	X	X	X	X	X			
de Albuquerque (1973,1974)*	X	X	X		X	X		X	X
Bancroft & Brown (1975)					X				
Simonen & Yorma (1969)	X	X	X	X	X				
Christofides & Sapountzis (1983)	X	X	X						
Baker et al. (1975)	X	X	X	X					
Fiala et al. (1976)	X	X	X	X					
Tanaka (1975)	X	X	X						
Kato & Onuki (1977)	X	X	X						
Fershtater et al. (1970)		X	X					X	X
Fershtater (1973)								X	

*Also includes data for Ba, Co, Cr, Ga, Ni, Sc, Sr, V, Zr, Li, Y, and Yb.

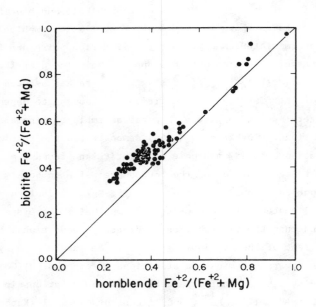

Figure 4. Distribution of $Fe^{2+}/(Fe^{2+}+Mg)$ in igneous hornblendes and biotites. Data from mineral pairs where ferrous/ferric iron ratio was determined: Larsen and Draisin (1948), Best and Mercy (1967), Haslam (1968), Simonen and Vorma (1969), Hietanen (1971), Dodge and Ross (1971), Ewart (1971), Kato (1972), Kanisawa (1972), Kato and Tanaka (1973), de Albuquerque (1973; 1974), Honma (1974), Tanaka (1975), Barker et al. (1975), Kato et al. (1977).

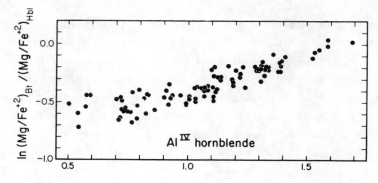

Figure 5. Relation between the natural logarithm of $(Mg/Fe^{2+})_{biotite}/(Mg/Fe^{2+})_{hornblende}$ and the tetrahedral content in hornblende for coexisting biotite and hornblende in igneous rocks. Data from Haslam (1968), Kanisawa (1969, 1972, 1974), Dodge and Ross (1971), Hietanen (1971), Kato (1971), de Albuquerque (1973, 1974), Kato and Tanaka (1973), Tanaka (1975), Christofides and Sapountzis (1983).

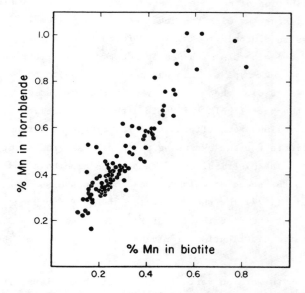

Figure 6. Distribution of Mn between coexisting hornblende and biotite in igneous rocks. Data are from various granitoids and volcanic rocks in the United States (Greenland et al., 1968) and the central Bohemian Massif, Czechoslovakia (Fiala et al., 1976).

Kanisawa (1972) found no evidence of a systematic variation in Fe-Mg distribution between amphibole and biotite with metamorphic grade, a conclusion similarly reached by Stephenson (1977) who found that the rock's composition was more important. By analogy, the Fe-Mg distribution between biotite and amphibole in igneous rocks is also largely independent of physical conditions. In only one instance has a different interpretation been made. Heitanen (1971) concluded that the difference between the distribution coefficient in a tonalite of the southern California batholith, $K_D = 0.83$, and a quartz diorite in the Sierra Nevada batholith, $K_D = 0.65$, was a result of different conditions of pressure and temperature during crystallization. The Sierra Nevada quartz diorite crystallized at 700°C and 4 kbar; the southern California tonalite crystallized at 700-800°C and 2-3 kbar. The opposite was found by de Albuquerque (1974), however, who pointed out that the distribution coefficients for low-pressure igneous rocks such as in the Aregos region, $K_D = 0.60$, or the Ben Nevis, $K_D = 0.71$ (Haslam, 1968), are less than for the deeper-seated igneous rocks of the Scottish Caledonian, $K_D = 0.79$-0.87 (Nockolds and Mitchell, 1948). It would appear that the influence of physical conditions is less important than the effect of the compositions of the amphibole and biotite, which in turn depend on rock chemistry.

The distribution of manganese between amphibole and biotite has received some attention, starting with the detailed study by Greenland et al. (1968) for occurrences in plutonic rocks. They found an equilibrium distribution of Mn between the two minerals with nearly identical distribution coefficients for a variety of rock types. This has been the case found by a variety of investigators (Fig. 6). The similarity of the distribution coefficients suggests that the partitioning of Mn is either insensitive to temperature or a result of subsolidus reequilibration. The slight difference of the distribution coefficients for the extrusive and hypabyssal rocks suggest the latter for most rocks. The distribution of Mn between amphibole and biotite and its relation to the speed at which a rock cools was also suggested by Czamanske et al. (1977) for the granitoids of the Pliny Range, New Hampshire. However, they felt the distribution of Mn between biotite and ilmenite retains its magmatic distribution, implying that only the amphibole reequilibrates.

In addition to work on various aspects of chemical partitioning between biotite and amphibole, some work has also been done on isotopic partitioning. Suzuoki and Epstein (1976) have shown that the hydrogen isotopic equilibria between water and hydrous silicates depend on temperature and chemical composition, especially Fe-Mg content. The hydrogen isotopic compositions of coexisting biotite and amphibole in granitoid rocks of Japan have been reported by Kuroda et al. (1974, 1976, 1977), who recognized two situations: rocks which have the expected D-H distribution of $\delta D°/oo$ hornblende greater than $\delta D°/oo$ biotite and those rocks that do not. The equilibrium mineral pairs

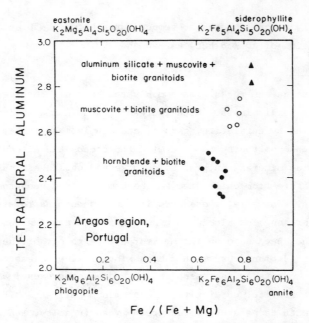

Figure 7. Biotites from the Aregos region, Portugal, projected onto the phlogopite-annite-siderophyllite-eastonite quadrilateral showing the increase in Al and Fe/(Fe+Mg) contents in biotites coexisting with hornblende, muscovite, and aluminum silicate. Data from de Albuquerque (1973).

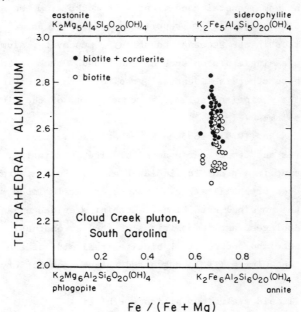

Figure 8. Biotites from the Clouds Creek pluton, South Carolina, projected onto the phlogopite-annite-eastonite-siderophyllite quadrilateral showing increased tetrahedral aluminum content of biotites coexisting with cordierite. Data from Speer (1981).

have isotopically reequilibrated at temperatures between 450 and 1000°C with a fluid composition of $\delta D °/oo$ between -40 and -90.

Aluminous minerals

Biotites in igneous rocks are commonly the host of the excess aluminum, making the rocks peraluminous. With increasing aluminum activity, biotite may be accompanied by cordierite, garnet, and aluminum silicate. Such assemblages present an opportunity to determine the compositional equilibria among the ferromagnesian minerals in granitoid rocks and obtain estimates of pressure and temperature from calibrated geothermometers and geobarometers or petrologic grids. Normally, these are difficult phenomena to study in igneous rocks, but the presence of aluminous minerals allows application of the extensive work that has been done on the metamorphic petrology of these minerals.

Biotite compositions vary with the coexisting mineral assemblage. The aluminum content generally increases in the sequence pyroxene, amphibole, biotite alone, muscovite, cordierite, and aluminum silicate. The chemistry will also vary with temperature, pressure, oxygen fugacity, and water fugacity and plots of all literature data will include those complications as well. It is best to examine analyses from one pluton or a cotemporal group that crystallized under the same general conditions. In rocks from the Aregos region of Portugal, de Albuquerque (1973) reports increasing Al as well as Fe/(Fe+Mg) contents in biotites in the sequence hornblende, muscovite, aluminum silicate (Fig. 7). Within a single pluton such as the Clouds Creek in South Carolina (Speer, 1981), where otherwise identical granitoids can be divided into biotite and cordierite + biotite facies, the compositions of the biotites reflect the mineral assemblages (Fig. 8).

Abbott and Clarke (1979) proposed hypothetical liquidus relations at differing pressures and temperatures among biotite, cordierite, aluminum silicate, and garnet in a granitic liquid (Fig. 9). Such diagrams allow understanding of the crystallization sequences of peraluminous melts as well as estimates of the physical conditions. Combining evidence based on textures, phase assemblages, and compositions of the ferromagnesian minerals of the garnet + biotite and andalusite + biotite granitoids of the South Mountain batholith, Allan and Clarke (1981) suggested the sequence of crystallization reactions:

$$\text{liquid} = \text{aluminum silicate} + \text{biotite}$$
$$\text{liquid} + \text{aluminum silicate} + \text{biotite} = \text{garnet}$$
$$\text{liquid} = \text{biotite} + \text{garnet}$$

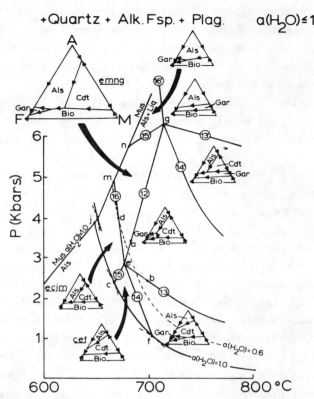

Figure 9. Regions of distinct AFM liquidus topologies in pressure-temperature space as a function of a(H$_2$O). After Abbott and Clarke (1979).

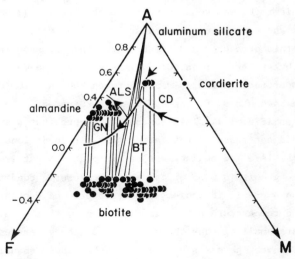

Figure 10. AFM projection from muscovite for the magmatic occurrences of aluminum silicate, biotite, cordierite, and garnet in peraluminous granites. After Clarke (1981).

This sequence of crystallization reactions corresponds to the pressure-temperature area *emng* in Figure 9 with estimated conditions of 675-725°C and 3-6 kbars. Using comparable arguments for the cordierite + biotite granitoids of the Clouds Creek pluton, South Carolina, Speer (1981) estimated a pressure of emplacement between 1 and 2.5 kbars. Clarke (1981) plotted the mineral chemistries from a number of aluminum silicate-, muscovite-, cordierite-, garnet-, and biotite-bearing granitoids (Fig. 10) and obtained cotectic paths consistent with topology in region *emng* of Abbott and Clarke (1979).

Several studies have used experimentally determined Fe-Mg distribution coefficients to determine pressures and temperatures of acidic rocks. Wyborn et al. (1981) estimated a pressure of 5-6 kbars and temperatures of 750-800°C for a biotite + garnet + cordierite + orthopyroxene-bearing dacite in the Silurian volcanics of the Lachian Fold belt, Australia. They concluded that these were the conditions of the source region and that the melt was extruded directly from the source area without reequilibration. By comparison, other volcanic rocks in the area were concluded to have reequilibrated in shallow-level magma chambers before extrusion. Wood (1974) used garnet-biotite geothermometry to estimate temperatures of 700-825°C for the Mt. Somers and Mt. Misery garnet + biotite rhyolites of New Zealand. These temperatures are comparable to the temperatures of 690-765°C found for the garnet + biotite + cordierite + orthopyroxene-bearing acid volcanics of the Cerberan Cauldron, Australia, reported by Birch and Gleadow (1974). An example of a similar study for plutonic rocks was done by Pattison et al. (1982), who combined biotite + garnet and anorthite + grossular + aluminum silicate + quartz equilibria to obtain pressures and temperatures for the Hepburn and Wentzel batholiths, Northwest Territories, Canada, as well as the wall rocks. The pressure and temperature estimates range between 800 and 1100°C and 5 and 10 kbars, which are in contrast to the 650°C and 3 kbars estimates for the adjacent metapelites (Fig. 11). This contrast suggests that the minerals in the granitoids did not reequilibrate at submagmatic conditions, nor did they crystallize at the level of emplacement of the pluton. Pattison et al. (1982) conclude that the granitoid mineral assemblages preserve conditions of the source area. Those granitoid bodies in the Hepburn and Wentzel batholiths with pressure and temperature estimates consistent with the level of emplacement (Fig. 11) are small plutons thought to have been generated *in situ* by anatexis. Other investigators have reported subsolidus reequilibration of such mineral pairs. Kistler et al. (1981) found subsolidus temperatures of

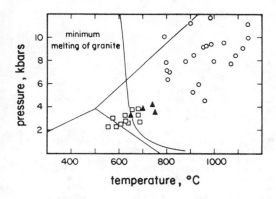

Figure 11. Pressure and temperature estimates for the Hepburn and Wentzel batholiths (open circles) and their enclosing Epworth Group metapelites (open squares) from garnet-biotite and garnet-plagioclase-sillimanite-quartz equilibria. Data from Pattison et al. (1982). The aluminum silicate triple point is from Holdaway (1971). The dark triangles are from small granitoid bodies believed to have formed *in situ* by anatexis.

365-505°C for garnet + biotite pairs from the muscovite + biotite granitoids of the Ruby Mountains, Nevada. The garnets are considered magmatic minerals; thus, reequilibration must have occurred.

The aluminous, ferromagnesian minerals of these igneous rocks are usually interpreted as restite phases, as suggested by the higher pressure estimates. The work of Abbott and Clarke (1979) indicates that a series of crystallization reactions should occur with drop in pressure and temperature during emplacement and cooling; such sequences were used to obtain pressure and temperature conditions for the South Mountain batholith and Clouds Creek pluton mentioned above.

Apatite

The coexistence of F-OH minerals such as muscovite, amphibole, topaz or apatite with biotite provides a possible geothermometer. Fluorine-OH partitioning between muscovite and biotite has been shown by Munoz and Ludington (1977) to be largely independent of temperature. The distribution of these elements between amphibole and topaz is unstudied. The apatite-biotite geothermometer was quantified by thermodynamic data when proposed by Stormer and Carmichael (1971). It gave temperatures in the subsolidus range. The geothermometer was revised by Ludington (1978) who included the effects of the octahedral site composition, found by Munoz and Ludington (1974) to greatly influence the F-OH distribution coefficients in biotite, and more recent thermodynamic data for apatite. Ludington's (1978) apatite - biotite geothermometer, T (Kelvins) = 1100/log K, where

$$\log K = \log(X_F/X_{OH})_{apatite} + \log(X_{OH}/X_F)_{biotite} - 1.107(X_{annite})$$
$$- 1.444(X_{siderophyllite}),$$

gives temperatures in the magmatic range. Blattner (1980) examined the

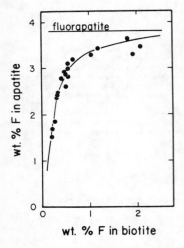

Figure 12 (left). Distribution of fluorine between coexisting apatite and biotite from the Neogene granitoid rocks of Kyushu, Japan, reported by Nedachi (1980).

Figure 13 (below). Distribution of F/(F+OH) between coexisting apatite and biotite from the Skaergaard intrusion, Greenland (Nash, 1976), the Lucerne pluton, Maine (Wones, 1980), and a quartz latite porphyry at Bingham, Utah (Parry et al., 1978). Solid lines are the 1100°C and 600°C isotherms for phlogopite; dashed lines are isotherms calculated for a biotite composition of the Lucerne pluton. All calculations are based on the work of Ludington (1978).

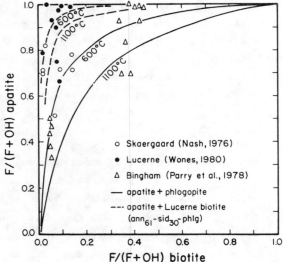

available F analyses of coexisting apatite + biotite pairs and concluded that the partitioning generally corresponds to the distribution expected of a primary igneous event with temperatures near 700°C. Nedachi (1980) found an even more regular distribution of F between apatite and biotites of the Neogene granitoid rocks of Japan (Fig. 12). The temperatures given by these mineral pairs according to Ludington's method are reasonable but rather scattered, Nedachi believes, because of the uncertainty in the F content of the apatite. As Ludington (1978) notes, this geothermometer is sensitive to analytical precision; uncertainties of ±5% in F/(F+OH) represents a 330°C range. Nevertheless, Nedachi feels that the F distribution between apatite

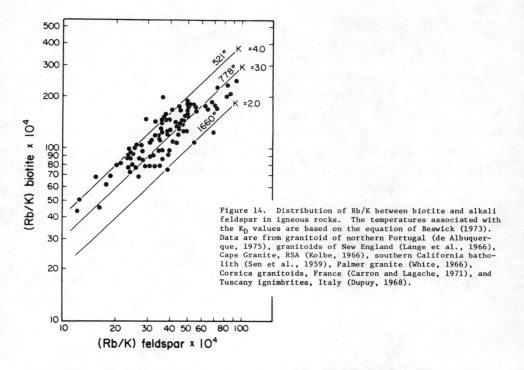

Figure 14. Distribution of Rb/K between biotite and alkali feldspar in igneous rocks. The temperatures associated with the K_D values are based on the equation of Beswick (1973). Data are from granitoid of northern Portugal (de Albuquerque, 1975), granitoids of New England (Lange et al., 1966), Cape Granite, RSA (Kolbe, 1966), southern California batholith (Sen et al., 1959), Palmer granite (White, 1966), Corsica granitoids, France (Carron and Lagache, 1971), and Tuscany ignimbrites, Italy (Dupuy, 1968).

and biotites in these Japanese rocks represents igneous partitioning. In contrast, Nash (1976) concluded that while the apatites in the Skaergaard intrusion had their original, igneous compositions, the biotites, interpreted as subsolidus reaction products, had subsequently lost F to migrating meteoric groundwaters. As seen in Figure 13, the distribution of F between the two minerals in the Skaergaard is irregular, just as it is for the granitoids of the Lucerne pluton reported by Wones (1980). Figure 13 also illustrates the sensitivity of the geothermometer to analytical uncertainty and to biotite composition; the results of Nash and Wones highlight the problem of continued exchange in the subsolidus.

Feldspars

The distribution of the alkalis between alkali feldspar and biotite has been explored as a possible geothermometer. Fershtater (1973) studied the distribution of sodium between natural occurrences and calibrated the geothermometer with the two-feldspar geothermometer. Aside from the difficulties of calibration, Fershtater concluded that reequilibration in the subsolidus had occurred. Beswick (1973) experimentally studied the distribution of Rb between phlogopite and sanidine and calculated that $K_D^{phlogopite/sanidine}$

increased from 2.9 at 800°C to 4.1 at 500°C. He suggested that this distribution could serve as a criterion for equilibrium as well as a geothermometer, but that it was limited by analytical precision and subsolidus reequilibration. Application to natural examples showed volcanic rocks had temperatures greater than 860°C whereas plutonic rocks range between 1100 and 550°C with no correlation with the rock type. In studies of the Rb distribution between alkali feldspar and biotite (Lange et al., 1966; de Albuquerque, 1975), systematic distributions are found suggesting equilibrium. A summary of published values in Figure 14 also shows a rather regular relation and generally "igneous" temperatures, although in individual cases the temperatures are unrealistic or inconsistent with accepted petrologic criteria such as the temperatures of more mafic rocks being greater than the more leucocratic ones. In part, this could be a result of analytical uncertainty or post-magmatic alteration. de Albuquerque (1975) suggests that differing feldspar compositions and the larger temperature range of biotite crystallization as compared to feldspar could also play a role.

Magnetite

Fershtater (1973) found that the distribution of titanium between biotite and magnetite varied between low and high temperature rocks. The distribution coefficient

$$K = [100Ti/(Ti+Fe)_{biotite}]/[100Ti/(Ti+Fe)_{magnetite}]$$

is greater than 15 for hypabyssal granitoids and gabbroids and felsic extrusive rocks and less than 10 for plutonic granitoids. He attempted to calibrate this with the two-feldspar and magnetite-ilmenite geothermometers, but obtained a rather large scatter attributed to post-magmatic reequilibration.

Magnetite coexisting with biotite as well as alkali feldspar forms the basis of determining oxygen fugacity. This aspect of coexisting mineral pairs is covered in the section on oxygen and water fugacity.

Muscovite

Distribution of elements between the two micas have been explored from several aspects. de Albuquerque (1975) suggested that the regular distribution of Rb, Cs, and Ba between coexisting muscovite and biotite in granitoids of the Aregos region of Portugal indicated that the minerals crystallized in equilibrium and concluded that the muscovite was primary. Munoz and Ludington (1977) investigated the F = OH exchange equilibria between muscovite and biotite and found that, although F is selectively partitioned in biotite, the distribution is independent of temperature and fluid composition. Munoz and Ludington suggest that the distribution of F and OH between the two micas can be used instead as a measure of F loss from the micas, an equilibriometer. As an example, they use the muscovite-biotite pairs from two granitoids described by Muller (1966) and find that the F content appears to be in equilibrium. Gunow et al. (1980) found F-OH equilibration for biotite-sericite pairs in the Henderson molybdenite deposit, Colorado. Because these particular minerals are secondary in the deposit, the results indicate that F-OH exchange occurs between micas in igneous rocks at postmagmatic conditions, perhaps as low as 350°C, based on fluid inclusion homogenization temperatures.

ISOTOPIC COMPOSITION

Isotopic compositions of igneous biotites and muscovites have been investigated for the information they contain about geochronology, extent of the isotopic exchange between the pluton and the wall rock at the magmatic stage, interactions between the pluton and aqueous fluids during postmagmatic cooling, the probable source region of the magma, and geothermometry.

Geochronology

Muscovite and biotite both incorporate radiogenic elements ^{87}Rb and ^{40}K largely to the exclusion of the ^{87}Sr or ^{40}Ar daughter products and thus can be used to date the minerals isotopically. This "date" represents the time at which the mica became a closed system with respect to Rb-Sr or K-Ar. When, in the 1950's, isotopic dates first began to be widely obtained on micas, these dates were interpreted as ages of igneous crystallization. But as the number of such dates increased and comparisons were made with other techniques, it became obvious that the Rb-Sr and K-Ar dates had commonly been "reset," giving younger ages than that of the igneous rock. This was particularly true for plutonic rocks, and dating of igneous rocks by these techniques became restricted to volcanic or hypabyssal rocks. By the 1970's it was realized that these "dates" were ages since the micas dropped below the temperature at which Ar and/or Sr cease to diffuse on a relatively widespread scale and begin to accumulate in the minerals. This blocking temperature has been determined at 130-225°C for K-Ar in biotite (Damon, 1968; Turner and Forbes, 1976) and 250-311°C for Rb-Sr in biotite (Hofmann and Giletti, 1970). The actual temperature depends on grain size and cooling rate. The important petrologic aspect of all this is that the "date" represents the time of a temperature on a cooling curve. Combining a number of such dates of minerals with different blocking temperatures will define a

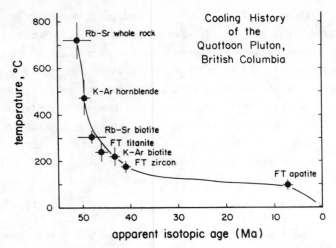

Figure 15. Cooling history of a sample from the Quottoon pluton, British Columbia, obtained by plotting the mineral K-Ar, Rb-Sr, or fission track (FT) ages against closure temperature estimates. Data from Harrison et al. (1979). The whole rock Rb-Sr age is interpreted as the igneous crystallization age.

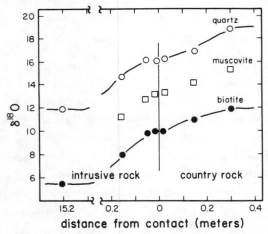

Figure 16. Variation of $\delta^{18}O$ across the contact of the Sawtooth stock trondhjemite, Nevada. Data from Shieh and Taylor (1969).

cooling curve for the host rock. An example of a cooling curve is given in Figure 15 for the Quottoon pluton, British Columbia. In this case the rock cools rapidly directly after crystallization, and more slowly as time progresses.

Wall-rock/igneous rock interaction

As a large number of isotopic age dates began to accumulate, it became evident that the isotopic compositions can record interactions between the magma, or cooling igneous rocks, and the enclosing rocks. This permits the study of the mechanism and duration of cooling; in addition, this exchange provides a tool for understanding the formation of ore deposits associated with igneous rocks.

An oxygen isotopic exchange zone was identified at the contact of the Sawtooth stock, Nevada, by Shieh and Taylor (1969). They determined the isotopic composition of the minerals in a traverse across the contact (Fig. 16) and found a smooth, systematic variation. The values at the contact were the arithmetic mean of the intrusive and country rock values. The isotopic fractionation between quartz and biotite in all rocks was constant and gave a temperature of 400-450°C using the quartz-biotite geothermometer of Shieh and Schwarz (1974). This was concluded to be the temperature of final equilibrium. Nagy and Parmentier (1982) considered these exchange zones in more detail and concluded that the oxygen isotopic variation is consistent with a diffusional transport of the oxygen isotopes. Other exchange zones, although not as well defined as that for the Sawtooth stock, are asymmetric, suggesting that either the diffusivities of the intrusive and country rocks differ or there has been transport of fluid out from the pluton. The width of the Sawtooth exchange zone is about 0.5 m which yields an estimate of the bulk diffusivity of oxygen in the rock at $10^{-9} cm^2/sec$. Nagy and Parmentier (1982) have calculated that solid state diffusion within mineral structures or along grain boundaries is too slow to explain the width of the exchange zone. However, the required diffusivity can be explained if transport is through a water-rich fluid phase occupying fractures.

By comparison with oxygen isotopic exchange, hydrogen isotopic exchange occurs on a truly regional scale, particularly in shallow intrusions. There are several reasons. In shallow environments, the rocks are sufficiently permeable to allow interaction of meteoric groundwaters with the melt or hot rock. The isotopic composition of the groundwater can have a greatly different value than the rocks, and the amount of hydrogen in the rocks is much less than the amount of oxygen. Therefore, much less water is required to change the hydrogen isotopic composition of the rock. These features make hydrogen isotopes sensitive to hydrothermal alteration. The study of the hydrogen isotopic composition of igneous rocks is largely the study of the hydrous minerals in the rock: biotite ± hornblende ± chlorite. A well-documented study of hydrogen isotope exchange is the work on the Idaho batholith in Idaho and Montana. The batholith comprises leucocratic, biotite ± hornblende or biotite ± muscovite granite to granodiorite plutons of Cretaceous age and contains a number of Au-Ag deposits. The eastern edge of

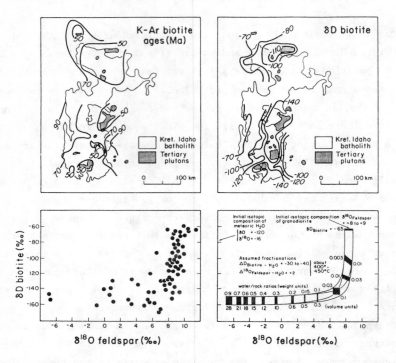

Figure 17. (a) Sketch map of the Idaho batholith contoured with the mica K-Ar ages (Ma). Data from Armstrong (1975) and Criss and Taylor (1983). (b) Sketch map of the Idaho batholith contoured with δD values of biotite ± chlorite. The deuterium depletions are centered on the younger, Tertiary plutons. Data from Taylor and Margaritz in Taylor (1978) and from Criss and Taylor (1983). (c) Plot of the δD values for biotite ± hornblende ± chlorite versus the $\delta^{18}O$ value of feldspar in the Idaho batholith. The original magmatic isotopic composition was about δD = -70 and $\delta^{18}O$ = +9. (d) Plot of calculated δD biotite and $\delta^{18}O$ feldspar values that would evolve during the meteoric-hydrothermal alteration of a granodiorite with increasing amounts of water assuming conditions on the diagram. After Taylor (1977).

the pluton was intruded by later (Tertiary) granitoid plutons. These later plutons had a profound effect on the Idaho batholith, systematically resetting the K-Ar biotite ages (Fig. 17) and causing widespread rock alteration. Taylor and Margaritz (1976, 1978) found that the alteration resulted in strong D and ^{18}O depletion. The isotope exchange was produced by a regional scale hydrothermal circulation induced by the crosscutting Tertiary plutons. The southern half of the batholith has been studied in much more detail by Criss et al. (1982) and Criss and Taylor (1983). The distribution of the K-Ar dates in the Idaho batholith (Fig. 17) shows a younging toward the Tertiary plutons over a broad area interpreted as reset (Armstrong, 1975). Contours of the δD values of the biotites ± hornblende ± chlorite show a systematic variation with the Tertiary plutons. The least disturbed isotopic compositions of the biotites are $\delta D = -70 \pm 5$; the feldspars have $\delta^{18}O$ values of $+9.3 \pm 1.5$. These values become -8.2 and -176, respectively, in intensely altered rocks. The mapping shows that anomalous isotopic compositions are centered about the later plutons, and cover 1500 km^2, indicating the lateral extent of the hydrothermal system. The circulating, meteoric water had an isotopic composition of $\delta^{18}O = 16$ and $\delta D = -120$.

A plot of $\delta^{18}O$ values of feldspars versus the δD values of biotites from the same rocks of the Idaho batholith (Fig. 17) shows an inverted L-shaped distribution of data. This distribution, discussed by Taylor (1977), is the result of rocks exposed to varying amounts of circulating water with different isotopic compositions. This causes the less abundant hydrogen to change faster at smaller water/rock ratios than the oxygen composition. At greater water/rock ratios, the hydrogen composition has essentially become that of the circulating water and the more abundant oxygen begins to change. The change in hydrogen isotopic composition is sensitive to minor hydrothermal alteration whereas oxygen isotopes are not. Thus the latter are better indicators of the extent of hydrothermal activity in areas of more intense alteration. Figure 17 suggests that an enormous volume of water was involved, but Criss and Taylor (1983) conclude such a volume could be supplied by a small percentage of the rainfall over a reasonable cooling time for the pluton. The Idaho batholith is an example where the later plutons established a system of overlapping meteoric-hydrothermal convective cooling systems. These convective systems caused widespread hydrothermal alteration, resetting the isotopic ages and isotopic reequilibration. This mechanism also answers the age-old question of how hydrothermal ore deposits can form when the likely soluble species are so insoluble: increase the volume of water.

Geothermometry

The distribution of two stable isotopes of an element between two coexisting minerals depends on the crystal structure of the minerals, the temperature of their last equilibration, and, to a lesser degree, elemental composition, and negligibly on pressure. This temperature effect is pronounced only for the light elements, and with suitable calibration the isotopic composition of a mineral pair can be used for geothermometry. The elements of isotopic thermometry have been reviewed by Javoy (1977) and Clayton (1981). The oxygen isotopic compositions of both biotite and muscovite have been investigated for their geothermometer potential. In practical

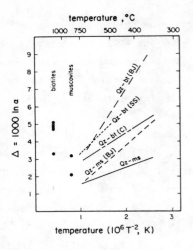
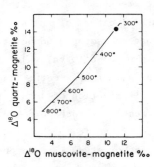

Figure 18 (left). Comparison of the experimental and empirical oxygen isotopic fractionation curves for quartz + biotite and quartz + muscovite pairs with temperature. The solid line for quartz + muscovite is obtained by combining the experimental work for muscovite + quartz (O'Neil and Taylor, 1969) and quartz + water (Matsuhisa et al., 1979). The remaining curves are empirical and are based on experimental determinations of other mineral pairs and the observed distribution between one of those minerals and the mica: SS, Shieh and Schwarcz (1974); BJ, Bottinga and Javoy (1975); and C, Clayton (1981). The dots are oxygen isotopic fractionation values of quartz-biotite and quartz-muscovite pairs from igneous rocks (Taylor and Epstein, 1962).

Figure 19. ^{18}O fractionation between quartz, muscovite, and magnetite (Deines, 1977). The dots are determined values from a trondhjemite from the Sawtooth stock (Shieh and Taylor, 1969).

terms, there are two ways to present the data, either as the difference in $^{18}O/^{16}O$ between two minerals with temperature, or the difference among three minerals.

The variation of δ quartz-muscovite has been derived by combining the experimental data on muscovite-water (O'Neil and Taylor, 1969) and quartz-water (Matsuhisa et al., 1979) which can be expressed as

$$\delta^{18}O = 10^6 T^{-2} + B .$$

Bottinga and Javoy (1975) propose the relation

$$\delta^{18}O = 2.20(10^6 T^{-2}) - 0.60$$

which differs because of theoretical considerations. A curve for quartz + biotite has been obtained from the observed distribution of oxygen isotopes among quartz + muscovite + biotite, calibrated by quartz + muscovite (Clayton, 1981) and quartz + biotite + magnetite, calibrated by quartz + magnetite (Shieh and Schwarcz, 1974). Bottinga and Javoy (1975) also suggest a curve. There is a fair difference among these curves (Fig. 18), and many of them yield temperatures well below magmatic conditions, suggesting that quartz + muscovite reequilibrate O isotopes well into the subsolidus. Rather than translating the calibrating mineral pair to temperature, the $\delta^{18}O$ value of each pair can be plotted to yield a smooth curve calibrated with temperature.

Table 2. Criteria used by various authors to recognize magmatic muscovite

texture
- grain size comparable to other magmatic minerals
- sharp terminations
- subhedral to euhedral shape
- no reaction - relation textures with other minerals
- relatively abundant
- the host rock is relatively unaltered

occurrence
- deep-seated, water-rich plutons emplaced under conditions within the experimentally determined field of muscovite stability

chemistry
- greater content of Ti and, depending on the individual setting, systematic difference in Mg, Fe, Na, Al, Si, and H_2O
- equilibrium distribution of elements between muscovite and the other magmatic minerals

mineral assemblage
- the mineral assemblage is indicative of a peraluminous magma composition with coexisting aluminous biotite, cordierite, garnet, aluminum silicate, topaz, and tourmaline

Figure 19 gives an example for quartz + magnetite and muscovite + magnetite from Deines (1977). These concordancy diagrams allow a test of how well the measured data satisfy the assumption that all three minerals reequilibrate at the same temperature. If this were true, a straight line would pass through the origin. Deines (1977) reviewed much of the available data among mineral triplets in igneous rocks and found that an approach to isotopic equilibrium is common but that only about half yield a concordant ^{18}O-derived temperature. Among the triplets of most promise is muscovite + biotite + magnetite. However, the problem of obtaining igneous temperatures when rocks reequilibrate in the subsolidus is still important. A trondhjemite analyzed by Shieh and Taylor (1969) was found to be concordant (Fig. 19) but yields a temperature of only 300°C.

MUSCOVITE IN IGNEOUS ROCKS

The major question concerning muscovite in igneous rocks is whether it is, in fact, an igneous mineral. The various criteria used to recognize an igneous origin are not definitive, but in combination they offer compelling reasons to believe the muscovite is a magmatic phase in some plutons (Table 2). The presence of muscovite as a magmatic phase can be combined with experimental data to yield estimates of temperature and pressure of crystallization. Recently, interest in magmatic muscovite has further intensified because peraluminous melts can be used as indicators of an aluminous or "sedimentary" source region.

Texture

The most important criterion for recognizing magmatic muscovite consists of textural evidence comparable to that traditionally used in petrography to decide if any mineral is an igneous phase. "Igneous features" include sharp grain boundaries, subhedral to euhedral shape, a grain size comparable to other magmatic minerals, absence of reaction relations with other minerals, absence of alteration in the rest of the host rock, and the mineral's relative abundance. The other criteria discussed below that suggest an igneous origin for muscovite rely on textural recognition of the magmatic muscovite as a first step. Nevertheless, the textural evidence is often equivocal, which is why other types of evidence are being investigated.

Chemistry

Miller et al. (1981) investigated the composition of igneous and secondary muscovites from 16 granitoid plutons, mostly from North America. They found that the "primary-looking" micas are richer in Ti, Al, and Na and poorer in Mg and Si than the "secondary-looking" muscovites. Investigation by Speer et al. (unpublished) of the igneous and secondary muscovites in 59 Acadian and Alleghanian granitoid plutons of the southern Appalachians found a systematically higher Ti content of the igneous muscovites (Fig. 20) but an overlapping of the Fe, Mg (Fig. 20), Na (Fig. 21), Al, and Si contents. The secondary muscovites can be more iron-rich and have much less than full occupancy in the interlayer site (Fig. 21), suggesting extensive illite substitution. Speer et al. (1980) had previously noted that secondary muscovites occurring as isolated saussuritization products in plagioclase can have higher Na contents than the matrix muscovites which include both magmatic and secondary muscovites. In these studies, the compositional overlap of the igneous and secondary muscovites can result from the limitations of the textural criteria used to recognize a muscovite's paragenesis, the subsequent equilibration of the mineral chemistry in the subsolidus, and the variety of physicochemical conditions represented by the samples which come from a large number of plutons spread over wide geographic areas. To overcome this last problem, detailed examination of one pluton would be instructive. Monier et al. (1984) did such a study of the Millevaches massif, France. They distinguished three generations of muscovite: magmatic, late- to postmagmatic, and hydrothermal; they were able to obtain excellent chemical distinction between the three types (Fig. 22). The magmatic muscovite is the most Ti-rich; the data are sufficiently detailed to show that decreasing Ti content coincides with the

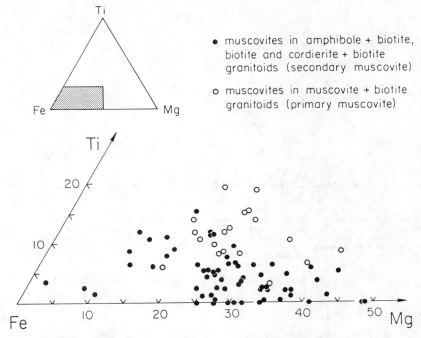

Figure 20. Ternary projection of the three major substituting cations for Al in muscovites: Fe, Mg, and Ti in the Acadian and Alleghanian granitoid plutons of the southern Appalachians. Data from Speer et al. (unpublished).

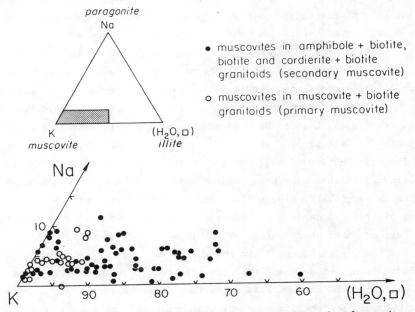

Figure 21. Ternary plot of the three major substitutions in the interlayer site of muscovites: K, Na, and H_2O, from the Acadian and Alleghanian granitoid plutons of the southern Appalachians. Data from Speer et al. (unpublished).

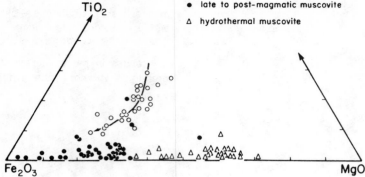

Figure 22. Compositions of white micas of the St. Julien leucogranite projected onto the TiO_2-(total Fe as)Fe_2O_3 - MgO compositional triangle. The data are from Monier et al. (1984) who recognize three generations of white micas: magmatic muscovites, late- to postmagmatic muscovite, and hydrothermal illite and smectite. The arrow is the changing compositions of the magmatic muscovites with the magmatic evolution of the granite.

magmatic evolution of the granite (Fig. 22). The late- to postmagmatic muscovite is more Fe-rich than the hydrothermal muscovite and both are much less Ti-rich than the magmatic muscovite. In addition to the Ti, Fe, and Mg contents, the different generations of the Millevaches' muscovites can be distinguished on the basis of the interlayer site composition. The magmatic muscovites are the most sodic, with Na/(Na+K) values between 0.06 and 0.12. The late- to postmagmatic muscovites have Na/(Na+K) values between 0.01 and 0.07, whereas the hydrothermal muscovites have less than 0.04. In addition, the hydrothermal muscovites contain Ca and significant amounts of water in the interlayer site. From these three studies, it is evident that muscovites texturally recognized as magmatic have elevated Ti contents, and that secondary muscovites can have appreciable interlayer water contents. Variations in Mg, Fe, Si, Al, Ca, and Na also occur, but the relationship depends on the individual situation.

The distribution of major or trace elements between muscovite and a magmatic mineral has also been used to determine whether the muscovite is igneous in origin. de Albuquerque (1975) reported the distribution of a number of trace elements among biotite, feldspar, and muscovite in some Portugese granites which suggests that these minerals are in equilibrium. The relatively high Cr and V but low Rb and Cs contents of the muscovites do not indicate a "metasomatic" replacement. For these reasons, de Albuquerque concludes that the muscovite is magmatic. Temperatures estimated from Rb-K distributions (see section on coexisting minerals) cover a large range between 521 and 1660°C, but an assumption of no subsolidus reequilibration is made. This may not be a good assumption because Monier et al. (1984) found a regular Rb-K distribution for both primary and secondary muscovites in the Millevaches massif, suggesting that Rb can be easily reequilibriated in the subsolidus.

Mineral assemblage

Muscovite is an aluminous mineral occurring in peraluminous igneous rocks, rocks where mol $Al_2O_3 > [Na_2O + K_2O + CaO]$ (Shand, 1947). The excess alumina goes into muscovite or such aluminous minerals as the aluminum silicates, aluminous biotite, cordierite, corundum, tourmaline, topaz, or garnet. Metaluminous and subaluminous rocks contain insufficient alumina to crystallize such minerals and are characterized by the presence of amphibole, pyroxene, or olivine. Muscovite occurring in rocks with these minerals is considered secondary.

Occurrence

Experimental work on the stability of muscovite is used to derive estimates primarily of pressure and, less commonly, of temperature of its crystallization in igneous rocks. These estimates are compared with other, independent estimates of the conditions to provide confirmation of the magmatic interpretation. For the biotite + muscovite granitoids of the Ruby Mountains, Nevada, Kistler et al. (1981) estimate a 3.4 kbar minimum pressure of crystallization on the basis of the intersection of the muscovite + quartz stability curve and the water-saturated granite solidus. A pressure estimate of 6.5 kbar in the wall rocks is compatible with this minimum pressure estimate and the muscovite is considered as magmatic.

The thermal stability of muscovite decreases with the substitution of Na for K (Chatterjee, 1972) and $(Mg,Fe)^{vi} + Si^{iv}$ for $Al^{vi} + Al^{iv}$ (Velde, 1965). Because the muscovite + quartz stability curve has a positive slope in P-T space and the granite solidus has a negative one, any of these compositional substitutions would decrease its thermal stability, thereby raising the pressure necessary for its crystallization from the magma. Thus, depending on the experimental work used, a pressure of about 3.5 kbar is usually considered the minimum pressure at which muscovite can crystallize in a melt. Miller et al. (1981) suggest that because "magmatic" muscovites can be found in igneous rocks which must have crystallized at pressures less than 3.5 kbar, the experimental work must be inadequate to assess accurately conditions of crystallization. They favor increased thermal stability of celadonitic muscovite to account for igneous occurrences, contrary to the work of Velde (1965). But this conclusion may be unnecessary because, as shown above, only Ti-substituted muscovites need have increased thermal stability. The granite solidus could also be at lower temperatures with addition of components such as boron, iron, or magnesium. Boron can lower the granite solidus at 1 kbar by 125°C (Chorlton and Martin, 1978), whereas Fe and Mg have a much smaller effect, estimated at about 20°C (Abbott and Clarke, 1979). Alternatively, Miller et al. (1981) suggest that much of the muscovite chemistry may reflect subsolidus reequilibration. As discussed throughout this chapter, subsolidus reequilibration of mica elemental and isotopic compositions is a widespread feature. Additional insight into this problem can be obtained from the two recent and convincing occurrences of magmatic muscovites in extrusive rocks in Germany (Schleicher and Lippolt, 1981) and Peru (Noble et al., 1984). Except for the high F content of the German muscovites, the muscovite and rock compositions for these extrusive rocks are comparable to most other occurrences of igneous muscovites. It is likely these muscovites are out of their stability fields and are preserved only because of rapid extrusion. A similar situation may account for a number of hypabyssal muscovite occurrences.

IGNEOUS MICAS AS METALLOGENIC INDICATORS

The major and trace element contents of biotites from mineralized porphyry copper deposits have been examined as indicators of the economic potential and evolution of the ore deposit. Elements most commonly studied have been Pb, Zn, and particularly Cu, as well as F and Cl. The behavior of the major elements as indicators of the economic potential of a pluton are less common.

The earlier work on the trace elements in biotites was based on analyses of bulk mineral separates (Parry and Nackowski, 1963; Al-Hashimi and Brownlow, 1970; Lovering, 1969; Graybeal, 1973; Kesler et al., 1975; Olade, 1979). This technique reflects the compositions of the biotites plus all their inclusions, and Al-Hashimi and Brownlow (1970) concluded that biotites with greater than 200 ppm Cu contained copper sulfide inclusions. Banks (1974) used an electron microprobe to overcome this problem but the limit of detection was 90 ppm. Banks did find that the Cu was undetectable in biotite but concentrated in the chloritized parts of the biotites. Hendry et al. (1981) analyzed biotites on a microscale using the more sensitive ion microprobe. They found that in the Koloula igneous complex, the Cu content of biotites from barren intrusions averaged 88 ppm (range 33 to 500 ppm) which is higher than the average of 23 ppm (range 6-85 ppm) of unaltered, mineralized rocks. They suggested that this difference arose because of Cu extraction during magmatic boiling.

The importance of F and Cl as transporting species of metals in ore deposits has led to increased interest in the occurrence and behavior of the halogens (see Chapter 11, this volume). Stollery et al. (1971) first suggested that the Cl content of igneous biotites could indicate the ore potential of an intrusive rock. They found that the Cl content of biotites from the mineralized Providencia stock, Mexico, averaged 3000 ppm (range 2500-5000 ppm) which is greater than the average Cl content of 1683 ppm reported from unmineralized German granites by Haack (1969). Subsequently, Parry (1972), Roegge et al. (1974), and Chivas (1981) determined the Cl contents of biotites for a number of mineralized and barren intrusions in the western U.S. and found an association between Cl-rich biotites and mineral deposits. Parry and Jacobs (1975) examined more biotites from Basin and Range plutons, but failed to confirm a difference in Cl content of biotites that would be useful in mineral exploration. Kesler et al. (1975) found that igneous biotites from North American plutons with significant Cu mineralizations have slightly greater F (av. 9535 ppm) and Cl (av. 863 ppm) contents than biotites from unmineralized intrusive rocks with lower F (av. 5365 ppm) and Cl (av. 827 ppm) contents. They felt that this magnitude of F and Cl enrichment is too small to be a useful exploration tool. Noting that the halogen content of biotites can be greater in more evolved rocks, Kesler et al. (1975) also performed an analysis of K-rich rocks only and found the same relationship. Van Loon et al.'s (1973) water leach analyses of several of these biotites indicate less than 10% Cl and 1% F are present as fluid inclusions. There is no assessment of the amount of halogens present in included apatite but an attempt was made to exclude chloritized biotites. Kesler et al. (1975) critically evaluated the use of biotite halogen composition in mineral exploration. For success it requires that the difference in F and Cl content between mineralized and unmineralized igneous rocks be greater than that caused by P, T, and biotite composition. They felt that this was unlikely, especially in view of the experimental studies of Munoz and Ludington (1974). Also threaded through such a discussion is an assumption that there is indeed a greater F or Cl content in intrusive rocks which will yield a mineral deposit.

Mason (1978) studied the major element compositions of the biotites and amphiboles of the porphyry copper-bearing and barren plutons of Papua

New Guinea. On the basis of the evolution of the Fe/(Fe+Mg) contents of the ferromagnesian minerals, he concluded that the barren plutons had "low" oxygen fugacities that possibly may increase with crystallization. By contrast, the mineralized plutons had "high" oxygen fugacities that increased with crystallization and alteration. Chivas (1981) found similar trends in the Koloula Igneous complex, Solomon Islands. In both cases the increase in oxygen fugacity is believed to result from separation of increasingly greater amounts of fluid with crystallization of the magma. The evolution of this abundant fluid is considered necessary for the formation of the economic Cu deposits in each case.

Most interest in biotites as metallogenic indicators centers on porphyry copper deposits, but biotites in Sn-bearing granitoids have also received some attention. The Sn content of biotites from granitoids associated with Sn mineralizations is greater than that of biotites from barren plutons (Bradshaw, 1967; Groves, 1972). Neiva (1976) noted that there was a positive correlation between the Sn contents of biotites and their host rocks but a negative correlation between Sn content of the rocks and modal abundance of biotite in the rock. This indicates that there are increasingly greater concentrations of Sn in increasingly smaller amounts of biotite but that only a small percentage of the Sn present in the rock is in the biotite, suggesting a potentially useful Sn-mineralization indicator. Tischendorf (1973) suggested that a condition necessary for the formation of economic Sn deposits is an elevated F content, which should be reflected in the chemistry of the contained biotites. In a study of the Nigerian Younger Granites, Imeokparia (1981) found that the average F content of biotites in barren plutons is 0.84 wt % (range 0.23-2.94%) whereas plutons with mineral deposits contain biotites with an average F content of 4.09 wt % (range 1.86-7.21%).

SUBSOLIDUS ALTERATION

Plutonic igneous micas, particularly biotite, commonly undergo postmagmatic autometamorphism. This is usually mentioned in a negative sense, as the cause why some "igneous" parameter cannot be obtained. It is an important feature to recognize in order to understand why some geothermometer, geobarometer, isotopic age, etc. is not giving the expected value. However, the alteration can yield the information on the postmagmatic history of the rock. Of particular interest are alterations associated with hydrothermal mineralizations.

Biotite most commonly alters to chlorite. Aside from the necessary change in Si, Al, K, and H_2O contents, the chlorites differ from their parent biotites in being more magnesian and slightly more oxidized, with a higher Fe^{3+} content (Dodge, 1973; Refaat and Abdallah, 1979). The chlorites contain little of the original Ti content of the biotite, which probably forms the commonly seen rutile or titanite intergrown with chlorite. The altering chlorite contains less Ba, V, and Cr but more Ga, Co, Ni, and Cu than the parent biotite (Dodge, 1973; Refaat and Abdallah, 1979). Dodge remarks that the process of chloritization is thus an unlikely mechanism to derive ore metals such as copper from the biotite.

Criss and Taylor (1983) studied the oxygen and hydrogen isotopic components of the biotite-to-chlorite alteration and found that the $^{18}O/^{16}O$ composition of biotites does not readily exchange with the deuteric fluid. The newly formed chlorite is, however, in isotopic equilibrium with the fluid, so that the isotopic composition of the "bulk" biotite samples is a linear mixture of the chlorite and biotite values and depends on the modal abundance of each. By contrast, the change in the δD during the chloritization of biotite suggests that the biotite itself undergoes some D-H exchange before alteration to chlorite. In a parallel study, O'Neil and Kharaka (1976) demonstrated that D-H exchange in clay minerals is faster than the oxygen isotopic exchange.

Altered biotite is generally accompanied by grains of alkali feldspar, suggesting that chloritization of biotite occurs by the reaction biotite + quartz + water = chlorite + alkali feldspar. A study by Chayes (1955) of the amount of model alkali feldspar produced by this reaction in a granitoid showed an insufficient amount present, suggesting that K had left the immediate site. Barriere and Cotten (1979) found a well-defined linear distribution of Rb/K between biotite and alkali feldspar in the Ploumanac'h biotites. These minerals show evidence of postmagmatic changes, and -- noting that Rb/K is temperature dependent -- Barriere and Cotten suggest that these minerals have reequilibrated at a lower temperature of 550°C.

Ferry (1979) studied the chloritization of a granitoid biotite as a part of the overall alteration which included sericitization of the alkali feldspar and saussuritization of the plagioclase. The reaction that takes place at the biotite is

$$\text{biotite} + H_2O + H^{1+} + Ca^{2+} + Fe^{2+} + Mg^{2+} + Mn^{2+} =$$
$$\text{chlorite} + \text{titanite} + H_2SiO_4 + K^{1+} + Na^{1+} .$$

This reaction assumes that Al and Ti are immobile and that 100-1000 cm^3 water passed through 1000 cm^3 rock at 425°C and 3.5 kbar. Veblen and Ferry (1983) studied the chloritized biotites from these same garnet + biotite + muscovite quartz monzonites by TEM. They found that a "talc" layer of biotites, comprising the sequence tetrahedral + octahedral + tetrahedral sheets, is replaced by a "brucite" sheet. The alteration of the brucite interlayers and talc layers comprises a chlorite sequence. The reaction is accompanied by a microprecipitate, interpreted as silica. Such a reaction is consistent with the chemistry observed in these rocks by Ferry (1979) and differs from a mechanism which involves growth of a "brucite" sheet into the biotites' interlayer site (Olives Banos et al., 1983).

The alteration of biotite in porphyry copper deposits has attracted the most attention. While alteration to chlorite is common, the actual alteration of interest is the appearance of a secondary biotite which is usually a phlogopite. Lanier et al. (1978) examined a traverse across the hydrothermal zones of the porphyry copper deposit associated with the Bingham and Last Chance stocks. Two types of textural and chemical variants of biotites were recognized: magmatic and secondary. The secondary biotites are either recrystallized magmatic grains or hydrothermal vein fillings and are modally more abundant than the original magmatic biotites. They are phlogopites with much smaller amounts of Ba, Ca, Mn, and Ti than the magmatic biotites. Le Bel (1979) found that in the porphyry copper deposits of Cerro Verde-Santa Rosa, Peru, the biotites' Al^{vi} content increases and Ti content decreases through several facies of hydrothermal alteration, whereas the Fe/(Fe+Mg) remains nearly constant. Le Bel (1979) interprets the Al^{vi} and Ti contents to reflect the temperature of formation based on the experimental stability of Ti-phlogopite (Robert, 1976). The constant Fe/(Fe+Mg) results from the external buffering of the oxygen fugacity at a constant value. In the Santa Rita porphyry copper deposit, New Mexico, Jacobs and Parry (1979) found that the magmatic biotites contain less F and Si and greater amounts of Fe/(Fe+Mg), Ti, Cl, and Al than secondary biotites. The actual amount can vary depending on the type of alteration, but magmatic biotites can invariably be distinguished by their higher Ba contents. With the compositional data, Jacobs and Parry (1979) attempted to determine conditions of both magmatic and secondary biotite crystallization using the techniques described at the beginning of this chapter in the section entitled OXYGEN AND WATER FUGACITIES. Plotting the T-f(H$_2$O) stability curve for the igneous biotite compositions and determining its intersection with the granodiorite solidus of Piwinskii and Wyllie (1968), Jacobs and Parry obtained a minimum temperature of 710°C and a maximum $f(H_2O)$ of 1270 bars for the magmatic conditions. Using the F-OH exchange equilibria of Munoz and Ludington (1974), magmatic temperatures were found to be 760-720°C. Assuming that the secondary biotites crystallized at the same $f(H_2O)/f(HF)$, the secondary biotites gave temperature estimates of 710-620°C and the vein biotites 645-570°C.

Prehnite has been noted as lens-shaped aggregates in biotites from a variety of igneous rocks ranging from gabbros (Moore, 1976) through diorites (Phillips and Rickwood, 1975) to granitoids (Tulloch, 1979). In addition, Tulloch (1979) reports the occurrence of andradite-grossular garnet, epidote and pumpellyite as well. These Ca-Al silicates are interpreted as subsolidus replacements of the biotite. In the case of prehnite the association is explained by replacement of an amphibole (Hall, 1965), noninvolvement of the biotite, which merely acts as a nucleation site (Moore, 1976: Phillips and Rickwood, 1975), or a subsolidus reaction of the form plagioclase + biotite + water = prehnite + chlorite + alkali feldspar + titanite ± muscovite (Tulloch, 1979).

HALOGENS IN BIOTITE

Fluorine

Between 70 and 90% of the F in muscovite- and fluorite-free granitoid rocks is contained in biotite (Grabezhev et al., 1979). The remainder is in apatite and titanite. The behavior of F in biotite is thus the key to the behavior of F in many igneous rocks. Experimental work by Munoz and Ludington (1974) has shown that the F content of biotite depends on the H_2O-HF composition and temperature of the fluid that last reequilibrated with the mica and on the composition of the mica itself (see Chapter 11, this volume). The effect of composition has long been known: chemical analyses of biotites show that Mg-biotites generally can contain much more F than Fe-biotites (Fig. 23). One aspect of Munoz and Ludington's work that is of petrologic interest is that the chemistry of a biotite can yield the fluid composition or temperature if the other is known (Fig. 24). Another useful view of this experimental work is an isobaric projection onto a log f(HG)-T plot. Figure 25 is a schematic construction for a biotite of arbitrary Fe-Mg-Al composition. During crystallization, most plutons act as closed systems and probably have a F-buffered system and only limited change in biotite chemistry. During cooling, biotites should decrease in F content. For open systems with externally buffered F fugacity, the F content of the biotites would increase

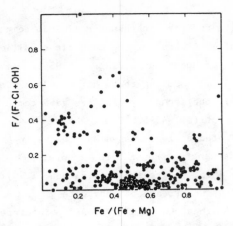

Figure 23. Variations of mole-fraction of F/(F+Cl+OH) and Fe/(Fe+Mg) in igneous biotites. The data are from Barker et al. (1975), Barriere and Cotten (1979), Best and Mercy (1967), Crecraft et al. (1981), Cross (1897), Czmanske et al. (1977), de Albuquerque (1973), Deer (1937), Dodge and Moore (1968), Dodge and Ross (1971), Erickson and Blade (1963), Fiala et al. (1976), Gunow et al. (1980), Hazen et al. (1980), Hietanen (1971), Huntington (1979), Kaspustin (1980), Larsen and Draisin (1948), Larsen et al. (1937), Lee et al. (1981), Lyons (1976), Nash (1976), Neiva (1976; 1983), Parry et al. (1978), Prider (1939), Simonen and Vorma (1969), van Kooten (1980), Vejnar (1971), Volkov and Gorbacheva (1980), and Wones (1980).

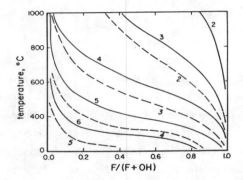

Figure 24. Fluorine content of synthetic biotites as a function of temperature in equilibrium with fluids of constant $f(H_2O)/f(HF)$ values. The solid lines are for phlogopite and the dashed lines are for a mean of siderophyllite and annite. From Munoz and Ludington (1974).

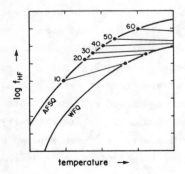

Figure 25. Schematic stability of biotites of differing F/(F+OH) compositions as a function of temperature and HF fugacity at a constant total pressure. The numbers are contours of constant 100 F/(F+OH) values for biotites in the assemblage biotite + fluid. The labeled curves are the fluorine buffer curves: WFQ wollastonite + fluorite + quartz + gas and AFSQ anorthite + fluorite + sillimanite + quartz + gas.

slightly. For cases with increasing F fugacity, the F contents of the biotites would increase rapidly.

Because there are other methods of estimating temperatures, estimates of the $f(H_2O)/f(HF)$ are normally sought. Estimating temperatures from fluid inclusions, Parry et al. (1978) found a ratio of 10^5 in a hypabyssal quartz latite plug at Bingham, Utah. In Japan, Kanisawa et al. (1979) found that the biotites from the Kitakami and Tabito granites have values of $f(H_2O)/f(HF)$ between values of 10^4 and 10^5; biotites from the Inagawa and Sanin granites values have somewhat lower values, at 10^3 and 10^4. The fluorite-bearing Naegi granites yield $f(H_2O)/f(HF)$ near 10^3. Similar Fe-rich and fluorite-bearing granites of the Pikes Peak batholith reported by Barker et al. (1975) also indicate values near 10^3. Barriere and Cotten (1979) estimated a value of 10^4 for the deuteric fluid of the Ploumanac'h complex in France, whereas Nash (1976) estimated a value of 10^5 for the secondary biotites in the Skaergaard intrusion although he concluded that these biotites lost F to heated, circulating meteoric groundwaters. These various estimates of $f(H_2O)/f(HF)$ suggest that only small amounts of F occur in the fluids associated with igneous rocks, even for those considered F-rich because of the presence of fluorite.

The effect of magmatic processes on the F content of biotite has been documented for several locations. The effects have not, however, been extensively investigated because of the complexity of the factors controlling the F content and the fact that they are continually varying. As the magma cools, temperature is decreasing and the chemistries of the melt and the crystallizing phases are changing. Several different trends have been noted. In porphyry copper deposits, such as the Santa Rita, there is an increase in the Mg and F contents from the magmatic biotites through the late- to post-magmatic and deuteric or hydrothermal biotites (Fig. 26a). These changes are covered in more detail by Munoz (Chapter 11, this volume). A decrease in the Mg and F contents of the biotites with differentiation is noted in the Ploumanac'h pluton (Fig. 26b) as well as in composite batholiths such as the central Bohemian massif (Fiala et al., 1976). The Ploumanac'h is believed to have a concentric cryptic layering caused by fractional crystallization, which produces a monzogranite cumulate and a residual syenogranite. As the temperature decreases, the biotites become increasingly Fe-rich. The slight decrease in the F content of the Ploumanac'h biotites could be entirely a result of the change in Fe content of the biotites which would suggest a constant $f(H_2O)/f(HF)$ ratio for the crystallizing pluton. Alternatively, or

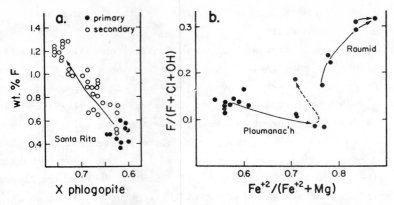

Figure 26. (a) Variation of wt % F with mole fraction phlogopite in primary and secondary biotites from the Santa Rita pluton and porphyry copper deposit (Jacobs and Parry, 1976). (b) Variation of the mole fraction $F/(F+Cl+OH)$ and $Fe^{2+}/Fe^{2+}+Mg)$ in biotites from the Ploumanac'h pluton, France (Barriere and Cotten, 1979) and the Raumid pluton, USSR (Volkov and Gorbacheva, 1980). The arrows are in the direction of falling temperatures and magmatic evolution.

in addition, the decrease in biotite F content could be a result of the drop in temperature (Fig. 25). The abrupt increase in the F content of the pegmatitic biotite shown by the dotted line in Figure 26b probably represents an increase in the F content of the melt. In the Raumid pluton, USSR (Fig. 26b), both Fe and F content of the biotites increases with fractional crystallization from both the bottom and top of the pluton. Similar trends are also noted in the Adamello massif, Italy (De Pieri and Jobstraibizer, 1983), although there is no recognizable sense of evolution. The Fe enrichment of the biotites with a presumed drop in temperature suggests a decrease in oxygen fugacity. The rapid increase in F content of the biotites suggests an increase in the F content of the evolving melt sufficient to overcome the decrease of F that would result from the decreasing temperature and increasing Fe content of the biotites.

Chlorine

The Cl content of igneous biotites is usually rather low. Parry and Jacobs (1975) found that biotites from 23 plutons in the Basin and Range Province of the U.S. contained between 0.45 wt % and less than 0.01 wt % Cl. Biotites with the highest Cl content coexist with pyroxene. While only a minor constituent, the Cl content is of interest because of its potential role as an indication of mineralization. This is based on the assumption that Cl is an important transporting species. While its usefulness in this regard remains unfulfilled as discussed above in the section on mineralization, interest in the Cl content of biotite remains high because biotites can help unravel the behavior of Cl in igneous rocks.

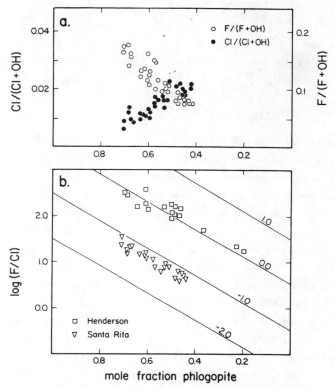

Figure 27. (a) Variation of Cl/(Cl+OH) and F/(F+OH) composition versus mole fraction of Mg in the octahedral sites (phlogopite) for the biotites of the Santa Rita, New Mexico porphyry deposit. Data from Jacobs (1976) in Munoz and Swenson (1981). (b) Log F/Cl plotted against mole fraction Mg in the octahedral sites for biotites of the Santa Rita and Henderson porphyry deposits. Contours of constant log f(HF)/f(HCl) composition of the equilibrating fluid. After Munoz and Swenson (1981).

Paralleling the work of F ⇄ OH exchange in biotites, Munoz and Swenson (1981) examined OH ⇄ Cl and Cl ⇄ F exchange in biotites in order to arrive at estimates of HCl fugacity and the behavior of Cl. They studied the Cl ⇄ OH exchange behavior of annite as part of the buffer assemblage KCl + muscovite + quartz = alkali feldspar + HCl and found that the exchange could be expressed as

$$\log K = \log (X_{Cl-Bt}/X_{OH-Bt})_{Fe-Bt} + \log [f(H_2O)/f(HCl)]_{fluid}$$
$$= 5151/T(Kelvin) = 5.01 , \qquad (3)$$

where ideal mixing of Cl ⇄ OH is assumed. Experiments with phlogopite were unsuccessful and the effect of Fe ⇄ Mg substitution on the Cl ⇄ OH exchange was investigated by examining natural biotites. In the biotites of the Santa Rita porphyry copper deposit in New Mexico, F increases with Mg substitution, whereas Cl decreases (Fig. 27a). Using this information, Munoz and Swenson expanded equation 1 to include intermediate composition biotites:

$$\log K_{OH-Cl} = 5151/T - 5.01 - 1.93 \, X_{Mg} \qquad [T \text{ in Kelvins}], \qquad (4)$$

where X_{Mg} is the mole fraction of octahedral Mg in the biotite, and ideal mixing of Mg ⇄ Fe is assumed. Combining this information with the F ⇄ OH exchange data of Ludington and Munoz (1975), an F ⇄ Cl exchange equation can be obtained

$$\log K_{F-Cl} = 3051/T - 5.34 + 3.13 \, X_{Mg} \qquad [T \text{ in Kelvins}] \qquad (5)$$

which allows calculation of isothermal contours of constant $\log[f(HF)/f(HCl)]$ in a fluid coexisting with biotites of differing Fe/(Fe+Mg) and F/Cl compositons (Fig. 27b). Plotting the compositions of biotites from the Santa Rita and Henderson porphyry deposits produce linear trends paralleling the contours (Fig. 27). Munoz and Swenson note that a linear trend can be produced only if the halogen composition of both the igneous and secondary biotites have reequilibrated at the same temperature with fluids of similar composition. For the Santa Rita biotites this would have been in the subsolidus at a temperature of about 310-370°C at $\log[f(HF)/f(HCl)]$ of -1.2 ± 0.8. The Henderson biotites (Wallace et al., 1978) have a parallel linear trend at a different F/Cl value as a result of either a different temperature or fluid composition. Munoz and Swenson prefer the latter because the Henderson deposit also reequilibrated in the subsolidus at a temperature of about 350°C (Gunow et al., 1980). The $f(HF)/f(HCl)$ difference between the two plutons is an order of magnitude with the Henderson being more F rich.

INTERLAYER SITES

The interlayer sites of the igneous micas are dominantly filled with K and small amounts of Na, Ca, Ba and H_2O and even lesser amounts of Rb, Cs, Sr, and NH_4.

Ammonium

Stevenson (1962) determined that the majority of nitrogen bound within an igneous rock is present as NH_4. Wlotzka (1961) found that micas are the minerals richest in ammonium followed by the feldspars. Subsequent infrared work confirmed that the ammonium substitutes for K in the interlayer site (Vedder, 1965; Yamamoto and Nakahira, 1966; Karakin et al., 1973; Higashi, 1978). Honma and Itihara (1981) and Itihara and Honma (1979) found a systematic distribution of ammonium in coexisting magmatic minerals suggesting that ammonium is a stable component in magmatic processes. For the igneous rocks they studied, the biotite contains between 5 and 149 ppm ammonium and muscovite 32 to 85 ppm, whereas feldspars and quartz contain less than 47 ppm.

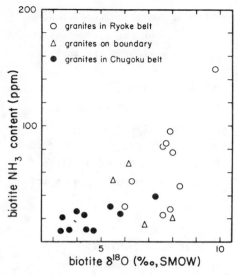

Figure 28. Relationship between the ammonium content and oxygen isotopic composition of biotites from granites emplaced in the high grade Ryoke belt rocks, granites emplaced in unmetamorphosed Chugoku belt rocks, and granites emplaced in the transition zone. The data for these rocks from Japan (Itihara and Honma, 1979) suggest that the ammonium content of the biotites is a result of magma-wall rock interaction.

For the biotites of plutons emplaced in unmetamorphosed Chugoku belt rocks, the ammonium content is between 5 and 53 ppm with an average of 22 ppm. Biotites in granitoid plutons emplaced in the high-grade metamorphic rocks of the Ryoke belt have ammonium contents from 23 to 149 ppm with an average of 67 ppm. Biotites in plutons emplaced at the boundary of these two belts contain between 15 and 67 ppm ammonium with an average of 39 ppm. This systematic variation in ammonium content correlates with the oxygen and initial strontium isotopic compositions of the rocks (Fig. 28). This suggested to Itihara and Honma (1979) that the enrichment of ammonium in these granitoid biotites is a result of magma-wall rock interaction. The ammonium content of biotite from migmatites is higher, with values between 391 and 534 ppm, indicating that the elevated ammonium content of a biotite can be indicative of the source rocks as suggested by Wlotzka (1961), Molovskiy and Volnets (1966), Urano (1971), and Itihara and Honma (1979).

Barium

Ba-rich micas from metamorphic rocks and skarns have been known for some time, a Ba-phlogopite was synthesized by Frondel and Ito (1967). The first Ba-rich igneous biotite, containing 7.32 wt % BaO, was reported by Thompson (1977) from a leucitite from the Alban Hills, Italy. This was followed by a Ba-phlogopite with up to 8.62 wt % BaO from a monticellite peridotite (Wendlandt, 1977) and a Ba-biotite with up to 20 wt % BaO from nephelinites from Hawaiian basalts (Mansker et al., 1979). The Ba-biotites from the feldspathoid-bearing volcanic rocks are interpreted as late-crystallizing phases. The substitution of K^{1+} by Ba^{2+} requires a charge compensation. Wendlandt (1977) suggested $Al^{3+} + Ba^{2+} \rightleftarrows Si^{4+} + K^{1+}$, but the extraordinary Ti content of the Hawaiian biotites, up to 14 wt %, suggested the substitutional scheme

$$K^{1+} + 3(Mg,Fe)^{2+} + 3Si^{4+} \rightleftarrows Ba^{2+} + 2Ti^{4+} + 3Al^{3+}.$$

A similar substitution scheme has been found by Velde (1979) in melilite-bearing olivine nephelinites. The Ba contents of most other igneous biotites is less than 1 wt % and reflects the low Ba content of igneous rocks, even though Higuchi and Nagasawa (1969) found that the Ba content of biotites in a dacite to be enriched by a factor of 10 over the Ba content of the matrix of the rock.

Calcium

Calcium substitutes in the interlayer site of igneous biotites up to about 27% of the cations. The Ca-rich micas occur in intermediate granitoid rocks with hornblende and pyroxene. Olesch (1979) suggested a substitutional scheme of $K^{1+} + Mg^{2+} + 2Si^{4+} \rightleftarrows Ca^{2+} + 3Al^{3+}$ for clintonite. He determined experimentally that the extent of solid solution is less than 15 mol % and that it varies little with pressure and temperature. Natural micas generally agree with these substitutional schemes. Strong support comes from kimberlites which are Al-poor, Ca-rich, biotite-bearing rocks. Olesch (1979) found that except in one case, the kimberlite micas contain less than 0.34 wt % CaO.

Sodium

The dioctahedral Na-mica is paragonite. Although widespread in metamorphic rocks, there are no reported igneous occurrences. Trioctahedral Na-micas are rare, but sodic biotites have been reported from K-poor alkali granites. These are older analyses and are summarized by Franz and Althaus (1976). The micas are intermediate Fe-Mg biotites and Na is 32 to 97% of the interlayer cations but only 39 to 58% of the site is filled. The charge balance because of the vacancy may be made up by the Fe^{3+} substituting in the tetrahedral site. The analyses of these sodian igneous biotites do not appear comparable to either preiswerkite, Na-phlogopite, or ephesite.

RARE EARTH ELEMENTS IN BIOTITES

Igneous biotites contain low amounts of rare earth elements, which show a flat distribution pattern clustered about a value of one (Fig. 29). The

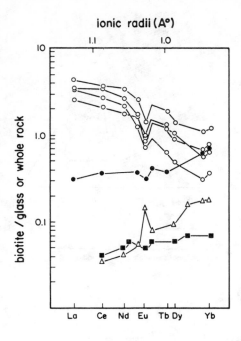

Figure 29. Biotite/glass or whole rock coefficients versus ionic radii as a measure of mineral-melt partitioning of the rare earth elements. Data are from the Bishop Tuff, open circles (Mahood and Hildreth, 1983), a biotite rhyodacite (Higuchi and Nagasawa, 1969; 1977), biotite + garnet dacite (Schnetzles and Philpotts, 1970), and a biotite granodiorite (Gromet and Silver, 1983).

biotite/liquid distribution of the REE summarized in Figure 29 ranges between 0.02 and 4.0, well within the limits of 0.10 to 14.7 found by Crecraft et al. (1981) in the Twin Peaks volcanic complex, Utah. Various workers have tried to model the crystallization history of melts using REE. Bender et al. (1984) completed such a study for the Cortlandt complex. The low REE content of the biotites and flat pattern both suggest, however, that biotite is not a critical mineral in such modelling. Gromet and Silver (1983) found that biotite contains less than 1% of the REE content in a granitoid.

PETROLOGY

Occurrence

An earlier interest in the geologic occurrence of biotite was correlation of its chemical composition with different types of host rocks. Heinrich (1946) examined the variation with rock type by means of MgO, $(Fe_2O_3 + TiO_2)$, FeO + MnO plots (Fig. 30). He found that biotites from different rock types have restricted but not unique compositional fields. Foster (1960) found considerable overlap of biotite compositions from different rock types (Fig. 30) in plots of octahedral cations on triangular $Al^{vi} + Fe^{3+} + Ti^{4+}$, $Fe^{2+} + Mn^{2+}$, Mg^{2+} diagrams. Because the rock names were not rigorously applied in these previous studies, Neilson and Haynes (1973) made a plot comparable to

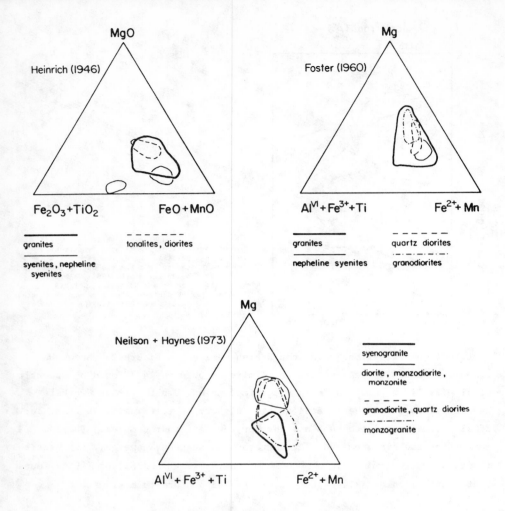

Figure 30. Relation between the composition of biotites and the rock type of the host.

Foster's for biotite compositions where there was also a modal analysis available for the host rock. The distribution of biotite compositions are similar to Foster's, with distinctive fields for syenogranites and the more plagioclase-rich diorites, monzodiorites, monzonites, granodiorites, and quartz diorites (Fig. 30). However, biotites from the intermediate monzogranites overlap all the other fields.

Variations of biotite composition with rock composition, rather than modal composition as in the above studies, is less often attempted. Vejnar (1971) reported on such a study for the West Bohemian plutons, Czechoslovakia. He found that in the series from lherzolite through muscovite + biotite

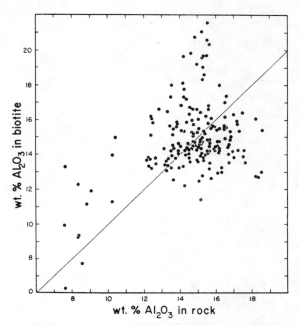

Figure 31. Variation of wt % Al_2O_3 between biotite and its host rock. Data from Carmichael (1967), Dodge and Moore (1968), Dodge and Ross (1971), Edgar (1979), Fiala et al. (1976), Haslam (1968), Kabesh and Aly (1980), Kabesh and Ragab (1974), Kabesh et al. (1975), Kanisawa (1972), Kuehner et al. (1981), Lee et al. (1981), Mahmood (1983), Murakami (1969), Neiva (1976; 1983), Sapountzis (1976), and Velde (1975).

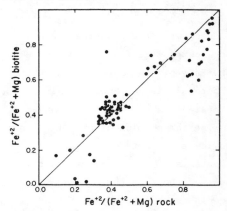

Figure 32. Variation of the $Fe^{2+}/(Fe^{2+}+Mg)$ ratio of biotite and its host rock. Data from Carmichael (1967), Dodge and Moore (1968), Dodge and Ross (1971), Edgar (1979), Erickson and Blade (1963), Flood and Shaw (1979), Haslam (1968), Kanisawa (1972), Kuehner et al. (1981), Lee et al. (1981), Lyons and Krueger (1976), Mahmood (1983), Murakami (1969), Neiva (1976; 1983), and Velde (1975).

granite, the biotites increase in Si, Ti, Mg, and Na content and decrease in Al, Fe^{3+}, Fe^{2+}, Mn, Li, and F. Potassium initially increases, reaching a maximum content in the biotites where the host rocks have the chemistry of diorites and granodiorites, then decreases. This decrease in K content in

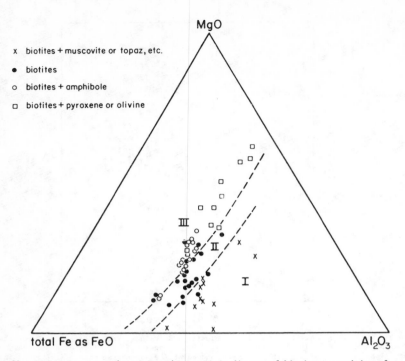

Figure 33. Triangular MgO - (total Fe as)FeO - Al_2O_3 diagram of biotite compositions from igneous rocks after Nockolds (1947). Roman numerals refer to the fields of biotites associated with muscovite or topaz (I), unaccompanied by other mafic minerals (II), and biotites associated with hornblende, pyroxene, or olivine (III).

the biotites of the leucocratic rocks during the deuteric stage. Variation in the biotite and rock compositions within a single body was reported by Mahmood (1983) for the Zaër pluton, Morocco. With increasing differentiation index, the biotites increased in Rb, MnO, and Al_2O_3 contents, decreased in MgO, CaO, Cr, and Ni contents and remained unchanged with respect to SiO_2, TiO_2, K_2O, and Na_2O. The composition of biotites generally follows that of the host rock as is evident in plots of the alumina (Fig. 31) and $Fe^{2+}/(Fe^{2+}+Mg)$ (Fig. 32) contents of the biotites and their host rocks. The poor correlation is a result of the fact that, in addition to rock chemistry, biotite composition is dependent on the physical conditions of crystallization, the coexisting minerals, and the extent of secondary alteration.

Following a slightly different approach, Nockolds (1947) examined the composition of biotites in relation to paragenesis. Using a triangular MgO - FeO - Al_2O_3 plot (Fig. 33), he was able to distinguish biotites associated with muscovite or topaz (I), unaccompanied by other mafic minerals (II), or associated with hornblende, pyroxene, or olivine (III). These biotites differ primarily in their aluminum content and much less so for Mg and Fe. This is more evident for the biotites of the Aregos region, Portugal, which coexist with the increasingly aluminous minerals hornblende, muscovite, and muscovite + aluminum silicate (Fig. 7). An example from a single pluton is the difference in tetrahedral Al contents of biotites from the biotite and cordierite + biotite facies of the Clouds Creek pluton, South Carolina (Fig. 8).

In an attempt to characterize chemically biotites from different kinds of igneous rocks, Lovering (1969) reported on the content of 36 minor elements of over 200 biotites. While the median values varied from rock type to rock type, the range of values generally overlap. The data were not considered significant because, among other reasons, the samples were bulk samples containing impurity minerals and alteration products.

The occurrence of biotite in a rock reflects the K and water content of the magma and ultimately its source region, for in many rock types biotite is the only phase containing major amounts of these elements. An interesting application of this idea is Sakuyama's (1979) study of the lateral occurrence of biotite (and amphibole) in the Quaternary volcanic rocks of northeastern Japan. He found that the volcanoes in the East Japan volcanic zone can be classified into three groups: (I) those without biotite or hornblende phenocrysts; (II) those with hornblende phenocrysts; and (III) those with biotite phenocrysts. These groups have an east-to-west distribution and indicate a westerly increase in the water contents of the andesitic to dacitic magmas. Quartz phenocrysts occur in the volcanoes without hydrous ferromagnesian phenocrysts, pointing to a gradual increase in the water content and decrease in the temperature of the parent magma toward the west. This conclusion is based on experimental work on variation of the liquidus temperature of quartz with water content. Sakuyama concludes that these lateral variations result from the extent of partial melting decreasing to the west which is also in agreement with the increase in the K, Rb, Sr, Th, and REE contents to the west.

Indicators of magmatic evolution

Controls on the composition of igneous biotite are only imperfectly known and, in most cases, knowledge of the magmatic evolution of a pluton is used to understand the behavior of the biotites. The converse was attempted by Leake (1974) to understand the crystallization of the Galway Granite, Ireland, which has an alkali granite at its margin, followed inward by an adamellite and then a granodiorite. This sequence conflicts with an interpretation of a pluton crystallizing from the outside inwards and having the most basic rocks at its margin. Leake found that the Fe/(Fe+Mg+Mn) content of the biotite increases from the margin of the granodiorite to its center, suggesting that it did crystallize from its periphery. The biotites of the alkali granite are Fe- and F-rich, suggesting they have evolved from the other rocks. Leake concludes that they are differentiates, drawn out from the residuum of the main part of the pluton by blocks of country rocks falling into the pluton.

Hybrid rocks

Biotites of hybrid igneous rocks (rocks that have assimilated the wall rock) have been used to document assimilation processes. Lee and Van Loenen (1970) found that the biotites in the hybrid granitoid rocks of the Snake Creek - Williams Canyon area, Nevada, have major and trace element compositions that resemble a classic differentiation sequence as does the composition of the host rocks. Neiva (1981) studied biotites from muscovite + biotite granitoids contaminated by metapelites and marbles in the Franzilhal, S. Lourenco, and Pombal areas, Portugal, and also found that the biotite compositions show an apparent calc-alkaline trend of differentiation. However, the compositional trends of some elements, e.g., Rb, K, Li, Cs, K/Rb, cannot be explained by fractionation of biotites and muscovites with compositions such as are found in the rocks, and Neiva concludes that the compositions of the igneous rocks can be explained by the assimilation of the wall rocks by the magma.

Walsh (1975) found that the biotites in the Centre III igneous complex of Ardnamurchan, Scotland, increase in Fe/(Fe+Mg) from gabbros through dolerites, but there is a small decrease in continuing to the tonalites and quartz monzonite. The compositional trend in the mafic rocks is believed to reflect a decrease in oxygen fugacity with falling temperatures. The more magnesian trend in the intermediate rocks indicates an increase in the oxygen fugacity. This is not the expected result if the intermediate rocks are fractionated from the gabbro, which further suggests that there was contamination of the basic magma by the wall rocks to form the intermediate rocks.

REFERENCES

Abbott, R.N., Jr. and Clarke, D.B. (1979) Hypothetical liquidus relationships in the subsystem Al_2O_3-FeO-MgO projected from quartz, alkali feldspar, and plagioclase for $a(H_2O) \leqslant 1$. Canadian Mineral. 17, 549-560.

Al-Hashimi, A.R.K. and Brownlow, A.H. (1970) Copper content of biotites from the Boulder Batholith, Montana. Econ. Geol. 65, 985-992.

Allan, B.D. and Clarke, D.B. (1981) Occurrence and origin of garnets in the South Mountain batholith, Nova Scotia. Canadian. Mineral. 19, 19-24.

Armstrong, R.L. (1975) The geochrononometry of Idaho. Isochron West 14, 1.

Bancroft, G.M. and Brown, J.R. (1975) A Mössbauer study of coexisting hornblendes and biotites: quantitative Fe^{3+}/Fe^{2+} ratios. Am. Mineral. 60, 265-272.

Banks, N.G. (1976) Halogen contents of igneous minerals as indicators of magmatic evolution of rocks associated with the Ray porphyry copper deposit, Arizona. U.S. Geol. Surv. J. Res. 4, 91-117.

Barker, F., Wones, D.R., Sharp, W.N. and Desborough, G.A. (1975) The Pikes Peak Batholith, Colorado Front Range, and a model for the origin of the gabbro-anorthosite-syenite-potassic granite suite. Precambrian Res. 2, 97-160.

Barriere, M. and Cotten, J. (1979) Biotites and associated minerals as markers of magmatic fractionation and deuteric equilibration in granites. Contrib. Mineral. Petrol. 70, 183-192.

Bateman, R. (1982) The zoned Bruinbun granitoid pluton and its aureole. J. Geol. Soc. Australia 29, 253-265.

Bender, J.F., Hanson, G.N. and Bence, A.E. (1984) Cortlandt Complex: differentiation and contamination in plutons of alkali basalt affinity. Am. J. Sci. 284, 1-57.

Best, M.G. and Mercy, E.L.P. (1967) Composition and crystallization of mafic minerals in the Guadalupe Igneous Complex, California. Am. Mineral. 52, 436-474.

Beswick, A.E. (1973) An experimental study of alkali metal distributions in feldspars and micas. Geochim. Cosmochim. Acta 37, 183-208.

Birch, W.D. and Gleadow, A.J.W. (1974) The genesis of garnet and cordierite in acid volcanic rocks: evidence from the Cerberean Cauldron, central Victoria, Australia. Contrib. Mineral. Petrol. 45, 1-13.

Blattner, P. (1980) Chlorine-enriched leucogabbro in Nelson and Fiordland, New Zealand. Contrib. Mineral. Petrol. 72, 291-296.

Bottinga, Y. and Javoy, M. (1975) Oxygen isotope partitioning among the minerals in igneous and metamorphic rocks. Rev. Geophys. Space Phys. 13, 401-408.

Bradshaw, P.M.D. (1967) Distribution of selected elements in feldspars, biotites and muscovites from British granites in relation to mineralization. Trans. Inst. Mining Metall. B76, B137-B148.

Carmichael, I.S.E. (1967) The mineralogy and petrology of the volcanic rocks from the Leucite Hills, Wyoming. Contrib. Mineral. Petrol. 15, 24-66.

Carron, J.P. and Lagache, M. (1971) La distribution des éléments alcalins Li, Na, K, Rb dans les minéraux essentiels des granites et granodiorites du sud de la Corse. Bull. Soc. franc. Minéral. Cristallogr. 94, 70-80.

Chappell, B.W. and White, A.J.R. (1974) Two contrasting granite types. Pacific Geol. 8, 173-174.

Chatterjee, N.D. (1972) The upper stability limit of the assemblage paragonite + quartz and its natural occurrences. Contrib. Mineral. Petrol. 34, 288-303.

Chayes, F. (1955) Alkali feldspar as a by-product of the biotite-chlorite transformation. J. Geol. 63, 75-82.

Chivas, A.R. (1981) Geochemical evidence for magmatic fluids in porphyry copper mineralization. Part I. Mafic silicates from the Koloula igneous complex. Contrib. Mineral. Petrol. 78, 389-403.

Chorlton, L.B. and Martin, R.F. (1978) The effect of boron on the granite solidus. Canadian Mineral. 16, 239-244.

Christofides, G. and Sapowitzis, E. (1983) Compositional dependence of the $Mg-Fe^{2+}$ distribution coefficient in biotite-hornblende pairs from the Xanthi (N. Greece) granitic rocks. N. Jahrb. Mineral. Monatsh. 1983, 1-12.

Clarke, D.B. (1981) The mineralogy of peraluminous granites: a review. Canadian Mineral. 19, 3-17.

Clayton, R.N. (1981) Isotopic Thermometry. In: Newton, R.C., Navrotsky, A., and Wood, B.J., Thermodynamics of Minerals and Melts, Ch. 5, Vol. 1, Advances in Physical Geochemistry, Springer-Verlag, 85-109.

Crecraft, H.R., Nash, W.P., and Evans, S.H., Jr. (1981) Late Cenozoic volcanism at Twin Peaks, Utah: geology and petrology. J. Geophys. Res. 86, 10303-10320.

Criss, R.E., Lanphere, M.A., and Taylor, H.P., Jr. (1982) Effects of regional uplift, deformation, and meteoric-hydrothermal metamorphism on K-Ar ages of biotites in the southern half of the Idaho batholith. J. Geophys. Res. 87, 7029-7046.

_____ and Taylor, H.P., Jr. (1983) An $^{18}O/^{16}O$ and D/H study of tertiary hydrothermal systems in the southern half of the Idaho batholith. Geol. Soc. Am. Bull. 94, 640-663.

Cross, W. (1897) Igneous rocks of the Leucite Hills and Pilot Butte, Wyoming. Am. J. Sci. 4, 115-141.

Czamanske, G.K. and Wones, D.R. (1973) Oxidation during magmatic differentiation, Finnmarka Complex, Oslo Area, Norway: Part 2, The mafic silicates. J. Petrol. 14, 349-380.

_____, _____, and Eichelberger, J.C. (1977) Mineralogy and petrology of the intrusive complex of the Pliny Range, New Hampshire. Am. J. Sci. 277, 1073-1123.

Damon, P. (1968) Potassium-argon dating of igneous and metamorphic rocks with applications to the basin ranges of Arizona and Sonora. In: Hamilton, E.I. and Farquhar, R.M., eds., Radiometric Dating for Geologists, Interscience Publishers, 1-71.

de Albuquerque, C.A.R. (1973) Geochemistry of biotites from granitic rocks, northern Portugal. Geochim. Cosmochim. Acta 37, 1779-1802.

_____ (1974) Geochemistry of actinolitic hornblendes from tonalitic rocks, Northern Portugal. Geochim. Cosmochim. Acta 38, 789-883.

_____ (1975) Partition of trace elements in co-existing biotite, muscovite and potassium feldspar of granitic rocks, northern Portugal. Chem. Geol. 16, 89-108.

Deer, W.A. (1937) The composition and paragenesis of the biotites of the Carspairn igneous complex. Mineral. Mag. 24, 495-502.

Deines, P. (1977) On the oxygen isotope distribution among mineral triplets in igneous and metamorphic rocks. Geochim. Cosmochim. Acta 41, 1709-1730.

De Pieri, R. and Jobstraibizer, P.G. (1977) On some biotites from Adamello massif (northern Italy). N. Jahrb. Mineral. Monatsh. 1977, 15-24.

_____ and _____ (1983) Crystal chemistry of biotites from dioritic to granodioritic rock-types of Adamello massif (northern Italy). N. Jahrb. Mineral. Abh. 148, 58-82.

Dodge, F.C.W. (1973) Chlorites from granitic rocks of the central Sierra Nevada batholith, California. Mineral. Mag. 39, 58-64.

_____ and Moore, J.G. (1968) Occurrence and composition of biotites from the Cartridge Pass pluton of the Sierra Nevada batholith, California. U.S. Geol. Surv. Prof. Paper 600-B, B6-B10.

_____ and Ross, D.C. (1971) Coexisting hornblendes and biotites from granitic rocks near the San Andreas fault, California. J. Geol. 79, 158-172.

_____, Smith, V.C. and Mays, R.E. (1969) Biotites from granitic rocks of the central Sierra Nevada batholith, California. J. Petrol. 10, 250-271.

Dupuy, C. (1968) Rubidium et caesium dans biotite, sanidine et verre des ignimbrites de Toscane (Italie). Chem. Geol. 3, 281-291.

Edgar, A.D. (1979) Mineral chemistry and petrogenesis of an ultrapotassic-ultramafic volcanic rock. Contrib. Mineral. Petrol. 71, 171-175.

Erickson, R.L. and Blade, L.V. (1963) Geochemistry and petrology of the alkalic igneous complex at Magnet Cove, Arkansas. U.S. Geol. Surv. Prof. Paper 425, 95 p.

Ewart, A. (1971) Notes on the chemistry of ferromagnesian phenocrysts from selected volcanic rocks, central volcanic region. New Zealand J. Geol. Geophys. 14, 323-340.

Ferry, J.M. (1979) Reaction mechanisms, physical conditions, and mass transfer during hydrothermal alteration of mica and feldspar in granitic rocks from south-central Maine, USA. Contrib. Mineral. Petrol. 68, 125-139.

Fershtater, G.B. (1973) Distribution of titanium and sodium between minerals of granitoids as a geological thermometer. Geochem. Int'l 10, 57-65.

_____, Bushlyakov, I.N., Borodina, N.S. (1970) Distribution of petrogenic elements between biotites and hornblendes in the granitoids of Urals. Geochem. Int'l 7, 903. (abstract)

Fiala, J., Vejnar, Z., Kucerova, D. (1976) Composition of the biotites and the coexisting biotite-hornblende pairs in granitic rocks of the central Bohemian pluton. Krystalinikum 12, 79-111.

Flood, R.H. and Shaw, S.E. (1979) K-rich cumulate diorite at the base of a tilted granodiorite pluton from the New England Batholith, Australia. J. Geol. 87, 417-425.

Foster, M.D. (1960) Interpretation of the composition of trioctahedral micas. U.S. Geol. Surv. Prof. Paper 354-B, 1-49.

Franz, G. and Althaus, E. (1976) Experimental investigation on the formation of solid solutions in sodium-aluminum-magnesium micas. N. Jahrb. Mineral. Abh. 126, 233-253.

Gillberg, M. (1964) Halogen and hydroxyl contents of micas and amphiboles in Swedish granitic rocks. Geochim. Cosmochim. Acta 28, 495-516.

Gorbatschev, R. (1969) Element distribution between biotite and Ca-amphibole in some igneous or pseudoigneous plutonic rocks. N. Jahrb. Mineral. Abh. 111, 314-342.

_____ (1970) Distribution of tetrahedral Al and Si in coexisting biotite and Ca-amphibole. Contrib. Mineral. Petrol. 28, 251-258.

_____ (1977) The influence of some compositional relations on the partition of Fe and Mg between biotite and Ca-amphibole. N. Jahrb. Mineral. Abh. 130, 3-11.

Grabezkev, A.I., Vigorova, V.G., and Chashukhina, V.A. (1979) Behavior of fluorine during crystallization of granites (in connection with validation of the criteria of granite specialization). Geochem. Int'l. 1979, 23-33.

Graybeal, F.T. (1973) Copper, manganese, and zinc in coexisting mafic minerals from Laramide intrusive rocks in Arizona. Econ. Geol. 68, 785-798.

Greenland, L.P., Gottfried, D. and Tilling, R.I. (1968) Distribution of manganese between coexisting biotite and hornblende in plutonic rocks. Geochim. Cosmochim. Acta 32, 1149-1163.

Gromet, L.P. (1983) Rare earth element distributions among minerals in a granodiorite and their petrogenetic implications. Geochim. Cosmochim. Acta 47, 925-939.

Groves, D.F. (1972) The geochemical evolution of tin-bearing granites in the Blue Tier batholith, Tasmania. Econ. Geol. 67, 443-457.

Gunow, A.J., Ludington, S., and Munoz, J.L. (1980) Fluorine in micas from the Henderson molybdenite deposit, Colorado. Econ. Geol. 75, 1127-1137.

Haack, U.K. (1969) Spurenelemente in Biotiten aus Graniten und Gneisen. Contrib. Mineral. Petrol. 22, 83-126.

Hall, A. (1965) The occurrence of prehnite in appinitic rocks from Donegal, Ireland. Mineral. Mag. 35, 235-236.

Harrison, T.M., Armstrong, R.L., Naeser, C.W., and Harakal, J.E. (1979) Geochronology and thermal history of the Coast Plutonic Complex, near Prince Rupert, British Columbia. Canadian J. Earth Sci. 16, 400-410.

Haslam, H.W. (1968) The crystallization of intermediate and acid magmas at Ben Nevis, Scotland. J. Petrol. 9, 84-104.

Hazen, R.M., Finger, L.W., and Velde, D. (1981) Crystal structure of a silica- and alkali-rich trioctahedral mica. Am. Mineral. 66, 586-591.

Heinrich, E.W. (1946) Studies in the mica group. J. Sci. 244, 836-848.

Hendry, D.A.F., Chivas, A.R., Reed, S.J.B., and Long, J.V.P. (1981) Geochemical evidence for magmatic fluids in porphyry copper mineralization. Part II. Ion-probe analysis of Cu contents of mafic minerals, Koloula Igneous Complex. Contrib. Mineral. Petrol. 78, 404-412.

Hietanen, A. (1971) Distribution of elements in biotite-hornblende pairs and in an orthopyroxene-clinopyroxene pair from zoned plutons, northern Sierra Nevada, California. Contrib. Mineral. Petrol. 30, 161-176.

Higashi, S. (1978) Dioctahedral mica minerals with ammonium ions. Mineral. J. (Japan) 9, 16-27.

Higuchi, H. and Nagasawa, H. (1969) Partition of trace elements between rock-forming minerals and the host volcanic rocks. Earth Planet. Sci. Letters 7, 281-287.

Hofmann, A. and Giletti, B.J. (1970) Diffusion of geochronologically important nuclides under hydrothermal conditions. Eclogae Geol. Helvetiae 63, 141-150.

Honma, H. (1974) Chemical features of biotites from metamorphic and granitic rocks of the Yanai district in the Ryoke belt, Japan. J. Japan. Assoc. Mineral., Petrol. and Econ. Geol. 69, 390-402.

_____ and Itihara, Y. (1981) Distribution of ammonium in minerals of metamorphic and granitic rocks. Geochim. Cosmochim. Acta 45, 983-988.

Huntington, H.D. (1979) Kiglapait mineralogy I: apatite, biotite, and volatiles. J. Petrol. 20, 625-652.

Imeokparia, E.G. (1981) Fluorine in biotites from the Afu younger granite complex (central Nigeria). Chem. Geol. 32, 247-254.

Itihara, Y. and Honma, H. (1979) Ammonium in biotite from metamorphic and granitic rocks of Japan. Geochim. Cosmochim. Acta 43, 503-509.

Ishihara, S. (1977) The magnetite-series and ilmenite-series granitic rocks. Mining Geol. 27, 293-305.

Jacobs, D.C. and Parry, W.T. (1976) A comparison of the geochemistry of biotite from some Basin and Range stocks. Econ. Geol. 71, 1029-1035.

_____ and _____ (1979) Geochemistry of biotite in the Santa Rita porphyry copper deposit, New Mexico. Econ. Geol. 74, 860-887.

Javoy, M. (1977) Stable isotopes and geothermometry. J. Geol. Soc. 133, 609-636.

Kabesh, M.L. and Aly, M.M. (1980) The chemistry of biotites as a guide to the petrogenesis of some Precambrian granitic rocks, Yemen Arab Republic. Chemie Erde 39, 313-324.

_____ and Ragab, A.I. (1974) The chemistry of biotites as a guide to the evolution trends of El Atawi granitic rocks, Eastern Desert, Egypt. N. Jahrb. Mineral. Monatsh., 307-316.

_____, _____ and Refaat, A.M. (1975) On the chemistry of biotites and variation of ferrous-ferric ratios in the granitic rocks of Umm Naggat Stock, Egypt. N. Jahrb. Mineral. Abh. 124, 47-60.

_____ and Refaat, A.M. (1972) The chemical composition of biotites as a guide to the petrogenesis of the granitic rocks of Wadi El Mellaha, Northern Eastern Desert, Egypt. N. Jahrb. Mineral. Abh. 117, 85-95.

Kanisawa, S. (1972) Coexisting biotites and hornblendes from some grantic rocks in southern Kitakami Mountains, Japan. J. Japan. Assoc. Mineral., Petrol., Econ. Geol. 67, 332-344.

_____, Tanaka, H. and Nakai, Y. (1979) Behaviour of fluorine in granitic rocks from the Abukuma Plateau, the Ryoke Zone, and the San-in Zone, Japan. J. Geol. Soc. Japan 85, 123-134.

Kapustin, Yu.L. (1980) Mineralogy of Carbonatites, Amerind Publishing Co., Put. Ltd., New Delhi, India, 259 p.

Karyakin, A.V., Volynets, V.F. and Kriventsova, G.A. (1973) Investigation of nitrogen compounds in micas by infrared spectroscopy. Geochim. Int'l. 10, 326-329.

Kato, K. (1972) Petrology of the Orikabe granitic body, Kitakami Mountainland. J. Japan. Assoc. Mineral., Petrol., Econ. Geol. 67, 50-59.

Kato, Y., Onuki, H. and Tanaka, H. (1977) Compositional dependence of the Mg/Fe^{2+} distribution coefficient between biotite-hornblende pairs from calc-alkaline granitic rocks. J. Japan. Assoc. Mineral., Petrol., Econ. Geol. 72, 252-258.

Kesler, S.E., Issigonis, M.J., Brownlow, A.H., Damon, P.E., Moore, W.J., Northcote, K.E. and Preto, V.A. (1975) Geochemistry of biotites from mineralized and barren intrusive systems. Econ. Geol. 70, 559-567.

Kistler, R.W., Ghent, E.D. and O'Neil, J.R. (1981) Petrogenesis of garnet two-mica granites in the Ruby Mountains, Nevada. J. Geophys. Res. 86, 10591-10606.

Kolbe, P. (1966) Geochemical investigation of the Cape granite, South-western Cape Province, South Africa. Trans. Geol. Soc. South Africa 69, 161-199.

Kuehner, S.M., Edgar, A.D. and Arima, M. (1981) Petrogenesis of the ultrapotassic rocks from the Leucite Hills, Wyoming. Am. Mineral. 66, 663-677.

Kuroda, Y., Suzuoki, T., Matsuo, S. and Kanisawa, S. (1974) D/H fractionation of coexisting biotite and hornblende in some granitic rock masses. J. Japan. Assoc. Mineral., Petrol., Econ. Geol. 69, 95-102.

_____, _____ and Tanaka, H. (1976) D/H fractionation of coexisting biotite and hornblende in Tabito composite mass, Abukuma plateau, Japan. J. Japan. Assoc. Mineral., Petrol., Econ. Geol. 71, 1-15.

_____, _____, _____, Murakami, N., Kanisawa, S. and Kinugawa, T. (1977) D/H ratios of biotites and hornblendes from some granitic rocks in the Chugoku district, southwest Japan. J. Geol. Soc. Japan 83, 719-724.

Lalonde, A.E. and Martin, R.F. (1983) The Baie-des-Moutons syenitic complex, La Tabatiere, Quebec. II. The ferromagnesian minerals. Canadian Mineral. 21, 81-91.

Lange, I.M., Reynolds, R.C. and Lyons, J.B. (1966) K/Rb ratios in coexisting K-feldspars and biotites from some New England granites and metasediments. Chem. Geol. 1, 317-322.

Lanier, G., Raab, W.J., Folsom, R.B., Cone, S. (1978) Alteration of equigranular monzonite, Bingham Mining District, Utah. Econ. Geol. 73, 1270-1286.

Larsen, E.S., Jr., Irving, J., Gonyer, F.A. and Larsen, E.S. III (1937) Petrologic results of a study of the minerals from the Tertiary volcanic rocks of the San Juan Region, Colorado. Am. Mineral. 22, 889-905.

_____ and Draisin, W.M. (1950) Composition of the minerals in the rocks of the southern California batholith. 18th Int'l. Geol. Congress, London, 1948, Rpt. Pt. 2, 66-79.

Leake, B.E. (1974) The crystallization history and mechanism of emplacement of the western part of the Galway Granite, Connemara, western Ireland. Mineral. Mag. 39, 498-513.

LeBel, L. (1979) Micas magmatiques et hydrothermaux dans l'environnement du porphyre cuprifère de Cerro Verde-Santa Rosa, Pérou. Bull. Minéral. 102, 35-41.

Lee, D.E. and Loenen, R.E.V. (1970) Biotites from hybrid granitoid rocks of the Southern Snake Range, Nevada. U.S. Geol. Surv. Prof. Paper 700D, 196-206.

_____, Kistler, R.W., Friedman, I. and van Loenen, R.E. (1981) Two-mica granites of northeastern Nevada. J. Geophys. Res. 86, 10607-10616.

Lovering, T.G. (1969) Distribution of minor elements in samples of biotite from igneous rocks. U.S. Geol. Surv. Prof. Paper 650-B, B101-B106.

Ludington, S. (1978) The biotite-apatite geothermometer revisited. Am. Mineral. 63, 551-553.

Lyons, P.C. and Krueger, H.W. (1976) Petrology, chemistry, and age of the Rattlesnake pluton and implications for other alkalic granite plutons of southern New England. Geol. Soc. Am. Memoir 146, 71-102.

Mahmood, A. (1983) Chemistry of biotites from a zoned granitic pluton in Morocco. Mineral. Mag. 47, 365-369.

Mansker, W.L., Ewing, R.C. and Keil, K. (1979) Barian-titanian biotites in nephelinites from Oahu, Hawaii. Am. Mineral. 64, 156-159.

Mahood, G. and Hildreth, W. (1983) Large partition coefficients for trace elements in high-silica rhyolites. Geochim. Cosmochim. Acta 47, 11-30.

Mason, D.R. (1978) Compositional variations in ferromagnesian minerals from porphry copper-generating and barren intrusions of the Western Highlands, Papua, New Guinea. Econ. Geol. 73, 878-890.

Matsuhisa, Y., Goldsmith, J.R. and Clayton, R.N. (1979) Oxygen isotope fractionation in the system quartz-albite-anorthite-water. Geochim. Cosmochim. Acta 43, 1131-1140.

Miller, C.F., Stoddard, E.F., Bradfish, L.J. and Dollase, W.A. (1981) Composition of plutonic muscovite: genetic implications. Canadian Mineral. 19, 25-34.

Milovskiy, A.V. and Volynets, V.F. (1966) Nitrogen in metamorphic rocks. Geochem. Int'l. 3, 752-758.

Monier, G., Mergoil-Daniel, J. and Labernardière, H. (1984) Générations successives de muscovites et feldspaths potassiques dans les leucogranite du massif de Millevaches (Massif Central francais). Bull. Minéral. 107, 55-68.

Moore, A.C. (1976) Intergrowth of prehnite and biotite. Mineral. Mag. 40, 526-529.

Muller, G. (1966) Die Beziehungen zwischen der chemischen Zusammensetzeng, Lichtbrechung, und Dichter einiger Koexistierender Biotit, Muskovit, und Chlorit aus granitischen Tiefengesteinen. Contrib. Mineral. Petrol. 12, 173-191.

Munoz, J.L. and Ludington, S.D. (1974) Fluoride-hydroxyl exchange in biotite. Am. J. Sci. 274, 396-413.

_____ and _____ (1977) Fluorine-hydroxyl exchange in synthetic muscovite, with application to muscovite-biotite assemblages. Am. Mineral. 62, 304-308.

_____ and Swenson, A. (1981) Chloride-hydroxyl exchange in biotite and estimation of relative HCl/HF activities in hydrothermal fluids. Econ. Geol. 76, 2212-2221.

Murakami, N. (1969) Two contrastive trends of evolution of biotite in granitic rocks. Ganseki Kobutsu Kosho Gakkaishi (J. Japan. Assoc. Mineral., Petrol., Econ. Geol.) 62, 223-247.

Nagy, K.L. and Parmentier, E.M. (1982) Oxygen isotopic exchange at an igneous intrusive contact. Earth Planet. Sci. Letters 59, 1-10.

Nash, W.P. (1976) Fluorine, chlorine, and OH-bearing minerals in the Skaergaard intrusion. Am. J. Sci. 276, 546-557.

Nedachi, M. (1980) Chlorine and fluorine contents of rock-forming minerals of the Neogene granitic rocks in Kyushu, Japan. Mining Geol. Spec. Issue 8, 39-48.

Neilson, M.J. and Haynes, S.J. (1973) Biotites in calc-alkaline intrusive rocks. Mineral. Mag. 39, 251-253.

Neiva, A.M.R. (1976) The geochemistry of biotites from granites of northern Portugal with special reference to their tin content. Mineral. Mag. 40, 453-466.

_____ (1981) Geochemistry of hybrid granitoid rocks and of their biotites from central northern Portugal and their petrogenesis. Lithos 14, 149-163.

_____ (1983) Geochemistry of granitic rocks and their micas from the west border of the Alvão Plateau, northern Portugal. Chemie Erde 43, 31-44.

Noble, D.C., Vogel, T.A., Peterson, P.S., Landis, G.P., Grant, N.K., Jezek, P.A. and McKee, E.H. (1984) Rare-element-enriched, S-type ash-flow tuffs containing phenocrysts of muscovite, andalusite, and sillimanite, southeastern Peru. Geol. 12, 35-39.

Nockolds, S.R. (1947) The relation between chemical composition and paragenesis in the biotite micas of igneous rocks. Am. J. Sci. 245, 401-420.

_____ and Mitchell, R.L. (1948) The geochemistry of some Caledonian plutonic rocks: a study in the relationship between the major and trace elements of igneous rocks and their minerals. Trans. Royal Soc. Edinburgh 61, 535-575.

Olade, M.A. (1979) Copper and zinc in biotite, magnetite and feldspar from a porphyry copper environment, Higland Valley, British Columbia. Canadian Mining Eng. 31, 1363-1370.

Olesch, M. (1979) Ca-bearing phlogopite: synthesis and solid solubility at high temperatures and pressures of 5 and 10 kilobars. Bull. Minéral. 102, 14-20.

Olives Baños, J., Amouric, M., de Fouquet, C. and Baronnet, A. (1983) Interlayering and interlayer slip in biotite as seen by HRTEM. Am. Mineral. 68, 754-758.

O'Neil, J.R. and Kharaka, Y.K. (1976) Hydrogen and oxygen isotope exchange reactions between clay minerals and water. Geochim. Cosmochim. Acta 40, 241-246.

_____ and Taylor, R.P., Jr. (1969) Oxygen isotope equilibrium between muscovite and water. J. Geophys. Res. 74, 6012-6022.

Parry, W.T. (1972) Chlorine in biotite from Basin and Range plutons. Econ. Geol. 67, 972-975.

_____ and Nackowski, M.P. (1963) Copper, lead and zinc in biotites from Basin and Range quartz monzonites. Econ. Geol. 58, 1126-1144.

_____ and Jacobs, D.C. (1975) Fluorine and chlorine in biotite from Basin and Range plutons. Econ. Geol. 70, 554-558.

_____, Ballantyne, G.H. and Wilson, J.C. (1978) Chemistry of biotite and apatite from a vesicular quartz latite porphyry plug at Bingham, Utah. Econ. Geol. 73, 1308-1314.

Parsons, I. (1979) The Klokken gabbro-syenite complex, South Greenland: cryptic variation and origin of inversely graded layering. J. Petrol. 20, 653-694.

_____, I. (1981) The Klokken gabbro-syenite complex, South Greenland: quantitative interpretation of mineral chemistry. J. Petrol. 22, 233-260.

Pattison, D.R.M., Carmichael, D.M. and St.-Onge, M.R. (1982) Geothermometry and geobarometry applied to Early Proterozoic "S-type" granitoid plutons, Wopmay Orogen, Northwest Territories, Canada. Contrib. Mineral. Petrol. 79, 394-404.

Phillips, E.R. and Rickwood, P.C. (1975) The biotite-prehnite association. Lithos 8, 275-281.

Piwinskii, A.J. and Wylle, P.H. (1968) Experimental studies of igneous rock series: a zoned pluton in the Wallowa batholith. J. Geol. 76, 205-234.

Potap'yev, V.V. (1964) Decrease in the index of refraction of biotite in late-phase granite of the Kolyvan Massif (Altai). Dokl. Akad. Nauk SSSR 155, 141-143.

Prider, R.T. (1939) Some minerals from the leucite-rich rocks of the west Kimberley area, Western Australia. Mineral. Mag. 25, 373-387.

Ramberg, H. (1952) Chemical bonds and the distribution of cations in silicates. J. Geol. 60, 331-355.

Refaat, A.M. and Abdallah, Z.M. (1979) Geochemical study of coexisting biotite and chlorite from Zaker Granitic rocks of Zanjan area, northwest Iran. N. Jahrb. Mineral. Abh. 136, 262-275.

Rehrig, W.A. and McKinny, C.N. (1976) The distribution and origin of anomalous copper in biotite. Soc. Mining Eng., A.I.M.E. preprint 76-L-64, 34 p.

Robert, J.-L. (1976) Titanium solubility in synthetic phlogopite solid solutions. Chem. Geol. 17, 213-227.

Roegge, J.S., Logsdon, M.J., Young, H.S., Barr, H.B., Borcsik, M. and Holland, H.D. (1974) Halogens in apatites from the Providencia area, Mexico. Econ. Geol. 69, 229-240.

Rutherford, M.J. (1973) The phase relations of aluminous iron biotites in the system $KAlSi_3O_8$-$KAlSiO_4$-Al_2O_3-Fe-O-H. J. Petrol. 14, 159-180.

Sakuyama, M. (1979) Lateral variations in H_2O contents in Quaternary magmas of northeastern Japan. Earth Planet. Sci. Letters 43, 103-111.

Sapountzis, E.S. (1976) Biotites from the Sithonia igneous complex (North Greece). N. Jahrb. Mineral. Abh. 126, 327-341.

Schleicher, H. and Lippolt, H.J. (1981) Magmatic muscovite in felsitic parts of rhyolites from southwest Germany. Contrib. Mineral. Petrol. 78, 220-224.

Schnetzler, C.C. and Philpotts, J.A. (1970) Partition coefficients of rare-earth elements between igneous matrix material and rock-forming-mineral phenocrysts. II. Geochim. Cosmochim. Acta 34, 331-340.

Sen, N., Nuckolds, S.R. and Allen, R. (1959) Trace elements in minerals from rocks of the southern California batholith. Geochim. Cosmochim. Acta 16, 58-78.

Shand, S.J. (1947) Eruptive Rocks. 488 p.

Shieh, Y.N. and Taylor, H.P. (1969) Oxygen and hydrogen isotope studies of contact metamorphism in the Santa Rosa Range, Nevada and other areas. Contrib. Mineral. Petrol. 20, 306-356.

_____ and Schwarcz, H.P. (1974) Oxygen isotope studies of granite and migmatite, Greenville province of Ontario, Canada. Geochim. Cosmochim. Acta 38, 21-45.

Simonen, A. and Vorma, A. (1969) Amphibole and biotite from rapakivi. Bull. Comm. Géol. Finlande 238, 28 p.

Speer, J.A. (1981) Petrology of cordierite- and almandine-bearing granitoid plutons of the southern Appalachian Piedmont, U.S.A. Canadian Mineral. 19, 35-46.

_____, Becker, S.W. and Farrar, S.S. (1980) Field relations and petrology of the postmetamorphic, coarse-grained granitoids and associated rocks of the Southern Appalachian Piedmont. In: D.R. Wones (ed.), Proc. Caledonides in the USA, I.G.C.P. Project 27: Caledonide Orogen, Virginia Polytechnic Inst. & State Univ., Geol. Sci. Memoir 2, 137-148.

Stephenson, N.C.N. (1977) Coexisting hornblendes and biotites from Precambrian gneisses of the south coast of western Australia. Lithos 10, 9-27.

Stevenson, F.J. (1962) Chemical state of nitrogen in rocks. Geochim. Cosmochim. Acta 26, 797-809.

Stollery, G., Borcsik, M. and Holland, H.D. (1971) Chlorine in intrusives: a possible prospecting tool. Econ. Geol. 66, 361-367.

Stormer, J.C. and Carmichael, I.S.E. (1971) Fluorine-hydroxyl exchange in apatite and biotites: a potential igneous geothermometer. Contrib. Mineral. Petrol. 31, 121-131.

Suzuoki, T. and Epstein, S. (1976) Hydrogen fractionation between OH-bearing silicate minerals and water. Geochim. Cosmochim. Acta 40, 1229-1240.

Tanaka, H. (1975) Magnesium-iron distribution in coexisting biotite and hornblende from granitic rocks. J. Japan. Assoc. Mineral., Petrol., Econ. Geol. 70, 118-124.

Taylor, H.P., Jr. (1977) Water/rock interactions and the origin of H_2O in granitic batholiths. J. Geol. Soc. 133, 509-558.

_____ (1978) Oxygen and hydrogen isotope studies of plutonic granitic rocks. Earth Planet. Sci. Letters 38, 177-210.

Thompson, R.N. (1977) Primary basalts and magma genesis III. Alban Hills, Roman comagmatic province, central Italy. Contrib. Mineral. Petrol. 60, 91-108.

Tischendorf, G. (1973) Metallogenic basis of tin exploration in the Erzgebirge. Inst. Mining Metall., Sect. B82, B7-B24.

Tulloch, A.J. (1979) Secondary Ca-Al silicates as low-grade alteration products of granitoid biotite. Contrib. Mineral. Petrol. 69, 105-117.

Turner, D.L. and Forbes, R.B. (1976) K-Ar studies in two deep basement drill holes: a new geologic estimate of argon blocking for biotite. Trans. Am. Geophys. Soc. EOS 57, 353.

Urano, H. (1971) Geochemical and petrological study on the origins of metamorphic rocks and granitic rocks by determination of fixed ammoniacal nitrogen. J. Earth Sci., Nagoya Univ. 19, 221-228.

van Kooten, G.K. (1980) Mineralogy, petrology, and geochemistry of an ultrapotassic basaltic suite, Central Sierra Nevada, California, U.S.A. J. Petrol. 21, 651-684.

van Loon, J.C., Kesler, S.E. and Moore, C.M. (1973) Analysis of water-extractable chlorite in rocks by use of a selective ion electrode. In: Jones, M.J. (ed.) Geochem. Exploration 1972, 4th Int'l. Geochem. Exploration Symp., London, 1972, Inst. Mining Metall., 429-434.

Veblen, D.R. and Ferry, J.M. (1983) A TEM study of the biotite-chlorite reaction and comparison with petrologic observations. Am. Mineral. 68, 1160-1168.

Vedder, W. (1965) Ammonium in muscovite. Geochim. Cosmochim. Acta 29, 221-228.

Vejnar, Z. (1971) Trioctahedral micas of west Bohemian pluton and their petrogenetic significance. Krystalinikum 7, 149-164.

Velde, B. (1965) Phengite micas: synthesis, stability and natural occurrence. Am. J. Sci. 263, 886-913.

Velde, D. (1975) Armalodite-Ti-phlogopite-analcite-bearing lamproites from Smoky Butte, Montana. Am. Mineral. 60, 566-573.

_____ (1979) Trioctahedral micas in melilite-bearing eruptive rocks. Carnegie Inst. Wash. Yearbook 78, 468-475.

Vlasov, K.A., Kuz'menko, M.V. and Es'Kova, E.M. (1966) The Lovozero Alkali Massif, Hafner Publishing Co., New York, 627 p.

Volkov, V.N. and Gorbacheva, S.A. (1980) Composition of rock-forming biotite and the variation in crystallization conditions in a vertically exposed intrusion. Geochem. Int'l. 1980, 75-79.

Volynets, O.N., Kutyev, F.Sh., Koloskov, A.V., and Erlich, E.N. (1979) Rock-forming and accessory minerals of acid andesite-rhyolite rock series. In: Erlich, E.N. and Gorshkov, G.S. (eds.) Quaternary Volcanism and Tectonics in Kamchatka. Bull. Volcano. 42, 185-211.

Wallace, S.R., MacKenzie, W.B., Blair, R.G. and Muncaster, N.K. (1978) Geology of the Urad and Henderson molybdenite deposits, Clear Creek County, Colorado, with a section on a comparison of these deposits with those at Climax, Colorado. Econ. Geol. 73, 325-368.

Walsh, J.N. (1975) Clinopyroxenes and biotites from the Centre III igneous complex, Ardnamurchan, Argyllshire. Mineral. Mag. 40, 335-345.

Waldbaum, D.R. and Thompson, J.B., Jr. (1969) Mixing properties of sanidine crystalline solutions. IV. Phase diagrams from equations of state. Am. Mineral. 54, 1274-1298.

Wentlandt, R.F. (1977) Barium-phlogopite from Haystack Butte, Highwood Mountains, Montana. Carnegie Inst. Wash. Yearbook 76, 534-539.

Whelan, J. (1980) Aspects of Granites and Associated Mineralization. Ph.D. dissertation. Australia National Univ., Canberra, A.C.T., Australia.

White, A.J.R. (1966) Genesis of magmatites from the Palmer region of South Australia. Chem. Geol. 6, 133-213.

Wiotzka, F. (1961) Untersuchungen zur geochimie des Stickstoffs. Geochim. Cosmochim. Acta 24, 106-154.

Wones, D.R. (1972) Stability of biotite: a reply. Am. Mineral. 57, 316-317.

_____ (1980) Contributions of crystallography, mineralogy, and petrology to the geology of the Lucerne pluton, Hancock County, Maine. Am. Mineral. 65, 411-437.

_____ (1981) Mafic silicates as indicators of intensive variables in granitic magmas. Mining Geol. 31, 191-212.

_____ and Eugster, H.P. (1965) Stability of biotite: experiment, theory and applications. Am. Mineral. 50, 1228-1272.

_____, Burns, R.G. and Carroll, B.M. (1971) Stability and properties of synthetic annite. Trans. Am. Geophys. Union 53, 550.

Wood, C.P. (1974) Petrogenesis of garnet-bearing rhyolites from Canterbury, New Zealand. New Zealand J. Geol. Geophys. 17, 759-787.

Wyborn, D., Chappell, B.W. and Johnston, R.M. (1981) Three S-type volcanic suites from the Lachlan Fold Belt, southeast Australia. J. Geophys. Res. 86, 10335-10348.

Yamamoto, T. and Nakahira, M. (1966) Ammonium ions in sericites. Am. Mineral. 51, 1775-1778.

10. MICAS in METAMORPHIC ROCKS
Charles V. Guidotti

INTRODUCTION

It is evident from Chapters 1 and 2 of this volume that the list of true micas is quite extensive and that there is much chemical variability among these micas. Fortunately for metamorphic petrologists, only a few micas are common in metamorphic rocks and their chemical variability is fairly limited. Nonetheless, the writer (albeit biased) believes that the micas may be the most useful group of minerals for gaining information on the petrogenesis of those metamorphic rocks containing them.

Table 1 in Chapter 1 shows that many micas involve elements such as Li, Ba, V, Zn, etc., which are scarce in most common metamorphic rocks -- hence such micas are virtually absent in metamorphic rocks. Rock-forming micas are composed of elements common in the crust of the earth, but not all micas are abundant, rock-forming minerals (e.g., Na-phlogopite, wonesite, and clintonite). Their scarcity probably results from their occurrence in rock types such as metamorphosed evaporites, Ca-deficient volcanic rocks, etc., which are not common bulk compositions for metamorphic rocks. However, it is also possible that they have been overlooked or misidentified. We will not discuss these relatively rare micas.

The micas which are common rock-forming minerals include the white micas, muscovite, margarite, and paragonite, and the dark micas which may be broadly described as the biotites. These micas are extremely useful as petrogenetic indicator phases because:

(a) They are abundant in several of the bulk compositions that make up the most common metamorphic rocks. Mica schists (i.e., metapelites) come to mind immediately. Micas are also common in the metamorphosed equivalents of impure limestones, greywacke, various igneous rocks, etc. However, it is well recognized that it is in the metapelites where the micas put on their best "show".

(b) The rock-forming micas occur over a wide range of metamorphic conditions. In the metapelites they are actively involved in most of the important isogradic reactions, especially the biotite isograd. The upper part of the amphibolite facies is also commonly defined in terms of a reaction related to the upper stability limit of muscovite (the K-feldspar + sillimanite isograd). Or again, some workers have defined the granulite facies in terms of a reaction in which biotite breaks down to hypersthene + sillimanite.

(c) All of the micas are hydrous phases, hence their phase relations provide information about $\mu(H_2)$, as well as P and T.

(d) The micas exhibit extensive compositional variation but in contrast with amphiboles and pyroxenes, only a few end members suffice to describe all of the variation. To a very great extent the compositional variation involves Na = K and Fe^{2+} = Mg. Other substitutions are relatively minor in the micas of metamorphic rocks, although petrologically they are extremely important and interesting.

(e) The observed compositional variation of micas is strongly controlled by metamorphic grade and it usually can be distinguished from compositional variation due to bulk composition factors. Hence, one can use the composition of micas to gain information about metamorphic processes in general or information about grade of metamorphism, even in areas that have not been studied in detail.

General groupings of the rock-forming micas

We have grouped the micas of metamorphic rocks as white and dark micas. Most earth scientists would think of these as muscovite and biotite and this would be correct in the great majority of cases. For petrologic purposes a more complete listing, split into two groups, is required. The micas within each group are intimately related in terms of their composition, but only minor solid solution occurs between the micas of the two groups.

Group I: The dioctahedral white micas

This group includes muscovite, margarite, and paragonite. Each is separated from the other by a miscibility gap. Muscovite is usually thought of as one of the most common and abundant minerals in metamorphic rocks, in part because it is very common in metashales (i.e. metapelites) -- one of the most common rocks in the geologic record -- and in part because it is so easily identified, even in hand specimen. In fact, it is truly abundant only in this relatively narrow, though very important, bulk composition range. Muscovite does occur in other rock types but to a much lesser extent than in the metapelites.

Margarite and paragonite were once thought to be rare. Margarite was not even thought of as a rock-forming mineral. Instead, it was considered as one of the "funny minerals" found in unusual rocks that sometimes were subjected to metamorphic processes, e.g. emery deposits. We now know that these two white micas are actually fairly common, though still more rare than muscovite. The petrologic significance and relative scarcity of paragonite (compared to muscovite) is now fairly well understood. The significance of the phase relations involving margarite are only now starting to be understood.

Two other dioctahedral white micas listed in Table 1 of Chapter 1 are phengite and celadonite. Both will be considered in detail later. For the present purposes we can merely state that they represent an important, commonly observed compositional deviation, i.e., $(Mg,Fe^{2+}),Si^{4+} = Al^{vi},Al^{iv}$, of muscovite from its ideal formula, $KAl_2(AlSi_3O_{10})(OH)_2$. Although one might anticipate analogous compositional variation for margarite and especially paragonite, observations show little of this substitution in either one.

Group II: The trioctahedral dark micas

Viewed petrologically these are referred to as biotites. Although one of this group (phlogopite) is not dark in color, it belongs in the group, because it is trioctahedral, and it is one of the end members of the general biotite group.

Most of the compositional variation in the biotite group can be described in terms of four end members:

phlogopite, $KMg_3(AlSi_3O_{10})(OH)_2$; annite, $KFe_3^{2+}(AlSi_3O_{10})(OH)_2$; eastonite, $K(Al_1Mg_2)(Al_2Si_2O_{10})(OH)_2$; and siderophyllite, $K(Al_1Fe_2^{2+})(Al_2Si_2O_{10})(OH)_2$.

Several modest though petrologically interesting compositional deviations do occur from the composition plane defined by these end members.

Although the dark micas have a much wider range of occurrence in metamorphic rocks than do the white micas, they are still most common and most abundant

Table 1 Nomenclature used by Different Workers for Various Real and Hypothetical Compositions Relevant to the Common, Rock-forming Dioctahedral K-White Micas

Formula in terms of VI & IV sites/22 Ox	Winchell (1927)(1951)	Schaller (1950)	Yoder & Eugster (1955)	Foster (1956)(1967)	Kanehira & Banno (1960)	Ernst (1963)	Wise & Eugster (1964)	Velde (1965)	Zen (1981)	Zen (in press)	This Discussion
$(Al_4)^{VI}(Al_2Si_6O_{20})^{IV}$	Muscovite	Muscovite	Muscovite	Muscovite	Muscovite	Muscovite	Muscovite	Muscovite	Muscovite	Muscovite	Muscovite
$(MgAl_3)(AlSi_7O_{20})$	Picro-Phengite	High Silica Sericite		Phengite	Phengite	Phengite		Phengite	Phengite		Phengite
$(Fe^{+2}Al_3)(AlSi_7O_{20})$	Ferro-Phengite										
$(Fm^{(a)}_2Al_2Fe^{+3})(AlSi_7O_{20})$					Ferri-Phengite	Ferri-Phengite			Ferri-Phengite		Ferri-Phengite
$(Mg_2Al_2)(Si_8O_{20})$		Leuco-Phyllite	Al-Celadonite				Leuco-Phyllite			Leuco-Phyllite	Leuco-Phyllite
$(Fe^{+2}_2Al_2)(Si_8O_{20})$			Celadonite	Celadonite			Celadonite	Celadonite		Celadonite	Celadonite
$(Mg_2Fe^{+3}_2)(Si_8O_{20})$		Celadonite					Most Important end Member				
$(Fe^{+2}_2Fe^{+3}_2)(Si_8O_{20})$											
$(Fe^{+3}Al_3)(Al_2Si_6O_{20})$					Ferri-Muscovite	Ferri-Muscovite		"Iron Muscovite"	Ferri-Muscovite	Ferri-Muscovite	*Ferri-Muscovite
$(Fe^{+3}_2Al_2)(Al_2Si_6O_{20})$	Ferri-Muscovite		Fe Muscovite				Ferri-Muscovite			Ferri-Muscovite	Ferri-Muscovite
$(Fe^{+3}_4)(Al_2Si_6O_{20})$											

(a) $Fm = Mg + Fe^{+2}$ Combined

in pelitic schists coexisting with the white micas. For example, biotite and muscovite are the pair which characterizes the "mica schists". The biotite group also exhibits a wider range of compositional variation than do the white micas. None of the miscibility gaps so prominent in the white micas seem to be present in the biotite group. Moreover, especially in the high-T ranges, the biotite group seems to have a markedly larger stability field than any of the white micas. This aspect of the biotite group probably explains its wider range of occurrence in comparison with the white micas.

DIOCTAHEDRAL WHITE MICAS

We will discuss the white micas in three independent categories, but each discussion will be in the context of the micas as they occur in metamorphic rocks. The categories are (I) White micas from a mineralogic perspective. (II) Petrologic consideration of the white micas. [This will deal with their petrogenetic implications in terms of the phase relations among the micas themselves and with respect to other coexisting minerals, the latter stressing the extreme importance of "mineral assemblage" when using micas as petrogenetic indicators.] (III) Determinative methods for recognizing these white micas and ascertaining their composition. [To avoid overlap with the material presented in Chapter 7, our discussion will be strictly from the viewpoint of petrologic problems.]

(I) White micas from a mineralogic perspective

Here we shall concentrate on: (A) the observed chemical variation of white micas, (B) white mica lattice spacings and polytypes, and (C) crystallochemical aspects of the observed chemical variation.

(I.A) Observed chemical variation of the white micas.

The observed chemical variation of the white micas in metamorphic rocks is most easily discussed in terms of the end-member subsystem muscovite (Mu), paragonite (Pg), and margarite (Marg) which can be defined in the ideal system $NaAlO_2$ - $KAlO_2$ - $CaAl_2O_4$ - Al_2O_3 - SiO_2 - H_2O. The most important compositional deviation of the white micas from the ideal subsystem Mu-Pg-Marg is the solid solution between Mu and celadonite (Cd -- see Table 1). This, and other lesser compositional deviations are discussed below.

The system $NaAlO_2$ - $KAlO_2$ - CAl_2O_4 - Al_2O_3 - SiO_2 - H_2O. This system is shown on Figure 1, with SiO_2 and H_2O considered to be in excess so that they do not need to be shown graphically. On this diagram the white mica plane is shown in terms of the end members, Mu-Pg-Marg. Similarly, the feldspar plane is shown in terms of the end members, Or-Ab-An.

Figure 2a is a schematic approximation (based on natural observations) of the white mica plane extracted from Figure 1. It shows the solution fields and connecting tie lines at some arbitrary, moderate T. On this diagram the various miscibility gaps and 1, 2, and 3 phase regions are readily apparent. Figure 2b is a familiar schematic representation of the feldspar plane from Figure 1. For Figure 2b, a temperature is assumed (e.g., 600°C) such that the alkali feldspar solvus is cut and such that at least some of the complications in the plagioclase series (e.g., the peristerite gap) will be avoided. For simplicity, we will ignore such complications. The main purpose of Figures 2a and 2b is to show that the white mica plane is similar to the more familiar feldspar plane. Basically, the following similarities can be noted from these two diagrams.

(a) Two prominent exchange solutions are present; a single site exchange,

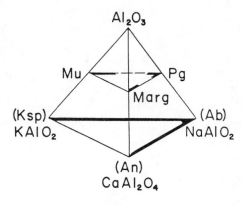

Figure 1 (left). The ideal system $NaAlO_2$ - $KAlO_2$ - $CaAl_2O_4$ - Al_2O_3 with quartz and H_2O in excess. The feldspar and white mica planes are shown in this ideal system.

Figure 2 (below). (a) Schematic representation of the white mica plane at some arbitrary, moderate T. (b) Schematic representation of the feldspar plane at about 600°C. Miscibility gaps in the plagioclase series are not considered.

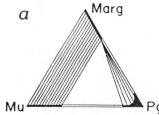 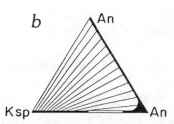

Na = K between Mu and Pg, and a coupled exchange Na,Si = Ca,Al between Pg and Marg. A third exchange K,Si = Ca,Al is very limited for both the feldspars and micas. In all cases the mica solutions are nonideal, a point readily apparent from the prominent miscibility gaps.

(b) The tie line configurations are generally similar on both planes with Na-poor Mu or Or coexisting with Ca-rich Marg or Pg, respectively. Even in the case of bulk compositions lying near the axial side of systems, the orientation of the tie lines shows that as Ca increases in the sodic phase (Pg or Ab), the coexisting potassic phase becomes less sodic. Later, we will show that the same relationship prevails between coexisting plagioclase and Mu and that it has considerable petrologic significance.

(c) In contrast with at least the alkali feldspar solvus, none of the white mica solvi close. Intrinsic instability intervenes for at least one limb of the three solvi before such closing can occur. From Figure 2a it is evident that at least qualitatively the Mu-Pg solvus has an asymmetry similar to the alkali feldspar solvus. The Pg-Marg solvus is asymmetric in the same manner. In the case of the Mu-Pg solvus a question arises as to whether the degree of asymmetry changes as a function of P. Data for the cell volume of K-bearing Pg by Chatterjee and Froese (1975) and d(002) data in the study of Guidotti and Crawford (1967) suggest that the volume of mixing on the Pg side of the solvus is (-). This contrasts with the alkali feldspar solution series which has a (+) volume of mixing over its whole range. As discussed later, other authors argue for a (+) volume of mixing for K-bearing Pg, and so the effect of P on the solvus limbs would be similar to that on the limbs of the alkali feldspar solvus. At present this question remains open, but its significance will be evident in our discussion of the Mu-Pg geothermometer.

(d) As seen on Figure 2a, Marg may have slightly greater solution toward Pg than does Mu. And Pg may have a somewhat greater degree of solution toward Mu than toward Marg. However, these small differences in degree of solution could be due to inadequate sample density. The main point of petrologic

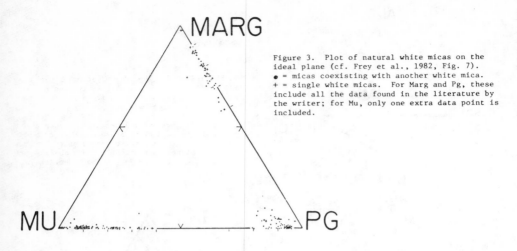

Figure 3. Plot of natural white micas on the ideal plane (cf. Frey et al., 1982, Fig. 7).
● = micas coexisting with another white mica.
+ = single white micas. For Marg and Pg, these include all the data found in the literature by the writer; for Mu, only one extra data point is included.

significance is that Pg coexisting with Mu is able to incorporate significant amounts of Ca whereas the Mu will contain virtually none.

Figure 3 is a plot of the natural data from which Figure 2a was generalized. Many of the data are from coexisting white mica pairs. In the case of Marg and Pg all the data the writer could find in the literature are included. For Mu only a few data in addition to samples containing two white micas are included. These extra Mu's are merely to enlarge the solution range that can be detected. In addition to the generalizations drawn from Figure 2a, Figure 3 permits additional conclusions; these are:

(a) Because several of the specimens with only one mica present show a greater degree of solution toward the opposite end member than do the coexisting pairs, it suggests that the available data from the white mica pairs have not provided us with the maximum solution limits.

(b) A close inspection of the distribution of points for Marg suggests (as noted by Frey et al., 1982) that as Na increases in Marg, it can take in more K. This is not unexpected.

(c) In detail it appears that there is very little solution of K-free Pg toward Marg. However, this may merely be reflecting the rarity of (Marg + Pg)-bearing bulk compositions which are devoid of K.

In summary, based on observation of natural materials, the above discussion covers most of what can be said about white micas in the context of the ideal system (Fig. 1) we have been considering. For this system we can consider the white micas as restricted to the plane shown on Figure 2a or Figure 3. Only Velde (1969) suggests that the white micas *in the ideal system* might deviate from the mica plane. He has presented some tentative arguments for minor solid solution between Mu and pyrophyllite. This question remains open; however, in the context of the suggested similarity between the mica and feldspar plane, it is interesting to note that similar deviations from the ideal feldspar plane were discussed earlier (see Thompson and Waldbaum, 1969, p. 827).

Figures 2a and 3 include specimens from over a wide range of T. Obviously, the degree of closing of each solvus will be a function of T and to a lesser extent P. Because these functional relationships relate to use of the solvi for geothermometry (and less so geobarometry), brief discussion will be given now and then more fully later in the section on petrogenetic aspects.

Theoretical and experimental studies bearing upon the solvi of the ideal white mica plane: Comparison with natural data. Numerous experimental and theoretical studies have now been made of the phase relations between Mu and Pg: Eugster and Yoder (1955), Iiyama (1964), Zen and Albee (1964), Nicol and Roy (1965), Fujii (1966), Popov (1968), Eugster et al. (1972), A.B. Thompson (1974), Blencoe (1974), and Chatterjee and Froese (1975). Some of these studies are primarily experimental, some primarily theoretical, and some are a combination. All of them suggest a solvus between Mu and Pg and that the solvus never closes due to truncation by other phase assemblages. The first study suggested that the solvus was symmetrical; however, almost all subsequent studies, whether experimental, theoretical, or a combination, argued for a solvus asymmetric toward Pg (i.e., less K in Pg than Na in Mu). Most workers have accepted an argument for this asymmetry which is based on crystallographic studies by Radoslovich (1960) and Burnham and Radoslovich (1964). These studies suggested that replacing K by Na in Mu would produce much less structural distortion than replacing Na by K in Pg. The same suggestion has routinely been advanced for the asymmetry of the alkali feldspar solvus. Hence, for both the Mu-Pg solvus and the Or-Ab solvus, it is not unexpected that the K-phase shows greater solution toward the Na-end member than vice versa.

Considerable deviation exists among the above studies with respect to the exact shape of the solvus, critical composition, critical T (both metastable), degree of asymmetry, effect of P on the solvus limbs, effect of P on the critical T, etc. Some of these points will be considered again below in our discussion of white-mica phase relations, but for now it is of interest to note that the data from natural samples formed over a wide range of PT conditions (plotted in Fig. 3) are fully consistent with the notion of a truncated solvus between Mu and Pg. However, none of the experimental or theoretical studies indicate a degree of solvus closure equal to that of the natural data. This point is even more remarkable considering that the effects of extra components in the natural Mu and Pg (celadonite in Mu and Ca in Pg) should work toward opening the solvus.

Much less work has been done on the Marg-Pg solvus. Nonetheless, the results of Franz et al. (1977) show an asymmetric solvus grossly similar to that of Mu-Pg. Thus general compatibility exists between the experimental data and the natural data plotted on Figure 3.

White mica compositional deviation from the ideal system. Having now considered most of the chemical variation that can take place among the white micas in the ideal system $KAlO_2 - NaAlO_2 - CaAl_2O_4 - Al_2O_3 - SiO_2 - H_2O$, we now ask about compositional variation involving other elements. Even a casual inspection of white mica analyses suggests several significant deviations by Mu from the ideal system. However, similar inspection shows surprisingly minor deviation by Marg and Pg. Some means of documenting these suggestions is requred.

One approach is to make various plots (e.g., histograms of ΣSi^{iv}, $\Sigma(Mg+Fe)^{vi}$, etc.) for large numbers of analyses of each of the three white micas. Frey et al. (1982; their Fig. 5, based on 60 analyses) have done this for Marg. Their histograms show that the ideal system accounts for all but a very small fraction of the composition of Marg. For example, the $(Mg+Fe^{2+})$ accounts for less than 0.2 atoms/4 octahedral sites, the ΣXII sites form a sharp peak at the ideal 2 per formula unit, and the ΣAl^{vi} form a sharp peak cenetered at 4 atoms per 4 octahedral sites. Inspection by the writer of 45 analyses of Marg from the literature shows that other elements such as Ti and Mn are present in only trace amounts. Hence, despite the small amounts of Mg and Fe noted above, Marg in *common* rock types can be almost wholly described by our ideal system. The only exceptions known to the writer occur in rather unusual settings such as at Chester, Massachusetts, where Langer et al. (1981) reported a significant

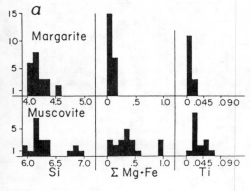
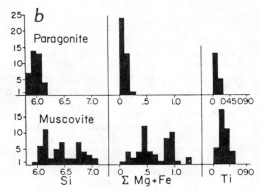
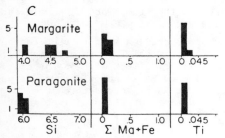

Figure 4. (a) A comparison of the extent of compositional deviation by coexisting Mu and Marg from the ideal white mica plane. (b) A comparison of the extent of compositional deviation by coexisting Mu and Pg from the ideal white mica plane. (c) A comparison of the extent of compositional deviation by coexisting Marg and Pg from the ideal white mica plane. The wide variation of Si in Marg does not represent a deviation from the ideal plane. Cations per 22 oxygens.

amount of ephesite, $Na(LiAl_2)(Al_2Si_2O_{10})(OH)_2$, in Marg.

Probably a more efficient and meaningful way of assessing deviations by Mu, Pg, and Marg from the ideal system is to make histograms for coexisting white mica pairs. This enables comparative as well as absolute assessment of the deviation by each white mica from the ideal system. For purposes of comparing the relative deviation of each white mica from the ideal system, this approach reduces the effects of PT variations.

Figures 4a, 4b, and 4c show histograms for ΣSi, $\Sigma(Mg+Fe)$, and ΣTi. Because most of the analyses used in these diagrams were done by electron probe, Fe^{2+} and Fe^{3+} are indistinguishable. Thus, $\Sigma(Mg+Fe)$ involves total Fe ($\equiv Fe_T$). Moreover, almost none of the data for Marg or Pg include elements such as F, Cl, or Ba. Nonetheless, because of the "reasonable" anhydrous analytical totals obtained, plus the seemingly "reasonable" mica formulas calculated therefrom, it appears that beside the components of the ideal system, only Fe_T, Mg, and Ti are consistently petrologically meaningful, extra elements in rock-forming white micas. The extent of these deviations is shown on Figures 4a-4c. To the extent that Fe is divalent, the $\Sigma(Mg+Fe)$ and ΣSi are both measures of solution toward a tetrasolicic "celadonite-type" of end member [e.g., $K_2(Mg_2Al_2)(Si_8O_{20})(OH)_4$]. It would be represented by an exchange of the type $(Mg,Fe^{2+})Si = Al^{iv}Al^{vi}$. In the case of Mu this exchange is also referred to as a phengite substitution and as a Tschermak substitution. For our purposes, it is a measure of the deviation from the $(Al^{iv}+Al^{vi})$ values permissible on the ideal mica plane.

In the context of Figure 4, the following can be said about deviations from the ideal system:

(a) Based on ΣSi, Pg is very close to being an ideal, trisilicic white mica and thus as expected the $\Sigma(Mg+Fe)$ is low in all of the Pg considered.

And although very little data are available on the Fe^{2+}/Fe^{3+} ratio in Pg, the fact that Fe_T is so low indicates that we can ignore any Fe^{3+} substitution for Al. It is also clear that Ti is negligible in Pg. Thus, for petrologic purposes, natural rock-forming Pg can be described by the ideal system.

(b) The Marg data on Figures 4a and 4c fully confirm the conclusions based upon Frey et al. (1982, Fig. 5). Obviously, the very nature of the Na, Si = Ca,Al exchange so prominent in Marg necessitates that it shows a wide range of Si. Nonetheless, we see that $\Sigma(Mg+Fe)$ is both low and restricted in its range, thereby indicating an absence of any significant Tschermak exchange.

(c) Figures 4a and b show a drastic change for Mu as compared with Marg and Pg. Marked Tschermak exchange can occur in Mu. By far, this is the most important of the several compositional deviations by Mu from the ideal system. The figures also show that significantly greater amounts of Ti can occur in Mu than in Marg or Pg. Thus, Mu commonly deviates significantly from plotting wholly within the white-mica plane discussed above. Moreover, this deviation has considerable petrologic significance.

We will return to a more detailed discussion of the *extent* to which Mu deviates from the ideal system, including data available for substitutions in addition to those considered via Figures 4a and 4b, but before leaving the present discussion, it is instructive to investigate how the Tschermak substitution in Mu might influence the phase relations among the three white micas. It will become evident that this provides some information on the crystallochemical controls of white mica compositions which bears heavily on formulating activity models for the micas. It will be especially important for studies trying to use the compositions of coexisting white micas (Mu-Pg in particular) to obtain petrologic insights about rocks.

Figures 5a and 5b show the extent of opening of two of the three solvi involved in the ideal white mica plane. These plots have been constructed so as to show (to the extent feasible from the data provided in the literature) the influence of T on the solvi. On each diagram the data are split into two groups, one with celadonite-rich Mu and one with celadonite-poor Mu. Inspection of the figures shows that, other things being equal, the Na/(Na+K) ratio of the Mu is influenced markedly by the Cd content of the Mu. Specifically, it appears that high Cd content in Mu inhibits substitution of Na for K. Katagas and Baltatzis (1980) made a similar suggestion, but it was based on only two data points. More recently Grambling (1984) has provided support for this suggestion and shown how the Cd content of Mu might also have a small effect on any coexisting Pg, making it less K-rich. Moreover, even a casual inspection of analyses from the literature of the celadonitic muscovites of high P terranes suggests the same conclusion. But, in such cases it is not a simple matter to determine whether it is a bulk composition effect that controls Na/(Na+K) or whether it is a crystallochemical control related to Cd content. In contrast, using the coexisting white mica pairs (especially the Mu-Pg pairs as on Fig. 5a), eliminates any such uncertainty. Ignoring the small effect of Ca in Pg on the Na/(Na+K) ratio of the coexisting Mu, the Mu coexisting with Pg must be the most sodic possible for the given P, T, and Cd content of the Mu.

Especially from Figure 5a it can be seen that a series of Mu-Pg solvi exist, depending upon the Cd content of the Mu -- with the most open solvi occurring in rocks with Cd-rich Mu. The two samples reported by Grambling (1984) are not plotted, but they are worthy of special mention. Grambling's specimen 77-244C would seem to plot out of sequence if ΣSi is used as the measure of Cd content. However, it does contain high Fe (2.99 wt % FeO). Because the rock contains hematite, it seems likely that the Fe in Mu may be mostly Fe^{3+}. If so, this sample would suggest that it is the $\Sigma(Mg+Fe_T)$ that influence

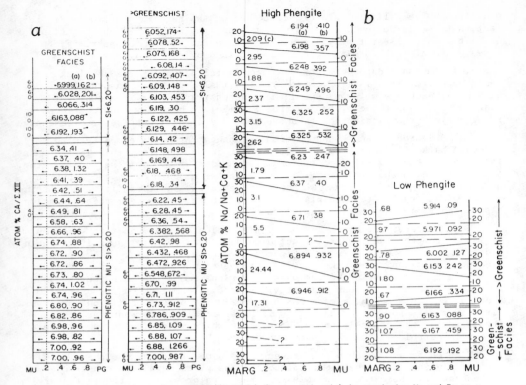

Figure 5. (a) Comparison of atom % Na/(Na+K+Ca) (horizontal axis) in coexisting Mu and Pg as a function of phengite content in Mu and metamorphic grade. Phengite content indicated by (a) = Si per 22 oxygens, and (b) = Σ(Mg+Fe$_T$) per 22 oxygens. (b) Comparison of atom % Na/(Na+Ca+K) in coexisting Marg and Mu as a function of phengite content in the Mu and metamorphic grade. Key to numbers: (a) = Si per 22 oxygens, (b) = sum of (Mg+Fe$_T$) per 22 oxygens, (c) = K$_D$, atom % Na for Marg/Mu.

the Na/(Na+K) ratio of Mu rather than the Tschermak exchange alone.

Recently Mohr and Newton (1983), as well as Grambling (1984), have recorded observations on the Mu-Pg solvus that may be related to the effects just described. Specifically, they suggested (their Fig. 2) an abrupt narrowing of the solvus of the isograd that brings in kyanite. Moreover, they noted some indications that the Cd content of Mu decreased from 15-18 mol % to 12 mol % at this isograd. Such a change for Mu could be due to the incoming of kyanite to serve as an Al-saturing phase.

It would appear that the effect of Cd on the Na/(Na+K) ratio of Mu can be quite marked, in fact more so than the effect that the Ca content of Pg has on "opening" the solvus. The potential effect of this "opening" of the solvus on Mu-Pg geothermometry is quite obvious. Finally, if we consider this "Cd-effect" in terms of Figure 2a, an interesting point emerges. The tie lines connecting Marg and Mu will change orientation as the Mu-Pg solvus opens. The net result is that the three phase field, Mu-Marg-Pg, will become significantly enlarged.

Details of the deviation of muscovite from the ideal white mica plane. The preceding discussion has shown that, for petrologic purposes, it is only for Mu that we must consider deviations from the ideal white-mica plane. The focus is on the Tschermak exchange but also includes several other substitutions of potential petrologic significance. Our approach involves consideration of the deviations in natural samples and then comparison with the available experimental data. At this point the substitutions thought to be involved in producing the deviations will merely be stated. In a later section the crystallochemical aspects of some of these substitutions will be considered. An extensive search through the literature on white micas indicates that the following compositional variations of Mu should receive at least some discussion.

(a) The phengite = celadonite = Tschermak substitution.
(b) The Fe^{3+} or ferrimuscovite content.
(c) The substitution of Ti into octahedral sites.
(d) The substitution of F and Cl for (OH).
(e) Deviation of ideal Mu from dioctahedral to trioctahedral.
(f) Substitutions involving the 12-coordinated (XII) sites.
(g) Miscellaneous compositional variations.

Our treatment focuses on Mu but, where appropriate, Marg and Pg will also be considered. Moreover, it should be noted that several factors may influence our ability to assess the nature and extent of these substitutions. Insofar as we cannot assess the effects of these factors, ambiguity exists in the generalizations made. These include the facts that

(i) Mica analyses usually include only the 10 or so elements common as major constituents of rock-forming minerals. Excluded are elements such as Zn and Ba. Moreover, since the advent of analysis via the electron probe, distinction between Fe^{2+} and Fe^{3+} has usually been neglected.

(ii) The nature and extent of (a)-(g) may be strongly influenced by metamorphic conditions, mineral assemblage, and bulk composition. By considering a very large sampling of analyses from the literature, it is hoped that the effects of these influences have been minimized so that we can obtain an idea of the maximum extent of (a)-(g) as they occur in metamorphic environments.

(iii) Closely related to the mineral assemblage influence in (ii) are the effects to be expected depending upon whether muscovite coexists with a so-called "saturating phase". For example, Mu coexisting with magnetite or hematite will, for a given P, T, etc. contain the maximum amount of Fe^{3+} possible. Guidotti et al. (1977) have discussed the Ti saturation limits of Mu and Bio in the context of ilmenite or rutile serving as a Ti-saturating phase. Unfortunately, very little information is available which enables one to ascertain the presence or absence of saturating phases.

(a) Phengite = celadonite = Tschermak substitution. Treatment of this substitution in Mu requires discussion of it first in terms of the appropriate solid solution end members because considerable confusion exists in the nomenclature of the various end members. Thus, we must define our nomenclature. The same point also arises in the subsequent discussion of the Fe^{3+} substitution in Mu.

Mu is known to deviate markedly from the ideal composition $K_2Al_4(Al_2Si_6O_{20})(OH)_4$ both by substitution of Fe^{3+} for Al^{vi} and by the Tschermak exchange $(Mg,Fe^{2+})^{vi}, Si^{iv} = Al^{vi}, Al^{iv}$. In the former, the absolute end-member composition would be $K_2Fe_4^{3+}(Al_2Si_6O_{20})(OH)_4$. If both substitutions occur together to the maximum extent, we could get $K_2Fe_2^{3+}(Mg,Fe^{2+})_2(Si_8O_{20})(OH)_4$. Among these compositional extremes a considerable array of names have been assigned to various composition ranges, and commonly the nomenclature employed has not been consistent from author to author. Rather than discuss the pros and cons of this terminology we have tabulated some of the more important terms that have been used (Table 1). The names employed herein are used largely for convenience in achieving our immediate goals, which are to gain some indication of the extent by which Mu deviates via the Tschermak exchange and Fe^{3+} substitution from the

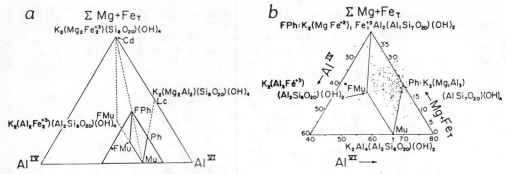

Figure 6. (a) Plot of various K-white micas in terms of amount of Al^{vi}, Al^{iv}, and $\Sigma(Mg+Fe_T)$. See Table 1 for mineral compositions. F MU = ferrimuscovite. *F Mu = ferrimuscovite of Kanehira and Banno (1960). Mu = muscovite. Ph = phengite. F Ph = ferriphengite. Lc = leucophyllite. Cd = celadonite. (b) phengitic Mu from the literature plotted on an enlargement of the darkened portion of Figure 6a. Only specimens with Si ≥ 6.50 atoms per 22 oxygens are plotted.

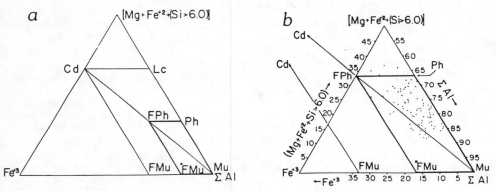

Figure 7. (a) Plot of various K-white micas in terms of Fe^{3+} versus $\Sigma(Al^{iv}+Al^{vi})$ versus $\Sigma[(Mg+Fe_T) + (Si\ in\ excess\ of\ 6.00)]$. Enlargement of darkened portion is shown on Figure 7b. Formulas and symbols as on Figure 6a,b. (b) Phengitic Mu from the literature plotted on an enlargement of the darkened portion of Figure 7a. Only specimens with Si ≥ 6.50 are plotted.

ideal white mica plane, and to gain some idea of the interdependence of these two substitutions. The main obstacle is that many analyses do not distinguish Fe^{2+} and Fe^{3+}. Because of this, the two goals are approached by means of graphical devices. Figure 6a shows the compositional relationship of a number of the end members, and Figure 6b shows the four end members described by Kanehira and Banno (1960) and subsequently used by workers such as Ernst (1963) and Zen (1981). Noting that ideal, end member, Mu ($K_2Al_4(Al_2Si_6O_{20})(OH)_4$ has four Al^{vi} and two Al^{iv} and plots on the bottom edge of the diagram, then points plotting progressively higher in the diagram deviate further and further from ideal Mu. However, because this diagram lumps Fe^{2+} and Fe^{3+} as total Fe, it makes no overt distinction between the substitutions discussed above. Figure 6b shows the location of many Mu analyses taken from the literature. Inasmuch as we are trying to determine the extent of deviation from ideal Mu, only analyses in which $Si^{iv} > 6.5$ are plotted. Consideration of the Tschermak exchange shows that ΣSi can serve as one measure of Mu deviation from its ideal end-member composition.

Figure 7a shows another view of the end members which bears directly upon both of our intended goals. It enables assessment of the extent and interdependence of the two substitutions listed above. As pointed out by Zen (in press), the end members shown on Figure 7b form a reciprocal ternary system. On Figure 7b are plotted a large number of wet chemical analyses, most of which have Si ⩾ 6.5.

The following are more detailed implications of Figure 6b. (1) Although Fe^{2+} and Fe^{3+} are not distinguished on this diagram, the absolute and relative amounts of Al^{iv} compared to $\Sigma(Mg+Fe_T)$ enable some assessment of the extent of Ph versus *FMu (i.e., the ferrimuscovite of Kanehira and Banno, 1960). Only a relatively few samples fall essentially on the Mu-Ph join. The majority show a significant deviation from ideal Mu, involving Fe^{3+} replacing Al^{vi}. In fact, the main deviation is approximately between Ph and FPh. (2) There is little evidence of an $Fe^{3+} = Al^{vi}$ substitution which is not also accompanied by the Ph substitution. (3) The points plotting above the Ph-FPh join are almost exclusively those with $Si^{iv} > 7.0$. Obviously such points are few, and they do not plot much above the join. Moreover, it is possible that some of them may be due to analytical error. In contrast, the density of points just below this join is fairly high. The implication emerges that the Ph-FPh join may serve as an approximate limit on the extent to which, starting from Mu, Al^{iv} can be replaced by Si, i.e., only up to $\sim(AlSi_7)$. Radoslovich (1963) and Foster (1956) have discussed the existence of such a limit. Complete solid solution, starting from Mu and going toward the Si_8 end members, does not seem to exist, at least not in metamorphic rocks (but see later discussion).

Because on Figure 7b Fe^{2+} is distinguished from Fe^{3+}, and Si in excess of that in ideal Mu is shown explicitly, it enables refinement of the comments made in the context of Figure 6b. Fewer data points are available for Figure 7b than Figure 6b, because probe analyses cannot be plotted on it. On Figure 7b the data points are scattered fairly evenly between the joins connecting Mu with Ph and Mu with FPh. Thus the points also lie between the joins connecting Mu with the tetrasilicic end members, Lc and Cd. In contrast with Figure 6b, virtually no points fall exactly on the Mu-Ph join but some are on the Mu-FPh join. Again it is evident that very little solid solution exists from Mu directly to *FMu. Finally, because Figure 7b explicitly includes Si, it shows more clearly the extent of solution from ideal Mu toward the tetrasilicic end members, Lc and Cd. The Ph-FPh join is the boundary between K-white mica with Si < 7 and Si > 7. Again, very few points fall on the Si > 7 side (and they may be due to analytical error). In fact as discussed later, the works of Foster (1956) and later Velde (1965a) suggest that the Lc end member (Table 1) may not even be possible.

The substitution along the Mu-Lc join is equivalent to the Tschermak exchange. It is of interest to consider the extent to which the $\Sigma(Mg+Fe^{2+})$ is due to this exchange. If the $\Sigma(Mg+Fe^{2+})$ in Mu were solely due to the Tschermak exchange a plot of Si versus $\Sigma(Mg+Fe^{2+})$ should produce a 45° sloping line connecting ideal Mu and ideal Ph. Figure 8 shows such a line plus a plot of Si versus $\Sigma(Mg+Fe^{2+})$ for analyses obtained from the literature. Despite considerable scatter in the data, it is still evident that a best fit line for the data points would have a 45° slope, but the line would be systematically above the ideal Tschermak line. Fe^{3+} could produce the observed effect (Laird and Albee, 1981), but use of only wet chemical data for Figure 8 has eliminated that possibility. Thus, although the bulk of the $(Mg+Fe^{2+})$ substitution is related to charge balance necessitated by substitution of Si for Al^{iv}, a small amount of the $\Sigma(Mg+Fe^{2+})$ represents charge balance for other substitutions such as Ti = Al^{vi}, or vacancies, or deviation toward a trioctahedral end member. These possibilities will be considered again below.

To increase our understanding of the compositional aspects of the Cd substitution, the histograms shown in Figures 9-11 have been constructed. Figure 9 confirms the implications of Figures 6b and 7b concerning a boundary at Si =

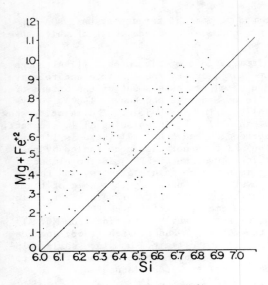

Figure 8. Plot of Si versus $\Sigma(Mg+Fe^{2+})$ with respect to the line for an ideal Tschermak substitution. If all the $(Mg+Fe^{2+})$ in Mu were charge-balanced by Si replacing Al^{iv}, the points should cluster along the ideal Tschermak substitution line.

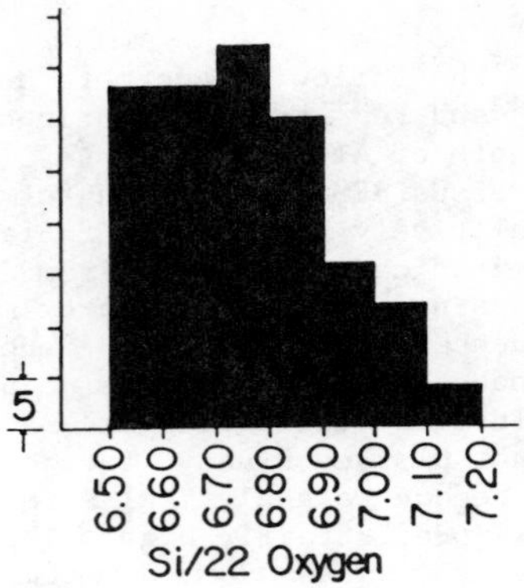

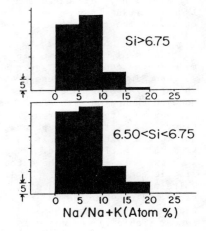

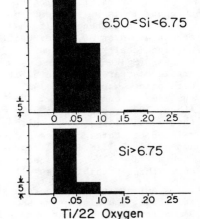

Figure 9 (top right). Data bearing on the suggested upper limit near Si = 7 for the Tschermak exchange in Mu.

Figure 10. Relative abundance of various ranges of Na/(Na+K) ratio for highly phengitic Mu samples.

Figure 11. Relative abundances of various ranges of Ti content in highly phengitic Mu samples.

7 for solid solution between ideal Mu and Lc. Indeed, the first big drop in frequency of Si values (6.80 - 6.90) coincides quite closely with the 6.80 value of Si postulated by Radoslovich (1963) as the limit of Si above which the Mu-type structure becomes unstable.

In discussing coexisting Mu and Pg pairs it was noted that the Cd content of the Mu influenced its Na/(Na+K) ratio. The data shown on Figure 10 are, at the very least, consistent with the earlier assertion that increase of Cd content favors a decrease of Na/(Na+K) ratio. However, because of factors related to grade and bulk composition, the implications of Figure 10 are less unequivocal than those of Figures 5a and 5b.

Some information on the relationship of Cd content and Ti content are given in Figure 11. To the extent that the data are meaningful, it appears that as Cd increases, Ti decreases. Once again, the influence of metamorphic grade and bulk composition cannot be easily ascertained, but (as discussed later) in some cases there may be crystallochemical reasons to expect Ti to decrease as Cd increases in Mu.

Finally, it should be noted that in the vast majority of cases one finds Mg > Fe^{2+}, commonly by about a 2:1 ratio. In those cases in which Fe^{2+} > Mg or Mg >> Fe^{2+}, it would seem that bulk composition is the controlling factor. This is shown clearly by the very high Mg/Fe ratios recorded in the Mu of the sulfidized rocks studied by Guidotti (1970a), Guidotti et al. (1977), and Mohr and Newton (1983). The more common 2:1 range for Mg:Fe^{2+} is consistent with the experimental results of Velde (1965a) which suggest that solution toward a Mg-Lc is more extensive than solution toward the Fe^{2+}-Lc.

(b) *The substitution of Fe^{3+} for Al^{vi}*. In discussing the Cd substitution in Mu, the most important aspects of the Fe^{3+} for Al^{vi} substitution have also been covered. Basically, this substitution does not occur by itself in metamorphic rocks but is accompanied by an even greater amount of the Ph substitution. Unfortunately distinction between the $Fe^{3+} \rightleftarrows Al^{vi}$ substitution and the Ph substitution requires distinguishing between Fe^{2+} and Fe^{3+}, but with the advent of electron probe analysis this distinction cannot be readily made. However, in view of the earlier discussion on "saturating phases" and a consideration of mineralogic indicators of oxidation state one might expect that (a) solid solution from ideal Mu directly toward Lc is most likely to occur in graphite-bearing rocks, and that (b) solid solution from ideal Mu directly toward FPh (on Figure 7b) is most likely to occur in magnetite- or hematite-bearing rocks. Unfortunately, the data in the literature often lack information on the presence or absence of accessory minerals such as graphite, magnetite, and hematite. However, it is encouraging to point out that Brown (1967) noted specifically that the most Fe^{3+}-rich Mu encountered in his study coexisted with magnetite. Also, the extremely Fe^{3+}-rich Mu (9.10 wt % Fe_2O_3; 0.95 atoms per 4 octahedral sites), reported by Kanehira and Banno (1960), coexists with hematite.

Finally, it should be noted that not all Fe^{3+} in Mu may be a result of Fe^{3+} replacing Al^{vi}. For example, Zen (in prep.) has mentioned the possibility of $(Fe^{2+})^{vi}$ + H = Fe^{3+vi} + □. This substitution involves an oxidation as well as a simple replacement. The writer has no information bearing on the significance of such a substitution.

(c) *The substitution of Ti into octahedral sites*. Substitution of Ti into Mu has been discussed by Saxena (1966), Kwak (1968), Tracy (1978), Ruiz et al. (1980), Guidotti (1973, 1978), and Guidotti et al. (1977). Some of these studies focused on the influence of metamorphic grade on the Ti content of Mu. Various substitution models by which Ti might enter an ideal Mu will be discussed later, along with their crystallochemical implications, but here our concern is with the analytical data bearing on the observed extent of Ti content in Mu. Several factors must be considered in ascertaining the upper limit for the Ti content in metamorphic Mu. These include (a) metamorphic grade, (b) the presence or absence of a Ti saturating phase such as rutile or ilmenite, and (c) the effects of bulk composition, particularly with regard to whether or not a given Mu is from an Al-saturated assemblage (cf. Guidotti, 1973).

With due consideration of (a)-(c), an extensive search of the published analyses of metamorphic Mu shows that (1) on a formula basis of 22 oxygens, the highest reliable values of Ti are approximately 0.16 per 4 octahedral sites, (2) the vast bulk of the Ti values are actually less than 0.10 per 4 octahedral sites, and (3) most Ti values greater than 0.10 per 4 octahedral sites are from Mu of high grade metamorphic rocks (e.g., Evans and Guidotti, 1966; Ashworth, 1975; Tyler and Ashworth, 1982; Boak and Dymek, 1982; Fletcher and Greenwood, 1978; Tracy, 1978).

The extensive compilation provided by Cipriani et al. (1971) is especially useful. Of the 358 specimens for which Ti data are listed, only 9 have Ti > 0.16 per 4 octahedral sites. Of these, the three highest are still only 0.22, 0.22, and 0.26. Moreover, all 9 of these are wet chemical analyses. In view of these facts, it seems likely that these 9 can be disregarded. No modern probe analyses of Ti in Mu exceed $\sim$0.16 Ti.

Although the Ti content of Mu can be systematically related to metamorphic grade and as discussed later is systematically partitioned with respect to coexisting Bio, the fact that it is present in such small amounts suggests that it will not result in any major effects on the thermodynamic solution properties of ideal Mu. Hence, although it is a petrologically interesting substitution, its effects are minor compared with those discussed above for the phengite substitution.

(d) Substitution of F and Cl for (OH). Deer et al. (1962, p. 14) note that the average F content of natural Mu is about 0.6 wt %. However, this would include muscovites from various igneous and vein type environments as well as metamorphic paragenesis. A search of the literature suggests that Mu of a clearly metamorphic origin usually has F in amounts up to about 0.25 wt % (Lee et al., 1969). Few studies have made a concerted attempt at gaining information about the halogen content in common metamorphic Mu. In one such study, Evans (1969) found the wt % of F in 22 high grade Mu samples was consistently less than 0.10, and the Cl content was less than 0.01 wt %. Henry (1981), working on staurolite zone rocks, found Cl and F in Mu similar to those noted by Evans. As the rationale for making his study, Evans pointed out that significant Cl and/or F in metamorphic Mu could have a significant effect on the thermal stability limit of the mineral, but because of the low amounts of halogens, he concluded that this effect would be minimal.

Because no F-saturating phase is present in the samples from common rock types considered above, it is unlikely that these muscovites are F-saturated. In fact, an extremely interesting study by Nêmec (1980) shows the F-contents considered above are well below saturation levels. He described "phengites" from the tin-bearing orthogneisses of the Bohemian-Moravian Heights, Czecholslovakia. Although called "phengites", his samples are not very phengite-rich inasmuch as they all have <6.41 Si per 22 oxygens. In spite of nomenclature problems, these metamorphic micas are of great interest because they are very rich in F -- up to 2.65 wt %. Moreover, Nêmec showed a close correlation between bulk F in the rocks and the F-content of the Mu. Most importantly, the three Mu samples with the highest F-contents (1.68, 2.45, and 2.65 wt %) coexist with syngenetic fluorite. In these fluorite-bearing specimens, it seems

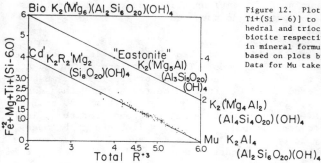

Figure 12. Plot of total R^{3+} versus $[Fe^{2+}+Mg+Ti+(Si-6)]$ to illustrate the ideal dioctahedral and trioctahedral substitutions in Mu and biotite respectively. $R^{3+} = Al + Fe^{3+}$ and 'Mg' in mineral formulas = $Mg + Fe^{2+} + Ti$. Diagram based on plots by Brown (1968) and Tracy (1978). Data for Mu taken from the literature.

likely that the F-contents in the coexisting silicates are at saturation levels. Based upon the mineral assemblages in pelitic schists associated with the orthogneisses, Němec suggests that the metamorphic grade is medium to high subfacies of the almandine-amphibolite facies (Mu + biotite + garnet + plagioclase + kyanite + sillimanite in the pelitic schist). Hence, these rocks are probably roughly intermediate in grade between those described by Henry (1981) and those described by Evans (1969). The important point is that the F-contents described in common rock types are well below saturation levels.

Obviously the tin-bearing orthogneisses described by Němec (1980) are rather unusual bulk compositions. Nonetheless, they provide us with some measure of the saturation level of F in Mu. In summary, it appears that in common rock types the F content in Mu is well below saturation levels, and deviation of ideal Mu by substitution of F or Cl for (OH) will be minor and so can be ignored for the purposes of metamorphic petrology.

(e) Deviation of ideal Mu from dioctahedral to trioctahedral. Several authors have discussed the nature and extent by which Mu deviates from being purely dioctahedral toward some trioctahedral mica such as biotite, including Foster (1956), Radoslovich (1963), Brown (1968), and Tracy (1978). This deviation is readily seen by the fact that, based on a formula with 22 oxygens, the total octahedral sites for Mu usually exceed the ideal of 4 by a small amount, e.g., 4.02 to 4.10. This is particularly true for Mu from metamorphic parageneses. Later some discussion will be addressed to the possibility that some of the deviation may be related to metamorphic grade.

Figure 12 shows the manner and extent by which Mu deviates from being solely dioctahedral and biotite from being trioctahedral. The trend along the line from Mu to Cd is controlled by substitutions which maintain a dioctahedral structure. These would include the phengite substitution, exchanges such as $(Mg,Fe^{2+}) + Ti^{4+} = 2Al^{vi}$ (Tracy, 1978; Guidotti, 1978), and $Ti^{vi} + Al^{iv} = Al^{vi} + Si^{iv}$ (Guidotti, 1978). In contrast, substitutions such as $2R^{3+} = 3R^{2+}$ (Brown, 1968) or $Ti + \square = 2(Mg,Fe^{2+})$ lead to a deviation toward a trioctahedral mica, resulting in points plotting between the two lines shown on Figure 12. To illustrate these generalizations, Mu analyses have been taken from the literature with as wide a range of $\Sigma(Mg+Fe^{2+})$ as obtainable and plotted on Figure 12. Arbitrary assignment of total Fe to Fe^{2+} would artificially move points off the Mu-Cd line toward the biotite line. In order to assign Fe properly to R^{3+} or R^{2+}, only wet chemical analyses have been used. Figure 12 confirms the small positive deviation of the total number of octahedral sites mentioned above, in that Mu typically plots just slightly off (toward Bio) the Mu-Cd line. Inasmuch as Ti contents are always low, Figure 12 also confirms the earlier assertion that most of the Mg and Fe^{2+} in Mu is due to the phengite substitution.

Because only very small amounts of the various trioctahedral substitutions occur, they can be ignored for petrologic purposes.

(f) Substitutions involving 12-coordinated (XII) sites. The sum of the cations reasonably assigned to the 12-coordinated (XII) site is commonly less than the theoretical 2 atoms per 22 oxygens. This deficiency was noted at least as far back as Lambert (1959), and, as implied by Zen (1981), most subsequent workers have considered the deficiency to be real rather than due to analytical error. Typically, the XII sites of Mu contain from 1.80 to 1.95 atoms per formula unit. Several substitutions might account for this, including (i) Ca^{2+} in the XII site accompanied by vacancies (for charge balance), (ii) Ba^{2+}, acting similarly to Ca^{2+}, (iii) $H_3O^+ = K^+$, and (iv) deficiencies in XII coupled with deviation by the octahedral sheet from dicotahedral to trioctahedral.

(i) Ca^{2+}: Although many wet chemical analyses show small amounts of Ca^{2+} present in Mu, very few probe analyses show it at much above background levels. Indeed, Evans and Guidotti (1966) made a concerted effort to determine if Ca^{2+} could enter Mu in petrologically meaningful amounts and concluded that it was essentially absent. Thus, Ca does not relate to the deficiency in the XII site. However, it is worth recalling that in the case of Pg, Ca^{2+} *is* present in the XII site in petrologically meaningful amounts.

(ii) Ba^{2+}: Generally this element is present in only small amounts for the Mu typical of common metamorphic rocks, but it is often omitted during chemical analysis. Thus, at least some of the XII site deficiency may be due to not reporting Ba. A reconnaissance of papers dealing with low to high metamorphic grades and which report analyses of Ba in Mu (Karamata et al., 1970; Tracy, 1978; Butler, 1967; Boak and Dymek, 1982; Evans and Guidotti, 1966; Guidotti, 1973, 1978; Ashworth, 1975; Tyler and Ashworth, 1982) shows that for a 22 oxygen formula base, Ba commonly ranges from 0.01 to 0.03 atoms per two XII sites. Occasionally it ranges up to 0.04 atoms (Abraham and Schreyer, 1976), and Ernst (1963) records two phengites with 0.08 and 0.10 Ba. Also, Ramos and Gallego (1979) report an amphibolite with a Mu containing 0.10 Ba. However, these seem to be highly unusual cases for *common* rock types.

Each substitution of Ba^{2+} for a K^+ could potentially be accompanied by the formation of a vacancy, assuming the exchange $K + Si = Ba + Al^{iv}$ is not operative. But this would account for only 0.01–0.03 of the 0.05–0.20 atom deficiency usually reported for the XII site of Mu. On the other hand, for analyses which omit Ba, any deficiency mentioned should take into account the probable presence of 0.01 to 0.03 Ba atoms plus, possibly, the presence of a similar number of vacancies resulting from a 1:1 replacement of K^+ by Ba^{2+}. Barium is not abundant in Mu in part because no Ba-saturating phase is present, a reflection of the fact that Ba is not an abundant element in common rock types. Nonetheless, crystallochemical reasons are probably the best explanation for the low Ba in Mu. These probably involve the large Ba in the XII site of Mu preventing the tetrahedral rotation required in order to have the tetrahedral and octahedral sheets in Mu fit together (Zussman, 1979). Such factors fall outside the scope of the implications that can be made from data obtained in the course of studying metamorphic rocks.

(iii) H_3O^+ replacing K^+: Numerous authors have also discussed the possibility that H_3O^+ might be able to replace some K^+, resulting in or contributing to a deficiency in the XII site of Mu. This suggestion goes back at least to Brown and Norrish (1952) and White and Burns (1963). However, to my knowledge, no one has yet demonstrated that H_3O^+ is present in metamorphic Mu.

(iv) Deficiencies in the XII site may be coupled to deviation by the octahedral sheet from dioctahedral to trioctahedral. This coupling might be expressed as $2K^{xii} = (R^{2+}) + 2[\,]^{xii}$. Because the numbers involved in the XII site deficiency and the octahedral excess are very small, it seems likely that analytical scatter would obscure any relationship predicted on the basis of the above exchange equation. Moreover, the fact that probe analyses do not distinguish Fe^{2+} from Fe^{3+} will also obscure such a relationship. In fact, inspection of a tabulation of $\Sigma XII, \Sigma (Mg+Fe^{2+})$, the total number of octahedral sites, and ΣSi from 31 wet chemical analyses of Mu (from Ernst, 1963, and Butler, 1967) failed to reveal the relationship expected on the basis of the above exchange reaction. Either such a relationship does not exist or, as suggested above, analytical variability obscures it.

In summary, any or all of (ii), (iii), or (iv) might contribute to the XII site deficiency, and, of course, some mechanism not yet considered may exist. Regardless, inasmuch as the XII site deficiency is almost always quite small (<0.20 atoms per 2 sites, and usually less than 0.15) it would again seem that for petrologic purposes it can be ignored. Finally it should be noted that deficiency in the XII site is observed for both probe and wet chemical analyses. Hence, volatilization of alkalis by the probe beam is probably not an important factor.

(g) Miscellaneous compositional variations in muscovite from metamorphic rocks. Several percent of Cr_2O_3 sometimes replaces Al_2O_3 in Mu from certain specialized petrologic environments. But this substitution is not significant in any of the common rock types that one finds on a broad scale in a typical orogenic belt. Characteristically, when Cr exceeds a few tenths of the 4 octahedral sites, the Mu has a bright green color, and the mineral is then called fuchsite. Chatterjee (1968) described a green phengite containing 1.75 wt % of Cr_2O_3 from an ankeritic marble, and Dymek et al. (1983) described some metamorphic green muscovites which are not only Cr- but also Ba-rich, (0.436 Ba per two XII sites!).

Manganese is usually present in only very small amounts in metamorphic Mu, e.g., 0.01 atom per 4 octahedral sites. However, in a few unusual bulk compositions such as piemontite-bearing schists, the coexisting Mu can contain markedly greater amounts of Mn. For example, Dal Piaz et al. (1979) report Mu with up to 0.081 Mn, Abraham and Schreyer (1976) report Ph with 0.07 Mn, and Smith and Albee (1967) report Mu with 0.06 Mn. In some viridine- (Mn_2SiO_5-) bearing hornfels, Abraham and Schreyer (1975) report Ph with up to 0.067 Mn. Following Winchell (1951), they point out that such Mn-rich Ph would be called by the varietal name alurgite. All of these cases represent uncommon bulk compositions such that an intrinsic Mn-phase (piemontite or viridine) is present. Nonetheless, these cases show that the negligible amount of Mn usually found in Mu is a result of bulk composition rather than crystallochemical control. Because of the presence of an Mn phase in the above cited studies, the Mn found in those Mu's may represent the maximum solution of Mn^{2+} into Mu *for the prevailing PT conditions*. Obviously it is much less than that of Mg^{2+} or Fe^{2+}, and this is not unexpected inasmuch as Mn^{2+} is a larger cation than Fe^{2+} or Mg^{2+}, or especially Al^{3+}.

(I.B) White mica lattice spacings and polytypes

Later the interrelationships between lattice spacings and composition will be considered in terms of indirect methods for composition determination. Here, we consider these interrelationships in terms of the crystallochemical aspects of white micas to be discussed in the next section. Our focus is on the $2M_1$ polymorph of Mu because (a) very little relevant information exists for Marg, (b) only modest amounts of information exist for Pg, and (c) considerable amounts of data are now available for the $2M_1$ type of Mu. The compositional variation to be interrelated with lattice spacings will be the ratio Na/(Na+K) and the Cd

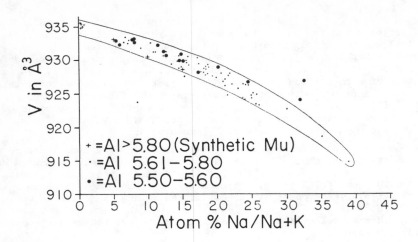

Figure 13. Plot of Mu cell volume versus Na/(Na+K) for Al-rich Mu (i.e., Al ⩾ 5.50 Al per 22 oxygens).

content. The latter will be indicated as suits the case in terms of $\Sigma(Mg+Fe_T)$, ΣSi, or Al. All other compositional variations discussed previously are not important. The relationships considered have long been recognized based upon crystallographic studies such as Radoslovich (1960, 1962, 1963), Radoslovich and Norrish (1962), Burnham and Radoslovich (1964), and Güven (1967). Here we will merely elaborate upon the relationships based upon data from metamorphic Mu. It is expedient to treat the compositional data selectively such that the two compositional parameters are varied only one at a time. All lattice spacing data employed are based upon single crystal structural determinations or cell refinements obtained from complete diffraction scans. It is convenient to begin with the relationship between Na/(Na+K) and unit cell volume and thus consideration of the most important compositional variable of Mu in terms of its aggregate effect on lattice spacings.

<u>Composition versus lattice spacings and cell volumes.</u> Figure 13 shows a large number of data points based upon synthetic and natural Mu. For those plotted, the Na/(Na+K) ratios range from 0 to 0.38 (or 0 to 38 "atom %") and the ΣAl has a restricted range. For the natural samples (based on a 22 oxygen formula) all have $\Sigma Al > 5.50$ and most have $\Sigma Al > 5.60$. Hence, they all fall in the range of the maximum Al contents found in natural Mu. Of course the synthetic Mu has $\Sigma Al = 6.00$. The point to note is that in all of these Mu, $\Sigma(Mg+Fe_T)$ is low and quite uniform. It is evident that the cell volume decreases as the Na(Na+K) increases. Decrease of V is due mainly to a decrease in the c-cell dimension as the smaller Na$^+$ replaces K$^+$. Reflecting this decrease in c is the concomitant decrease of ½$c$sinβ or the d(002) spacing for the $2M_1$ polymorph. Inspection of the a and b cell parameters *for this group of samples* shows that the former remains essentially constant and the latter shows only minor changes. Figure 14 is a plot of ½$c$sinβ versus Na/(Na+K) (see also Kotov et al., 1969, for a similar curve) for the same samples, and the nature of the envelope enclosing the data points shows that this diagram is quite similar to Figure 13. This similarity is also reflected by the form of the regression equations for each data set. For Figure 13 we have the equation:

$$V = 934.92 - 0.23 X - 0.0069 X^2 , \qquad (1)$$

in which V = cell volume in Å^3 and X = atom % Na/(Na+K). For Figure 14 the

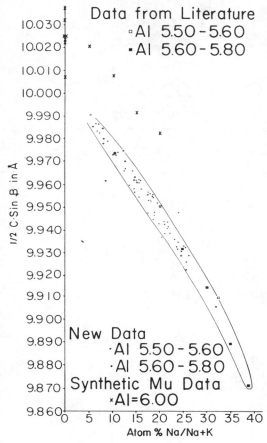

Figure 14. Plot of ½ $c\sin\beta$ versus atom % Na/(Na+K) for specimens with ΣAl ≥ 5.50 per 22 oxygens.

equation is:

$$\tfrac{1}{2}c\sin\beta = 10.00 - 0.00217\,X - 0.0000259\,X^2 \,, \quad (2)$$

in which ½$c\sin\beta$ = d(002) in Å and X = atom % Na/(Na+K).

Figure 13 and Equation 1 show that V versus Na/(Na+K) ratio is nonlinear, and the volume of mixing is (+). From them one finds that end-member Mu (i.e., Na/(Na+K) = 0 from natural rocks has a volume of ~935 Å^3, essentially the same as that determined from synthetic end-member Mu.

Figure 15 is similar to Figure 13 and includes the envelope of data points for Al-rich Mu. However, the only specific data points (+) *repeated* on Figure 15 are those for the synthetic Mu of Blencoe (1978); all of the others are for Mu with the ΣAl progressively less than 5.50, as seen in the key to the diagram. Figure 15 shows that as ΣAl decreases (i.e., Σ(Mg+Fe$_T$) increases) the volume of Mu increases more or less progressively for a given Na/(Na+K) ratio. Of interest is the fact that the synthetic muscovites fall near the lower boundary of the envelope that enclosed the Al-rich samples on Figure 13. Considering that such samples contain no Mg or Fe, they plot as would be expected, based upon the general contrast between the data in Figure 13 and that in Figure 15.

Figure 16 is a plot of ½$c\sin\beta$ versus Na/(Na+K) for the same specimens plotted on Figure 15. Observe that a decrease in ΣAl [increase in Σ(Mg+Fe$_T$)] causes a decrease in the magnitude of d(002) and hence c. This decrease occurs even though Na/(Na+K) is held constant. Ernst (1963), Brown (1967), Guidotti and Sassi (1976a), and others have since noted this inverse relationship. Thus, when we reconsider the increase of V [for constant Na/(Na+K)] shown by the samples in Figure 15, it is clear that the increase in V is solely due to increase in the lateral cell dimensions, inasmuch as the c dimension -- as we have just shown -- actually decreases! The relationship between b and Σ(Mg+Fe$_T$) is shown on Figure 17. From the envelope enclosing the data points, the relationships between b and Σ(Mg+Fe$_T$) appears to be nearly linear and this is confirmed by the equation

$$b = 8.990 + 0.04\Sigma(\text{Mg}+\text{Fe}_T) \,, \quad (3)$$

in which b is in Å and Σ(Mg+Fe$_T$) is in atoms per 22 oxygens.

More scatter is seen in Figure 17 than in Figure 13. Some of this may be due to the influence of the Na/(Na+K) ratio on b. The data on Figure 17 exclude samples with Na/(Na+K) > 12.5. Although the influence of Na/(Na+K) ratio will

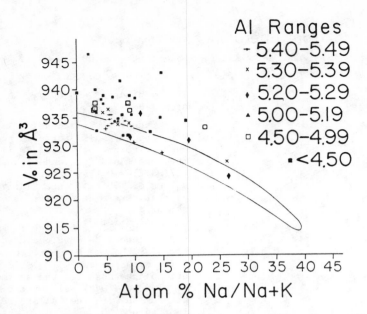

Figure 15. Plot of Mu cell volume versus atom % Na/(Na+K) for Al-poor Mu (ΣAl < 5.50 atoms per 22 oxygens).

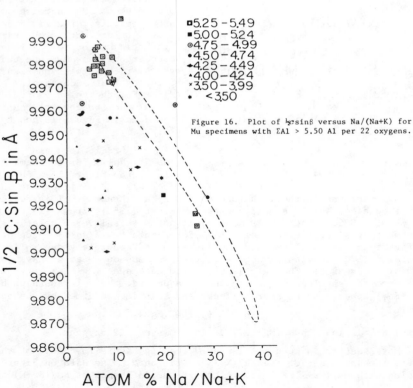

Figure 16. Plot of $\frac{1}{2}c\sin\beta$ versus Na/(Na+K) for Mu specimens with ΣAl > 5.50 Al per 22 oxygens.

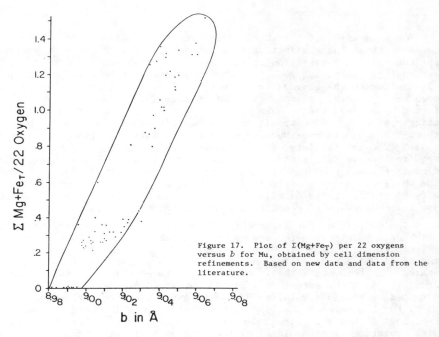

Figure 17. Plot of $\Sigma(Mg+Fe_T)$ per 22 oxygens versus b for Mu, obtained by cell dimension refinements. Based on new data and data from the literature.

not be pursued in depth here, inspection of data presently available indicates that increase of this ratio results in a fairly systematic small decrease of b.

Inspection of the data base used for developing the above relationships shows that a increases as a function of phengite content of Mu. However, because no measurements have been devised for relating a to the phengite content of Mu, no effort was made at this time to examine this relationship further.

General summary of the composition versus lattice spacing relationships. (1) At essentially constant $\Sigma(Mg+Fe_T)$, increasing Na/(Na+K) causes Mu cell volume to decrease, largely because c decreases. (2) For a given value of Na/(Na +K), decrease of ΣAl [i.e., increase of $\Sigma(Mg+Fe_T)$] causes the cell volume to increase, solely due to increase of the lateral dimensions, inasmuch as c actually decreases. (3) To a close approximation, b value is a linear function of $\Sigma(Mg+Fe_T)$. The Na/(Na+K) ratio has a small but as yet unquantified effect on b.

White mica polymorphs in metamorphic rocks. In Table 3 of Chapter 1, Bailey listed the types of polymorphs that have been reported for the micas, including those germane to the present discussion: Mu, Pg, Ph, and Marg. There are only five polytypes in which these white micas occur ($2M_1$, $1M$, $1Md$, $3T$, $2M_2$), but Marg has only one and Pg two polytypes.

Petrologic factors suggested as controls for the occurrence of a given polymorph will be discussed later, but for now only two points need be made. (1) The $1M$ and $1Md$ polymorphs are largely restricted to rocks that have not been influenced by more than diagenetic effects. The experimental work of Velde (1965b) suggests that they are limited to temperatures below 300°C, and (2) the $2M_1$ and $3T$ polymorphs are by far the most abundant in metamorphic rocks, with $2M_1$ being the most common for Mu, Pg, and Marg. Only in the case of Ph does $3T$ become dominant.

(I.C) Crystal chemistry of the white micas

Consideration of the white micas from a crystallochemical viewpoint focuses on those factors intrinsic to a given phase which influence or control the compositional variation it can undergo. These would contrast with factors external to the phase, such as metamorphic grade or bulk composition, which also influence the composition of micas. Obviously, crystal-chemical factors such as (i) charge balance, (ii) cation sizes, and (iii) the types of bonds formed by a given cation when associated with a particular anion constrain the effects of P, T, and composition. Ultimately, these factors are manifest through the limits observed for the range of solid solution exhibited by a phase. The intermediate effects include (i) the specific substitution model required for a given variation of composition, and (ii) the physical and thermodynamic effects produced by a given variation of composition. In the case of the white micas these would include ΔV_{mix}, ΔH_{mix}, effects on cell parameters, effects of the steric details of the crystal structure (such as the bond angles), the size and shape of coordination polyhedra, the rotation of tetrahedra, etc.

The context of our treatment is limited to that compositional variation which occurs in the white micas of common metamorphic rocks and has general petrologic significance. We will treat first substitutions occurring within the ideal white mica system and then those involving deviations from the ideal system. In both cases the focus is on crystal-chemical factors which (at given P and T) limit the extent of a given substitution. Particular attention will be directed at the interrelationship between the two types of compositional variation. For the category dealing with variations within the *ideal* white mica system, most emphasis is directed at the Na = K exchange. For the category dealing with deviation from the ideal system, the emphasis will be on the petrologically important, well documented celadonite plus $Fe^{3+} = Al^{vi}$ substitutions.

<u>Chemical variation within the ideal white mica systems</u>. As seen on Figure 3, miscibility gaps separate all three of the white mica end members. Hence, it is obvious that all of the solution series shown are nonideal.

(i) The Mu-Marg solution: $K^+, Si^{4+} = Ca^{2+}, Al^{3+}$. The large gap between Mu and Marg on Figure 3 shows that very little solution occurs between these end members. This parallels the behavior of the solution between K-feldspar and An-rich plagioclase. In both cases the wide gap probably reflects the very different size of K^+ and Ca^{2+} as well as their difference in charge.

(ii) The Pg-Marg solution: $Na^+, Si^{4+} = Ca^{2+}, Al^{3+}$. Natural and experimental data (Franz et al., 1977) indicate a miscibility gap which is asymmetric such that the Ca limb of the solvus is wider. Considering that Ca^{2+} and Na^+ are nearly the same size and a simple coupled substitution maintains charge balance, it may seem a bit surprising that Pg and Marg do not form a more complete solid solution. Even more surprising is the fact that the Pg limb of this solvus is quite similar to the Pg limb of the Mu-Pg solvus. Although the Marg-Pg solution is similar to the Ab-An solution in the form of the coupled substitution, the parallelism is limited in that the XII site of the micas is obviously different than the alkali site of a framework silicate. In fact, the siting of Na^+ or Ca^{2+} in the XII site of Pg and Marg involves considerable rotation of the tetrahedra (Zussman, 1979, Table III). Moreover, ordering occurs in the tetrahedral sites of Marg (Guggenheim and Bailey, 1975). And finally, a recent structural refinement of Pg (Lin and Bailey, 1984) shows that two of the Na-O bonds are shorter than the others and form a pincer to hold in the small Na ion. Hence, it can at least be suggested that some differences between divalent Ca and monovalent Na in the context of influence on tetrahedral rotation, tetrahedral ordering, and XII bond lengths results in the two cations not being fully equivalent in the sense of one substituting for the other.

(iii) The Mu-Pg solution: $Na^+ = K^+$. The miscibility gap is also asymmetric toward the Na side and is not known to close under any PT conditions, although it does narrow at elevated T. The cause for the miscibility gap is the difference in size between Na^+ and K^+ (Radoslovich, 1960). The asymmetry of the solvus results from greater distortion caused when K replaces Na in Pg than when Na replaces K in Mu (Radoslovich, 1960; Burnham and Radoslovich, 1964).

Data for the effects of Na = K on lattice spacings are poor for the Pg limb of the solvus but quite good for the Mu limb. Figure 13 shows that the effect of increased Na/(Na+K) on the cell volume of Mu is nonlinear and that ΔV_{mix} is clearly (+). In this sense, the shape of the volume curve based on natural samples is similar to that of Blencoe (1978) based on synthetic materials. However, as discussed in a previous section, the "synthetic curve" is systematically shifted to slightly lower values of V. Moreover, the previous discussion showed that the change in V as a function of Na/(Na+K) ratio is due almost wholly to change in c (i.e., $\sim d(002)$. On the Pg side of the solvus the V data are poor. It has generally been assumed (e.g., Eugster et al., 1972) that ΔV_{mix} is (+) for this limb of the solvus too. However, Guidotti and Crawford's (1967) study of d(002) and a later volume plot of Chatterjee and Froese (1975) have suggested a slightly (-) ΔV_{mix}. Based upon his experimental data Blencoe (1978) favored a (+) ΔV_{mix}. However, a raw plot of his data with no regard for error bars is in fact slightly concave upward. At this point it would seem that the question of (+) versus (-) ΔV_{mix} for the Pg limb remains unsettled. For petrologic purposes this question is of some importance inasmuch as it bears on how P variation will affect the limbs of the Mu-Pg solvus.

<u>Deviations from the ideal white mica plane</u>. In terms of cyrstallochemical effects the celadonite and $Fe^{3+} = Al^{vi}$ substitutions are by far the most important deviations from the ideal white mica plane. Our earlier discussion, centered around Figures 6 and 7, showed that in rocks these two substitutions always occur together. Because of the sizes of the cations involved, the effects of Fe^{2+}, Mg^{2+}, and Fe^{3+} on lattice spacings, coordination polyhedra, etc. will be fairly similar. In discussing Figures 6b and 7b, it was noted that the solid solution of the Tschermak exchange reaction ends fairly abruptly near Si_7, about halfway between the trisilicic and tetrasilicic end members. If there is truly a limit near Si_7 for the amount of solid solution from Mu toward the Si_8 end members, it would seem to be a crystal-chemical one. In this context it is instructive to compare Figure 9 with the fact that Radoslovich (1963) has noted that the Mu structure will be unfavorable if Si exceeds 6.8. Moreover, both Foster (1956) and Velde (1965a) pointed out that the Al-bearing tetrasilicic end members

$$K_2(Al_2Mg_2)(Si_8O_{20})(OH)_4 \quad \text{and} \quad K_2(Al_2Fe_2^{2+})(Si_8O_{20})(OH)_4$$

are nonexistent. Instead, the tetrasilicic end members have Fe^{3+} as the trivalent ion in the octahedral sites. In the context of this observation, Velde (1972) called attention to the large size difference for the pairs (Mg_2Al_2) and $(Fe_2^{2+}Al_2)$.

Inasmuch as we are viewing the question from the perspective of a solution starting at pure Mu, the possibility must be considered that the proposed limit at $\sim Si_7$ could be a reflection of natural rocks being restricted to bulk compositions ($Al_2O_3 > Fe_2O_3$) for which no tetrasilicic end members exist. However, Velde's experiments also suggested an absence of solid solution from Mu to the $(Mg_2Fe^{3+})^{vi}$ or $(Fe_2^{2+}Fe_2^{3+})^{vi}$ tetrasilicic end members. In fact, in his Figure 1, Velde (1965a) showed less solid solution from Mu toward these two Fe^{3+}-bearing end members than toward the two nonexistent, hypothetical Al-bearing end members. In summary, in order to maintain a true Mu structure, there may be a crystallographic constraint of Si < 6.8 (Radoslovich, 1963) which serves to limit the extent of solution from Mu toward the tetrasilicic end members.[1]

(a) *Effects of celadonite substitution on lattice dimensions.* We have noted that for a given Na/(Na+K) ratio in Mu, increases in the amounts of celadonite and/or Fe^{3+} contents produced increases in b and a but a decrease in c with a net increase in cell volume. As $\Sigma(Mg+Fe_T)$ increases in Mu, several effects have been observed (summarized by Zussman, 1979) which allow an understanding of the inverse behavior between the lateral dimensions (a and b) and c. For the increase of a and b crystallographic studies show that as the $\Sigma(Mg+Fe_T)$ increases; (1) the octahedra become larger and tend to flatten, and (2) concomitantly, there is less tetrahedral rotation (α) required in order to get the tetrahedral sheets to fit with the octahedral sheets. For example, in Zussman (1979, Table 1) the following values are given for α: Mu ($\sim13°$), Ph ($\sim6-11°$), and Cd ($\sim2°$). With regard to the effect on c, it would seem likely that decrease of α will enlarge or open the XII site. This will enable the K^+ to sink deeper into the site and thus lead to a decrease of $\frac{1}{2}c\sin\beta$. Inasmuch as V increases as $\Sigma(Mg+Fe_T)$ increases [at constant Na/(Na+K), it is evident that increases in the lateral dimensions more than compensate for the decrease of c.

(b) *Interrelationships between substitutions in XII and substitutions in the octahedral and tetrahedral sites.* It has been noted that as the phengite content of Mu increases, the Na/(Na+K) ratio decreases. This can be understood from a crystallochemical viewpoint as a result of recent structural studies by S.W. Bailey (pers. comm.). Based on refinements of a $2M_1$ Pg and a $2M_1$ Ph, Bailey concludes that (i) in part, the larger XII sites, due to the decrease of α, favor K over Na, and (ii) more importantly, it is necessary in Pg to distort and corrugate the interlayer site to hold the small Na^+ in place. Hence, as the phengite (and Fe^{3+}) content increases, the rotation and corrugation required to distort the XII site to hold in Na becomes impossible if the tetrahedral sheet is to be fitted together with the octahedral sheet. In essence, the pincer which was required to hold in Na^+ becomes unclamped so that the XII site enlarges and becomes less favorable for holding Na^+.

It is interesting to note that these arguments can be inverted so that they provide a rationale for the very minor amounts of Mg, Fe^{2+}, and Fe^{3+} in Pg. If one is dealing with Pg ($\alpha \sim 16°$), the rotation and corrugation of the tetrahedra is such that adding Mg, Fe^{2+}, or Fe^{3+} to the octahedral sites would enlarge the XII site as well as "unlock the pincers" and thus destabilize the hold on Na. In the case of Marg, the "pincer effect" does not occur (Bailey, pers. comm.), but enlargement of the XII site as Mg, Fe^{2+}, or Fe^{3+} increase would probably decrease the hold on the highly charged but relatively small Ca^{2+} ion. Hence, for Pg or Marg to remain stable, it is probably necessary that very little Mg, Fe^{2+}, or Fe^{3+} enters the octahedral sites. The experimental work of Franz and Althaus (1976) correlates well with these conclusions:

---

[1] It should be noted that some experimental work not yet fully available to the writer (Massone, 1981) does suggest that at P > 10 kbar, the Si content of Mu can exceed 7.0 (See Fig. 2 in Hoinkes and Thoni, 1982). Thus, the general rarity of phengitic Mu with Si > 7 may reflect lack of sampling in the appropriate geological settings. Indeed, two of the studies reporting Mu with Si > 7.0 are from the high P rocks of the Alps (Saliot and Velde, 1982; Chopin, 1979). It appears that the final verdict on the degree of solid solution between trisilicic and tetrasilicic K-white mica is still uncertain.

(i) very little Mg could be substituted into Pg, and
(ii) it was not possible to synthesize a Na-phengite.

(c) *Other substitutions*. Radoslovich (1963) suggested that deviation by Mu from being purely dioctahedral toward being trioctahedral should be minor, and data from natural materials support this. Moreover, the experimental study of Green (1981) indicates that even at very high P and T (20-30 kbar and 800-1000°C) there is still only minor deviation from dioctahedral K-white micas toward the trioctahedral micas. However, a phase intermediate between the dioctahedral and trioctahedral K-micas [$K_2Mg_3Al_2(Al_2Si_6O_{20})(OH)_4$] could be synthesized. Although no solution occurs between this phase and the dioctahedral micas, it appears to have complete solid solution with the trioctahedral micas.

As for the "other substitutions", because all of them are so small in magnitude, it is quite difficult to detect crystallochemical controls in the context of data obtained from metamorphic studies. The extent of substitutions of Mn, Cr, Ba, Ti may be largely controlled by metamorphic grade and especially by bulk composition effects. For example, the absence of a Ba-rich phase (to serve as a Ba-saturating or -buffering phase) coexisting with Mu makes it highly unlikely that the observed Ba content of Mu is crystallochemically controlled. Nonetheless, Ti can enter octahedral sites as Ti^{4+} and substitute for Al, Fe^{3+}, Mg, or Fe^{2+}. Some of the exchange reactions could involve the formation of vacancies. From our earlier discussion we have seen that Mu is the only white mica in which Ti enters to the extent that any relationship with petrologic factors can be detected. Because a Ti-saturating phase (ilmenite or rutile) usually coexists with Mu, the Ti content in the Mu should be at saturation levels for the PT conditions prevailing during its formation. Hence, the amount of Ti present is probably crystallochemically controlled. Evidence that Ti is at saturation levels in Mu is seen from the fact that the Ti content is closely related to metamorphic grade. The increase of Ti with increase of grade implies that the Ti saturation boundary has moved as a function of T, certainly not an unexpected effect.

Inasmuch as the amount of Ti in Mu is never very large, it would seem likely that charge balance factors rather than ionic size control the saturation limit. The amount of Ti that can enter Mu at various grades might be closely interrelated (via charge balance) with other compositional variations, primarily in octahedral sites. This is well illustrated by the different Ti contents in isograde Mu in western Maine where Guidotti (1974) found that the Ti content was higher in the low-Al Mu of the Al-poor rocks than in the higher-Al Mu found in the Al-richer assemblages. Thus, it seems likely that the specific substitution reactions by which Ti enters Mu will reflect some of the effects metamorphic grade has on Mu composition. In fact, the data discussed by Guidotti (1978) suggest the following *in Al-saturated rocks*. (i) From lower garnet zone into staurolite zone the exchange reaction is (a): $2(Mg,Fe^{2+}) \rightleftarrows Ti + \square$, and (ii) from lower sillimanite zone to upper sillimanite zone the exchange reaction is mainly (b): $2Al^{vi} \rightleftarrows Ti + (Mg,Fe^{2+})$ and possibly to a minor extent (c): $Al^{vi} + Si^{iv} \rightleftarrows Ti^{vi} + Al^{iv}$. Tracy (1978) has also provided evidence for (b) in similar grade rocks. It is interesting to note that the left hand sides of (a) and (c) either singly or combined would, on a casual inspection, seem like an increase in celadonite content. The point to recall is that earlier it was noted that an inverse relationship might exist between the $\Sigma(Mg+Fe^{2+})$ and Ti and between Si and Ti. Equations (a) and (c) could explain such a suggestion. Finally, an exchange like (b) might explain the fact that on Figure 8 the data points fall on a line which lies above the ideal Tschermak line.

(II) Petrologic aspects of white micas

The petrologic aspects of the phase relations of the white micas relative to bulk compositions in which they occur and relative to the minerals with which they coexist will be discussed in three categories. (A) Whether a given bulk composition range can be expected to contain a particular white mica and, if so, the composition range of the white mica for a particular bulk composition range. For example, one might consider the nature or possibility of white micas occurring in metabasites versus a similar consideration for metagreywacke. This would be dealing with white micas in a very general petrologic way. (B) The detailed chemistry of white micas relative to specific coexisting phases and to the effects of P, T, and other environmental conditions. This is a more rigorous petrologic approach in that it relates to experimental work and thermodynamic considerations. The white mica phase relations considered are those which have particular petrologic significance. Hence, to a significant extent the treatment will emphasize white micas in pelitic schists. And (C) inversely, using white micas to gain information on petrogenesis.

(II.A) White micas in the context of gross lithologic types

It should be noted at the outset that white micas are Al-rich phases, although Ph represents a slight exception. Thus, the relative Al contents of various rock types will bear strongly on the possibility of white micas being present and on their relative abundance. The term "relative Al content" takes into account the manner in which Al is tied up in the various other phases present and, therefore, bears on whether the Al will be available to form micas. The latter can be strongly influenced by metamorphic conditions, but, in general, the higher the Al_2O_3 content of a rock the more likely it is that white micas will be present and abundant. This general point is well illustrated by the fact that the white micas are so abundant and such characteristic phases of the Al-rich rocks known as pelitic schists.

To a significant degree this relation to Al content is an extension of the Al-saturation concept so commonly used in igneous petrology when distinguishing among per-aluminous, calc-alkaline, and per-alkaline compositions in the sense of Shand (1969). This applies in the gross petrologic sense considered here and also in the more specific sense when dealing with comparisons between different mineral assemblages within a narrower range of bulk composition. However, in metamorphic rocks there needs to be some tempering of this approach due to the influence of the P, T, and X (fluid) on the actual phases making up a given bulk composition and thus controlling the availability of Al to form white micas. At the extreme, this metamorphic control would be illustrated by rocks subjected to PT conditions such that the white micas were intrinsically unstable. A particularly important aspect of the influence of metamorphism is that for a general bulk composition range (e.g., metabasalt), even when considered in a gross petrologic sense, there are quite significant differences of mineralogy in rocks affected by "Blue Schist Style" metamorphism, as opposed to "Green Schist Style" metamorphism.[2] In turn, this is reflected by variation in the occurrence,

---

[2]This term is used in a very general sense to mean the sort of metamorphism characterized by high P, low T conditions and commonly involving mineral assemblages that contain phases like lawsonite, glaucophane, jadeitic pyroxenes, aragonite, etc. Most typically this style of metamorphism is found in Mesozoic and Cenozoic orogenic belts. In contrast, the term "Green Schist Style" is used in a very general sense to include metamorphism taking place at the generally lower P, higher T conditions than the "Blue Schist Style". Minerals like lawsonite, etc. would not be present in any assemblages.

abundance and composition of the white micas. Inasmuch as the "Blue Schist Style" of metamorphism is now generally believed to reflect higher P conditions, the gross differences in mineralogy (including the white micas) for a given bulk composition appear to reflect the influence of P on how the various elements are tied up in different phases.

In our treatment of the gross petrologic aspects, we will take each white mica separately and discuss it in terms of the various general bulk compositions in which it has been observed to occur. The reader should be aware that information presented reflects the perception of this reviewer as to the stage of our knowledge at present. To the extent that a reader disagrees, it is hoped that it will spur discussion and thus result in refinement by the general petrologic community of our overall perception about the white micas.

Margarite occurrence. Many reports have been published since 1970 bearing on general and specific petrologic aspects of Marg, but there is little question that the bulk of our knowledge about this mineral is due to work of Martin Frey and coworkers (Frey and Orville, 1974; Frey et al., 1982; and Bucher et al., 1983). Indeed, Frey and Niggli (1972) appear to be the first to have emphasized the truly general significance of Marg as a rock-forming phase. References to virtually all of the literature on Marg can be found in Frey et al. (1982) and Bucher et al. (1983). A survey of these shows that there are two main categories of Marg occurrences: (a) as a "normal prograde rock-forming mineral interspersed and intergrown with other minerals, and (b) as a mineral forming pseudomorphs after other minerals.

(a) *"Normal", rock-forming occurrences.* In such cases Marg occurs at grades ranging from the lower greenschist facies to well up into the amphibolite facies. It has also been found in a few rocks formed at blue schist conditions. The general rock types listed by Frey et al. (1982) in which Marg has been observed exclusive of the Alps include: metapelites (12), metamarls (6) (similar to metapelites but also having calcite as a coexisting phase), metabauxites (3) (i.e., the emery deposits mentioned earlier), metabasites (3) (i.e., metabasalts), and meta-anorthosites (3) [numbers in parentheses indicate the number of localities listed]. The first two rock types are clearly the dominant "home" for Marg. Frey et al. (1982) also list 47 Marg-bearing Alpine localities, and the bulk of these involve metapelites or metamarls. For the 271 Alpine specimens studied, they list the following minerals as also present (on the basis of the percentage of the 271 samples containing the mineral) quartz (98.5), Mu (97.8), graphite (86.7), chlorite (80.8), plagioclase (64.2), calcite (59.8), epidote-group (53.5), biotite (45.4), garnet (31.0), dolomite (31.0), Pg (25.1), chloritoid (10.5), kyanite (8.9), staurolite (6.3), hornblende (4.1), and corundum (1.1). A scan of the assemblages listed for the localities, exclusive of the Alps, suggests that a similar listing of coexisting minerals would apply to them also.

There are some general points worth mentioning with regard to the minerals observed to coexist with Marg. (i) Marg is commonly associated with Ca-rich phases such as zoisite, plagioclase, and calcite -- certainly an expectable observation. (ii) In the majority of cases Marg is associated with graphite; however, there are exceptions, and in some of the metabauxites it occurs with magnetite and/or hematite. (iii) If an Al-silicate is present, it is kyanite and never andalusite or sillimanite. (iv) Marg is commonly associated with quartz, the main exception being the metabauxites, and in these rocks the Marg is associated with corundum. (v) Marg is commonly associated with biotite, and in this respect it differs from Pg which is common with biotite only at higher grades. (vi) Most, but not all, of the occurrences of Marg with hornblende are in the metabasites.

(b) *Pseudomorph-forming occurrences*. Although Frey et al. (1982) did not emphasize such occurrences of Marg, they noted that the minerals it replaces so commonly in this manner include Al-rich phases such as andalusite, kyanite, sillimanite, or corundum, and more rarely minerals like chloritoid, staurolite, or even Mu. The papers by Guidotti and Cheney (1976) and Guidotti et al. (1979) describe such pseudomorphs in considerable detail and cite references to numerous other such occurrences. Some question arises as to whether the Marg is in equilibrium with the minerals in the groundmass, or in the case of partial pseudomorphs with the remnant of the original phase, e.g., andalusite or sillimanite. The latter is of particular interest, because in the more straightforward prograde occurrences of Marg described above, kyanite was the only Al-silicate that seemed to occur with Marg. Nonetheless, as suggested by Baltatzis and Katagas (1981), even if equilibrium has been attained for the Marg in the various pseudomorphs, it seems likely that it results from polymetamorphism rather than a simple prograde event. Because Marg which forms pseudomorphs typically replaces Al-rich phases, this type of occurrence is largely restricted to pelitic schists. Hence, the associated suite of minerals is similar to that expected in typical low to medium grade metapelites and metamarls. Particularly notable is that graphite is a characteristic (though not ubiquitous) accessory for this type of Marg occurrence also.

Paragonite occurrence. Previous to routine use of X-ray diffraction for petrologic studies, Pg was considered to be a rare mineral, but Rosenfeld (1956), Rosenfeld et al. (1958), and Zen and Albee (1964) clearly illustrated the general occurrence and significance of Pg. Because most occurrences of Pg involve very fine-grained material and intergrowth with other minerals, it was not until electron microprobe analyses became routine that data on the chemistry of Pg became readily available. Since 1970 many data have been accumulated on Pg and the nature of its occurrences, and some picture is starting to emerge with respect to the general petrologic framework in which it occurs.

The possibility of Pg occurring at sub-greenschist conditions has been questioned by several authors (see discussion in Chatterjee, 1973), but for our purposes we can say that Pg (coexisting with quartz) can occur throughout the greenschist facies conditions and persist into the amphibolite facies at least into the sillimanite zone (Warner and Al-Mishwt, 1968; Grambling, 1984). In metamorphic terranes affected *only* by "Green Schist Style" metamorphism Pg is largely restricted to pelitic schists and metamarls (calc-schists). In the greenschist and lower amphibolite facies of such rocks Pg is largely restricted to very Al-rich bulk compositions such that it coexists with highly aluminous minerals such as pyrophyllite, chloritoid, kyanite, and staurolite and is incompatible with minerals such as biotite. At higher grades the range of compositions in which Pg can occur expands to somewhat lower Al contents such that it can then coexist with minerals such as biotite. Only relatively few exceptions to these constraints occur and in most of them there is usually some question as to whether equilibrium prevails. Morever, in many of the seeming exceptions there is clear evidence that the rocks are polymetamorphic, having first been affected by a "Blue Schist Style" of metamorphism. Hence, in general the prime control on the occurrence of Pg in "Green Schist Style" terrane is the relative availability of Al_2O_3. Relative availability of Na_2O is only of minor importance.

In a very few areas (e.g., Henley, 1970) hornblendes or actinolites have been found coexisting with Pg in the above described rock types. In a few cases Pg has also been reported for greenschist facies metabasites (e.g., Ernst and Dal Piaz, 1978) but invariably such occurrences also involve an earlier "Blue Schist Style" metamorphism such that it is unclear whether the Pg is truly stable. It does not appear that Pg occurs in metabasites which have been affected only by a "Green Schist Style" metamorphism. In "Blue Schist Style" terranes the range of rock types in which Pg can occur enlarges markedly to include pelites, metamarls, greywacke, and metabasites (including eclogites, e.g., Black, 1977; Krogh, 1980). Hence, it occurs in some rocks with markedly lower bulk content of Al_2O_3. In the pelites and metamarls it occurs with many of the same minerals mentioned above with the exception of biotite because biotite tends to be absent in "Blue Schist Style" rocks. In some cases typical blue schist minerals such as glaucophane and lawsonite will be present. In the metabasites Pg can be found coexisting with Ca-amphiboles, epidotes, etc.

In general, it seems that Pg is considerably more common in "Blue Schist Style" than in "Green Schist Style". As will be developed in a later section, the expansion of the Pg stability field to include lower ranges of Al_2O_3 bulk compositions is probably just reflecting changes in tie line orientation in composition space in response to higher P. Also possibly contributing to the increased abundance is the opening of the Mu-Pg solvus due to the high phengite content that typifies Mu in "Blue Schist Style".

<u>Muscovite occurrence</u>. Mu has long been recognized as an important rock forming mineral, and much has been written about it from a petrologic viewpoint (e.g., see a summary by Guidotti and Sassi, 1976a). Mu occurs over a wider range of metamorphic conditions than either Marg or Pg, and it occurs from lower greenschist into upper amphibolite facies. Its range of rock types is considerably greater than that of Pg or Marg, occurring in all those in which Pg or Marg can occur, plus a number of others in which Pg and Marg are absent. An indication of the wider composition range in which Mu can occur is seen from the fact that it can coexist with K-feldspar, whereas Marg and Pg cannot. In discussing the range of rock types in which Mu can occur, it is again useful to distinguish between Mu in "Green Schist Style" terranes versus "Blue Schist Style" terranes.

In "Green Schist Style" terranes, Mu is most abundant in Al-rich rocks such as metapelites and metamarls. It also occurs in metagreywacke, impure limestones at low grades of metamorphism, metamorphosed acid to intermediate intrusive and extrusive igneous rocks, and in small amounts even in low-grade metabasites, where it is associated with calcic amphiboles, epidotes, etc. According to Laird (1982) Mu can persist in metabasites into the epidote amphibolite and amphibolite facies, if the rocks are rich enough in K_2O and Al_2O_3. It is apparent that Mu occurs over a wider range of bulk Al_2O_3 content than Pg or Marg. Obviously rocks with only minor K_2O are not likely to contain Mu, e.g., meta-ultramafics. About the only minerals with which Mu tends not to occur are some of the Ca-rich phases like hornblende, tremolite, diopside, etc., especially at grades exceeding greenschist facies. At these somewhat elevated temperatures the mineralogy starts to adhere more closely to the Shand Al-saturation principle such that Al in Mu and Ca of the calcic phases starts to have an affinity to combine and form calcic plagioclase. However, some occurrences of Mu coexisting with hornblende have been recorded (e.g., Henley, 1970).

In "Blue Schist Style" terranes Mu occurs in the same range of bulk compositions as described above. However, in the appropriate rocks it can also be found together with the typical "Blue Schist Style" minerals, glaucophane, lawsonite, jadeitic pyroxene, etc. Mu is probably more abundant in the metabasites of "Blue Schist Style" terranes than in "Green Schist Style" terranes. For example, it is commonly reported in eclogites although in some cases it may be the result of post eclogite formation, e.g., Ernst and Dal Piaz (1978). Laird's (1982) discussion of "Blue Schist Style" terranes contains much relevant information about Mu in metabasites.

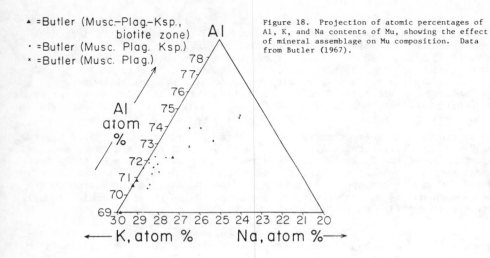

Figure 18. Projection of atomic percentages of Al, K, and Na contents of Mu, showing the effect of mineral assemblage on Mu composition. Data from Butler (1967).

White mica composition in the context of gross lithologic differences. The specific contents of the XII, octahedral and tetrahedral sites of the white micas are controlled by a combination of bulk composition (mineral assemblage) and metamorphic conditions. The only compositional aspect worth special note in this section pertains to the Al content of Mu because it is particularly responsive to variation in the bulk Al content of rocks. Earlier it was seen that the amount of Al^{vi} and Al^{iv} in Mu is a measure of its celadonite content. Data accumulated from many studies have shown that for given P and T, the Cd content is very strongly related to bulk composition (mineral assemblage). In highly aluminous rocks (as indicated by the presence of Al-rich minerals like chloritoid, pyrophyllite, staurolite, etc.), the Cd content of Mu will be greatly decreased. In rocks from the same locality which have a low Al content (as indicated by the presence of low Al minerals like biotite and K-feldspar) the Mu will have a much higher Cd content (see Fig. 18). However, as developed in the next section, for a given bulk composition (mineral assemblage) P seems to have a large effect on Cd content so that Mu from "Blue Schist Style" terranes is typically quite Cd-rich.

(II.B) Metamorphic controls on white mica occurrence and composition

Here we are concerned with the manner in which metamorphic conditions -- P, T, X (fluid) -- determine the specific reactions that produce or destroy the white micas, and also how they control the composition of a given white mica. This involves the specific phase relations of the white micas and is a more detailed treatment than that given above, in which the white micas were discussed merely in terms of the type of rocks and range of metamorphic grades in which they have been found. We shall review the phase relations of white micas in common rocks based on data from numerous workers. The discussion is strongly slanted toward the Al-rich metapelites and metamarls because these are the rocks in which: (i) micas are most abundant, (ii) the most data exist, and (iii) there is the best evidence that the white micas are in equilibrium with the other phases present. It is expedient to discuss each of the white micas separately except when reviewing data bearing on the solvus relationships. Then the micas are considered in pairs. In order to describe the phase relations of white micas in the context of metamorphic controls, it is necessary

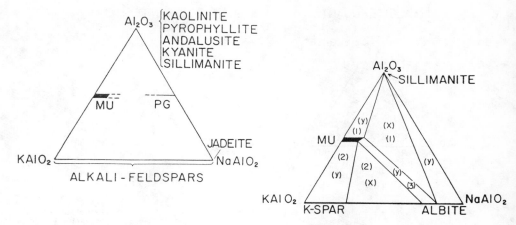

Figure 19 (left). AKNa projection of the phases in the system Al_2O_3-K_2O-Na_2O-SiO_2-H_2O (Thompson, 1961). Projected from quartz at a given P, T, and activity of H_2O.

Figure 20 (right). Schematic AKNa projection of the mineral facies of sillimanite-grade metapelites. X = Limiting assemblages and Y = non-limiting assemblages (see text). Assemblages (1) have Al-saturated Mu and assemblages (2) have Mu with the lowest Al for the given conditions. The Al-content of Mu in assemblage (3) can vary with bulk composition change.

to review briefly a theoretical framework for understanding the interrelationships among bulk composition, P, T, X (fluid) and the occurrence and composition of a given white mica.

Theoretical framework for treating the phase relations of white micas. A theoretical framework for discussing the phase relationships of white micas has been reviewed in some detail by Guidotti and Sassi (1976a). A virtually identical, though more brief, approach will be followed here. For our purposes only the following aspects of a theoretical framework need to be reviewed.
(a) A chemical system (or systems) must be chosen such that it covers most of the chemical variation possible by a given white mica. It should be capable of describing the most important reactions that can affect the mica and thus also capable of describing all of the phases which participate in the reactions. Figure 19 illustrates a system capable of describing many of the phase relations pertinent to either Mu or Pg or both. (b) With respect to the system(s) chosen, one should have an understanding of notions listed below.

(i) Limiting and non-limiting assemblages: Limiting assemblages are those in which the number of phases equals the number of components. Because of Phase Rule considerations, the composition of phases in such assemblages will be independent of bulk composition and will be a function of metamorphic conditions only. In non-limiting assemblages the number of phases is less than the number of components. Hence, the composition of the solution phases can reflect variation of bulk composition as well as metamorphic conditions. On Figure 20 assemblages marked with an (x) are limiting and those with a (y) are non-limiting.

(ii) Two types of saturating phases: One type would involve a phase that is outside the system of concern, e.g. ilmenite or rutile, for the system being considered. The presence of such a Ti-rich phase insures that the phases of the ideal system contain the maximum amount permissible (for the metamorphic conditions) of some component (TiO_2 in this case) outside of our system. In the second usage the term would apply to phases intrinsic to our system. An example best serves our attempt to illustrate this usage. Consider the

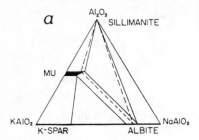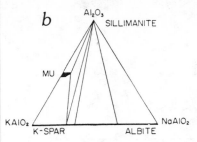

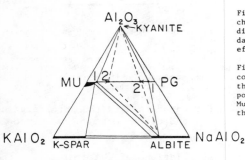

Figure 21 (above). Schematic drawings showing changes in the AKNa topology accompanying the discontinuous reaction 17. On Figure 21a the dashed-line three-phase assemblage shows the effect of continuous reaction 28.

Figure 22 (left). Schematic AKNa showing the composition of coexisting Mu + Pg at T(1) versus the composition at higher T(2). The movement of points (1) and (2) reflect the closing of the Mu-Pg solvus due to the exchange Na = K between the two phases.

schematic diagram shown in Figure 20 and the solution field of Mu. Because Mu in assemblage (1) coexists with an Al-silicate it contains the maximum amount of Al_2O_3 for the prevailing metamorphic condition, (i.e., it is Al-saturated). It should also be noted that the Mu in assemblage (2) would have the lowest Al content allowed by the prevailing metamorphic conditions. And in assemblage (3) the Al content of Mu is a function of bulk composition plus metamorphic conditions.

(iii) Discontinuous, continuous, and exchange reactions: Discontinuous reactions involve a fundamental change in the compatibilities shown on a phase diagram (see Figs. 21a and 21b). Continuous reactions merely involve the shifts in the position of a limiting assemblage as seen on a phase diagram (Fig. 21a). An example of an exchange reaction would be that taking place due to closing of a solvus (see Fig. 22 for the case of Mu-Pg solvus).

(iv) Compositional deviation from the system being considered: Based upon earlier discussion it is apparent that for the white micas this applies only to Mu and primarily with regard to its Cd content. If Mu were a fully trisilicic phase and thus fully describable by the system represented in Figure 19, it would plot along a single line at the top of the solution field shown for Mu. In fact, as discussed before, via the Tschermaks substitution it deviates to less aluminous compositions. On Figures 19-22 this is shown by that portion of the Mu solution field that plots below the ideal trisilicic line. As muscovites plot progressively below this line they contain progressively greater amounts of FeO, Fe_2O_3, and MgO, three components which are not part of the designated system. Hence, a consideration of where a Mu plots on the Mu solution field gives us some measure of its Cd content. Obviously Mu coexisting with K-feldspar will, for the prevailing metamorphic conditions, have the maximum permissible Cd content.

The preceding outline of a framework for discussion of mica phase relations should suffice for our purposes. A much more complete treatment of the theoretical aspects can be found in Thompson (1961) and Thompson and Thompson (1976). We can now treat the following in a systematic and rational fashion: (a) the occurrence of a white mica as a function of grade, including the reactions marking the lower and upper stability limits; (b) the effect of bulk composition (mineral assemblage) on the occurrence and composition of a given white mica; and (c) composition of the white micas as a function of metamorphic conditions with bulk composition effects factored out.

Margarite phase relations. The most complete treatments of the phase relations of Marg as a prograde, rock forming mineral are those of Frey and Orville (1974) and Bucher et al. (1983) (cf. Perkins et al. (1979) for a purely theoretical treatment of the phase relations of Marg in the system $CaO - Al_2O_3 - SiO_2 - H_2O$). Much of their study is based on the data accumulated over the years by Frey (1969, 1978). Frey and Orville (1974) used the system $CaO - Na_2O - Al_2O_3$ with SiO_2, H_2O, and CO_2 in excess, and Bucher et al. (1983) used the system $CaO - Al_2O_3 - SiO_2 - (C-O-H)$ to treat the phase relations of Marg in the calcareous, graphitic metapelites and metamarls of the Central Alps. Hence, their efforts were directed at the most important type of occurrence of Marg. Both studies merged observations of natural assemblages (including mineral compositions) with theoretical calculations (based on thermodynamic data and activity models) of the P, T, X (fluid) controls of mineral assemblages to obtain an understanding of the metamorphism forming the Marg-bearing assemblages. Here, we briefly review the main reactions producing the observed assemblages and the compositions of minerals found in these assemblages, following Bucher et al. (1983).

At metamorphic conditions below the incoming of Marg, the precursor assemblage is believed to be pyrophyllite + calcite + quartz. Margarite comes in at lower greenschist facies conditions by the reaction:

$$2 \text{ pyrophyllite} + 1 \text{ calcite} = 1 \text{ Marg} + 6 \text{ quartz} + 1 \text{ CO}_2 + 1 \text{ H}_2\text{O} . \quad (4)$$

Thenceforth the assemblage Marg + quartz + calcite is common and characteristic of the greenschist facies, and additional reactions subsequently bring in clinozoisite and/or plagioclase. In the case of clinozoisite the reaction is:

$$3 \text{ Marg} + 5 \text{ calcite} + 6 \text{ quartz} = 4 \text{ clinozoisite} + 5 \text{ CO}_2 + 1 \text{ H}_2\text{O} . \quad (5)$$

In the case of plagioclase the reaction is:

$$1 \text{ Marg} + 1 \text{ calcite} + 2 \text{ quartz} = 2 \text{ anorthite} + 1 \text{ CO}_2 + 1 \text{ H}_2\text{O} . \quad (6)$$

Reaction 6 is supported by textural observations (plagioclase replacing Marg), as well as the mineral assemblage data. The existence of Equilibria 5 and 6 is shown by the stable coexistence of Marg + calcite + quartz + clinozoisite and of Marg + calcite + quartz + plagioclase. The existence of these four-phase assemblages over an area can be explained either in terms of buffered gas composition or the presence of extra components (e.g., Na in Marg and Plag). However, the five-phase assemblage, Marg + calcite + quartz + plagioclase + clinozoisite, also occurs over an area along the prograde traverse. In the system $CaO - Al_2O_3 - SiO_2 - (C-O-H)$ the simultaneous occurrence of 5 and 6 takes place at an isobaric invariant point, (see Fig. 4 of Bucher et al., 1983). Hence, the occurrence of this assemblage over an area is presumed to result from the variable plagioclase composition.

Regardless of complications caused by fluid buffering and extra components, Bucher et al. (1983) found that the highest grade occurrence of Marg + quartz +

calcite could be mapped as a sharp isograd such that at higher grades either calcite or quartz but not both were present in the Marg-bearing assemblages. For the assemblage Marg + quartz + clinozoisite + plagioclase the following reaction equilibrium was suggested:

$$1 \text{ Marg} + 2 \text{ clinozoisite} + 2 \text{ quartz} = 5 \text{ anorthite} + 2 \text{ H}_2\text{O} . \qquad (7)$$

The existence of the isobarically invariant (in the ideal system without Na_2O) assemblage Marg + quartz + clinozoisite + plagioclase + kyanite above the isograd marking the end of the Marg + quartz + calcite assemblage requires in addition to Reaction 7 the simultaneous existence of reaction equilibrium:

$$3 \text{ quartz} + 4 \text{ Marg} = 5 \text{ kyanite} + 2 \text{ clinozoisite} + 3 \text{ H}_2\text{O} . \qquad (8)$$

The existence of this five-phase assemblage over an area along the traverse is again explainable in terms of extra components in Marg, plagioclase, and clinozoisite.

The suggestion that Reactions 7 and 8 take place progressively agrees well with the observed progressive decrease in the modal amount of Marg noted by the authors. And the occurrence of 6 and 7 (and 7 simultaneously with 8) coincides with the progressive increase in the anorthite (An) content of plagioclase noted by the authors for many rocks above the occurrence of Reactions 5 and 6. On their Figure 7, they show (with grade) a progressive change from An_{30} to about An_{60}. At the high grade end of the rocks they considered, Marg persisted only in quartz-absent rocks. It has been removed from all quartz-bearing rocks by Reaction 8 and from all calcite bearing rocks by Reaction 6. The final breakdown of all Marg occurred by the reaction:

$$1 \text{ Marg} = 1 \text{ anorthite} + 1 \text{ corundum} + 1 \text{ H}_2\text{O} . \qquad (9)$$

The theoretical calculations of Bucher et al. (1983) provided $T-X_f$ diagrams for the system $CaO - Al_2O_3 - SiO_2 - (C-O-H)$ at 4 kbar and 7 kbar. Considering only assemblages containing *both* calcite and quartz, two distinctly different $T-X_f$ topologies were calculated. In the 4 kbar topology Marg has a wide stability field and can coexist with anorthite. At 7 kbar the Marg stability field has shrunk and the Marg + anorthite compatibility is replaced by kyanite + clinozoisite. Although Na substitution into plagioclase causes some blurring of the distinction between the two $T-X_f$ topologies, e.g., enabling Marg to coexist with Na-bearing plagioclase at 7 kbar, the general implication still prevails with regard to kyanite + clinozoisite coexisting (regardless of T) only at the higher pressure ranges. Bucher et al. calculated that the topology change occurs at 6 kbar but that the kyanite + clinozoisite compatibility would be manifest only if the plagioclase contained more than 60 mol % An.

In the case of the study of Bucher et al. (1983), the aspect of the topology calculations which is significant for the Marg-bearing reactions discussed above is that the high grade ones do involve the coexistence of kyanite + clinozoisite, thereby suggesting that the metamorphic gradient they described involved an increase of P as well as T. The authors note that such a suggestion is corroborated from the results of studies on other parageneses associated with the Marg-bearing rocks. Hence, it lends credibility to the significance of their calculated T-X grids. Based upon their theoretical calculation of the T, P, X (fluid) controls on the reactions involving Marg, Bucher et al. believe the formation of the Marg + calcite + quartz assemblage occurred in the lower greenschist facies. The last occurrence of this assemblage forms a sharp isograd which they believe occurs at near 480°C at 4 kbar and 520°C at 7 kbar. They suggest that in quartz-saturated rocks, Marg persists above the Marg + calcite + quartz isograd for about 70°C. And Marg in quartz-free and calcite-free rocks

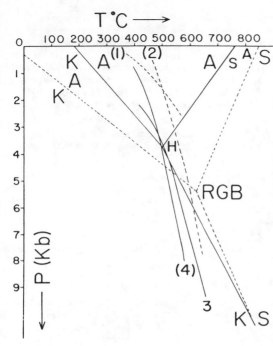

Figure 23. Experimental and theoretically calculated curves for Marg (dashed lines) and Marg + quartz (full lines). Al-silicate curves, from Richardson et al. (1969) (RGB) and Holdaway (1971) (H), are shown for reference purposes. (1) Velde (1971), (2) Chatterjee (1974), (3) Storre and Nitsch (1972), (4) Chatterjee (1976) calculated. A = andalusite, S = sillimanite, K = kyanite.

finally breaks down at about 620°C at 7 kbar. In general they found that these P and T estimates based on Marg-bearing assemblages coincide well with data obtained from mineral assemblages in other quite different rock types. The suggestions of Bucher et al. (1983) can be compared with the experimental curves for Marg stability (Fig. 23).

Finally it is appropriate to comment about the phase relations of Marg where it is involved in the formation of pseudomorphs. As mentioned earlier, some question arises about the spatial extent of equilibrium in such cases. Inasmuch as these pseudomorphs most typically form after highly aluminous phases in ordinary metapelites, they would appear to result from transfer of the components Na_2O and CaO (e.g., from groundmass plagioclase) and H_2O (from pore fluids) to combine with the Al and Si of the original aluminosilicate and thereby form Marg (Guidotti et al., 1979). Presumably some newly imposed metamorphic conditions differing from those which formed the Al-silicate serve to drive the required transfer of components.

Paragonite phase relations. Numerous occurrences are recorded for Pg in non-pelitic bulk compositions (especially in "Blue Schist Style" terranes), but there is relatively little discussion of its phase relations in such rocks. Only a few papers have proposed specific mineralogic reactions involving Pg and minerals such as calcic amphiboles or "Blue Schist" minerals, including those by Goffe (1977), Black (1977), and Laird and Albee (1981) in which reactions involving both Pg and glaucophane are discussed. Others, such as Holland (1979), discuss occurrence of Pg and jadeite or omphacite. By far, most observational and theoretical considerations of Pg phase relations involve bulk compositions like pelites and metamarls. Moreover, most of the observational studies involve rocks best considered as belonging to a "Green Schist Style" of metamorphism. Our discussion will reflect these facts.

(a) *Theoretical models of Pg phase relations.* The papers of Thompson (1961) and Thompson and Thompson (1976) provide a good model for treating the phase relations of Pg. The AKNa system considered by them allows a nearly complete chemical description of the phases involved in the reactions bringing in Pg at low T and those removing it at high T. Although their model is referred to as theoretical, it incorporates many observations based on natural assemblages. Considering only the summary PT grid of Thompson and Thompson (1976, their Fig. 11) several generalizations can be suggested for the reactions forming or destroying Pg.

(i) For the general P range in which the metamorphism would be of the "Green Schist Style". At low T the first Pg to form is K-bearing and the reaction is:

$$\text{kaolinite + Mu + albite = Pg + quartz + H}_2\text{O} . \tag{10}$$

This is followed by the reaction forming end-member Pg

$$\text{kaolinite + albite = Pg + quartz + H}_2\text{O} . \tag{11}$$

At higher T, end-member Pg disappears by the reaction

$$\text{Pg + quartz = albite + Al-silicate + H}_2\text{O} . \tag{12}$$

This is followed at slightly higher T by the breakdown of a potassic Pg:

$$\text{Pg + quartz = Mu + Al-silicate + albite + H}_2\text{O} . \tag{13}$$

(ii) For the pressure range in which jadeite is shown as present on Figure 11 of Thompson and Thompson (i,e., corresponding to "Blue Schist Style"), the reactions which form or destroy Pg are similar in form but somewhat more complex, as they involve jadeite in some cases and albite in others. Because there are still only a few natural observations with which these proposed reactions can be compared, we will not pursue the matter further at this point. The set of Pg-bearing mineral facies diagrams shown on Figure 11 of Thompson and Thompson (1976) enables recognition of several continuous reactions that will affect the composition of Pg. However, there are few observations of natural materials which enable one to determine if they really do occur. This probably results from the fact that the range of Na/(Na+K) shown by Pg is quite restricted ($\sim$12 atom %). By far the most interesting compositional variation of Pg is when it coexists with Mu. This will be treated as part of our discussion of Mu phase relations.

(b) *Pg phase relations based on natural parageneses.* Observations of natural rocks suggest several reactions for the first production of Pg at the low T range of "Green Schist Style" metamorphism. Zen (1960) suggested initial formation of Pg in metapelites by the reaction:

$$1 \text{ albite + 1 kaolinite = 1 Pg + 2 quartz + 1 H}_2\text{O} . \tag{14}$$

This is identical with Reaction 11 above.

Chatterjee (1973) discussed the possibility of Pg occurring in sedimentary rocks but Pg + quartz requiring the onset of some metamorphism. Based upon natural parageneses and his experimental results he suggested that Pg + quartz resulted from the reaction:

$$3 \text{ Na-montmorillonite + 2 albite = 3 Pg + 8 quartz} . \tag{15}$$

Frey (1969) suggested that a mixed-layer intergrowth of Ph and Pg in very low grade rocks was an intermediate step in the formation of Pg from an Na-bearing illite. He suggested the reaction:

$$\text{Na-bearing illite = Pg + Ph + H}_2\text{O} . \tag{16}$$

Later, Frey (1978) suggested that the mineralogic sequence from unmetamorphosed rocks to low grade metamorphism involved: irregular mixed-layer montmorillonite to rectorite to mixed-layer Pg/Mu to discrete Pg and Mu.

Studies at the high T end of the Pg stability range seem to be uniform regarding the reactions terminating Pg (coexisting with quartz). They are similar to the theoretical reactions of Thompson and Thompson (1976). For example, Purtscheller et al. (1972) suggest the breakdown of Pg by:

$$Pg + quartz = Al\text{-}silicate + albite + H_2O \ . \tag{17}$$

Hoffer (1978) using quite extensive mineral assemblage data combined with probe data suggested the reaction:

$$Pg(Ms_7Marg_{4.5}) + Mu(Pg_{22}Marg_{0.2}) + oligoclose(Ab_{73}An_{25.4}Or_{0.6}) + quartz =$$

$$Mu(Pg_{11}Marg_{0.1}) + oligoclose(Ab_{84}An_{15}Or_{1.0}) + andalusite + H_2O \ . \tag{18}$$

Reaction 18 is identical with 13.

Data bearing on the phase relations of Pg in the T range between its lower and upper stability limits are not very extensive, but Hoffer (1978) suggested the following reaction for the formation of Pg at the boundary between his low and medium grade of metamorphism.

$$Chlorite + phengitic \ Mu(Pg_{14}) =$$

$$staurolite + biotite + Pg(Mu_7) + Mu(Pg_{22}) + quartz + H_2O \ . \tag{19}$$

Most intermediate T reactions involving Pg also involve Fe-Mg phases and are considered below in the context of discussing bulk composition controls on the occurrence of Pg.

(c) *Experimental and theoretical approaches to Pg stability conditions*. Chatterjee (1973) has discussed experimental studies bearing on the low T stability of Pg. For example, the study of Hemley et al. (1961) suggests the stabilization of Pg + quartz relative to albite + Na-montmorillorite at T > 330°C at 1 kbar. From rate studies, Chatterjee (1973) claims that at 2, 4, and 7 kbar this transformation takes place at T's between 335 to 315°C (from low to high P). At higher P, Chatterjee suggests that the Na-montmorillonite is unstable and then Pg forms by reaction (11). Because the reaction involving Na-montmorillonite has a (-) PT slope, it is evident that the metastable extension of the Reaction 11 actually lies at lower T than Reaction 15. Hence, the T at which Pg + quartz first become stable depends largely upon whether it comes in via Reaction 11 or 15, a matter still open to question.

The question of the upper stability limit of Pg is more tractable, because the reaction by which Pg is terminated is fairly well recognized (see above). Aside from the actual P and T of this limit, the most important question is whether this stability curve passes through part of the sillimanite PT field. The observations of Warner and Al-Mishwt (1968), Grambling (1984), and Holdaway (pers. comm.) plus the arguments of Guidotti (1968) and Guidotti and Sassi (1976a) suggest that Pg + quartz stability does extend into the sillimanite field. In contrast, the observations of Hoffer (1978) suggest that it lies below the Al-silicate triple point. It should be noted that the occurrence described by Hoffer (1978) seems to be the only definite occurrence of Pg with andalusite as an equilibrium assemblage. Moreover, it may be the only occurrence recorded of Pg + cordierite. The Pg + andalusite occurrence described by Harder (1956) involves partial pseudomorphs of andalusite, and so attainment

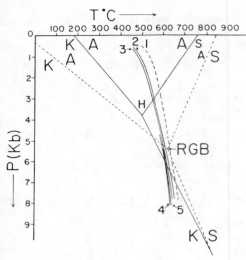

Figure 24. Experimental and theoretically calculated stability curves for Pg (dashed lines) and Pg + Quartz (full lines). Al-silicate curves as on Figure 23. (1) Chatterjee (1970). (2) Chatterjee & Froese (1975), curve calculated for Reaction 10 of text. (3) Chatterjee (1972). (4) A. B. Thompson (1974) curve calculated for Pg + quartz. (5) A. B. Thompson (1974) curve calculated for Reaction 10 of text. At P < 5 kbar, curves (4) and (5) merge with (2) and (3) and thus are not shown.

of equilibrium cannot be assumed. Reasons were suggested by Guidotti (1968) to explain the general rarity of the assemblage Pg + quartz + andalusite, and the same reasons would apply to the assemblage Pg + quartz + sillimanite (without biotite) described recently by Grambling (1984, discussed below). Unfortunately the various experimental and theoretical calculations for the upper stability of Pg do not provide confirmation of the question of overlap by the Pg and sillimanite stability fields, because the question also involves making a choice of whose Al-silicate triple point to use. Morever, the influence of $a(H_2O) < 1$ during metamorphism could shift the Pg stability field such that it no longer overlaps with that of sillimanite. However, based on the natural parageneses noted above, it would appear that this question is now settled.

Estimates of the PT conditions at which Reaction 12 and 13 occur are available from either/or both experimental work and theoretical calculations. Reaction 12 has been done experimentally by Ivanov and Guaynin (1970) and by Chatterjee (1972), the latter doing an extensive job in terms of the PT range covered. The PT curves for Reaction 12 and 13 have been calculated theoretically by A.B. Thompson (1974) and 13 by Chatterjee and Froese (1975). These curves and the experimental curve of Chatterjee (1972) for Reaction 12 are shown on Figure 24. Also plotted are the Al-silicate triple points of Holdaway (1971) and Richardson et al. (1969). The position of the various Pg curves relative to the location of the two triple points should be noted in the context of the previous discussion.

(d) *Miscellaneous aspects of Pg phase relations.*
(i) Pg compatibility with calcite. Numerous workers have reported this compatibility, but few have discussed its PT extent or controls. One exception is the study by Frey (1978), who observed that Pg + calcite occurs at extremely low grades, becoming less common with increase in grade and extremely rare by the time staurolite was stable. At still higher grade the Pg + calcite pair is replaced by assemblages involving phases like plagioclase, Marg, clinozoisite, etc. Several reactions were suggested to explain the progressive decrease in the occurrence of the Pg + calcite pair. Chatterjee (1972) pointed out that at low grades of metamorphism Pg + quartz is restricted to highly aluminous bulk compositions. But if calcite is present, Pg can occur in rocks

not especially Al-rich. The latter situation has been recorded in glaucophane-lawsonite schist facies, zeolite facies, and greenschist facies. Chatterjee suggests that at high $\mu(CO_2)$ the assemblage calcite + Pg + quartz is stabilized relative to albite + lawsonite or albite + zoisite. However, as T increases into the upper greenschist facies, this three-phase assemblage becomes incompatible and is converted to albite + zoisite. Thus, Chatterjee believes the calcite + Pg + quartz assemblage is restricted to low T and high $\mu(CO_2)$. At higher T, Pg and quartz remain stable but only in more Al-rich bulk compositions. Hence, at higher T, metamarls would no longer contain Pg + calcite -- a suggestion in general agreement with the observation of Frey (1978).

(ii) Well-defined bulk compositional controls on Pg occurrence. At least since Harder (1956), petrologists have been aware of the influence of bulk Al_2O_3 on the occurrence of Pg. A few have also suggested that bulk Na_2O can influence the occurrence of Pg. Guidotti (1968) summarized much of this and related it to assemblages in which Pg can occur. Particular emphasis was given to the incompatibility of Pg and biotite in greenschist facies metamorphism. It was pointed out that to have Pg present required bulk compositions sufficiently Al-rich that on an AFM diagram they plotted above a join connecting garnet and chlorite. Only after grade increased enough to break joins like garnet-chlorite and form AFM joins like staurolite - biotite or Al-silicate - biotite could one expect to find Pg compatible with biotite. With very few exceptions, these suggestions do seem to hold true for metapelites and metamarls.

However, a much more fundamental understanding of the interrelationship between Pg occurrence and bulk composition has been developed by J.B. Thompson; this arises from the fact that he relates observations to basic physicochemical theory rather than remaining at a purely empirical level. It will be seen that the control on Pg occurrence is basically a phase-rule control. This approach was presented indirectly by Thompson and Norton (1968) and more directly by Thompson (1972), who introduced the idea of the presence of an Na-saturating phase (Pg or albite) with respect to the mineral assemblages portrayed on an AFM mineral facies diagram. Unfortunately the paper by Thompson (1972) was not published in one of the "standard journals" and was primarily concerned with applying the "saturation phase" concept to the oxide and Fe-sulfide phases that can occur with AFM mineral assemblages. Hence, its significance for the controls on the occurrence of Pg has not generally been appreciated. Guidotti and Sassi (1976b) showed in a simple graphical way the most general aspects of this approach for understanding the controls on the occurrence of Pg.

Consideration of the Na-saturating phases for an AFM mineral facies diagram can be accomplished by means of simple diagrams. For the sake of simplicity the effects of CaO will be ignored. Any convenient AFM mineral facies diagram can be used. Ideally, it will consist of a number of one-, two-, and three-phase fields of AFM minerals. As shown on Figure 25, another composition axis leading to $NaAlO_2$ can be attached to the Al_2O_3 corner of the original AFM triangle, resulting in a tetrahedron. Pg and albite plot along the Al_2O_3 - $NaAlO_2$ line and for a given P, T, and $a(H_2O)$ one must consider the orientation of the tie lines between the two sodic phases and the various AFM phase assemblages. The tie line orientation will control which AFM assemblages have Pg and which have Ab.

In terms of the Phase Rule, it is evident that at arbitrary P, T, $a(H_2O)$ conditions, the AFM three-phase assemblages can have only one sodic phase added to them (i.e., to maintain the condition of P = C). Only the AFM two-phase assemblages can have two sodic phases present. This, in conjunction with the fact that tie lines cannot cross, suggests that the AFM two-phase field(s) containing two sodic phases will form a boundary separating portions of the AFM

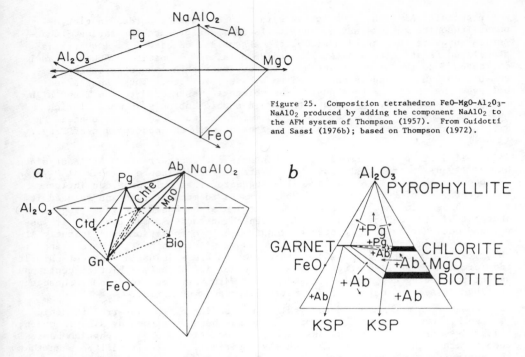

Figure 25. Composition tetrahedron FeO-MgO-Al$_2$O$_3$-NaAlO$_2$ produced by adding the component NaAlO$_2$ to the AFM system of Thompson (1957). From Guidotti and Sassi (1976b); based on Thompson (1972).

Figure 26 (above). (a) Schematic representation of the lines between the sodic phases on Figure 25 and some of the AFM phases at garnet-grade metamorphism. The existence of the four-phase volume Pg + albite + chlorite + garnet precludes the coexistence of biotite with Pg. From Guidotti and Sassi (1976b). (b) AFM garnet-grade mineral facies diagram with the Na-saturation phases labeled so as to be consistent with (a). Dashed line in the garnet + chlorite two-phase field forms the Pg - Albite boundary. Quartz in all assemblages and uniform P, T, a(H$_2$O) assumed.

Figure 27 (below, left). A possible tie line configuration between sodic phases and AFM phases at staurolite grade metamorphism. Biotite and Pg would still be incompatible. From Guidotti and Sassi (1976b).

Figure 28 (below). A possible tie line configuration between sodic phases and AFM phases at kyanite grade metamorphism. This configuration would allow Pg and biotite to be compitable. From Guidotti and Sassi (1976b).

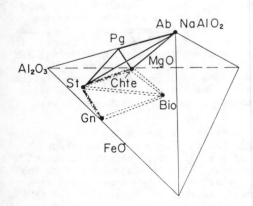
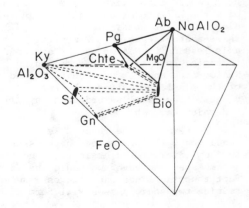

topology which contain Pg from those which contain albite. Figures 26a and 26b show the manner in which this boundary arises for garnet-grade pelites and provide an explanation for the incompatibility at low grades of Pg and biotite. Biotite occurs in less Al-rich bulk compositions and the sodic phase with which it coexists is albite. On Figure 26b the AFM diagram has the Pg versus albite bearing fields labeled, i.e., showing which is the Na-saturating phase. Albee (1965a) showed a similar labeling of AFM phase fields.

As grade arises, reactions of the AFM assemblages will in some cases involve the two-phase field through which this boundary runs. The result will be systematic changes in the AFM two-phase fields through which this boundary runs (see Fig. 27). Thompson and Norton (1968) and Thompson et al. (1977) have considered the nature of some of these reactions which involve AFM phases and the two sodic phases. In fact, most of the discontinuous reactions involving Pg between its lower and upper stability limits will be of this type. In general these reactions are such that the AFM two-phase fields with the Pg-albite boundary progressively re-orient so that they connect relatively aluminous phases with relatively low alumina minerals. Ultimately the AFM two-phase field(s) containing the Pg-albite boundary are oriented such that they permit the coexistence of biotite + Pg + Al-silicate. Figure 28 is a schematic possibility for the latter, but it must be emphasized that it remains to be determined as a function of grade the actual sequence(s) for the orientation of the AFM two-phase fields containing the Pg-albite boundary. Unfortunately, there has been very little recognition of this aspect of the interrelationship between the AFM mineral assemblages and the sodic phases present (but see Thompson et al., 1977).

In summary, it should be emphasized that this approach to understanding the compositional controls on the occurrence of Pg involves a Phase Rule constraint. Hence, when violations (e.g., biotite + Pg at low grades) occur, serious consideration should be given with regard to whether equilibrium prevailed. Alternatively, consideration should be directed at the possibility of other components stabilizing a given phase outside of its expected stability field. Finally, it should be noted that this principle of saturating phases can also be used to understand the compatibility relations between Pg (and other AKNa phases) with various calcic phases such as calcite, Marg, zoisite, etc. Although Thompson and Thompson (1976) discuss this only briefly, it is apparent that the phase relations between Pg and the various calcic minerals will be better understood when treated in this manner.

Guidotti (1968) noted that the assemblage andalusite + biotite + Pg is extremely rare (see previous discussion), whereas kyanite + biotite + Pg is fairly common. It was suggested that these resulted from the position of the upper stability curve of Pg relative to the PT curve for the AFM tie line flip that brings in the Al-silicate + biotite pair (see Fig. 29). In the context of the above discussion, the Pg + Al-silicate + biotite compatibility would involve some changes in the AFM two-phase fields containing the Pg - albite boundary, such that the newly formed Al-silicate and biotite assemblages will have Pg as the Na-saturating phase. Figure 29 suggests that at low P, the biotite - Al-silicate compatibility usually occurs above the upper stability limit of Pg and so precludes the coexistence of Pg + andalusite + biotite. Figure 28 shows hypothetical locations for the Pg-albite boundary which would allow the Pg + kyanite + biotite assemblage. Based upon observations Thompson et al. (1977, Fig. 2) show a more realistic location of the Pg-albite boundary which still permits the assemblage kyanite + Pg + biotite.

(iii) Pg in "Blue Schist Style" rocks. It has been noted that in "Blue Schist Style" rocks Pg seems to be more common and to occur in a wider range of bulk compositions than in "Green Schist Style" rocks. This appears to be a direct

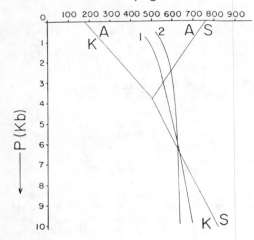

Figure 29. Suggested schematic relationship between the upper stability limit of Pg (curve 1) from Chatterjee and Froese (1975) and the AFM discontinuous reaction (Curve 2) which first produces the compatibility of Al-silicate + biotite (Reaction 17 of text). See Figure 28 for the AFM-NaAlO$_2$ tie line configuration at P greater than the intersection of Curves 1 and 2 (after Guidotti, 1968). Al-silicate curves from Holdaway (1971).

response to the manner in which tie lines are oriented or what minerals serve to express a given bulk composition in response to high P. This is best shown by the recent recognition of Pg occurring in metabasites of blueschist and eclogite facies rocks. Holland (1979b) noted that such parageneses have been described in recent years by Bearth (1973), Abraham et al. (1974), Miller (1974, 1977), Holland (1979a), and Raith et al. (1977). By way of contrast, Pg would not occur in the metabasites of a "Green Schist Style" terrane. Indeed, it is not even especially common in metapelites as higher than average bulk Al$_2$O$_3$ is required to form Pg. In eclogites of "Blue Schist Style" terranes the mineralogy of metabasites is such that the components Na$_2$O and Al$_2$O$_3$ are available to form the minerals kyanite, Pg, and omphacite ($\sim$Jd$_{50}$). Of particular interest is the fact that because K$_2$O can be relatively low in metabasite composition, the phase assemblages of the AKNa system lie over near the NaAlO$_2$-Al$_2$O$_3$ side of the system, thereby enabling observation of AKNa bulk compositions not seen in "Green Schist Style" rocks.

Holland (1979b) has experimentally determined that the breakdown of endmember Pg to jadeite + kyanite has a low negative slope ($\sim$10 bars/°C) passing through 500°C, 24-26 kbar and 700°C, 23-24.6 kbar, and intersecting the extension of the Pg-breakdown curve of Chatterjee (1970) (see Fig. 24) at about 22 kbar and 780°C. The low slope enables this reaction to serve as a good geobarometer. Moreover, the presence of omphacite ($\sim$Jd$_{50}$) rather than pure jadeite lowers the P of the reaction by only about 5 kbar. Thus, it would appear that the Pg + kyanite + omphacite and kyanite + omphacite parageneses recently recorded in eclogites have formed at very high P. Considered as lithostatic P, the depths at which such metamorphism occurred would be on the order of 60 to 70 km!

It is worth calling attention to several other effects which might contribute to an enlargement of the composition range in which Pg can occur in "Blue Schist Style" rocks. These include: (a) The high celadonite content in Mu of such rocks appears to favor an Na-poor Mu. This would widen the Mu-Pg solvus and thus increase the composition range that would contain both Mu and Pg. And (b) if the relationship between cell volume and Na/(Na+K) ratio of Pg does involve a (−) ΔV as discussed earlier, then the higher P of "Blue Schist

Style" rocks would probably increase the amount of K (for a given T) that can enter Pg. This would have the effect of enlarging the Pg composition field and so enlarging the range of bulk composition that contains Pg. At the same time, the effect of increased P on the Mu limb of the Mu-Pg solvus would move it to a more K-rich position, thereby further contributing to an enlargement of the bulk composition range that can contain Pg.

Points (a) and (b) are obviously speculative but should be investigated in future studies of Pg-bearing parageneses.

Muscovite phase relations. Mu (including Ph) is much more abundant and widespread in its occurrence than the other white micas and much more is known about its phase relations. Cipriani et al. (1971), Korikovski (1973), Seki (1973), and Guidotti and Sassi (1976a) have summarized its occurrence and phase relations in pelitic schists, and because these are available, our treatment here will emphasize general points pertinent to pelitic schists.

(a) *Occurrence*. Mu can occur in virtually all pelitic schists, regardless of whether they are Al-rich varieties or relatively low Al such that they contain K-feldspar at low to medium grades; it seems to be present even at the lowest grades of metamorphism, at which it appears to form by recrystallization of illites and clays. Subsequently, the vast majority of mineralogic reactions occurring in pelitic schists (including those involving Fe-Mg phases) have Mu as a participating phase. Mu generally decreases modally as grade increases, but it usually persists as a stable phase into the upper amphibolite facies. At this grade there is fairly general agreement among theoretical and experimental work and studies of natural parageneses that two reactions occur which result in the complete disappearance of Mu (in quartz-bearing rocks). (See Guidotti and Sassi (1976a) for some of the many references on these reactions.) Ignoring for now the CaO that will be present in plagioclase of natural parageneses, the first reaction is:

Na-bearing Mu + plagioclase + quartz = Al-silicate + Na-bearing KsP + H_2O. (20)

As seen on Figure 21b, this reaction severely limits the compositional range available to Mu. Continued increase of T causes a progressive change in Mu composition toward the Na-free end member. Upon reaching the T at which Mu is free of Na, the final, terminal reaction occurs for Mu.

$$\text{Mu + quartz = Al-silicate + K-feldspar} + H_2O \ . \qquad (21)$$

Numerous experimental and theoretical studies have been done on this reaction, as summarized in Figure 30 which also shows that the theoretically calculated PT curves for Reaction 20 lie only a small amount below the curve for 21. The PT curves for Reaction 21 lie at fairly high T, thus if $P(H_2O)$ is high, these curves may encounter some of the first melt curves of the granitic system. In fact, rocks formed at conditions where Reactions 20 and 21 have occurred often are migmatites. Extended treatment of the phase relations of Mu under conditions where a melt phase occurs can be found in Kerrick (1972), Tracy (1978), A.B. Thompson and Algor (1977), and Tyler and Ashworth (1982), etc.

(b) *Compositional variation of Mu in response to metamorphic conditions*. Previous discussion has covered Mu composition as affected by bulk composition in general and by specific mineral assemblage when dealing with a relatively restricted bulk composition range, e.g., pelites. Moreover, theoretical considerations permit the discrimination of Mu compositional variations due to bulk composition effects from those due to metamorphic effects. When the former are factored out, it is found that the major variations are Cd content and Na/(Na+K) ratio, and the minor variations include Ti, Ba, Mg/Fe ratio, and the number of vacancies, among others.

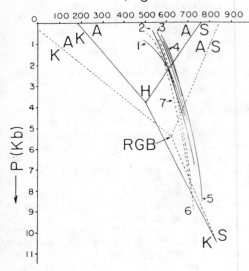

Figure 30. Selected post-1964 experimental and theoretically calculated stability curves for Mu + quartz and for Reaction 17 of the text (dotted). Al-silicate curves as on Figure 23. (1) Evans (1965). (2) Chatterjee and Johannes (1974). (3) Althaus et al. (1970). (4) Day (1973). (5) Thompson (1974) calculated. (6) Thompson (1974) calculated for Reaction 17 of text. (7) Chatterjee and Froese (1975) calculated for Reaction 17 of text.

(c) *Variation of celadonite content*. As summarized in Guidotti and Sassi (1976a), numerous studies have shown that the Cd content of Mu increases markedly as the P of formation increases. This is clearly evident from the celadonitic Mu routinely found in "Blue Schist Style" rocks, regardless of whether or not the Mu coexists with an Al-saturating phase. Coleman (1967, Fig. 7) showed the clear contrast in Cd content of Mu from "Blue Schist Style" areas versus "Green Schist Style" areas. Moreover, the experimental work of Velde (1965a) also suggests that increase of P is probably responsible for the formation of celadonitic Mu[3]. In contrast, numerous workers have noted that increase in the T of formation of Mu generally results in a progressively less celadonitic Mu (cf. review by Guidotti and Sassi (1976a)).

Ernst (1963) was probably the first to discuss in a detailed manner the effect of P at producing celadonitic Mu. He suggested that the P-sensitive, controlling reaction was:

$$8 \text{ phengite} + \text{chlorite} = 5 \text{ Mu} + 3 \text{ biotite} + 7 \text{ quartz} + 4 \text{ H}_2\text{O} . \quad (22)$$

Ernst discussed the ΔV aspect of this equation and suggested that if $P(\text{fluid}) = P(H_2O) = P(\text{total})$, the left hand side of Equation 22 would be favored by increase of P. Other reactions have been discussed for effecting changes in the celadonite content of Mu in response to changes in P, T, and $P(H_2O)$. For rocks at low grades these would include:

$$\text{Ph} = \text{Mu} + \text{biotite} + \text{K-feldspar} + \text{quartz} + H_2O \quad (23)$$

(Velde, 1965a, 1967; Plas, 1959).

$$\text{Ph} + \text{Al-rich chlorite} + \text{quartz} = \text{Mu} + \text{chlorite} + \text{K-feldspar} + H_2O \quad (24)$$

(Velde, 1965a).

---

[3] As noted in Schreyer (1981), a refined experimental calibration of the Cd content of Mu as a function of $P(H_2O)$ (to 30 kbar) has been carried out by Massone (1981) and soon will be published.

$$Ph + chlorite + K\text{-}feldspar = biotite + Mu + quartz + H_2O \qquad (25)$$

(Mather, 1970).

$$Ph + chlorite = Mu + chlorite + biotite + H_2O \qquad (26)$$

(Mather, 1970).

Many papers considering Reactions 22 to 26 show diagrams with a fairly extensive two-phase field in which celadonitic Mu and chlorite exhibit variations in $\Sigma(Mg+Fe^{2+})$ and Al contents. Rotation of the tie lines implies the exchange reaction

$$celadonitic\ Mu + Al\text{-}chlorite = less\ celadonitic\ Mu + Al\text{-}poorer\ chlorite. \qquad (27)$$

A detailed discussion of this reaction has recently been given by Miyashiro and Shido (in prep.). At medium grades the following reactions have been discussed in terms of their effect on the Cd content of Mu.

$$chlorite + phengitic\ Mu(Pg_{14}) =$$
$$staurolite + biotite + Pg(Mu_7) + Mu(Pg_{22}) + quartz + H_2O \qquad (28)$$

(Hoffer, 1978).

$$Fe\ chlorite + Mu = garnet + biotite + Mg\text{-}chlorite + Na\text{-}richer\ Mu + H_2O \qquad (29)$$

(Guidotti and Sassi, 1976a). In the case of Reactions 29 and 30, the Mu on the right hand side contains less celadonite than that on the left.

A particularly important consideration in Reactions 22-27 is whether ΔV_R is such that the side containing the Cd-richer Mu has a lower total volume and thus is consistent with the common observation that the Cd content of Mu increases with P of formation (especially in the lower metamorphic grades). The effect of Cd substitution into Mu considered as an isolated phase (see earlier discussion of cell-volume relations) clearly shows that P increase on Mu alone would favor a decrease in its Cd content. Hence, the observed increase of the Cd content of Mu in response to P increase must involve reaction with other phases such that ΔV_R is (-). In only a few cases has this question been addressed with regard to the above reactions.

Reaction 27 is especially interesting in this context. As noted by Guidotti (1969), early work by Fawcett and Yoder (1966) suggested that increase of P causes an increase of the Al of pure Mg-chlorite. Consistent with this observation is the work of Jenkins and Chernosky (in prep.) on synthetic Mg-chlorites, showing that the cell volume decreases as the Al content increases. The few data available (Hock, 1974b) for natural chlorites with intermediate Mg/Fe ratio also show a decrease in cell volume as Al content rises. In addition, Kotov (1974) calculated ΔV for Reaction 27 and did find a (-) ΔV for the reaction going to the left. However, his volume data were not of high quality. These observations are at least suggestive that progress to the left by Reaction 27 could be accompanied by a $(-\Delta V)$ and so favored by increase of P. In essence the tie lines connecting Mu and chlorite might rotate in response to increase in P. It would seem that some such effect must occur to be consistent with the high celadonite content so routinely observed in Mu in rocks that have the assemblage Mu + chlorite and for which other information implies high P during recrystallization. It is suggested that this effect is worth serious investigation inasmuch as Reaction 27 might serve as a useful geobarometer.

Reactions 22-30 involve a Tschermak exchange between Mu and the coexisting AFM phases. Thompson (1979, in Russian!) has developed a very compact and complete system for treating such exchanges between Mu and AFM phases, but Cheney

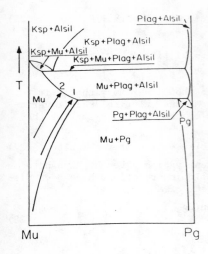

Figure 31 (left). Schematic pseudobinary section along the Mu-Pg tie line of the AKNa system. From Guidotti and Sassi (1976).

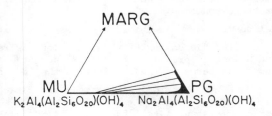

Figure 32 (above). Schematic illustration of how Ca entering into Pg results in a widening of the Mu-Pg miscibility gap.

(1980) uses this approach directly.

(d) *Variation of Na/(Na+K) ratio*. Figure 31 illustrates the most important phase relations bearing on variation in Na/(Na+K) ratio. In the T range below the final stability of Pg, one of the most important controls on Na/(Na +K) in Mu is the closing of the Mu-Pg solvus. For specimens containing both Mu and Pg natural and experimental data show that Pg does not change much as a function of T. However, Mu can undergo a marked change in Na/(Na+K) ratio in response to increasing T. The actual extent of this variation is unknown: neither the experimental data nor data on natural coexisting pairs have shown the *maximum* amount of the Pg end member which can dissolve in Mu. This is shown by the observation of Mu (not coexisting with Pg) that has a higher Na/(Na+K) ratio than any Mu coexisting with Pg (see early discussion of a Mu with Na/(Na+K) > 38 atom %. Nonetheless, the existence of a solvus between Mu and Pg cannot be doubted. Chemical and X-ray data accumulated since 1950 (cf. Zen and Albee, 1964) for coexisting Mu + Pg show that in general as the T of formation (based upon other data) rises, the coexisting Mu and Pg approach each other compositionally. Much of this data is summarized in Figure 5a (compare the greenschist and > greenschist columns), in which it appears that when Mu and Pg coexist and as T increases, the Na/(Na+K) ratio of Pg can vary from 95.43 (Hock, 1974a) to at least 78.06 (Rumble, 1978), and similarly for Mu from 3.16 (Goffe, 1977) to at least 36.37 (Guidotti, unpub. data). However, as mentioned above, the maximum attainable Na/(Na+K) ratio for Mu is probably somewhat greater than that indicated by available data from coexisting Mu + Pg pairs.

In discussing the extent of closing of the Mu-Pg solvus, our attention so far has been directed at the effect of T. At least three other factors can influence the degree of closure of this solvus:

(a) Pressure: Because the Mu side of the Mu-Pg solvus has a (+) ΔV_{mix}, the effect of high P should shift the Mu limb to a less sodic position. In contrast, because some question exists as to whether the Pg side of the solvus has a (+) or (-) ΔV_{mix}, the effect of P on the Pg limb is ambiguous.

(b) Ca replacing Na in Pg coexisting with Mu. This long recognized effect (e.g., Zen and Albee, 1964) can be illustrated graphically, and Figure 32 shows that Mu coexisting with progressively more calcic Pg has a progressively lower Na/(Na+K) ratio. From Figure 5a it is evident that, although this effect is real, it never is very large because Pg found to date does not contain large amounts of Ca replacing Na. With few exceptions, less than 6% of Na is replaced by Ca. Indeed, it has been noted earlier that the Marg-Pg solvus is strongly asymmetric toward Pg.

(c) The effect of Cd content on Na/(Na+K) in Mu. Using Figures 5a and 5b, it was shown that as the Cd content of Mu increases there is a crystallochemical effect such that Mu cannot contain as much Na in place of K. From Figure 5b, in particular, it is evident that this effect can be significant. If the proposed effect of P on the Mu limb of the solvus is correct, it would work in tandem with the effect of Cd on the Na/(Na+K) ratio of Mu. Thus, P could influence the position of the Mu limb directly because of the (+) ΔV and indirectly because high P favors a Cd-rich Mu. (The implications of (a)-(c) for Mu-Pg geothermometry will be discussed further below.)

The arrows shown in the Mu solution field of Figure 31 reflect another prominent change observed in the Na/(Na+K) ratio of Mu at T's below the final breakdown of Pg. The details of the T-driven reactions affecting this variation of Na/(Na+K) are discussed in Guidotti and Sassi (1976a). In brief, many of the continuous reactions among the limiting assemblages of Fe-Mg minerals of metapelites (e.g., Reactions 29 and 30) involve Mu, and the net effect is that, as shown by the arrows, Mu becomes progressively enriched in Na. A given arrow would apply to groups of rocks with a relatively narrow range of bulk composition, e.g., as might occur within a particular field area. This enrichment of Mu with Na at T's below the incoming of any Al-silicate has been recorded many times by Lambert (1959), Guidotti and Sassi (1976a), Bocchio et al. (1980), Cipriani et al. (1968, 1971), Henry (1974), Ruiz et al. (1980), among others.

In rocks which contain an Al-silicate + plagioclase + Mu the following continuous reaction takes place in the AKNa system:

$$\text{Na-Mu} + \text{quartz} = \text{Al-silicate} + \text{albite} + \text{K-richer Mu} + H_2O \ . \tag{31}$$

As implied from the shape of the Al-silicate + plagioclase + Mu field on Figure 31, this reaction can produce a significant change in the Na/(Na+K) ratio of Mu. This has been described in many parts of the world by Bocchio et al. (1980), Ruiz et al. (1980), Guidotti and Sassi (1976a), Cipriani et al. (1968, 1971), Cheney (1975), Henry (1974), Fletcher and Greenwood (1979), etc., suggesting that in response to Reaction 31 the Na/(Na+K) ratio of Mu can change in response to T from ~ 30 to ~ 6 atom %. Although variation of intensive parameters are the prime control on Reaction 31, it has been shown that the An content of the coexisting plagioclase also affects the Na/(Na+K) ratio of Mu. It is considered in a quantitative fashion in the section on geothermometry presented later.

At still higher grades where the Al-silicate + K-feldspar compatibility has been formed, the following continuous reactions (starting with Mu at ~ 6 atom % Na/(Na+K)) continue to produce a less sodic Mu. In the limit they produce a pure K-Mu and thereupon the terminal reaction of Mu occurs. In a Ca-free system the continuous reaction is:

$$\text{Na-Mu} + \text{quartz} = \text{Al-silicate} + \text{Na-Ksp} + H_2O \ . \tag{32}$$

In a Ca-bearing system the four-phase assemblage can occur and have the following continuous reaction take place:

$$\text{Na-Mu} + \text{plagioclase} + \text{quartz} =$$
$$\text{Al-silicate} + \text{Na-bearing orthoclase} + \text{Ca-richer plagioclase} + H_2O \ . \tag{33}$$

Because of bulk composition controls in natural parageneses, it appears that neither Reaction 32 or 33 ever proceeds to the extent that end-member Mu is produced. Instead, in Reaction 33, Mu, which still contains a few atom % Na/(Na+K), becomes used up, and the bulk composition is expressed in the AKNaCa system by the assemblage Al-silicate + Na-bearing K-feldspar + plagioclase. Evans and Guidotti (1966) have discussed this reaction and its relation with bulk composition in considerable detail.

(e) *Summary of (c) and (d)*. Summarizing the two main compositional changes that can affect Mu in response to metamorphic conditions, the following should be noted. (1) The ranges of Na/(Na+K) ratio and Cd content are extensive (and easily monitored) for Mu and thus contrast markedly with the ranges permissible in Pg. (2) For both variations, straightforward considerations discussed briefly here and in detail by Guidotti and Sassi (1976a) enable separation of effects due to bulk composition and those due to metamorphic conditions. (3) Based upon a large mass of natural observations it appears that P and secondarily T control the Cd content of Mu. (4) In contrast, the Na/(Na+K) ratio of Mu seems to be influenced most by T (possibly a(H_2O) also as discussed later) and less so by P. All these will be important in our subsequent discussion of how Mu can be used to gain information about the paragenesis of rocks.

(f) *Other compositional variations of Mu in response to metamorphism*. Our early discussion of Mu compositional variation included several things besides Cd content and Na/(Na+K) ratio. Here, our concern is with those that can be related to metamorphic grade. These include:

(1) Ti-content. Increase of Ti in Mu with increase of metamorphic grade has been described in numerous reports, e.g., Kwak (1968), Guidotti (1970a, 1973, 1978), Guidotti et al. (1977), Ruiz et al. (1980), Cipriani et al. (1968, 1971). Because some of these studies did not specify whether or not a Ti-saturating phase was present, it has not always been possible to ascertain if the Ti content reported is an indication of the saturation limit for the existing PT conditions. Also the possible effects of Na/(Na+K), Cd content, etc. on the Ti content in Mu are not yet clear. However, it appears in general that over the range from greenschist facies to upper amphibolite facies where Mu finally breaks down, the Ti content increases fairly steadily from about 0.02 atoms per 22 oxygens to about 0.12.

(2) The ratio of Mg/Fe in Mu has been observed to change by Guidotti (1978) and Cheney (1975) in a systematic way in response to the manner in which the Mg/Fe ratio of coexisting biotite varies in response to metamorphic grade. However, this variation can be detected only in the context of specific mineral assemblages which are capable of undergoing the required continuous reactions. Indeed, this would be an extreme example of mineral assemblage control combined with metamorphic control on Mu composition.

Figure 33 summarizes some of the variations of Mu chemistry in response to metamorphic change. As emphasized by Guidotti (1978), care was taken to eliminate bulk composition effects on the Mu samples used for Figure 33, and most of the variations shown therein have been discussed above. Obviously some of them, e.g., ΣAl and Si^{iv}, are included in the Cd content of Mu. The details of how the variations shown on Figure 33 are interrelated are given in Guidotti (1978; cf. Tracy, 1978).

<u>Polymorphs of Mu -- metamorphic aspects</u>. As noted before, different polymorphs occur among the white micas in metapelites. The greatest variation in polytypes occurs among Mu and Ph, and it is regarding them that we now discuss controls on

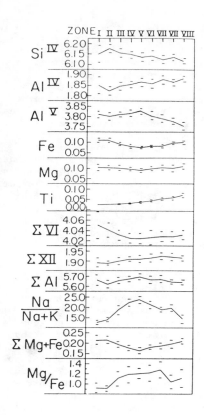

Figure 33. Illustration of Mu compositional variation in high Al specimens as a function of metamorphism. Zone 1 = lower garnet zone, Zone VIII = upper sillimanite zone. From Guidotti (1978).

the occurrence of a given polytype. The question is of interest because understanding the controls might provide information about petrogenetic conditions when a particular polytype is present. The question of what controls the occurrence of a given polymorph has been approached from both an epxerimental and observations point of view. Frey et al. (1983) reviewed the distribution of polymorphs of Mu and Ph in the central Alps, and they provide a valuable, general review of the question of Mu polymorphs in pelitic schists on which the following is based.

From observational studies on natural parageneses (e.g., Maxwell and Hower, 1967) and experimental studies (e.g., Eugster and Yoder, 1955; and Velde, 1965b), it is apparent that the $2M_1$ polymorph is the most common K-white mica in pelitic schists. The $1M$ and $1Md$ polymorphs are restricted to rocks formed at sedimentary conditions. The $3T$ polytype is rarely reported, and the controls on its occurrence in preference to $2M_1$ have been of particular interest. Frey et al. showed that the $3T$ variety has been observed throughout the world in rocks seemingly formed over a wide range of P and T. They review (p. 185) the controls that have been suggested for the occurrence of this polymorph, including: (1) compositional control, (2) an unknown physical factor furnished by a hydrothermal or metamorphic environment, (3) high-P, low-T metamorphic conditions, and (4) significant partial pressure of CO_2 in the fluid phase. They note that although most $3T$ K-white micas are quite phengitic, there are some data in the literature in which $3T$ polymorphs are not phengitic. Of particular significance is their observation that the distribution of the $3T$ polymorphs suggests that these samples were formed by the Eo-Alpine high pressure/"low temperature" metamorphism. Based on their data, it is evident that much of this event would qualify as "Blue Schist Style".

In contrast with the distribution of the $3T$ polymorph, rocks affected by the high grade portion of the Meso-Alpine metamorphism (Lepontine) and rocks affected only by the pre-Alpine Hercynian Metamorphism (Low Pressure type) contain the $2M_1$ polymorphs. In some of the low to middle grade Meso-Alpine metamorphic rocks the $2M_1$ and $3T$ polymorphs are both present but in such cases there is evidence that these rocks have also been affected by the earlier Eo-Alpine event, and it is suggested that the $3T$ polymorph is a remnant from that earlier event. Although Frey et al. (1983) discuss some minor problems with their conclusions, they nonetheless conclude that in the Alps the $3T$ polymorph is a result of the high-P, low-T conditions of the Eo-Alpine event. They suggest that an effort should be made to see if the $3T$ polymorph is present in other "Blue Schist Style" terranes such as the Franciscan complex.

(II.C) White micas as petrogenetic indicators

We have now discussed the mineralogic aspects and phase relations of the white micas, with the main emphasis on Mu. Particular attention was given to the effect of P, T, and a(H_2O) on the composition of the white micas. Also discussed was a theoretical framework by which the effects of P, T, a(H_2O) on the mica compositions could be distinguished from effects due to variation of bulk composition or mineral assemblage. Included as part of the considerations of the phase relations were the various theoretical and experimental studies bearing on the upper and lower stability limits, solvi, and reactions with other phases. To a considerable extent these phase relations could be summarized in terms of the pseudobinary shown in Figure 31. These phase relations provide the basis for using white micas to ascertain the P, T, P(H_2O) conditions at which rocks containing micas were re-crystallized. In the case of the experimental and theoretical calculations of the white mica stability curves, solvi, etc., the results are used to subdivide P, T, P(fluid) space in order to develop a constrained petrogenetic grid. Comparison of natural parageneses with the stability ranges shown for similar parageneses on the constrained grid enable some estimation of the conditions under which the natural white mica-bearing assemblages formed.

If the natural assemblages are quite similar to those considered in the experimental and theoretical studies, one presumes that the comparative approach has some validity. This is a fairly standard approach for estimating the conditions at which metamorphism occurred. It has been used by petrologists at least since Bowen (1940) first described the concept of a petrogenetic grid. Many grids applicable to white mica-bearing parageneses are now in the literature, so this approach is not pursued except to list the main difficulties encountered in using it. These are: (1) if more than one stability curve, solvus, etc. has been determined for a given equilibrium, a choice has to be made as to which one will be used. (2) Care must be taken that the natural assemblages are as close as possible to those used to establish the petrogenetic grid. This will include taking into account the presence of components in the natural phases that are absent in the studies used to establish the petrogenetic grid. For our purposes, the Cd content in Mu would be of particular significance in this respect. (3) Many of the lines on a typical petrogenetic grid were determined in the context of a pure gas end member, usually P(H_2O) = P (Total). To the extent that any gas present during metamorphism deviates from such an end-member composition, conditions of formation of the natural parageneses will differ from those inferred from the grid. Kerrick (1972) has shown quantitatively the effects of deviation from a pure H_2O gas on the Mu breakdown curve.

<u>Use of white mica composition data for petrogenetic purposes</u>. Those aspects of the phase relations of the white micas that relate to compositional variation will be our main concern for discussing petrogenetic uses of the white micas as geothermometers, geobarometers, and for gaining information on the composition of metamorphic fluids. The question can be approached both qualitatively and quantitatively (see Guidotti and Sassi, 1976a, for a review of the latter). To date, Mu has been most important in terms of gaining petrogenetic information. We will provide an overview of how the compositions of white micas can be used as geothermometers, etc.

For the above purposes several assumptions and conditions must be met (see review by Essene, 1982). These follow from the fact that the compositions of the white micas involve some sort of chemical reactions with the coexisting phases. For our purposes the only point requiring further comment is the nature of the activity model used for Mu in the various equilibria that form the basis for the different geothermometers, etc. Of particular importance is the influence of Cd content on the Na/(Na+K) ratio of Mu. In the qualitative approaches to geothermometry this effect can be taken into account on an empirical basis but for the quantitative approaches it requires an appropriate mathematical formulation.

The form for an activity model for Mu in natural parageneses has been discussed by Ghent (1975), Chatterjee and Froese (1975), Ferry (1976), Fletcher and Greenwood (1978), Cheney and Guidotti (1979), Yardley et al. (1980), and Holdaway (1980), some of whom introduce a term to reflect the presence of a Cd content in Mu. However, at best these models assume ideal mixing over the octahedral and tetrahedral sites and no interaction between substitution in the octahedral (and tetrahedral) and XII sites. Data from natural parageneses indicate that substitution in the octahedral sites does affect the Na/(Na+K) ratio of the XII sites. Hence, a more accurate activity model for Mu must be based on at least some sort of ternary solution in which all of the site substitutions are interdependent. Unfortunately, as pointed out by Saxena (1973), activity models for such solutions are much more difficult to formulate than those for binary solutions.

In summary, it would seem that the various attempts at geothermometry using the composition of Mu need to have the activity term for Mu modified to take into account the above considerations. Similarly, theoretical calculations of the Mu-Pg solvus by Eugster et al. (1972), A.B. Thompson (1974), and Chatterjee and Froese (1975) need to have such a term added, if the calculated solvi are to have direct relevance to natural parageneses. Conceivably the failure of the Mu-Pg solvus to serve as a quantitative geothermometer is due to the Cd effect on the Na/(Na+K) ratio of Mu.

Geothermometry using white micas

(a) *The Mu-Pg solvus*. As long ago as the 1950's petrologists were trying to use the solvus relationship between Mu and Pg as a geothermometer. In principle its use is fairly straightforward. From a qualitative point of view it seems to work, as shown by Rosenfeld et al. (1958) and Zen and Albee (1964), who used an indirect approach for monitoring the solvus closure as a function of T. Taking into account the effect of variation of the Na/(Na+K) ratio in both Mu and Pg, these authors used the difference between the basal spacing of Mu and that for Pg [d(002)Mu - d(002)Pg] as a measure of how much the solvus had closed and thus as a measure of T.

The Mu-Pg solvus works well as a qualitative geothermometer, especially if used on samples all collected in the same field area, probably because factors that might influence the solvus are relatively similar in the specimens of a given region. Many authors have documented the qualitative closing of the Mu-Pg solvus as a function of T, primarily by means of the compositional change of the Mu inasmuch as the coexisting Pg shows relatively little change (cf. Frank (1983) for a well documented example).

As noted by Guidotti and Sassi (1976a) and by others since, use of the Mu-Pg solvus as a quantitative geothermometer has not been very successful at all. Very likely this difficulty results from using experimental or theoretical solvi which fail to take into account important factors that influence the natural solvi, as noted by Laird and Albee (1981) and Feininger (1980). These were reviewed in our discussion of phase relations, and only one additional comment is made: If increase of P causes both limbs of the solvus to move apart (in a fashion similar to the effect of P on the Or-Ab solvus), it would imply that as a function of P the composition of Pg at its final breakdown changes (assuming no melt phase intercedes first) from being a K-bearing Pg to a pure Na-Pg. In essence, at high P, a different set of topologies would be required on Figure

11 of Thompson and Thompson (1976) to indicate the termination of Pg. In turn, this would imply that the distribution of Na/(Na+K) between Pg and feldspar changes as a function of P. Although theoretical in nature these arguments would seem to lend some support to the suggestion that V versus Na/(Na+K) ratio of Pg involves a negative volume of mixing.

(b) *The assemblage Mu + plagioclase + Al-silicate*. Numerous petrologists have recognized that the Mu in rocks with this assemblage becomes K-enriched with increase in grade and so can serve as at least a qualitative geothermometer. Influencing factors can be identified with a fair degree of certainty; they are: (a) a P effect roughly proportional to the slope of the experimental PT curve for the breakdown of pure Mu; (b) an effect due to Ca in the plagioclase (see discussion on phase relations), although as pointed out in Cheney and Guidotti (1979, p. 424), this effect is actually a relatively minor one; (c) an effect due to the Cd content of the Mu; and (d) an effect due to $a(H_2O)$, because the K-enrichment of Mu via Reaction 31 involves dehydration. Within a given area (a), (b), and (c) will probably be fairly uniform and thus not obviate this qualitative approach. In the case of (d) it appears that in some areas variation in $a(H_2O)$ can occur such that the effect of a T increase on the Na/(Na+K) ratio is neutralized (see Evans and Guidotti, 1966; Guidotti, 1970a). However, this should not be looked upon with despair because it is still providing information about a very important parameter in petrology.

In summary, the Na/(Na+K) ratio of Mu in the assemblage Mu + plagioclase + Al-silicate will usually serve as a good qualitative geothermometer within a given area. In those cases where it seems not to work, it will usually provide information about other important factors, such as $a(H_2O)$, or whether the rocks are not at equilibrium. However, because factors (a)-(d) might vary significantly in widely separated areas, it would seem that "caution" is the watchword for using this qualitative geothermometer for interregional comparisons.

Attempts have now been made to relate Reaction 31 to experimental and thermodynamic data such that it will serve as a quantitative geothermometer. With the appropriate activity terms in the equations presented by Cheney and Guidotti (1979) and Yardley et al. (1980), it is possible to remove the effects of Ca in the plagioclase and the Cd content of Mu (see previous discussion on the need for a new activity term for the Cd in Mu). This results in an equation with P, T, and $f(H_2O)$ as variables. Pressure can usually be estimated by considering the mineral assemblages occurring in the various bulk compositions present. Hence, it is necessary only to arrive at some estimate of $f(H_2O)$ and then solve for T. [See Cheney and Guidotti (1979) for a discussion of the ramifications of estimating $f(H_2O)$.] Because one usually has to assume or estimate some value for $f(H_2O)$, the T values obtained using the thermodynamic formulation of Cheney and Guidotti (1979) will be numerical but still relative in nature. This is no real problem because the general assemblage data associated with rocks containing Al-silicate + plagioclase + Mu usually allow a pretty good estimate of the actual T's involved. Hence, the relative T's calculated enable calculation of *absolute T gradients* between isograds. For example, Cheney and Guidotti (1979) used this approach to argue that a thermal gradient of 60° to 80°C/km existed during the metamorphism of the Puzzle Mountain area in western Maine.

(c) *The assemblage Mu + plagioclase + K-feldspar + Al-silicate*. Guidotti and Sassi (1976a) discussed the composition of Mu in this assemblage serving as a geothermometer and noted that in some cases buffering of $a(H_2O)$ by the four-phase assemblage seems to counteract the effect of T such that the Mu maintains a constant composition. Moreover, even in those cases in which the Mu of this assemblage does change composition in response to T, the range is quite small; e.g., for the Mu reported by Tracy (1978), Na/(Na+K) ranges only from 5.1 to 2.5 atom %.

It would appear that the qualitative approach to geothermometry using the composition of Mu in this assemblage may not be realistic. However, as pointed out by Cheney and Guidotti (1979, p. 424), the same equation employed as a basis for the Mu + plagioclase + Al-silicate geothermometer also applies to the four-phase assemblage. Hence, assuming an appropriate activity model for the Cd in Mu can be devised, it would seem that quantitative geothermometry would be more readily used for the Mu in the four-phase assemblage. It should be noted that in this case the Ca content of the plagioclase will have no effect on the composition of Mu because one is dealing with a limiting assemblage.

(d) *Other geothermometers involving the composition of Mu.* Kotov et al. (1969) have proposed a geothermometer using Na/(Na+K) of Mu coexisting with another Na-bearing phase, either albitic plagioclase or paragonite. However, our discussion of the phase relations of Mu has shown that the Mu + plagioclase and Mu + Pg assemblages will not be equivalent for purposes of geothermometry. Subsequently Talantsev (1971) proposed a similar but more elaborate geothermometer based upon the distribution of Na (molecular proportions) between coexisting Mu and plagioclase. Although avoiding the problems of the Kotov et al. geothermometer, no attempt was made to account for bulk compositional effects or deviations from ideality by Mu or plagioclase. Basically it is a geothermometer calibrated on a purely empirical basis. Kotov (1975) proposed another geothermometer based on coexisting Mu + chlorite. Inasmuch as it involves the Cd content of Mu, it is significantly influenced by P. Kotov tried to calculate that influence, but unfortunately the volume data available to him were of very poor quality.

Geobarometry using white micas. A vast literature has accumulated since Ernst (1963) dealing with the relationship between pressure and the Cd content of Mu. The experimental work of Velde (1965a) provided some idea of the quantitative aspects of this relationship, but most efforts to calibrate it for geobarometry have been qualitative in nature. In most cases the attempt has involved calibration in terms of baric type or in terms of the facies series as defined by Miyashiro (1973). In a few cases specific effort has been made to consider such calibrations in terms of the geothermal gradient prevailing in the region during metamorphism. Many of these attempts at calibration have used the b cell dimension of Mu, thereby making use of the now well recognized relationship between b and the Cd content of Mu. Hence, we have the term "b geobarometer". The earliest attempt at b calibration was that by Sassi (1972) and Sassi and Scolari (1974a), who showed that in greenschist facies pelites at the lowest P ranges believed to occur, the b value was $\sim$8.990 Å, and as P increased to that for the glaucophane schist facies the b value increased to $\sim$9.055 Å (see also Ritter, 1981). At present it would appear that b serves as a workable, qualitative geobarometer, but further calibration is required. Guidotti and Sassi (1976a) reviewed the precautions and requirements for its successful use. Particularly important is the need to account for the effects of bulk composition and mineral assemblage and to carry out the x-ray work in a fashion that will avoid interference of the 060 peak by the $\overline{3}31$ peak. This will be discussed in a subsequent section.

The b geobarometer is of prime use in greenschist to blue schist facies rocks, i.e. low T parageneses such that the Cd content has not been depleted due to T (see previous discussion of Mu phase relations). The fact that the b technique works mainly in greenschist facies rocks is of considerable value because in them none of the other phase assemblages provide much information about P. This contrasts with medium and high grade rocks in which one can readily employ the bathograd technique of Carmichael (1978) or some of the numerous quantitative geobarometers (see review by Essene, 1982). It would seem that something like the b approach is the only way to obtain geologically important information about P (and hence depth of burial, structural level, etc.) of the vast greenschist terranes occurring in some orogens, because it is impractical to carry out vast numbers of probe analyses on such fine grained rocks. Numerous studies have incorporated b measurements to gain useful insights bearing on the tectonic aspects of the area being studied, including Geyssant and Sassi (1972), Sassi and Zanferrari (1972), Sassi et al. (1974a, b,c;1976), Fettes et al. (1976), Iwasaki et al. (1978), Schimann (1978), and Briand (1980). There have also been some attempts to employ this approach for gaining information on P for sub-greenschist facies rocks (Padan et al., 1982).

Many other workers have chosen to make direct analyses of the Cd content of Mu to gain information on the relative variation of P within a given region. Recent examples are Saliot and Velde (1982) and Frey et al. (1983). The latter are noteworthy because they used the Mu + plagioclase + K-feldspar assemblage in their attempt to monitor P variation via Cd content of Mu. Hence, the Cd contents they observed were maximized for the given PT values. This work is a fine example of the care that should be given to what mineral assemblages are employed.

<u>Information on the fluid phase via white micas</u>. The discussion of Mu geothermometry related to the Na/(Na+K) ratio made evident the manner in which this ratio could be used to ascertain qualitative information about the behavior of H_2O in the fluid phase during metamorphism. The review of Guidotti and Sassi (1976a) provides details of such usage. They describe observations which have led some authors to propose buffering of $f(H_2O)$ in some regions during metamorphism. Interested readers should consult Evans and Guidotti (1966). Here, we will review some attempts to use white mica phase relations to gain quantitative insights as to the nature of the fluid phase.

Ghent (1975), employing the equilibrium,

Pg (solid sol'n in MU) + quartz =

albite (solid sol'n in plagioclase) + kyanite + H_2O , (34)

calculated the $f(H_2O)$ for the assemblage Na,K-Mu + quartz + plagioclase + kyanite for a given P and T. His approach was basically the same as that used in Cheney and Guidotti (1979) except that his activity term for a Pg in Mu ignored any terms for the Cd content of the Mu.

In another study Ghent et al. (1970) combined thermochemical data for one reaction with experimental data for another reaction to obtain an equilibrium constant expression in terms of P and T for the $f(H_2O)$ and $f(CO_2)$ of the equilibrium

Pg + calcite + quartz = albite + anorthite + CO_2 + H_2O . (35)

This expression was then used as a starting point for calculating the fluid composition of some rocks with the assemblage calcite + quartz + paragonite + graphite. In calculating the gas composition it was necessary to obtain values of P and T from other means and to assume $P_f = P_T$. Ghent et al. noted that even if their calculations are only approximate, the assemblage calcite + Pg + plagioclase + graphite enables at least an estimate of the gas composition. Similar calculations of gas compositions can be made for various graphitic, Marg-bearing assemblages as shown by Bucher et al. (1983).

The approach used in Cheney and Guidotti (1979) for geothermometry based on the assemblage Mu + plagioclase + Al-silicate can also be used to calculate the $f(H_2O)$ required for the Mu and plagioclase compositions found in a given equilibrium. As with the similar approach of Ghent (1975), estimates of P and

T must be obtained by other means. Obviously the $f(H_2O)$ calculated is then dependent upon the validity of the P and T estimates. In their study, however, they showed that such calculations can be extremely useful even if only relative values are obtained. They applied their calculation to rocks from the K-feldspar + Sillimanite zone of western Maine. On an outcrop scale, some of these rocks contain sillimanite + plagioclase + Mu and some contain sillimanite + plagioclase + orthoclase + Mu. Inasmuch as both assemblages formed at the same T and P, one can use the approach of Cheney and Guidotti (1979) to calculate the difference in $f(H_2O)$ between those rocks containing the four-phase assemblage which is typical of the K-feldspar + Sillimanite zone and the three-phase assemblage which is typical of the next lower metamorphic grade, i.e., the Upper Sillimanite zone. Their calculation showed that $f(H_2O)$ was 30% higher in the three-phase assemblage rocks. This confirmed in a quantitative way the purely qualitative suggestion made earlier by Evans and Guidotti (1966) that the three-phase assemblage persisted into the K-feldspar + Sillimanite zone because of higher H_2O in such rocks.

(III) White micas: Identification and determination of composition

Here the focus is on mineralogical methods commonly used on white micas when determining the metamorphic conditions of formation of white mica-bearing rocks. These methods include: (A) optical, (B) x-ray, (C) electron microprobe, (D) ways to distinguish Fe^{2+} from Fe^{3+}, and (E) staining. Our treatment is not exhaustive or "cookbook" in nature but merely calls attention to particularly useful aspects as well as limitations of the various methods.

(III.A) Optical methods

Chapter 6 covers the details of optical techniques for studying micas. In most cases a petrologist views minerals in thin section and thus uses optics mainly to establish the presence or absence of minerals. Mu and Pg are easily recognized (but not distinguished from each other) by their high birefringence, platy habit, cleavage, parallel extinction, and the nature of the associated minerals. Only rarely will a petrographer have to do more than simple sight identication to establish the presence of Mu ± Pg and to distinguish them from other minerals. However, in a few cases Mu ± Pg can occur in rocks also containing talc, phlogopite, or pyrophyllite. Then sight identification may be inadequate for establishing their presence or distinguishing them from the talc, etc. It may be necessary to use optics beyond sight identification (see Chapter 6) or some of the x-ray techniques discussed below.

Obviously it would be prohibitive to scrutinize in detail every Mu ± Pg-bearing rock for the presence of talc, etc. Thus, it is useful to consider what auxiliary petrographic information serves as an alert to check further for the presence of minerals like talc or for the presence of Pg as well as Mu. Such information is an outgrowth of the experience of metamorphic petrologists over many decades which shows that the combinations of minerals in a rock are a good clue as to whether minerals easily confused optically with Mu ± Pg are also present. A few examples serve to illustrate this point.

(1) A Mu-bearing schist containing biotite plus minerals such as garnet + chlorite is not likely to contain pyrophyllite (as an equilibrium phase) and is also not likely to contain Pg. On the other hand, if biotite is absent and the Mu coexists with garnet + chlorite + chloritoid one should suspect the presence of Pg and possibly pyrophyllite. If the Mu coexists with chlorite + chloritoid (no garnet), experience shows that the presence of Pg is highly probable and with almost as high probability that pyrophyllite is also present. A chemographic understanding of these observations was provided in our discussion of

the phase relations of Pg (see Figs. 26a and 26b).

(2) In a Mu-bearing schist containing Al-silicates it is unlikely that pyrophyllite is also present (as an equilibrium phase), the reason being that Al-silicates form above the upper stability limit of pyrophyllite.

(3) As a final example, consider a rock with Mu + sillimanite + abundant pyrrhotite (e.g. 10% modally) ± cordierite. Experience shows (e.g. Robinson and Tracy, 1977; Guidotti et al., 1977) that almost end-member phlogopite is likely to be present also, and it could easily be mistaken for Mu. [This phlogopite (as opposed to biotite) results from sulfide-silicate reactions.]

With experience, recognition of Marg in thin section is fairly easy. Mu, Pg, talc, etc., all have high birefringence whereas Marg has only low to moderate birefringence. Thus, if after taking into account the thickness of the petrographic slide, one notes the presence of a micaceous phase with systematically lower birefringence, the presence of Marg should be considered. About the only micaceous mineral with a similar birefringence is chlorite with compositions nearly at the Mg-end member. Final confirmation of Marg presence can be done by x-ray or probe methods.

(III.B) X-ray methods

From a petrologic viewpoint, only two aspects of x-ray methods are germane. These are (1) using them to identify and distinguish among the various layer silicates, and (2) using them to gain compositional information about a particular mica. Our focus for (2) is on Mu and to a minor extent Pg.

(1) The necessity to use x-ray to identify and distinguish various layer silicates usually results from the difficulty of distinguishing in thin section Mu, Pg, end-member phlogopite, talc, and pyrophyllite. Commonly one employs a diffractometer scan on a partial concentrate and then uses the low angle basal reflections to identify the presence of a given layer silicate. Details of such procedures can be found in the appropriate textbooks.

(2) The routine x-ray work that can provide compositional information about Mu (and to some extent Pg) is based upon several observations covered previously, although in a different context. (a) As the Na/(Na+K) ratio of Mu or Pg increases, then the c cell dimension decreases. Hence, $\frac{1}{2}c\sin\beta$ or d(002) (assuming, as is the common case, a $2M_1$ polymorph) also decreases. (b) As the $\Sigma(Mg+Fe_T)$ in the octahedral sites increases, the b cell dimension increases and thus so does the d(060) spacing.

Numerous studies by petrologists and crystallographers have confirmed these observations. Thus, careful determination of d(002) and d(060) can provide information on the two most important compositional aspects of Mu and Pg. However, as developed previously, the two interrelationships between composition and d-spacing cannot be treated independently of each other. For a given Na/(Na+K) ratio, increase of the $\Sigma(Mg+Fe_T)$ in octahedral sites causes d(002) to decrease markedly (see Figs. 14 and 16) -- an effect first noted by Ernst (1963). Possibly it is in part the failure to take into account this effect that led Naef and Stern (1982) to question the existence of any relationship whatsoever between x-ray spacing and composition for Mu and Pg. The interrelationships between composition and lattice spacings have already been discussed, and the specific x-ray laboratory techniques are treated in the literature. Here we will focus on several limitations or precautions that have sometimes been ignored when using white mica lattice spacings to determine composition.

(1) Use of an internal standard. Although this may seem ridiculously obvious, it is dismaying to note that there is no mention of internal standards

having been used in many studies reporting d-spacings.

(2) Determination of Na/(Na+K) in Mu and Pg. From earlier discussion it is evident that use of the curve shown on Figure 14 or Equation 2 to determine the Na/(Na+K) ratio of Mu involves an important constraint. The curve or equation can be used only for Mu samples having ΣAl near the saturation limit. This is no great handicap, because Mu in the vast majority of medium to high grade pelitic schists would be of an Al-rich variety. Most importantly, the curve of Figure 14 or Equation 2 should not be applied to any Mu suspected to be Cd-rich. Although the relationship between d(002) and Na/(Na+K) in Pg is generally similar, it has not yet been worked out in detail. Because there is very little Cd substituted into Pg the problem discussed above for Mu will be obviated. On the other hand, because the Na/(Na+K) ratio of Pg does not have a very wide range, it may be fairly difficult to calibrate its relationship with d(002). Nonetheless, if one takes into account the effect of Cd content on d(002) of Mu, it should be possible to use the difference in d(002) for coexisting Mu + Pg to serve as a qualitative measure of the closing of the Mu-Pg solvus in a given sequence of metamorphic rocks (cf. Zen and Albee, 1964).

(3) Determination of $\Sigma(Mg+Fe_T)$ in Mu. It was noted previously how the relationships shown by Figure 17 or Equation 3 could enable one to determine $\Sigma(Mg+Fe_T)$ in Mu by measuring d(060) or b. It was also noted that the Na/(Na+K) ratio of Mu has a small effect on b. However, this effect will not be any problem because it is a small effect, and it is only for greenschist grade rocks that one would be interested in using b as a monitor of Cd content, and at such grades the Mu is invariably relatively Na-poor. It is also obvious that concern will exist over whether an observed b value is reflecting a Tschermak substitution or the substitution Fe^{3+} = Al in the octahedral sites. Inasmuch as these two substitutions have quite different petrologic meaning, it is necessary to get some idea of the relative contribution of each to the observed b value. Fortunately, a number of petrographic considerations (discussed in detail in Guidotti and Sassi, 1976a, and summarized in the next section) enable one to assess fairly well the relative contribution in a given sample of each of these substitutions to b, and thus a petrologic significance can usually be attached to it.

Ideally one should be able to determine b merely by obtaining the d(060) by means of a diffractometer scan or by a powder camera pattern. However, as emphasized recently by Frey et al. (1983), the 060 peak of $2M_1$ Mu cannot be clearly separated from the $\bar{3}31$ peak. Moreover, the calculated patterns of Borg and Smith (1969) indicate that the $\bar{3}31$ peak has an intensity approximately twice that of the 060 peak. Indeed, in the process of doing cell refinements of 75 Mu samples based on diffractometer scans from 8° to 70°, the writer was never able to obtain adequate resolution of the 060 peak such that it could be used directly in the refinement.

On Figure 17 all of the b values were determined by single-crystal methods or are calculated values obtained by unit cell refinement. Although not shown, a similar plot using b values determined by simple measurement of d(060) (diffraction or powder camera) had so much scatter that one would despair of using d(060) as an indirect measure of Mu composition. This supports the skepticism expressed by Frey et al. (1983) with regard to using d(060) or b to monitor variation in the Cd content of Mu. As an alternative, Frey et al. developed a plot (for $2M_1$ Mu) of a "combined value" of d(060,$\bar{3}$31) versus RM (where RM = the molar proportions of 2 Fe_2O_3 + FeO + MgO -- Cipriani et al., 1971). Their plot results in a linear equation:

$$d(060,\bar{3}31) = 1.4976 + 0.0819 \text{ RM} . \qquad (36)$$

They also made a similar plot of d(300) versus RM for $3T$ Mu, but the data showed considerably greater scatter and the authors considered it as not yet satisfactory as a means of indirectly obtaining composition data.

Although determination of b by simple measurement of the 060 peak by means of conventional x-ray methods seems fraught with difficulties, there is a simple and rapid way to circumvent these problems. Unfortunately, this technique has been used by only a few workers thus far. It was developed by Sassi and Scolari (1974a) and involves cutting a thin chip of schist perpendicular to the foliation of the rock and then lightly polishing it. The slice of rock is mounted in a diffractometer and scanned over the 2θ range of 060 and that of the nearby 211 peak of quartz (to serve as an internal standard). Because of experimental constraints, large numbers of Mu flakes are in a preferred orientation which greatly enhances the intensity of the 060 reflection, and thus diminishes the intensity of $\bar{3}31$ and 331 peaks. Preliminary comparison of the b thus obtained with that determined for the same sample by means of a full cell refinement has been quite encouraging. However, further work is needed to ascertain quantitatively how much uncertainty is present in the b values obtained by this technique. The geologically meaningful results noted earlier that have been obtained thus far by Sassi and his colleagues and others using this technique clearly show that even at the present time this approach can be used at least qualitatively.

At a time when so many petrologic studies present massive amounts of probe data, it may seem quaint to some to still consider using indirect (x-ray) methods in order to obtain compositional information about rock-forming minerals. Be that as it may, it must be remembered that relatively few medium and smaller size geology departments have an electron probe, but many do have an x-ray laboratory. Moreover, maintaining an x-ray laboratory is a relatively easy and low cost matter, and the technical "know how" for generating petrologically useful data on micas by means of x-ray is markedly less than that required for using a probe. Of particular importance is the fact that large numbers of samples can be processed by x-ray techniques at low cost in terms of time and money. Indeed, because much of the x-ray work is rote or repetitious in nature, after only modest instruction it can be carried out by relatively unskilled helpers. Although it might seem that aside from the time and expense required, making use of Cd contents obtained directly by chemical analysis would be desirable over using the indirect b value as a monitor of Cd content, this may not always be the case. Because probe analyses do not enable distinguishing Fe^{2+} from Fe^{3+} (see next section) most studies using probe analyses use the ΣSi as the measure of Cd content. Unfortunately, it is now becoming recognized that there can be considerable variation in the Si values obtained from probe lab to probe lab depending upon the standard used for Si. Hence, one may not always be able to compare with confidence data for Si in Mu obtained in different labs. For example, in discussing the solution between Mu and Cd, it was noted that very few phengitic Mu have Si > 7.0. This was particularly true for analyses done by wet chemical techniques. Yet a (very) few studies using the probe have reported Si values as high as 7.4 to 7.6. Obviously the problem of differences between labs should not be a problem for the b technique, because the same internal standard (quartz) is used in all cases.

(III.C) Electron microprobe analysis of white micas

Probably the first broad-scale use of the probe for analysis of white micas in a petrologic context was the study by Evans and Guidotti (1966). Since then the veritable explosion in the literature of probe generated white mica analyses attests to the extreme importance of the microprobe. About the only drawbacks of this method are the expense and time involved in maintaining a probe-lab, the fact that routine probe analyses cannot distinguish Fe^{2+} from Fe^{3+}, and the need for careful standardization so that results from different labs can be interrelated. In addition to these factors, most workers using the probe for white mica analyses have suggested using a slightly enlarged electron beam so as to avoid possible volatilization of the alkali elements, especially Na.

(III.D) Problems of Fe^{2+} versus Fe^{3+} in white micas

Because Marg and Pg contain only negligible Fe, distinguishing Fe^{2+} from Fe^{3+} is important only with regard to Mu, in these respects: (1) for considerations of the extent of and crystallochemical features of Mu compositional variation, and (2) for determining the petrologic implications of Mu, because, as noted, useful relationships exist between the phengite content of Mu and the conditions under which it formed. Below we shall discuss some ways of minimizing the difficulties which arise for (1) and (2) due to deficiencies of analytical methods at distinguishing Fe^{2+} from Fe^{3+}. These include both indirect approaches (b) and direct ones (electron microprobe). It will be expedient to carry out this discussion in terms of (1) and (2) listed above.

In the case of (1) most substitution models, exchange reactions, etc. which depend upon knowledge of the ionization states of Fe in Mu will probably require wet chemical analysis. For those relationships which may not have such a direct tie to the Fe^{2+}/Fe^{3+} ratio (e.g., the extent of variation of the Na/(Na+K) ratio) then the absence of these data is not so important. However, because much of our understanding of the crystal chemistry of a solid solution mineral involves charge balance and ionic size consideration it is evident that direct and accurate determination of Fe^{2+} and Fe^{3+} is essential for truly complete crystallochemical understanding of substitution models, exchange reactions, etc., which affect Mu. The various methods for approximating or rationalizing the Fe^{2+}/Fe^{3+} question when considering (2) will probably be inadequate for the considerations involved in (1). In summary, wet chemical analysis, at least for Fe, is still highly essential for crystal-chemical considerations.

In considering (2), the absence of data on Fe^{2+} versus Fe^{3+} can be considered in terms of microprobe data and b data. In the case of probe analyses, inability to distinguish Fe^{2+} from Fe^{3+} necessitates using Si content as the indicator of the extent of phengite substitution. Using the $\Sigma(Mg+Fe_T)$ or ΣAl as the sole indicators of phengite content would encounter problems if much Fe^{3+} were present. Various authors (e.g., Brown, 1967; Zen, 1981) have attempted to devise methods by which, given total Fe as part of a microprobe analysis, one could calculate in an approximate fashion the amounts of Fe^{2+} and Fe^{3+}. Such schemes involve charge balance considerations, comparison with Mg/Fe ratio in phases not containing significant Fe^{3+}, etc. However, as concluded by Zen (1981, p. 9) "... the problem of partition of oxidation states of iron in microprobe analyses of muscovite must remain unresolved".

When considering the phengite content in terms of b measurements, several factors must be taken into account. It will be evident that these factors will also apply to those studies in which the phengite content has been determined by probe analyses. As emphasized above the value of b is related to the $\Sigma(Mg+Fe_T)$. However it is the $\Sigma(Mg+Fe^{2+})$ which reflects the phengite substitution in Mu, i.e., the substitution which is supposed to reflect the PT during formation of the Mu. Thus, it is desirable to find some way to circumvent the possibility that the observed b is monitoring Fe^{3+} rather than $\Sigma(Mg+Fe^{2+})$. With regard to specifically minimizing the effect of Fe^{3+} on the b value of Mu, Guidotti and Sassi (1976a, p. 107-108), based on data recorded in studies by Chinner, 1960, and Hounslow and Moore, 1967, noted that the assemblage of opaque minerals in a rock will provide a means for estimating the relative abundances of Fe^{2+} and Fe^{3+}. Specifically:

(a) Rocks containing hematite or magnetite formed under relatively high

$f(O_2)$ conditions and thus are likely to contain fairly high Fe^{3+}/Fe^{2+} ratios. In fact, because hematite and to a great extent magnetite are Fe^{3+} phases they will set $a(Fe_2O_3)$ at 1 for any rock containing them. Thus, for a given PT, all other Fe^{3+}-bearing phases will be saturated with respect to that cation. From the perspective of trying to use b of Mu to measure the $\Sigma(Mg+Fe^{2+})$, hematite or magnetite-bearing rocks would be inappropriate.

(b) Rocks with ilmenite or ilmenite + Fe sulfide could contain Mu with significant Fe^{3+}, but most likely the Fe^{3+}/Fe^{2+} ratio will be in some low to moderate range. Thus the b value will to a large extent probably reflect the $\Sigma(Mg+Fe^{2+})$.

(c) Rocks containing graphite ± ilmenite ± Fe-sulfide will have formed under low $f(O_2)$ conditions. Thus the minerals in such rocks should have relatively low and uniform values for the Fe^{3+}/Fe^{2+} ratio. Discussion of the control (including buffer capacity) that graphite exerts on $f(O_2)$ are given by Miyashiro (1964), French (1966), and French and Eugster (1965). Moreover, Miyashiro (1964) has pointed out that graphite is quite common in metamorphic rocks. In contrast, hematite is quite rare and magnetite is not common.

By taking these assemblages into account it should be feasible to eliminate Mu samples which will have unusually high Fe^{3+} contents and focus on those with only low to moderate amounts of Fe^{3+}. Most importantly, we should be able to select samples with relatively uniform Fe^{3+} contents. Certainly it would be unwise to try to use b data as an estimate of phengite content if our population of specimens includes some with hematite and some with graphite. As seen in Figure 7, most natural phengites contain a moderate amount of Fe^{3+} and represent solid solution to the celadonite end member rather than the leucophyllite end member. Thus, in using b as a monitor of phengite content, what we really want to avoid is Mu with a marked solution toward the ferri-Mu end member. This can be avoided by not using hematite- or magnetite-bearing rocks.

It would be highly desirable to obtain wet chemical data for Fe^{3+} and Fe^{2+} of Mu formed in rocks with different $f(O_2)$ conditions as indicated by the opaque minerals present. Unfortunately the wet chemical analyses of Mu presently reported in the literature are not useful for this purpose inasmuch as only rarely do the authors make mention of the opaque minerals with which the Mu was associated.

(III.E) Staining techniques

Laduron (1971) has described a staining technique by which Mu and Pg can be distinguished from each other in thin section. A particular value of such an approach would be the opportunity to observe any difference in the textural aspects of the two white micas.

TRIOCTAHEDRAL "DARK MICAS"

In the context of metamorphic rocks, the term trioctahedral "dark micas" can be considered largely under the common name biotite. The compositional range covered by this term was discussed in the beginning of this chapter. Other trioctahedral micas can occur in metamorphic rocks (e.g., wonesite and clintonite), but they are either not common or not well studied to date. Hence, our treatment is mainly restricted to biotite, the one that has general petrologic significance. As in the case of the white micas, our discussion of biotite will be carried out under three headings: (I) biotite from a mineralogic perspective, (II) petrologic considerations, and (III) determinative methods for recognizing biotites and ascertaining their compositions.

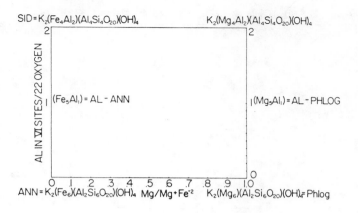

Figure 34. An "ideal biotite plane" in which the only compositional variation allowable is the Mg $\rightleftharpoons$ Fe^{2+} exchange and the Tschermak exchange, (Mg,Fe^{2+}) + Si^{iv} = Al^{iv} + Al^{vi} -- plotted as Al^{vi} versus Mg/(Mg+Fe^{2+}). Ann = annite; Phlog = phlogopite; Al-Phog = aluminous phlogopite; Al-Ann = aluminous annite; Sid = siderophyllite; the composition at the $K_2(Mg_4Al_2)(Al_4Si_4O_{20})$ $(OH)_4$ corner would be eastonite in the old nomenclature (see text).

(I) Biotite from mineralogic perspective

(I.A) Observed chemical variation of biotite

To a very large extent the compositional variation shown by biotite in metamorphic rocks can be described in terms of the end members listed at the beginning of this chapter. Figure 34 shows these end members graphically and defines them relative to the nomenclature given on Table 1 of Chapter 1 of this volume. In the interest of simplifying the terminology used herein, the compositional points shown on Figure 34 will be designated as Ann, Phlog, Al-Ann, Al-Phlog, and Sid. The space enclosed by the end members Ann-Phlog-(Al-Phlog)-(Al-Ann) will be designated the *biotite plane* and the term biotite (Bio) will be used to indicate any general point on the plane. The end-member terms will be used when trying to indicate relative position on the plane. For the sake of those familiar with the older literature, it should be noted that what is here designated as Al-Ann and Al-Phlog used to be called siderophyllite and eastonite respectively (see Deer et al., 1962, p. 57). Although the term siderophyllite is still used, it is now applied to a more Al-rich composition (see Fig. 34). The term eastonite is no longer used.

The composition of biotite from metamorphic paragenesis can undergo several minor to moderate deviations from being fully represented by the Bio plane of Figure 34, namely, (1) Ti^{4+} substituting into octahedral sites, (2) Fe^{3+} substituting into octahedral sites, (3) Cl and F replacing (OH), (4) vacancies in octahedral sites and other substitutions therein, and (5) substitutions in XII sites. We will consider variations within the Bio plane first and then (1)-(5) in sequence.

<u>Compositional variation within the biotite plane.</u> In order to gain some idea of the extent of compositional variation within the biotite plane by specimens obtained from metamorphic parageneses, it will be expedient to treat the matter in a manner which will bear on subsequent petrologic considerations. This can be accomplished by making a survey of biotite analyses from the various bulk compositions that can be found in metamorphic rocks, e.g. metapelites, calc-silicates, metabasites, etc., and also by considering them over a range

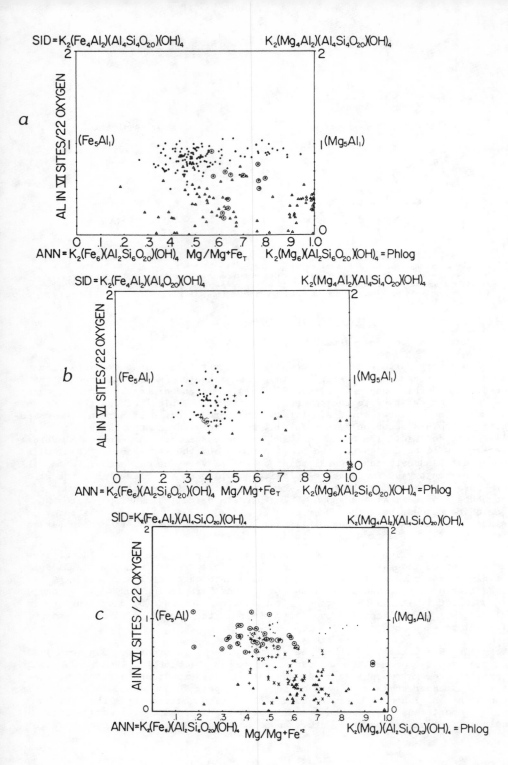

of metamorphic grades. At the outset we shall ignore the problem of Fe^{2+} versus Fe^{3+} and consider only total Fe.

Figure 35a includes a large number of wet and probe analyses of biotites obtained from the literature. All of these samples are from amphibolite grade rocks. Almost all of them have Ti contents of less than 0.3 atoms per six octahedral sites. Hence, the plot accounts for the vast bulk of the contents of the octahedral sites. To a first approximation, such a diagram also illustrates the extent to which metamorphic biotite deviates from having the ideal tetrahedral Si:Al ratio of 6:2 as in Ann or Phlog. This follows from the fact that Al > 2 in tetrahedral sites is the primary way of maintaining charge balance for Al replacing Mg or Fe^{2+} in octahedral sites. Thus, the total substitution is identical with the Tschermak exchange described for Mu. The samples plotted on Figure 35a come from a wide diversity of rock types -- pelitic schists, metabasites, Fe-formations, calc-silicates, and talc-phlogopite rocks. Hence, virtually all of the bulk compositions and amphibolite-grade assemblages are represented.

With this in mind, the following generalizations can be made. (1) Amphibolite-grade biotites are almost wholly restricted to lying between Al^{vi} of 0 and 1 and $Mg/(Mg+Fe_T)$ of 0.3 and 1. (2) Biotites from typical Mu-bearing pelites and other Mu-bearing parageneses have Al between 0.7 and 1.0 per six octahedral sites and $Mg/(Mg+Fe_T)$ ranges from 0.3 to close to 1.0. In contrast biotites from pelitic and other Al-rich rocks *that are lacking in Mu* have a tendency to contain somewhat lower Al^{vi} contents. This point was noted recently by Rabotka (1983) and the more extensive sampling shown on Figure 35a confirms his observation. (3) Biotites from amphibolites, calc-silicates, etc. typically have Al^{vi} < 0.6 and some have Al^{vi} approximately that of the Ann-Phlog join.

Several additional important features are not discernible from Figure 35a. These include: (1) At $Mg/(Mg+Fe_T)$ values less than 0.65, the Al to Si ratio in tetrahedral sites is typically in the range of 2.70/5.30 to 2.60/5.40. However, at higher values of $Mg/(Mg+Fe_T)$, this ratio drops more or less progressively to about 2.20/5.80 and Labotka (1983) reports a specimen with 2.08/5.92 and $Mg/(Mg+Fe_T)$ of 0.934. In essence, as Mg increases, these biotites are nearly approximate to the ideal Al:Si = 1:3 ratio of Ann or Phlog. However, it is evident from Figure 35a that the Al^{vi} content does not change in the manner expected from a simple Tschermak exchange as Si^{iv} increases. Guidotti et al. (1977) have discussed this relationship in some detail and Atzori et al. (1973) noted it for igneous and metamorphic biotites in general. (2) Those

Figure 35. (a) Biotites from amphibolite grade rocks (except for K-feldspar + sillimanite grade) plotted in terms of the "ideal biotite plane." Data sources from the literature. Symbols are: ● = biotite from Mu-bearing rocks, mainly ordinary pelitic schists but a few with coexisting amphibole also. Double circles = biotite from Mu-free pelitic schists and other rocks with coexisting, Al-rich phases. Δ = biotite from Mu-free rocks not containing any Al-rich phases, mainly calc-silicates and amphibolites.

(b) Biotites from sub-amphibolite grade rocks plotted in terms of the "ideal biotite plane". Data sources from the literature. Symbols are: ● = biotite from Mu-bearing rocks, mainly pelitic schists. + = biotite from rocks containing Mu + amphibole. Double circles = biotite from Mu-free rocks but otherwise fairly aluminous specimens as indicated by the presence of phases like garnet, cordierite, etc. Δ = biotite from clearly low-Al rocks such as metabasites, calc-silicate, etc.

(c) Biotites from upper amphibolite and granulite grade rocks plotted in terms of the "ideal biotite plane". Data sources from the literature. Symbols are: ● = biotite from Al-silicate + Mu-bearing rocks from just down grade of the Al-silicate + K-spar zone. Double circles = biotite from rocks containing Mu + Al-silicate + K-spar. + = biotite from rocks containing Al-silicate + K-spar but no Mu. x = biotites from granulite-grade pelitic schists (most with an Al-silicate present but a few with only cordierite ± garnet. Δ = biotite from non-pelitic rocks of Al-silicate + K-spar or granulite grade, mainly amphibolites.

specimens on Figure 35a which have $Mg/(Mg+Fe_T) > 0.950$ are mainly from calc-silicates (i.e., metamorphosed, impure siliceous marbles; cf. Rice, 1977). Biotite from pelites with $Mg/(Mg+Fe_T) > 0.7$ are mainly from sulfide-rich rocks (Guidotti et al., 1975, 1977; Mohr and Newton, 1983). Robinson (pers. comm.) has found essentially end-member Al-Phlog in similar parageneses in central Massachusetts.

Figure 35b shows biotite data for sub-amphibolite facies rocks again from a wide range of bulk compositions and Fe is lumped as Fe_T, and Ti amounts to less than 0.25 atoms per six octahedral sites. Although the data base is smaller, it would appear that the same generalizations apply here as with the amphibolite grade biotites. At most, one might suggest that the upper limit of Al^{vi} is slightly lower.

Once again the same relationship between $Mg/(Mg+Fe_T)$ and tetrahedral Al/Si occurs. In this case, however, several parageneses are present which illustrate this relationship in a spectacular fashion. These are the talc-Phlog rocks described by Abraham and Schreyer (1976), Chopin (1981), and Schreyer et al. (1980). Although these biotites still appear to have some Al in octahedral sites (see points near $Mg/(Mg+Fe) = 1$, many have such high Si contents that the Al to Si ratio in the tetrahedral sites *is less than 2:6*. In this context it should be noted that the three biotites from *Mu-bearing* rocks plotting over near the pure Mg side of the diagram have Al:Si ratios < 2:6. Nonetheless, they still have significant Al in octahedral sites. These biotites coexist with talc and highly phengitic Mu. In contrast, the biotites plotting almost at the Phlog corner coexist with talc, but not with phengitic-Mu. Hence, again it is seen that the presence or absence of Mu has a marked effect on the Al^{vi} content of biotite.

Figure 35c is a plot of biotite from high-grade metapelites (upper sillimanite zone-granulite grade) and biotites from non-pelitic granulite grade rocks. The upper sillimanite zone biotites were also plotted on Figure 35a but the higher grade biotites were not. Excluding the non-pelitic, granulite-grade rocks, all of the other biotites plotted on Figure 35c are from Al-saturated parageneses. However, these biotites have been subdivided to show the effect of the presence or absence of Mu on the Al^{vi} of biotite. The following gross generalizations emerge: (1) Biotite from Mu-bearing rocks of the upper sillimanite zone and K-feldspar + Al-silicate zone plot in much the same fashion as the Mu-bearing rocks on Figure 35a, i.e. Al^{vi} between 0.7 and 1.0. For those K-feldspar + Al-silicate zone rocks *without Mu*, the Al^{vi} content starts to decrease. (2) For the granulite grade K-feldspar + Al-silicate-bearing rocks, Al^{vi} is markedly lower, usually < 0.55. (3) Biotite from non-pelitic granulite grade rocks has, as expected, relatively low Al^{vi} contents.

Not apparent from Figure 35c is the fact that as Al^{vi} decreases in the biotites from Al-saturated rocks, the Ti content rises markedly from about 0.25 atoms per 22 oxygens in the amphibolite grade biotites, to more than 0.5 atoms per 22 oxygens in the granulite-grade biotites. It would appear that with the demise of Mu, the Al^{vi} of biotite is not held at high values and is at least in part replaced by Ti^{vi} by some type of substitution mechanism (see later sections) as grade rises further.

Although the pattern of Al^{vi} and Ti^{vi} variation as a function of grade just discussed is based on a gross overview of many data points, it can be clearly seen in several homogeneous data sets given in the literature. Examples would include Schmid and Wood (1978) covering the range from just above the stability limit of Mu into granulite-grade, Ruiz et al. (1978) covering the range from lower sillimanite zone to above the stability limit of Mu, Dymek (1983) covering the range from upper sillimanite zone to granulite grade, and Hunziker and Zingg (1980) covering the range from upper sillimanite zone to well above the upper stability limit of Mu.

Also not apparent from Figure 35c is the fact that, although possibly less pronounced, the same relationship noted earlier between $Mg/(Mg+Fe_T)$ and Al^{iv}/Si^{iv} exists for biotite in the K-feldspar + Al-silicate grade rocks and granulite grade rocks.

A feature not considered yet but present for the biotites on Figures 35a, b, and also c to a fair extent is a strong tendency for an inverse relationship between $Mg(Mg+Fe_T)$ ratio and the amount of Ti^{vi}. Inspection of the biotite analyses plotted on the three figures shows that such a relationship is quite universal. High Mg-biotites from granulite-grade rocks have fairly high Ti contents compared to virtually all biotites from lower grade rocks, but relatively low Ti contents when compared with lower Mg-biotites from other granulite grade rocks. It should also be pointed out that although the above generalizations are based upon an amalgamation of a large amount of analytical data, many of them have previously been noted based upon smaller, homogeneous data sets of specific studies. For example: The interrelationship among Si, $Mg/(Mg+Fe_T)$, and Ti was noted by Guidotti et al. (1977). Honma (1974, his Fig. 2) showed a clear separation of Al^{iv} and Al^{vi} (hence total Al) in amphibolite grade biotites depending upon whether the biotite was associated with Mu, hornblende, or neither. In essence, these associations reflect different degrees of Al-saturation. On an even finer scale, many workers have shown how the Al content of biotite responds to variation in assemblage within a given general rock type, e.g. pelites (Boak and Dymek, 1982). And Labotka (1983) showed the effect of the presence or absence of Mu on the Al^{vi} content of biotite from a suite of pelitic schists.

Although the general patterns described have been described to varying extents by previous studies, e.g., Korikovskiy (1965), it should be emphasized that the compositional ranges discussed above apply to common metamorphic parageneses and may not indicate the most general constraints for chemistry of biotite. Examples of how the above described patterns do not represent the case for biotites in general are: (1) The near absence of metamorphic biotites with $Mg/(Mg+Fe_T)$ less than 0.3 may be a function of bulk composition controls of metamorphic controls but not crystallochemical controls. For example, Foster (1960) reports biotites lying very close to the Ann-Sid join. And (2) Dawson et al. (1970) describe phlogopites from ultrabasic xenoliths which are both Ti and Si rich. However, in other cases (Smith et al., 1978; Exley et al., 1982), phlogopites from ultramafic bulk compositions do follow the pattern for $Mg/(Mg+Fe_T)$, Si, and Ti noted for common metamorphic parageneses.

Titanium substitution in octahedral sites. Some discussion has occurred in the literature concerning the possibility of Ti in biotite being in the trivalent state (Engel and Engel, 1960; Gorbatschev, 1972; Evans and Raftery, 1980), or the Ti substituting for Si in tetrahedral sites (Kunitz, 1936; Farmer and Boettcher, 1981). However, the general consensus is that Ti in biotite is quidrivalent and substitutes into octahedral sites, and we will follow this consensus view. Workers as long ago as Engel and Engel (1960) and Oki (1961) have noted that in general the Ti content of biotite increases with metamorphic grade. Virtually all of the numerous studies since have agreed on this. Strictly speaking, this assertion should be made in the context of biotite coexisting with an intrinsically Ti phase such as ilmenite or rutile. The presence of such a phase ensures that the coexisting biotite is saturated with Ti, and thus the Ti content is a function of the prevailing intensive parameters. Guidotti et al. (1977) discussed this in detail, and it is seen diagramatically on Figure 36. The fact that so many studies have confirmed the original suggestion of Engel and Engel (1960) reflects the nearly ubiquitous appearance of ilmenite, rutile, or sphene in all kinds of metamorphic parageneses.

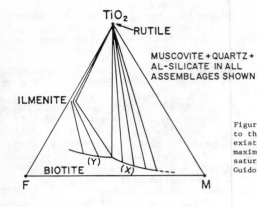

Figure 36. Schematic projection from Al_2SiO_5 on to the TiFM plane showing how the biotite coexisting with ilmenite or rutile will contain the maximum amount of Ti possible (i.e. be Ti-saturated) for given external conditions. From Guidotti et al. (1977).

In greenschist grade metapelites and semi-pelites the Ti content of biotite ranges from about 0.15-0.25 to about 0.2 to 0.30 atoms per six octahedral sites in upper sillimanite zone rocks. The same general trend occurs in other rock types like metabasites, etc., but the specific values attained are not yet well established. In Mg-rich bulk compositions (impure siliceous dolomite) the tendency is less marked because of the interrelationships among Ti, Si, and $Mg/(Mg+Fe_T)$ in biotite which were discussed above.

At grades exceeding the upper sillimanite zone, the Ti content of biotite from metapelites increases quite markedly to about 0.35 to 0.65 atoms per six octahedral sites. And as noted previously, the initiation of this sharp rise appears to coincide with the breakdown of Mu. Concomitantly Al^{vi} decreases, although the biotite is Al-saturated as indicated by its coexistence with an Al-silicate such as sillimanite. The Ti-content of biotite from other rock types (e.g. meta-igneous rocks, etc.) at granulite grade also increases to high values, commonly 0.4-0.6 atoms per six octahedral sites.

The inverse relationship between Ti content of biotite and its $Mg/(Mg+Fe_T)$ ratio plus Si content seems to be quite universal for most of the common metamorphic parageneses. Moreover, it seems to be at least qualitatively in agreement with the experiments of Robert (1976). A particularly extreme example of this phenomenon is illustrated by the pure Mg biotites described by Schreyer et al. (1980). Despite the presence of a Ti-saturating phase (rutile), these phlogopites contain no Ti. In some rocks it can be shown that this inverse relationship takes place while Al^{vi} remains constant (Guidotti et al., 1977). However, it is important to note that data presented recently by Whitney and McLelland (1983) seem to contradict some of the above. They describe biotite forming reaction coronas around ilmenite in granulite grade rocks. This biotite has quite high $Mg/(Mg+Fe_T)$ ratios (~0.7 atom %) but it is also Ti rich (0.7 to 1.07 Ti atoms per six octahedral sites; 9.45 wt % TiO_2) and only modestly enriched in Si. At present, one can only speculate that the direct reaction relationship between ilmenite and biotite is responsible for this seemingly isolated violation of the above generalizations.

Finally it should be noted that a considerable literature has grown during the last 10 years on phlogopites in lamprophyres, potassic rocks, kimberlites, and mantle xenoliths, as summarized by Bachinski and Simpson (1984). For our purposes it is interesting to note that the common metamorphic phlogopites we have been considering are similar to the primary, primary-metasomatic, and MARID-suite phlogopites in the sense of high $Mg/(Mg+Fe_T)$ and Si and low Ti.

In contrast, the biotites described by Whitney and McLelland (1983) are more similar to those from certain types of lamprophyres and potassic rocks (moderately high $Mg/(Mg+Fe_T)$, variable Si, and very high Ti). Bachinski and Simpson suggest that differences in $f(O_2)$ were important in controlling the Ti, etc., of the phlogopites considered in their review with the ones having high $Mg/(Mg+Fe_T)$, high Si, and low Ti corresponding to lower $f(O_2)$ conditions. Because of the much lower PT conditions prevailing for common metamorphic parageneses, it is difficult to determine if the same arguments would apply to the similar phlogopites in them.

Fe^{3+} substitution into octahedral sites. As with Ti, some discussion exists in the literature regarding the possibility of Fe^{3+} in tetrahedral sites, especially in cases for which the $\Sigma(Si+Al)$ is < 8, (Dawson and Smith, 1977; Delaney et al., 1980). Again, we shall adhere to the consensus view that it substitutes into octahedral sites only (but see Chapter 5). Various authors have tried to devise schemes based largely upon charge balance arguments for estimating Fe^{3+} when an analysis is done by the probe, e.g., Dymek (1983). However, as with the situation discussed earlier for Mu, such procedures remain unsatisfactory. Hence, the only real information on Fe^{3+} in biotite is based on pre-microprobe wet chemical analyses. From the viewpoint of charge balance it is obvious that one must consider Fe^{3+} substituting for Al^{vi}. As noted by Dymek (1983) this could lead to consideration of a Ferri-Tschermak substitution:

$$(R^{+2})^{vi} + (Si^{4+})^{iv} = (Fe^{3+})^{vi} + (Al^{3+})^{iv} \quad . \tag{37}$$

Hence, in terms of Figures 35a, b, and c, one might replot them with $\Sigma(Al^{3+}+Fe^{3+})$ substituted on the vertical axis and $Mg/(Mg+Fe_T)$ plotted on the horizontal axis. Such a plot would give a more complete picture of the extent to which Al:Si > 2:6 in tetrahedral sites is directly balanced by X^{3+} cations in octahedral sites. In fact, inspection of biotite formulas shows that Al^{vi} alone usually more than balances the Al^{iv} in "excess" of that for the 2:6 ratio in the tetrahedral sites. Only in biotites from granulite grade rocks is Al^{vi} insufficient to balance the "excess" Al^{iv}. At the other extreme, in biotites from pelitic rocks the Al^{vi} commonly exceeds the "excess" Al^{iv} by 0.2 to 0.3 atoms. Hence, it is clear that Fe^{3+} in octahedral sites is usually not required for the purpose of balancing the Al^{iv} which is in "excess" of the 2:6 ratio.

In order to get some idea of how much Fe^{3+} is present in common metamorphic biotites one must consider the wet chemical analyses provided by the pre-microprobe literature. Although many such analyses can be found, only a part of them are set in the context of well characterized petrologic suites. The data used herein include only that for which a fairly complete assemblage is stated and for which there is a reasonable certainty that the biotite is part of an equilibrium assemblage.

From Figure 37 it is apparent that on the basis of six octahedral sites, the Fe^{3+} content of biotite from pelites and semi-pelites tends to lie between 0.1 and 0.3 atoms. Although the data base is fairly small, Figure 37 suggests that biotite from amphibolites tends to have a somewhat greater amount of Fe^{3+} (0.2-0.5 atoms). It must be stressed that these values are general ranges based upon data regarded as acceptable in the context of the above defined criteria. In fact, one can find a few values in the literature that markedly exceed the ranges shown on Figure 37, thus they should not be interpreted as providing limits on the amount of Fe^{3+} that can enter biotite from metamorphic parageneses -- and certainly not from igneous parageneses (see also Miyano and Miyano (1982) for very low grade Fe-formations). Rather, the ranges indicated probably reflect the degree of oxidation that commonly prevails during metamorphism.

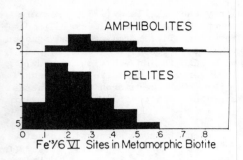
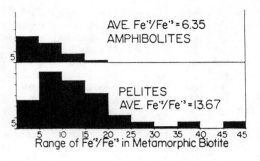

Figure 37 (left). Relative frequency of various ranges of Fe^{3+} contents (per six octahedral sites) in biotites from common metamorphic rocks. Based on wet-chemical data from the literature.

Figure 38 (right). Relative frequency of various ranges of Fe^{2+}/Fe^{3+} ratios in biotites from common metamorphic rocks. Same data base as for Figure 37.

Figure 38 is a histogram of the range and abundance of the Fe^{2+}/Fe^{3+} ratios for the same biotites considered in Figure 37. For the pelites + semi-pelites the average Fe^{2+}/Fe^{3+} ratio is 13.67 and for the amphibolites the ratio is 6.35 (the latter based on only 26 data points). From Figure 38 it is apparent that very few biotites from pelites and semi-pelites have $Fe^{2+}/Fe^{3+} > 20$.

Ideally one should find a close relationship between the Fe^{3+} content of biotite (especially its Fe^{2+}/Fe^{3+} ratio) and the nature of the coexisting suite of opaque minerals. As pointed out long ago by Eugster and Skippen (1967) and Eugster (1972), the suite of opaque minerals coexisting with the silicates in a rock gives direct information on the oxidation conditions prevailing during formation of the rock. Thus, a relationship is to be expected between the Fe^{3+}-content of a mineral like biotite and the nature of the coexisting opaque minerals. Indeed, Hounslow and Moore (1967) show a close correlation between biotite Fe^{3+}/Fe^{2+} ratio and rock oxidation ratio, which in turn is closely correlated with the nature of the opaque mineral suite. As an example of the above, rocks containing graphite + ilmenite should have formed at markedly lower oxidation conditions than rocks containing hematite + magnetite. Thus, at the same PT, biotite coexisting with graphite should have systematically less Fe^{3+} than biotite in rocks containing hematite. In principle, information on biotite composition in such a context should enable one to estimate, at least qualitatively, the Fe^{2+}/Fe^{3+} ratio for biotite on which total Fe is obtained by microprobe analysis. Unfortunately the data on assemblages present are usually too poor to enable rigorous establishment of such a relationship.[4]

A few data are available which at least enable one to see how the expectation outlined above does seem to hold true. These would include the studies of Chinner (1960) and Hounslow and Moore (1967), from which it can be determined that at least four of the pelitic biotites in the column with Fe^{2+}/Fe^{3+} between 0 and 5 do, as expected, coexist with hematite. No data are available for the coexisting opaque phases of the other five specimens in this range of Fe^{2+}/Fe^{3+}. Another very informative data set in this context is that for a metamorphosed iron formation in northern Sweden. Annersten (1968) reports three biotites coexisting with magnetite which have Fe^{2+}/Fe^{3+} ratios ranging from 5.18 to 31.8, averaging 15.8. In contrast, eight biotites which coexist with hematite have

---

[4]Unfortunately journal editors will permit authors to list only partial mineral assemblages.

Fe^{2+}/Fe^{3+} ratios ranging from 1.57 to 4.57, averaging 2.50. It seems apparent that a larger data base of this type would enable petrologists to make fairly reasonable estimates of what portion of the Fe they obtain via probe analysis is Fe^{+3} and what portion Fe^{+2}.

<u>Substitution of F and Cl for (OH) in biotite</u>. Consideration of F and Cl substituting for (OH) in biotite has received marked attention since about 1970. In part, this is due to the recent emphasis on the nature of metamorphic fluids and thus recognition that substitution of significant Cl or F into (OH)-bearing phases might influence their stability relations. Moreover, in recent years microprobe techniques have been adapted so that the halogens can be determined on an "almost" routine basis (see Peterson et al., 1982, for further discussion). To date most of the emphasis has been on F and very little consideration has been given to Cl. As noted by Peterson et al., however, a surprisingly small portion of the biotite analyses in the literature provide data on the halogens -- regardless of whether the analyses were done by probe or wet chemical techniques. This is most unfortunate, because, especially for F, there have been a number of theoretical and experimental studies dealing with its significance in minerals like biotite and the phases with which it coexists. Munoz and Ludington (1974, 1977), Peterson et al. (1982) and Valley et al. (1982) provide an especially valuable review of what is known from theoretical, experimental, and mineralogic studies about the substitution of F for (OH) in biotite. Moreover, they provide references to much of the relevant literature. Hence, we shall draw heavily from these papers for those aspects having direct relevance to metamorphic biotites (also see Chapter 11).

These authors emphasize that, based upon theoretical and experimental considerations, knowledge about the F content of micas in metamorphic parageneses may be of considerable importance in the context of displacement of stability limits of the micas (a point noted even earlier by Evans, 1969) and the nature of the fluid phase during metamorphism. Indeed, Peterson et al. (1982), Valley et al. (1982), as well as Kearns et al. (1980) discuss the possibility of minerals such as F-rich phlogopite and tremolite being stabilized to higher T's in granulite-grade rocks than would be the case for (OH)-rich phlogopite and tremolite. In a similar context, Peterson et al. (1982) suggest that F-rich micas might be expected to occur in high grade marbles, skarns, and granulite-grade rocks.

From the restricted viewpoint of metamorphic parageneses, a number of points concerning natural biotites can be extracted from the above mentioned literature plus some other less detailed studies cited in these papers. However, as a prelude to reviewing the F-contents observed in natural biotites, two points require comment. First, the biotites from natural metamorphic parageneses probably are not fully saturated with Fe for the prevailing PT conditions. This statement probably applies even to some of the extremely F-rich (X = 0.96) phlogopites from marbles described by Peterson et al. (1982). This probable lack of saturation arises because none of the parageneses appear to have a coexisting, intrinsically F-bearing phase such as fluorite. In contrast, some of the biotites and phlogopites from skarns (e.g. Kwak and Askins, 1981) and mineral deposits (Gunow et al., 1980) do coexist with fluorite and so are probably saturated with respect to F. Consistent with this suggestion is the fact that the highest f(F) that Valley et al. (1982) calculated for any Adirondack rocks was from a fluorite-bearing orthogneiss. Second, from the descriptions and discussions in the literature it seems evident that the F-content in hydrous, metamorphic minerals is directly inherited from the amount of F present in the pre-metamorphic protolith. Valley et al. (1982) emphasized and discuss this point. Hence, in this sense the F-content in the minerals of common metamorphic rocks probably differs from that in the minerals of skarns, greisens, hydrothermal mineral deposits, etc.

Considering that the hydrous phases are the only ones capable (in most cases) of taking in F and Cl in place of (OH), the likelihood that the F content of the hydrous minerals of the common metamorphic parageneses is inherited from the pre-metamorphic protolith has an important implication. That is, as grade rises and progressively decreases the amount of hydrous minerals in a rock, the remaining hydrous phases should have a progressively greater amount of F (and Cl) in them. The hydrous phases act as sinks for F (and Cl) and so should behave much as garnet does for Mn or staurolite for Zn. Obviously one should expect the F- and Cl-content of a hydrous phase, such as biotite, *in a given rock type*, to show a continuous increase of F as grade rises. However, because common metamorphic parageneses are not saturated with F, one should not expect the correlation between grade and F- and/or Cl-content of biotite to be as regular as for example the case for Ti which as we have noted before is commonly at saturation levels in biotite.

Considering first the data for biotite in pelitic bulk compositions, we find that they conform fairly well to the above suggestions. Biotite from low to middle grade rocks typically contains less than 0.2 wt % F. Biotite from sillimanite zone rocks commonly contains 0.2-0.4 wt % F, and in the granulite facies the values range up to and exceed 1.0 wt %. However, it should be re-emphasized that a great deal of variability exists in these limits because the amount of F in biotite depends greatly on the amount of F in the original bulk composition and upon the amount of other hydroxyl-bearing phases (usually Mu) that are present. For example, Dymek (1983) reports biotite from granulite grade meta-pelites with only about 0.30 wt % F. Nonetheless, the generally observed increase of F in pelitic, granulite-grade biotites probably reflects the demise of Mu and thus the concentration of all of the F into biotite. In turn, this marked concentration of F in biotite only after Mu is gone suggests that at lower grades the F-content of the hydrous phases will be insufficient to offset significantly any equilibria involving Mu or biotite.

In other rock types the data on F in biotite are too poor to attempt any generalizations, but they are of interest nevertheless. Němec (1980) reports F-rich biotites from tin-bearing orthogneiss (amphibolite grade) containing 0.32 to 1.45 wt % F. As noted in our discussion of Mu, this orthogneiss is F-rich and some specimens contain fluorite as a phase. Unfortunately none of the analyzed biotites were from the fluorite-bearing rocks. Nonetheless the coexisting phengitic Mu is also quite F-rich and comparison of the Mu and biotite data suggests that F concentrates in biotite by a ratio averaging at least somewhat greater than 2:1. By way of contrast, in high grade pelites with low phengite Mu, Evans (1969) found this ratio averaging 4.8. In still another type of paragenesis, Leelanandam (1969a) found biotites in various types of charnockites (ultrabasic, mafic, acid or intermediate, and biotite gneiss) with F-contents ranging from 1.22 to 3.99 wt %. Moreover, he found a positive correlation between F-content and Mg(Mg+Fe) ratio of the biotite. Ekstrom (1972a) reports F-rich biotites from several amphibolite-grade iron formations in Sweden (mostly > 1.0 wt %, but ranging as high as 4.9 wt % F). His observations suggest that F fractionates into biotite relative to Ca-amphibole (K_D = 0.5 to 1.1). He also reports a positive correlation between biotite F-content and Mg/(Mg+Fe) ratio.

In most cases the Cl-content of biotite is much less than the F-content. However, the data base for Cl in biotite from metamorphic rocks is extremely poor and so it seems appropriate only to cite the results of several studies. Evans (1969) found Cl mainly in the range of 0.01 to 0.04 wt % in biotite from sillimanite to sillimanite + K-feldspar grade biotites. Dallmeyer (1974) found Cl values as high as 1.19 wt % in some biotites from pelitic granulite grade rocks. Leelanandam (1969b) reports biotites (from the charnockites noted above) with Cl-contents ranging from 0.19 to 0.62 wt %. Moreover, his data suggest that Cl is somewhat fractionated in favor of coexisting hornblendes. In contrast with the correlation between F and Mg/(Mg+Fe$_T$) of biotite, Leelanandam (1969a) shows that biotite Cl-content appears to be independent of the Mg/(Mg+Fe$_T$) ratio. Finally, Ekstrom (1972b) reports on the Cl-content of biotite coexisting with Cl-bearing scapolite from amphibolite grade Fe-formations in Sweden. Biotite Cl-contents in the range of 0.3 to 0.5 wt % appear to be common in these rocks. However, inspection of his data for the Cl-content of coexisting Ca-amphibolites and biotites suggests, in contrast with Leelanandam (1969b), that Cl is fractionated in favor of biotite instead of Ca-amphibole.

Because the comments made above concerning the F-content of biotite as a function of grade pertained mainly to metapelites, they apply with a degree of rigor only to biotites with intermediate Mg/(Mg+Fe$_T$) ratio. However, that data noted for the other parageneses (charnockites and Fe-formations) revealed a correlation between biotite F-content and Mg/(Mg+Fe$_T$) ratio also. Most of the biotites reported on in Ekstrom (1972a) have Mg/(Mg+Fe) ratios of 0.7 or greater. The discussion of Munoz and Ludington (1974), Peterson et al. (1982), and Valley et al. (1982) further emphasizes that a distinct relationship exists between high Mg/(Mg+Fe$_T$) ratio in biotite and replacement of (OH) by F. Moreover, virtually all of the common metamorphic biotites with almost complete replacement of (OH) by F are from the phlogopite side of the biotite plane (from marbles and calc-silicates). Valley et al. (1982) discuss the relationship between high Mg and high F in terms of an Fe-F avoidance and cite evidence for such a relationship based upon crystal structure analysis, experimental studies, and crystal-field theory as well as mineral composition studies like those noted above. It would appear that for a given ratio of H$_2$O/F in the fluid phase that an Mg-rich biotite will have a higher percentage of its (OH) replaced by F and an Fe-rich biotite a smaller percentage. Probably some of the Fe-rich biotites that also have high F are from rocks with unusually low H$_2$O/F ratios in the fluid phase. The F-rich, Fe-rich biotites reported by Nêmec (1980) would be an example of such an occurrence. Although the analyzed biotites were not from fluorite-bearing specimens, some associated rocks do contain fluorite.

Valley et al. (1982) calculated the H$_2$O/F ratio for the Adirondack marbles containing the extremely F-rich phlogopites. Because the ratios obtained fall in a range that implies only minor F present in the fluid, they suggest that many of the phlogopites found in high grade marbles will probably be F-rich. The data of Rice (1977a, 1979), Bucher-Nurminen (1981), etc. seem to confirm their suggestion.

Other factors which the above authors discuss as having an influence on the F-content of biotite are the Al-content of the biotite, temperature, and the ratio of H$_2$O/F in the fluid phase. Based upon their experiments with annite and siderophyllite Munoz and Ludington (1974) claimed that the Al^{vi} content does not influence the F = (OH) exchange but Valley et al. (1982) suggested the contrary. Figure 39 would seem to suggest the Al^{vi} content may have an inverse correlation with F-content.

The effect of T will have two aspects: an innate control of the F = (OH) exchange -- as shown by the experiments of Munoz and Ludington (1974), and the control it exerts on the modal amounts of other F-bearing phases and thus its control on how much F is available to get concentrated in biotite. The H$_2$O/F ratio in the fluid phase obviously depends upon the equilibria with other volatile species (CO$_2$, CH$_4$, etc.) which can affect the percentage of H$_2$O in the fluid and thus the ratio of H$_2$O to F in the fluid. Valley et al. (1982) discuss the controls on the distribution of F between biotite and amphibole or Mu and Munoz and Ludington (1977) the distribution of F between biotite and Mu. In

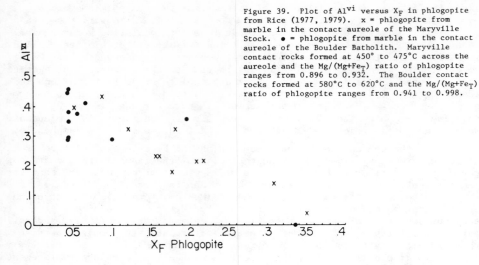

Figure 39. Plot of Al^{vi} versus X_F in phlogopite from Rice (1977, 1979). x = phlogopite from marble in the contact aureole of the Maryville Stock. ● = phlogopite from marble in the contact aureole of the Boulder Batholith. Maryville contact rocks formed at 450° to 475°C across the aureole and the $Mg/(Mg+Fe_T)$ ratio of phlogopite ranges from 0.896 to 0.932. The Boulder contact rocks formed at 580°C to 620°C and the $Mg/(Mg+Fe_T)$ ratio of phlogopite ranges from 0.941 to 0.998.

general they conclude that mineral chemistry and not grade is the dominant control on the partitioning of F and (OH) between biotite and other hydrous phases. The reader is referred to the cited papers for details on the factors controlling the F-content in biotite and the partitioning of F between biotite and other hydrous phases in metamorphic parageneses. In summary of the petrologic implications of F in biotite, it should be noted that Valley et al. (1982) present calculations on the quantitative effects of F at stabilizing phlogopite in granulite facies marbles. They conclude that, especially in such rocks, application of devolatilization equilibria without taking into account F, can lead to large errors in estimates of P, T, and X (fluid). From a purely mineralogic viewpoint, it is useful to note that virtually all of the Mg-rich (and F-rich) biotites we have been discussing conform to the Si-rich, Ti-poor pattern discussed earlier. In one case the $Si:Al^{iv}$ ratio is even > 6:2. Moreover, many of these biotites also have substantial Al^{vi} and so do not plot at the phlogopite end member but about half-way between phlogopite and Al-phlogopite (see Fig. 35a).

<u>Vacancies in octahedral sites and other substitutions</u>. Consideration of several hundred biotite analyses in the literature (wet and probe analyses) confirms an assertion routinely made by petrologists dealing with metamorphic biotites. That is, formulas calculated on a basis of 22 oxygens typically have total number of octahedral cations between 5.6 and 5.9, rather than 6.0 as implied by an ideal formula for biotite. This deficiency is universal in that it includes biotites from all grades and bulk compositions. Indeed, only rarely does it equal 6.0. In most cases a small fraction of this deficit is due to analytical procedures, mainly failure to analyze for minor constituents like Cr, Zn, etc. However, the bulk of the deficit cannot be explained in this fashion and is thus attributed by most workers to vacancies in the octahedral sites. Because this would appear to represent a deviation toward a dioctahedral sheet, most workers have likened the deficit to a solution between biotite and muscovite. Considered from this viewpoint the range between 5.6 and 5.9 would amount to a 20-5% solution range toward a dioctahedral mica like Mu. However, Hazen and Burnham (1973) suggest that such a description is misleading because the vacancies (at least for annite) in the octahedral positions of biotite are random. This contrasts with the case for dioctahedral micas which have the M1

site vacant and the M2 site filled. Considered on the basis of the percentage of vacancies relative to an ideal six octahedral sites, we obtain a range of 6.6 to 1.6%. These vacancies are important in formulating exchange substitutions in biotite and especially in the formulation of activity models for use in thermodynamic calculations (e.g. Holdaway, 1980; Labotka, 1983).

Several other substitutions are commonly recorded in small amounts for the octahedral sites of biotite. Typical analyses show a few thousandths to a few hundredths of an atom of Mn per six octahedral sites. However, these amounts are almost surely not at saturation levels, as indicated by the fact that biotite from assemblages with Mn-rich parageneses typically contains much more Mn. For example, Abraham and Schreyer (1976) report 0.17 Mn in a phlogopite (coexisting with piemontite) and Brown et al. (1978) 0.5 Mn in phlogopite (coexisting with piemontite, braunite, etc.). Zinc is also sometimes reported, typically at levels ranging from 0.002 to 0.004 atoms. Interest in Zn arises in the context of its relationship with the high Zn contents sometimes found in coexisting phases, staurolite in particular. And Dietvorst (1980) found the Zn in biotite coexisting with gahnite ranged as high as 0.015 to 0.026 atoms per six octahedral sites. Tracy (1978) found levels as high as 0.012 atoms of Cr in biotite from high grade rocks, and Boak and Dymek (1982) found values as high as 0.025 atoms per six octahedral sites.

<u>Substitutions in the XII sites</u>. The 12-coordinated sites of biotite typically appear to contain less than the ideal 2 atoms expected for a formula based on 22 oxygens, ranging from 1.65 to 1.90 with values between 1.80 and 1.85 being most common. Of the XII atoms, K is always the vastly predominant cation in metamorphic biotites, Na is usually present as several hundredths of an atom (occasionally ranging to > 0.1 atom), and Ca is rarely present in amounts greater than a few thousandths of an atom, even in Ca-rich parageneses. Rice (1977a,b) reported Ca of a few hundredths of an atom for the phlogopites present in metamorphosed marbles; however, these amounts may be just analytical fluctuation, because Rice (1979) reports phlogopite coexisting with clintonite with less than 0.001 Ca (the clintonite contained < 0.001 K). Barium is the only other XII cation reported on a fairly frequent basis, especially for high grade parageneses. In such parageneses, where biotite coexists with K-feldspar, Ba has been recorded in amounts ranging up to 0.01 to 0.018 atoms (Lonker, 1975; Tracy, 1978). A particularly interesting study of the Na content of biotite is that by Schreyer et al. (1980), who report phlogopite in rocks coexisting with Na-phlogopite that contains 0.10 to 0.52 atoms of Na (average of 9 specimens is 0.248). This suggests that the smaller amounts of Na usually reported are real; they are not impurities but represent values far removed from saturation levels.

(I.B) Crystallochemical aspects of biotite compositional variation

In comparison with white micas, biotite from metamorphic rocks shows much more compositional variation of elements in major proportions. These include Fe^{2+}, Mg, Fe^{3+}, Ti, Tschermak exchange, vacancies, and F (and Cl) for (OH). This compositional variation occurs "within and from" the ideal biotite plane, in response to bulk composition and/or mineral assemblage variation, and in response to variation of metamorphic conditions. The last involve changes in assemblage, changes in saturation limits of a given element, and changes in oxidation state.

With regard to the chemical variation occurring in metamorphic biotites, two crystallochemical aspects need to be considered. (1) Are there any absolute composition limits on the general solution field of biotite that would set boundaries on the extent of solid solution that can occur for the biotite that occurs in metamorphic rocks? The possibility of miscibility gaps would be included in this question. (2) What is the nature of the substitutions that occur in the four sites (12-, 6-, and 4-fold and the (OH) sites) of biotite? This would include consideration of whether a substitution occurs in a single site, is coupled between sites, etc.

To a large extent the basic controls bearing on these questions will be the need for charge balance and the requirements for fitting together the structural units of biotite (i.e., the tetrahedral and octahedral sheets). The latter aspect will involve consideration of tetrahedral rotation (α) and octahedral flattening. Some understanding of (1) and (2) has importance concerning any deviations by biotite from being an ideal solution and for formulation of activity models.

Many different substitutions and schemes have been proposed for biotite, some operating within a single site, some coupled between and among sites. Because of the complex interaction of the effects of bulk composition and metamorphism on biotite composition, it is probably rare that a substitution scheme operates by itself. In most cases, two or more substitution schemes are probably in operation, e.g., see Abrecht and Hewitt (1980, 1981) or Labotka (1983).

Dymek (1983) listed a number of theoretical substitutions that could be operating in a singular fashion within biotite. His list includes:

$$(R^{2+})^{vi} + (Si^{4+})^{iv} = (Al^{3+})^{vi} + (Al^{3+})^{iv} \quad \text{(Tschermak exchange)} \tag{38}$$

$$3(R^{2+})^{vi} = 2(Al^{3+})^{vi} + (\square)^{vi} \tag{39}$$

$$(R^{2+})^{vi} + 2(Si^{4+})^{iv} = (Ti^{4+})^{vi} + 2(Al^{3+})^{iv} \tag{40}$$

$$(Al^{3+})^{vi} + (Si^{4+})^{iv} = (Ti^{4+})^{vi} + (Al^{3+})^{iv} \tag{41}$$

$$2(Al^{3+})^{vi} = (Ti^{4+})^{vi} = (R^{2+})^{vi} \tag{42}$$

$$2(R^{2+})^{vi} = (Ti^{4+})^{vi} + (\square)^{vi} \tag{43}$$

$$(R^{2+})^{vi} + 2(OH)^- = (Ti^{4+})^{vi} + 2(O^{2-}) + H_2 \tag{44}$$

$$(R^{2+})^{vi} + (Si^{4+})^{iv} = (Fe^{3+})^{vi} + (Al^{3+})^{iv} \tag{45}$$

$$Fe^{2+} + (OH)^- = Fe^{3+} + O^{2-} + \tfrac{1}{2}H_2 \tag{46}$$

$$(K^+)^{xii} + (Al^{3+})^{iv} = (\square)^{xii} + (Si^{4+})^{iv} \tag{47}$$

Others that obviously should be mentioned include:

$$K^+ = H_3O^+ \tag{48}$$

$$Fe^{2+} = Mg^{2+} \tag{49}$$

$$K^+ = Na^+ \tag{50}$$

$$Fe^{3+} = Al^{3+} \tag{51}$$

$$F^- \text{ or } Cl^- = (OH)^- \tag{52}$$

We shall discuss some of these which seem to be well supported by experimental or observational data. For natural samples there are several reasons for the difficulty in identifying the substitution schemes that occur in biotite.

(1) Biotite has a wide range in composition, and the compositional variation may take place by means of several substitution schemes working in conjunction. (2) In the past investigators have attempted to design substitution models by combining analytical data on biotite from widely disparate parageneses -- e.g. biotites from igneous and metamorphic rocks, biotites from Al-saturated and -undersaturated rocks, and biotites from reduced and oxidized parageneses. (3) Poor quality data sets have been merged with good quality data sets, and in some cases all of the data used were poor quality. (4) The effects of grade change have not been separated from bulk composition effects. (5) The site and valence assignment of Fe and Ti have been incorrect.

Because some of the possible substitution schemes have direct petrologic importance, attempts must be made to sort out and identify those that can actually occur as biotite changes composition in response to petrologic factors (grade, bulk composition, etc.). Two approaches that have distinct promise are: (1) experimental studies on homogeneous and controlled bulk compositions that are relevant to metamorphic rocks [examples -- Hewitt and Wones (1975), Abrecht and Hewitt (1980, 1981), Robert (1976a,b), Hazen and Wones (1972)], and (2) studies of natural parageneses that form a homogeneous population chosen so that the number of potential compositional variation is constrained by phase equilibrium considerations [examples -- Guidotti et al. (1975, 1977) and Labotka (1983)]. We shall concentrate on a few crystallochemical aspects of biotite that can be recognized with a fair degree of certainty. To a large extent this will be based on observed variations described in the previous section, and our discussion will be structured in terms of the crystal chemistry of the compositional changes "within and from" the ideal biotite plane shown on Figure 34.

Compositional variation within the biotite plane. Within the ideal biotite plane, the only compositional limits of concern are the Al Tschermak exchange and $Fe^{2+} = Mg^{2+}$. Within the context of these, the solution range of the ideal plane has been studied experimentally and theoretically by Hazen and Wones (1972, 1978), Hewitt and Wones (1975), and Robert (1976), who provide references and discussion of most of the relevant earlier literature. From the compositional limits they discussed, it is apparent that natural, metamorphic biotites do not "bump against" the Al or Fe^{2+} limits possible for biotite. This is true even if one decides to consider Fe^{3+} as equivalent to Al^{3+} and to designate the $\Sigma(Al^{3+}+Fe^{3+})^{vi}$ as part of the Tschermak exchange. Experimental studies show Al^{3+} can be as high as 1.50 in Fe biotites (Hewitt and Wones, 1975), and as high as 1.625 in Mg biotites (Robert, 1976). From Figures 35a-c it is evident that Al^{3+} (even if added with Fe^{3+}) is usually less than 1.0. Similarly, for Al-bearing biotite, Hewitt and Wones (1975) were able to synthesize end-member Fe^{2+}-biotite. Figure 35 also shows that natural metamorphic biotites rarely have $Mg/(Mg+Fe_T) < 0.3$. Thus one can conclude that the composition *limits* of the *common* natural metamorphic biotites are determined by petrologic factors and not crystal-chemical factors.

However, for the observed range of composition variation of metamorphic biotite several crystallochemical factors can be recognized. (1) Especially for biotites from pelitic schists, the amount of Al^{vi} is commonly near 0.9 atoms. Comparison with the Al^{iv} present in these same biotites immediately reveals that the Al^{vi} is much higher than that which would be required (via a Tschermak substitution) to balance the Al^{iv} in excess of 2.0. In this context the Al^{vi} is commonly in excess by about 0.2 to 0.4 atoms, and it would appear that this excess is involved in some substitutions that involve a deviation from the ideal biotite plane. (2) As noted by Guidotti et al. (1975), there appears to be no relationship between $Mg/(Mg+Fe_T)$ ratio and Al^{vi}. In contrast there is a strong inverse relationship between $Mg/(Mg+Fe_T)$ and Al^{iv}, with only a few exceptions. Guidotti et al. (1977) argued that the increase of $Si:Al^{iv}$ is a means of decreasing the amount of tetrahedral rotation required to maintain fit between the tetrahedral sheet and an octahedral sheet that has a decreased b cell dimension due to having octahedra filled with Mg. This argument would also include the considerations of Hewitt and Wones (1975) that requiring too much tetrahedral rotation would shorten the $K-O_T$ bonds to the point of destabilizing the structure. An observation possibly interrelated with the above suggestions is that made by Mohr and Newton (1983), Cheney (1975), and Guidotti et al. (1977), who noted that, at least for the high Al phlogopites they studied, the number of vacancies in the XII site seemed to increase as $Mg/(Mg+Fe_T)$ and Si^{iv} increase to higher values.

The question of whether biotite acts as an ideal solution within the ideal biotite plane has been addressed by various authors. Mueller (1972) presented arguments that the $Fe^{2+} = Mg^{2+}$ exchange approximates ideality but that it is greatly perturbed by "extraneous" ions like Al^{3+} and Si^{4+}. In essence, it would be perturbed by the Tschermak exchange. The experimental data of Wones (1972) and Wones et al. (1971) also suggest that the annite-phlogopite exchange is ideal. Based upon an experimental study Hewitt and Wones (1975) suggest that

$$Fe^{2+} = Mg^{2+} \text{ and } (Mg^{2+})^{vi} + (Si^{4+})^{iv} = (Al^{3+})^{vi} + (Al^{3+})^{iv}$$

approach ideality but that

$$(Fe^{2+})^{vi} + (Si^{4+})^{iv} = (Al^{3+})^{vi} + (Al^{3+})^{iv}$$

has a positive volume of mixing and so deviates markedly from ideality.

Compositional variation deviating from the biotite plane

(a) *Experimental data*. Several experimental studies bear on the crystal-chemical aspects of those compositional variations which represent deviations from the ideal biotite plane. Robert (1976) studied the extent of Ti solubility in synthetic phlogopite as a function of P and T. He suggested that Ti enters the phlogopite via the exchange $2Si^{iv} + Mg^{vi} = 2Al^{iv} + Ti^{vi}$, and he found that the solubility of Ti in phlogopite was very small at 600°C and 1000 bars (i.e., only 0.07 Ti per 22 oxygens) and it increased only up to 0.2 Ti at 800° and 1000°C. Pressure seemed to have an opposite effect. Comparison of these observations with the Ti contents of natural phlogopites suggests a fairly close agreement between the solubility limit obtained (but not in terms of T's required) by the two approaches. Recalling that the increase of Si as $Mg/(Mg+Fe_T)$ increases is accompanied by a decrease of Ti also suggests that the distinct correlation between Mg and Si noted in the previous section involves Ti as part of the total exchange taking place.

Seifert and Schreyer (1965, 1971) and Green (1981) have discussed the synthesis of micas which are compositionally intermediate between the common dioctahedral and trioctahedral series. At a pressure range of 20-35 kbar and T's of 800-1000°C, Green (1981) suggests a complete range of octahedral occupancy from 6 and 5 octahedral sites filled. Consideration of our previous discussion on the number of vacancies observed in natural biotite suggests that the natural mineral does not "bump against" the one full vacancy per six octahedral sites suggested by Green's experiments. (The reader is referred to Zussman (1979) for a further review of micas intermediate between the dioctahedral and trioctahedral series.) Moreover, the intermediate phase of Green has 5 occupied octahedra and Si = 4, but the natural biotites (phlogopites) with the ratio of $Si:Al > 3:1$ show no tendency to have a greater deficiency of occupied octahedral sites than other biotites. In fact, the data of Schreyer et al. (1980) and Chopin (1981) suggest that these phlogopites have a higher than average number of octahedral sites filled.

Over the past ten years, various authors have synthesized the Na-equivalents of essentially all phases shown along the Mg side of Figure 34. These include Carmen (1974), Hewitt and Wones (1975), and Franz and Althaus (1976), reviewed by Keusen and Peters (1980). Moreover, there seems to be evidence that phases with intermediate Na/K ratio can also be synthesized. Natural analogs have now also been found for the Na end members; e.g. preiswerkite with the formula close to $Na_2(Mg_4Al_2)(Al_4Si_4O_{22})(OH)_4$ (Keusen and Peters, 1980), wonesite with a formula

$$(Ca_{0.004}Na_{0.79}K_{0.145})_{0.939}(Mg_{4.39}Fe_{0.778}Mn_{0.004}Cr_{0.008}Ti_{0.074}Al_{0.62})_{5.874}$$

$$(Al_{1.534}Si_{6.466})_8O_{20}(OH,F)_4$$

(Spear et al., 1981), and Na-phlogopite with a formula approximating to $Na_2Mg_6(Al_2Si_6O_{20})(OH)_4$ (Schreyer et al., 1980). Although Weidner and Carmen (1968) and Franz and Althaus (1976) were able to synthesize Na-annite, there is little other evidence from experiments or natural parageneses for Fe-rich Na-biotites (e.g., Na-siderophyllites).

From our viewpoint of the common biotites found in metamorphic parageneses, the rarely-discussed questions which are germane are: how much K^+ can be replaced by Na^+ in a continuous solution series, and, does a limit exist in the form of a miscibility gap between the various K-biotites and the Na-analogues? Several factors suggest that in metamorphic rocks the extent of Na^+ replacing K^+ (starting at the K end of the biotite solutions) is quite limited. (1) Natural biotites rarely contain > 0.10 Na atom per six octahedral sites. (2) In the unusual case of the rocks described by Schreyer et al. (1980) which contain both Na and K phlogopite, the Na in the K-phlogopite still only reaches a maximum of 0.52 Na atoms per six octahedral sites (Na/Na+K = 0.28).

If one considers the arguments of Hewitt and Wones (1975) relating size of the XII sites and the α tetrahedral rotation required to fit the tetrahedral sheet with an Mg-rich or Mg+Al-rich octahedral sheet, one might expect Na^+ to be most successful at replacing K^+ in the various phlogopites. The data of Schreyer et al. (1980) seem to bear this out. On the other hand, in the Fe-richer biotites there would be less need for α rotation and so the XII site might remain enlarged to the extent that it would be unfavorable to accepting the small Na (or Ca) ion. In essence, this suggestion parallels that made earlier for the observed low Na contents in celadonitic Mu. Hence, one is led to at least a tentative speculation that there may be crystal-chemical reasons (as well as petrologic reasons, see Spear et al., 1981) why very little Na (or Ca) occurs in the common metamorphic biotites. This suggests that with the possible exception of the Mg end members, wide miscibility gaps exist between the common K-biotites and their Na equivalents.

(b) *Substitution of Ti.* Turning now to the purely natural data that bear on the deviations from the ideal biotite plane and considering only those that seem fairly unequivocal, let us first consider Ti. A problem that has hampered unequivocal recognition of the Ti substitution schemes involves the merging of data sets from Al-saturated and non-Al-saturated parageneses. Labotka (1983) discussed aspects of this problem as well as the more general case of merging data from dissimilar assemblages. Another difficulty has been lack of analyses of Fe^{2+} and Fe^{3+}. Despite these drawbacks, at least three types of Ti substitution in biotite can be discussed with some degree of certainty -- mainly in the context of pelitic parageneses. These include: (1) the increase of Ti with grade from biotite zone to upper sillimanite zone, (2) the increase of Ti in biotite of granulite-grade rocks, and (3) Ti content as a function of biotite Mg/(Mg+Fe) ratio at constant metamorphic grade.

(1) Many workers have documented the increase of biotite Ti content as a function of grade, and many substitution reactions have been proposed as the means by which the Ti enters the biotite. There is a common point in many of the substitution schemes which involves the formation of vacancies concomitant with increase of Ti. In some cases the models involve replacement of only divalent cations or only trivalent cations, and in some cases both. In a few of the substitution models the XII sites and (OH) site are also involved. Quite likely, a number of these schemes are correct for the rocks that were studied. Hopefully, future efforts to work up data sets in the context of narrowly defined assemblages will permit meaningful comparisons of schemes suggested by different authors.

(2) The biotite composition data discussed earlier have shown that the Ti content of biotite rises markedly in granulite-grade rocks. This is true for all rock types, but here we will focus on the enrichment of Ti in biotite from pelitic rocks. In general, it seems to start with the demise of Mu. The most distinctive aspect of this Ti increase is that it takes place to a considerable extent at the expense of Al^{vi} -- despite the fact that the rocks are Al-saturated, as shown by the manner in which biotite from granulite-grade metapelites plots on Figure 35c. The exact nature of this substitution remains to be determined. For example, it might involve Ti replacing Al^{3+} alone with concomitant formation of vacancies or alternatively take place along with other substitutions (e.g., R^{2+} for Al^{3+}, etc.) such that vacancies are not formed. It is beyond the scope of this review to provide such a determination, but, it is germane to suggest that the substitution scheme will become evident by restricting attention to Ti and Al-saturated pelites over the range from upper amphibolite facies through granulite facies. Moreover, it will be very important to focus on biotites with a relatively narrow range of $Mg/(Mg+Fe_T)$ in order to avoid the effect to be described next. Finally, it would be well to choose a suite of rocks with a uniform set of opaque minerals so that one can avoid potentially radical variation of Fe^{2+}/Fe^{3+} ratios.

(3) As noted in several previous sections, the Ti content of biotite is also strongly influenced by the $Mg/(Mg+Fe_T)$ ratio of the biotite. Because this effect shows up even if metamorphic conditions are held constant, it is clearly a crystal-chemical effect. In fact, it involves two substitutions, one being the replacement of Fe^{2+} by Mg^{2+}. In turn this reaction facilitates the reaction: $2Al^{iv} + Ti^{vi} = Si^{iv} + Mg^{vi}$. It should be recalled that the experimental work of Robert (1976b) showed that increase of grade (T) drove this exchange to the left. Here, we are considering the exchange in terms of constant grade and examining the influence of $Mg/(Mg+Fe_T)$ ratio. Guidotti et al. (1977) discussed in detail this interrelationship among $Mg/(Mg+Fe_T)$, Si, and Ti, and in an earlier section the crystallochemical aspects of the Mg and Si covariance were considered. Here, we see the additional aspect involving Ti as part of the overall substitution. It would appear that the Ti decrease is necessitated for charge balance reasons to balance out the increase of Si^{iv}. Moreover, replacing Ti^{vi} by Mg^{vi} as $Mg/(Mg+Fe_T)$ increases would also help to minimize the disparity of fit between the tetrahedral and octahedral sheets.

(c) *Substitution of F for (OH) and its relation to $Mg/(Mg+Fe_T)$*. Our discussion of the F content of biotite encompassed metamorphic as well as compositional constraints on this substitution. It is this latter aspect for which crystallochemical interactions seem to occur. Specifically, we have the well documented relationship between the Mg/(Mg+Fe$_T$) ratio of biotite and its F content. As noted earlier, Valley et al. (1982) discuss it in terms of an Fe-F avoidance. The reader should consult their paper for further details.

(d) *Al^{vi} in excess of that required for the Tschermak exchange*. This excess has already been commented on. Numerous suggested substitution schemes involved only octahedral sites and so would have a direct bearing on this excess Al^{vi}. However, there seems to be little in the way of unambiguous arguments for its origin. Possibly it is involved with the formation of vacancies (e.g., 2Al = $3R^{2+}$ + ☐) and thus the so-called deviation toward the dioctahedral micas (see earlier discussion).

(e) *Substitution of Fe^{3+}*. This substitution is very poorly understood because (especially with microprobe analysis) one commonly has no Fe^{3+} analyses and little attention is given to the oxidation state of the rocks being studied. Thus, it is rare that one has any idea of whether Fe^{3+} is participating in a Tschermak exchange or is merely substituting for Al^{vi}, or is part of a redox exchange such as Fe^{2+} + $(OH)^-$ = Fe^{3+} + O^{2-} + $\frac{1}{2}H_2$.

(f) *Substitutions in the XII site*. Obviously some of this substitution involves simple replacement of K by Na. Suggestions were made earlier as to why this substitution might be favored in Mg-rich biotites. Substitution of Ca or Ba for K is very minor in metamorphic biotites and can be presumed to occur via a typical brittle mica substitution. One of the most ambiguous aspects of substitutions in the XII site is the ΣXII atoms or vacancies. Obtaining meaningful values for the XII atoms is plagued by poor quality analytical data and the possibility that, as discussed for Mu, H_3O^+ might substitute for K^+ to a significant extent. Obtaining a meaningful value for ΣXII in biotite is important because some of the substitution schemes in the literature make use of vacancies or an increase in the total atoms in the XII site to balance out substitutions in octahedral and tetrahedral sites. Judging the validity of these proposals is beyond the scope of this review. The only proposal that seems to recur is the decrease of ΣXII atoms as Mg/(Mg+Fe$_T$) increases, as discussed above.

(II) Petrologic aspects of biotite

As with the white micas, the concern here is with the relationship of biotite to the coexisting phases in metamorphic parageneses. Hence, the questions addressed are similar to those with the white micas, although a few are unique to biotite. These include (1) the nature of the biotite isograd, (2) the general absence of biotite in "Blue Schist Style" terranes, (3) the granulite grade reactions involving biotite -- including its termination as a stable phase, and (4) the effects on biotite of sulfidation and oxidation. It should also be noted that, in comparison with the white micas, biotite has a wider stability field in terms of both metamorphic conditions and bulk composition range in which it can occur, because it becomes involved in phase assemblages not possible for the white micas. The question whether biotite is in equilibrium with the other phases present in a given rock arises because biotite is so easily altered to chlorite. In many cases this alteration may be part of a reaction approaching a new equilibrium state -- something that is common in polymetamorphic terranes. In other cases the alteration to chlorite may represent only deviation away from an original equilibrium with no approach to a new equilibrium state. There appears to be no formula for quickly distinguishing the two possibilities from microscopic study alone. Each case must be considered on its own merits in the context of extensive analytical data on the biotite and the "co-existing" phases -- especially the chlorite. Probably the only valid generalization to make is that when a petrographer observes biotite and chlorite intergrown in a fashion that suggests alteration, then caution is advised about taking the visual appearance at face value.

(II.A) Biotite occurrence and composition in the context of gross lithology

Our discussion of the compositional variation occurring in metamorphic biotites has touched on most of the important aspects of the interrelationship between biotite composition and gross litholigic type. In addition we have just noted that in general biotite occurs in a wider range of bulk compositions than the white micas. In particular this applies to biotite occurring in rocks of markedly lower bulk Al_2O_3 than would be the case for the white micas. Hence, it is commonly found in metagreywackes, metabasites, high-grade calc-silicates, etc., as well as most of the more aluminous rocks in which the white micas commonly occur. In fact it is in some of the more Al-rich parageneses that biotite tends not to occur. (See previous discussion of the phase relations of Pg.) In general it appears that about all that is required to insure the presence of at least some biotite in a metamoprhic rocks is a little K2O and sufficient Al_2O_3 to at least form a biotite on the annite-phlogopite join and, of course, the appropriate metamorphic conditions. With regard to general interrealtionships between biotite composition and gross lithologic type, our previous considerations permit the following generalizations in the context of the ideal biotite plane.

<u>Biotite in pelitic schists</u>. The Al^{vi} is higher than in other rock types and usually lies between 0.7 and 0.9 atoms per six octahedral sites. Only in granulite-grade rocks does this value commonly drop below 0.7. The $Mg/(Mg+Fe_T)$ ratio can vary widely from about 0.3 to 1.0. The higher values of $Mg/(Mg+Fe_T)$ for biotite (> 0.65) tend to occur in rocks that have been strongly affected by sulfidation or oxidation reactions. In such biotites Al^{iv} decreases.

<u>Biotite in metagreywacke and impure quartzites</u>. The main difference of biotite in pelites versus that in metagreywackes and impure quartzites concerns the Al^{vi} content which generally is systematically lower in the latter rock types.

<u>Biotite in calc-silicates and marbles</u>. Not uncommonly, biotites in such parageneses are Mg-rich enough to be described as phlogopites. In most cases Al^{vi} is distinctly lower than for biotites from pelites, but some do reach values comparable to those from pelites. Al^{iv} tends to be quite low, with Al:Si approaching 1:3.

<u>Biotite in metabasites</u>. Such rocks typically have Ca-amphiboles present. The biotites in them have a wide range of $Mg/(Mg+Fe_T)$ ratio, but they tend to have values > 0.5. The Al^{vi} content ranges up to values seen in biotites from metapelites, especially if Mu is also present. Probably these occur in rocks from middle to lower grades of metamorphism. Much more commonly the Al^{vi} content is distinctly lower than in biotite from pelites and in a few cases (especially from granulite-grade rocks) the Al^{vi} content plots along the annite-phlogopite line. As with the previously described gross lithologic types, the biotites with very high $Mg/(Mg+Fe_T)$ ratios tend to have low Al^{iv} contents.

<u>Biotite in metamorphosed granitic rocks</u>. This group would include metamorphosed plutonic and volcanic rocks of intermediate to true granite compositions. Biotite from the metamorphosed equivalents of such bulk composition are quite similar to biotites found in metagreywacke. Their $Mg/(Mg+Fe_T)$ ratios tend to be somewhat less than 0.5.

Miscellaneous. Although not qualifying as common metamorphic parageneses it is interesting to comment on the biotites found in the talc-phengite parageneses described by Chopin (1981) (pelites subjected to "Blue Schist Style" metamorphism) and by Schreyer et al. (1980) (metamorphosed evaporites). These biotites tend to approach end-member phlogopite and are so Si-rich that the $Si:Al^{iv}$ ratio actually exceeds 3:1. Nonetheless, some of them still contain Al^{vi}, thereby serving as a spectacular example of Al^{vi} in excess of that needed for the Tschermak exchange.

A general thread running through the above discussion is the manner in which Al^{vi} in biotite maintains a fairly close correlation with the Al content of a rock type as reflected by the gross mineralogy. Honma (1974), Labotka (1983), and Guidotti et al. (1975) have commented on this relationship.

Finally, it has been noted that biotite is rare or absent in "Blue Schist Style" terranes, and the explanation advanced usually makes use of the fact that chlorite + Cd-rich Mu are abundant and nearly ubiquitous in "Blue Schist Style" metamorphic rocks (cf. Ernst, 1963). Hence, the suggestion made is that this Cd-rich Mu + chlorite take the place of biotite at high P, low T conditions. In the context of the biotite isograds to be discussed in the next section, it would appear that the high P of "Blue Schist Style" terranes prevents the biotite-producing reactions from proceeding to the right. Ernst (1963) discussed this question in terms of the ΔV involved and the influence of high $P(H_2O)$ and low T. The only seeming exception regarding the occurrence of biotite in "Blue Schist Style" rocks involves the occurrences of phlogopite + talc + phengite recently described by Chopin (1981).

(II.B) Metamorphic controls on the occurrence and composition of biotite

The approach taken here is similar to that used for the white micas, i.e., a consideration of the manner in which metamorphic conditions -- P, T, and X (fluid) -- determine the specific phase relations of biotite. We now go beyond the previous section and consider the detailed relationship between the occurrence and composition of biotite and specific mineral assemblages. Our treatment will cover only the more interesting, yet common rock types such as calc-silicates, metabasites, and metapelites. Emphasis will be on the last group because it is the one in which the phase relations of biotite are best known. The theoretical framework required for considering the phase relations of biotite (especially in metapelites) is essentially identical with that used for the white micas, except that one uses the AFM system rather than the AKNa system. To the extent possible we shall consider the reactions bringing in and terminating biotite plus a few of the more important intermediate T reactions. Also discussed are those continuous reactions that are especially important for controlling the composition of biotite. Moreover, in order to illustrate an example of the kind of biotite compositional variation that can occur in a prograde sequence of pelitic schists, the focus for the continuous reactions will be on a single, well studied group of rocks from western Maine.

Biotite in marbles and calc-silicates. The terms marble and calc-silicate are used here to distinguish between the following biotite-bearing parageneses: (1) metamorphosed siliceous limestones and dolomites that contain only minor K_2O and Al_2O_3 such that, in addition to abundant carbonate minerals, most of the other minerals are phases like tremolite, diopside, etc., with only small amounts of K-feldspar ± phlogopite as additional phases, and (2) metamorphosed siliceous limestones and dolomites which contain abundant impurities that include components such as K_2O, Al_2O_3, FeO, and Na_2O. As expected, the phase relations for biotite are quite different in (1) and (2) and so must be considered separately.

Type (1) can be thought of as starting out with a siliceous dolomite that contains some detrital K-feldspar. Rice (1977a) used the system $CaO - MgO - SiO_2 - KAlO_2 - CO_2 - H_2O$ to describe such rocks; he, DeBethune (1976), and others have proposed the low T reaction as follows:

$$3 \text{ dolomite} + \text{K-feldspar} + H_2O = \text{phlogopite} + 3 \text{ calcite} + 3 CO_2 , \qquad (53)$$

a reaction that has also been studied by Puhan and Johannes (1974). According to Rice (1977a) it is followed by two reactions which result in an incompatibility of phlogopite and dolomite. These are:

$$5 \text{ dolomite} + 8 \text{ quartz} + H_2O = \text{tremolite} + 3 \text{ calcite} + 7 CO_2 , \text{ and} \qquad (54)$$

$$5 \text{ phlogopite} + 6 \text{ calcite} + 24 \text{ quartz} =$$
$$3 \text{ tremolite} + 5 \text{ K-feldspar} + 6 CO_2 + 2 H_2O . \qquad (55)$$

Reaction 55 was studied experimentally by Hewitt (1975) and Hoschek (1973). Phlogopite is still present at grades higher than Reactions 54 and 55, but it is restricted to the assemblages tremolite + quartz + K-feldspar + phlogopite and tremolite + calcite + K-feldspar + phlogopite. The next reactions bring in diopside in a bulk composition not involving phlogopite by means of the well known reactions

$$\text{dolomite} + 2 \text{ quartz} = \text{diopside} + 2 CO_2 , \qquad (56)$$

$$\text{tremolite} + 3 \text{ calcite} + 2 \text{ quartz} = 5 \text{ diopside} + 3 CO_2 + H_2O . \qquad (57)$$

However, the tie line configuration resulting still precludes phlogopite from coexisting with diopside. At still higher grade, the following reaction occurs and makes phlogopite compatible with diopside.

$$3 \text{ tremolite} + 6 \text{ calcite} + \text{K-feldspar} =$$
$$\text{phlogopite} + 12 \text{ diopside} + 6 CO_2 + 2 H_2O . \qquad (58)$$

The sequence of phlogopite-bearing reactions described above has occurred in the contact aureole of the Marysville stock in Montana. Based on thermal modelling, Rice (1977a) estimates the T at the contact was 600°-700°C. Although phlogopite occurred in the innermost metamorphic zone, it did not seem to persist right up to the contact. Hence, one can put an upper T limit of the phlogopite *in this sequence of rocks* as below the 600°-700°C believed to have occurred directly on the contact. In addition Rice suggested that P(total) was about 1 kbar. Hence, based on his P estimate and the experimental work mentioned above, the maximum T possible for Reaction 55 is below 500°C.

It is evident from the nature of the reactions and the experimental studies that fluid composition will strongly influence the T's of some of these phlogopite-forming reactions. Presumably at sufficiently high T the phlogopite will break down to an Mg phase such as enstatite with concomitant production of K-feldspar. However, it should be recalled from our discussion of the F content of biotite, the phlogopite in marbles is a prime candidate for taking in significant F in place of (OH) and thus having its stability range extended to higher T's. Finally, the fact that the biotite in siliceous dolomitic marbles approaches the Mg-side of Figure 34 clearly reflects the bulk composition involved. As seen below, as soon as FeO becomes an abundant component, the trioctahedral mica approaches "garden variety" biotite.

Turning now to type (2), the siliceous carbonates with abundant extra components in addition to $KAlO_2$, we find that in these "dirty carbonates" the reactions involving biotite are more complex. Various authors have discussed specific reactions that can occur in such bulk compositions but the papers by Ferry (1976a,b; 1983) are among the more complete in the sense of covering the reactions occurring over a wide range of grades. Moreover, in the area he studied it was possible to relate the reactions involving biotite in the calc-silicates with the metamorphic grades recognized in the associated pelitic rocks. The first mineralogic change he observed involved formation of biotite + chlorite + calcite -- at approximately the same grade as the biotite isograd in the adjacent pelites. The proposed reaction is

$$\text{Mu} + \text{quartz} + \text{ankerite} + H_2O = \text{calcite} + \text{chlorite} + \text{biotite} + CO_2 \ . \qquad (59)$$

Ferry (1976b) noted that this reaction differs from that proposed by other workers in that chlorite rather than Ca-plagioclase (Hewitt, 1973) or zoisite (P.H. Thompson, 1973) takes up the Al released from Mu. Moreover, the reactions of Hewitt and Thompson occur in the staurolite zone. Subsequent to Reaction 59, a continuous reaction occurs which facilitates a modal increase of biotite, increase in the An content of plagioclase, and a modal decrease of Mu and its final elimination. Ferry (1976b) gives this continuous reaction as

$$\text{Mu} + \text{calcite} + \text{chlorite} + \text{quartz} + \text{albite}$$
$$= \text{biotite} + \text{intermediate plagioclase} + H_2O + CO_2 \ . \qquad (60)$$

As a result of Reaction 60, plagioclase changes from near An_0 to about An_{60}, as the next discontinuous reaction is encountered. This reaction defines an amphibole-anorthite isograd that does not involve biotite directly, and thus is not given here. Several other reactions not involving biotite then take place. Just downgrade of the sillimanite isograd as mapped in the associated pelites, another discontinuous reaction that involves biotite occurs in the carbonate-bearing rocks. Ferry (1976b) refers to this reaction as the microcline-amphibole isograd. It is written as

$$\text{biotite} + \text{calcite} + \text{quartz} = \text{calcic-amphibole} + \text{microcline} + H_2O + CO_2. \quad (61)$$

Ferry notes that this reaction has also been recorded by Hewitt (1973), P.H. Thompson (1973), and Carmichael (1970) in rocks at staurolite grade. Subsequent to the occurrence of Reaction 61 as a discontinuous change, the same reaction then describes the continuous shift of the three-phase assemblage biotite + calcic-amphibole + microcline as a function of T, $f(H_2O)$, and $f(CO_2)$.

The next reaction recorded by Ferry (1976b) is the diopside isograd, but it does not involve biotite as part of the reaction. Nonetheless, biotite still persists as a stable phase in the impure meta-carbonates in the sillimanite + K-feldspar zone as reported by Guidotti (1963, 1965) in the areas about 30 km to the west of the area studied by Ferry. Guidotti (1963) noted that comparison of the sillimanite grade impure carbonates just to the south of the area studied by Ferry with the same units to the west in the K-feldspar + sillimanite zone showed a progressive decrease (to just minor amounts) of biotite and an increase in diopside (concomitant with a darkening of its color). Based upon these observations, the following continuous reaction was proposed:

$$\text{quartz} + \text{calcite} + \text{biotite} = \text{diopside} + \text{microcline} + H_2O + CO_2. \qquad (62)$$

Presumably this reaction involved progressive migration of the assemblage calcite + biotite + diopside to more Fe-rich compositions for the biotite and diopside concomitant with the modal changes noted above.

The reactions in impure carbonates which have been discussed above should not be considered as unique for all impure carbonate sequences. They merely reflect that set of reactions appropriate for the bulk composition range of the impure carbonates that occur in central and western Maine. In other regions different reactions are to be expected. DeBethune (1976) suggests biotite forming by the reaction

$$\text{dolomite} + \text{Mu} = \text{biotite} + \text{chlorite} + \text{calcite} + CO_2 \ . \tag{63}$$

Hoinkes (1983) suggests a more complete form for this reaction based upon rocks in the Eastern Alps:

$$\text{Mu} + \text{dolomite} + \text{quartz} + H_2O = \text{biotite} + \text{chlorite} + \text{calcite} + CO_2 \ . \tag{64}$$

Melson (1966) suggests a sequence of reactions in impure carbonates (and involving phlogopite in several of the reactions) which appears to represent a sequence roughly intermediate between that described by Rice (1977a) and Ferry (1976). Here intermediate is used in the sense of extent of deviation due to impurities (K_2O, $NaAlO_2$, Al_2O_3, etc.) of the system from the ideal "end-member, pure" siliceous dolomite.

Finally it should be noted that in the siliceous carbonates containing only $KAlO_2$ as an impurity, the resulting biotite was essentially at the Mg end member and so the mineralogic reactions involved phases approximating pure substances. As a result, the mineralogic reactions were similar in style to those in ideal siliceous dolomites. In the impure carbonates, such as described by Ferry (1976), the trioctahedral mica is dark, "garden variety" biotite with intermediate Mg/Fe ratios. This biotite commonly participates in continuous reactions as well as discontinuous reactions. In this sense the phase relations are more similar to those to be described below for pelitic schists. However, the sparse chemical data available indicate that biotite in the highly impure meta-carbonates is systematically less aluminous than that from typical pelitic schists. It is more akin to the biotite found in metabasites.

<u>Biotite in metabasites</u>. Biotite is commonly mentioned as occurring in metabasites but usually only as "being present" and rarely in terms of participating in any of the major continuous or discontinuous reactions that occur as a function of grade. To some extent, one can consider the phase relations of biotite in metabasites in terms of which phase contains the relatively small amount of $KAlO_2$ present. Drawing upon the reviews of the literature given by Miyashiro (1973) and Laird (1982) (and realizing that exceptions are not unknown) the following generalizations seem to emerge. In low and medium pressure parageneses of metabasites the K-phase (if present) tends to be Mu ± biotite in the greenschist and epidote amphibolite facies. As grade increases into the amphibolite facies muscovite tends to react with calcic amphibole to form biotite. Miyashiro (1973) suggests the reaction

$$\text{Mu} + \text{Ca-amphibole} = \text{biotite} + \text{anorthite} + \text{quartz} + H_2O \ . \tag{65}$$

Biotite can then persist into the granulite facies and eventually breaks down to pyroxene and K-feldspar.

As pointed out by Miyashiro (1973) and Laird (1982), many authors have noted a general tendency for the biotite of metabasites in the greenschist or epidote-amphibolite facies to have green or greenish-brown colors. At higher grades it becomes more brownish in color. Laird (1982) suggests that the color change may be related to increase of Ti-content and cites relevant optical absorption studies of Robbins and Strens (1972).

Metabasites that have been affected by high pressure metamorphism (e.g. approaching and in "Blue Schist Style"), do not have biotite. Instead the $KAlO_2$ appears to reside in phengitic Mu and stilpnomelane. Moreover, although the occurrence of stilpnomelane is strongly influenced by rock chemistry, it tends to be rare in metabasites of low and medium pressure parageneses. Brown (1971, 1975) has discussed reactions interrelating stilpnomelane and biotite parageneses (in bulk compositions not quite fitting in the metabasite category), and it would seem likely that some of these reactions would explain the absence of biotite and presence of stilpnomelane in metabasites formed in high pressure rocks.

Biotite in pelitic schists: Considerable controversy has arisen with regard to what reactions first give rise to biotite in pelitic schists -- hence the reactions responsible for the biotite isograd. All of the standard textbooks provide reviews of this controversy. In part it has arisen because workers have commonly blurred the distinction between "true" pelites and bulk compositions grading toward calc-schists, metagreywacke, or reworked pyroclastic volcanic rocks. The reactions bringing in biotite as discussed in the previous sections would cover the three bulk compositions just listed, especially if one includes the reactions involving stilpnomelane (see Brown, 1971, 1975).

Here we shall restrict the term pelite to those meta-shales which at low grade consist of quartz + chlorite + Mu + plagioclase ± K-feldspar and which on metamorphism to higher grades involve phases like garnet, staurolite, Al-silicates, etc. In some cases small amounts of a carbonate (dolomite, ankerite, etc.) might be present. In the case of the carbonate-bearing pelites, Thompson and Norton (1968) suggest that biotite might appear by the reaction

$$3 \text{ dolomite (or ankerite)} + \text{K-feldspar} + H_2O = \text{biotite} + 3 \text{ calcite} + 3 CO_2. \quad (66)$$

In the case of the carbonate-free pelites, Thompson (1979) notes that the first appearance of biotite must involve reactions among the minerals quartz, Mu, biotite, chlorite, and K-feldspar. He developed a procedure by which a complete set of five net transfer reactions (essentially continuous reactions as used earlier in this discussion) among these phases can be derived. His treatment focused especially on the Mg-Fe exchanges (Mg Fe_) and Tschermak exchanges (Fe Si Al$_{-2}$) among these phases. Three of the reactions which are especially relevant to our purposes are

$$3 \text{ celad} = 2 \text{ K-feldspar} + \text{biotite} + 3 \text{ quartz} + 2 H_2O , \quad (67)$$

$$3 \text{ celad} + 2 \text{ chlorite} = 3 \text{ biotite} + 7 \text{ quartz} + 4 H_2O , \quad (68)$$

$$\text{K-feldspar} + 3 \text{ chlorite} = \text{biotite} + 2 \text{ quartz} + H_2O . \quad (69)$$

[See Thompson (1979) for appropriate stoichiometry of mineral formulas.]

It is especially important to recognize that these are continuous reactions and thus involve progressive changes of mineral compositions as a function of grade increase. Hence, any rock containing the assemblage given by one of the equations will be subjected to the appropriate changes in modal amounts and compositions of the minerals. The equations in Thompson's treatment are written in terms of Fe-end members. Hence, he notes that in terms of Mg-Fe exchange and Tschermak exchange, these reactions cover much of what is essential for the appearance of biotite in pelitic schists. Specifically, these reactions enlarge the AKFM composition range in which biotite-bearing assemblages can occur and enlarge the composition field of biotite itself. Concomitantly Mu decreases modally and becomes less celadonitic and chlorite decreases modally but becomes more Al-rich.

In essence, it appears that the *discontinuous* reaction that first brings in biotite occurs in bulk compositions not represented by natural parageneses. Hence, the "incoming" of biotite is by means of one of the net-transfer reactions listed above -- the particular one being a function of bulk composition. All of these take place at grade somewhat above the initial, intrinsic stability of biotite. As noted by Thompson (1979), many other reactions have been written for the biotite isograd but none is independent of the five net transfer reactions derived in his treatment of this isograd. Indeed, consideration of the biotite-forming reactions for pelites as discussed in Mather (1970), McNamara (1976), Brown (1971, 1975), Ernst (1963), etc., shows them to have obvious similarity to one or more of Reactions 67 to 69.

Subsequent to biotite's appearance in pelites it is involved in a vast number of continuous and discontinuous reactions. It is not feasible to consider very many of these reactions here, so for the continuous reactions we will merely cover the sets that have been observed in a single, well studied metamorphic sequence. Our goal is to provide an illustration of the approach used for identifying the continuous reactions that occur in a particular metamorphic sequence. Other metamorphic sequences may display different patterns depending upon the bulk compositions available and the PT path followed during metamorphism.

For the discontinuous reactions involving biotite our attention will be on those that markedly affect the range of bulk compositions (in the AKFM system) in which biotite can occur. To accomplish this goal it is first necessary to consider the bulk composition of pelitic schists with regard to the AKFM projection of Thompson (1957). Albee (1952) showed that with regard to this system, the majority of pelitic schists have bulk compositions that plot below a line that would connect the points at which garnet and chlorite plot on the projection. Only a relatively few pelitic schists are sufficiently Al-rich that they plot above this line. Moreover, the fact that in the vast majority of pelitic metamorphic sequences we find that the chlorite zone is followed by the biotite zone and then garnet zone (rather than zones(s) involving chloritoid) strongly supports the suggestion that the majority of pelitic schists plot below the garnet-chlorite line of an AKFM diagram. Thus, our consideration of the discontinuous reactions involving biotite will be focused on bulk compositions plotting below the garnet-chlorite line. Moreover, we will ignore reactions occurring in bulk compositions lying significantly removed from a 1:1 mole ratio of FeO:MgO. These restrictions will enable us to concentrate on the most important reactions that involve biotite. Moreover, these reactions also constitute the most important changes (from the viewpoint of mapping metamorphic grades) that occur in pelitic schists up through the amphibolite facies.

Upon attaining garnet grade, the majority of pelitic schists will consist of AKFM mineral assemblages involving garnet, chlorite, and biotite in some combination. As long as the garnet + chlorite compatibility exists, biotite is unable to coexist with any of the more aluminous AKFM phases (chloritoid, staurolite, Al-silicate, etc.) [see extended discussion Guidotti (1968) and Albee (1972)]. However, with further increase in grade, a series of tie line changes occur which systematically bring biotite into compatibility with progressively more Al-rich phases. In essence, the range of AKFM bulk compositions (in terms of Al_2O_3 content) that will include biotite as a coexisting phase becomes enlarged.

As developed in detail by Albee (1972), starting from the garnet zone compatibilities discussed above, two sequences of reactions appear to be possible, depending upon the external conditions (P, T, etc.) that prevail in a given metamorphic sequence. Judging from the assemblages and isogradic reactions described in the literature, it appears that one sequence is markedly more common that the other. In the context of the bulk composition range we are considering, the more common sequence involves

$$\text{garnet + Mg-chlorite + Na-Mu =} \tag{70}$$
$$\text{staurolite + biotite + Na-richer Mu + albite + quartz + } H_2O \;.$$

In many areas this reaction forms the staurolite isograd. Then

$$\text{staurolite + Mg-chlorite + Na-Mu =}$$

$$\text{Al-silicate + biotite + K-richer Mu + albite + quartz + H}_2\text{O} \ . \tag{71}$$

This reaction commonly is the basis for the andalusite, kyanite, or sillimanite isograd. Then

$$\text{staurolite + Na-Mu + quartz =}$$
$$\text{Al-silicate + biotite + garnet + K-richer Mu + albite + H}_2\text{O} \ . \tag{71}$$

This reaction (studied experimentally by Hoschek, 1969) is the basis for the isograd marking the upper kyanite or upper sillimanite zones. On an AKFM projection the resulting Al-silicate + garnet + biotite three-phase field occupies an extensive part of the bulk composition projected. The metamorphic grade would be well up in the amphibolite facies.

In the more common sequence, chloritoid became unstable before the garnet-chlorite join was broken. In the less common sequence this is not so. Hence, the reactions are (modified from Albee, 1972, Fig. 14)

$$\text{garnet + chlorite + Mu = biotite + chloritoid + Na-richer Mu + H}_2\text{O} \ , \tag{73}$$

$$\text{chloritoid + chlorite + Mu = staurolite + biotite + Na-richer Mu + H}_2\text{O}, \tag{74}$$

(another possible staurolite isograd), and then

$$\text{chloritoid + Mu = garnet + staurolite + biotite + Na-richer Mu + H}_2\text{O} \ . \tag{75}$$

Thereafter, Reactions 71 and 72 follow as in the more common sequence.

None of the above reactions have involved cordierite because (at least in regional metamorphic parageneses) it requires more Mg-rich compositions than those we have chosen to consider. However, at grades exceeding the upper sillimanite (kyanite) zone as described above, reactions occur which set the stage for a biotite-cordierite compatibility in rocks of intermediate Mg-Fe ratio. The preliminary reaction involves the breakdown of Mu to K-feldspar + sillimanite. At still higher grade -- essentially merging into granulite grade conditions -- the Al-silicate-biotite join becomes unstable and reorients to form a garnet-cordierite join. The specific reaction is

$$\text{Al-silicate + biotite + quartz = garnet + cordierite + K-feldspar + H}_2\text{O} \tag{76}$$

From our perspective of biotite-bearing parageneses, this reaction *(ideally)* results in biotite coexisting with garnet and cordierite and no longer with an Al-silicate. In contrast with the effects of the lower grade reactions, this higher grade reaction reverses the trend of enlarging the range of bulk compositions in terms of Al_2O_3 content that can include biotite as a coexisting phase. Briefly recalling our previous discussion of biotite composition, it should be noted that Reaction 76 roughly coincides with the metamorphic conditions at which the Ti content of biotite starts to increase dramatically. Ultimately biotite becomes intrinsically unstable and breaks down by reactions such as

$$\text{biotite + quartz = orthopyroxene + K-feldspar + H}_2\text{O} \ . \tag{77}$$

Turning now to an example that illustrates the nature of some of the continuous reactions that involve biotite we will make use of a data set from a study in progress (C.V. Guidotti, and J.T. Cheney). These data will also provide a detailed picture of biotite compositional variation as a function of grade. The example involves biotite from the same composition range as used in our treatment of the important discontinuous reactions. All of the specimens used for this data set are from the well studied, continuous metamorphic sequence

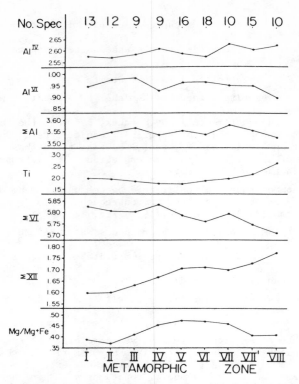

Figure 40. Compositional variation of biotite as a function of metamorphic grade in the pelitic schists in Western Maine. Zones I-III are lower to upper garnet zone and the AFM assemblage is biotite + garnet + chlorite. Zones IV-V are lower and upper staurolite zone, and the AFM assemblage is biotite + garnet + staurolite + chlorite. Zone VI is the transition zone, and the AFM assemblage is biotite + garnet + staurolite + sillimanite ± chlorite. Zones VII and VII' are the lower sillimanite zone of the Rangeley and Oquossoc areas, respectively, and the AFM assemblage is biotite + garnet + staurolite + sillimanite. Zone VIII is the upper sillimanite zone, and the AFM assemblage is biotite + garnet+ sillimanite. Mu + quartz + plagioclase are present in all rocks.

of grades that occurs in western Maine (Guidotti, 1966, 1970a,b, 1974; Henry, 1981; Wood, 1981). The grades range through the lower to upper parts of the garnet zone into the staurolite zone and the lower and upper sillimanite zones, all of which are separated by true isogradic reactions in the sense of the reactions involving distinct AKFM topology changes. The garnet zone has been subdivided into lower, middle, and upper portions based upon progressive changes in the composition of various solution phases (Mu, chlorite, and biotite). In a similar fashion the staurolite and lower sillimanite zones have been subdivided into lower and upper portions. Finally a thin zone exists between the staurolite zone and lower sillimanite zone in which the rocks have mineral assemblages intermediate between the staurolite and lower sillimanite zones. Hence, it is termed the "Transition Zone". Additional details on the designation of zones are available in the references cited above and also in Guidotti (1973, 1978).

Figure 40 shows the average values for various sites, ratios, etc. of biotite as a function of the grades that have been distinguished in the rocks of western Maine. The data which have been plotted are solely from the limiting assemblage produced in the most abundant bulk composition in response to discontinuous Reactions 70, 71, and 72. Hence, Zone I-III specimens contain garnet + chlorite + biotite, Zone IV-V specimens contain staurolite + garnet +

chlorite + biotite, Zone VI specimens contain sillimanite + garnet + staurolite + biotite ± chlorite, Zone VII and VII' specimens contain sillimanite + garnet + staurolite + biotite, and Zone VIII specimens contain sillimanite + garnet + biotite. Our discussion of Figure 40 is mainly limited to aspects related to the continuous reactions that occur as a function of grade in the limiting assemblages listed above. Guidotti and Cheney (in preparation) will discuss in detail the other biotite compositional variations that occur in these samples.

For our consideration of biotite participating in continuous reactions, the most important feature on Figure 40 is the pattern shown by the $Mg/(Mg+Fe_T)$ ratio. In Zones I-V, biotite (in the limiting assemblages) as well as the coexisting phases become progressively enriched in Mg via the reactions

$$\text{Fe-chlorite} + \text{Mu} = \text{garnet} + \text{Mg-richer biotite} + \text{Mg-richer chlorite} + \text{Na-richer Mu} + H_2O , \quad (78)$$

$$\text{Mg-chlorite} + \text{Mu} = \text{staurolite} + \text{Mg-richer biotite} + \text{Mg-richer chlorite} + \text{Na-richer Mu} + H_2O . \quad (79)$$

Subsequent to the formation of the four-phase limiting assemblage of zones VII-VII', via discontinuous Reaction 71, it is evident that biotite then reverses to enrichment in Fe as grade increases. The continuous reaction responsible is:

$$\text{staurolite} + \text{Na-Mu} + \text{quartz} = \text{Al-silicate} + \text{Fe-richer biotite} + \text{Fe-richer staurolite} + \text{K-richer Mu} + \text{albite} + H_2O \quad (80)$$

The Fe-enrichment of biotite occurring in the assemblage Al-silicate + staurolite + garnet + biotite was recognized by Chinner (1965), but has rarely been documented since. Comparison with Figure 33 shows that the muscovites coexisting with the biotites plotted on Figure 40 show the same pattern with respect to $Mg/(Mg+Fe_T)$ ratio. Several points worthy of further note are: (1) Because all of the specimens used in Figure 40 have essentially the same degree of Al-saturation, Al^{iv}, Al^{vi}, and ΣAl all remain at the high levels typical of biotite from Al-saturated parageneses. And (2) many people have tried to make gross generalizations about how biotite varies compositionally in response to grade -- it is obvious that such statements must be couched precisely in terms of a specific assemblage and thus a specific continuous reaction.

Finally, a brief comment should be made concerning the pattern for TiO_2 on Figure 40. Previously it was noted that in general Ti in biotite increases as a function of grade. However, it appears that in detail there may be deviations from this generalization. From Zone I to V it appears that Ti is constant to slightly decreasing: only after sillimanite comes in does Ti start rising. No detailed attempt will be made here to provide an understanding for the pattern of Ti in Zones I-V, but note that $Mg/(Mg+Fe_T)$ rises markedly over this same range, and previous discussion has focused on an antipathy between Ti and high $Mg/(Mg+Fe_T)$ ratio in biotite. However, that antipathy usually is manifest at $Mg/(Mg+Fe_T)$ ratios in a range > 0.6.

(II.C) Petrogenetic aspects of biotite

Several aspects of biotite phase relations have direct implications for discussions about the petrogenesis of metamorphic rocks. Here, we shall review them briefly under the following categories: (1) isogradic reactions, (2) continuous changes in biotite composition in response to continuous reactions, (3) relationships with volatile constituents, and (4) exchange reactions and geothermometry. These will be discussed mainly in the context of pelitic bulk compositions, because only for them is adequate information available for our purposes.

Isogradic reactions. Obviously the biotite isograd will come to mind immediately. Despite some controversy over the exact nature and PT conditions for the occurrence of this isograd, it is well accepted as being an important indicator for metamorphism rising above chlorite zone. Our previous discussion has considered some aspects of the reactions that bring in biotite. It was noted that the reactions bringing in biotite in pelitic bulk compositions are probably continuous rather than discontinuous types. Moreover, the reactions bringing in biotite in pelitic rocks are quite distinct from those producing biotite in other bulk compositions. Although the biotite isograd is the most renowned isograd involving biotite, it is important to note that several other very commonly mapped isograds have biotite participating in the reaction that defines the isograd. For example Reaction 70 is probably the essence of the most commonly mapped staurolite isograd, and Reaction 71 is probably the manner in which andalusite, sillimanite, or kyanite is formed in the majority of pelites. Hence, it forms the basis for many andalusite, sillimanite, or kyanite isograds. Other examples could be cited, but our purpose here is only to emphasize how biotite is a fundamental participant in many of the most important metamorphic reactions used as major markers of metamorphic grade.

Continuous variation of biotite composition in response to continuous reactions. Our previous discussion has described some of the ways in which biotite changes composition in response to metamorphic grade changes. Here we will discuss only those that can be related directly with major AKFM continuous reactions. Referring to Figure 40, we saw that due to Reactions 78 and 79 biotite became progressively enriched in Mg and then, due to Reaction 80, it became Fe-enriched. The possibility thus arises that within a given grade (i.e. mineral facies in the sense of Thompson, 1957), one might be able to contour progressive changes in grade on the basis of a continuous change in composition of a solution phase like biotite. Indeed, the distinction between Zone II and III in the garnet zone of western Maine (see Fig. 40) is largely based on variation in the Mg/(Mg+Fe$_T$) ratio of biotite and similarly for Zone IV and V in the staurolite zone.

Relationships with volatile constituents. Obviously most of the continuous and discontinuous reactions involving biotite depend on a(H_2O), as well as P and T. Hence, the phase relations shown by biotite can be used to infer information on the behavior, activity, etc., of H_2O during a metamorphic event. Here we shall comment mainly on the nature of biotite phase relations in terms of other volatile constituents, in particular $f(O_2)$ and $f(S_2)$. Fluorine can also have an important influence on biotite phase relations, but see our earlier treatment.

A vast literature has grown up concerning the effects of $f(O_2)$ and $f(S_2)$ on silicate phase relation. In terms of the minerals present in rocks, this has involved detailed consideration of the phase relations among the opaque minerals (oxides, sulfides, and graphite) and the silicates. Some of the more important theoretical starting points for dealing with the phase relations between the opaque and silicate minerals might include Eugster and Skippen (1967), Eugster (1972), Thompson (1972), and Miyashiro (1964). References to much of the more recent literature are given in Mohr and Newton (1983) and Nesbitt and Essene (1982). Experimental aspects have been considered also, e.g., see Tso et al. (1979). Consideration of the phase relations between opaques and silicates can be done in the context of reactions controlled by variation of $f(O_2)$ and $f(S_2)$, although Thompson (1972) has shown that the same reactions can be written in terms of dehydration equations. One of the more important aspects of such considerations is as a means of calculating the composition of the fluid phase during metamorphism. An early example of such a usage was given by Guidotti (1970a), and many such calculations have since been done.

For our purposes of considering biotite in the context of sulfide-silicate reactions or oxidation reactions we need only consider the consequences of high $f(O_2)$ or high $f(S_2)$. The former is generally manifest by the presence of magnetite and especially hematite, and the latter is manifest by high modal pyrrhotite and especially pyrite. In a very simplified way one can visualize the results of high $f(O_2)$ producing Fe^{3+} from Fe^{2+} and so depleting the amount of Fe^{2+} available to form silicates. A similar simplistic model resulting in the formation of iron sulfides can be considered for the effects of high $f(S_2)$. Considered in the context of biotite-bearing parageneses, the net results are series of AKFM assemblages in which biotite coexists with progressively more Mg-rich minerals and the biotite becomes more Mg-enriched.

For example, in the case of progressive increase of $f(O_2)$ in a sequence of medium high-grade pelitic schists as indicated by rock oxidation ratio, Hounslow and Moore (1967) found that the mineral assemblages progressively involved more Mg-rich phases and the Mg/(Mg+Fe) ratio of the biotite participating in the assemblages rose from ~ 0.4 to ~ 0.6. The rock oxidation ratio was measured in terms of the ratio (molar $2Fe_2O_3 \times 100/2Fe_2O_3 + FeO$), but it was generally reflected closely in terms of the opaque mineral assemblages present in each rock as well as with the modal amounts of oxide phases present. In an earlier study, Chinner (1960) found very similar relationships regarding rock oxidation state, silicate mineral assemblage, and biotite Mg/(Mg+Fe) ratio in pelitic gneisses from Scotland. Annersten (1968), though working on a quite different bulk composition (Fe-formation), also demonstrated a relationship between the silicate chemistry -- Mg/(Mg+Fe$_T$) ratio -- and oxidation state, $f(O_2)$, as implied by the oxide assemblages present. However, in the study conducted by Annersten, the biotite Mg/(Mg+Fe$_T$) ratios attain values > 0.8 in the rocks containing hematite and, as a result, the inverse effect on the Ti content becomes quite pronounced.

Comparable relationships between high $f(S_2)$ (as indicated by modal amount of sulfides and pyrite versus pyrrhotite) and the silicate assemblages and biotite Mg/(Mg+Fe$_T$) ratio have been recorded by Guidotti (1970), Guidotti et al. (1977), Robinson and Tracy (1977), and Mohr and Newton (1983).

In addition to the obvious petrogenetic significance for gaining information on the nature of the fluid phase during metamorphism, the sulfide-silicate and oxide-silicate phase relations just reviewed have several other very significant aspects bearing directly on characterizing biotite in metamorphic rocks.

(1) By looking at biotites formed in a series of samples subjected to different degrees of sulfidation or oxidation, one is able to study the effects of a wide variation of biotite Mg/(Mg+Fe$_T$) ratio in terms of crystal-chemical effects. Most importantly, by appropriate choices one can conduct such a study while holding Ti, Al, etc. at saturation levels. The reports of Guidotti et al. (1975, 1977) illustrate exploitation of such an approach to study of the substitutions that can occur in biotite. For example, using this approach, they showed, contrary to previous ideas, that increase of biotite Mg/(Mg+Fe$_T$) ratio did not result in a decrease of Al^{vi} in biotite.

(2) The studies of Hounslow and Moore (1967), Chinner (1960), etc., have shown how the Fe^{3+} content of biotite increases as rock oxidation state increases and how this correlates with the oxide mineral assemblages and modal amounts of oxides. Obviously such results imply that in suites of rocks for which the mineral analyses do not distinguish Fe^{3+} from Fe^{2+}, a petrologist can use the

opaque mineralogy to obtain a fairly good idea of the relative amount of Fe^{3+} in minerals like biotite. At the very least, a wide variation in the opaque mineralogy in a suite of rocks should caution a petrologist against merging the chemical data for a mineral like biotite from such a suite when developing petrologic or crystallochemical models. Finally, studying biotite chemistry from a suite of rocks exhibiting different oxidation states as indicated by the opaque minerals should provide insights into the influence of varying Fe^{3+} on other substitutions that can occur in biotite.

Exchange-reactions and geothermometry. Many studies have been conducted on $Mg-Fe^{2+}$ exchange reactions between biotite and other Mg,Fe^{2+} phases. For example, Albee (1965) considered the $Mg-Fe^{2+}$ exchange between biotite and chlorite. Hounslow and Moore (1967) considered the distribution of Mg and Fe^{2+} between biotite and staurolite, etc. However, by far most consideration has been given to the distribution of Mg and Fe^{2+} between garnet and biotite. Goldman and Albee (1977) provide an extensive list of references to earlier studies on this distribution. The culmination of such studies has been the use of the Mg-Fe exchange between garnet and biotite as the basis of a quantitative geothermometer. The three commonly considered calibrations include A.B. Thompson (1976) -- a combined field and thermodynamic calibration; Goldman and Albee (1977) -- a calibration relative to oxygen isotope temperatures; and Ferry and Spear (1978) -- an experimental calibration. Subsequently, numerous studies have used one or more of these geothermometers (Ferry and Spear, most commonly) to determine temperatures of metamorphism. A multitude of papers have also been directed at testing and comparing these three geothermometers and discussing the constraints on their use (Ghent et al., 1982, is a good example of such a paper).

We will not review the garnet-biotite geothermometer again. However, it is notable that the vast majority of the literature dealing with the constraints on this geothermometer have concentrated on problems arising due to the fact that garnets commonly show compositional zoning. Relatively few authors have dealt extensively with the implications of other substitutions in garnet or biotite that might influence their Mg/Fe^{2+} ratios; Goldman and Albee (1977) and Essene (1982) are exceptions. And of course most studies have also assumed that Mg and Fe^{2+} mix ideally in both garnet and biotite. Here, our concern is only with other substitutions that occur in biotite.

Goldman and Albee (1977) point out how substitution of cations beside Fe^{2+} and Mg in octahedral sites of biotite will affect the K_D of the exchange reaction between garnet and biotite. This is apparent by considering the reaction they wrote:

$$8[Fe_3Al_2(SiO_4)_3] + 9[KMg_{2\ 2/3}Al_{1/3}(Al_{1\ 1/3}Si_{2\ 2/3}O_{10})(OH)_2] =$$
$$8[Mg_3Al_2(SiO_4)_3] + 9[KFe_{2\ 2/3}Al_{1/3}(Al_{1\ 1/3}Si_{2\ 2/3}O_{10})(OH)_2] \ . \tag{81}$$

The equilibrium constant would be

$$K = \frac{(a^G_{Mg-G})^8 (a^B_{Fe-B})^9}{(a^G_{Fe-G})^8 (a^B_{Mg-B})^9} \tag{82}$$

Substitution of activity coefficients and mole fractions gives:

$$K = \left[\frac{\gamma^G_{Mg-G}}{\gamma^G_{Fe-G}} \cdot \frac{\gamma^B_{Fe-B}}{\gamma^B_{Mg-B}}\right] \cdot \left[\frac{x^G_{Mg}}{x^G_{Fe}} \cdot \frac{x^B_{Fe}}{x^B_{Mg}}\right]^{24} \tag{83}$$

where the second term in brackets on the right = K_D. Assuming Mg and Fe^{2+} mix ideally in both garnet and biotite then:

$$K = K_D = \left[\frac{X_{Mg}^G}{X_{Fe}^G} \cdot \frac{X_{Fe}^B}{X_{Mg}^B} \right]^{24} \tag{84}$$

The exponent for the term on the right hand side of 84 is specific for the formulas of biotite as written in Equation 81. It would be different if the biotite were exactly on the annite-phlogopite join, $K(Fe,Mg)_3(AlSi_3O_{10})(OH)_2$, because such a biotite has a different number of exchangeable Mg-Fe sites. Indeed, the Ferry and Spear (1978) calibration is for annite-phlogopite and not for the more common biotites that have significant Tschermak exchange present.

Although Goldman and Albee (1977) tried to take into account the influence of Al^{vi}, Ti^{vi}, etc. in biotite on the geothermometer, they, like virtually all other researchers, ignored the problem of some of the Fe being trivalent. This is obviously a problem since the exchange reaction involves only Fe^{2+} and Mg. Moreover, their approach was basically statistical in nature and made little attempt at identifying how the other substitutions that can occur in biotite influenced its Mg/Fe^{2+} ratio and thus its exchange with garnet.

It would appear that for the biotite part of this geothermometer the following efforts should be made in an attempt to achieve a more rigorous calibration. (1) Ascertain the extent of any influence on the geothermometer resulting from lumping Fe^{3+} with Fe^{2+}. Determining the importance of such an influence might enable petrologists to make some assessment on whether the Fe^{3+} content in biotite as inferred from the coexisting opaque minerals will cause problems or not. [For example, it might be found that the Fe^{3+} content in biotite coexisting with graphite is sufficiently low that no problems arise, but if coexisting with magnetite the Fe^{3+} might be sufficiently great in the biotite to significantly offset the inferred T. To date, no attempts have been made to assess such an influence.] (2) Attempt to determine if the high Ti^{vi} that occurs in upper sillimanite zone to granulite zone biotite influences the Mg/Fe^{2+} ratio as well as affecting the number of Mg-Fe exchange sites available. (3) Most pelitic schists are Al-saturated and so there will be relatively little variation in the Al^{vi} total as a function of grade or $Mg/(Mg+Fe_T)$ ratio. The effect of Al^{vi} in such cases will be a uniform decrease in the number of Mg-Fe exchange sites available. In this context, however, it should be recalled that the Ferry and Spear calibration is for biotite with the maximum number of octahedral exchangeable sites, in contrast with biotites from pelites which have $\sim$ 1 per six octahedral sites occupied by other cations -- primarily Al^{3+}. Only when considering rocks like metagreywackes or especially amphibolites will the biotites start to approach closely those used in the calibration of Ferry and Spear.

The fact that Guidotti et al. (1975) found that in graphite-bearing rocks the Mg/Fe_T ratio of biotite did not seem to have any effect on the ΣAl^{vi} suggests that the Tschermak exchange will not influence the Mg/Fe^{2+} *ratio* in biotite. However, this would contrast with the case in Mu where it was seen that Mg was favored over Fe in the Tschermak exchange when written in the direction of Si^{4+} increase.

In summary, garnet-biotite geothermometry has become one of the most important geothermometers used by metamorphic petrologists. However, it appears that the influence of several substitutions, besides $Mg-Fe^{2+}$, that can occur in biotite may have a bearing on further refinement of the calibration of this geothermometer.

(III) Biotite identification and determination of composition

The approach taken here for identification and determination of composition of biotite is the same as followed in the treatment of white micas.

(III.A) Optical methods

The only optical aspect of biotite of particular use from the viewpoint of metamorphic petrology and petrography concerns the color of biotite as seen in thin section. Here it is important to emphasize that the following discussion is restricted only to biotite as found in common metamorphic rocks, amphibolites, metagreywackes, etc., and especially metapelites. It is not meant to apply to biotite in igneous parageneses, veins, ore deposits, etc. Optical absorption studies, etc. (e.g., Robbins and Strens, 1972) have shown that Ti should have a strong effect on the color of minerals like biotite (see Chapter 5). It was noted earlier that in general the Ti content of biotite increases with grade. In cases for which there is reason to believe that Fe^{3+} remains fairly constant (and low) as in a sequence of graphite-bearing rocks, it has been found in many studies that the biotite becomes progressively darker in shades of orange to reddish-brown. Commonly such biotite with more than 2.0 wt % TiO_2 is strongly pleochroic, and in the appropriate optical orientation it has a dark chestnut, reddish-brown color.

Suggestions of relationships between biotite composition and color were made as long ago as Grout (1924), Tilley (1926), and Harker (1932). Hall (1941) provided a more comprehensive overview of the question and suggested, among other things, that the reddish colors noted above were, as implied above, due to TiO_2. Hayama (1959) reviewed the question in considerable detail and concluded that the color of biotite depends mainly on its TiO_2 content and $Fe_2O_3/(Fe_2O_3+FeO)$ ratio. Basically he showed that in cases for which $Fe_2O_3/(Fe_2O_3+FeO)$ was less than about 0.2 that the color of biotite was controlled by its TiO_2 content, becoming progressively darker in shades of reddish brown as TiO_2 increased. In contrast, biotite with relatively low TiO_2 contents (< 2 wt %) was yellowish to greenish-brown to green as the Fe_2O_3 ratio increased from $\sim$ 0.2 to 0.6. Biotite with intermediate contents of both TiO_2 and Fe_2O_3 tended to have intermediate colors but in general it appears that in many cases TiO_2 and Fe_2O_3 are not both abundant in the same biotite specimen.

In the reviewer's experience the suggestions of Hayama appear to be valid at least for the biotite that occurs in common pelitic metamorphic rocks. Although one might consider the color of biotite as a rough monitor of grade, this is probably much too risky. The most important use of biotite colors is that they can serve as an indicator of what sort of opaque minerals to anticipate when doing petrography of the rocks containing the biotite. Shades of red in the biotite suggest low Fe_2O_3 contents and so one might (at least as a starter) anticipate any opaque oxide is ilmenite rather than magnetite or hematite. Obviously shades of greenish-brown should alert one to look especially carefully for the presence of magnetite +/or hematite. And of course it would also serve as a warning that Fe^{3+} may form a significant fraction of the total Fe in biotite and so might influence things like garnet-biotite geothermometry. Consult Chapter 5 for details.

(III.B) Chemical analyses of biotite

Without question, the electron microprobe is the most useful and important approach to obtaining chemical analysis of biotite. Some relatively minor problems arise for getting elements like Cl and F and, of course, H_2O and the very light elements like Li and B cannot be obtained by microprobe analysis, but the main difficulty is the inability to distinguish Fe^{2+} and Fe^{3+}. Our previous discussion of biotite in a petrogenetic context has noted the serious problems that might arise from this limitation. At present the only way to avoid these difficulties is to carry out analyses that distinguish Fe^{2+} from Fe^{3+}, and the most obvious approach is a wet chemical determination of the Fe^{2+}/Fe^{3+} ratio. Alternatively, Mössbauer techniques also hold promise for solving this problem.

In the absence of directed termination of Fe^{2+}/Fe^{3+}, it is still possible for a petrologist to reduce the problems arising from mixing data based upon assemblages containing "Fe^{3+}-poor" and "Fe^{3+}-rich" biotite. The approach for achieving this highly desirable end is the same as that discussed earlier for Mu. As discussed in that section, one can use the opaque mineral assemblages as a guide to the relative Fe^{3+} contents that will be present in silicates like biotite.

SUGGESTIONS FOR FUTURE RESEARCH

From a petrologic point of view, a major goal in using micas is to gain information on the external conditions (P, T, etc.) prevailing during formation of the micas. This requires information on the mineral assemblages and the composition of the micas *in specific assemblages*. Below are considered four approaches to studying micas that should contribute to the goals of a metamorphic petrologist.

(1) Improvement of solution models of micas: Improvement of the solution models of micas would lead directly to more accurate activity models for use in the ever more sophisticated thermodynamic formulations of mineral equilibria. Included in such attempts at improving the solution models would be:

(a) The various substitution models should be determined in detail and examined for interrelationships. This would enable one to ascertain whether substitutions in one site influence the substitutions that can occur concommitantly in other sites, such as the effect that the Cd-substitution has on the Na/(Na+K) ratio of Mu.

(b) An effort should be made to determine how the major substitutions influence the cell dimensions of the micas, thereby providing information on volumes of mixing. Such work has already been done for natural Mu and synthetic biotite, but it would be desirable to study both synthetic and natural materials for all of the important rock forming micas. It is evident from our earlier discussion that it would be especially useful to determine the relationship between Na/(Na+K) ratio and unit cell volume for Pg.

(c) An effort should be made to determine more rigorously the solution limits for the various micas. A particularly important case involves the extent of solution between Mu and Cd. Suggestions were made that, starting with ideal trisilicic Mu, the Tschermak exchange can occur to about halfway to the tetrasilicic end member (i.e., up to Si_7). However, it was also noted that some recent natural data suggest a greater degree of solution.

(2) Improvement of data on mica compositions in the context of mineral assemblage: Here the term mineral assemblage would include the opaque as well as silicate phases. Obtaining a more refined picture of mica compositions in the context of specific assemblages has obvious, direct

petrologic significance. However, in a less obvious way it bears on
the validity of the various activity models for micas used in the now
popular thermodynamic calculations of mineral equilibria. For example,
it may be found that the Fe^{3+} content of biotite from magnetite-bearing
assemblages is sufficient to invalidate straightforward use of the garnet-
biotite geothermometer if using total Fe for the biotite. On the other
hand, it may be found that such a problem does not arise in magnetite-
free rocks that contain graphite. Another simple example would be to
ascertain at what level the Cd content of Mu starts to influence its
Na/(Na+K) ratio. It may be found that the Cd content of Mu of medium
to high grade pelites from "Green Schist Style" terranes does not significantly affect the Na/(Na+K) ratio of the Mu. Then the simple activity
models employed in the past would probably be valid. With regard to the
question of Fe^{2+}/Fe^{3+} ratio in micas, there are two approaches that should
be followed in the context of this discussion: (a) direct determination
of the Fe^{2+}/Fe^{3+} ratio via wet chemical techniques, and (b) relating the
directly determined Fe^{2+}/Fe^{3+} ratios to the mineral assemblage, in particular the suite of opaque phases present.

Finally, it should be added, that studying the compositional variation
of micas in the context of specific mineral assemblages would greatly
aid in attempts to determine the substitution models in and between the
various structural sites of the micas. Aspects of this approach have been
covered in previous discussions of specific micas.

(3) Experimental and thermodynamic modelling studies: The basic stability
curves of the common micas have now been fairly well established via experiments or thermodynamic calculations. This would be with reference to the
equilibria marking the upper stability limits of various micas in the presence of quartz. Future work should concentrate on:

(a) Additional studies of the numerous reactions involving micas
which occur below the upper stability limits of the various micas.

(b) Calibration of the effects of external conditions (P, T, etc.)
on the compositional variation and solution limits of the micas in
the context of specific assemblages. The study of Massone (1981) on
the Cd content of Mu as a function of P is a good example.

(c) Attempts should be made via experiments or thermodynamic modelling to ascertain the influence of extra components (in micas) on
specific equilibria. A simple example for which such work would be
highly desirable would be to determine the effect of the Cd content
of Mu on the Mu-Pg solvus.

(4) Field investigations: In the course of preparing this chapter on
micas, it has become apparent to the writer that expansion of studies on
the phase relations and solution ranges of micas in the high P rocks of
"Blue Schist Style" areas would be highly informative. It would be particularly desirable to see more work on rocks affected by pressures exceeding 10-15 kbar. Holland (1979), Chopin (1981), and others have provided
some especially exciting new insights on mica phase assemblages (e.g.,
for the Pg + jadeite and talc + phengite + phlogopite assemblages). Such
studies will serve as guides on the effect of P on the phase assemblages
involving micas and the compositional response of the micas to variation
of P.

ACKNOWLEDGMENTS

The list of people to whom the writer is indebted is large and includes many of the authors cited in the references list. Some of them were especially helpful in providing preprints, unpublished data, etc. Among these would be included T. Labotka, P.M. Black, J.G. Blencoe, T. Feininger, M. Frey, S. Guggenheim, G. Hoinkes, A. Miyashiro, E-an Zen, J.T. Cheney, R. Dymek, F.P. Sassi, D.W. Mohr, J.W. Valley, J.M. McLelland, A. Zingg, M.T. Gomez-Pugnaire, M.J. Holdaway, D. Rumble, C. Chopin, O.D. Hermes, A.L. Albee, and J. Rice. I am particularly grateful to M. Ross for making available to me his data on the (to date) most Na-rich Mu that has been observed. People who have reviewed or commented upon portions of Chapter 10 include J.T. Cheney, W.E. Trzcienski, M. Ross, D.A. Hewitt, J. Chernosky, F.P. Sassi, and S.W. Bailey. To all of them I express my gratitude but absolve them of any blame for shortcomings (or longcomings) that might be present. I also acknowledge the support of NSF Grants GA 77-04521, EAR-7902597, and EAR-8200580, which have been extremely important for my research on micas as well as for preparation of Chapter 10.

Technical assistance has been provided by the following people: drafting, Jim Hayden and Alice Kelley; typing, Judy Polyot; proof reading and compilation of references, Gretchen L. Guidotti. To all of these people I am grateful for carrying out some of these time-consuming but essential tasks.

REFERENCES

Abraham, K., Hörmann, P.K. and Raith, M. (1974) Progressive metamorphism of basic rocks from the Southern Hohe Tauern area, Tyrol (Austria). N. Jahrb. Mineral. Abh. 122, 1-35.

_____ and Schreyer W. (1975) Minerals of the viridine hornfels from Darmstadt, Germany. Contrib. Mineral. Petrol. 49, 1-20.

_____ and _____ (1976) A talc-phengite assemblage in piemontite schist from Brezovica, Serbia, Yugoslavia. J. Petrol. 17, 421-439.

Abrecht, J. and Hewitt, D.A. (1980) Ti-substitution in synthetic Fe-biotites. Geol. Soc. Am. Abstr. with Progr. 12, 377.

_____ and _____ (1981) Substitution in synthetic Ti-biolites. Geol. Soc. Am. Abstr. with Progr. 13, 393.

Ackermand, D. and Morteani, G. (1973) Occurrence and breakdown of paragonite and margarite in the Greiner Schiefer series (Zillerthal Aps, Tyrol). Contrib. Mineral. Petrol. 40, 293-304.

Albee, A.L. (1952) Comparison of the chemical analyses of sedimentary and metamorphic rocks. Geol. Soc. Am. Bull. 63, 1229 (abstr.).

_____ (1965) Phase equilibrium in three assemblages of kyanite-zone pelitic schists, Lincoln Mountain Quadrangle, Central Vermont. J. Petrol. 6, 246-301.

_____ (1972) Metamorphism of pelitic schists: Reaction relations of chloritoid and staurolite. Geol. Soc. Am. Bull. 83, 3249-3268.

Althaus, E., Karotke, E., Nitsch, K.H. and Winkler, H.G.F. (1970) An experimental re-examination of the upper stability limit of muscovite plus quartz. N. Jahrb. Mineral. Monatsh. 7, 325-336.

Annersten, H. (1968) A mineral chemical study of a metamorphosed iron formation in Northern Sweden. Lithos, 1, 374-397.

Ashworth, J.R. (1975) The sillimanite zones of the Huntley-Portsoy area in Northeastern Dalradian, Scotland. Geol. Mag. 112, 113-136.

Atzori, P., LoGiudice, A., Pezzino, A. and Rittmann, L. (1973) Analisis fattoriale della correlazione fra le variabili chimiche di biotiti di diversi ambianti genetici. Rivista mineraria Siciliana, Anno XXIV, 171-190.

_____ and Sassi, F.P. (1973) The barometric significance of the muscovite from the Savoca Phyllites (Peloritani, Sicily). Schweiz. Mineral. Petrogr. Mitt. 53, 243-253.

Bachinski, S.W. and Simpson, E.L. (1984) Ti-phlogopites of the Shaw's Cove Minette: A comparison with micas of other lamprophyres, potassic rocks, kimberlites, and mantle xenoliths. Am. Mineral. 69, 41-56.

Baltatzis, E. and Katagas, C. (1981) Margarite pseudomorphs after kyanite in Glen Esk, Scotland. Am. Mineral. 66, 213-216.

_____ and Wood, B.J. (1977) The occurrence of paragonite in chloritoid schists from Stonehaven, Scotland. Mineral. Mag. 41, 211-216.

Bearth, P. (1973) Gesteins und Mineralparagenesen aus den Ophiolithen von Zermatt. Schweiz. Mineral. Petrogr. Mitt. 53, 299-334.

Beran, A. (1969) Beiträge zur Verbreitung und Genesis Phengit-fuhrender Gesteine in den Ostalpen. Tscherm. Mineral. Petrog. Mitt. 13, 115-130.

Black, P.M. (1975) Mineralogy of New Caledonian metamorphic rocks; IV Sheet silicates from the Ouégoa District. Contrib. Mineral. Petrol. 49, 269-284.

_____ (1977) Regional high-pressure metamorphism in New Caledonia: Phase equilibrium in the Ouégoa District. Tectonophys. 43, 89-107.

Blasi, A. and Blasi DePol, C. (1973) $2M_1$ e $3T$ polimorfi delle miche diottaedriche coesistenti nei granite del Massiccio dell'Argentera (Alpi Marittime). Rend. Acc. Naz. Lincei 55m, 528-545.

_____ and _____ (1974) Condizioni di cristallizzazione del polimorfo $3T$ delle miche diottaedriche. Rend. Acc. Naz. Lincei 8/55, 1-7.

Blencoe, J.G. (1974) An experimental study of muscovite-paragonite stability relations. Ph.D. dissertation, Stanford Univ., Palo Alto, California.

_____ (1977) Molar volumes of synthetic paragonite-muscovite micas. Am. Mineral. 62, 1200-1215.

Boak, J.L. and Dymek, R.F. (1982) Metamorphism of the ca. 3800 Ma supracrustal rocks at Isua, West Greenland: Implications for early Archean crustal evolution. Earth Planet. Sci. Let. 59, 155-176.

Bocchio, R. (1977) $3T$ nuscovite from a staurolite-zone South-Alpine gneiss Carmeledo, Italy. Mineral. Mag. 41, 400-402.

_____, Crespi, R., Liborio, G. and Mottana, A. (1980) Variazioni composizionali delle miche chiare nel metamorfismo progrado defli sciste sudalpini dell'alto Lago di Como. Mem. 1st Geol. Mineral Univ. Padova 34, 153-176.

Bohlen, S., Peacor, D. and Essene, E. (1980) Crystal chemistry of a metamorphic biotite and its significance in water barometry. Am. Mineral. 54, 55-62.

Borg, I.Y. and Smith, D.K. (1969) Calculated x-ray powder patterns for silicate minerals. Geol. Soc. Am. Mem. 122, 896 p.

Bowen, N.L. (1940) Progressive metamorphism of siliceous limestone and dolomite. J. Geol. 48, 225-274.

Briand, B. (1980) Geobarometric application of the b_0 value of K-white mica to the Lot Valley and Middle Cevennes metapelites. N. Jahrb. Mineral. Montsh. 12, 529-542.

Brown, E.H. (1967) The greenschist facies in part of Eastern Otago, New Zealand. Contrib. Mineral. Petrol. 14, 259-292.

_____ (1968) The Si^{+4} content of natural phengites: A discussion. Contrib. Mineral. Petrol. 17, 78-81.

_____ (1971) Phase relations of biotite and stilphomelane in the greenschist facies. Contrib. Mineral. Petol. 31, 275-299.

_____ (1975) A petrogenetic grid for reactions producing biotite and other Al-Fe-Mg silicates in the greenschist facies. J. Petrol. 16, 258-271.

Brown, G. and Norrish, K. (1952) Hydrous micas. Mineral. Mag. 29, 929-932.

Brown, P. Essene, E.J. and Peacor, D.R. (1978) The mineralogy and petrology of manganese-rich rocks from St. Marcel, Piedmont, Italy. Contrib. Mineral. Petrol. 67, 227-232.

Bucher-Nurminen, K. (1981) Petrology of chlorite-spinel marbles from Northwestern Spitsbergen (Svalbard). Lithos 14, 203-213.

_____, Frank, E. and Frey, M. (1983) A model for the progressive regional metamorphism of margarite-bearing rocks in the Central Alps. Am. J. Sci. 283A, 370-395.

Burnham, C.W. and Radoslovich, E.W. (1964) Crystal structure of coexisting muscovite and paragonite. Carnegie Inst. Wash. Yearbook 63, 232-236.

Butler, B.C.M. (1967) Chemical study of minerals from the Moine schists of the Ardnamurchan area, Argyllshire, Scotland. J. Petrol. 8, 233-267.

Carmen, J.H. (1974) Synthetic sodium phlogopite and its two hydrates: Stabilities, properties, and mineralogic implications. Am. Mineral. 59, 261-273.

Carmichael, D.M. (1970) Intersecting isograds in the Whetstone Lake area, Ontario. J. Petrol. 11, 147-181.

_____ (1978) Metamorphic bathozones and bathograds: A measure of the depth of post-metamorphic uplift and erosion on the regional scale. Am. J. Sci. 278, 769-797.

Chatterjee, N.D. (1968) Chromian phengite in an ankertie marble from the Susa Valley, Western Italian Alps. N. Jahrb. Mineral. Monatsh. 3, 103-110.

_____ (1970) Synthesis and upper stability of paragonite. Contrib. Mineral. Petrol. 27, 244-257.

_____ (1971) Phase equilibria in the Alpine metamorphic rocks of the environs of the Dora-Maira-Massif, Western Italian Alps. N. Jahrb. Mineral. Abh. 114, 181-210.

_____ (1972) The upper stability limit of the assemblage paragonite and quartz and its natural occurrences. Contrib. Mineral. Petrol. 34, 288-303.

_____ (1973) Low-temperature compatibility relations of the assemblage quartz-paragonite and the thermodynamic status of the phase rectorite. Contrib. Mineral. Petrol. 42, 259-271.

_____ (1974) Synthesis and upper thermal stability limit of $2M$-margarite, $CaAl_2(Al_2Si_2O_{10})(OH)_2$. Schweiz. Mineral. Petrogr. Mitt. 52, 753-767.

_____ (1976) Margarite stability and compatibility relations in the system $CaO-Al_2O_3-SiO_2-H_2O$ as a pressure-temperature indicator. Am. Mineral. 61, 699-709.

_____ and Froese, E. (1975) A thermodynamic study of the pseudobinary join muscovite-paragonite in the system $KAlSi_3O_8-NaAlSi_3O_8-Al_2O_3-SiO_2-H_2O$. Am. Mineral. 60, 985-993.

_____ and Johannes, W. (1974) Thermal stability and standard thermodynamic properties of synthetic $2M_1$ muscovite $KAl_2(AlSi_3O_{10})(OH)_2$. Contrib. Mineral. Petrol. 48, 89-114.

Cheney, J.T. (1975) Mineralogy and petrology of lower sillimanite through sillimanite + K-feldspar zone pelitic schists, Puzzle Mountain area, Northwestern Maine. Ph.D. dissertation, Univ. Wisconsin, Madison, Wisconsin.

_____ (1980) Chloritoid through sillimanite zone metamorphism of high-alumina pelites from the Hoosac Formation, Western Massachusetts. Geol. Soc. Am. Abstr. with Progr. 12, 401.

_____ and Guidotti, C.V. (1979) Muscovite-plagioclase equilibria in sillimanite + quartz-bearing metapelites, Puzzle Mountain area, Northwest Maine. Am. J. Sci. 279, 411-434.

Chenhall, B.E. (1976) Chemical variation of almandine and biotite with progressive regional metamorphism of the Willyama Complex, Broken Hill, New South Wales. J. Geol. Soc. Australia 23, 234-242.

Chiesa, S., Liborio, G., Mottana, A. and Pasquarè, G. (1972) La paragonite nei calcescisti delle alpi: Distribuzione e interpretazione geopetrologica. Mem. Soc. Geol. Ital. 11, 1-30.

Chinner, G.A. (1960) Pelitic gneisses with varying ferrous/ferric ratios from Glen Clova, Angus, Scotland, J. Petrol. 1, 178-217.

_____ (1965) The kyanite isograd in Glen Clova, Angus, Scotland. Mineral. Mag. 34 (Tilley Vol.) 1322-213.

_____ (1974) Dalradian margarite: A preliminary note. Geol. Mag. 111, 75-78.

Chopin, C. (1977) Une paragenèse à margarite en domaine métamorphique de haute pression-base temperature, massif du Grand Pardis, Alpes Francaises. C. R. Acad. Sci. Paris 285, Serie D., 1383-1386.

_____ (1981) Talc-phengite: A widespread assemblage in high-grade pelitic blueschists of the Western Alps. J. Petrol. 22, 628-650.

Cipriani, C., Sassi, F.P. and Scolari, A. (1971) Metamorphic white micas: Definition of paragenetic field. Schweiz. Mineral. Petrogr. Mitt. 51, 259-302.

_____, _____ and Viterbo-Bassani, C. (1968) La composizione delle miche chiare in rapporto con le constanti reticolari e col grado metamorfico. Rend. Soc. Ital. Mineral. Petr. 24, 153-187.

Coleman, R.G. (1967) Glaucophane schists from California and New Caledonia. Tectonophysics 4, 479-498.

Cooper, A.F. (1980) Retrograde alteration of chromian kyanite in metachert and amphibolite whiteschist from the Southern Alps, New Zealand, with implications for uplift on the Alpine Fault. Contrib. Mineral. Petrol. 75, 153-164.

Dahl, O. (1970) Octahedral titanium and aluminum in biotite. Lithos 3, 161-166.

Dal Piaz, G.V., Battistini, G. Di., Kienast, J.R. and Venturelli, G. (1979) Managaniferous quartzitic schists of the piemonte ophiolite nappe--in the Valsesia-Valtournanche area (Italian Western Alps). Mem. Sci. Geol.: già Mom. degli 1st. Geol. Miner. dell'Univ. di Padova XXXII, 24.

Dallmeyer, R.D. (1974a) The role of crystal structure in controlling and partitioning of Mg and Fe between coexisting garnet and biotite. Am. Mineral. 59, 201-203.

_____ (1974b) Metamorphic history of the northeastern Reading Prong, New York and northern New Jersey. J. Petrol. 15, 325-359.

Davidson, L.T. and Mathison, C.I. (1974) Aluminous orthopyroxenes and associated cordierties, garnets and biotites from granulites of the Quairading District, Western Australia. N. Jahrb. Mineral. Monatsh. 6, 272-287.

Dawson, J.B., Powell, D.G. and Reid, A.M. (1970) Ultrabasic xenoliths and lava from the Lashaine Volcano, Northern Tanzania. J. Petrol. 11, 519-548.

_____ and Smith, J.V. (1977) The MARID (mica-amphibole-rutile-ilmenite-diopside) suite of xenoliths in kimberlites. Geochim. Cosmochim. Acta 41, 309-323.

Day, H.W. (1973) The upper stability limit of muscovite plus quartz. Am. Mineral. 58, 255-262.

De Bethune (1976) Formation of metamorphic biotite by decarbonation. Lithos 9, 309-318.

Deer, W.A., Howie, R.A. and Zussman, J. (1962) Rock Forming Minerals, Vol. 3, Sheet Silicates. John Wiley, New York.

Delaney, J.S., Smith, J.V., Carswell, D.A. and Dawson, J.B. (1980) Chemistry of micas from kimberlites and zenoliths. II. Primary- and secondary-textured micas from peridotite zenoliths. Geochim. Cosmochim. Acta 44, 857-872.

Dietvorst, Eugene, J.L. (1980) Biotite breakdown and the formation of gahnite in metapelite rocks from Kemiö, Southwest Finland. Contrib. Mineral. Petrol. 75, 327-337.

Dymek, R.F. (1983a) Margarite pseudomorphs after corundum, Qôrqut area, Godthåbsfjord, West Greenland. Rapp. Grønlands geol. Unders. 112, 95-99.

_____ (1983b) Titanium, aluminum and interlayer cation substitutions in biotite from high-grade gneisses, West Greenland. Am. Mineral. 68, 880-899.

_____, Boak, J.L. and Kerr, M.T. (1983) Green micas in the Archaean Isua and Malene supracrustal rocks. Southern West Greenland, and the occurence of a barian-chromian muscovite. Rapp. Grønlands geol. Unders. 112, 71-82.

Ekstrom, T.K. (1972a) The distribution of fluorine among some coexisting minerals. Contrib. Mineral. Petrol. 34, 192-200.

_____ (1972b) Coexisting scapolite and plagioclase from two iron formations in Northern Sweden. Lithos 5, 175-185.

Engel, A.E.J. and Engel, C.S. (1960) Progressive metamorphism and granitization of the major paragneiss, Northwest Adirondacks Mountains, New York. Part II: Mineralogy. Geol. Soc. Am. Bull. 71, 1-58.

Ernst, W.G. (1963) Significance of phengitic micas from low grade schists. Am. Mineral. 48, 1357-1373.

_____ and Dal Piaz, G. (1978) Mineral parageneses of eclogite rocks and related mafic schists of the Piemonte ophiolite mappe, Breuil-St. Jacques area, Italian Western Alps. Am. Mineral. 63, 621-640.

Essene, E.J. (1982) Geologic thermometry and barometry. In: Ferry, J.W., ed., Reviews in Mineralogy 10, 153-206.

Eugster, H.P. (1972) Reduction and oxidation in metamorphism (II). 24th Int'l. Geol. Congress, Sec. 10, 3-11.

_____, Albee, A.L., Bence, A.E., Thompson, J.B. and Waldbaum, D.R. (1972) The two-phase region and excess mixing properties of paragonite-muscovite crystalline solutions. J. Petrol.13, 147-179.

_____ and Skippen, G.B. (1967) Igneous and metamorphic reactions involving gas equilibria. In: Abelson, P.H., ed., Researches in Geochemistry. John Wiley & Sons, New York, v. 2, 492-520.

_____ and Yoder, H.S., Jr. (1955) The join muscovite-paragonite. Carnegie Inst. Wash. Yearbook 54, 124-126.

Evans, B.W. (1965) Application of a reaction-rate method to the breakdown equilibria of muscovite and muscovite plus quartz. Am. J. Sci. 263,647-667.

_____ (1969) Chlorine and fluorine in micas of pelitic schists from the sillimanite-orthoclase isograd, Maine. Am. Mineral. 54, 1209-1211.

_____ and Guidotti, C.V. (1966) The sillimanite-potash feldspar isograd in Western Maine, U.S.A. Contrib. Mineral. Petrol.12, 25-62.

Evans, S. and Raftery, E. (1980) X-ray photoelectron studies of titanium in biotite and phlogopite. Clays & Clay Minerals 15, 209-218.

Exley, R.A., Sills, J.D. and Smith, J.V. (1982) Geochemistry of micas from the Finero Spinel-lherzolite, Italian Alps. Contrib. Mineral. Petrol. 81, 63-69.

Farmer, G.L. and Boettcher, A.L. (1981) Petrologic and crystal-chemical significance of some deep-seated phlogopites. Am. Mineral. 66, 1154-1163.

Fawcett, J.J. and Yoder, H.S., Jr. (1966) Phase relations of chlorites in the system $MgO-Al_2O_3-SiO_2-H_2O$. Am. Mineral. 51, 353-380.

Feininger, T. (1980) Eclogite and related high-pressure regional metamorphic rocks from the Andes of Ecuador. J. Petrol. 21, 107-140.

Ferry, J.M. (1976a) Metamorphism of calcareous sediments in the Waterville-Vassalboro area, South-Central Maine: Mineral reactions and graphical analysis. Am. J. Sci. 276, 841-882.

_____ (1976b) P, T, fCO_2, and fH_2O during metamorphism of calcareous sediments in the Waterville area, South-Central Maine. Contrib. Mineral. Petrol. 57, 119-143.

_____ (1983) Regional metamorphism of the Vassalboro Formation, South-Central Maine, U.S.A.: A case study of the role of fluid in metamorphic petrogenesis. J. Geol. Soc. London 140, 551-576.

_____ and Spear, F.S. (1978) Experimental calibration of the partitioning of Fe and Mg between biotite and garnet. Contrib. Mineral. Petrol. 66, 113-117.

Fettes, D.J., Graham, C.W., Sassi, F.P. and Scolari, A. (1976) The lateral spacing of potassic white micas and facies series variation across the Calondonides. Scott. J. Geol. 12, 227-236.

Fiorentini, P.M. and Morelli, G. (1968) La paragenesi delle metamorfite a fengiti $3T$ e muscovite $2M$, in Val Chiusella-Zona Sesia-Lanzo. Atti. Soc. Ital. Sci. Nat. Mus. Civ. Stor. Nat. Milano 107, 5-36.

Fletcher, C.J.N. and Greenwood H.J. (1979) Metamorphism and structure of Penfold Creek area near Quesnel Lake, British Columbia. J. Petol. 20, 743-794.

Foster, M.D. (1956) Correlation of dioctahedral potassium micas on the basis of their charge relations. U.S. Geol. Surv. Bull. 1036-D, 57-67.

_____ (1960) Interpretation of the composition of trioctahedral micas. U.S. Geol. Surv. Prof. Paper 354-B, 11-46.

_____ (1967) Tetrasilicic dioctahedral micas-celadonite from near Reno, Nevada. U.S. Geol. Surv. Prof. Paper 575-C, C17-C22.

Fox, J.S. (1974) Petrology of some low variance meta-pelites from the Lukmanier Pass area, Switzerland. Ph.D. dissertation, Univ. of Cambridge, Cambridge, England.

Frank, E. (1983) Alpine metamorphism of calcareous rocks along a cross-section in the Central Alps: Occurrence and breakdown of muscovite, margarite, and paragonite. Schweiz. Mineral. Petrogr. Mitt. 63, 37-93.

Franz, G. and Althaus, E. (1976) Experimental investigation on the formation of solid solutions in sodium-aluminum-magnesium micas. N. Jahrb. Mineral. Abh. 126, 233-253.

_____, Hinrichsen, T. and Wannemacher, E. (1977) Determination of the miscibility gap on the solid solution series paragonite-margarite by means of the infrared spectroscopy. Contrib. Mineral. Petrol. 59, 307-316.

French, B.M. (1966) Some geological implications of equilibrium between graphite and a C-O-H gas phase at high temperatures and pressures. Rev. Geophys. 4, 223-253.

_____ and Eugster, H.P. (1965) Experimental control of oxygen fugacities by graphite-gas-equilibriums. J. Geophys. Res. 70, 1529-1593.

Frey, M. (1969) A mixed-layer paragonite/phengite of low grade metamorphic origin. Contrib. Mineral. Petrol. 24, 63-65.

_____ (1978) Progressive low-grade metamorphism of a black shale formation, Central Swiss Alps, with special reference to pyrophyllite and margarite-bearing assemblages. J. Petrol. 19, 93-135.

_____, Bucher, K., Franks, E. and Schwander, H. (1982) Margarite in the Central Alps. Schweiz. Mineral. Petrogr. Mitt. 62, 21-45.

_____, Hunziker, J.C., Jäger, E. and Stern, W.B. (1983) Regional distribution of white K-mica polymorphs and their phengite content in the Central Alps. Contrib. Mineral. Petrol. 83, 185-197.

_____ and Niggli, E. (1972) Margarite, an important rock-forming mineral in regionally metamorphased low-grade rocks. Naturwiss. 59, 214-215.

_____ and Orville, P.M. (1974) Plagioclase in margarite-bearing rocks. Am. J. Sci. 274, 31-47.

Froese, E. (1978) The graphical representation of mineral assemblages in biotite-bearing granulites. Current Res., Part A, Geol. Surv. Canada, Paper 78-1A, 323-325.

Fujii, T. (1966) Muscovite-paragonite equilibria. Ph.D. dissertation, Harvard Univ., Cambridge, Massachusetts.

Geyssant, J. and Sassi, F.P. (1972) I lembi di filladi e quarziti del Padauner Kögel (Brennero) ed il loro significato. Mem. Acc. Patav. SS. I.L. AA 84, 15 p.

Ghent, E.D. (1975) Temperature, pressure, and mixed-volatile equilibria attending metamorphism of staurolite-kyanite-bearing assemblages, Esplanade Range, British Columbia. Geol. Soc. Am. Bull. 86, 1654-1660.

_____, Jones, J.W. and Nicholls, J. (1970) A note on the significance of the assemblage calcite-quartz-plagioclase-paragonite-graphite. Contrib. Mineral. Petrol. 28, 112-116.

_____, Knitter, C.C., Raeside, R.P. and Stout, M.Z. (1982) Geothermometry and geobarometry of pelitic rocks, upper kyanite and sillimanite zones, Mica Creek area, British Columbia. Canadian Mineral. 20, 295-305.

Gibson, G.M. (1979) Margarite in kyanite and corundum-bearing anothosite, amphibolite and hornblendite from Central Fiordland, New Zealand. Contrib. Mineral. Petrol. 68, 171-179.

Goffe, B. (1977) Succession de subfacies metamorphiques en Vanoise Meridionale (Savoie). Contrib. Mineral. Petrol. 62, 23-41.

Goldman, D.S. and Albee, A.L. (1977) Correlation of Mg/Fe partitioning between garnet and biotite with 18O/16O partitioning between quartz and magnetite. Am J. Sci. 277, 750-767.

Gorbatschev, R. (1972) Coexisting varicolored biotites in migmatitic rocks and some aspects of element distribution. N. Jahrb. Mineral. Abh. 118, 1-22.

Graeser, S. and Niggli, E. (1967) Zur Verbreitung der Phengite in den Schweizer Alpen. Ein Beitrag zur Zonegraphie der alpinen Metamorphose. Etages Tect.-Colloque de Neuchâtel, 18, 89-104.

Grambling, J.A. (1984) Coexisting paragonite and quartz in sillimanitic rocks from New Mexico. Am. Mineral. 69, 79-87.

Green, T.H. (1981) Synthetic high-pressure micas compositionally intermediate between the dioctahedral and trioctahedral mica series. Contrib. Mineral. Petrol. 78, 452-458.

_____ and Hellman, P.L. (1982) Fe-Mg partitioning between coexisting garnet and phengite at high pressure, and comments on a garnet-phengite geothermometer. Lithos 15, 253-266.

Grew, E.S. (1981) Granulite-facies metamorphism at Molodezhnaya Station, East Antarctica. J. Petrol. 22, 397-336.

Grout, F.F. (1924) Notes on biotite. Am. Mineral. 9, 159-165.

Guggenheim, S. and Bailey, S.W. (1975) Refinement of the margarite structure in subgroup symmetry. Am. Mineral. 60, 1023-1029.

Guidotti, C.V. (1963) Metamorphism of the pelitic schists in the Bryant Pond Quadrangle, Maine. Am. Mineral. 48, 772-791.

_____ (1965) Geology of the Bryant Pond Quadrangle, Maine. Maine Geol. Surv., Quadrangle Mapping Ser. 3, 1-116.

_____ (1966) Variation of the basal spacings of muscovite in sillimanite-bearing pelitic schists of Northwestern Maine. Am. Mineral. 51, 1778-1786.

_____ (1968a) On the relative scarcity of paragonite. Am. Mineral. 53, 963-974.

_____ (1968b) Prograde muscovite pseudomorphs after staurolite in the Rangeley-Oquossoc area, Maine. Am. Mineral. 53, 1368-1376.

_____ (1969) A comment on "chemical study of minerals from the Moine Schists of the Ardnamurchan area, Argyllshire, Scotland" by B.C.M. Butter, and its implications for the phengite problem. J. Petrol. 10, 164-170.

_____ (1970a) The mineralogy and petrology of the transition from the lower to upper sillimanite zone in the Oquossoc area, Maine. J. Petrol. 11, 277-336.

_____ (1970b) Metamorphic petrology, mineralogy, and polymetamorphism in a portion of Northwestern Maine. In: Boone, G.M. (ed.) 1970 New England. Int'l. Coll. Geol. Cong. 62nd Ann. Mtg. Field Trip B-1, 1-29.

_____ (1973) Compositional variation of muscovite as a function of metamorphic grade and assemblage in metapelites from Northwestern Maine. Contrib. Mineral. Petrol. 42, 33-42.

_____ (1974) Transition from staurolite to sillimanite zone, Rangeley Quadrangle, Maine. Geol. Soc. Am. Bull. 85, 475-490.

_____ (1978) Compositional variation of muscovite in medium- to high-grade metapelites of Northwestern Maine. Am. Mineral. 63, 878-884.

_____ and Cheney, J.T. (1976) Margarite pseudomorphs after chiastolite in the Rangeley area, Maine. Am. Mineral. 61, 431-434.

_____, _____ and Conatore, P.D. (1975) Coexisting cordierite + biotite + chlorite from the Rumford Quadrangle, Maine. Geology 3, 147-148.

_____, _____ and Guggenheim, S. (1977) Distribution of titanium between coexisting muscovite and biotite in pelitic schists from Northwestern Maine. Am. Mineral. 62, 438-448.

_____, _____ and Henry, D.J. (1977) Sulfide-silicate phase relations in metapelites of Northwestern Maine. Trans. Am. Geophy. Union 58, 524.

_____ and Crawford, K.E. (1967) Determination of Na/Na+K in muscovite by x-ray diffraction and its use in the study of pelitic schists in Northwestern Maine. Geol. Soc. Am. Abstr. with Progr. (New Orleans).

_____, Post, J.L. and Cheney, J.T. (1979) Margarite pseudomorphs after chiastolite in the Georgetown area, California. Am. Mineral. 64, 728-732.

_____ and Sassi, F.P. (1976a) Muscovite as a petrogenetic indicator mineral in pelitic schists. N. Jahrb. Mineral. Abh. 127, 97-142.

_____ and _____ (1976b) L'Utilita' dell' analisi grafica delle compatibilita' fra minerali di rocce metamorfiche: Un esempio riquardante la paragonite E I silicati di Fe-Mg negli scisti pelitici. Per. Mineral. Roma 45, 65-80.

Gunow, A.J., Ludington, S. and Munoz, J.L. (1980) Fluorine in micas from the Henderson Molybdenite Deposit, Colorado. Econ. Geol. 75, 1127-1131.

Güven, N. (1967) The crystal structure of $2M_1$ phengite and $2M_1$ muscovite. Carnegie. Inst. Wash. Yearbook 66, 4487-4492.

Hall, A.J. (1941) The relation between color and chemical composition in the biotites. Am. Mineral. 26, 29-33.

Harder, H. (1956) Untersuchungen an Paragoniten und an natrium haltigen Muskowiten. Heidelb. Beitr. Mineral. Petrogr. 5, 227-272.

Harker, A. (1932) Metamorphism. Methuen, London.

Harris, N.B.W., Holt, R.W. and Drury, S.A. (1982) Geobarometry, geothermometry, and late Archean geotherms from the granulite facies terrain of South India. J. Geol. 90, 509-527.

Hayama, Y. (1959) Some considerations of the color of biotite and its relation to metamorphism. J. Geol. Soc. Japan 65, 21-30.

Hazen, R.M. and Burnham, C.W. (1973) The crystal structures of one-layer phlogopite and annite. Am. Mineral. 58, 889-900.

_____ and Wones, D.R. (1972) The effect of cation substitution on the physical properties of trioctahedral micas. Am. Mineral 57, 103-129.

_____ and _____ (1978) Predicted and observed compositional limits to trioctahedral micas. Am. Mineral. 63, 885-892.

Hemley, J.J., Meyer, C. and Richter, D.H. (1961) Some alteration reactions in the system $Na_2O-Al_2O_3-SiO_2-H_2O$. U.S. Geol. Surv. Prof. Paper 424-D, 338-340.

Henley, K.J. (1970) Application of the muscovite-paragonite geothermometer to a staurolite-grade schist from Sulitjelma, North Norway. Mineral. Mag. 37, 693-704.

Henry, D.J. (1981) Sulfide-silicate relations of the staurolite grade pelitic schists, Rangeley Quadrangle, Maine. Ph.D. dissertation, Univ. Wisconsin-Madison, Wisconsin.

Henry, W.E. (1974) Metamorphism of the pelitic schists in the Dixfield Quadrangle, Northwestern Maine. Ph.D. dissertation, Univ. Wisconsin-Madison, Wisconsin.

Hewitt, D.A. (1973) The metamorphism of micaceous limestones from South-Central Connecticut. Am. J. Sci. 273-A, 444-469.

_____ (1975) Stability of the assemblage phlogopite-calcite-quartz. Am. Mineral. 60, 391-397.

_____ and Wones, D.R. (1975) Physical properties of some synthetic Fe-Mg-Al trioctahedral biotites. Am. Mineral. 60, 854-862.

Höck, V. (1974a) Coexisting phengite, paragonite and margarite in metasediments of the Mittlere Hohe Tauern, Austria. Contrib. Mineral. Petrol. 43, 261-273.

_____ (1974b) Zur metamorphose Mesozoischer metasedimente in den mittleren Hohen Tauern (Österreich). Schweiz. Mineral. Petrogr. Mitt. 54, 567-593.

Hoffer, E. (1978) On the "late" formation of paragonite and its breakdown in pelitic rocks of the Southern Damara Orogen (Namibia). Contrib. Mineral. Petrol. 67, 209-219.

Hoinkes, G. (1983) Cretaceous metamorphism of metacarbonates in the Austroalpine Schneeberg Complex, Tirol. Schweiz. Mineral. Petrog. Mitt. 63, 95-114.

_____ and Thöni, M. (1983) Neue geochronologische und geothermobarometrische daten zum ablauf und zur verbreitung der Kretazischen metamorphose im Ötztalkristallin. Jber. 1982 Hochschulschwerpkt. 15, 73-83.

Holdaway, M.J. (1971) Stability of andalusite and the aluminosilicate silicate phase diagram. Am. J. Sci. 271, 97-131.

_____ (1980) Chemical formulae and activity models for biotite, muscovite, and chlorite applicable to pelitic metamorphic rocks. Am. Mineral. 65, 711-719.

Holland, T.J.B. (1979a) Experimental determination of the reaction paragonite = jadeite + kyanite + H_2O, and internally consistent themodynamic data for part of the system $Na_2O-Al_2O_3-SiO_2-H_2O$, with applications to ecolgites and blueschists. Contrib. Mineral. Petrol. 68, 293-301.

_____ (1979b) High water activities in the generation of high pressure kyanite eclogites of the Tauern Window, Austria. J. Geol 87, 1-27.

Honma, H. (1974) Chemical features of biotite from metamorphic and granitic rocks of the Yanai Distict in the Ryoke Belt, Japan. J. Japan. Assoc. Mineral. Petrol. Econ. Geol. 69, 390-402.

Hoschek, G. (1969) The stability of staurolite and chloritoid and their significance in metamorphism of pelitic rocks. Contrib. Mineral. Petrol. 22, 208-232.

_____ (1973) Die reaktion plogopit + calcit + quarz = tremolit + kalifeldspat + H_2O + CO_2. Contrib. Mineral. Petrol. 39, 231-237.

Hounslow, A.W. and Moore, J.M., Jr. (1967) Chemical petrology of Grenville schists near Fernleigh, Ontario. J. Petrol. 8, 1-28.

Hunziker, J.C. and Zingg, A. (1980) Lower Palaeozoic amphibolite to granulite facies metamorphism in the Ivrea Zone (Southern Alps, Northern Italy). Schweiz. Mineral. Petrogr. Mitt. 60, 181-213.

Iiyama, J.T. (1964) Étude des reactions d'échange d'ions Na-K-dans la serie muscovite-paragonite. Bull. Soc. franc. Minéral. Cristallogr. 87, 532-541.

Ivanov, I.P. and Guaynin, V.F. (1970) Stability of paragonite in the system $SiO_2-NaAlSi_3O_8-Al_2O_3-H_2O$. Geochem. Int'l. F, 578-587.

Iwasaki, M. Sassi, F.P. and Zirpoli, G. (1978) New data on the K-white micas from the Sanbagawa Metamorphic Belt, and their petrologic significance. J. Japan. Assoc. Mineral. Petrol. Econ. Geol. 73,274-280.

Jones, J.W. (1971) Zoned margarite from the Badshot Formation (Cambrian) near Kaslo, British Columbia. Canadian J. Earth Sci. 8, 1145-1147.

Kanehira, K. and Banno, S. (1960) Ferriphengite and aefirinejadeite in a crystalline schist of the Iimori District, Kii Peninsula. J. Geol. Soc. Japan 66, 654-659.

Karamata, S.I., Keesmann, and Okrusch, M. (1970) Ein paragonite-fuhrender Granatquarzite in Raum Brezonvica, Subserbien. N. Jahrb. Mineral. Monatsh. 1, 1-19.

Katagas, C. and Baltatzis, E. (1980) Coexisting celadonitic muscovite and paragonite in chlorite-zone metapelites. N. Jahrb. Mineral. Monatsh. 5, 206-214.

Kearns, L.E., Kite, L.E., Leavens, P.B. and Nelen, J.A. (1980) Fluorine distribution in the hydrous silicate minerals of the Franklin Marble, Orange County, New York. Am. Mineral. 65, 557-562.

Kerrick, D.M. (1972) Experimental determination of muscovite + quartz stability with $P(H_2O) < P(total)$. Am. J. Sci. 272, 946-958.

Keusen, H.R. and Peters, T.J. (1980) Preiswerkite, an Al-rich trioctahedral sodium mica from the Geisspfad ultramafic complex (Penninic Alps). Am. Mineral. 65, 1134-1137.

Korikovski, S.P. (1965) Biotite from metamorphic rocks of the greenschist and amphibolite facies. Translated from: Biotity iz porod zelenoslantsevoy i amfibolitovoy fatsiy metamorfizma. Dokl. Akad. Nauk SSSR 160, 189-192.

_____ (1973) Changes in the composition of muscovite-phengite micas during metamorphism. In: Phase Equilibrium and Processes of Mineral Formation. Moscow, Nauka (in Russian), 71-95.

Kotov, N.V. (1974) Muscovite-chlorite paleothermometer. Dokl. Akad. Nauk SSSR 222, 701-704.

_____ (1975) Muscovite-chlorite paleothermometer. Dokl. Akad. Nauk SSSR 222, 174-176.

_____, Mil'kevich, R.I. and Turchenko, S.I. (1969) Paleothermometry of muscovite-bearing metamorphic rocks based on X-ray and chemical analysis of muscovite. Dokl. Akad. Nauk SSSR 184, 1180-1182.

Krogh, E.J. (1980) Geochemistry and petrology of glaucophane-bearing eclogites and associated rocks from Sunnfiord, Western Norway. Lithos 13, 355-380.

Kunitz, W. (1936) Beitrage zur Kenntis der magmatischen Assoziation. III. Die Rolle des Titans und Zirkoniums in den gesteinbildenden Silikaten. N. Jahrb. Mineral. Geol. Paläont. 70, 385-416.

Kwak, T.A.P. (1980) Ti in biotite and muscovite as an indication of metamorphic grade in almandine facies rocks from Sudbury, Ontario. Geochim. Cosmochim. Acta 32, 1222-1229.

_____ and Askins, P.W. (1981) Geology and genesis of the F-Sn-W (-Be-Zn) skarn (Wrigglite) at Moina, Tasmania. Econ. Geol. 76, 439-467.

Labotka, T.C. (1980) Petrology of a medium-pressure regional metamorphic terrane, Funeral Mountains, California. Am. Mineral. 65, 670-689.

_____ (1983) Analysis of the compositional variations of biotite in pelitic hornfelses from Northeastern Minnesota. Am. Mineral. 68, 900-914.

Laduron, D.M. (1971) A staining method for distinguishing paragonite from muscovite in thin section. Am. Mineral. 56, 1117-1119.

Laird, J. (1982) Amphiboles in metamorphosed basaltic rocks: Greenschist facies to amphibolite faces. In: Veblen, D.R. and Ribbe, P.H., eds., Reviews in Mineralogy 9B, 113-135.

_____ and Albee, A.L. (1981) High-pressure metamorphism in mafic schist from Northern Vermont. Am. J. Sci. 281, 97-126.

Lambert, R.St.J. (1959) The mineralogy and metamorphism of the Moine schists of the Morar and Knoydart Districts of Inverness-shire. Trans. Roy. Soc. Edin. 63, 553-588.

Langer, K., Chatterjee, N.D. and Abraham, K. (1981) Infrared studies of some synthetic and natural $2M_1$ diocathedral micas. N. Jahrb. Mineral. Abh. 142, 91-110.

Lee, D.E. and Van Loenen, E. (1969) Phengite micas from the Cambrian Prospect Mountain Quartzite of Eastern Nevada. U.S. Geol. Surv. Prof. Paper 650-C, C455-C458.

Leelanandem, C. (1969a) H_2O+, F and Cl in the charnockitic biotites from Kondapalli, India. N. Jahrb. Mineral. Monatsh. 10, 461-468.

_____ (1969b) Electron microprobe analyses of chlorine in hornblendes and biotites from the charnockitic rocks of Kondapalli, India. Mineral. Mag. 37, 362-365.

Liborio, G. and Mottana, A. (1975) White micas with $3T$ polymorphs from the calcescisti of the Alps. N. Jahrb. Mineral. Monatsh. 11, 546-555.

Lin, Cheng-Yi and Bailey, S.W. (1984) The crystal structure of paragonite-$2M_1$. Am. Mineral. 69, 122-127.

Lonker, S.W. (1975) Geology of the Mt. Zircon Quadrangle, Northwestern Maine. M.S. thesis, Univ. Wisconsin-Madison, Wisconsin.

McNamara, M. (1965) The lower greenschist facies in the Scottish Highlands. Geol. Foren. Stockholm Forhand. 87, 347-389.

Major, V. and Mason, R. (1983) High pressure metamorphism between the Pelagonian massif and Vardor ophiolite belt, Yugoslavia. Mineral. Mag. 47, 139-141.

Massone, H.-J. (1981) Phengite: Eine experimentelle Untersuchung ihres Druck-Temperatur-Verhaltens im System $K_2O-MgO-Al_2O_3-SiO_2-H_2O$. Ph.D. dissertation, Ruhr Univ., Bochum, West Germany.

Mather, J.D. (1970) The biotite isograd and the lower greenschist facies in the Dalradian Rocks of Scotland. J. Petrol. 11, 253-275.

Maxwell, D.T. and Hower, J. (1967) High-grade diagenesis and low-grade metamorphism of illite in the Precambrian Belt Series. Am. Mineral. 52, 843-857.

Melson, W.G. (1966) Phase equilibria in the calc-silicate hornfels, Lewis and Clark County, Montana. Am. Mineral. 51, 402-421.

Miller, C. (1974) On the metamorphism of eclogites and high grade blueschists from the Penninic Terrain of the Tauern Window, Austria. Schweiz. Mineral. Petrogr. Mitt. 54, 371-384.

_____ (1977) Mineral parageneses recording the P, T history of Alpine eclogites in the Tauern Window, Austria. N. Jahrb. Mineral. Abh. 130, 69-77.

Miyano, T. and Miyano, S. (1982) Ferri-annitefrom the Dales Gorge Member iron formations, Wittenoom area, Western Australia. Am. Mineral. 67, 1179-1194.

Miyashiro, A. (1961) Evolution of metamorphic belts. J. Petrol. 2, 277-311.

_____ (1964) Oxidation and reduction in the earth's crust with special reference to the rôle of graphite. Geochim. Cosmochim. Acta 28, 771-729.

_____ (1973) Metamorphism and Metamorphic Belts. Halsted Press, John Wiley & Sons, New York, 492 p.

Mohr, D.W. and Newton, R.C. (1983) Kyanite-staurolite metamorphism in sulfidic schists of the Anakeesta Formation, Great Smoky Mountains, North Carolina. Am. J. Sci. 283, 97-134.

Mueller, R.F. (1972) Stability of biotite: A discussion. Am. Mineral. 57, 300-316.

Munoz, J.L. and Ludington, S.D. (1974) Fluoride-hydroxyl exchange in biotite. Am. J. Sci. 274, 396-413.

_____ and _____ (1977) Fluorine-hydroxyl exchange in synthetic muscovite and its application to muscovite-biotite assemblages. Am. Mineral. 62, 304-308.

Naef, U. and Stern, W.B. (1982) Some critical remarks on the analysis of phengite and paragonite components in muscovite by x-ray diffractometry. Contrib. Mineral. Petrol. 79, 355-360.

Němec, D. (198) Fluorine phengites from tin-bearing orthogneisses of the Bohemian-Moravian Heights, Czechoslovakia. N. Jahrb. Mineral. Abh. 139, 155-169.

Nesbitt, B.E. and Essene, E.J. (1982) Metamorphic thermometry and barometry of a portion of the Southern Blue Ridge Province. Am. J. Sci. 282, 701-729.

_____ and _____ (1983) Metamorphic volatile equilibria in a portion of the Southern Blue Ridge Province. Am. J. Sci. 283, 135-165.

Nicol, A.W. and Roy, R. (1965) Some observations on the system muscovite-paragonite. Canadian J. Earth Sci. 2, 401-405.

Nitsch, K.-H. and Storre, B. (1972) Zur Stabilität von Margarite in H_2O-CO_2-Gasgemischen: Fortschr. Mineral. 50, 71-73.

Oki. Y. (1961) Biotites in metamorphic rocks. Contrib. Geol. Inst. Tokyo, 497-506.

Padan, A., Kisch, H.J. and Shagam, R. (1982) Use of the lattice parameter b_0 of dioctahedral illite/muscovite for the characterization of P/T gradients of incipient metamorphism. Contrib. Mineral. Petrol. 79, 85-95.

Perkins, D., III, Westrum, E.R., Jr. and Essene, E.J. (1980) The thermodynamic properties and phase relations of some minerals in the system CaO-Al_2O_3-SiO_2-H_2O. Geochim. Cosmochim. Acta 44, 61-84.

Petersen, E.V., Essene, E.J., Peacor, D.R. and Valley, J.W. (1982) Fluorine end-member micas and amphiboles. Am. Mineral. 67, 538-544.

Plas, L. van der (1959) Petrology of the Northen Adula region, Switzerland. Leidse-Geol. Med. 24, 415-602.

Popov, A.A. (1968) Composition of muscovites and paragonites synthesized at temperatures of 350°C to 500°C. Geokhimiya 2, 131-144.

Puhan, D. and Johannes, W. (1974) Epperimentelle Untersuchung der Reaktion Dolomite + Kalifeldspat + H_2O = Phlogopite + Calcite + CO_2. Contrib. Mineral. Petrol. 48, 23-31.

Purtscheller, F., Hoernes, S. and Brown, G.L. (1972) An example of occurrence and breakdown of paragonite. Contrib. Mineral. Petrol. 35, 34-42.

Radoslovich, E.W. (1960) The structure of muscovite $KAl_2(AlSi_3O_{10})(OH)_2$. Acta Crystallogr. 13, 919-932.

_____ (1962) The cell dimensions and symmetry of layer-lattice silicates. II. Regression relations. Am. Mineral. 47, 617-636.

_____ (1963) The cell dimensions and symmetry of layer-lattice silicates. V. Composition limits. Am. Mineral. 48, 348-367.

_____ and Norrish, K. (1962) The cell dimensions and symmetry of layer lattice silicates. I. Some structural considerations. Am. Mineral. 47, 599-616.

Raith, M., Hormann, P.K. and Abraham, K. (1977) Petrology and metamorphic evolution of the Penninic ophiolites in the Western Tauern Window (Austria). Schweiz. Mineral. Petrogr. Mitt. 57, 187-232.

Raju, R.D., Satyanarayana, B. and Rao, J.S.R.K. (1977) Variation of MnO and TiO_2 contents of biotite in relation to metamorphic grade. Indian J. Earth Sci. 4, 13-19.

Ramos, J.D.M. and Gallego, M.R. (1979) Sobre una Moscovita con bario de Sierra Nevada (Cordillera Betica). Soc. Esp. Mineral. 1, 103-109.

Rice, J.M. (1977a) Progressive metamorphism of impure dolomitic limestone in the Marysville Aureole, Montana. Am. J. Sci. 277, 1-24.

—— (1977b) Contact metamorphism of impure dolomitic limestone in the Boulder Aureole, Montana. Contrib. Mineral. Petrol. 59, 237-259.

—— (1979) Petrology of clintonite-bearing marbles in the Boulder Aureole, Montana. Am. Mineral. 64, 519-526.

Richardson, S.W. Gilbert, M.C. and Bell, P.M. (1969) Experimental determination of kyanite-andalusite and andalusite-sillimanite equilibria: The aluminum silicate triple point. Am. J. Sci. 267, 259-272.

Ritter, U. (1981) The baric conditions of mid- and Late Proterozoic metamorphic terrains in the lower Orange River region, South Africa. N. Jahrb. Mineral. Monatsh. 3, 133-144.

Robbins, D.W. and Strens, R.G.J. (1972) Charge-tranfer in ferromagnesian silicates: The polarized electronic spectra of trioctahedral micas. Mineral. Mag. 38, 551-563.

Robert, J.-L. (1967a) Phlogopite solid solutions in the system $K_2O-MgO-Al_2O_3-SiO_2-H_2O$. Chem. Geol. 17, 195-212.

—— (1967b) Titanium solubility in synthetic phlogopite solid solutions. Chem. Geol. 17, 213-227.

Robinson, P. and Tracy, R.J. (1977) Sulfide-silicate-oxide equilibria in sillimanite-K-feldspar grade pelitic schists, central Massachusetts. Trans. Am. Geophys. Union 58, 524.

Rosenfeld, J.L. (1956) Paragonite in the schist of Glebe Mountain, Southern Vermont. Am. Mineral. 41, 144-147.

——, Thompson, J.B. and Zen, E-an (1958) Data on coexistent muscovite and paragonite. Bull. Geol. Soc. Am. 69, 1637 (abstr.).

Ruiz, J.L., Aparicio, A. and García Cacho, L. (1978) Chemical variations in biotites during prograde metamorphism, Sierra De Guadarrama, Sistema Central, Spain. Chem. Geol. 21, 113-129.

——, —— and —— (1980) Chemical variations of muscovites from the Sierra de Guadarrama Metamorphic area, Sistema Central, Spain. Geol. Rund. 69, 94-106.

Rumble, D., III (1978) Mineralogy, petrology, and oxygen isotopic geochemistry of the Clough Formation, Black Mountain, Western New Hampshire, U.S.A. J. Petrol. 19, 317-340.

Saliot, P. and Velde, B. (1982) Phengite compositions and post-nappe high-pressure metamorphism in the Pennine Zone of the French Alps. Earth Planet. Sci. Lett. 57, 133-138.

Sassi, F.P. (1972) The petrologic and geologic significance of the b_0 value of potassic white micas in low-grade metamorphic rocks. An application to the Eastern Alps. Tschermaks Mineral. Petrogr. Mitt. 18, 105-113.

——, Krautner, H.G. and Zirpoli, G. (1976) Recognition of the pressure character in greenschist facies metamorphism. Schweiz. Mineral. Petrogr. Mitt. 56, 427-434.

—— and Scolari, A. (1974) The b_0 value of the potassic white micas as a barometric indicator in low-grade metamorphism of pelitic schist. Contrib. Mineral. Petrol. 45, 143-152.

—— and Zanferrari, A. (1972) Il significato geologico del Complesso del Turntaler (Pusteria) con particolare riguardo allá successione di eventi metamorfica pre-alpini nel basamento austridico delle Alpi Orientali. Bull. Soc. Geol. Ital. 91, 533-557.

——, —— and Zirpoli, G. (1974) Some considerations on the South-Alpine basement of the Eastern Alps. N. Jahrb. Geol. Paläont. Mh. 10, 609-624.

——, ——, ——, Borsi, S. and Moro, A. (1974) The Austrides to the south of the Tauern Window and the Periadriatic Lineament between Mules and Mauthen. N. Jahrb. Geol. Paläont. Mh. 7, 421-434.

Saxena, S.K. (1966) Distribution of elements between coexisting muscovite and biotite and crystal chemical role of titanium in the micas. N. Jahrb. Mineral. Abh. 105, 1-17.

—— (1973) Thermodynamics of Rock-Forming Crystalline Solutions. Minerals, Rocks and Inorganic Materials 8, Springer-Verlag, New York.

Schaller, W.T. (1950) An interpretation of the composition of high-silica sericites. Mineral. Mag. 29, 407-415.

Schimann, K. (1978) On regional metamorphism in the Wakeham Bay area, New Quebec. In: Metamorphism in the Canadian Shield. Geol. Surv. Canada Paper 78-10, 245-248.

Schmid, R. and Wood, B.J. (1976) Phase relationships in granulitic metapelites from the Ivrea-Verbano zone, Northern Italy. Contrib. Mineral. Petrol. 54, 255-279.

Schreyer, W. (1981) High pressure research relevant to geodynamics. Terra Cognita, 119-127.

——, Abraham, K. and Kulke, H. (1980) Natural sodium phlogopite coexisting with potassium phlogopite and sodian aluminian talc in a metamorphic evaporite sequence from Derrag, Tell Atlas, Algeria. Contrib. Mineral. Petrol. 74, 223-233.

Seifert, F. and Schreyer, W. (1965) Synthesis of a new mica, $KMg_{2.5}[Si_4O_{10}](OH)_2$. Am. Mineral. 50, 1114-1118.

_____ and _____ (1971) Synthesis and stability of micas in the system $K_2O-MgO-SiO_2-H_2O$ and their relations to phlogopite. Contrib. Mineral. Petrol. 30, 196-215.

Seki, Y. (1973) Basal spacings of metamorphic white micas and type of metamorphism. J. Geol. Soc. Japan 79, 611-620.

Shand, S.J. (1969) Eruptive Rocks. Hafner Publ. Co., N.Y. and London, 488p.

Smith, D. and Albee, A.L. (1967) Petrology of a piemontite-bearing gneiss. San Gorgonio Pass, California. Contrib. Mineral. Petrol. 16, 189-203.

Smith, J.V., Brennesholtz, R. and Dawson, J.B. (1978) Chemistry of micas from kimberlites and xenoliths--I. Micaceous kimberlites. Geochim. Cosmochim. Acta 42, 959-971.

Spear, F.S., Hazen, R.M., and Rumble, D., III (1981) Wonesite: A new rock-forming silicate from the Post Pond Volcanics, Vermont. Am. Mineral 66, 100-105.

Stephenson, N.C.N. (1977) Coexisting hornblendes and biotites from Precambrian gneisses of the South Coast of Western Australia. Lithos 10, 9-28.

Storre, B. and Nitsch, K.-H. (1974) Zur Stabilität von Margarit in System $CaO-Al_2O_3-SiO_2-H_2O$. Contrib. Mineral. Petrol. 43, 1-24.

Talantsev, A.S. (1971) The plagioclase-muscovite geothermometer. Dokl. Akad. Nauk SSSR 196, 1193-1195.

Thompson, A.B. (1974) Calculation of muscovite-paragonite-alkali feldspar phase relations. Contrib. Mineral. Petrol. 44, 173-194.

_____ (1976) Mineral reactions in pelitic rocks: I. Prediction of P-T-X (Fe-Mg) phase relations. Am. J. Sci. 276, 401-424.

_____ and Algor, J.R. (1977) Model systems for anatexis of pelitic rocks. I. Theory of melting reactions in the system $KAlO_2-NaAlO_2-Al_2O_3-SiO_2-H_2O$. Contrib. Mineral. Petrol. 63, 247-269.

_____, Lyttle, P.T. and Thompson, J.B. (1977) Mineral reactions and A-Na-K and A-F-M facies types in the gassetts schists, Vermont. Am. J. Sci. 277, 1124-1151.

Thompson, J.B., Jr. (1957) The graphical analysis of mineral assemblages in pelitic schists. Am. Mineral. 42, 842-858.

_____ (1961) Mineral facies in pelitic schist (in Russian with English summary). In: Sokolov, G.A., ed., Physico-chemical Problems of the Formation of Rocks and Ores. Moscow Akad. Nauk SSSR, 3131-325.

_____ (1972) Oxides and sulfides in regionals metamorphism of pelitic schists. 24th Int'l. Geol. Cong. (Montreal) Section 10, 27-35.

_____ (1979) Tschermak replacement and reactions in pelitic schists. In: Problems of Physical and Chem. Petrol. 1, 146-159 (in Russian).

_____ and Norton, S.A. (1968) Paleozoic regional metamorphism in New England and adjacent areas. In: Zen, E-an et al., eds., Studies of Appalachian Geology, Northern and Maritime, John Wiley & Sons, New York, 319-327.

_____ and Thompson, A.B.(1976) A model system for mineral facies in pelitic schists. Contrib. Mineral. Petrol. 58, 243-277.

_____ and Waldbaum, D.R. (1969) Mixing properties of sanidine crystalline solutions. III. Calculations based on two-phase data. Am. Mineral. 54, 811-838.

Thompson, P.H. (1973) Mineral zones and isograds in "impure" calcareous rocks, and alternative means of evaluating metamorhpic grade. Contrib. Mineral. Petrol. 42, 63-80.

Tilley, C.E. (1926) On some mineralogical transformations in crystalline schist. Mineral. Mag. 21, 34-46.

Tracy, R. (1978) High grade metamorphic reactions and partial melting in pelitic schist, West-Central Massachusetts. Am. J. Sci. 278, 150-178.

Tso, J.L., Gilbert, M.C. and Craig, J.R. (1979) Sulfidation of synthetic biotites. Am. Mineral. 64, 304-316.

Tyler, I.M. and Ashworth, J.R. (1982) Sillimanite-potash feldspar assemblages in graphitic pelites, Strontian Area, Scotland. Contrib. Mineral. Petrol. 81, 18-29.

Valley, J.W., Petersen, E.V., Essene, E.J. and Bowman, J.R. (1982) Fluorphlogopite and fluortremolite in Adirondack marbles and calcualted C-O-H-F fluid compositions. Am. Mineral. 67, 545-557.

Velde, B. (1965a) Phengite micas: Synthesis, stability and natural occurrence. Am. J. Sci. 263, 886-913.

_____ (1965b) Experimental determination of muscovite polymorph stabilities. Am. Mineral. 50, 436-449.

_____ (1967) Si^{+4} content of natural phengites. Contrib. Mineral. Petrol. 14, 250-258.

_____ (1969) The compositional join muscovite-pyrophyllite at moderate pressures and temperatures. Bull. Soc. franc Minéral. Cristallogr. 92, 360-368.

_____ (1971) The stability and natural occurrence of margarite. Mineral. Mag. 38, 317-323.

_____ (1972) Celadonite mica: Solid solution and stability. Contrib. Mineral. Petrol. 37, 235-247.

Warner, J. and Al-Mishwt, A. (1968) Variations of the basal spacings of muscovite and metamorphic grade in Southeastern Pennsylvania. Proc. Pennsylvania Acad. Sci. 42, 193-202.

Weidner, J.R. and Carmen, J.H. (1968) Synthesis and stability of sodium annite, $NaFe_3AlSi_3O_{10}(OH)_2$. Ann. Meet. Geol. Soc. Am. Abstr. with Progr. 315.

White, J.L. and Burns, A.F. (1963) Infra-red spectra of hydronium ion in micaceous minerals. Science 141, 800-801.

Whitney, P.R. and McLelland, J.M. (1983) Origin of biotite-hornblende-garnet coronas between oxides and plagioclase in olivine metagabbros, Adirondack Region, New York. Contrib. Mineral. Petrol. 82, 34-41.

Winchell, A.N. (1927) Further studies in the mica group. Am. Mineral. 12, 267-279.

_____ and Winchell, H. (1951) Elements of Optical Mineralogy. Part II. Description of Minerals. John Wiley & Sons, New York.

Wise, W.S. and Eugster, H.P. (1964) Celadonite: Synthesis, thermal stability and occurrence. Am. Mineral. 49, 1031-1083.

Wones, D.R. (1972) Stability of biotite: A reply. Am. Mineral. 57, 316-317.

_____, Burns, R.G. and Carroll, B.M. (1971) Stability and properties of synthetic annite. Trans. Am. Geophys. Union 52, 369 (abstr.).

Wood, C. (1981) Chemical and Textural Zoning in Metamorphic Garnets, Rangeley Area, Maine. Ph.D. dissertation, Univ. Wisconsin—Madison, Wisconsin.

Yardley, B.W.D., Leake, B.E. and Farrow, C.M. (1980) The metamorphism of Fe-rich pelites from Connemara, Ireland. J. Petrol. 21, 365-399.

Yoder, H.S. and Eugster, H.P. (1955) Synthetic and natural muscovites. Geochim. Cosmochim Acta 8, 225-280.

Zen, E-an (196) Metamorphism of Lower Paleozoic rocks in the vicinity of the Taconic Range in West-Central Vermont. Am. Mineral. 45, 129-175.

_____ (1981) Metamorphic mineral assemblages of slighty calcic pelitic rocks in and around the Taconic Allochthon, Southwestern Massachusetts and adjacent Connecticut and New York. U.S. Geol. Surv. Prof. Paper 1113, 1-128.

_____ and Albee, A.L. (1964) Coexistent muscovite and paragonite in pelitic schist. Am. Mineral. 49, 904-925.

Zingg, A. (1980) Regional metamorphism in the Ivrea Zone (Southern Alps, Northern Italy): Field and microscopic investigations. Schweiz. Mineral. Petrogr. Mitt. 60, 153-179.

Zussman, J. (1979) The crystal chemistry of the micas. Bull. Minéral. 102, 5-13.

11. F-OH and Cl-OH EXCHANGE in MICAS with APPLICATIONS to HYDROTHERMAL ORE DEPOSITS

J.L. Munoz

INTRODUCTION

Fluorine and chlorine are two of the most important and interesting of the trace elements present in rocks. The phyllosilicate mineral group in general, and the micas in particular, provide one of the most common mineral residences for these halogens. Both halogens occur in micas by replacing hydroxyl anions, and thus are found in the mica structure at the same level as the apical oxygens of the tetrahedral sheets where they are bonded to octahedral cations (see Bailey, Chapter 1, Fig. 1). In synthetic trioctahedral micas, OH groups may be completely replaced by F, and even in natural biotites the molal ratio F/(OH+F+Cl) may approach unity. In strong contrast, the amount of Cl substitution in the OH site is much less -- a few exceptional biotites have Cl/(OH+F+Cl) greater than 0.1, but generally that ratio is much less than 0.05. The maximum permissible halogen contents of the dioctahedral micas are considerably less than those for trioctahedral micas, rarely exceeding 0.2 for F/(OH+F+Cl), and 0.01 for Cl/(OH+F+Cl).

The extent of halogen replacement of hydroxyl in micas is governed by a number of independent factors. The most important of these are (1) the activity of halogen ion or halogen acid present during crystallization, (2) the cation population of the octahedral sheet, (3) the temperature of hydroxyl-halogen exchange, and finally (4) the effects, if any, of post-crystallization leaching or enrichment due to hydrothermal fluids or groundwater. This review will discuss the systematics of F=OH and Cl=OH exchange with particular attention to the use of the halogen contents of micas to obtain information concerning the halogen activities in the magma or hydrothermal fluid which was assumed to be in exchange equilibrium with the micas. Most examples will be drawn from hydrothermal ore deposits and related rocks, but the principles discussed are valid for igneous and metamorphic rocks as well. The first subject is a brief review of the thermodynamic model for halogen-hydroxyl exchange.

THERMODYNAMICS OF HALOGEN=HYDROXYL EXCHANGE

The model equation assumes exchange equilibrium between a mica and either a silicate melt or aqueous fluid phase, and is written

$$mica(OH) + HX = mica(X) + H_2O \qquad (1)$$

where X stands for either F or Cl. The log of the equilibrium constant for the exchange is thus

$$\log K = \log(a_X/a_{OH}) + \log(a_{H_2O}/a_{HX}) \quad (2)$$

where a_X and a_{OH} are the activities of halogen and hydroxyl components present in mica solid solution, and a_{H_2O} and a_{HX} are the activities of H_2O and HF or HCl present in the magma or fluid. By assuming a standard state of pure H_2O and HX, respectively, at 1 bar and the temperature of interest, the fluid activities may be replaced by fugacities. A considerable simplification is obtained by assuming ideal mixing of OH, F, and Cl ions in the hydroxyl site, which allows the substitution of mole fractions for activities in the mica activity term. The propriety of this assumption will be discussed later. The log of the equilibrium constant is now written as

$$\log K = \log(X_X/X_{OH})_{mica} + \log(a_{H_2O}/a_{HX}). \quad (3)$$

If the equilibrium constant is known, it is possible to use the analyzed halogen/hydroxyl ratio in the mica to calculate the fugacity ratio of H_2O to halogen acid which prevailed during the exchange. It is important to remember that these exchange equations cannot define an absolute activity or fugacity of halogen acid, but only a ratio relative to H_2O -- in this review the terms "relative activity" or "relative fugacity" may be used as a shorthand substitute notation for the log of that fugacity ratio.

The equilibrium constant for Equation 1 is strongly dependent on the composition of the mica, and especially on the cationic occupancy of the octahedral sheet. This is important, because it means that two micas with comparable F or Cl contents do not necessarily imply the existence of comparable relative halogen acid activity during exchange. The dependence of log K for X=OH exchange on mica composition, and the calculation of relative halogen activities from chemical analyses of micas are the principal subjects of this chapter.

SYSTEMATICS OF F=OH EXCHANGE

Effect on physical properties

The substitution of F for OH has a marked effect on the physical properties of the trioctahedral micas. For example, in F-phlogopite significant increases in hardness, specific resitivity, and dielectic constant relative to the F-free composition have been noted (Bloss et al., 1959; Shell and Ivey, 1969, p. 188). In both phlogopite and annite, the interlayer distance ($c\sin\beta$ or d_{001}) is reduced considerably by F=OH substitution (Table 1). Note that the amount of reduction is about the same for both compositions. Noda and Ushiro (1964) studied synthetic (F,O,OH)-phlogopite compositions, and showed that the c-dimensions of intermediate solid solutions fell on a straight line between the pure F- and OH-end members. Thus, X-ray measurements of the interlayer

Table 1. Effect of F=OH substitution on the interlayer spacing of synthetic phlogopite (Ph) and annite (An)

composition	$c \sin \beta$ (Å)
OH-Ph	10.149 (Wones, 1963)
F-Ph	9.978 (Takeda and Morosin, 1975)
OH-An	10.135 (Wones, 1963)[a]
F-An	9.981 (Munoz and Ludington, 1974)[b]

a. synthesized in equilibrium with the magnetite-wuestite buffer
b. synthesized in an evacuated silica tube at unknown, but presumably low oxygen fugacity

spacing of *synthetic* (F,OH)-phlogopites can be used to determine their F contents. Data of comparable quality do not exist for muscovite because of the inability to synthesize a pure F-muscovite of known composition -- such synthesis attempts are invariably contaminated with topaz, sanidine, or other phases. Nonetheless, Yoder and Eugster (1955) reported no discernible differences between the X-ray patterns of "F-muscovite" synthesized in the presence of additional phases and that of pure OH-muscovite. Thus it appears that the F=OH substitution in muscovite does not reduce the interplanar spacing to the same extent as noted for phlogopite and annite.

This difference in behavior between trioctahedral and dioctahedral micas has been explained by presumed differences in the orientation of the hydroxyl group (e.g. McCauley et al., 1973; Bailey, Chapter 2). In trioctahedral mica structures, the O-H bond is nearly perpendicular to the tetrahedral sheets, which places the proton relatively closer to the interlayer cation. This orientation is thought to weaken the strength of the interlayer bond. Substitution of F for OH would presumably cancel this effect, increasing the interlayer bond strength and slightly collapsing the layers. In contrast, in muscovite the proton is directed towards the vacant M1 octahedron, and thus cannot affect the interlayer bond strength. Giese (1975, and Chapter 4) proposed an electrostatic model which makes use of these differences in OH group orientation to calculate the effect of F⇌OH substitution on the strength of interlayer bonding. The calculations agree with the qualitative predictions discussed above, and show a negligible effect for muscovite and a large effect for phlogopite.

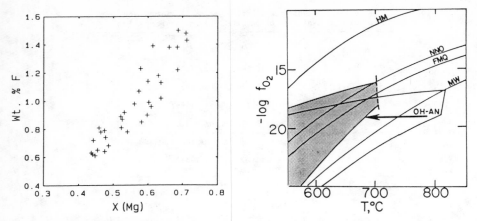

Figure 1 (left). Fe-F avoidance shown by biotites from overlapping zones of phyllic-potassic alteration at the Santa Rita porphyry copper deposit. X(Mg): mole fraction Mg in biotites. Data from Jacobs (1976).

Figure 2 (right). Comparison of stability field of (F,OH)-annite buffered by anorthite-fluorite-sillimanite-quartz (shaded area) with that of OH-annite (unshaded area) as a function of temperature and $f(O_2)$. The dashed line represents the solidus for (F,OH)-annite. Total fluid pressure was 2 kbar. Adapted from Munoz and Ludington (1974).

Fe-F avoidance

It has been observed that hydroxyl-bearing ferromagnesian silicate minerals with high Mg/Fe ratios tend to incorporate more F than comparable minerals with lower Mg/Fe ratios. Because those phases strongly enriched in Fe show relatively low F contents, this general principle is commonly referred to as "Fe-F avoidance". The effect has been shown by chemical analyses of coexisting amphiboles and micas (Ekström, 1972), and has been extensively documented for natural biotite compositions of widely varying Mg/Fe ratios (e.g., Jacobs and Parry, 1979; Zaw and Clark, 1978; Valley and Essene, 1980). Figure 1 illustrates the principle by plotting the mole fraction of the phlogopite component (X_{Mg}) versus the wt % F for biotites from the porphyry copper deposit at Santa Rita, New Mexico (Jacobs, 1976). The positive correlation between high Mg/Fe ratios and high F is obvious. Such empirical observations of Fe-F avoidance in minerals have been substantiated by both theoretical and experimental studies.

Stability relations. It has been widely noted that the substitution of F for OH markedly increases the thermal stability of phlogopite: fluorophlogopite melts at about 1390°C at 1 bar (Shell and Ivey, 1969, p. 160-164), whereas hydroxyl phlogopite dehydrates at 905°C for $P(H_2O)$ = 100 bars (Wones, 1967). A similar increase is not observed for annite compositions, however. In fact, under hydrothermal conditions at a total pressure of 2 kbar, synthetic (OH,F)-annites containing 2.3 to 3.7 % F break down at temperatures well below

Table 2. Equilibrium constants for the exchange reaction
OH-mica + HF = F-mica + H_2O

Composition	Log K
Phlogopite	2100/T + 1.52
Annite[a]	2100/T + 0.41
Siderophyllite[a]	2100/T + 0.20[b]
Muscovite	2100/T − 0.11

Total fluid pressure was either 1 or 2 kbar, and the temperature range was 673 to 1000 kelvins (Munoz and Ludington, 1974, 1977; Ludington and Munoz, 1975).

a. synthesized at oxygen fugacities defined by either methane-graphite (hydrogen fugacity in membrane equilibrium with an O-H gas), or fayalite-magnetite-quartz.
b. log K intercept for siderophyllite revised in 1982.

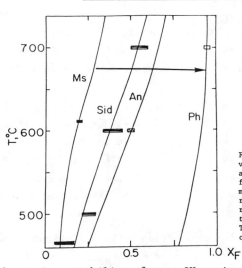

Figure 3. Fluorine contents of synthetic muscovite (Ms), siderophyllite (Sid), Annite (An), and phlogopite (Ph) buffered by anorthite-fluorite-sillimanite-quartz at 1 kbar. X_F = mole fraction F in hydroxyl site of mica. The rectangles show the width of the experimental reversal brackets, and the curves drawn through the brackets follow the equations in Table 2. The arrow emphasizes the strong effect of cation occupancy of F OH partitioning.

the maximum stability of pure OH-annite, when compared at comparable oxygen fugacities (Fig. 2). At very low total pressures in the absence of H_2O (in evacuated silica capsules), pure F-annite melts incongruently to fayalite + liquid at the relatively low temperature of 825±20°C (Munoz, unpub.). These data show that the substitution of F for OH has a somewhat destabilizing influence on the annite structure, in marked contrast to the very strong increase in stability shown for phlogopite compositions. This behavior may be interpreted as additional evidence for Fe-F avoidance.

F-OH exchange experiments. Munoz and Ludington (1974, 1977) used the F buffer assemblage anorthite-fluorite-sillimanite-quartz (AFSQ) to control the f(H_2O)/f(HF) ratio in hydrothermal experiments, and thus were able to measure the equilibrium constants of F=OH exchange for synthetic phlogopite, annite, siderophyllite, and muscovite. The results of reversal experiments are shown in Figure 3. Note that for constant relative HF activity and temperature the amount of F present in the micas increases in the series muscovite - siderophyllite - annite - phlogopite, with the greatest difference noted between phlogopite and annite. From these data, and from similar experiments which used the wollastonite - fluorite - quartz buffer assemblage, it was possible to calculate both the temperature and composition effect on the equilibrium constant (Table 2). The isopleths shown on Figure 3 were calculated from the equations in Table 2 and from fugacity ratios for the AFSQ buffer which were calculated from thermodynamic data.

Thermodynamic models. Ramberg (1952) explained Fe-F avoidance in silicates in terms of the strongly exothermic model reaction,

$$FeF_2 + Mg(OH)_2 = Fe(OH)_2 + MgF_2 \quad . \quad . \tag{4}$$

The equilibrium constant for the reaction is very positive - log K is 9.57 at 298 K (Wagman et al., 1982) which attests to the preference of Fe-OH bonds over Fe-F bonds. Using the same reasoning, one can write an analogous model reaction for Al:

$$FeF_2 + 2/3\ Al(OH)_3 = Fe(OH)_2 + 2/3\ AlF_3, \tag{5}$$

which has a free energy change much closer to zero (log K_{298} = -0.38). This reaction suggests that Al and Fe should behave in very similar fashion with respect to (OH,F) bonding, but that Al should be somewhat less likely than Fe to seek F bonds over OH bonds. This qualitative idea appears to be confirmed by the somewhat smaller equilibrium constant for siderophyllite as compared to annite (Table 2).

Crystal field theory. An alternative explanation of Fe-F avoidance was proposed by Rosenberg and Foit (1977). They pointed out that the crystal field splitting parameter (Δ_o) of Fe^{2+} octahedrally coordinated to F^- is considerably smaller than the same value for Fe^{2+} octahedrally coordinated to OH (7660 cm^{-1} to 6990 cm^{-1} for F^- as opposed to about 10000 cm^{-1} for OH^-). Thus, Fe^{2+} will not be preferred in those sites where coordination by F is possible -- rather, ions with higher bond strengths should be favored in order to maximize the crystal field stabilization energy.

Cation ordering. Perhaps one of the most direct expressions of Fe-F avoidance comes from spectral investigations of natural biotites. Sanz and Stone (1979) obtained NMR spectra of phlogopites with Mg/(Mg+Fe) ratios ranging between 0.93 and 0.99 and F contents between 0.7 and 5.2%. They showed that the OH groups were coordinated to Fe^{2+}, which implies that the F ions specifically avoided those sites which would require coordination by Fe^{2+}.

Thermodynamics of (OH,F) biotite solutions

Relatively little information is available which would allow an accurate description of the thermodynamics of mixing halogens in the hydroxyl site of biotites. As long as the mole fraction of halogen is very small relative to hydroxyl, an ideal mixing model is probably satisfactory. For less dilute solutions, the assumption of ideality may be inappropriate. For example, Duffy and Greenwood (1979) studied the unit cell dimensions and phase relations of synthetic (OH,F)-brucite as well as clinohumite, chondrodite, and norbergite, and preferred a non-ideal model to describe (OH,F) mixing in all these phases. These data may be relevant to biotite solutions because of the presence of brucite-like octahedral sheets in biotite. The situation becomes more complicated when simultaneous mixing is considered in both octahedral sites and anion sites. The principle of Fe-F avoidance in natural biotites argues forcefully against the assumption of ideality for at least certain compositional ranges. Nonetheless, it may still be worthwhile to begin with very simple assumptions for very simple solution compositions, with the understanding that these assumptions may be modified as more data become available.

Thus, with the foregoing restrictions in mind it is possible to consider the simultaneous mixing of (Mg, Fe^{2+}) and (F, OH) in a homogeneous biotite solution in terms of the exchange reaction

$$\text{OH-phlogopite} + \text{F-annite} = \text{OH-annite} + \text{F-phlogopite}. \quad (6)$$

Wood and Nichols (1978) showed that any reciprocal solution whose components may be written in terms of an exchange reaction such as this cannot be ideal unless ΔG for the exchange reaction is equal to zero. From the data in Table 2, it is clear that this is not the case for annite-phlogopite (OH,F) solutions -- for Reaction 6, $\log K = \log K_{Ph} - \log K_{An} = 1.107$. The equilibrium constant for this exchange reaction is independent of temperature, but this independence is probably not valid outside the limited temperature range of experimental data.

By taking OH-phlogopite as the standard state composition,

$$\log K = 2100/T + 1.523 = \log(a_{FPh}/a_{OHPh}) + \log[f(H_2O)/f(HF)] \quad . \quad (7)$$

The simplest possible solution model assumes ideal mixing on each of the sublattices, the compositions of which are defined by the four reciprocal components in Equation 6. The activity terms in Equation 7 are then expanded to

$$\log\left[\frac{a_{FPh}}{a_{OHPh}}\right] = \log\left[\frac{X_{Mg}^{3/2} X_F}{X_{Mg}^{3/2} X_{OH}}\right] + \log\left[\frac{\gamma_{FPh}}{\gamma_{OHPh}}\right] , \quad (8)$$

where γ_{FPh} and γ_{OHPh} are the activity coefficients of F-phlogopite and OH-phlogopite in the reciprocal solution. The activity coefficients are given by

(cf. Wood and Nichols, 1978, Eqn. 19)

$$\log \gamma_{OHPh} = -X_{Fe} X_F \log K \tag{9}$$

$$\log \gamma_{FPh} = X_{Fe} X_{OH} \log K \tag{10}$$

which by substitution of $X_{OH} = 1 - X_F$ leads to the simplification

$$\log(\gamma_{FPh}/\gamma_{OHPh}) = X_{Fe} \log K \tag{11}$$

The equilibrium constant is now written

$$\log K = 2100/T + 1.523 - 1.107 X_{Fe} \tag{12}$$

which is equivalent to

$$\log K = 2100/T + 1.523 X_{Mg} + 0.416 X_{Fe}. \tag{13}$$

In analogous fashion, it is possible to incorporate octahedral Al into the biotite solution by including the equilibrium constant for siderophyllite exchange. The complete equation then becomes

$$\log K = 2100/T + 1.523 X_{Mg} + 0.416 X_{An} + 0.200 X_{Sid} = \log[f(H_2O)/f(HF)] + \log[X_F/X_{OH}]. \tag{14}$$

Considerable ambiguity exists in selecting the most appropriate method for calculating the mole fractions of annite (An) and siderophyllite (Sid) in Equation 14. Gunow et al. (1980) defined

$$X_{Sid} = [(3 - Si/Al)/1.75][1 - X_{Mg}] \text{ and } X_{An} = 1 - (X_{Mg} + X_{Sid}),$$

which accurately describe synthetic phlogopite-annite-siderophyllite solutions. In natural biotite solutions, however, sizable amounts of other cations (most notably Fe^{3+}, Ti, Mn, and Li) may be present in octahedral sites. The preceding recipes for calculating X_{An} and X_{Sid} imply that all "extra" octahedral cations will behave like Fe^{2+}, because they will be counted as part of the annite component. This may be reasonable for Mn^{2+} because of the geochemical similarity of Fe and Mn. For instance, the crystal-field splitting parameters for octahedrally coordinated Mn^{2+} in MnF_6^{4-} is very similar to FeF_6^{4-} (8000 cm^{-1} versus 7600 cm^{-1}, Gerloch and Slade, 1973). However, it is almost certainly an unreasonable assumption for Li because of the strong positive correlation between Li and F in micas (Foster, 1960). More evidence of this correlation can be obtained from a study of zinnwaldite compositions, and will be presented later in this chapter. Equation 14 should not be used if substantial Li in biotite is either present or suspected. As for Ti, no data exist which permit prudent speculation. For this reason, the same warning may be warranted for high-Ti biotites.

Fluorine intercept value and fluorine index

It is convenient for comparative purposes to define a single numerical value which expresses the relative degree of F enrichment in a mica. Because of Fe-F avoidance, the wt % F by itself is an inadequate measure of relative enrichment except for micas of very similar Mg/Fe ratio. By separating the compositional variables from Equation 14, it is useful to define a fluorine intercept value (IV(F)) for biotite as

$$IV(F)_{bio} = 1.52\, X_{Mg} + 0.42\, X_{An} + 0.20\, X_{Sid} - \log(X_F/X_{OH}) \ . \quad (15)$$

The complementary term for a muscovite - phengite solid solution is

$$IV(F)_{ms} = 1.52\, X_{Mg} + 0.42\, X_{Fe} - 0.11\, X_{Al} - \log(X_F/X_{OH}) \quad (16)$$

where X_{Al} and X_{Fe} are the mole fractions of Al and Fe (total) in dioctahedral mica (cf. Gunow et al., 1980, and Table 2). By substitution of Equation 15 into a rearranged version of 14, it is apparent that

$$\log[f(H_2O)/f(HF)] = 2100/T + IV(F) \ , \quad (17)$$

which shows that the intercept value is equal to the log of the fugacity ratio at infinite temperature (1/T = 0). It may also be thought of as a measure of the F content of mica which has been corrected for Mg/Fe ratio, and thus can be used to compare relative degrees of F enrichment in different micas without necessarily implying differences in relative halogen activity. Note from Equations 15 and 16 that *smaller* intercept values are to be correlated with *higher* degrees of F enrichment.

Because of the overwhelming effect of Mg on the fractionation of F into the mica structure, Munoz and Gunow (1982) showed that a graphical method using only the wt % MgO and the log wt % F in biotite or phengite as coordinate axes could be used to interpolate a value called the fluorine index (FI). The fluorine index is related to the intercept value by the equation FI = 12.8 - 4 IV(F). For most micas, the FI which is interpolated graphically is within 0.2 units of the FI which is calculated from a measured intercept value. Thus, the principal benefit of the FI graph is that a measure of relative F enrichment in micas can be obtained from incomplete chemical analyses.

Uncertainties relating to anion site occupancy

The calculation of either intercept values or relative HF activities from biotite analyses is obviously dependent on the F/OH ratio of the anion site. Unfortunately, sizable uncertainties -- both analytical and crystallographic -- exist in the accurate determination of this ratio. If both H_2O+ and F have been measured then the calculation of F/OH is straightforward, although it does depend to some degree on the method chosen for converting the chemical analysis to a structural formula. Ludington (1974) and Stern (1979) recommended normalization of the biotite formula with reference to a constant number of tetrahedral + octahedral cations, rather than with reference to a fixed anionic charge. This method does not permit octahedral vacancies. The reason for this recommendation was based on the observation that substantial analytical uncertainties are encountered in the measurement of both F and H_2O+, and any structural formula which is based on a constant anionic charge which includes F^- and Cl^- will reflect this uncertainty. Nonetheless, most subsequent workers have preferred to base biotite formulae on either fixed anionic or cationic charge, and are willing to accept the octahedral vacancies which are commonly implied by such calculations.

The advent of the electron microprobe as the major analytical instrument for minerals has added a new dimension to the anion site problem. Because hydrogen cannot be measured, it is common practice to report hydroxyl in microprobe analyses by difference, i.e., $OH = 2.0 - F - Cl$, for structural formulae based on $12(O + OH + F + Cl)$. The latter approach assumes that the hydroxyl site is filled only with OH groups or with halogens. However, it is well known from studies of natural biotites that the sum of $(OH + F + Cl)$ is commonly less than a completely filled site. For instance, Levillain (1980) found in a survey of many published biotite analyses that the mean value for $(OH + F + Cl)$ was 3.59, assuming 4.0 represented full occupancy. These hydroxyl site deficiencies have been attributed by many to the oxidation of Fe^{2+} (with concomittant loss of H_2), resulting in the conversion of OH groups to O^{2-} anions (Eugster and Wones, 1962; Rimsaite, 1970; Levillain, 1980). This reaction may be written in terms of the ideal annite formula

$$KFe_3^{2+}Si_3AlO_{10}(OH)_2 = K(Fe_2^{3+}Fe^{2+})Si_3AlO_{12} + H_2 \tag{18}$$

Hazen and Wones (1972) used both crystallographic and Mössbauer spectroscopic evidence to demonstrate that synthetic annite contains at least 10 mol % of octahedral Fe^{3+}. However, Foster (1964) concluded that the oxidation of Fe^{2+} will not explain all hydroxyl site deficiencies in biotite. In an alternative interpretation which does not depend on oxidation, Forbes (1972) found a negative correlation between the net cationic charge of the octahedral and tetrahedral sheets and the hydroxyl site occupancy in 100 biotites, and concluded that the OH/O ratio varies in response to cationic substitutions such as $R^{3+} + H^+ = Si^{4+}$.

Regardless of mechanism, the determination of the hydroxyl content of biotites by difference is likely to result in a value for X_{OH} which is too large. Such an error will propagate to an intercept value which is also too large. Based on a survey of several hundred complete biotite analyses, the positive error in IV(F) which results from calculating OH by difference may be as large as 0.25, but is more commonly in the range of 0.05 to 0.10.

SYSTEMATICS OF Cl = OH EXCHANGE

Unfortunately, much less information is available regarding Cl=OH exchange in micas. The main reason for this difference is that, with few notable exceptions, the Cl content of micas is considerably less than the F content, and pure (i.e. OH-free) Cl-micas cannot be synthesized for study. This may reflect mainly differences in ionic radii -- the ionic crystal radius of Cl^- in micas is 1.81 Å, which is significantly larger than either F^- (1.31 Å) or OH^- (1.38 Å). Because of this, relatively few mica analyses in the literature report Cl. Nonetheless, the information which is available is sufficient to emphasize the markedly different behavior of Cl=OH exchange as opposed to F=OH exchange.

Mg-Cl avoidance

If the model halogen exchange reaction (cf. Eqn. 4) is written in terms of Cl rather than F, i.e.

$$FeCl_2 + Mg(OH)_2 = Fe(OH)_2 + MgCl_2 \tag{19}$$

then the equilibrium constant calculated from thermodynamic data is very negative (log K = -10.1 at 298.15 K), indicating that Fe-Cl bonds should be strongly preferred over Mg-Cl bonds. By analogy with the effect of Mg/Fe ratio on F=OH exchange, this tendency may be referred to as "Mg-Cl avoidance", and has been confirmed in many analyses of natural biotites. In Figure 4, the same biotites

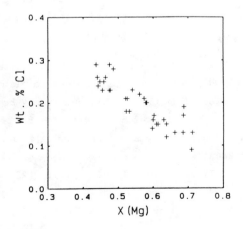

Figure 4. Mg-Cl avoidance shown by biotites from overlapping zones of phyllic-potassic alteration at the Santa Rita porphyry copper deposit. X(Mg): mole fraction Mg in biotite. Data from Jacobs (1976); see also Fig. 1.

which were plotted previously in Figure 1 show a negative correlation between the amount of Mg in the octahedral sheet (X_{Mg}) and the wt % Cl. Note that this relationship is the opposite of Fe-F avoidance.

Munoz and Swenson (1981) measured the equilibrium constant for reaction (1) when X = Cl, and found that synthetic annite contained from 0.27 to 0.75 % Cl when equilibrated hydrothermally with the buffer assemblage muscovite – sanidine – quartz – KCl – HCl (ΣCl = 2 molal) at 1 kbar and for temperatures ranging from 445° to 575°C. In contrast, synthetic phlogopite equilibrated under similar conditions contained less than 0.1 % Cl, a value too low to allow calculation of the equiilbrium constant for Cl=OH exchange. Nonetheless, by combining the annite measurements with data on natural Cl-rich biotites of different Mg/Fe ratios, a semi-quantitative model for Cl=OH exchange in biotite was developed. The resulting expression for the equilibrium constant (written in accordance with Eqn. 1 when X = Cl) is

$$\log K = 5151/T - 5.01 - 1.93\, X_{Mg} \ . \tag{20}$$

Chlorine intercept value

The intercept value for Cl=OH exchange [IV(Cl)] is a measure of relative Cl-enrichment in biotite corrected for Mg-Cl avoidance. By analogy with Equation 15, and assuming the validity of Equation 20,

$$IV(Cl) = -5.01 - 1.93\, X_{Mg} - \log(X_{Cl}/X_{OH}) \ . \tag{21}$$

The chlorine intercept value is related through the equilibrium constant to the relative HCl activity by

$$\log [f(H_2O)/f(HCl)] = 5151/T + IV(Cl) \ . \tag{22}$$

The IV(Cl) is readily calculated from a biotite structural formula, and is always negative; more negative numbers imply higher degrees of Cl enrichment. No comparable equation for IV(Cl) exists for muscovite, reflecting the extremely low Cl contents of most dioctahedral micas.

F/Cl intercept value

Equilibrium constants for F=OH and Cl=OH exchange may be combined to eliminate OH, resulting in the F/Cl intercept value [IV(F/CL)]. This number may be used either as a measure of the F/Cl ratio in biotite which is corrected for the opposing effects of Fe-F and Mg-Cl avoidance, or as a measure of $\log[f(HCl)/f(HF)]$. By subtracting Equation 22 from Equation 17,

$$\log[f(HCl)/f(HF)] = -3051/T + IV(F) - IV(Cl) = -3051/T + IV(F/Cl), \tag{23}$$

where

$$IV(F/Cl) = 3.45\, X_{Mg} + 0.41\, X_{An} + 0.20\, X_{Sid} - \log(X_F/X_{Cl}) + 5.01 \ . \tag{24}$$

Table 3. Mean halogen intercept values of micas from porphyry copper deposits and related igneous rocks
Values in parentheses represent 1 standard deviation

locality (number of analyses)	IV(F)	-IV(Cl)	IV(F/Cl)
1a. Santa Rita (120)	2.02 (0.13)	4.27 (0.07)	6.30 (0.17)
1b. Santa Rita (3)	1.92,1.90,2.00		
2a. Bingham (5)	1.55 (0.07)	4.48 (0.17)	6.03 (0.14)
2b. Bingham (11)	2.11 (0.17)	4.02 (0.41)	6.14 (0.38)
2c. Bingham (5)	1.77 (0.21)		
3. British Columbia (11)	2.14 (0.36)		
4. Arizona (8)	2.12 (0.24)		
5. Boulder batholith (17)	2.18 (0.18)		
6. Tintic (3)	1.63,1.61,1.47		
7. Guadalcanal (9)	2.65,2.82	4.17 (0.11)	6.55,7.10
8. Turkey (6)	2.50 (0.15)	4.31 (0.07)	6.80 (0.16)
9. Chile (5)	2.00 (0.05)	4.64 (0.05)	6.64 (0.03)

1a. biotites from hydrothermally altered granodiorite porphyry at Santa Rita, New Mexico (Jacobs, 1976, Jacobs and Parry, 1979)
1b. hydrothermal sericites (average of 25 analyses) from Santa Rita, New Mexico (Parry et al., 1984)
2a. biotites from latite porphyry dike at Bingham, Utah (Parry et al., 1978)
2b. biotites from equigranular quartz monzonite at Bingham, Utah (Lanier et al., 1978)
2c - 5. biotites from various intrusive systems associated with economically significant mineralization in the western U.S. and Canada (Kesler et al., 1975). Analytical data were reported as ppm F, ppm Cl, and Mg/(Mg+Fe) which allowed use of the fluorine index graph (Munoz and Gunow, 1982) to calculate an approximate IV(F). The following localities are represented: British Columbia- Highland Valley, Copper Mountain, Cat Face, Brenda, Island Copper, Granisle; Arizona- Twin Buttes, Copper Hill, Safford, Florence, Vulture Mountains, Rare Metals, Silverbelle; Utah- Bingham; Montana- Boulder batholith, including the intensely mineralized area around Butte.
6. hydrothermal sericites (averages of 47 analyses) from Southwest Tintic, Utah (Parry et al., 1984)
7. biotites from various phases of the Inamumu zoned pluton (tonalite), Guadalcanal, Solomon Islands (Chivas, 1981)
8. biotites from porphyry copper prospect at Bakircay, Turkey (based on averages of 119 analyses of granodiorite porphyry and associated alteration; Taylor, 1983)
9. biotites from potassic and potassic-phyllic alteration at Los Pelambres, Chile (based on averages of 109 analyses; Taylor, 1983)

Implicit in Equation 23 is the definition of the F/Cl intercept value:

$$IV(F/Cl) = IV(F) - IV(Cl) . \qquad (25)$$

Because Equation 24 does not contain a term for X_{OH}, the calculation of IV(F/Cl) is insensitive to uncertainties in hydroxyl anion occupancy. For this reason, F/Cl intercept values are likely to be more accurate than either IV(F) or IV(Cl) alone.

HALOGEN COMPOSITIONS OF HYDROTHERMAL MICAS RELATED TO ORE DEPOSITS

Based primarily on the abundance of published data, it is convenient to subdivide the chemical analyses of hydrothermal micas and micas from associated ore deposits and igneous rocks into two major groups -- porphyry copper systems and Mo-Sn-W-Be systems (including porphyry deposits, skarns, and pegmatites and

Table 4. Mean halogen intercept values of micas from Mo-Sn-W-Be deposits and related igneous rocks
Values in parentheses represent 1 standard deviation

locality (number of analyses)	IV(F)	-IV(Cl)	IV(F/Cl)
10a. Henderson (39)	0.68 (0.21)	3.96 (0.40)	4.67 (0.40)
10b. Henderson (40)	0.76 (0.14)		
11. N.W.T., Canada (8)	1.58 (0.19)		
12. Portugal (12)	1.62 (0.45)		
13a. Tasmania (12)	1.53 (0.28)	4.07 (0.22)	5.60 (0.18)
13b. Tasmania (12)	1.07 (0.23)	4.06 (0.32)	5.15 (0.51)
14. Alaska (2)	1.49, 1.68		
15a. Nigeria (13)	1.69 (0.32)		
15b. Nigeria (23)	0.34 (0.22)		
16. Japan (11)	1.50 (0.37)	3.81 (0.22)	5.33 (0.45)

10a. biotites associated with hydrothermal alteration and mineralization in the porphyry molybdenum deposit at Henderson, Colorado (Gunow, 1978, Gunow et al., 1980)
10b. hydrothermal sericites from quartz-sericite-pyrite alteration at Henderson, Colorado (ibid.)
11. biotites from Catung E-zone scheelite orebody, Tungsten, N.W.T., Canada (Zaw and Clark, 1978)
12. muscovites from aplites, pegmatites, greisenized granite, and hypothermal quartz veins associated with Sn-W mineralization in northern Portugal (Neiva, 1982)
13a. sector-zoned biotite crystals from the Mt. Lindsay Sn-W-Be deposit, Tasmania (Kwak, 1981). The ranges shown reflect variations within single crystals.
13b. biotites and phengites from Sn-W-Be skarns at Mt. Lindsay and Moina Tasmania (Kwak, 1983, Kwak and Askins, 1981)
14. biotites from Sn-W skarn at Lost River, Alaska (Dobson, 1982)
15a. biotites from the Afu Younger granite complex, central Nigeria (Imeokparia, 1982)
15b. biotites associated with Sn mineralization at the Afu Younger granite complex, Nigeria (Imeokparia, 1982)
16. biotites from granitic rocks associated with Sn-W-Mo mineralization in Kyushu, Japan. Calculated from averages of 37 analyses. (Nedachi, 1980)

Table 5. Mean halogen intercept values of micas from other hydrothermal systems
Values in parentheses represent 1 standard deviation

locality (number of analyses)	IV(F)	-IV(Cl)	IV(F/Cl)
17. Salton Sea (6)	2.54 (0.18)	4.09 (0.24)	6.52 (0.26)
18. Japan (3)	2.03, 1.75, 2.06	4.23, 4.45, 4.31	6.26, 6.20, 6.38
19. France (5)	0.98 (0.17)		

17. authigenic biotites in sandstone from borehole Elmore 1, Salton Sea geothermal system, California. Temperatures ranged from 335° C to 361° C, corresponding to depths of 1219 to 2169 m. Averages of 13 analyses (McDowell and Elders, 1980)
18. biotites in hydrothermally altered granitic rocks from the Taishu district (Pb-Zn), Kyushu, Japan. Averages of 9 analyses. (Nedachi, 1980)
19. phengites from two-mica Hercynian granites associated with hydrothermal uranium deposits. Averages of 36 analyses. (Leroy and Cathelineau, 1982)

Table 6. Mean halogen intercept values of biotites from selected igneous rocks
Values in parentheses represent 1 standard deviation

locality (number of analyses)	IV(F)	-IV(Cl)	IV(F/Cl)
20. Sierran granites (22)	2.14 (0.25)	3.63 (0.50)	5.78 (0.52)
21. Pikes Peak (4)	1.47 (0.11)	3.37 (0.16)	4.83 (0.05)
22. Bishop tuff (37)	2.04 (0.14)	3.61 (0.10)	5.64 (0.18)
23. Peruvian rhyolites (13)	2.30 (0.20)	3.90 (0.21)	6.17 (0.30)

20. biotites from plutonic rocks of the central Sierra Nevada and Inyo Mountains, California (Dodge et al., 1969)
21. biotites from fayalite-bearing granites and quartz syenites of the Pikes Peak batholith (Barker et al., 1975)
22. biotite phenocrysts from the Bishop tuff, California (Hildreth, 1977)
23. biotite phenocrysts from glassy rhyodacites at Julcani, Peru (Drexler, 1982)

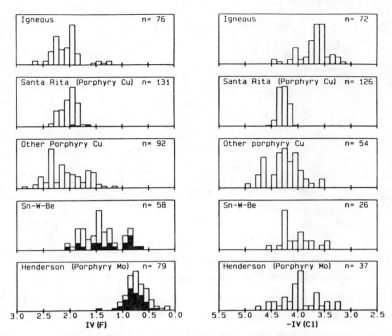

Figure 5. Histograms of F intercept values for micas listed in Tables 3-6. The number in the upper right corner is the number of analyses on which the histogram was based. Relative F enrichment in micas increases to the right. Shaded areas represent sericite or muscovite, unshaded areas are biotite.

Figure 6. Histograms of negative Cl intercept values - [IV(Cl)] for biotites listed in Tables 3-6. The number in the upper right corner is the number of analyses on which the histogram was based. Relative Cl enrichment increases to the left.

related granitic rocks). Halogen data from these two groups are presented in tabular format in terms of intercept values rather than weight % F or Cl, so that relative halogen enrichment can be compared directly without needing to consider differences in the Mg/Fe ratios of the biotites (Tables 3-4). The mean, standard deviation, and number of analyses are given for each set of data. Table 5 includes comparable data for three other hydrothermal systems — a hydrothermal uranium deposit, a Pb-Zn deposit, and the Salton Sea geothermal system. Finally, a small sampling of biotite analyses from volcanic and plutonic rocks are included for comparison (Table 6). All these data are summarized as histograms of both F and Cl values (Figs. 5 and 6).

Fluorine

The F intercept values of most natural biotites and muscovites range from about 3.0 (equivalent to ~0.2 % F for a Mg-rich biotite and ~0.05 % F for an Fe-rich biotite) to near 0.0 for very F-rich micas (a natural F-phlogopite from Grenville marbles near Balmat, New York has 8.53 % F (Valley et al., 1982), corresponding to an intercept value of 0.28). The data in Tables 3 and 6 and Figure

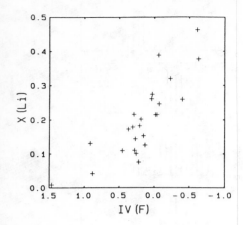

Figure 7. Correlation of high F content [low IV(F)] with mole fraction of octahedral Li [X(Li)] for zinnwaldites from the Erzegebirge region Data from Rieder (1970).

5 suggest that the range in F intercept values of many common igneous biotites substantially overlap the range of intercept values of hydrothermal biotites associated with porphyry copper mineralization. Thus, F contents alone cannot distinguish micas in mineralized copper systems from those in barren systems, a fact noted previously by Kesler et al. (1975). On the other hand, a marked separation of more than 1.0 IV(F) unit is apparent for the F-rich biotites and sericites from Henderson, Colorado (Table 4), a strongly differentiated porphyry molybdenum deposit of the Climax type (White et al., 1981). The other Mo-W-Sn-Be deposits listed in Table 4 are not nearly as F-enriched as the Henderson micas, but have significantly lower F intercept values than the micas from porphyry copper deposits.

The F intercept data from Mo-W-Sn-Be deposits must be viewed with some reservation because of the strong positive correlation of Li with these elements, as revealed by the common presence of zinnwaldites (Li-rich Fe biotites) in many of these ore deposits and in their related igneous rocks. The complicating effect of Li on the interpretation of the F chemistry of biotite can be assessed by calculating F intercept values for a number of zinnwaldites with widely varying Li content. Figure 7 presents results of such calculations for 26 Li-biotites and zinnwaldites from hydrothermally altered granites associated with Sn-W deposits of the Erzegebirge region in central Europe (Rieder, 1970), and shows a positive correlation between IV(F) and X_{Li}. The slope of the trend suggests that Li may have a stronger effect on the partitioning of F into biotite than does Mg. If the zinnwaldites shown in Figure 7 equilibrated with a fluid of approximately constant relative HF activity, then a fluorine intercept value calculated from a mica with as little as 2000 ppm Li_2O would be 0.1 IV(F) units too low (i.e., too F-rich), because the calculation of IV(F) does not include a term to correct for the Li content. As an example of the potential impact of this problem, undetected Li is almost certainly present in significant amounts in the "mineralized" biotites from the Afu Younger complex in Nigeria (Table 4, no. 15b), because the Al_2O_3 contents in these micas are much higher than normal for biotite (they average from 18-20% Al_2O_3), and the sum of all reported octahedral occupancies calculated on the basis of 12 (O+OH+F) averages about 2.6, much too low for normal biotite but typical for zinnwaldite. It is interesting to note that these micas have extremely low F intercept values. If substantial Li_2O is present in these micas, these intercept values are likely to be substantially depressed, and should not be compared with other intercept values in Tables 3-6.

Chlorine

The chlorine intercept values for most biotites range from about -3.0 (corresponding to less than 0.05 % Cl for a very F-rich biotite) to around -4.5 (corresponding to about 0.4 to 0.8 % Cl, depending on Mg/Fe ratio). There are a few noteworthy exceptions; for example, a Cl-rich biotite from an Indian charnockite has 2.07 % Cl (X_{Mg} = 0.44), and a Cl intercept of -5.23 (Kamineni et al., 1982).

The available data suggest that chlorine intercept values from micas in all the mineralized or hydrothermally altered rocks tend to be more negative (more Cl-enriched) than comparable values from micas in igneous rocks. For example, mean IV(Cl)'s for the five porphyry copper deposits in Table 3 are substantially less than -4.0, whereas the mean igneous IV(Cl)'s are more positive than -4.0. Because of significant overlap, conclusions based on single biotite analyses are unwarranted (cf. Fig. 6).

Fluorine/chlorine ratio

F/Cl intercept values for biotites from porphyry copper deposits, Mo-W-Sn-Be deposits (excluding Henderson) and the Henderson porphyry molybdenum deposit are shown as histograms in Figure 8, and are plotted against IV(F) in Figure 9. Because IV(F/Cl) is directly related to the fugacity ratio f(HCl)/f(HF) (cf. Eqn. 23), lower IV numbers correspond to higher F/Cl ratios (with reference to constant temperature). The trend of increasing F/Cl ratio from porphyry copper biotites to porphyry Mo biotites is striking but is due more to increasing F enrichment than to declining Cl enrichment (cf. Figs. 5 and 6). Separation of mineralization type based on F/Cl ratio in biotite may also be shown by plotting F/Cl in biotite directly against X_{Mg} (Munoz and Swenson, 1981; Gilzean and Brimhall, 1983).

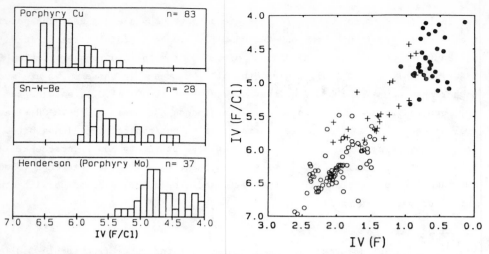

Figure 8. Histograms of F/Cl intercept values for biotites from porphyry copper deposits in Table 3 (including potassic alteration zone at Santa Rita, and all micas from Bingham, Bakircay, and Los Pelambres), and for biotites from Mo-Sn-W-Be deposits in Table 4 (including Henderson, Tasmanian skarns, and Japanese Sn deposits).

Figure 9. F/Cl intercept value [IV(F/Cl)] plotted against F intercept value [IV(F)] for biotites shown in Figure 8. Open circles: porphyry copper deposits, crosses: Sn-W-Be deposits, closed circles: porphyry molybdenum deposits (Henderson, Colorado).

INTERPRETATION OF HALOGEN INTERCEPT DATA IN TERMS OF FLUID COMPOSITIONS

The data presented in Tables 3-6 are of limited value because they show only large differences between unrelated groups of micas, and it is commonly not possible to interpret such differences without more detailed information. It is more rewarding to study the halogen geochemistry of a small population of micas which have been carefully characterized in terms of their textural relations, mineral assemblages, and whole-rock chemistry. For instance, the averaged data from the Santa Rita porphyry copper deposit (Table 3, no. 1a) reveal differences when separated according to alteration type (Table 7). Note a systematic decrease in IV(F) from the least-altered biotites (2.22) to the biotites from zones of overlapping phyllic-potassic alteration (1.90). In contrast, no meaningful differences in IV(Cl) are apparent for any of the alteration types. These data can also be shown on a graph which plots IV(F/Cl) for all biotites in the deposit against IV(F) (Fig. 10). Note the complete separation in IV(F) between the fresh biotites (open circles) and the biotites from phyllic-potassic alteration (crosses). on the other hand, biotites associated with potassic alteration (dark circles) overlap the entire range of IV(F).

The diagonal line drawn on Figure 10 has a slope of +1, and fits the trend of the data reasonably well. This linear trend arises because of the definition of IV(F/Cl) (Eqn. 25, cf. the form $y = ax + b$), and the small variation in IV(Cl)

Table 7. Mean halogen intercept values for biotites from the Santa Rita porphyry copper deposit, classfied according to alteration type.

Alteration type	IV(F)	-IV(Cl)	IV(F/Cl)
Fr	2.22 (0.06)	4.25 (0.06)	6.47 (0.07)
PrPot	2.20 (0.09)	4.25 (0.09)	6.45 (0.14)
Pot	2.06 (0.12)	4.32 (0.08)	6.38 (0.15)
ArgPot	1.95 (0.07)	4.30 (0.08)	6.25 (0.10)
PhPot	1.90 (0.06)	4.23 (0.09)	6.12 (0.11)

Key to alteration symbols:
Fr: least-altered biotite porphyry
PrPot: propylitic alteration (chlorite ± siderite ± epidote ± sericite ± clay minerals) superimposed to varying degrees on potassic alteration
Pot: potassic alteration (montmorillonite/kaolinite) superimposed on potassic alteration
PhPot: phyllic alteration (quartz + sericite/muscovite + pyrite + chalcopyrite + rutile/anatase) superimposed on potassic alteration

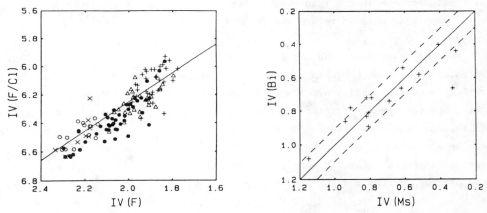

Figure 10. Correlation of F/Cl intercept value [IV(F/Cl)] with F intercept value [IV(F)] for biotites from the Santa Rita porphyry copper deposit, plotted according to alteration type. Open circles: least altered biotite porphyry, filled circles: potassic alteration, X's: overlapping propylitic-potassic alteration, triangles: overlapping argillic-potassic alteration, crosses: overlapping phyllic-potassic alteration. See Table 7 and text for discussion.

Figure 11. Comparison of fluorine intercept values of hydrothermal sericites [IV(Ms)] with those of primary biotites [IV(Bi)] from the Henderson molybdenite deposit. Each cross represents a sericite rim on a biotite crystal. The solid diagonal line represents perfect correspondence (sericite and biotite in F OH exchange equilibrium), and the dashed lines define an envelope of ±0.1 IV. The figure illustrates hydrothermal overprinting of primary F content in micas.

for all alteration types. It is this restricted variation in IV(Cl) which limits the vertical spread in the data. Thus, the y-intercept of the line in Figure 10 is 4.27, which is equal to the mean -IV(Cl) for all alteration types listed in Table 7.

The most interesting question remaining is how to correlate intercept values in micas such as those at Santa Rita with some more fundamental geochemical parameter such as magma or fluid composition. Neglecting analytical errors, differences in mica intercept values can be related either to differences in relative halogen acid activity present during halogen=OH exchange, or to differences in the exchange temperature. When confronted with the problem of estimating an exchange temperature, it is always important to decide whether the halogen content in an igneous mica represents a "pristine" high-temperature value, or whether the original halogen content in the mica may have been overprinted subsequently by low-temperature hydrothermal fluids. Such overprinting may occur without substantially changing the textural appearance of the mica. For instance, in the Henderson Mo deposit, primary igneous biotite is commonly replaced by hydrothermal sericite. F intercept values calculated separately for the biotite core and the sericite rim are very nearly equal (within ±0.1 IV(F)) in 11 out of 14 biotite-sericite pairs (Fig. 11). Such correspondence is possible only if the high-temperature biotites continued to exchange F with the hydrothermal fluid throughout the hydrothermal stage. The range in intercept values shown in the figure was interpreted by Gunow et al. (1980) as reflecting gradients in relative HF activity which existed in the hydrothermal system.

On the other hand, the rhyolites from Julcani, Peru (Table 6, no. 23) are of interest because they are glassy and clearly unaltered, and thus presumably reflect magmatic conditions. In such rocks where low-temperature alteration is not suspected, the temperature of crystallization can be assumed to be the same as the temperature of halogen=OH exchange. If the crystallization temperature can be estimated by some convenient geothermometer, then magmatic fugacity ratios can be calculated directly from Equations 17 and 22. This procedure was used by Drexler and Noble (1983) for the Julcani biotites, where they calculated values for $\log [f(H_2O)/f(HF)]$ and $\log [f(H_2O)/f(HCl)]$ values ranging from 3.9 to 4.4 and 3.6 to 4.2, respectively, for assumed temperatures between 830° and 850°C. Other examples of the calculation of halogen fugacity ratios based on halogen contents of micas and specific assumed temperatures are Zaw and Clark (1978), Jacobs and Parry (1979), Gunow et al. (1980), Valley et al. (1982), and Parry et al. (1984).

In hydrothermal systems, defining a meaningful exchange temperature may be difficult. Commonly employed geothermometers such as fluid inclusion filling temperatures or stable isotopic ratios may be unrelated to the latest temperature of halogen=OH exchange. Moreover, it is impossible to predict a generally applicable $T-a_{HF}$ or $T-a_{HCl}$ path for fluids descending from magmatic temperatures to hydrothermal temperatures. Burt (1981) has outlined the most important factors which may result in either increases or decreases in the chemical potential of HF in a magmatic/hydrothermal system. As temperature falls, relative F enrichment as measured by mineral assemblages (e.g. andalusite versus topaz) may either increase or decrease, depending on the relative importance of boiling, mineral-fluid buffering, fluid-wall rock interaction, and dilution of the

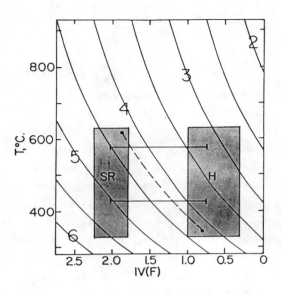

Figure 12. Relation between F intercept value [IV(F)] temperature, and fluid composition for mica suites at the Santa Rita porphyry copper deposit (SR) and the Henderson porphyry molybdenum deposit (H). The diagonal lines are contoured in $\log [f(H_2O)/f(HF)]$. See text for discussion.

hydrothermal fluids by meteoric waters.

The effect of temperature uncertainty on fugacity ratio is shown for F⇌OH exchange in Figure 12, which plots mica composition in terms of IV(F) against temperature. The contours represent the fluid composition (in $\log [f(H_2O)/f(HF)]$), and were calculated by solving Equation 17 for a number of temperatures and intercept values. The shaded boxes show the ranges in IV(F) for both the Santa Rita porphyry Cu biotites (Table 3, no. 1a) and the Henderson micas (Table 4, no. 10a and 10b), extended over a temperature range of 327° to 627°C (600 - 900 Kelvins). The first thing to notice is that the inferred difference in log fugacity ratio based on the means of the two groups of data is about 1.4 log units at any constant temperature (cf. the equal lengths of the lines on the 580° and 420° isotherms, for instance). This difference in log fugacity ratio could be diminished substantially only if the Santa Rita biotites stopped exchanging F at a higher temperature than the Henderson biotites. A 300° difference between the exchange temperatures of the two suites of biotites is required to make the fugacity ratios in both systems of comparable magnitudes (cf. the dashed line on Fig. 12). On the other hand, if the Henderson biotites stopped exchanging at a higher temperature than the Santa Rita biotites, the implied difference in the log fugacity ratio would be even greater than the 1.4 log units calculated for isothermal conditions. Because both suites of micas are taken from hydrothermal systems, it is just as likely that they both continued to exchange halogens to about the same temperature, and the differences in their IV(F) numbers accurately reflect a substantial difference in the relative HF activities between the two magmatic/hydrothermal systems.

CONCLUSIONS

A complete chemical analysis of a mica should include both F and Cl. This is particularly true for micas from ore deposits and hydrothermally altered rocks, in view of the importance of halogens in ore-forming processes.

The halogen content of mica depends on temperature, the relative halogen acid activity (log $[f(H_2O)/f(HF)]$ and log $[f(H_2O)/f(HCl)]$) which prevailed during halogen=OH exchange, and the Mg/Fe ratio in biotite or X_{Mg} in muscovite - phengite. At constant relative a_{HF} and temperature, F is increasingly fractionated into micas according to the series muscovite - siderophyllite - annite - phlogopite. In contrast, at constant relative a_{HCl} and temperature, Cl is increasingly fractionated into biotites with decreasing Mg/Fe ratio. Halogen intercept values are measures of relative F or Cl enrichment in micas. Intercept values are calculated from a chemical analysis, and include corrections for the opposing effects of Mg/Fe ratio on F=OH and Cl=OH exchange.

The F/Cl intercept value in biotite increases from porphyry copper deposits (and related granitic rocks) to Mo-Sn-W-Be deposits (including skarns, porphyry deposits, granites, and pegmatites) to the highly differentiated porphyry Mo deposits of the Climax type. This trend is due primarily to the very high F contents in the micas from the Climax-type deposits.

In hydrothermal systems, the halogen contents of primary (high temperature) micas are commonly reset by lower temperature fluids. Intercept values of micas can be used as a quantitative measure of relative halogen acid activity only if the temperature of halogen=OH exchange is known. Nonetheless, intercept values of micas in hydrothermal systems can be related to differences in relative halogen acid activity if it is assumed that the micas exchanged halogens to some uniformly low temperature.

REFERENCES

Barker, F., Wones, D.R., Sharp, W.N., and Desborough, G.A. (1975) The Pikes Peak batholith, Colorado Front Range, and a model for the origin of the gabbro- anorthosite- syenite- potassic granite suite. Precambrian Res. 2, 97-160.

Bloss, F.D., Shekarchi, E., and Shell, H.R. (1959) Hardness of synthetic and natural micas. Am. Mineral. 44, 33-48.

Burt, D.M. (1981) Acidity-salinity diagrams - Application to gresien and porphyry deposits. Econ. Geol. 76, 832-843.

Chivas, A.R. (1981) Geochemical evidence for magmatic fluids in porphyry copper minerilization. Part I. Mafic silicates from the Koloula igneous complex. Contrib. Mineral. Petrol. 78, 389-403.

Dobson, D.C. (1982) Geology and alteration of the Lost River Tin-tungsten-fluorine deposit, Alaska. Econ. Geol. 77, 1033-1052.

Dodge, F.C.W., Smith, V.C., and Mays, R.E. (1969) Biotites from granitic rocks of the central Sierra Nevada Batholith, California. J. Petrol. 10, 250-271.

Drexler, J.W. (1982) Mineralogy and Geochemistry of Miocene Volcanic Rocks Genetically Associated with the Julcani; Ag-Bi-Pb-Cu-Au-W Deposit, Peru: Physicochemical Conditions of a Productive Magma Body. Ph.D. dissertation, Michigan Technological University, Houghton, Michigan.

_____ and Noble, D.C. (1983) Highly oxidized, sulfur-rich dacite-rhyolite magmas of the Julcani silver district, Peru: Insight into a "productive" magma system. Geol. Soc. Am. Abstr. with Programs 15, 325.

Duffy, C.J. and Greenwood, H.J. (1979) Phase equilibria in the system $MgO-MgF_2-SiO_2-H_2O$. Am. Mineral. 64, 1156-1172.

Ekström, T.K. (1972) The distribution of fluorine among some coexisting minerals. Contrib. Mineral. Petrol. 34, 192-200.

Eugster, H.P. and Wones, D.R. (1962) Stability relations of the ferruginous biotite, annite. J. Petrol. 3, 82-125.

Forbes, W.C. (1972) An interpretation of the hydroxyl contents of biotites and muscovites. Mineral. Mag. 38, 712-720.

Foster, M.D. (1960) Interpretation of the composition of lithium micas. U.S. Geol. Surv. Prof. Paper 354-E, 115-147.

_____ (1964) Water contents of micas and chlorites. U.S. Geol. Surv. Prof. Paper 474-F, 1-15.

Gerloch, M. and Slade, R.C. (1973) Ligand Field Parameters. Cambridge Univ. Press, Cambridge, 235 pp.

Giese, R.F., Jr. (1975) The effect of F/OH substitution on some layer-silicate minerals. Z. Kristallogr. 141, 138-144.

Gilzean, M.N. and Brimhall, G.H. (1983) Alteration biotite chemistry and nature of deep hydro- thermal system beneath Silverton district, Colorado. Geol. Soc. Am. Abstr. with Programs 15, 581.

Gunow, A.J. (1978) The geochemistry of fluorine-rich micas at the Henderson molybdenite deposit. M.S. Thesis, Univ. of Colorado, Boulder, Colorado.

_____, Ludington, S., and Munoz, J.L. (1980) Fluorine in micas from the Henderson molybdenite deposit, Colorado. Econ. Geol. 75, 1127-1137.

Hazen, R.M. and Wones, D.R. (1972) The effect of cation substitutions on the physical proper- ties of trioctahedral micas. Am. Mineral. 57, 103-129.

Hildreth, E.W. (1977) The magma chamber of the Bishop Tuff: Gradients in temperature, pressure, and composition. Ph.D. dissertation, Univ. of California, Berkeley, California.

Imeokparia, E.G. (1982) Tin content of biotites from the Afu Younger Granite Complex, Central Nigeria. Econ. Geol. 77, 1710-1724.

Jacobs, D.C. (1976) Geochemistry of biotite in the Santa Rita and Hanover-Fierro stocks, Central mining district, Grant County, New Mexico. Ph.D. dissertation, Univ. of Utah, Salt Lake City, Utah.

_____ and Parry, W.T. (1979) Geochemistry of biotite in the Santa Rita porphyry copper deposit, New Mexico. Econ. Geol. 74, 860-887.

Kamineni, D.C., Bonardi, M., and Rao, A.T. (1982) Halogen-bearing minerals from Airport Hill, Visakhapatnam, India. Am. Mineral. 67, 1001-1004.

Kesler, S.E., Issigonis, M.J., Browlow, A.H., Damon, P.E., Moore, W.J., Northcotte, K.E., and Preto, V.A. (1975) Geochemistry of biotites from barren and intrusive systems. Econ. Geol. 70, 559-567.

Kwak, T.A.P. (1981) Sector-zoned annite$_{8.5}$-phlogopite$_{15}$ micas from the Mt. Lindsay Sn-W-F(-Be) deposit, Tasmania, Australia. Canadian Mineral. 19, 643-650.

_____ (1983) The geology and geochemistry of the zoned Sn-W-F-Be skarns at Mt. Lindsay, Tasmania, Australia. Econ. Geol. 78, 1440-1465.

_____ and Askins, P.W. (1981) Geology and genesis of the F-Sn-W-(Be-Zn) skarn (wrigglite) at Moina, Tasmania. Econ. Geol. 76, 439-467.

Lanier, G., Raab, W.J., Folsom, R.M., and Cone, S. (1978) Alteration of equigranular monzonite, Bingham mining district, Utah. Econ. Geol. 73, 1270-1286.

Leroy, J. and Cathelineau, M. (1982) Les minéraux phylliteux dans les gisement hydrothermaux d'uranium I. Cristallochimie des micas hérités et néoformés. Bull. Minéral. 105, 99-109.

Levillain, C. (1980) Étude statistique des variations de la teneur en OH et F dans les micas. Tscher. Mineral. Petrol. Mitt. 27, 209-223.

Ludington, S.D. (1974) Application of fluoride-hydroxyl exchange data to natural minerals. Ph.D. dissertation, Univ. of Colorado, Boulder, Colorado.

_____ and Munoz, J. (1975) Application of fluor-hydroxyl exchange data to natural micas. Geol. Soc. Am. Abstr. with Programs 7, 1179.

McCauley, J.W., Newnham, R.E., and Gibbs, G.V. (1973) Crystal structure analysis of synthetic fluorphlogopite. Am. Mineral. 58, 249-254.

McDowell, S.D. and Elders, W.A. (1980) Authigenic layer silicate minerals in borehole Elmore 1, Salton Sea geothermal field, California, U.S.A. Contrib. Mineral. Petrol. 74, 293-310.

Munoz, J.L. and Gunow, A.J. (1982) Fluorine index: A simple guide to high fluorine environments. Geol. Soc. Am. Abstr. with Programs 14, 573.

_____ and Ludington, S.D. (1974) Fluorine-hydroxyl exchange in biotite. Am. J. Sci. 274, 396-413.

_____ and _____ (1977) Fluoride-hydroxyl exchange in synthetic muscovite and its application to muscovite-biotite assemblages. Am. Mineral. 62, 304-308.

_____ and Swenson, A. (1981) Chloride-hydroxyl exchange in biotite and estimation of relative HCl/HF activities in hydrothermal fluids. Econ. Geol. 76, 2212-2221.

Nedachi, M. (1980) Chlorine and fluorine contents of rock-forming minerals of the Neogene granitic rocks in Kyushu, Japan. Mining Geol. Spec. Issue 8, Tokyo, 39-48.

Neiva, A.M.R. (1982) Geochemistry of muscovite and some physico-chemical conditions of the formation of some tin-tungsten deposits in Portugal. In: Metallization associated with Acid Magmatism, A.M. Evans, ed., John Wiley & Sons, New York, 243-259.

Noda, T. and Ushiro, M. (1964) Hydrothermal synthesis of fluorine-hydroxyl phlogopite. II. Relationship between the fluorine content, lattice constants, and conditions of synthesis of fluorine-hydroxyl phlogopite. Geochem. Int'l. 1, 96-104.

Parry, W.T., Ballantyne, G.H., and Wilson, J.C. (1978) Chemistry of biotite and apatite from a vesicular quartz latite porphyry plug at Bingham, Utah. Econ. Geol. 73, 1308-1314.

_____, _____, and Jacobs, D.C. (1984) Geochemistry of hydrothermal sericite from Roosevelt hot springs and the Tintic and Santa Rita porphyry copper systems. Econ. Geol. 79, 72-86.

Ramberg, H. (1952) Chemical bonds and the distribution of cations in silicates. J. Geol. 60, 331-355.

Rieder, M. (1970) Chemical composition and physical properties of lithium-iron micas from the Krušné hory Mtns. (Erzgebirge) Part A: Chemical composition. Contrib. Mineral. Petrol. 27, 131-158.

Rimsaite, J.H.Y. (1970) Structural formulae of oxidized and hydroxyl-deficient micas and decomposition of the hydroxyl group. Contrib. Petrol. 25, 225-240.

Rosenberg, P.E., and Foit, F.F., Jr. (1977) Fe^{2+}-F avoidance in silicates. Geochim. Cosmochim. Acta 41, 345-346.

Sanz, J. and Stone, W.E. (1979) NMR study of Micas. II. Distribution of Fe^{2+}, F^-, and OH^- in the octahedral sheet of phlogopites. Am. Mineral. 64, 119-126.

Shell, H.R. and Ivey, K.H. (1969) Fluorine Micas. U.S. Bur. Mines Bull. 647, 291 pp.

Stern, W.B. (1979) Zur Strukturformelberechnung von Glimmermineralien. Schweiz. Mineral. Petrogr. Mitt. 59, 75-82.

Takeda, H. and Morosin, B. (1975) Comparison of observed and predicted structural parameters of mica at high temperature. Acta Crystallogr. B31, 2444-2452.

Taylor, R.P. (1983) Comparison of biotite geochemistry of Bakircay, Turkey, and Los Pelambres, Chile. Trans. Inst. Min. Metall. Sec. B92, B16-B22.

Valley, J.W. and Essene, E.J. (1980) Calc-silicate reactions in Adirondack marbles: The role of fluids and solid solutions: Summary. Geol. Soc. Am. Bull. Part I, 91, 114-117, Part II, 91, 720-815.

_____, Petersen, E.U., Essene, E.J., and Bowman, J.R. (1982) Fluorphlogopite and fluortremolite in Adirondack marbles and calculated C-O-H-F compositions. Am. Mineral. 67, 545-557.

Wagman, D.D., Evans, W.H., Parker, V.B., Schumm, R.H., Halow, I., Bailey, S.M., Churney, K.L., and Nuttall, R.L. (1982) The NBS tables of chemical thermodynamic properties. Selected values for inorganic and C1 and C2 organic substances in SI units. J. Phys. Chem. Ref. Data 11, supp. no. 2, 392 pp.

White, W.H., Bookstrom, A.A., Kamilli, R.J., Ganster, M.W., Smith, R.P., Ranta, D.E., and Steininger, R.C. (1981) Character and origin of Climax-type molybdenum deposits. Econ. Geol. 75th Anniv. Volume, B.J. Skinner, ed., 270-316.

Wones, D.R. (1963) Physical properties of synthetic biotites on the join phlogopite-annite. Am. Mineral. 48, 1300-1321.

_____ (1967) A low pressure investigation of the stability of phlogopite. Geochim. Cosmochim. Acta 31, 2248-2253.

Wood, B.J. and Nicholls, J. (1978) The thermodynamic properties of reciprocal solid solutions. Contrib. Mineral. Petrol. 66, 389-400.

Yoder, H.S. and Eugster, H.P. (1955) Synthetic and natural muscovites. Géochim. Cosmochim. Acta 8, 225-280.

Zaw, U.K. and Clark, A.H. (1978) Fluoride-hydroxyl ratios of skarn silicates, Catung E-zone scheelite orebody, Tungsten, Northwest Territories. Canadian Mineral. 16, 207-221.

12. ILLITE
Jan Środoń & Dennis D. Eberl

INTRODUCTION

The term "illite" was introduced by Grim et al. (1937) as a general name for clay-size minerals of the mica group, which are commonly found in argillaceous sediments. Used in this sense, the term is synonymous with hydromica, a name common in the Russian literature. Since the original paper by Grim et al., however, many varieties of micaceous clay have been recognized, and thus there is a need for more specific nomenclature. In the present paper, the term "illite" refers to a nonexpanding, dioctahedral, aluminous, potassium mica-like mineral which occurs in the clay-size (less than 4 μm) fraction. This definition distinguishes illite from similar species such as mixed-layer illite/smectite (I/S), which expands, trioctahedral illite, which is not dioctahedral, glauconite and celadonite, which are ferruginous, brammallite and ammonium illite, which are Na and NH_4 species, muscovite that is ground to clay-size particles in the laboratory, which is not naturally occurring, and coarse muscovite, which is not clay-size. This definition is more specific than the original definition, but is general enough to be useful petrologically in that parameters difficult to determine in clay mixtures, such as polytype and K content, need not be specified.

Illite by this definition can be specified operationally by X-ray diffraction (XRD) techniques. It gives d(001) = 10 ± 0.05 Å (measured using the 002 reflection and higher orders), showing that it is micaceous and suggesting that K is the interlayer cation; it has a d(060) = 1.50 ± 0.01 Å, demonstrating that it is dioctahedral; its intensity for the 002 reflection, I(002), is greater than approximately I(001)/4, suggesting that it is aluminous because an Fe-rich octahedral sheet would result in a relatively smaller I(002); and, finally,

$$Ir = \frac{I(001)/I(003) \quad \text{(air dried)}}{I(001)/I(003) \quad \text{(ethylene glycol-treated)}} = 1, \qquad (1)$$

showing that it is nonexpanding, because this ratio will be greater than one if an expanding component is present.

Two additional definitions are required. An "illite layer" is an approximately 10 Å, nonexpanding component of dioctahedral, mixed-layer clay which contains interlayer K that is nonexchangeable upon treatment with 1 M NaCl, or upon similar treatment. This layer is centered on the interlayer plane of K ions. "Illitic material" is a general petrologic term for the approximately 10 Å component of the clay-size fraction and is synonymous with the term "illite" as defined by Grim et al. (1937). It may exhibit some expandability. This term is used if a more detailed identification is impossible, or has not been made. A more detailed identification may be impossible by XRD, for example, if randomly interstratified illite/smectite (random I/S) and illitic material occur together, because the expandability of the latter cannot be determined due to interference by random I/S peaks.

Illitic material, together with kaolinite, chlorite, and I/S, is one of the four major clay components of sedimentary rock. In addition, I/S contains a significant proportion of illite layers. Thus a large portion of K found in the sedimentary column is fixed in illite layers. Illitic material also occurs in hydrothermal zones, where both illite and I/S can be abundant. Statistical data show an increase in illitic material in older rocks. It is unclear whether this trend reflects an evolving chemistry for the hydrosphere (Weaver, 1967), or a higher average degree of diagenesis for older rocks (Grim, 1968; Weaver and Beck, 1971; Hower et al., 1976; Nadeau and Reynolds, 1981b). Because illitic material is relatively stable under Earth surface conditions, a significant proportion of illite in sedimentary rocks is considered to be recycled from older sedimentary formations. This material can form initially by neoformation and by transformation in the weathering environment and in the higher temperature diagenetic, metamorphic, and hydrothermal environments (Eberl, 1984), as will be discussed.

X-RAY DIFFRACTION (XRD) ANALYSIS

XRD identification of illitic materials by basal reflections

Illitic materials analyzed by XRD in oriented preparations fall into three categories: (1) pure mineral illite, as defined previously, (2) highly illitic I/S, and (3) mixtures of (1) and (2). Examples of each category are given in Figure 1. All three categories were described traditionally as illites, and the "crystallinity indices" (Weaver, 1960; Kubler, 1964) were used to describe their evolution during geologic processes such as diagenesis. Środoń (1984a), using the computer program of Reynolds (1980) and samples of diagenetic origin, developed methods for differentiating between these types of illitic materials and for identifying precisely the mixed-layer component. These techniques require an analysis of both air-dried and ethylene glycol solvated preparations.

The first step in the identification procedure is to plot the positions of the 002 and 003 reflections from glycolated preparations onto Figure 2.

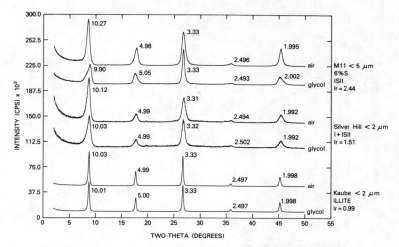

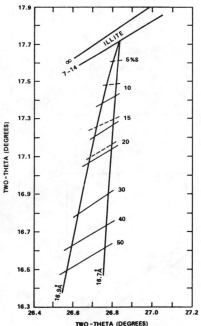

Figure 1 (above). XRD patterns (oriented preparations) of three types of illitic materials: M11 = highly illitic I/S; Silver Hill = a mixture of illite and I/S; Kaube = pure illite. Ir is a parameter for detecting expandability, as is explained in the text. In this figure, and in all subsequent figures containing XRD patterns, samples were run with Cu $K\alpha$ radiation with a step size of 0.02° 2θ and a counting time of 0.5 sec per step using a Siemens D500 automated X-ray diffraction system. Peak positions were measured at a more sensitive scale than that shown here.

Figure 2 (left). A plot of XRD 002 and 003 peak positions (Cu $K\alpha$ radiation) and expandability for oriented, glycolated preparations. Figure was prepared from computer simulated XRD patterns (Reynolds, 1980) using spacings of 16.9 and 16.7 Å for expanding layers, and 9.97 Å for illite layers. IS ordering = continuous lines; ISII ordering = dashed lines. This figure is used to distinguish between illite (or mixtures dominated by illite), and ordered I/S (or mixtures dominated by ordered I/S). Expandabilities can be determined from this figure for pure I/S. Środoń (1984a).

This step differentiates between mineral illite or mixtures dominated by illite, both of which plot in the illite field, and ordered I/S or mixtures dominated by ordered I/S, both of which plot in the I/S field. In the former case, illite can be distinguished from a mixture because XRD peak shapes, positions and intensities for pure illite are not affected by glycolation. The following indicators of expandability for mixtures are apparent after glycolation:

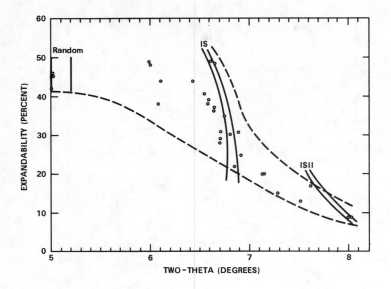

Figure 3. Low-angle XRD peak positions (Cu $K\alpha$ radiation) for glycolated I/S plotted versus expandability. This figure is used to distinguish between ordering types. Solid lines = theoretical plots for random, IS, and ISII-ordered I/S. Calculations used spacings of 16.9 Å (left line), 16.7 Å (right line), and 9.97 Å, with a crystallite size of 7 to 14 layers. Dashed lines define a field in which natural I/S from bentonites (open circles) have been found to plot. Samples that plot between pure ordering types are mixtures (R/IS or IS/ISII). Modified from Środoń (1984a). See also Figure 25.

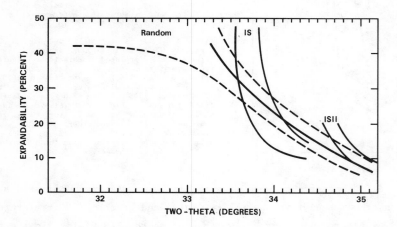

Figure 4. A plot of experimental XRD peak position (glycolated I/S; Cu $K\alpha$ radiation) versus expandability. This figure is used to determine expandability in the presence of discrete illite by using a weak I/S reflection that does not overlap an illite reflection. The dashed lines give measurement error, as inferred from the spread of experimental data. Theoretical fields for pure ordering types are shown also (left curves were calculated using spacings of 16.9 and 9.97 Å; right curves used spacings of 16.7 and 9.97 Å). Modified from Środoń (1984a).

(1) The 001 reflection becomes more asymmetric and features a low-angle tail or shoulder. A tail or shoulder also appears on the low-angle side of the 004 peak.

(2) The 001 reflection is displaced to a higher angle, and the 002 reflection is displaced to a lower angle.

(3) The intensity of the 001 reflection decreases relative to 003. As a result, the parameter Ir, as defined in Equation 1, is greater than one. All three of these indicators can be observed in Figure 1 for the M11 and Silver Hill samples.

If the peak positions plot in the I/S field in Figure 2, the percent smectite layers in I/S can be determined tentatively from the figure. Figure 3 is consulted for differentiating between IS (R1) and ISII (R3) types of ordering and for selecting the appropriate lines to be used in Figure 2. Distinction between pure I/S and mixtures containing I/S and minor amounts of discrete illite is made by comparing percent smectite obtained from Figure 2 with that obtained from Figure 4. Figure 4 uses the weak I/S reflection between 32° and 35° 2θ (Cu $K\alpha$ radiation). The latter determination is less accurate than that obtained from the angles used in Figure 2, but is not affected by the presence of discrete illite. For pure I/S, the value obtained from Figure 2 equals approximately that obtained from Figure 4, and the value from Figure 2 is accepted as the more accurate. For a mixture, the value obtained from Figure 2 is more illitic than that obtained from Figure 4, and the I/S composition to be reported is that obtained from Figure 4.

Because Figure 4 uses a very weak reflection, careful sample preparation is required. Using the <0.2 µm fraction of a sample usually eliminates quartz (a reflection which interferes with the 003 illite peak), concentrates expandable clay, and yields a high degree of preferred orientation, thereby enhancing basal reflections. Separation of adequate amounts of the <0.2 µm fraction (about 100 mg) from shales typically requires application of ultrasonics, sodium acetate buffer treatment (Jackson, 1975), and dialysis. Removal of iron-coatings and organic matter (Jackson, 1975) may be necessary also.

Studies of pure, diagenetic I/S indicate that the change from IS to ISII-type ordering takes place around 15% smectite layers. This change in ordering can be detected in patterns in which the 6°–8° (Fig. 3) and 32°–35° (Fig. 4) peaks cannot be measured (e.g., in samples in which the expandable component is very minor, or in samples in which the expandable phase is very illitic) by using parameters called BB1 and BB2 (Środoń, 1984a), as shown in Figure 5. These parameters are defined as the joint breadth of the 001 (BB1) or 004 (BB2) illite and adjacent I/S reflections, measured in degrees 2θ from where the tails of the peaks join the X-ray background. The BB parameters are independent of the illite:I/S ratio and change abruptly when the type of ordering changes. If a BB parameter is >4°, then IS ordering is present; if it is <4°, then the I/S is ISII ordered, and expandability is approximately <15%. Examples of this type of analysis are given in Środoń (1984a).

Identification techniques presented in this section are summarized in a flow diagram (Fig. 6). An alternative method for determining types of interstratification has been proposed by Watanabe (1981) and will be presented in a later section on hydrothermal clays. If the sample contains random I/S or smectite, which is often the case in surface or near-surface sediments, then identification of the illitic ecomponent cannot proceed beyond stating that illitic material is present.

Characterization of "standard illites" by basal reflections

It has been recognized since the work of Gaudette et al. (1966) that some of the well-known and widely-used "illites" contain expandable layers. Techniques described in the previous section were used to analyze eight illite standards: Morris (Ward's H36), Tazewell (Ward's H41), Fithian (Ward's H35), Strasbourg (Ward's metabentonite No. 37), Madison (from a shale break in the Madison Limestone, Wyoming), Marblehead (Gaudette, 1965), Beavers Bend (Mankin and Dodd, 1963), and Silver Hill (CMS, IMt-1). All of them are mixtures of illite and small amounts of ordered I/S (IS ordered in the Morris sample, and ISII ordered in the others). The relative amounts of I/S and illite in these samples cannot be determined with existing techniques; nor can expandabilities of the I/S components be determined, with the exception of the Fithian sample in which the ISII component is sufficiently abundant. I/S in this sample contains 9% smectite layers (Środoń, 1984a). Thus, experiments using different "standard illites" may differ considerably in their results. All of these illitic materials come from shales and differ from illitic materials of bentonitic and hydrothermal origin in that the proportion of the expandable component increases in the finer fractions, an increase that is accompanied by a decrease in the $2M_1/1M$ polytype ratio (see section on polytypes), thereby confirming that these "illites" are mixtures.

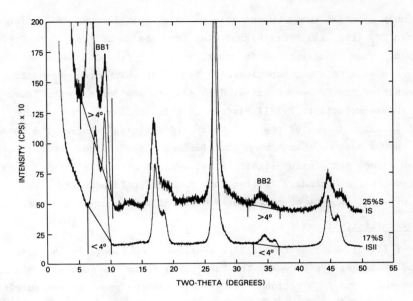

Figure 5. XRD patterns of glycolated illite/smectites, one having IS and the other ISII types of ordering, showing the method for measuring BB1 and BB2 parameters.

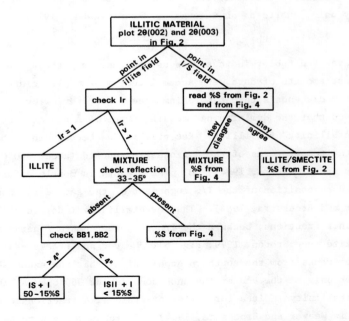

Figure 6. Flow sheet for XRD identification of illitic materials from their basal reflections. Modified from Šrodoń (1984a).

The only material known to the authors which fulfills our definition of illite is a $2M_1$ (Fig. 11) hydrothermal clay from the Kaube Mine, Nara Prefecture, Japan. When disaggregated in water, this clay produces large amounts of <2 μm, and even <0.2 μm, monomineralic fractions which are nonexpandable. XRD patterns for the Kaube and Silver Hill samples are compared in Figure 1. Features characteristic to I+ISII mixtures are found in patterns for the Silver Hill, which is one of the best "standard illites," whereas the Kaube patterns show almost no change upon glycolation. In addition, the 001 peak of the Kaube does not display significant low-angle scatter. This feature is displayed by the "standard illites" and could be related both to the presence of small amounts of mixed-layer material and to crystallite size.

Illite crystallinity index

Weaver (1960) observed that illite peaks vary considerably in their shapes and breadths. He proposed the "sharpness ratio" to quantify this phenomenon and found in a regional study that peak sharpness increased with increasing diagenetic and metamorphic grade. Kubler (1964) introduced another, more convenient, shape parameter called the "illite crystallinity index," which is simply 001 peak breadth, measured in mm or degrees 2θ, at half peak height (Hh). The measurement was made less dependent on recording conditions by using quartz as an internal standard (Weber, 1972):

$$Hh_{rel} = \frac{Hh \text{ illite (mm)}}{Hh \text{ quartz (mm)}} \times 100 .$$

Both Weaver and Kubler indices were used extensively throughout the world, and vast amounts of data were accumulated, most of it related to the transition from diagenesis to metamorphism (see recent review by Kisch, 1983). It was believed that the diminishing breadth of the 10 Å peak reflects a growing "crystallinity" for illite. Recently it has become clear, however, that before the temperatures of the epizone are reached (about 360°C according to Weaver and Broekstra, 1984), 001 peak breadth is mostly a function of the amount and composition of the I/S component of the sample (Środoń, 1979, 1984a; Weaver and Broekstra, 1984). Thus crystallinity index is significantly larger for finer fractions, because the I/S component has a finer particle size than illite (see Środoń, 1979, Fig. 5). Weaver and Broekstra (1984) have found for rocks from the southern Appalachians that the expandable component is lost only at the end of the anchizone (about 360°C). At this temperature crystallinity indices for different grain-size fractions converge (see Fig. 64 in Weaver and Broekstra, 1984) and probably begin to be controlled by crystallite size rather than by the presence of I/S. Students of

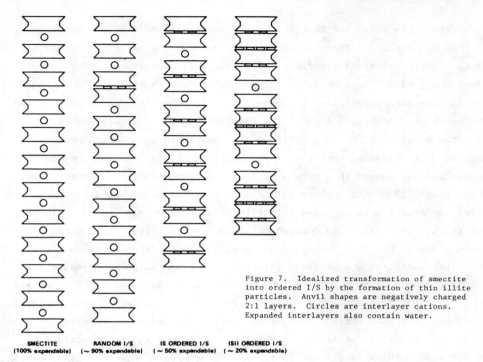

Figure 7. Idealized transformation of smectite into ordered I/S by the formation of thin illite particles. Anvil shapes are negatively charged 2:1 layers. Circles are interlayer cations. Expanded interlayers also contain water.

the crystallinity index advocate using the <2 μm-size fractions to standardize measurements made in diagenetic and metamorphic grades below the epizone. Experience indicates, however, that in most instances, mixed-layer clay in the diagenetic zone can be identified by the techniques outlined previously if the <0.2 μm fraction is analyzed, giving a diagenetic index that is independent of sample composition (I/S:illite weight ratio) and grain size (e.g., Środoń, 1984a, Fig. 9).

Interparticle diffraction

The question arises as to the physical nature of clays that diffract X-rays in illitic materials. In 1979 Frey and Lagaly showed that mechanical mixing of very fine fractions of low- and high-charge smectites in very dilute suspensions results in the formation of mixed-layer low/high-charge smectites. Thus the arrangement of silicate layers in Na-smectite crystallites, crystallites which were known previously for their ability to swell infinitely (Norrish, 1954), is unstable, and the fundamental stable particles are single 2:1 silicate layers. In 1982 McHardy et al. found that an authigenic, fibrous illite from sandstone pores displayed XRD characteristics of 20% expandable I/S, although transmission electron microscopy (TEM) showed that the illite laths averaged only three 2:1 layers thick. To reconcile these data, they suggested that the illite laths swelled at their interfaces, and that a stacking of these laths on top of each other gave rise to the diffraction effect of mixed layering. Following this suggestion, Nadeau et al. (1984) compared XRD data obtained for I/S with electron micrographs obtained from disaggregated samples of the same material. They found that whereas XRD gave particle thicknesses greater than 90 Å based on X-ray peak widths (Scherrer equation), TEM gave particle thicknesses that were much smaller, based on a Pt shadowing technique. TEM analysis of clays having a range of expandabilities showed that pure smectite is composed of particles that are approximately 10 Å thick, these particles corresponding to one 2:1 layer, whereas ordered I/S is composed primarily of particles that are approximately 20 Å thick, these particles corresponding to two 2:1 layers coordinated by a single plane of K ions. These "fundamental" clay particles stack on top of each other and glycolate at their interfaces in glass slide XRD preparations, thereby giving rise to the thicker mixed-layer crystals found in XRD analysis, and to IS ordering. Similarly, ISII-ordered I/S is composed of still thicker illite crystals that contain an average of four 2:1 layers coordinated to each other by three planes of fixed K ions, thereby forming illite crystals that are about 40 Å thick (see Figs. 7 and 8). Measurements made for illite having no detectable expanding component gave thicknesses that averaged 70 Å. Thus the evolution of I/S towards pure illite can be perceived as a growth of illite crystals parallel to c^*, beginning from the most elementary illite particle which is two 2:1 layers, or 20 Å thick. I/S with expandabilities greater than 50% must contain single 2:1 layers in addition to thin illite crystallites.

By this theory, "illitic material" is a mixture of illite crystallites of different thicknesses, the thin ones showing interparticle diffraction, thereby exhibiting expansion, and the thicker ones being chiefly nonexpanding illite, because interfaces between illite crystallites make up too small a portion of the crystallites for the effect to be detectable. Thus very fine-grain illitic material, by its very nature, should show some expansion.

In support of theory of interparticle diffraction, Nadeau et al. (1984) produced a series of randomly interstratified I/S by mixing ordered I/S (about 50% expandable) with various proportions of smectite in very dilute suspensions. Thus Frey and Lagaly's finding about the instability of layer sequence in smectite also applies to I/S.

The formation of mixed-layer clays by mechanical mixing may turn out to be an important clue to understanding the behavior of expanding clays during

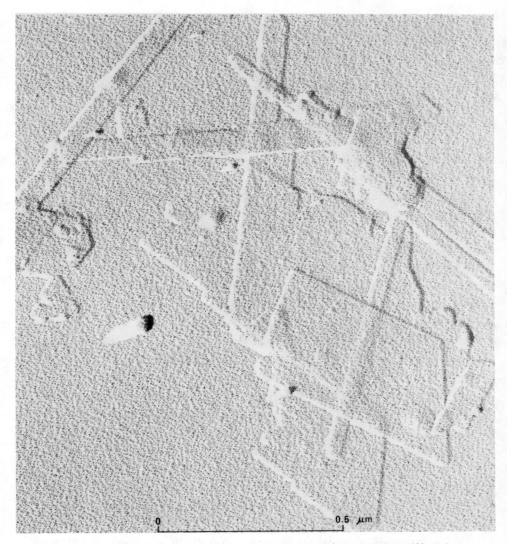

Figure 8. According to the theory of interparticle diffraction (McHardy et al. 1982; Nadeau et al., 1984), thin illite particles ("fundamental" clay particles), shown in this TEM photo of material from sandstones (and diagrammatically in Fig. 7), stack on top of each other, giving rise to interparticle expansion and to the XRD characteristics for I/S. Photo contributed by W.J. McHardy.

transportation and deposition, and during laboratory preparation, but the significance of this effect cannot be evaluated at present. The changes in interlayering were accomplished only in very dilute suspensions with Na-clay. Thicker suspensions gave mixtures. Mixtures of IS-ordered and random I/S are observed in the <0.2 μm fractions of natural clays (Fig. 19). One of the present authors (J.Ś.) failed to induce changes in XRD characteristics of component clays when trying to reproduce Frey and Lagaly's (1970) experiment with <0.2 µm fractions of Wyoming montmorillonite and various types of I/S.

Polytypes

Levinson (1955) was first to recognize in illitic materials all of the major polytypes previously known from muscovites: $1M$, $2M_1$, and $3T$, as well as the disordered $1M_d$ variety. This work later was confirmed by numerous studies, and rare occurrences of $2M_2$ were reported (Drits et al., 1966; Shimoda, 1970). A review of mineralogical work and XRD data on illite polytypes can be found in Brindley (1980).

Most petrological work on illite polytypism was inspired by the experimental work of Yoder and Eugster (1955), Yoder (1959), and Velde (1965). They found the sequence $1M_d -- 1M -- 2M_1$ with increasing hydrothermal run time and temperature. In regional geologic studies, Reynolds (1963) and Maxwell and Hower (1967) showed that the $2M/1M_d$ ratio increases with increasing metamorphic grade in limestones and shales. This approach, including their technique for measuring the polytype ratio, has been used widely in regional studies (see review in Kisch, 1983). In a recent study of a metamorphic sequence in the southern Appalachians, Weaver and Broekstra (1984) found a gradual increase in the $2M_1$ polytype to be characteristic of the anchizone (280°-360°C), the lower temperature diagenetic zone being characterized by 0%, and the higher temperature epizone by 100% of the $2M_1$ polytype in the 0.2-2 µm size fraction. In the epizone, all size fractions contain only $2M_1$ illite. In the lower metamorphic grades, the $2M_1/1M_d$ ratio is larger in the coarser fraction, reflecting the sedimentary input of detrital $2M_1$ mica. According to Weaver and Broekstra (1984), the appearance of 100% $2M_1$ illite in the epizone correlates with a complete loss of expandability for illitic materials (as detected by a heating test), and with crystallinity indices that become independent of grain size.

$3T$ illite has been synthesized by Warshaw (1959) from gels of illitic composition that contained 0.7 moles of K per $O_{10}(OH)_2$ at temperatures ranging between 375°-500°C. Its formation with respect to the $1M$ polytype was favored by lower temperatures and by a smaller layer charge: At 500°-600°C, the $1M$ polytype was synthesized from gels of the same composition, and gels of a muscovite composition yielded only the $1M$ polytype at 375°-600°C. Natural $3T$ illitic material has been reported for I/S associated with hydrothermal ore deposits by Horton (1983). He observed the $1M_d$ polytype for I/S having expandabilities greater than 20% and either $1M$, $1M$-$3T$ mixtures, or $3T$

for ordered I/S of lower expandabilities. $3T$, coarse-grained micas have been observed locally in regionally metamorphosed rocks from the Alps (see review in Frey et al., 1983), but this polytype has not been found in illitic material from diagenetic or low-grade metamorphic sequences.

Changes in polytypes during burial diagenesis have not been studied. The present authors investigated polytypes for a series of diagenetic I/S occurring in Silurian and Carboniferous bentonies. The chemistry and mixed layering of these clays have been characterized previously (Środoń, 1980, 1984a; Środoń et al., 1985). Fifteen samples were studied, and the results are summarized in Figure 9 and as follows:

(1) Unmodulated hk diffraction bands at 4.48, 2.55, and 1.63 Å, bands which are characteristic of completely disordered stacking (translational and rotational faults are so common that each 2:1 layer is diffracting independently, except for in the $c*$ direction; see Plancon et al., 1979), are found exclusively in pure smectites. As the percentage of illite layers increases, the hk bands become modulated, gradually developing reflections of the $1M$ polytype. The intensity of these peaks increases with increasing illite layers, and they become sharper. A complete set of well-developed, $1M$ reflections is observed in a 6% expandable ISII mineral (Fig. 9); but the most sensitive signs of increasing order, which are modulation of the (20,13) band at about 2.55 Å and the (15,24,31) band at about 1.63 Å, are observed in highly smectitic, randomly interstratified I/S.

(2) If the $1M_d$ polytype is defined as 100% turbostratic (0% ordering), then it is probable that there are no $1M_d$ illites or I/S, but that samples so described represent varying degrees of $1M$ ordering, with the $1M_d/1M$ ratio related to the percentage of illite layers. Detailed crystallographic work is needed to find a means for quantifying disorder; but modulation of the (15, 24,31) band provides a simple, empirical parameter, called α (Fig. 9), for characterizing $1M_d/1M$ structures. This parameter correlates well with the illite:smectite layer ratio (Fig. 10) and with the shape of the (20,13) band (Fig. 9). It was observed that the intensities of the $11\bar{2}$ reflection at 3.63 Å and the 112 reflection at 3.07 Å, which are the most intense reflections characteristic to $1M$, do not correlate well with α or with the illite:smectite interlayer ratio (compare the Kalkberg sample in Fig. 11 with the M11 sample in Fig. 9). Thus the pattern found here for $1M$ ordering has been overlooked previously.

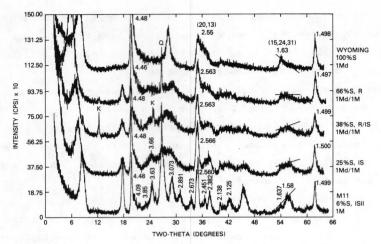

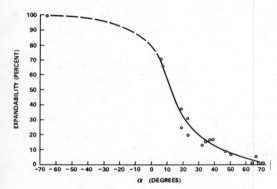

Figure 9 (above). Evolution of polytypes from $1M_d$ (turbostratic) to $1M$ for I/S from bentonites. K = kaolinite; Q = quartz. The change in the α parameter (Fig. 10), measured between 54° and 56° 2θ, with changing ordering and expandability, also is shown. α is the deviation from the horizontal, in degrees, of a line connecting maximum intensities in the region between 54° and 56° 2θ.

Figure 10 (left). Empirical parameter α versus expandability. $1M$ ordering, as defined by α, increases with % illite layers in I/S (circles). Pure $1M$ illites of different origins (Fig. 12) also are shown (squares).

Figure 11 (below). XRD patterns of random preparations of two size fractions of illite from a shale parting in the Madison limestone (Wyoming) compared with $1M$ and $2M_1$ illitic materials. The Madison sample is a mixture of these polytypes, (as seen especially in the 22°-33° 2θ range.

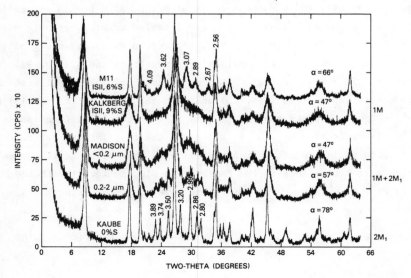

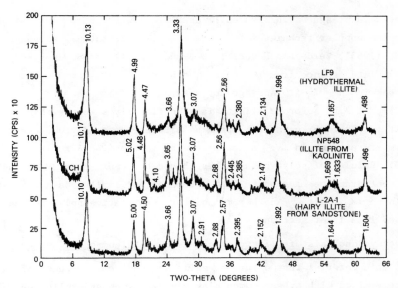

Figure 12. XRD patterns of 1M illites of different origins. Observe the difference in intensities for reflections in the 22° to 33° range, but the similarity in intensities for reflections in the 34° to 38°, and in the 54° to 56° (α parameter) ranges. K = kaolinite.

(3) Variations in polytype with respect to grain size for bentonitic I/S are not observed, in contrast to variations found for shales, as was discussed previously.

(4) The end product for the illitization of smectite is well-ordered, 1M illite. This polytype also is found for hydrothermal illites, for illites that originate from the alteration of kaolinite, and for hairy illites that precipitate in sandstone pores (Fig. 12; see also McHardy et al., 1982; Rochewicz, in preparation; Dopita and Králík, 1977). We know of no data on the polytypes of illites formed from feldspar. According to our data, illites formed from kaolinites and hairy illites formed in sandstone pores do not vary in polytype with respect to grain size. The same conclusion was drawn by Horton (1983) for hydrothermal illitic materials.

Variation of polytype with grain size for illitic material from a shale (Madison sample) is shown in Figure 11. This sample is a mixture of illite and ordered I/S, as was discussed previously. It also is a mixture of polytypes. The $2M_1$ component increases in the coarser fraction (Fig. 11). High values for the α parameter (Figs. 10 and 11) suggest that the dominant polytype is 1M rather than $1M_d$ (the amount of the $2M_1$ polytype is too small to influence α significantly). This conclusion would not have been reached by studying the weak 11$\bar{2}$ and 112 reflections, as is also the case for the bentonitic Kalkberg sample (Fig. 11).

The data presented in Figures 11 and 12 allow for some conclusions regarding the popular techniques for measuring $1M_d/2M_1$ ratios, as suggested by Reynolds (1963), Velde and Hower (1963), and Maxwell and Hower (1967):

(1) Interference from $1M$ peaks must be taken into account when choosing a $2M_1$ peak in the 2.6-4.0 Å range for analysis.

(2) The 2.56 Å for $2M_1$ illite is significantly stronger than that for $1M$ illite; therefore, using this peak as a reference tends to overestimate $2M_1$ in mixtures rich in $1M$. A better choice is the 4.48 Å reflection.

(3) Results of polytype analysis should be reported as $2M/1M$ instead of the current practice of report $2M/1M_d$, because our limited data suggest that there is no $1M_d$ illite. Another method for quantitative polytype analysis is described in a later section.

Quantitative analysis of illitic materials in multicomponent systems

Illitic material that occurs in soils and sedimentary rocks often is intimately mixed with other common clay minerals such as kaolinite, chlorite, and expandable clay (usually I/S). XRD analysis of oriented and random preparations can be used to determine semiquantitatively the ratios of these clays, and abundances in the bulk rock.

Analysis of oriented preparations. Two simple techniques often used for quantitative analysis are those of Biscaye (1965) and Schultz (1964). These authors provided "weighting factors" for the low angle peaks of four clays (e.g., intensity of the 17 Å smectite peak = 4 × intensity of the 10 Å illite peak, etc.). The weighted intensities are normalized to 100% to provide weight percentages in the clay fraction.

Several serious objections to techniques that use simple weighted intensities are:

(1) Mass attenuation coefficients of the component clays are not taken into account, a simplification that is justified only if iron-rich clays are not present.

(2) The 17 Å smectite peak represents not only smectite, but also random I/S, and its intensity is very dependent on expandability (an order of magnitude decrease between 100% and 50% expandable according to computer simulation).

(3) The degree of orientation of each clay mineral in a sample is different, thereby making quantitative analysis of oriented preparations difficult without orientation measurements (R.C. Reynolds, pers. comm.).

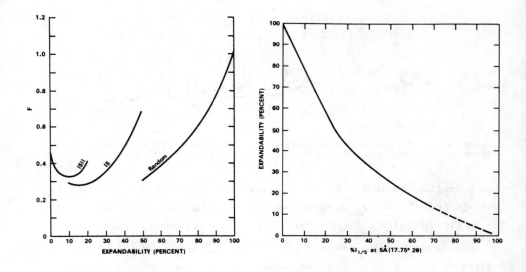

Figure 13 (left). Expandability versus the F parameter. The F parameter is used in semiquantitative analysis to correct intensities of the I/S reflection at 5.7 to 5.0 Å for variations due to mixed layering. Intensity measurements for I/S (Fig. 19) are divided by F. Modified from Środoń (1984b).

Figure 14 (right). Experimental plot for correcting the intensity of the 5 Å illite reflection for interference from I/S in semiquantitative analysis. % $I_{I/S}$ is read from the figure for a given expandability and then is multipled by the maximum intensity of the I/S peak that overlaps with the 002 illite peak. The resulting value then is subtracted from the maximum intensity of the illite peak.

If a homogeneous series of samples is investigated, it can be assumed that the degree of orientation for different clays does not vary considerably. With this assumption, semiquantitative, relative abundances for clays can be obtained from patterns of glycolated samples if corrections are made for variations in peak intensities related to I/S layer ratios, to overlapping of illite and I/S peaks, and to the absorption of X-rays by iron-rich samples.

Figure 19 (below) presents that portion of the XRD pattern most suitable for the analysis and the method for measuring intensities. Figure 13 gives the correction factor (F) for the intensity of the I/S peak as a function of expandability and type of ordering. The intensity of the I/S peak between 5.7 and 5.0 Å is divided by F to find the corrected intensity. Figure 14 presents a curve that can be used to correct the intensity of the 5 Å illite peak for interference from I/S of various expandabilities. The percent of maximum intensity of I/S at 5 Å (%$I_{I/S}$) is read from the figure using previously determined expandability and is multiplied by the intensity of the I/S peak. This value is then subtracted from the intensity of the 5 Å illite peak to give its corrected intensity. Corrected peak intensities are divided by mass attenuation coefficients if iron-rich clay is present, and

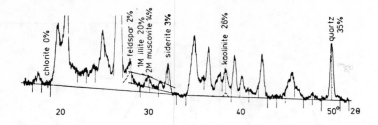

Figure 15. Examples of intensity measurements made for quantitative XRD analysis of illite polytypes in random preparations if kaolinite is present. From Środoń (1979).

then normalized to 100%. A more detailed description of the technique is given in Środoń (1984b). A technique for measuring the layer ratio of random I/S in mixtures with illitic materials is given in Środoń (1981).

Analysis of random preparations. Methods used by metamorphic petrologists to measure $2M/1M$ polytype ratios were discussed previously (see Polytypes). These techniques often are difficult to apply to lower-grade rocks, because kaolinite reflections interfere with illite peaks at 2.56 and 4.48 Å. An alternate approach is presented in Figure 15. The $1M$ contribution is measured at 29° 2θ (Cu $K\alpha$ radiation), even if no 112 reflection is present, as the height of the $1M$ illite band above the background. $2M_1$ illite is measured as the intensity of the 30° 2θ peak above the $1M$ illite band. Details for using this technique, including methods for computation and for random sample mount preparation, are given in Środoń (1984b). Pure $1M$ illite must be separated from the sample suite for use as a standard because the 112 reflection of $1M/1M_d$ illite varies significantly in intensity for different illites (see Polytypes). $1M$ illite can be concentrated by taking the <0.2 μm-size fraction.

CHEMICAL COMPOSITION

Composition of illite in relation to other 2:1 phyllosilicates

Several clay-size micas give XRD patterns that are similar to illite. Glauconite, for example, occurs in a glauconite/smectite (G/S) series of mixed-layer clays that is analogous to the I/S series (Thompson and Hower, 1975). Although similar in structure, the G/S and the I/S series are separated from each other compositionally. I/S contains an average of 5% Fe_2O_3, whereas G/S contains an average of 23%. Based on limited data, there appears to be a compositional break between the two series at 10-15% total Fe_2O_3, suggesting that there is no continuity between G/S and I/S. [Ed: See

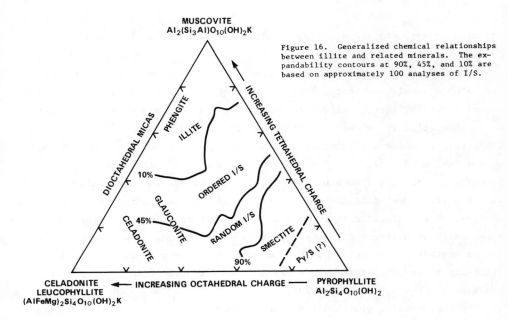

Figure 16. Generalized chemical relationships between illite and related minerals. The expandability contours at 90%, 45%, and 10% are based on approximately 100 analyses of I/S.

following chapter for a dissenting view.] This discontinuity probably results from environmental factors, rather than from crystallographic factors (Odin and Matter, 1981, Fig. 8). As shown in Figure 16, glauconite generally has a greater degree of octahedral, charge-building substitutions than does illite, and G/S has less K for a given expandability than does I/S (Thompson and Hower, 1975). Glauconite, which generally has 0.17 to 0.43 atoms of tetrahedral Al per $O_{10}(OH)_2$, can be distinguished from celadonite, which generally has less than 0.17 atoms of tetrahedral Al (Buckley et al., 1978; Bailey, 1980); also, glauconite has less than -0.7, and celadonite greater than -0.7 equivalents of layer charge developed in the octahedral sheet (Bailey et al., 1984). Phengite, like celadonite, has close to a full mica charge, but, unlike celadonite, has a lower octahedral charge, having 1.4 to 1.9 octahedral Al per $O_{10}(OH)_2$ (Bailey et al., 1979). Most illites are similar to phengite, except that they usually have less than a mica charge, having approximately 0.75 K per $O_{10}(OH)_2$ rather than the approximately 1.0 K of phengite (as will be discussed), although data from Beaufort and Meunier (1983a, Fig. 6e) indicate that hydrothermal micaceous samples display a complete range of K-contents, from 0.74 to 1.0 per $O_{10}(OH)_2$. The term "sericite" is a petrographic term used to indicate highly birefringent, fine-grain, micaceous material seen under the petrographic microscope.

The term is not used for materials studied by XRD because it is possible to be more specific.

Many of these compositional differences between clay-size, dioctahedral, micaceous minerals are represented in the "magic triangle" (Fig. 16). Starting with pyrophyllite, a structure in which 2:1 layers are electrically neutral, theoretically increasing octahedral charge by KR^{2+} for R^{3+} substitution leads to celadonite or leucophyllite (depending on whether $R^{3+} = Fe^{3+}$ or Al^{3+}) when a full mica charge of -1.0 equivalent per $O_{10}(OH)_2$ is developed. Increasing tetrahedral charge by KAl^{3+} for Si^{4+} substitution eventually leads to muscovite. Most dioctahedral clays contain both octahedral and tetrahedral charges, as is indicated in the figure. The smectite field in Figure 16 may be expanded or contracted in the direction of increasing layer charge depending on whether the smectite has been exposed to K, as will be discussed. Mixed-layer pyrophyllite/smectite has not been reported from nature, but it has been synthesized (Eberl, 1979).

Another species similar to illite is ammonium illite. Ammonium illite from black shales hosting a stratiform base metal deposit has been reported by Sterne et al. (1982). This illite has greater than 50% NH_4^+ for K^+ substitution. It can be distinguished from illite by a larger d(001), with values as large as 10.16 Å, by an infrared absorption band at 1430 cm^{-1}, and by ratios for I(001)/I(003) and I(002)/I(005) of about 2. Cooper and Abedin (1981) have reported the incorporation of NH_4^+ in I/S formed during burial diagenesis in the Texas Gulf Coast. The NH_4^+ occupies up to 7% of the K sites. Ammonium muscovite has been reported by Vedder (1965), and ammonium sericite by Yamamoto and Nakahira (1966) and Kozáč et al. (1977). Ammonium micas have been synthesized by Levinson and Day (1968) and by Tsunashima et al. (1975). Ammonium I/S has been synthesized by Wright et al. (1972).

Sodium illite (barmmallite) is a rare clay first found by Bannister (1943) in shales overlying coal measures. It also has been reported by Blokh et al. (1974) in Ordovician mudstones, where it is presumably of metamorphic origin. A related species, Na-rectorite, contains a regular interlaying of beidellite and brammallite layers (Kodama, 1966). Paragonite has been found in many metamorphic rocks (Kisch, 1983), but it is not clay-size.

Trioctahedral illite has been reported by Bodine (1979) from metamorphosed evaporites, by Weiss et al. (1956), by Ushizawa (1981) from talc mines, and by Walker (1950) for soil clays.

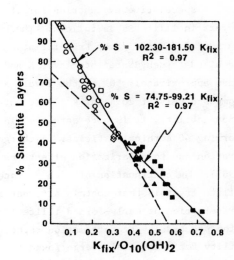

Figure 17. Expandability versus fixed interlayer K for I/S. Solid symbols represent specimens with ordered stacking, open symbols random stacking. From Środoń et al. (1985).

Chemistry of illite and illite layers

The chemistry of illitic materials, based on available published data, was summarized and analyzed statistically by Weaver and Pollard (1973). The obvious shortcoming of this approach is that illitic materials are heterogeneous, as was discussed previously. An alternative, indirect method for estimating the composition of end-member illite is by extrapolation of data for illite/smectites. This approach was attempted first by Hower and Mowatt (1966), and it yielded the widely accepted value of 0.75 K per $O_{10}(OH)_2$ for the interlayer occupancy of illite. Środoń et al. (1985) followed this approach, using a broader sample set and a more precise technique for determining expandability. This plot of expandability versus fixed interlayer K for I/S derived from bentonites is shown in Figure 17.

A statistical analysis of the data in Figure 17 showed that the data points can be fitted best by two straight lines which intersect at about 45% expandable. The upper line gives a fixed K content of about 0.55 per $O_{10}(OH)_2$ for each illite layer. The lower line gives a value of about 1.0 K for the formation of each additional illite layer when expandabilities are less than about 50%. Thus it appears that end-member illite is composed of at least two kinds of illite layers, with early-formed layers having approximately half the K content of those formed later. Because low- and high-K layers are almost equally abundant, the lower curve extrapolates to a value of 0.75 K per $O_{10}(OH)_2$, a value which is close to their arithmetic average.

A speculative explanation for the existence of two types of illite layers in illite is as follows. The upper line in Figure 17 represents formation of random I/S by transformation from smectite, and a charge of -0.55 is required to dehydrate interlayer K and fix it between adjacent 2:1 layers. Once approximately 50% of the layers are collapsed, the clay (ideally) is composed only of 20 Å thick illite layers. In order to further decrease expandability, it is necessary to join these 20 Å illites together, thereby forming 40 Å thick particles. A charge of -1.0 is required for this dehydration due to polarization effects caused by adjacent illite layers (Sawhney, 1967). The continuation of such a process would result in the formation of illite crystals that contain alternating low- and high-K interlayers (Fig. 7). Preservation of early-formed illite layers in illite suggests that the reaction proceeds by transformation rather than by neoformation, as does a stability for octahedral charge found throughout the mixed-layer series (Środoń et al., 1985). The fact that IIS-ordered I/S has not been found beyond doubt lends additional support to this mechanism, because this type of ordering would require a preponderance of 30 Å thick illite particles, according to the theory of interparticle diffraction discussed previously. Previous reports of IIS ordering are better explained as mixtures of IS and ISII-ordered I/S (Środoń, 1984a; see section on hydrothermal environment). A third type of illite layer also may be present in illite. This type of layer could be formed prior to burial diagenesis by wetting and drying of high-charge smectite layers in the presence of K^+ (Środoń et al., 1985, as discussed later). This type of layer could form up to 50% of the illite layers in illite, and its K content could vary between 0.55 and 1.0.

Plots of cation exchange capacity (CEC) and surface area versus expandability, similar to Figure 17, give a CEC for end-member illite of 15 meq per 100 g, a total layer charge of about -0.8 equivalents per $O_{10}(OH)_2$, and a surface area of 150 m^2 per g (Środoń et al., 1985).

All the above characteristics inferred from studying the illite/smectite series apply only to illites derived from the transformation of smectite in bentonites. Chemical analyses of <2 μm fractions of illitic materials from different origins, identified according to the XRD techniques presented previously, are given in Table 1. Structural formulae, calculated by assuming a negative layer charge of 22, are as follows:

(1) M11 (bentonite, ISII, 6% exp. (Fig. 1), $1M$ (Fig. 9)):

$(Al_{1.68}Fe_{0.04}Mg_{0.28})^{-0.28}(Si_{3.53}Al_{0.47})^{-0.47}O_{10}(OH)_2(K_{0.72}Na_{0.01}Ca_{0.01})^{+0.75}$

Table 1. Chemical analyses* of illitic materials of different origins (<2μm fraction).

Sample	Weight percent							
	SiO_2	Al_2O_3	Fe_2O_3	MgO	CaO	Na_2O	K_2O	TiO_2
Kaube	47.4	35.6	1.50	0.30	<0.02	0.53	9.12	0.23
NP 548	45.7	35.1	1.14	.35	.06	.24	7.95	.69
L-2A-1	49.2	25.7	2.49	2.73	.25	.29	8.73	.20
LF9	48.1	32.6	1.85	.75	.08	.34	8.38	.33
M11	53.6	27.0	.96	2.75	.24	.05	8.47	.33
Silver Hill	48.6	24.0	7.20	2.44	.35	<.15	8.02	.59

*Samples, except for M11, were analyzed by X-ray fluorescence spectroscopy (Taggart et al., 1981), K. Stewart and A. Bartel, analysts. Samples were Na-saturated, except for M11, which was Ca-saturated, and Silver Hill, which is natural. NP 548, L-2A-1, and LF9 were treated with sodium dithionite to remove iron coatings (Jackson, 1975). Sample NP 548 and L-2A-1 contain a chlorite and possibly a kaolinite impurity. The analysis for M11 is from Środoń et al. (1985).

(2) Silver Hill (shale I + ISII mixture, Ir = 1.51 (Fig. 1), $1M + 2M_1$ mixture):

$(Al_{1.38}Fe_{0.38}Mg_{0.25})^{-0.22}(Si_{3.40}Al_{0.60})^{-0.60}O_{10}(OH)_2(K_{0.72}Na_{0.03}Ca_{0.03})^{+0.81}$

(3) NP 548 (pseudomorphs after kaolinite, small admixture of chlorite, I + ISII (Ir = 1.56, ca. 2% exp.), $1M$ (Fig. 12)):

$(Al_{1.96}Fe_{0.06}Mg_{0.04})^{+0.14}(Si_{3.13}Al_{0.87})^{-0.87}O_{10}(OH)_2(K_{0.69}Na_{0.05})^{+0.74}$

Subtraction of 4.5% chlorite (see XRD pattern in Fig. 12) gives a more reasonable structural formula:

$(Al_{2.0})^{-0.0}(Si_{3.23}Al_{0.77})^{-0.77}O_{10}(OH)_2(K_{0.73}Na_{0.05})^{+0.78}$

(4) L-2A-1 (hairy illite from sandstone, small admixture of chlorite, I + ISII, ca. 1% exp. (Fig. 24), $1M$ (Fig. 12)):

$(Al_{1.57}Fe_{0.13}Mg_{0.28})^{-0.34}(Si_{3.45}Al_{0.55})^{-0.55}O_{10}(OH)_2(K_{0.78}Na_{0.06}Ca_{0.02})^{+0.88}$

(5) LF9 (hydrothermal, I + ISII (Ir = 1.88), ca. 3% exp., $1M$ (Fig. 12)):

$(Al_{1.84}Fe_{0.09}Mg_{0.08})^{-0.05}(Si_{3.25}Al_{0.75})^{-0.75}O_{10}(OH)_2(K_{0.72}Na_{0.07}Ca_{0.01})^{+0.81}$

(6) Kaube (hydrothermal, I (Fig. 1), $2M_1$ (Fig. 11)):

$(Al_{1.90}Fe_{0.08}Mg_{0.03})^{-0.0}(Si_{3.13}Al_{0.87})^{-0.87}O_{10}(OH)_2(K_{0.77}Na_{0.10})^{+0.87}$

The hydrothermal clays and the clay that is pseudomorphous after kaolinite are characterized by very low octahedral Mg and Fe; whereas, those from shales and sandstone pores resemble the composition of bentonitic clay, having more Mg and Fe and a larger octahedral charge. Layer charge for these clays seems to correlate with traces of expandability, as determined by Ir. Layer charge is higher than -0.75 and lower than -1.00 in all nonbentonitic samples. A systematic study of illite chemistry could result in a correlation between illite chemical composition and origin.

Illites contain "structural" water in excess of that required to complete a formula based on $O_{10}(OH)_2$. This extra water, which is not removed by ordinary drying, may be physically or chemically sorbed on crystal edges and faces, or may exist as H_3O^+ or H_2O in interlayer positions. It is interesting to note that when muscovite is finely ground, its water content is increased, and this excess water is not removed by drying at 110°C (Brindley, 1980).

ILLITE IN NATURAL, SYNTHETIC, AND THEORETICAL SYSTEMS

Weathering environment

Three major processes involving illite layers are encountered in the weathering zone:

(1) degradation of detrital illitic material by leaching, leading to formation of expandable layers ("opening" of illite);

(2) crystallization of illite in the bottom parts of weathering zones (sericitization);

(3) transformation of smectite into illite layers by wetting and drying.

Opening of illite layers. Acid leaching experiments (Scott and Smith, 1966) indicate that illitic material is more resistant to weathering than are coarse-grain micas such as muscovite and biotite. Nevertheless, numerous examples of illite opening in soil profiles have been published. The process usually is interpreted as an illite--vermiculite-smectite sequence (e.g., Flexor et al., 1975) with increasing degree of weathering. Robert et al. (1974) stated that true smectites (nontronite) are formed in soil profiles from glauconitic parent material; whereas, illite weathers to clay beidellitic composition, with mixed smectite and vermiculite characteristics.

The authors were unable to locate any published, high-quality XRD data on the weathering of illitic material. Our own example is atypical, because it is a buried weathering sequence (Pennsylvanian underclay described by

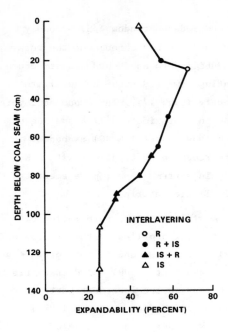

Figure 18. Relationship between depth below coal, expandability, and type of interlayering for underclays from Illinois (see core No. 1 in Rimmer and Eberl, 1982). These measurements were made on the <0.2 μm-size fraction using techniques described in the present paper. See Figure 19 below for sample XRD patterns.

Figure 19. The progressive opening of illite is shown in this weathering sequence for illitic material from a Pennsylvanian underclay from Illinois (<0.2 μm fractions from core no. 1; see Rimmer and Eberl, 1982). Percent smectite (%S) refers to expandabilities measured (using the reflections labeled with precise 2θ values) according to techniques described in Figures 2 and 3 in this paper and in Środoń (1981) for random I/S. Expandabilities given for the 90, 70, and 65 cm samples are average values because the coexistence of random (R) and IS-ordered I/S is shown clearly in the low-angle region. A method for measuring peak intensitites for semi-quantitative analysis for kaolinite, I/S, and illite is shown also.

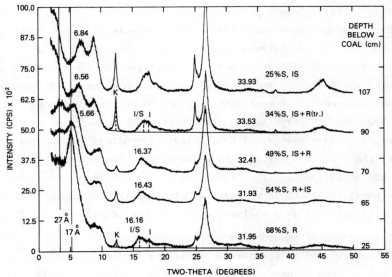

Rimmer and Eberl, 1982). Acid solutions generated in or near an overlying coal have opened illite layers, thereby forming I/S. Expandability increases towards the coal, where acid leaching was most intense (Fig. 18). The small reversal of this trend adjacent to the coal may represent a local increase in K^+ derived from vegetal matter in the coal. Unlike diagenetic materials,

XRD patterns (Fig. 19) show a mixture of ordered and random I/S, probably indicating that the process is proceeding nonuniformly throughout the rock. The opening of illite or I/S of low expandability to an IS-ordered structure may be related to the presence of alternating low- and high-charge illite layers in illite, as was discussed previously (Fig. 17). The expanded layers show characteristics typical of smectite: an approximately 17 Å spacing on glycolation, and a lack of irreversible collapse after 1 N KCl exchange. Layers which do collapse on K-exchange are reopened by treatment with 1 N NaCl. The profile suggests that I/S, and, in extreme cases, pure smectite, can form from the degradation of detrital illitic material.

Neoformation of illite in weathering profiles. The neformation of illitic material in the lower parts of weathering profiles by the process of sericitization is well known to optical microscopists, and is most often seen in Ca-rich portions of zoned plagioclase. Lapparent (1909) noted that this process is one of the first signs of weathering in the lowermost parts of weathering profiles. Stoch and Sikora (1976) and Gardener et al. (1978) observed a secondary, dioctahedral illite-like mineral in the base of a kaolinitic saprolite developed on granite. Konta (1972) separated illitic pseudomorphs after potassium feldspar from a kaolinized granite dyke and found that the illitic material shows slight expandability (<5%), 1M polytype, and highly aluminous composition. Eggleton and Buseck (1980) studied K-feldspar weathering by means of high-resolution electron microscopy and concluded that the first weathering product to develop formed on crystal defects throughout the crystal and was either illite, smectite, or I/S.

Despite a long history of sericitization studies, the volumetric importance of the weathering product to sedimentary rocks is still unclear. Older literature that reports illite formation in the weathering zone (Schlocker and van Horn, 1958) should be regarded with caution because of the possible misidentification of 10 Å halloysite, as has been discussed by Schultz (1963).

Illitization of smectite by wetting and drying. The transformation of smectite layers into illite layers by wetting and drying has been suggested as a cause for the phenomenon of K-fixation by soils (Volk, 1934; Truog and Jones, 1938). This process is not limited to vermiculites derived by the opening of micas, but also is displayed by bentonitic smectites. Further information on the process was provided by Mamy and Gaultier (1975), who showed by electron diffraction that K-fixation by wetting and drying leads to development of three-dimensional stacking order.

The present authors studied the wetting and drying mechanism in more detail (Eberl and Środoń, in preparation), and their conclusions are summarized as follows:

(1) K-exchange without wetting and drying results in complete or partial dehydration of a portion of the smectite layers, producing three-component mixed layering (17/14/10 Å with ethylene glycol). The percentage of illite-like layers so produced is small in low-charge smectites (e.g., Wyoming) but is quite large (up to 50%) in high-charge smectites (e.g., Cheto). Dehydration is not permanent, and original expandability is completely restored by simple exchange with 1 M NaCl.

(2) K-smectites run through wetting and drying cycles fix K irreversibly (as tested by 1 M NaCl exchange), transforming the clay into random I/S containing up to 50% illite layers for high-charge smectites (Cheto). In low-charge smectites (Wyoming), only a few percent of the layers collapse irreversibly. Most layers collapse irreversibly in less than 40 cycles.

(3) Na-smectite subjected to wetting and drying cycles with a K-bearing mineral (feldspar, muscovite) in distilled water strips K from the mineral and produces illite layers at a rate similar to that found for the K-smectite.

Because the land's surface is submitted to wetting and drying cycles, and because most smectites have charge densities higher than that of the Wyoming (Weaver and Pollard, 1973, Fig. 14), the present data suggest that wetting and drying is a geologically important mechanism that is responsible for forming significant amounts of illite layers. Although field evidence is needed, one can speculate that this mechanism is responsible for forming the vast amount of I/S found in rocks that have never experienced elevated temperatures (Schultz, 1978; Środoń, 1984b).

Sedimentary environment

Formation of illite in sea water. For many years it was thought that illite was formed in the marine environment by illitization of land-derived, expandable material (Dietz, 1941). K-Ar age dating has shown, however, that most marine illite is detrital (Hurley et al., 1963; Lisitzin, 1972; see Grim, 1968, for a summary). Other experimental approaches also suggest that illite does not form abundantly in the sea:

(1) Hoffman (1979), in a detailed XRD and K-Ar study of Mississippi river muds, and sediments and well-cuttings from the Gulf Coast, showed that sediments entering the Gulf of Mexico from the Mississippi River gain only

about 0.2-0.3% K_2O, and that their mineralogy remains unchanged until the onset of burial diagenesis at about 50°C. Weaver (1958) suggested that soil smectite, formed by the opening of mica layers in the weathering environment, is transformed back into illite by the fixation of K^+ from sea water. Hoffman's data suggest that the amount of illite formed by this process is minor. Our data, discussed previously, have shown that expanding clays, even expanding clays formed from the opening of illitic material, are reluctant to fix K irreversibly unless subjected to wetting and drying cycles. Other experiments (Eberl and Środoń, in preparation) have shown, however, that vermiculite and high-charge smectite layers in I/S can fix small amounts of K irreversibly without wetting and drying.

(2) Schultz (1978), in an extensive study of the Cretaceous Pierre shale, found no difference in expandability for I/S from sea water and fresh water parts of the sedimentary basin.

(3) Środoń (1984b) found no correlation between salinity and percent discrete illite or percent illite layers in I/S for profiles studied in a shallow, Miocene marine basin in southern Poland.

(4) Eberl and Hower (1976) extrapolated hydrothermal, kinetic data for the formation of illite layers from smectite to ocean bottom temperatures and found (if experimental results for their simple system are applicable to nature) that the illitization process is too slow to produce significant illite in the ocean. Subsequent experiments (Roberson and Lahann, 1981) with systems that more closely approximate nature support this conclusion. These studies suggest that land-derived, expandable material is relatively stable in the marine environment. Thus most illite layers are formed from smectite prior to entering the ocean by wetting and drying and, after burial, by higher temperature diagenesis, as will be discussed.

Although these studies suggest that illite does not form abundantly in the sea, it is well established that two closely related species, glauconite and celadonite, do form on the ocean bottom. Glauconite forms on outer continental shelves and slopes where sedimentation rates are low (Odin and Matter, 1981; Ireland et al., 1983), and celadonite forms from the alteration of oceanic basalts (Kastner and Gieskes, 1976). High-charge smectites have been reported as being neoformed on the ocean bottom by alteration of volcanic rocks and glass, by reaction of biogenic silica with Fe-oxhydroxides, and by direct precipitation from hydrothermal fluids. Chemical analyses of these "smectites" (Cole and Shaw, 1983, Table 1) show that K is often a major or even the dominant interlayer cation, but precise XRD data

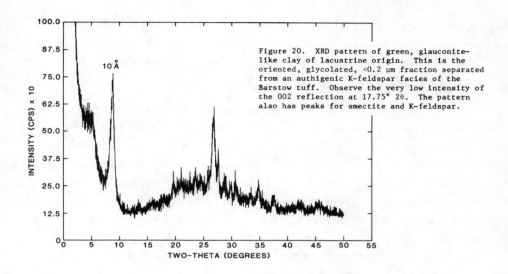

Figure 20. XRD pattern of green, glauconite-like clay of lacustrine origin. This is the oriented, glycolated, <0.2 μm fraction separated from an authigenic K-feldspar facies of the Barstow tuff. Observe the very low intensity of the 002 reflection at 17.75° 2θ. The pattern also has peaks for smectite and K-feldspar.

are lacking. The possibility exists that these "smectites" are I/S formed in the marine environment.

Formation of illite in lakes. Mineralogical data on illitic material occurring in lakes are not abundant, and detrital origin is postulated most often (review in Jones and Bowser, 1978). On the other hand, it is well established that a micaceous clay similar to glauconite does form authigenically under rare conditions (summary in Porrenga, 1968). This clay can be distinguished from illite by a relatively low intensity for the 002 reflection, as was discussed previously. We observed such a green, iron-rich mineral (Fig. 20) occurring with authigenic K-feldspar in Miocene tuff from the lacustrine Barstow Formation of California (Sheppard and Gude, 1969). Pure smectites occur with this glacuonite and with the various zeolites. Thus the illitization of smectite does not occur even under the alkaline (and sometimes K-rich) environment required for zeolite formation.

These findings are contrary to those of Singer and Stoffers (1980) who report illite neoformation in a saline lake. Their observed variation could be due to changes in detrital input rather than due to authigenic reaction. Because only the low-angle portions of their XRD patterns are presented, the possible presence of a glauconite-like mineral cannot be determined. Jones and Weir (1983), however, have shown convincingly that illite layers constitute a significant portion of mixed-layer clay in saline lake sediments; whereas, pure smectites are found in the sediment source area. We suspect that these illite layers formed during transportation to the lake by wetting and drying, because there is no significant difference between the lake samples and those found in the adjacent playa. The possible formation of illite or illite layers in saline lakes deserves further study.

Diagenetic/metamorphic environment

Illitization of smectite. This process was first recognized in Gulf Coast Tertiary sediments (Burst, 1959). Shutov et al. (1969) and Perry and Hower (1970) studied this reaction in more detail and discovered that ordered I/S appears with depth. Numerous subsequent studies have shown that illitization is the typical path for smectite alteration in the sedimentary column, although chloritization (formation of corrensites) does occur in volcanogenic sediments (Hoffman and Hower, 1979) and in arkoses (Helmold and van de Kamp, 1984).

The illitization reaction becomes detectable only at temperatures significantly higher than surface temperatures (50°C, according to Perry and Hower, 1970). For example, in a Miocene marine shale profile representing a time span of a few million years but exposed only to surface temperatures (less than 200 m depth), no detectable trend in mixed-layering can be measured (Środoń, 1984b). In a high-potassium environment of first-cycle desert alluvia, authigenic potassium feldspar is crystallizing along with essentially pure smectite during low-temperature diagenesis (Walker et al., 1978).

The reaction that occurs in Gulf Coast sediments is (Hower et al., 1976):

$$\text{smectite} + \text{K-feldspar} (+ \text{mica?}) \rightarrow \text{I/S} + \text{chlorite} + \text{quartz} .$$

This reaction ceases in the Gulf Coast at about 20% expandable with an IS-ordered structure (Perry and Hower, 1970; Weaver and Beck, 1971; Hower et al., 1976; Boles and Franks, 1979), but this cessation is a local feature, perhaps related to the complete dissolution of K-feldspar according to the previous reaction. In other areas the reaction proceeds further towards illite through ISII interstratification (Eslinger et al., 1979; Hoffman and Hower, 1979; Środoń, 1979, 1984a; also Dunoyer de Segonzac, 1969; if his X-ray data are reinterpreted). The appearance of IIS (R2) ordering has been suggested by Drits and Sakharov (1976) and Nadeau and Reynolds (1981a), but, as was mentioned previously, Środoń (1984a) suggested that the data are better explained as a mixture of IS and ISII ordering.

When data from the Gulf Coast were the only ones available, it was thought that expandability of I/S could be used as a geothermometer (Hoffman and Hower, 1979). When diagenetic profiles from different basins are plotted together (Fig. 21), however, it is clear that expandability is not solely a

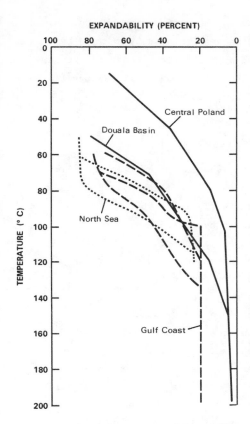

Figure 21. Smectite to illite transformation in shales from different sedimentary basins, plotted as a function of temperature. Central Poland from Środoń (1984a); Douala Basin from Dunoyer de Segonzac (1969) as interpreted by the present authors; North Sea from Pearson et al. (1982); Gulf Coast from Hower et al. (1976) and from Boles and Franks (1979). Dotted lines represent two extreme profiles for profiles from the North Sea. The three Gulf Coast profiles are Pliocene-Pleistocene (lower line), Eocene-Oligocene (middle line), and Lower Eocene (upper line).

function of temperature. Figure 21 suggests that reaction time plays an important role, as does rock and pore fluid composition, particularly with regard to the supply of K. The effect of time perhaps can be seen for profiles from the Gulf Coast: The older profiles have I/S of a lower expandability for a given temperature. The effect of K availability is demonstrated by cessation of reaction for the Gulf Coast when all of the K-feldspar has been dissolved and by a more illitic curve for the Polish basin. The Polish basin has a greater supply of K for the reaction than does the Gulf Coast because clays and mudstones in this basin are richer in detrital micas and feldspars, and expandable clay is a minor component.

Other evidence for the importance of K-availability to the reaction comes from studies of bentonites and sandstones. Bentonites often contain more highly expandable I/S than surrounding shales (Środoń, 1979; Nadeau and Reynolds, 1981b); and thick bentonite beds are frequently zoned, with expandability increasing towards the center (Parachoniak and Środoń, 1973; (Środoń, 1976; Altaner et al., 1984). Both phenomena may be related to a greater availability of K from K-minerals in the surrounding shales. Higher expandabilities found for I/S from sandstones surrounded by shales (Boles and Franks, 1979; Howard, 1981) is attributed by the latter author to the same effect. Środoń (unpublished data) observed that expandabilities for I/S in sandstones from the Carboniferous of the Upper Silesian Coal Basin lag one kilometer in depth behind the shales.

The question arises as to why K^+, rather than Na^+, is fixed in illite layers during burial diagenesis. The mechanism of how Na^+ is excluded from illite interlayers is a puzzle because the reaction of smectite to illite proceeds in a Na-rich pore water in which interlayer exchange sites are dominated by Na^+. It has been suggested (Eberl, 1980) that K^+ is preferred during illitization because it has the lower hydration energy and therefore dehydrates first in response to increasing layer charge. Dehydration of K^+ by interlayer electrical forces leads to increased selectivity by the interlayer for the smaller, dehydrated K^+ ion compared to hydrated Na^+ (according to Coulomb's law) and to collapse of the smectite interlayer during K^+ for Na^+ exchange to form illite by expulsion of Na^+ and associated water.

The illitization reaction slows considerably after reaching the ISII structure (Fig. 21); the last expandable layers are persistent. Close to 10% expandability was observed for Carboniferous rocks at the bottom of the profile from central Poland (Fig. 21) where temperatrues are 200°C and vitrinite reflectance has reached 3%. At this level, the noncoincident illite/smectite peak can still be recorded in the finest fraction (Środoń, 1984a, Fig. 9).

The further evolution of illitic material with increasing metamorphic grade has been studied extensively by means of illite crystallinity index, $2M_1/1M$ polytype ratio, and other parameters. For example, a distinction between diagenesis and the first metamorphic zone (anchizone) in pelitic rocks is based on a value of 0.42° 2θ (Cu Kα) for the Kubler index; and the border between the anchizone and epizone is defined at a value of 0.25° (Weaver and Broekstra (1984), These transitions have been studied extensively by Weaver and Boekstra (1984), and their results are summarized as follows:

(1) Temperatures for diagenesis--anchizone, and anchizone--epizone boundaries are 280° and 360°C, respectively. Epizone rocks are completely recrystallized (including the micas, as shown by K-Ar age determinations). The occurrence of illitic material is confined to the diagenesis zone and anchizone.

(2) Kubler and Weaver indices (measured for <2 μm fraction) evolve gradually throughout the sequence, but there are significant changes in the

slope of the curve at both boundaries, thereby suggesting an abrupt mineralogical change.

(3) Expandability of illitic material, estimated at 10% at the end of diagenesis, decreases gradually to 0% at the onset of the epizone (no contraction of the 10 Å reflection on heating). At this point, the yield of <0.2 μm clay decreases to 0, crystallinity indices for different size fractions converge to one value, and the $2M_1/(2M_1 + 1M)$ ratio becomes the same for all size fractions and approaches 1. The complete conversion to $2M_1$ was observed to have occurred in the Montana Disturbed Belt at 400°C, based on oxygen isotopic evidence (Eslinger and Savin, 1973a).

(4) MgO/Fe_2O_3 ratios and 002/001 peak intensity ratios increase systematically in the anchizone, and these ratios seem to be related.

(5) Phengite is the stable phase in the anchizone, and illitic material is metastable: The latter persists as a matrix material until recrystallization occurs in the epizone, but coarse-grained phengite crystallizes in the anchizone in porphyroblasts and fossil fragments.

The data of Weaver and Broekstra (1984) suggest that illite derived from the transformation of smectite does not exist in these types of rocks: Complete disappearance of expandability is coincident with complete crystallization of illitic matrix into coarse-grained micas at the onset of the epizone. The complete diagenetic-metamorphic process is accompanied by an increasing perfection for illite layers, as has been shown in TEM studies by Lee et al. (1985).

Illitization of kaolinite. The illitization of kaolinite was described by Kulbicki and Millot (1960), who also observed that the presence of oil in pore spaces inhibits illitization. Numerous later studies (review in Kisch, 1983) confirm that the process is common, that kaolinite usually is illitized completely before smectite is illitized completely, and that there is no strict relationship between this reaction and temperature. Illitization of kaolinite in the Upper Silesian Coal Basin begins when I/S in shales is at least 15% expandable, with a vitrinite reflectance of 0.8% (Środoń, 1979). Completely illitized tonstein from this basin has been described by Dopita and Králík (1977). It contains almost pure, very aluminous illite, with a minute admixture of ISSI I/S (see sample NP 548, described previously). The I/S probably is mixed-layer clay that was originally present in the tonstein; whereas, the pseudomorphs are pure illite. These illites preserve the characteristic morphology of the preceeding kaolinite and are the 1M polytype (Fig. 12). SEM studies of other illitized

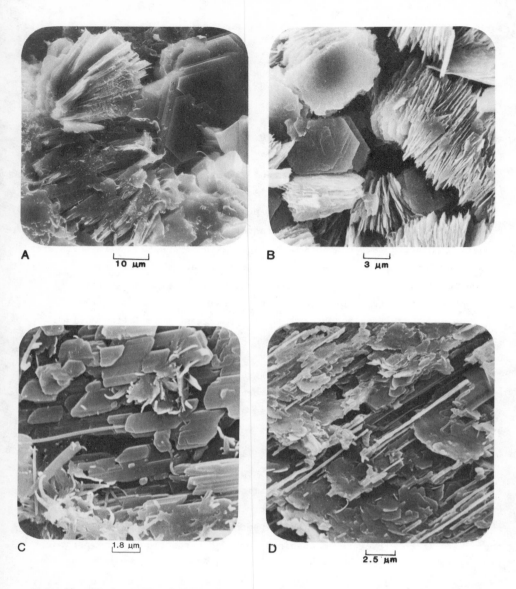

Figure 22. SEM photographs of diagenetic illites in sandstones. (A,B) Initial and advanced stages of illitization of vermicular aggregates of kaolinite. (C,D) Initial and advanced stages of illitization of feldspar. From Rochewicz (in preparation).

kaolinites reveal that the shapes of kaolinite books are preserved, but that the sharp edges are lost (Figs. 22a and 22b). Chemical analysis (Table 1) yields structural formulae that are very low in Fe and Mg compared with illitic materials of other origins. The fate of illitized kaolinite with increasing degree of metamorphism has not been studied.

Illitization of feldspar. Feldspar usually is altered to kaolinite or albite during diagenesis; but under some conditions it may alter to illite, as has been shown by Rochewicz (in preparation) in an SEM study of Rotliegend well samples from Poland (Figs. 22e and 22d). Little is known about this process, and no XRD data are available.

Illitization of muscovite. McDowell and Elders (1980) reported that detrital muscovite alters to sericite (illite or I/S) with depth in the Salton Sea Geothermal Field at temperatures of 280°C or less. At greater temperatures this sericite becomes rare and reacts to form idioblastic phengite. In general, the reaction sequence with increasing temperature is: detrital muscovite → sericite (I/S or illite) → phengite → muscovite. These data suggest that I/S and/or illite have a stability field with respect to muscovite. Some of the sericite is Na-rich, suggesting that it has a brammallite or paragonite component. Biotite reacts analogously, having a low interlayer K-content when it first appears at the biotite isograd, and increasing in K-content with increasing temperature. Similar results for muscovite illitization have been reported by Nicot (1981) for the anchi-epizone transition for metamorphic, Precambrian Belt Series rocks from northwestern Montana, and by Jasmund et al. (1969) for hydrothermally altered, Triassic sandstones from southeast of Cologne.

Neoformation of illite in sandstone pores (hairy illite). Weaver (1953) first observed lath-shaped illites in clay separated from sandstones. Subsequent SEM and TEM studies have documented the development of hairy (filamentous) illites in sandstone pore spaces (Stalder, 1973; review in McHardy et al., 1982). Illite laths are up to 30 µm in length, typically 0.1-0.3 µm in diameter, and up to 200 Å thick (Güven et al., 1980), although some samples are dominated by particles only 20 to 30 Å thick (McHardy et al., 1982). Some illite laths possess perfect morphology, displaying (110) prism forms at terminations (Fig. 23a). Microprobe data (Güven et al., 1980; Rochewicz, in preparation) indicate compositions low in Mg and Fe, but our XRF analysis (Table 1) gives a composition that is similar to illites derived from smectites. Diffraction data show exclusively the 1M polytype (Fig. 12).

SEM photographs by Rochewicz and Bakun (1980) and Rochewicz (in preparation) indicate compositions low in Mg and Fe, but our XRF analysis (Table 1) gives a composition that is similar to illites derived from smectites. Diffraction data show exclusively the 1M polytype (Fig. 12).

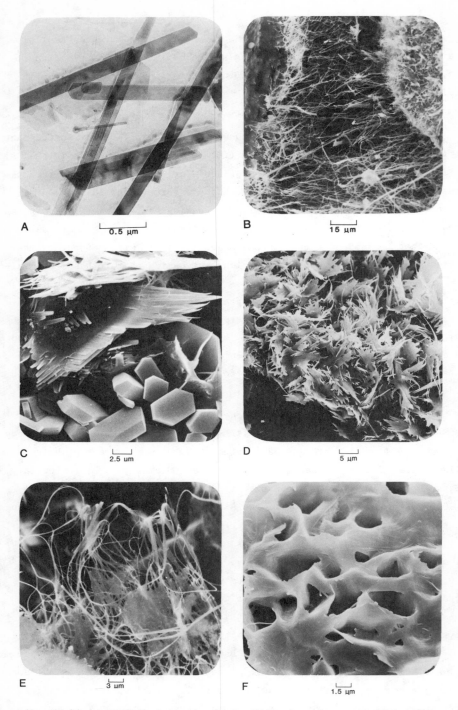

Figure 23. TEM (A) and SEM (B-F) photographs of hairy illites in sandstones. A = hairy illite crystals showing perfect terminations. B-F = stages in development of diagenetic illite in sandstone pores (see text). From A. Rochewicz (in preparation).

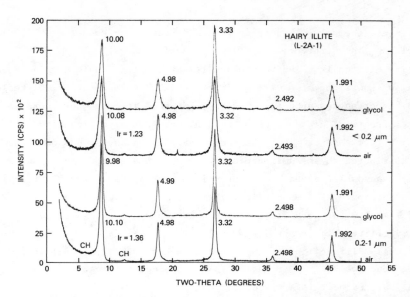

Figure 24. XRD patterns (oriented) of two size fractions of a hairy illite from the Rotliegend sandstone, central Poland (about 3,500 meters depth). Traces of expandability (001 shift after glycolation, and Ir > 1) is similar in both fractions, but peak breadth is wider for the finer fraction. CH = chlorite.

SEM phogotraphs by Rochewicz and Bakun (1980) and Rochewicz (in preparation) trace the gradual development of filamentous illites in cores from Carboniferous and Permian sandstones from central Poland. Lath growth begins from the surfaces of detrital grains (Fig. 23b), or from dark, Fe-rich material (an observation also made by Güven et al., 1980). The fibers then aggregate either by regular stacking at 120° (Fig. 23c), or by irregular stacking (Fig. 23d), or by being bound by another substance of similar chemical composition, but richer in Fe (Fig. 23e). Irregular stacking leads finally to the development of honeycomb-like pore filling, with single laths often visible on the surfaces (Fig. 23f). No relationships have been established between morphology, XRD characteristics, and geological parameters.

Are hairy illites truly illite, or can they also be composed of I/S? This is a difficult question because hairy illite may be mixed with other clay when separated from sandstones for XRD analysis. Of the three filamentous illites studied by Güven et al. (1980), the deepest was reported as being pure illite, and the others were mixtures of illites and I/S. XRD data reported by McHardy et al. (1982) also are interpreted (according to the XRD criteria discussed previously) as a mixture of illite and approximately 20% expandable IS-ordered I/S. The expanding component found with these filamentous illites may come from detrital grain coatings, rather than from the filaments themselves, as is suggested by data from Rossel (1982). This I/S would be concentrated in the finest size fractions.

Hairy illites from central Poland (eight samples provided by A. Rochewicz) also show signs of swelling (Ir > 1, with displacement of 001 reflection to higher angles upon glycolation), but swelling does not increase in the finer fraction, although peak breadth increases significantly due to decreasing particle size. Thus the samples are homogeneous, and swelling cannot be attributed to an admixture of I/S. The sample shown in Figure 24 is about 1% expandable, based on an analysis of the Ir parameter. (If Ir is taken as a measure of expandability, Ir = 2.44 for 6% expandable clay (Fig. 1), and Ir = 1.0 for pure illite; interpolation gives an expandability for the hairy illite which has an Ir = 1.2). This swelling may be caused by interparticle diffraction, as suggested by McHardy et al. (1983). It may be that this one swelling layer in 100 is not a smectite layer in a large illite crystal, but, rather, is a glycolated interface between illite laths that are stacked on top of each other in the XRD slide (Fig. 8).

Hydrothermal environment

Most of the illitic materials found in the diagenetic/metamorphic environment also are found in the hydrothermal environment. Hydrothermal illitic materials that have been studied most are those associated with active geothermal fields (Steiner, 1968; Muffler and White, 1969; Eslinger and Savin, 1973b; McDowell and Elders, 1980), and with ore deposits (Inoue and Utada, 1983; Beaufort and Meunier, 1983a,b; Duba and Williams-Jones, 1983; Cathelineau, 1983; Horton, 1983; Parry et al., 1984). The progressive alteration of smectite into illite with increasing temperature has been observed, for example, by Steiner (1968) for clays from the Wairakei Geothermal Field (New Zealand). According to our interpretation of his data, IS ordering first appears at about 50% expandable, and ordering becomes complete at about 35% expandable; ISII ordering begins at about 18% expandable. Similar data are reported by Inoue and Utada (1983) for clays from a hydrothermal envelope surrounding a Kuroko-type ore deposit at Shinzan, Japan. They plotted their data on Watanabe's (1981) diagram to distinguish between random (R0), IS (R1), and IIS (R2) types of ordering. An apparent discrepancy between the R2 theoretical curve and the experimental data leads them to suggest that longer range ordering must be present. A theoretical plot for ISII (R3) interstratification fits the data well (Fig. 25), offering additional evidence that IIS ordering either does not exist, or is rare (Środoń, 1984a).

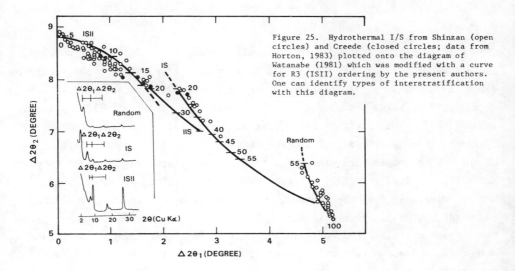

Figure 25. Hydrothermal I/S from Shinzan (open circles) and Creede (closed circles; data from Horton, 1983) plotted onto the diagram of Watanabe (1981) which was modified with a curve for R3 (ISII) ordering by the present authors. One can identify types of interstratification with this diagram.

Samples from a hydrothermal ore deposit at Creede, Colorado (Horton, 1983) also are plotted on this figure and also support this conclusion.

There seems to a better correlation between expandability and temperature for hydrothermal I/S than for diagenetic I/S. The appearance of ordering ("disappearance of montmorillonite" in the older literature) was observed at 98°–135°C in the Salton Sea Geothermal Field (Muffler and White, 1969) and at 85°–110°C in the Wairakei Geothermal Field (Steiner, 1968). The appearance of illite (meaning the disappearance of I/S peaks that are noncoincident with illite, which happens at 5% expandable) takes place at 203°–217°C in the Salton Sea, 230°–240°C in Wairakei, and somewhere around 200°C in Shinzan (Inoue and Utada, 1983). In the Wairakei section, illitization seems to be complete below 240°C (based on no change in 001 spacing after glycolation).

Polytypic relationships found for hydrothermal clays seem more complex than those found for diagenetic clays, although little work has been done. Horton (1983) documented a $1M_d$––$1M$––$1M/3T$––$3T$ evolution with decreasing expandability in clays associated with a hydrothermal ore deposit. Shirozu and Higashi (1972) found a $1M_d$––$1M$––$2M_1$ set in a similar geologic situation, $1M_d$ being restricted to clays with measurable expandability. Beaufort and Meunier (1983b) reported that $1M$ illite had a lower K content than did $2M_1$ illite for clays from a hydrothermally altered porphyry copper deposit. The hydrothermal synthesis of illite polytypes has been discussed previously (see Polytypes section).

Synthesis

Illitic materials have been synthesized hydrothermally by many workers, including Noll (1936), Gruner (1939), Yoder and Eugster (1955), Yoder (1959), Velde (1965, 1969, 1977), Eberl and Hower (1976, 1977), Eberl (1978), Eberl et al. (1978), Velde and Weir (1979), Frank-Kamenetskii et al. (1979), and Inoue (1983). The clay that forms initially from gels in hydrothermal experiments frequently reflects the composition of the starting gel (e.g., gel of phlogopite composition initially forms phlogopite; beidellite gel forms beidellite). This initially formed clay may not be thermodynamically stable, but may disappear over the course of weeks or months of reaction, yielding new phases as equilibrium is approached gradually in a series of steps. Reactions that follow this pattern are said to react according to Ostwald's step rule. Thus a more complete understanding of illite synthesis is derived from kinetic, rather than single run, hydrothermal experiments.

The kinetics of illite formation from a smectite precursor have been studied in laboratory hydrothermal experiments and in nature. Experiments starting with gels of K-beidellite composition (Eberl and Hower, 1976) showed that K-beidellite is the first phase to crystallize, and that this smectite evolves towards illite through random and ordered I/S. The appearance of illite layers in this chemically simple system could be described by a first order rate equation, with an activation energy of about 20 kcal/mole. It is difficult to apply these data to diagenetic sequences, however, because natural systems are so much more complex. By using natural smectites as starting material, by varying solution composition, and by assuming first order kinetics, Roberson and Lahann (1981) calculated an activation energy of about 30 kcal/mole and found that Na, Ca, and Mg inhibit the reaction to illite in the proportion 1:10:30. This study, although based on sketchy data, is closer to modelling natural systems and suggests that a solution term must appear in the rate equation. Such an equation was suggested by Pytte (1982), based on studies of I/S suites in shales adjacent to basaltic dikes. His data could be modelled by fourth, fifth, and sixth order rate equations, but the fifth order equation, using an activation energy of 30 kcal/mole, gave the best empirical fit for both dike alterations and burial diagenetic sequences. The equation is:

$$-dS/dt = k([K^+]/[Na^+])^1 S^4 , \qquad (2)$$

where S = the mole fraction of smectite in the mixed layer clay, t is time; $[K^+]/[Na^+]$ is the ratio of the activities of these ions in solution; and k

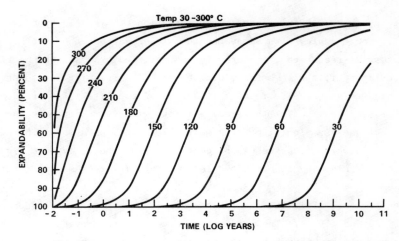

Figure 26. Plots of fifth-order rate equation (see text) for various temperatures for the reaction of smectite into illite. The calculation assumes an activation energy of 30 kcal/mole, and that $[K^+]/[Na^+]$ in solution is a function of microcline and albite dissolution equilibria for a given temperature. From Pytte (1982).

is the rate constant, where $k = Ae^{(-U/RT)}$. A is the frequency factor, U is the activation energy, R is the gas constant, and T is the temperature in Kelvins. Equation 2 is graphed in Figure 26 for several temperatures using $[Na^+]/[K^+]$ for a solution composition that is controlled by albite-microcline dissolution equilibria. This equation, however, also falls short of modelling many natural systems (for example, the effects of $[Mg^{2+}]$ and $[Ca^{2+}]$ are not included); but the equation does suggest an approach for explaining quantitatively differences in expandabilities found for a given temperature in burial diagenetic sequences (Fig. 21).

Other complexities not addressed in this equation are the effects of pressure and of initial chemistry for the solid (Eberl et al., 1978). Increasing pressure in closed systems tends to inhibit the reaction of smectite to illite. A starting material composed of trioctahedral smectite is much less reactive than is dioctahedral smectite, and a starting material of illite or muscovite composition may react directly to form illite or I/S of low expandability, without forming more expandable I/S as a precursor, as is suggested by the experiments of Velde (1969) and Warshaw (1959). Experiments by Howard and Roy (1983) show that hydrothermal treatment of a low-charge montmorillonite (Na-Wyoming) in a K-deficient hydrothermal system at 150° and 250°C for 30 to 180 days can increase layer charge without leading to collapse of the layers. Later K-saturation of this clay caused collapse of up to 40% of the layers in the most extreme case. Hydrothermal treatment of montmorillonite saturated with cations other than K^+ at higher temperatures (e.g., 400°C) leads to the formation of mixed-layer clays other than I/S. For example, with Na as the interlayer cation, the smectite reacts to form Na-rectorite (regularly interstratified brammallite/smectite); with interlayer Li, the reaction yields Li-tosudite (regularly interstratified chlorite/smectite). Thus the reaction of smectite → I/S (random) → I/S (IS ordered) → I/S (ISII ordered) → illite with increasing diagenetic grade is but one example (albeit the most common one) of a more general trend in reaction in which smectite can be converted hydrothermally into a variety of mixed-layer clays by changing the initial interlayer chemistry (Eberl, 1978).

The synthesis of illite and I/S at surface temperatures has been reported by Harder (1974). The products were very poorly crystallized, requiring analysis by Debye-Scherrer camera techniques. Very special conditions were required for this synthesis, including solutions low in Si and Na, and relatively high in K and Mg.

Stability diagrams

In attempting to construct stability diagrams for illitic materials, one is confronted with two unanswered and related questions: (1) What free energies of formation should be used for illite and smectite, both of which vary in composition and structure, and (2) should I/S be treated as a solid solution of smectite and illite, or is I/S a mixture of these two end members? One approach to the first problem has been to measure free energies for specific illites and smectites (e.g., Routson and Kittrick, 1971; Merino and Ransom, 1982); but a problem with this approach is that vast numbers of measurements would have to be made to cover the range of natural compositions. A second approach is to use thermochemical approximations for the phases; these free energies can be calculated from chemical compositions (Tardy and Garrels, 1974; Nriagu, 1975; Mattigod and Sposito, 1978; Aagaard and Helgeson, 1983; Stoessell, 1979, 1981; Tardy and Fritz, 1981).

In an attempt to answer the second question, Garrels (1984) constructed two theoretical stability diagrams using idealized compositions for the phases. One diagram was calculated by assuming that I/S is a solid solution, and the other was calculated by assuming that I/S is a mixture. Water analyses of solutions in contact with clays were plotted on these diagrams, and it was found that the mixture model, in which smectite and illite behave as two discrete phases, better fits the data (Fig. 27).

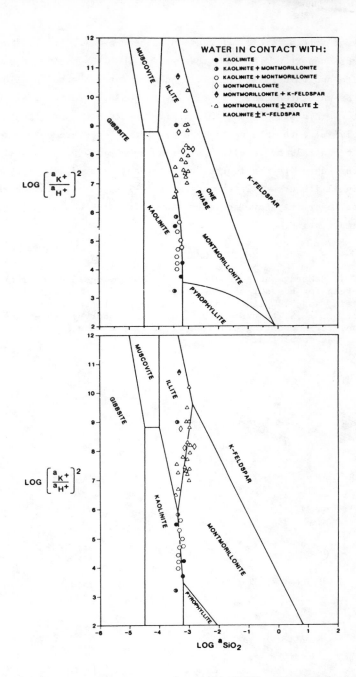

Figure 27. Ion activity diagrams from Garrels (1984). The upper diagram assumes that I/S is a solid solution of illite and montmorillonite. The lower diagram assumes that I/S is a mixture of illite and montmorillonite. Water analyses fall along phase boundaries in the latter diagram, thereby suggesting that montmorillonite and illite in I/S behave as two discrete phases.

The theory of interparticle diffraction (Nadeau et al., 1984), discussed previously, lends support to the idea that random I/S is a mixture because discrete smectite and thin illite particles can be interlayered in dilute suspensions of Na-saturated clay. By similar reasoning, ordered I/S could be considered to be a single phase (thin illite), were it not for the edge effects which give rise to interparticle expansion. Edges are not a problem in coarser grain minerals because they comprise so small a portion of the crystals. If the edges are considered, then perhaps ordered I/S is a solid solution of thin illite crystallites and smectite-like edges. The observation that illite layers may have at least two different K contents (Fig. 17) is a further complication.

ACKNOWLEDGMENTS

We thank W.J. McHardy and A. Rochewicz for supplying electron micrographs, D.J. Morgan for his set of I/S samples, J. Králík for a sample of illitized tonstein, and G. Whitney and L.G. Schultz for reviewing the original manuscript. J.Ś. thanks the U.S. Geological Survey, Water Resources Division, for supporting him as a Visiting Exchange Scientist while this paper was completed.

REFERENCES

Aagaard, P. and Helgeson, H.C. (1983) Activity/composition relations among silicates and aqueous solutions: II. Chemical and thermodynamic consequences of ideal mixing of atoms on homological sites in montmorillonites, illites, and mixed-layer clays. Clays & Clay Minerals 31, 207-217.

Altaner, S.P., Hower, J., Whitney, G. and Aronson, J.L. (1984) Model for K-bentonite formation: evidence from zoned K-bentonites in the disturbed belt, Montana. Geology 12, 412-415.

Bailey, S.W. (1980) Summary and recommendations of AIPEA Nomenclature Committee. Clays & Clay Minerals 28, 73-78.

_____, Brindley, G.W., Fanning, D.S., Kodama, H. and Martin, R.T. (1984) Report of the Clay Minerals Society Nomenclature Committee for 1982 and 1983. Clays & Clay Minerals 32, 239-240.

_____, _____, Kodama, H. and Martin, R.T. (1979) Report of the Clay Mineral Society Nomenclature Committee for 1977 and 1978. Clays & Clay Minerals 27, 238-239.

Bannister, F.A. (1943) Brammallite (sodium-illite), a new mineral from Llandebie, South Wales. Mineral. Mag. 26, 304-307.

Beaufort, D. and Meunier, A. (1983a) A petrographic study of phyllic alteration superimposed on potassic alteration: the Sibert porphyry deposit (Rhone, France). Econ. Geol. 78, 1514-1527.

_____ and _____ (1983b) Petrographic characterization of an argillic hydrothermal alteration containing illite, K-rectorite, K-beidellite, kaolinite and carbonates in a cupromolybdenic porphyry at Sibert (Rhône, France). Bull. Minéral. 106, 535-551.

Biscaye, P.E. (1965) Mineralogy and sedimentation of recent deep-sea clay in the Atlantic Ocean and adjacent seas and oceans. Geol. Soc. Am. Bull. 76, 803-832.

Blokh, A.M., Sidorenko, G.A., Dubinchuk, V.T. and Kuznetsova, N.N. (1974) A find of brammallite (sodium hydromica). Dokl. Acad. Sci. USSR, Earth Sci. Sect. 208, 157-160.

Bodine, M.W. (1979) Contact metamorphism of salt clays, Kerr-McGee potash mine, Eddy County, New Mexico (abstract). 28th Annual Clay Minerals Conf., Macon, Georgia, p. 40.

Boles, J.R., and Franks, S.G. (1979) Clay diagenesis in Wilcox sandstones of Southwest Texas: implications of smectite diagenesis on sandstone cementation. J. Sed. Petrol. 49, 55-70.

Brindley, G.W. (1980) Order-disorder in clay mineral structures. In: Brindley, G.W. and Brown, G. (eds.), Crystal Structures of Clay Minerals and Their X-ray Identification. Mineralogical Society, London, 125-195.

Buckley, H.A., Bevan, J.C., Brown, K.M., Johnson, L.R. and Farmer, V.C. (1978) Glauconite and celadonite: two separate mineral species. Mineral. Mag. 42, 373-382.

Burst, J.F. (1959) Post diagenetic clay mineral-environmental relationships in the Gulf Coast Eocene in clays and clay minerals. Clays & Clay Minerals 6, 327-341.

Cathielineau, M. (1983) Les minéraux phylliteux dans les gisements hydrothermaux d'uranium. II. Distribution et évolution cristallochimique des illites, interstratifiés, smectites et chlorites. Bull. Minéral. 106, 553-569.

Cole, T.G. and Shaw, H.F. (1983) The nature and origin of authigenic smectites in some recent marine sediments. Clay Minerals 18, 239-252.

Cooper, J.E. and Abedin, K.Z. (1981) The relationship between fixed ammonium-nitrogen and potassium in clays from a deep well on the Texas Gulf Coast. Texas J. Sci. 33, 103-111.

Dietz, R.S. (1941) Clay Minerals in Recent Marine Sediments. Ph.D. dissertation, University of Illinois at Urbana-Champaign.

Dopita, M. and Králík, J. (1977) Coal Tonsteins of Ostrava-Karvina Region. Ostrava (in Czech.).

Drits, V.A. and Sakharov, B.A. (1976) X-ray Structural Analysis of Mixed Layer Minerals. Acad. Sci. USSR (in Russian).

_____, Zvyagin, B.B. and Tokmakov, P.P. (1966) Gümbelite, a dioctahedral $2M_2$ mica. Dokl. Acad. Sci. USSR 170, 156-159.

Duba, D. and Williams-Jones, A.E. (1983) The application of illite crystallinity, organic matter reflectance, and isotopic techniques to mineral exploration: a case study in southwestern Gaspé, Quebec. Econ. Geol. 78, 1350-1363.

Dunoyer de Segonzac, G. (1969) Les minéraux argileux dans la diagénèse. Passage au metamorphisme. Mém. Serv. Carte Géol. Alsace-Lorraine 29, 320 p.

Eberl, D. (1978) Reaction series for dioctahedral smectites. Clays & Clay Minerals 26, 327-340.

_____ (1979) Reaction series for dioctahedral smectite: the synthesis of mixed-layer pyrophyllite/smectite. Proc. Int'l. Clay Conf., Oxford, 375-383.

_____ (1980) Alkali cation selectivity and fixation by clay minerals. Clays & Clay Minerals 28, 161-172.

_____ (1984) Clay mineral formation and transformation in rocks and soils. Phil. Trans. Royal Soc. London, A 311, 241-257.

_____ and Hower, J. (1976) Kinetics of illite formation. Geol. Soc. Am. Bull. 87, 1326-1330.

_____ and _____ (1977) The hydrothermal transformation of sodium and potassium smectite into mixed-layer clay. Clays & Clay Minerals 25, 215-228.

_____, Whitney, G. and Khoury, H. (1978) Hydrothermal reactivity of smectite. Am. Mineral. 63, 401-409.

Eggleton, R.A. and Buseck, P.R. (1980) High resolution electron microscopy of feldspar weathering. Clays & Clay Minerals 28, 173-178.

Eslinger, E., Highsmith, P., Albers, D. and deMayo, B. (1979) Role of iron reduction in the conversion of smectite to illite in bentonites in the Disturbed Belt, Montana. Clays & Clay Minerals 27, 327-338.

_____ and Savin, S.M. (1973a) Oxygen isotope geothermometry of the burial metamorphic rocks of the Precambrian Belt Supergroup, Glacier National Park, Montana. Geol. Soc. Am. Bull. 84, 2549-2560.

_____ and _____ (1973b) Mineralogy and oxygen isotope geochemistry of the hydrothermally altered rocks of the Ohaki-Broadlands, New Zealand geothermal area. Am. J. Sci. 273, 240-267.

Flexor, J.M., de Oliveira, J.J., Rapaire, J.L. and Siefermann, G. (1975) La dégradation des illites en montmorillonite dans l'alios de podzols tropicaux humo-ferrugineux du recôncavo bahianais et du Pará. Cah. ORSTOM, sér. Pédol. 13, 41-48.

Frank-Kamenetskii, V.A., Kotelnikova, E.N., Kotov, N.V. and Starke, R. (1979) Influence of tetrahedral aluminum on the hydrothermal transformation of montmorillonite and beidellite to mixed-layer illite-montmorillonite and illite. Kristall Technik 14, 303-311.

Frey, M., Hunziker, J.C., Jager, E. and Stern, W.B. (1983) Regional distribution of white K-mica polymorphs and their phengite content in the Central Alps. Contrib. Mineral. Petrogr. 83, 185-197.

Frey, E. and Lagaly, G. (1979) Selective coagulation and mixed-layer formation from sodium smectite solutions. Proc. Int'l. Clay Conf., Oxford, 131-140.

Gardner, L.R., Kheoruenromne, I. and Chen, H.S. (1978) Isovolumetric geochemical investigation of a buried granite saprolite near Columbia, SC, USA. Geochim. Cosmochim. Acta 42, 417-424.

Garrels, R.M. (1984) Montmorillonite/illite stability diagrams. Clays & Clay Minerals 32, 161-166.

Gaudette, H.E. (1965) Illite from Fond du Lac County, Wisconsin. Am. Mineral. 50, 411-417.

_____, Eades, J.L. and Grim, R.E. (1966) The nature of illite. Clays & Clay Minerals 12, 33-48.

Grim, R.E. (1968) Clay Mineralogy. McGraw-Hill, New York.

_____, Bray, R.H. and Bradley, W.F. (1937) The mica in argillaceous sediments. Am. Mineral. 22, 813-829.

Gruner, J.W. (1939) Formation and stability of muscovite in acid solutions at elevated temperatures. Am. Mineral. 24, 624-628.

Güven, N., Hower, W.F. and Davies, D.K. (1980) Nature of authigenic illites in sandstone reservoirs. J. Sed. Petrol. 50, 761-766.

Harder, H. (1974) Illite mineral synthesis at surface temperatures. Chem. Geol. 14, 241-253.

Helmold, K.P. and van de Kamp, P.C. (1984) Diagenetic mineralogy and controls on albitization and laumontite formation in Paleogene arkoses, Santa Ynez Mountains, California. In: McDonald, D.A. and Surdam, R.C. (eds.), Clastic Diagenesis. AAPG Memoir 37 (in press).

Hoffman, J.C. (1979) An Evaluation of Potassium Uptake by Mississippi River Borne Clays Following Deposition in the Gulf of Mexico. Ph.D. dissertation, Case Western Reserve University, Cleveland, Ohio.

Hoffman, J. and Hower, J. (1979) Clay mineral assemblages as low grade metamorphic geothermometers: application to the thrust faulted disturbed belt of Montana, USA. In: Scholle, P.A. and Schluger, P.R. (eds.), Aspects of Diagenesis. SEPM Spec. Publ. 26, 55-79.

Horton, D. (1983) Argillitic Alteration Associated with the Amethyst Vein System, Creede Mining District, Colorado. Ph.D. dissertation, University of Illinois at Urbana-Champaign.

Hower, J., Eslinger, E.V., Hower, M.E. and Perry, E.A. (1976) Mechanism of burial metamorphism of argillaceous sediments: 1. Mineralogical and chemical evidence. Geol. Soc. Am. Bull. 87, 725-737.

_____ and Mowatt, T.C. (1966) The mineralogy of illites and mixed-layer illite-montmorillonites. Am. Mineral. 51, 825-854.

Howard, J.J. (1981) Lithium and potassium saturation of illite/smectite clays from interlaminated shales and sandstones. Clays & Clay Minerals 29, 136-142.

_____ and Roy, D.M. (1983) Hydrothermal reactivity of smectite in basaltic groundwaters (abstract). 32nd Annual Clay Minerals Conf., Buffalo, New York, p. 46.

Hurley, P.M., Hunt, J.M., Pinson, W.H. and Fairbairn, H.W. (1963) K-Ar age values on the clay fractions of dated shales. Geochim. Cosmochim. Acta 27, 279-284.

Innoue, A. (1983) Potassium fixation by clay minerals during hydrothermal treatment. Clays & Clay Minerals 31, 81-91.

_____ and Utada, M. (1983) Further investigations of a conversion series of dioctahedral mica/ smectites in the Shinzan hydrothermal alteration area, northeast Japan. Clays & Clay Minerals 31, 401-412.

Ireland, B.J., Curtis, C.D. and Whiteman, J.A. (1983) Compositional variation within some glauconites and illites and implications for their stability and origins. Sedimentology 30, 769-786.

Jackson, M.L. (1975) Soil Chemical Analysis - Advanced Course, 2nd ed. Published by the author, Madison, Wisconsin.

Jasmund, K., Riedel, D. and Keddeinis, H. (1969) Neubildung von leistenformigem Illit und von Dickit beider Zersetzung des Muskovits in Sandstein. Proc. Int'l. Clay Conf. 1, Jerusalem, 493-500.

Jones, B.F. and Bowser, C.J. (1978) The mineralogy and related chemistry of lake sediments. In: Lerman, A. (ed.), Lakes: Chemistry, Geology, Physics. Springer-Verlag, New York, 179-235.

_____ and Weir, A.H. (1983) Clay minerals in Lake Albert, an alkaline, saline lake. Clays & Clay Minerals. 31, 161-172.

Kastner, M. and Gieskes, J.M. (1976) Interstitial water profiles and sites of diagenetic reactions, Leg 35, DSDP, Bellingshausen Abyssal Plain. Earth Planet. Sci. Letters 33, 11-20.

Kisch, H.J. (1983) Mineralogy and petrology of burial diagenesis (burial metamorphism) and incipient metamorphism in clastic rocks. In: Larsen, G. and Chilingar, G.V. (eds.), Diagenesis in Sediments and Sedimentary Rocks, 2. Elsevier, Amsterdam, 289-493.

Kodama, H. (1966) The nature of the component layers of rectorite. Am. Mineral. 51, 1035-1055.

Konta, J. (1972) Secondary illite and primary muscovite in kaolins from Karlovy Vary area, Czechoslovakia. In: Serratosa, J.M. (ed.) Kaolin Symposium, 1972 Int'l. Clay Conf., Madrid, 143-157.

Kozáč, J., Očenáš, D. and Derco, J. (1977) The discovery of ammonium hydromica in the Vihorlat Mts. (Eastern Slovakia). Mineralia slovaca 9, 479-494 (in Slovak).

Kubler, B. (1964) Les argiles, indicateurs de métamorphisme. Rev. Inst. Franc. Petrole 19, 1093-1112.

Kulbicki, G. and Millot, G. (1969) L'évolution de la fraction argileuse des grès pétroliers cambro-ordoviciens du Sahara central. Bull. Serv. Carte Géol. Alsace-Lorraine 13, 147-156.

Lapparent, J. (1909) Étude comparative de quelques porphyroides francaises. Bull. Soc. franc. Minéral. 32, 174-304.

Lee, J.H., Ahn, J. and Peacor, D.R. (1985) Textures of layered silicates: progressive changes through diagenesis and low temperature metamorphism. J. Sed. Petrol. (in press).

Levinson, A.A. (1955) Studies in the mica group: polymorphism among illites and hydrous micas. Am. Mineral. 40, 41-49.

_____ and Day, J.J. (1968) Low temperature hydrothermal synthesis of montmorillonite, ammonium-micas and ammonium-zeolites. Earth Planet. Sci. Letters 5, 52-54.

Lisitzyn, A.P. (1972) Sedimentation in the World Ocean. Rodolfo, K.S. (ed.), SEPM Spec. Publ. 17.

McDowell, D.S. and Elders, W.A. (1980) Authigenic layer silicate minerals in borehole Elmore 1, Salton Sea Geothermal Field, California, USA. Contrib. Mineral. Petrol. 74, 293-310.

McHardy, W.J., Wilson, M.J. and Tait, J.M. (1982) Electron microscope and X-ray diffraction studies of filamentous illitic clay from sandstones of the Magnus Field. Clay Minerals 17, 23-39.

Mamy, J. and Gaultier, J.P. (1975) Étude de l'évolution de l'ordre cristallin dans la montmorillonite en relation avec la diminution d'échangeabilite du potassium. Proc. Int'l. Clay Conf., Mexico City. 149-155.

Mankin, C.L. and Dodd, C.G. (1963) Proposed reference illite from the Ouachita Mountains of southeastern Oklahoma. Clays & Clay Minerals 10, 373-379.

Mattigod, S.V. and Sposito, G. (1978) Improved method for estimating the standard free energies of formation ($\delta G_{f,298.15}$) of smectites. Geochim. Cosmochim. Acta 42, 1753-1762.

Maxwell, D.T. and Hower, J. (1967) High-grade diagenesis and low-grade metamorphism of illite in the Precambrian Belt series. Am. Mineral. 52, 843-857.

Merino, E. and Ransom, B. (1982) Free energies of formation of illite solid solutions and their compositional dependence. Clays & Clay Minerals 30, 29-39.

Muffler, P.L.J. and White, D.E. (1969) Active metamorphism of Upper Cenzoic sediments in the Salton Sea geothermal field and the Salton Trough, southeastern California. Geol. Soc. Am. Bull. 80, 157-182.

Nadeau, P.H. and Reynolds, R.C., Jr. (1981a) Burial and contact metamorphism in the Mancos Shale. Clays & Clay Minerals 29, 249-259.

_____ and _____ (1981b) Volcanic components in pelitic sediments. Nature 294, 72-74.

_____, Tait, J.M., McHardy, W.J. and Wilson, M.J. (1984) Interstratified XRD characteristics of physical mixtures of elementary clay particles. Clay Minerals 19, 67-76.

Nicot, E. (1981) Les phyllosilicates des terrains précambriens du Nord-Ouest du Montana (U.S.A.) dans la transition anchizone-épizone. Bull. Minéral. 104, 615-624.

Noll, W. (1936) Über die Bildungsbedingungen von Kaolin, Montmorillonit, Sericit, Pyrophyllit, und Analcim. Mineral. Petrogr. Mitt. 48, 210-246.

Norrish, K. (1954) The swelling of montmorillonite. Trans. Faraday Soc. 18, 120-134.

Nriagu, J.O. (1975) Thermochemical approximations for clay minerals. Am. Mineral. 60, 834-839.

Odin, G.S. and Matter, A. (1981) De glauconiarum origine. Sedimentology 28, 611-641.

Parachoniak, W. and Środoń, J. (1973) The formation of kaolinite, montmorillonite, and mixed-layer montmorillonite-illites during the alteration of Carboniferous tuff (The Upper Silesian Coal Basin). Mineralogia Polonica 4, 37-56.

Parry, W.T., Ballantyne, J.M. and Jacobs, D.C. (1984) Geochemistry of hydrothermal sericite from Roosevelt Hot Springs and the Tintic and Santa Rita porphyry copper systems. Econ. Geol. 79, 72-86.

Pearson, M.J., Watkins, D. and Small, J.S. (1982) Clay diagenesis and organic maturation in Northern North Sea sediments. Proc. Int'l. Clay Conf., Italy, 665-675.

Perry, E. and Hower, J. (1970) Burial diagenesis in Gulf Coast pelitic sediments. Clays & Clay Minerals 18, 165-177.

Plancon, A., Besson, G., Gaultier, J.P., Mamy, J. and Tchoubar, C. (1979) Qualitative and quantitative study of a structural reorganization in montmorillonite after potassium fixation. Proc. Int'l. Clay Conf., Oxford, 45-54.

Porrenga, D.H. (1968) Non-marine glauconitic illite in the Lower Oligocene of Aardebrug, Belgium. Clay Minerals 7, 421-430.

Pytte, A.M. (1982) The Kinetics of the Smectite to Illite Reaction in Contact Metamorphic Shales. M.A. thesis, Dartmouth College, Hanover, New Hampshire

Reynolds, R.C., Jr. (1963) Potassium-rubidium ratios and polymorphism in illites and microclines from the clay size fractions of proterozoic carbonate rocks. Geochim. Cosmochim. Acta 27, 1097-1112.

_____ (1980) Interstratified clay minerals. In: Brindley, G.W. and Brown, G. (eds.), Crystal Structures of Clay Minerals and Their X-ray Identification. Mineralogical Society, London, 249-303.

Rimmer, S.M. and Eberl, D.D. (1982) Origin of an underclay as revealed by vertical variations in mineralogy and chemistry. Clays & Clay Minerals 30, 422-430.

Roberson, H.E. and Lahann, R.W. (1981) Smectite to illite conversion rates: effect of solution chemistry. Clays & Clay Minerals 29, 129-135.

Robert, M., Tessier, D., Isambert, M. and Baize, D. (1974) Évolution des glauconites et illites; contribution a la connaissance des smectites des sols. Int'l. Congr. Soil Sci., Trans. 10, vol. 7, 97-106.

Rochewicz, A. (in preparation) Morphology of authigenic illites in sandstone reservoirs.

_____ and Bakun, N.N. (1980) Secondary minerals in the Rotliegend sandstones of western Poland. Archiwum Mineralogiczne 36, 47-54 (in Polish).

Rossel, N.C. (1982) Clay mineral diagenesis in Rotliegend aeolian sandstones of the southern North Sea. Clay Minerals 17, 69-77.

Routson, R.C. and Kittrick, J.A. (1971) Illite solubility. Soil Sci. Soc. Am. Proc. 35, 714-718.

Sawhney, B.L. (1967) Interstratification in vermiculite. Clays & Clay Minerals 15, 75-84.

Schlocker, J. and van Horn, R. (1958) Alteration of volcanic ash near Denver, Colorado. J. Sed. Petrol. 28, 31-35.

Schultz, L.G. (1963) Nonmontmorillonitic composition of some bentonite beds. Clays & Clay Minerals 11, 169-177.

_____ (1964) Quantitative interpretation of mineralogical composition from X-ray and chemical data for the Pierre Shale. U.S. Geol. Surv. Prof. Paper 391-C.

_____ (1978) Mixed-layer clay in the Pierre Shale and equivalent rocks, northern Great Plains Region. U.S. Geol. Surv. Prof. Paper 1064-A.

Scott, A.D. and Smith, S.J. (1966) Susceptibility of interlayer potassium in micas to exchange with sodium. Clays & Clay Minerals 14, 69-81.

Sheppard, R.A. and Gude, A.J. (1969) Diagenesis of tuffs in the Barstow Formation, Mud Hills, San Bernardino County, California. U.S. Geol. Surv. Prof. Paper 634.

Shimoda, S. (1970) A hydromuscovite from the Shakanai mine, Akita prefecture, Japan. Clays & Clay Minerals 18, 269-274.

Shirozu, H. and Higashi, S. (1972) X-ray examination of sericite minerals associated with the Kuroko deposits. Clay Science 4, 137-142.

Shutov, V.D., Drits, V.A. and Sakharov, B.A. (1969) On the mechanism of a postsedimentary transformation of montmorillonite into hydromica. Proc. Int'l. Clay Conf. 1, Tokyo, 523-532.

Singer, A. and Stoffers, P. (1980) Clay mineral diagenesis in two East African lake sediments. Clay Minerals 15, 291-307.

Środoń, J. (1976) Mixed-layer smectite/illites in the bentonites and tonsteins of the Upper Silesian Coal Basin. Prace Mineral., 49.

―――― (1979) Correlation between coal and clay diagenesis in the Carboniferous of the Upper Silesian Coal Basin. Proc. Int'l. Clay Conf., Oxford, 251-260.

―――― (1980) Precise identification of illite/smectite interstratifications by X-ray powder diffraction. Clays & Clay Minerals 28, 401-411.

―――― (1981) X-ray identification of randomly interstratified illite/smectite in mixtures with discrete illite. Clay Minerals 16, 297-304.

―――― (1984a) X-ray identification of illitic materials. Clays & Clay Minerals (in press).

―――― (1984b) Illite/smectite in low-temperature diagenesis: data from the Miocene of the Carpathian Foredeep. Clay Minerals (in press).

――――, Morgan, D.J., Eslinger, E.V., Eberl, D.D. and Karlinger, M.A. (1985) Chemistry of illite/smectite and end-member illite. Clays & Clay Minerals (in press).

Stalder, P.J. (1973) Influence of crystallographic habit and aggregate structure of authigenic clay minerals on sandstone permeability. Geol. Mijnb. 52, 217-220.

Steiner, A. (1968) Clay minerals in hydrothermally altered rocks at Wairakei, New Zealand. Clays & Clay Minerals 16, 193-213.

Sterne, E.J., Reynolds, R.C. and Zantop, H. (1982) Natural ammonium illites from black shales hosting a stratiform base metal deposit, Delong Mountains, northern Alaska. Clays & Clay Minerals 30, 161-166.

Stoch, L. and Sikora, W. (1976) Transformations of micas in the process of kaolinitization of granites and gneisses. Clays & Clay Minerals 24, 156-162.

Stoessel, R.K. (1979) A regular solution site-mixing model for illites. Geochim. Cosmochim. Acta 43, 1151-1159.

―――― (1981) Refinements in a site mixing model for illites: local electrostatic balance and the quasi-chemical approximation. Geochim. Cosmochim. Acta 45, 1733-1741.

Taggart, J.E., Lichte, F.E. and Wahlberg, J.S. (1981) Methods of analysis of samples using X-ray fluoresence and induction-coupled plasma spectroscopy. U.S. Geol. Surv. Prof. Paper 1250, 683-687.

Tardy, Y. and Fritz, B. (1981) An ideal solid solution model for calculating solubility of clay minerals. Clay Minerals. 16, 361-373.

―――― and Garrels, R.M. (1974) A method for estimating the Gibbs energies of formation of layer silicates. Geochim. Cosmochim. Acta 38, 1101-1116.

Thompson, G.R. and Hower, J. (1975) The mineralogy of glauconite. Clays & Clay Minerals 23, 289-300.

Truog, E. and Jones, R.J. (1938) Fate of soluble potash applied to soils. Indus. Eng. Chem. 30, 882-885.

Tsunashima, A., Kanamaru, F., Ueda, S., Kiozumi, M. and Matsushita, T. (1975) Hydrothermal syntheses of amino acid-montmorillonites and ammonium-micas. Clays & Clay Minerals 23, 115-118.

Ushizawa, N. (1981) Trioctahedral illites from two talc mines in southwestern Hokkaido. Clay Science 5, 299-304.

Vedder, W. (1965) Ammonium in muscovite. Geochim. Cosmochim. Acta 29, 221-228.

Velde, B. (1965) Experimental determination of muscovite polymorphs stabilities. Am. Mineral. 50, 436-449.

―――― (1969) Compositional join muscovite-pyrophyllite at moderate pressures and temperatures. Bull. Soc. franc. Minéral. Cristallogr. 92, 360-368.

―――― (1977) Clays and Clay Minerals in Natural and Synthetic Systems. Elsevier, Amsterdam.

―――― and Hower, J. (1963) Petrological significance of illite polymorphism in Paleozoic sedimentary rocks. Am. Mineral. 48, 1239-1254.

―――― and Weir, A.H. (1979) Synthetic illite in the chemical system $K_2O-Al_2O_3-SiO_2-H_2O$ at 300°C and 2kb. Proc. Int'l. Clay Conf., Oxford, 395-404.

Volk, N.J. (1934) The fixation of potash in difficultly available forms in soils. Soil Sci. 37, 267-287.

Walker, G. (1950) Trioctahedral minerals in soil clays. Mineral. Mag. 29, 72-84.

Walker, T.R., Waugh, B. and Crone, A.J. (1978) Diagenesis in first-cycle desert alluvium of Cenozoic age, southwestern United States and northwestern Mexico. Geol. Soc. Am. Bull. 89, 19-32.

Warshaw, C.M. (1959) Experimental studies of illite. Clays & Clay Minerals 7, 303-316.

Watanabe, T. (1981) Identification of illite/montmorillonite interstratifications by X-ray powder diffraction. J. Mineral. Soc. Japan, Spec. Issue 15, 32-41 (in Japanese).

Weaver, C.E. (1953) A lath-shaped non-expanded dioctahedral 2:1 mineral. Am. Mineral. 38, 279-289.

_____ (1958) The effects and geologic significance of potassium "fixation" by expandable clay minerals derived from muscovite, biotite, chlorite, and volcanic material. Am. Mineral. 43, 839-861.

_____ (1960) Possible use of clay minerals in the search for oil. Bull. Am. Assoc. Petr. Geol. 44, 1505-1518.

_____ (1967) The significance of clay minerals in sediments. In: Nagy, B. and Colombo, U. (eds.), Fundamental Aspects of Petroleum Geochemistry. Elsevier, Amsterdam, 37-75.

_____ and Beck, K.C. (1971) Clay water diagenesis during burial: how mud becomes gneiss. Geol. Soc. Am. Spec. Paper 134.

_____ and Broekstra, B.R. (1984) Illite-mica. In: Weaver, C.E. and Associates, Shale Slate Metamorphism in Southern Appalachians. Elsevier, Amsterdam, 67-199.

_____ and Pollard, L.D. (1973) The Chemistry of Clay Minerals. Elsevier, Amsterdam.

Weber, K. (1972) Notes on determination of illite crystallinity. N. Jahrb. Mineral. Abh. 6, 267-276.

Weiss, A., Scholz, A. and Hoffman, U. (1956) Zur Kenntnis von trioktaedrischen Illit. Z. Naturforsch. 11b, 249.

Wright, A.C., Granquist, W.T. and Kennedy, J.V. (1972) Catalysis by layer lattice silicates. I. The structure and thermal modification of a synthetic ammonium dioctahedral clay. J. Catalysis 25, 65-80.

Yamamoto, T. and Nakahira, M. (1966) Ammonium ions in sericites. Am. Mineral. 51, 1775-1778.

Yoder, H.S. (1959) Experimental studies on micas: a synthesis. Clays & Clay Minerals 6, 42-60.

_____ and Eugster, H.P. (1955) Synthetic and natural muscovites. Geochim. Cosmochim. Acta 8, 225-280.

13. GLAUCONITE and CELADONITE MINERALS
I. Edgar Odom

Glauconite and celadonite minerals are hydrous iron and magnesium aluminosilicate clay micas. In addition to a mica type of structure, their distinguishing features include a substantial amount of K in interlayer positions and a high content of Fe^{3+} in the octahedral sheet. Glauconites and celadonites have identical unit structures which are in turn similar to the unit structures of the illite and the interstratified illite-smectite clay groups, but glauconites and celadonites are easily differentiated from each other and from the illitic group on the basis of structural, chemical and morphological characteristics, and/or geological occurrence.

Glauconites and celadonites were first described in the early and middle parts of the 19th century. The term "glauconie" (Greek work meaning bluish or pale green) was apparently first used by Brongniart in 1823, and the word "glauconite" was proposed by Keferstein in 1828 (Millot, 1970). Although the suffix "ite" implies a mineralogical meaning, it is not known whether Keferstein used the work "glauconite" as a morphological or mineralogical term. The name "celadonite from the French word "celadon," (meaning pale or sea green) was first proposed by Glocker (1847) for a hydrous K-Mg-Fe silicate mineral.

NOMENCLATURE

As the science of mineralogy developed in the late 19th and early 20th centuries, glauconites especially attracted the attention of many investigators because of their widespread and problematic occurrences in sedimentary rocks, unusual chemical characteristics, and intriguing origins. Since all but a few investigators before about 1940 lacked the analytical instrumentation required to analyze the mineralogical nature of the fine-grained micas, the term "glauconite" came to have many connotations. The term was used interchangeably to describe almost all types of greenish-colored grains or pellets, almost all greenish-colored clay pigments in sediments, and as the name of a mineral species. Obviously, the multiple meanings of the term "glauconite" caused much confusion, and beginning in the late 1950's, as the mineralogical nature of "glauconite" became better understood, nomenclature changes were proposed by several investigators. Both Warshaw (1957) and Burst (1958) recognized the heterogeneous mineralogical as well as chemical composition of "glauconite," which led Burst to propose using it only as a rock morphological term. Warshaw (1957), Burst (1958b), and Hower (1961) suggested that four mineralogical classes of "glauconite" be recognized: (1) ordered (10 Å) structures, (2) disordered (10 Å) structures, (3) interlayered (interstratified) structures, and (4) mixed mineral (glauconite) pellets. Bentor and Kastner (1965) condensed the Burst and Hower Classes 1 and 2 into one class they called "mineral glauconite," then added a fourth class reserved for green pellets containing no "glauconite mineral" phase.

In 1978 the AIPEA Nomenclature Committee made the following recommendation on the nomenclature of "cleadonite and glauconite": "Celadonite is defined as a dioctahedral mica of ideal composition, $KMgFe^{3+}Si_4O_{10}(OH)_2$, but allowing a tetrahedral Al (or Fe^{3+}) range of 0.0 to 0.2 atoms per formula unit. Substantial octahedral variations from this formula can be described by adjective modifiers, such as aluminian celadonite or ferroan celadonite. Further characteristics of celadonite are d(060) < 1.510 Å and sharp IR spectra. There is an area of potential overlap of celadonite and glauconite analyses between 0.17 and 0.20 tetrahedral Al atoms. For compositions near this boundary and for cases where analytical errors or impurities are suspect, application of other identification criteria are especially important.

"Glauconite is defined as an Fe-rich dioctahedral mica with tetrahedral Al (or Fe^{3+}) usually greater than 0.2 atoms per formula unit and octahedral R^{3+} correspondingly greater than 1.2 atoms. A generalized formula is $K(R^{3+}_{1.33}R^{2+}_{0.67})(Si_{3.67}Al_{0.33})O_{10}(OH)_2$ with $Fe^{3+} \gg Al$ and $Mg > Fe^{2+}$ (unless altered). Further characteristics of glauconite are d(060) > 1.510 Å and usually broader IR spectra than celadonite. The species glauconite is single-phase and ideally is non-interstratified. Mixtures containing an iron-rich mica as a major component can be called glauconitic. Specimens with expandable layers can be described as randomly interstratified glauconite-smectite." Mode of origin is not a criterion, and a green mineral in a marine or continental sediment or in igneous or metamorphic rocks that meets the definition for celadonite or glauconite should be so defined.

In spite of many attempts to develop a common understanding of the meaning of the term "glauconite," this goal has not completely been achieved. As recently as 1981, Odin and Matter proposed that the term "glauconite" be abandoned and introduced the general term "glaucony" for all morphological forms and the terms "glauconitic smectite" and "glauconitic mica" as end members of the glauconitic mineral family.

In this paper, the nomenclature as recommended by the APIEA is adopted for the $1M$ polymorphic forms of glauconites and celadonites, but for clarity it is recommended that the $1M$ polymorphs be called ordered glauconite and ordered celadonite. Since there are crystallographic forms of glauconites and celadonites that are less ordered and interstratified with smectite, the terms disordered glauconite (celadonite) and interstratified glauconite-smectite (celadonite-smectite) are recommended. This mineralogical nomenclature is for all practical purposes harmonious with that proposed by AIPEA (Bailey, 1980; Burst, 1958b; Hower, 1961; Bentor and Kastner, 1965; Odin and Matter, 1981).

Terms such as glauconite pellets, glauconite films, glauconitic sediments, etc., should be used to describe various morphological forms and occurrences of mineral glauconites. Morphological forms of celadonites should be described in a similar manner. Detailed morphological classifications of glauconite pellets, glauconite replacements, and glauconite films have been described by Triplehorn (1966) and Odin and Matter (1981).

LITERATURE

Over the past century several hundred papers have been published on one or more aspects of the occurrence, morphology, mineralogy, chemistry, and/or origin of glauconites, but only a few papers specifically describe the morphology, mineralogy, and/or chemistry of celadonites. Several papers have been published that attempt to relate the morphological, mineralogical, and chemical difference between glauconites and celadonites. No attempt will be made in this paper to review the entire literature available on either glauconites or celadonites, as this has been done by other authors (Hendricks and Ross, 1941; Foster, 1969; McRae, 1972; Wise and Eugster, 1964). The literature review on glauconites by McRae is especially comprehensive.

The general picture that has emerged from the studies of both glauconites and celadonites is that both minerals, especially glauconites, are characterized by morphological, mineralogical, and chemical heterogeneities, yet there are certain mineralogical and chemical properties that along with variations in modes of occurrence serve to differentiate the two minerals.

GEOLOGICAL OCCURRENCES OF GLAUCONITES AND CELADONITES

For many years it has been recognized that glauconites and celadonites, because of their modes of origin, occur in distinctly different geological environments. Glauconites most commonly occur as an accessory mineral constituent of sandy and/or limey sediments, and these sediments in most cases contain paleontological or sedimentological evidence of marine origin. The marine origin of glauconites has been further documented in the past decade through numerous reports of its occurrence in Recent marine sediments on the shelves bordering

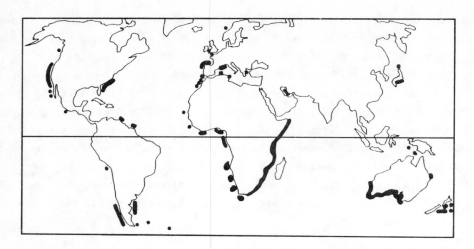

Figure 1. Distribution of glauconite minerals on the present sea floor. Modified from Odin and Matter (1981), Figure 10A.

several continents (Seed, 1965; Porrenga, 1966; Bell and Goodal, 1967; Giresse and Odin, 1973; Lamboy and Odin, 1975; Odin and Matter, 1981). The widespread distribution of various mineralogical forms of glauconites on the present seafloor is shown in Figure 1.

Glauconite pellets are known to occur extensively in sedimentary rocks ranging in age from Cambrian to Recent, but rocks belonging to the Cambrian, Cretaceous, early Tertiary, and Recent Systems seem to contain especially large concentrations. Russian (Polevaya et al., 1961) and American (Tyler and Bailey, 1961) investigators have also reported occurrences of glauconite in Precambrian sedimentary rocks, but radiometric ages suggest that the U.S. occurrences may have formed in post Precambrian time. Greenish, iron-rich nonpelletal clay micas having chemical and mineralogical characteristics intermediate between illites and glauconites have been described from non-marine sedimentary rocks by Keller (1958), Porrenga (1968), and Parry and Reeves (1966). The writer is not aware of any reported indigenous occurrences of ordered (mineral) glauconite in nonmarine sedimentary rocks.

Up to the present, all celadonites described in the literature are associated with igneous rocks, mainly basalts, where they occur as vesicular cavity fillings or less commonly as replacements of olivine or hypersthene.

These typical occurrences of glauconites and celadonites in no way imply that geological occurrence is a criterion for recognition of these minerals. For example, there are geological conditions where celadonites could be eroded from igneous rocks and become a component of either marine or nonmarine sediments.

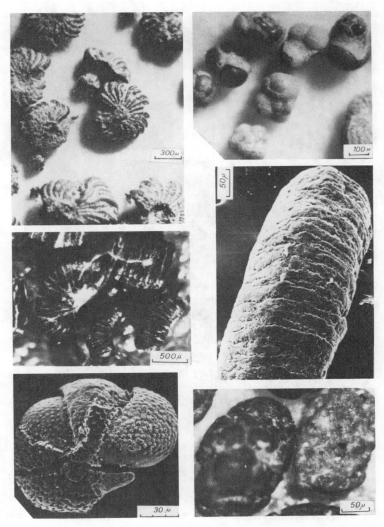

Figure 2. Common or morphological forms of glauconite pellets. Photos provided by G.S. Odin.

It is also conceivable that glauconites could be found in low rank metamorphosed sediments.

MORPHOLOGICAL FORMS OF GLAUCONITES AND CELADONITES

The most common morphological form of glauconites is pellets (Fig. 2) that usually occur in the size range of 100 to 500 micrometers, but glauconites also occur as pigments or films coating and penetrating various types of substrates (Fig. 3), as laminae within other minerals, and as a replacement of other mineral

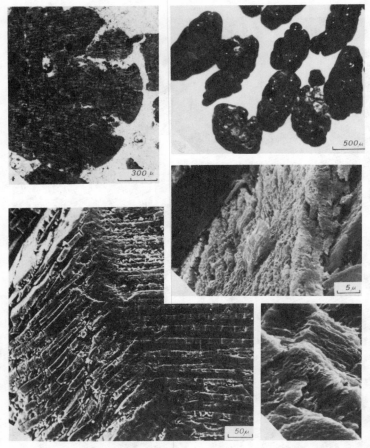

Figure 3. Examples of mineralogical substrates partially or completely replaced by glauconites. Photos provided by G.S. Odin.

such as biotite (Fig. 2), feldspars, etc. Triplehorn (1966) recognized the following morphological forms of glauconite pellets irrespective of mode of origin: (1) ovoidal or spheroidal, (2) tabular or discoidal, (3) mamillated or lobate, (4) ellipsoidal, (5) vermicular, (6) composite, and (7) fossil casts, internal molds or replacements. Odin and Matter (1981) recognized essentially two types of morphological glauconite, (1) granular and (2) film. According to them the granular form may represent internal molds or casts (Fig. 2), replaced fecal pellets, replaced biogenic carbonate debris (Fig. 3), or replaced mineral grains and rock fragments. Boyer et al. (1977) found that in the Marshalltown Formation of New Jersey, where glauconite pellets are very abundant, the pellet shapes could be classified into four forms: (1) ovoid, (2) accordion, (3) botryoidal, and (4) micaceous. Galliher (1935) described a large variety of glauconite pellets in sediments of Monterey Bay, California, all of which he correlated with variations in structural changes accompanying the alteration of biotite. Pratt (1963) found that glauconite pellets in sediments off of southern California displayed a wide range of shapes, and the origin of some of the shapes was readily identifiable. There have been many other descriptions of the possible origin of the morphological forms of glauconite pellets (Wermund, 1961; Ehlmann et al., 1963; Tapper and Fanning, 1968).

Studies done to date show that morphological criteria are not a reliable means for the characterization of either the mineralogical or chemical nature of glauconites. Color and magnetic susceptibility have been used with some degree of success to separate different mineralogical and chemical forms of glauconite pellets from the same sample. This author (Odom, 1976) used detailed thin section observations and electron microprobe analyses to separate glauconite pellets from several formations of Cambrian age (having nearly the same external morphology) into different chemical and crystallographic forms.

Unlike glauconites, the morphological forms of celadonites have not been emphasized, and to the writer's knowledge no classification of celadonites based on morphology or mode of occurrence has been proposed. Most occurrences of celadonites are associated with basalt or some other igneous rock types. Vein and vesicle fillings, crystals in various types of matrix, and pseudomorphs after certain ferromagnesian minerals are the most common occurrences of celadonites reported by both Buckley et al. (1978) and Hendricks and Ross (1941).

UNIT STRUCTURES OF GLAUCONITES AND CELADONITES

Structural characteristics

Glauconites and celadonites are classified as true micas on the basis of the physical and chemical composition of their unit structures. As with many other true micas, the basic unit structures of glauconites and celadonites consist of a 2:1 layer sequence bonded together by interlayer K cations. Glauconites and celadonites are recognized as separate mineral species on the basis of the differences in the chemical composition of their 2:1 layers with the most distinguishing chemical characteristic of both minerals being a large Fe^{+3} content in the octahedral sheet. Glauconites and celadonites are classified in the dioctahedral subgroup of the true mica group because it is generally assumed that among the three octahedral cation sites in each half unit cell, only two sites are occupied. There is, however, chemical evidence which shows that glauconites and to a lesser extent celadonites tend to be intermediate between dioctahedral and trioctahedral.

Glauconites and celadonites have certain chemical and structural heterogeneities that are not generally present in other true micas. Both celadonites and glauconites, like the illites, have smaller tetrahedral substitutions and contain less K than most other K-bearing true micas. In addition, both minerals show considerable complexity in the cation substitutions that occur within the octahedral sheet. The chemical variabilities within the unit structures are primarily responsible for the variations in mineralogical characteristics observed in glauconites and celadonites.

Mineralogy of glauconites

Most studies of the various morphological forms of glauconites prior to about 1955 were based on chemical analyses and/or on physical properties determined by microscopic methods. These studies showed that the chemical composition of mineral material called glauconites was highly variable (Hendricks and Ross, 1941; Smulikowski, 1954). Detailed X-ray diffraction studies of the mineralogy of various types of morphological glauconite were first published in the late 1950's (Warshaw, 1957; Burst, 1958). These studies provided insight into the nature of the mineralogical variability possible in glauconites, and also gave important data on the cause for the widely varying chemical composition of the unit structures. Succeeding studies by Hower (1961) and Bentor and Kastner (1965) provided additional detail regarding the interrelations of the mineralogy and chemical properties of mineral material called glauconite, including various types of glauconite pellets.

Burst (1958), Hower (1961), and Bentor and Kastner (1965) each proposed classifications of glauconite pellets based on X-ray diffraction characteristics (Table 1). These classifications are based on (1) the types of layers composing the unit structures, (2) the amount of K in interlayer positions, and (3) the total mineral composition, since some glauconite pellets contain other clay minerals, i.e., discrete illite, chlorite, smectite, etc. Three mineralogical types of glauconites having distinct unit structures are recognized, (1) ordered, (2) disordered, and (3) interstratified glauconite-smectite. In addition, Burst (1958) and Bentor and Kastner (1965) included in their classifications pellets in which one of the three mineralogical forms of glauconite may occur mixed with other clay and/or nonclay minerals, and green pellets containing none of the mineralogical forms of glauconites.

The three mineralogical forms of glauconite are defined on the basis of the percentage of glauconite (nonexpandable) and smectite (expandable) layers that compose the unit structures. The unit structure of ordered glauconite

Table 1. Classifications of the mineralogical form of glauconites as proposed by Burst (1958b), Hower (1961), and Bentor and Kastner (1965).

	Burst (1958a, 1958b)	Hower (1961)	Bentor and Kastner (1965)
Ordered Glauconites	Well-ordered - non-swelling, high in potassium and showing sharp symmetrical peaks characteristic of a micaceous 10Å unit structure. This constitutes the type mineral glauconite.	<10% expandable layers	Class 1: mineral glauconite (a) well ordered 1M with symmetric and sharp diffraction at 10.1, 4.53, 3.3Å. Reflections (112) and (11$\bar{2}$) are always present.
Disordered Glauconites	Disordered - non-swelling, low potassium, micaceous and monomineralic, but with subdued peaks, displaying broad bases and asymmetric sides.	10-20% expandable (smectite) layers	(b) disordered 1Md with asymmetric basal diffractions broadened at the base. Reflections (112) and (11$\bar{2}$) are absent.
Inter-Stratified Glauconite Smectite	Interlayered clay minerals - extremely disordered, expandable, low potassium, montmorillonite type layers abundant.	>20% expandable (smectite) layers	Class 2: interlayered glauconite, d(001) = 10.13 Å.
Mixed-Mineral Pellets	Mixed mineral - mixtures of two or more clay minerals with and without glauconites - the most frequent combinations being illite with montmorillonite, and illite with chlorite or kaolinite.		Class 3: mixed-mineral glauconites with: (a) two or more clay minerals (b) mixtures of clay with non-clay minerals.
Green Pellets	No separate classification given for pellets containing minerals which could be classed as impurities.		Class 4: green pellets containing no glauconites.

must meet the following criteria: (1) contains less than 10% expandable layers; (2) has a high K content; and (3) exhibits the 1M layer stacking sequence which is indicated by well defined 11$\bar{2}$ and 112 reflections. Also, the 00ℓ reflections at 10.1, 4.53 and 3.3 Å are usually symmetrical and sharp (Fig. 4). This is considered to be the species glauconite. There are, however, probably no instances where even ordered glauconite does not contain some interstratified smectite layers, but specimens containing less than 5% expandable layers are not uncommon (Fig. 4, Samples 1 and 2).

Disordered glauconite was defined by Hower to contain 10 to 20% expandable smectite layers, a lower K content than ordered glauconite, and usually a 1Md layer stacking sequence. The 11$\bar{2}$ and 112 reflections are poorly developed or absent (Fig. 4). After ethylene glycol solvation, basal reflections are broadened, especially at the base, and the diffraction effects may be spread over a broad range between 10.1 and 15 Å (Fig. 5). In a few cases the glauconite and smectite layers may be ordered (Thompson and Hower, 1975).

Interstratified glauconite-smectite contains from 20 to as much as 60% expandable layers (Fig. 6). This mineralogical form has the lowest potassium

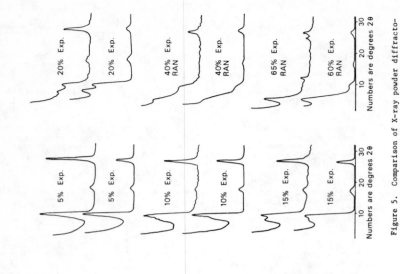

Figure 4. X-ray powder diffractograms (CuKα) of ordered (1M) (samples 1 and 2) and disordered (1Md) glauconite. The slight splitting of the 11$\bar{2}$ and 11$\bar{2}$ peaks in sample 3 is caused by approximately 10% expandable layers. Sharp diffraction peaks produced by spray drying solution containing dispersed glauconites. Modified from Odom (1976), Figure 5.

Figure 5. Comparison of X-ray powder diffractograms of oriented, glycol solvated glauconites (upper trace) with computer calculated diffraction profiles (lower trace). From Thompson and Hower (1975), Figure 1.

Table 2. X-ray powder data for glauconite (1M) and celadonite (1M). From Brindley and Brown (1980).

A. Glauconite-1M (Warshaw, 1957)			B. Celadonite-1M (Wise and Eugster, 1964)			
d(obs.)	I	hkl	hkl	d(obs.)	d(calc.)	I
10·1	100	001	001	9·97	9·94	47
4·98	*	002	020	4·53	4·53	85
4·53	80	020	$11\bar{1}$	4·35	4·36	42
4·35	20	$11\bar{1}$	021	4·14	4·123	37
4·12	10	021	$11\bar{2}$	3·635	3·638	80
3·63	40	$11\bar{2}$	022	$3·35^{1}$	3·349	60
3·33	60	003,022	003	$3·318^{1}$	3·314	70
3·09	40	112	112	3·087	3·081	80
2·89	5	$11\bar{3}$	$11\bar{3}$	2·90 B	2·901	10
2·67	10	023	023	2·678	2·675	75
2·587	100	$130,13\bar{1}$	130	2·604	2·605	70
		200	$13\bar{1}$	2·580	2·581	100
2·396	60	$132,20\bar{1}$	132	2·402	2·404	75
2·263	20	$040,22\bar{1}$	040	2·264	2·265	18
2·213	10	220,041	041	2·209	2·208	25
2·154	20	$13\bar{3},202$	$13\bar{3}$	2·148	(2·150)	31
1·994	20 B	005	$20\bar{2}$		(2·125)	
1·817	5	$22\bar{4}$	005	2·092	1·988	10
1·715	10	$311,24\bar{1}$	151	1·65 B	1·665	15
1·66	30 B	$240,31\bar{2}$	060	1·509	1·510	60
		$310,24\bar{1}$				
1·511	60	$060,33\bar{1}$				
1·495	10	330				
1·307	30	260,400				
1·258	10	$170,350$				
		420				

spacings in Å

A. * Obtained only with oriented sample.
$a = 5.234, b = 9.066, c = 10.16$ Å, $\beta = 100.5°$.
Specimen from Vidono Fm. Anzoategui, Venezuela.
[1] The exact location of these two peaks is difficult to obtain. B = broad peak.

B. $a = 5.23, b = 9.06, c = 10.13$ Å, $\beta = 100° 55'$.
Specimen from Wind River area, Washington.

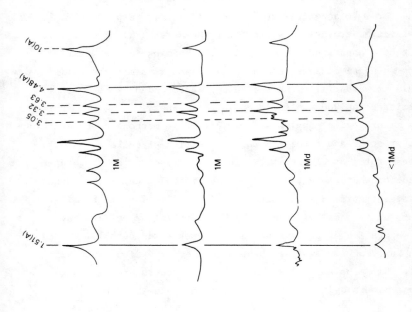

Figure 6. Examples of mica polymorphs in glauconites used by Burst (1958) to illustrate ordered, disordered, and interstratified mineralogical forms of glauconites. Modified from Burst (1958b), Figure 4.

content, and the structure is extremely disordered (1Md or <1Md) because the glauconite and smectite layers are usually randomly interstratified. After ethylene glycol solvation, diffraction effects may occur over a broad range between 10 and 17 Å (Fig. 5).

Table 2 shows X-ray powder data for ordered 1M glauconite (Warshaw, 1957) and 1M celadonite (Wise and Eugster, 1964). The data illustrate the absence or very low intensity of the 002 reflection typical of glauconite and celadonites having high octahedral Fe content. A 1M stacking sequence is indicated by the presence of well defined $11\bar{2}$ and 112 reflections.

Reynolds and Hower (1970) prepared an analog computer program to synthesize theoretical X-ray powder diffraction patterns for interstratified illite-smectite, and Thompson and Hower (1975) used this program to test the idea that the interlayering of glauconite-smectite layers is structurally similar to the illite-smectite series. Figure 5 shows the X-ray powder diffraction patterns of oriented glycol solvated natural glauconites having varying percentages of expandable smectite layers compared to computer calculated diffraction profiles, as presented by Thompson and Hower. These patterns well illustrate the differences in the characteristics of the basal reflections for the ordered, disordered, and interstratified glauconite-smectite mineralogical types.

In recent years, Buckley et al. (1978) and Odin and Matter (1981) have proposed that the terms "glauconite" or "glauconite mica," respectively, be used only in a strict mineralogical sense for the ordered (1M), Fe-rich, high-K form, that both constituents be specified in naming interstratified varieties, and that mineral mixtures containing an iron-rich mica be called glauconitic. This is essentially the same usage recommended by the AIPEA Nomenclature Committee in 1980 and followed in this paper.

While X-ray mineralogical analysis is an effective method for determining the layer characteristics and total mineral composition of pellets and of other morphological forms of glauconite, it does not resolve the frequently encountered situation where more than one of the mineralogical forms (ordered, disordered, or interstratified glauconite-smectite) as well as mixed mineral pellets may occur in the same bulk samples. Hand picking of pellets having similar colors, separation of pellets based on magnetic susceptibility, and sample selection based on XRD analysis have been used by various investigators to alleviate this problem at least partially. Readers are cautioned that more than one mineralogical form of glauconite may be present in a single bulk sample.

The mineralogical characteristics of the illite and interstratified illite-smectite clay group are discussed in Chapter 12. The writer believes that a continuous mineralogical and chemical series exists between the ordered and disordered forms of glauconites and illites, although several investigators (e.g., Odin and Matter, 1981) maintain that there is a chemical gap between glauconites and illites because of the predominance of Fe^{3+} in glauconites.

Mineralogy of celadonites

Although there have been fewer detailed X-ray diffraction studies published for celadonites, the data that are available suggest the possibility that the interstratification of smectite (expandable) layers is not as extensive in celadonites as occurs in glauconites. Buckley et al. (1978) showed that among 15 celadonite specimens studied, 14 specimens gave sharp 00ℓ and hkℓ reflections indicating a 1M structure. Buckley et al. do not provide X-ray diffraction data for the one celadonite specimen considered to contain more than 5% smectite layers. Table 2 shows X-ray powder diffraction data for a specimen of celadonite having a 1M structure.

Studies by Hendricks and Ross (1941), Wise and Eugster (1964), or Foster (1969) do not provide sufficient X-ray data to evaluate the extent of mixed-layering that might exist in celadonites. The only classifications of celadonites published to date are based on chemical criteria and are discussed in a later section. For the present, it seems necessary to presume that the same mineralogical heterogeneity shown in glauconites also may be found in celadonites, but the current data available show that the unit structure of celadonites tends as a rule to be more ordered.

CHEMICAL CHARACTERISTICS

Glauconites

A large number of chemical analyses of glauconite pellets have been reported in the literature, and several investigators have attempted to synthesize these data to show the range and average composition of glauconites (Hendricks and Ross, 1941; Smulikowski, 1954; Borchert and Braun, 1963; Weaver and Pollard, 1975). Unfortunately, many of the chemical analyses of glauconites included in these syntheses were done prior to the time that the large mineralogical variabilities possible in glauconite pellets were fully appreciated.

The following are average formulas of glauconite pellets.

Hendricks and Ross (32 analyses):
(KCaNa)$_{0.84}$(Al$_{0.47}$Fe$^{3+}_{0.97}$Fe$^{2+}_{0.19}$Mg$_{0.40}$)(Si$_{3.65}$Al$_{0.35}$)O$_{10}$(OH)$_2$.

Smulikowski (60 analyses):
(K$_{0.67}$Ca$_{0.08}$Na$_{0.08}$)$_{0.83}$(Al$_{0.40}$Fe$^{3+}_{1.05}$Fe$^{2+}_{0.17}$Mg$_{0.41}$)(Si$_{3.66}$Al$_{0.34}$)O$_{10}$(OH)$_2$.

Weaver and Pollard (82 analyses):
(K$_{0.66}$Ca$_{0.07}$Na$_{0.06}$)$_{0.78}$(Al$_{0.45}$Fe$^{3+}_{1.01}$Fe$^{2+}_{0.20}$Mg$_{0.30}$)(Si$_{3.65}$Al$_{0.35}$)O$_{10}$(OH)$_2$.

Considering that many of the above chemical analyses were done before the chemical differences between expandable and nonexpandable layers in glauconite unit structures were known in detail, the three average formulas are remarkably similar. These earlier formulas are also not greatly different from formulas of theoretical or known ordered (10 Å) glauconite as derived more recently by Cimbalnikova (1971) and Odom (1976).

Cimbalnikova (20 analyses -- corrected for % expandable layers):
K$_{0.78}$(NaCa)(Al$_{0.45}$Fe$^{3+}_{1.03}$Fe$^{2+}_{0.13}$Mg$_{0.34}$)(Si$_{3.69}$Al$_{0.31}$)O$_{10}$(OH)$_2$.

Odom (6 analyses):
(K$_{0.76}$Ca$_{0.03}$)$_{0.79}$(Al$_{0.39}$Fe$^{3+}_{1.10}$Fe$^{2+}_{0.18}$Mg$_{0.35}$)(Si$_{3.62}$Al$_{0.38}$)O$_{10}$(OH)$_2$.

In spite of the reasonably good agreement of chemical formula of glauconite pellets (when the analyses of numerous specimens are averaged), it is recognized that there is considerable chemical variability among glauconite pellets of different ages and from different areas. The possible range of chemical variability is illustrated in histograms (Fig. 7) compiled by Weaver and Pollard (1975) showing the distribution of 82 glauconite structural formulas. The ranges shown in the histograms may be due to cation variation in octahedral, tetrahedral, and interlayer sites, to percentage of expandable layers, to presence of extraneous materials such as free iron oxide (Bentor and Kastner, 1965), or to other clay minerals. For glauconite pellets free of impurities, the percentage of expandable layers within the unit structures appears to have the greatest effect on the contents of K, Al, Fe, Mg, and Si.

Burst (1958), Hower (1961), Bentor and Kastner (1965), Cimbalnikova (1971), and others have shown that interlayer K_2O content is directly related to the content of expandable layers, increasing with decreasing percentage of expandable layers (Fig. 8). Manghnani and Hower (1964), Cimbalnikova (1971), and Birch et al. (1976) have suggested that most ordered glauconite usually contains less than 10% expandable layers and at least 7.0 wt % K_2O. Manghnani and Hower (1964) show that pellets consisting predominantly of interstratified glauconite-smectite (40-50% expandable layers) contain in excess of 3.5 wt % K_2O. Odin and Matter (1981) suggest a lower limit of 5.0 wt % K_2O (Fig. 9)

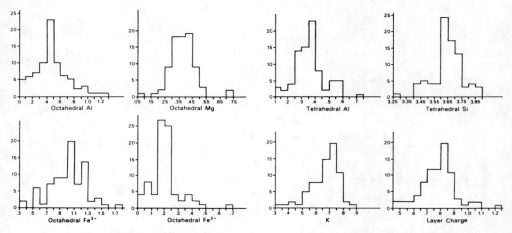

Figure 7. Histograms showing the distribution of the cations of 82 structural formulas of glauconites. From Weaver and Pollard (1975), Figure 5.

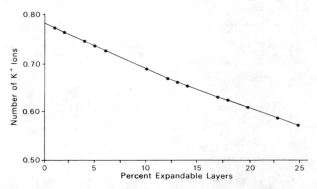

Figure 8. Relation between the percentage of expandable layers and the number of K ions in glauconites. Modified from Cimbalnikova (1971), Figure 1.

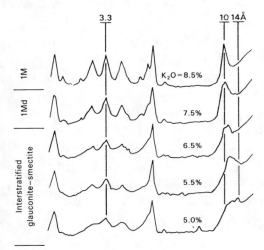

Figure 9. Powder diffractograms showing the relation of structural order to the K_2O content of the three mineralogical forms of glauconites. Modified from Odin (1975), Figure 12.

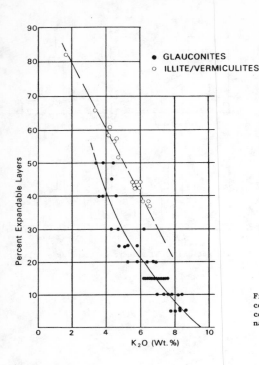

Figure 10. The relationship between potassium content and percent expandable layers in glauconites compared to some illites. From Manghnani and Hower (1964), Figure 4.

for interstratified glauconite-smectite. The wt % K_2O is related to the magnitude of the negative charge on the 2:1 structure, increasing with increasing charge (Manghnani and Hower, 1964; Foster, 1969).

Thompson and Hower (1975) showed that at the same K_2O content the percentage of expandable layers is greater in interstratified illite-smectite than in interstratified glauconite-smectite (Fig. 10), and they attribute this to difference in crystal structure and to iron in the octahedral sheets.

Within the 2:1 layer, the major chemical variation in the octahedral sheet is the ratio of Al to Fe^{3+}; Al increases as Fe^{3+} decreases with increasing percentage of expandable layers (Weaver and Pollard, 1975). Bentor and Kastner (1965) have shown that in wet chemical analyses of glauconite pellets, errors in the amount of Fe^{3+} can occur as nonstructural iron oxide are frequently present, and they caution that this iron oxide should be corrected to obtain an accurate structural Al:Fe^{3+} ratio.

The octahedral Fe^{2+} in glauconites shows a narrow range of variability (Fig. 7), and the Fe^{2+} content does not vary in a consistent manner with either the percent expandable layers or the percentage of divalent Mg ions. Mg also shows a narrow range of variability in glauconites, but in some cases there is a slight increase in Mg with increasing percentage of expandable layers (Cimbalnikova, 1971). The Mg content in ordered glauconite ranges from approximately 0.30 to 0.40 atoms per formula unit. Both Fe^{2+} and Mg decrease with increasing Al in glauconite-smectite with 40-50% expandable layers.

The tetrahedral sheet of all mineralogical forms of glauconites always shows substantial substitution of Al for Si. Although the range of Al for Si substitution (from many chemical analyses of glauconites) is from 0.1 to 0.7 atoms per formula unit, the average tetrahedral Al for Si substitution is approximately 0.35 atoms (Fig. 7). The ratio of tetrahedral Al to Si does not seem to correlate with percentage of expandable layers.

In summary, the variations in the chemical composition of the three mineralogical forms of glauconites are mainly related to the percentage of expandable layers. The chemical composition of the octahedral sheet of expandable layers is believed to be more aluminous than the composition of nonexpandable layers. Hower (1961) presented a generalization of the relationship between the compositional and structural variations of glauconites, which is as follows:

$1M$ glauconite (<5% expandable layers)

+1.00   −0.65   −0.35

$K_{0.90}(Al_{0.25}Fe^{3+}_{1.10}R^{2+}_{0.65})(Si_{3.65}Al_{0.35})O_{10}(OH)_2$.

Interstratified glauconite-smectite ($\sim$40% expandable layers)

+0.55   −0.20   −0.35

$K_{0.35}(Al_{1.30}Fe^{3+}_{0.50}R^{2+}_{0.20})(Si_{3.65}Al_{0.35})O_{10}(OH)_2$.

Foster (1969) prepared a twofold chemical classification of glauconites and celadonites based on the number of K atoms per formula unit. The glauconites and celadonites were further classified based on layer charge. This ordering arrangement permitted Foster to make interpretation of the unit structure chemical composition and to show interrelations between the abundance of various cations as related to K content and layer charge.

Celadonites

The chemical composition of celadonites has been studied by Hendricks and Ross (1941), Wise and Eugster (1964), and Foster (1969). Weaver and Pollard (1975) reviewed these earlier studies and summarized the apparent chemical differences between celadonites and glauconites (Fig. 11). Like glauconites, the published chemical data on celadonites show variations that are related to mineralogical characteristics of the specimen materials.

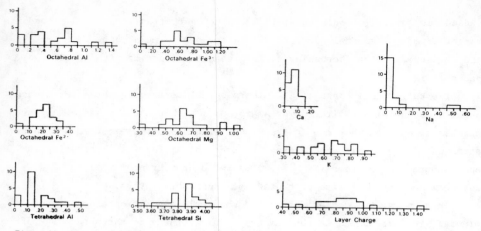

Figure 11. Histograms showing the distribution of the cations of 21 structural formulas of celadonites. From Weaver and Pollard (1975), Figure 10.

Also, Hendricks and Ross (1941) made a special point about the importance of other mineral impurities that are frequently present in celadonites.

Chemical differences between glauconites and celadonites occur in both the tetrahedral and octahedral sheets. For example, celadonites may have less than 0.30 tetrahedral Al (Fig. 11) and some specimens are wholly tetrasilicic (Foster, 1969), whereas most glauconites have between 0.25 and 0.60 tetrahedral Al. There is overlap in the Al composition of the tetrahedral sheets (Figs. 7 and 11) so that Al content cannot be considered entirely diagnostic. The octahedral sheet of celadonites tends to contain about the same range of Fe^{2+} but considerably more Mg than glauconites (Fig. 11). The larger octahedral divalent ion content is accompanied by a reduction in both Fe^{3+} and Al content.

In a relatively recent study of selected specimens consisting of 22 glauconites (1M and 1Md stacking order) and 15 celadonites (1M stacking order), Buckley et al. (1978) concluded that: (1) celadonites and glauconites form separate mineral species but within each group considerable chemical variation can occur; (2) celadonites usually have high Mg and K but lower Fe^{3+} than glauconites; (3) octahedral R^{3+} to R^{2+} ratio is much lower in celadonites (approximately 1:1, whereas in glauconites the ratio is approximately 2:1); (4) tetrahedral Al substitution in celadonites is much lower than in glauconites (Figs. 7 and 11); and (5) chemical differences between celadonites and glauconites make it possible to differentiate them readily by both IR and XRD. Figure 12 illustrates the difference in d(060) values between celadonites and glauconites as a function of octahedral Fe^{3+} content. For celadonite d(060) < 1.510 Å and for glauconite d(060) > 1.510 Å.

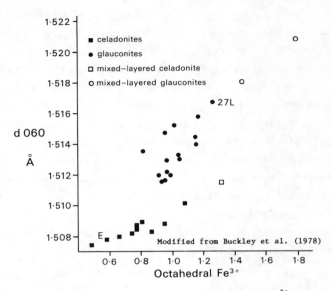

Figure 12. Relationship of d(060) spacing to Fe^{3+} ions.

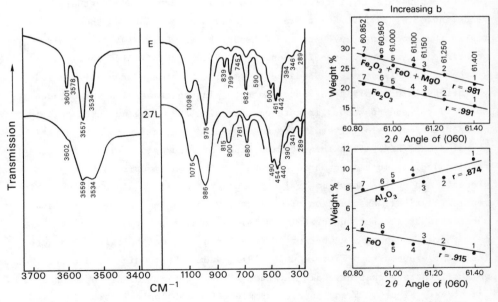

Figure 13 (to the left). Infra-red spectra of ordered celadonite (E) and glauconite (27L) illustrating the differences in and extremes of spectral sharpness. Modified from Buckley et al. (1978), Figure 4.

Figure 14 (to the right). Relationship of the b-cell dimension to Al, Fe, and Mg oxide contents in ordered (samples 2-7) and disordered (sample 1) glauconites of Cambrian age from Central United States. From Odom (1976), Figure 7.

Figure 13 shows the differences in the IR spectra of celadonites and glauconites (Buckley et al., 1978). The differences in and greater sharpness of the celadonite IR spectra reflect the more ordered structure associated with the low extent of Al for Si substitution in the tetrahedral sheet, and the regular distribution of divalent and trivalent ions in the octahedral sheet (Buckley et al., 1978). The cation content of the octahedral sheet of both celadonites and glauconites is reflected in the b dimension of the unit cell. Figure 14 shows the chemical influences on the d-spacing of the 060 reflections of six $1M$ and one $1Md$ glauconites. The d-spacing of 060 shows that celadonites are essentially dioctahedral, whereas glauconites tend to be intermediate (Fig. 12) between dioctahedral and trioctahedral (Buckley et al., 1978). Mössbauer spectra of some ordered glauconite of Cambrian age showed that Fe^{3+} strongly prefers the M_2 site, thus the octahedral cations are partially ordered (Rolf et al., 1977).

MISCELLANEOUS PHYSICAL AND CHEMICAL PROPERTIES OF GLAUCONITES AND CELADONITES

The refractive indices of glauconites range between 1.56 and 1.64 (Toler and Hower, 1959). The refractive indices are related to chemical composition and mineralogy, increasing with increasing Fe content and decreasing percentage of expandable layers (Toler and Hower, 1951; Bentor and Kastner, 1965; Velde and Odin, 1975). The refractive indices of celadonites show essentially the same range, 1.60 to 1.63 (Hendricks and Ross, 1941). In plane-polarized light both glauconites and celadonites exhibit different shades of green and greenish yellow. Because of small size of the crystallites and variations in crystallite orientation both glauconite pellets and celadonite spherulites tend to show aggregate polarization, but in glauconite pellets quite often masses of crystallites are oriented internally or around peripheral areas (Fig. 15). Thin-section study of glauconite pellets is strongly recommended, as much information can be obtained on impurities that might be present, and on the morphological nature of crystallites. In addition, thin-section (optical) examination of pellets may reveal the nature of replaced substrates or whether substrate material is still present.

There have been a number of SEM studies of the morphological characteristics of glauconite pellets (Zumpe, 1971; Odom, 1976; Odin and Matter, 1981), which illustrate that crystals may have a large variety of different shapes and sizes. Figures 15 and 16 are examples of morphological variations within single pellets and among different samples. Morphology alone is not a definitive method of identification of glauconites or celadonites. SEM combined with energy dispersive X-ray analysis is, however, a very effective method:

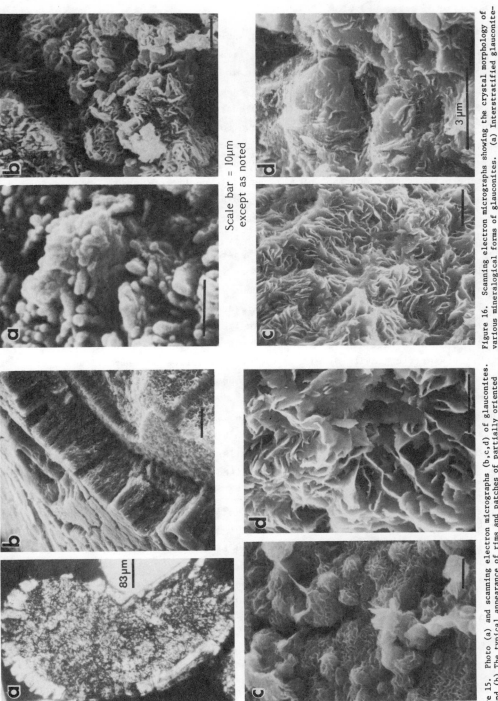

Figure 15. Photo (a) and scanning electron micrographs (b,c,d) of glauconites. (a) and (b) The typical appearance of rims and patches of partially oriented glauconites frequently present on and in glauconite pellets. (c) and (d) Some of the morphological forms of crystalline glauconites.

Figure 16. Scanning electron micrographs showing the crystal morphology of various mineralogical forms of glauconites. (a) Interstratified glauconite-smectite. (b,c,d) Disordered glauconites. The interior of the pellet shown in Figure 15b consists of ordered glauconite. Photos (a) and (b) provided by G.S. Odin.

565

(1) for identification of glauconites in whole rock samples; (2) for determining the presence of impurities (i.e., iron oxide); (3) for determining chemical variations within a single grain or among different pellets in a single sample; (4) for evaluation of the chemical nature of pellets from different outcrop locations; and (5) for qualitative and even quantitative chemical characterization of pellets known to have similar or dissimilar mineralogical characteristics. Certain computer-based energy dispersive X-ray systems have special spectral match programs (Odom and Culver, 1980) that permit direct comparison of the chemical spectra of unknown samples to standards (pellets) of known mineralogy and chemical composition.

Both glauconites and celadonites exhibit a relatively high degree of magnetic susceptibility which varies with the content of iron. Glauconite and celadonite grains may be effectively separated from other nonclay mineral grains using an isodynamic separator. When it is desired to magnetically separate glauconite grains having different susceptibilities, the magnetic field must be varied in a specific manner, and the grains should be repeatedly subjected to the individual magnetic fields. The magnetic separation does not, however, eliminate mineral impurities imbedded in the grains.

The specific gravity of glauconite pellets varies between 2.3 and 2.9; it is dependent on chemical and mineral composition, internal crystal microstructure, and degree of surface oxidation. There is limited information on the range of specific gravity of celadonites. Heavy liquids have been used to attempt to separate glauconite pellets having different values, but it has been found that pellets apparently similar in appearance have different specific gravities. Caution must be exercised in using heavy liquids, as glauconites may interact with the liquid.

Glauconites and presumably celadonites may exhibit a large range in cation exchange capacity (C.E.C.). Maghnani and Hower (1964) and Cimbalnikova (1971) have determined that C.E.C. is strongly related to percentage of expandable layers, but other factors such as total surface area and crystal edge characteristics also influence the C.E.C. The C.E.C. of glauconites may range from five to more than 40 meq/100 grams. There is no specific data known to the author on the C.E.C. of celadonites.

ORIGIN OF GLAUCONITE MINERALS

Prior to the late 1950's, the genesis of glauconites was considered to result primarily in one of three ways: (1) crystallization of Fe-Al-Mg-Si-rich gels (Twenhofel, 1936); (2) recrystallization of fecal pellets

(Takahashi, 1939); or (3) alteration of biotite grains (Galliher, 1935). Warshaw (1957) and Burst (1958a,b) showed through detailed X-ray diffraction studies that glauconite pellets of different geological ages and from several lithic associations may have a heterogeneous mineralogical composition. This finding was the basis for Burst's proposal that glauconite minerals result from the chemical upgrading of precursor, degraded layer silicate clays. This concept became known as the "layer lattice theory" and was popular with students of clay minerals until the mid-1960's when other investigations found it difficult to correlate certain properties and modes of occurrence of glauconite minerals with the existence of a precursor clay. Since about 1970 the "neoformation theory" of glauconite genesis has been gaining in popularity as more detailed chemical and microscopic studies have been possible with instrumentation such as the electron microprobe and the scanning electron microscope.

Layer lattice theory

As proposed by Burst and further developed by Hower (1961), the "layer lattice theory" considers that the formation of most glauconite minerals requires simply three conditions: (1) a precursor degraded layer silicate, preferably of the 2:1 type; (2) a plentiful supply of Fe and K; and (3) a reducing redox potential.

The organic material of soft-bodied animals (worms) in fecal pellets, and within foraminiferal tests and other shell debris, may provide the appropriate redox potential to initiate the glauconitization process in an overall oxidizing environment. The beginning stage in the development of glauconite minerals is considered to involve the substitution of iron into the octahedral sheet of the degraded layer silicate. Iron is considered to diffuse into the octahedral sheet whereby the layer charge slowly increases, which in turn is satisfied by adsorption of K ions in interlayer sites, if available.

The initial glauconite mineral is interstratified glauconite-smectite which progresses with time and increasing Fe and K adsorption to disordered glauconite, then to ordered $1M$ (mineral) glauconite. This sequence may be arrested at any stage by various physicochemical factors, but formation of ordered $1M$ glauconite could also occur very rapidly.

In discussing the layer lattice theory, Hower (1961) stressed the serial relation in composition and structure found in glauconites, with one end of this series being a 2:1 layer deficient in Fe and K and high in Al and percentage of expandable layers. Millot (1970) used the term "transformation" for the process of the chemical upgrading of preexisting micaceous layer silicates to glauconite minerals. It should be acknowledged that the major proponents of the "layer lattice theory" also recognized that other minerals such as biotite may be altered to glauconite.

Neoformation theory

Since about 1970 numerous investigators have reported new observations relating to the morphology, mineralogy, chemical composition, and modes of occurrence of glauconite minerals that are not easily reconcilable with the "layer lattice theory" concept requiring a degraded 2:1 layer silicate, other than biotite, as a precursor starting material. The foremost problems have been to explain the origin and occurrence of glauconite minerals where no reasonable precursor clay can be identified, i.e., within foraminiferal shells, and the morphological characteristics of glauconite minerals which indicate authigenic growth. Millot (1970), Odin (1975), Odom (1976), Odin and Matter (1981), and others have invoked the "neoformation theory" to reconcile these problematic situations. This theory considers that glauconite minerals initially form by precipitation directly or after solution of other minerals having appropriate chemistries. In both cases the initial glauconite mineral that is crystallographically recognizable is a low-K, high-Fe smectite-like clay which may have more than 50% expandable layers. Even the formation of glauconite minerals from biotite or fecal pellets is believed to involve solution of the precursor minerals and/or the simultaneous direct precipitation of interstratified glauconite-smectite or an Fe-smectite in pores combined with solution of preexisting mineral materials, the cations of which may or may not be part of the glauconite mineral formed. Odin and Matter (1981) used the term "nascent" to describe the low-K, high-Fe, high-layer charge, and poorly crystallized precipitated phase that constitutes the material from which better crystallized glauconite minerals are formed. Thus, the primary difference between the neoformation and the layer lattice theories involves the nature of the "starting" material. Other physico-chemical factors relating to the origin and chemical mineralogical evolution of glauconite minerals, including the requirement that organic material is probably required to develop a micro reducing environment, are essentially the same for both theories.

Odin, who along with several co-workers have made extensive studies of the occurrence, morphology, mineralogy, and chemistry of glauconite minerals, has presented a large body of evidence supporting the neoformation theory. This evidence shows that glauconite minerals result primarily from authigenic crystal growth that begins with the precipitation of a poorly crystallized Fe-rich smectite-like layer silicate that quickly evolves into a better crystallized glauconite-smectite phase. As additional precipitation (glauconitization) occurs, the previously formed glauconite-smectite evolves under appropriate chemical conditions into disordered glauconite, then to possibly ordered (mineral) glauconite. In the cases where other minerals are the source of the chemical elements for the initial glauconite mineral phase, solution followed by reprecipitation is required, i.e., formation of glauconite minerals from biotite. If K is readily available, ordered glauconite could rapidly form.

The salient characteristics of glauconite minerals that seem better explained by the "neoformation theory" than the "layer lattice theory" are: (1) the recent observation that the glauconitization of detrital biotite occurs by authigenic crystal growth between mica sheets (Fig. 2); (2) the frequently observed situation where nonmicaceous, porous materials, particularly carbonates, provide a substrate on or in which glauconite minerals crystallize with the substrate eventually being partially or entirely replaced by the glauconite mineral through solution (Fig. 3); (3) the patterns of growth of glauconite minerals and pellets especially within foraminiferal tests with the enclosing shell being broken by the pressure from glauconite crystal growth or replaced by solution (Fig. 2); (4) the solution of nonmicaceous mineral phases, i.e., kaolinite or berthierine, that may be initially present (Fig. 3); (5) the crystal morphology of glauconite crystals and the structure of some glauconite pellets that show characteristics of authigenic crystal growth with morphological features like authigenic illite, kaolinite and chlorite (Fig. 15); (6) growth of glauconite minerals in porous limestones to form what Odin and Matter (1981) called glauconite hardgrounds; (7) the presence of glauconite pellets in clay matrix that is very different chemically from the glauconite mineral composing the pellet and shows no trace of glauconitization; and (8) the thermodynamic and geochemical problems involved in low-temperature octahedral substitution of Fe for Al in a degraded illitic precursor clay -- these ions are geochemically very much alike.

In summary, based on current data available, the author strongly favors the neoformation theory over the layer lattice theory, and is in agreement with Odin and Matter (1981), who state "If the diagenetic changes from a degraded clay mineral to glauconite minerals should occur at all, it would represent a specific case of local importance compared with the large number of examples demonstrating authigenic growth of glauconite minerals." The author believes that further investigations, perhaps employing high resolution transmission electron microsocpy, will add further substantiating evidence for the neoformation theory.

ORIGIN OF CELADONITE MINERALS

All of the celadonite minerals studied by Hendricks and Ross (1941), Foster (1969), and Buckley et al. (1978) occur in basalts as vesicular fillings or as replacements of ferromagnesian minerals such as olivine and hypersthene. These associations suggest that the primary origin of celadonite minerals is by direct precipitation or by transformation. Both origins probably can be considered to be hydrothermal in nature. There is no information available on the specific environmental conditions under which the various chemical and mineralogical types of celadonites might form. The fact that a large percentage of celadonites have a $1M$ stacking sequence suggests formation under temperatures characteristic of hydrothermal conditions.

FUTURE RESEARCH

Students and others considering research on glauconites should review the numerous papers published by Odin and his co-workers. Their work on occurrence and mineralogy of Recent as well as Ancient glauconites is outstanding. There is still need for refined data on the relation between the major element composition and mineralogy of glauconites and celadonites. Electron microprobe and energy dispersive X-ray analysis as demonstrated by Buckley et al. (1978) appears to offer considerable promise for obtaining accurate chemical analysis of glauconites and celadonites on a pellet (particle) by pellet basis. Such analysis with appropriate sample preparation could largely overcome the problem of chemical analyses reflecting more than one mineralogical form of glauconite or celadonite in the same sample as well as the problem of enclosed nonclay mineral impurities.

The writer's experiences suggest that a major contribution on the mineralogy, origin, and growth mechanism of glauconites and possibly celadonites could be achieved with the aid of high resolution transmission electron microscopy. This research should focus on the nature of the layer composition and chemistry of disordered glauconite and interstratified glauconite-smectite rather than on ordered glauconite.

REFERENCES

Bailey, S.W. (1980) Summary of recommendations of AIPEA nomenclature committee. Clay Minerals 15, 85-93.

Bell, D.L. and Goodell, H.G. (1967) A comparative study of glauconite and associated clay fraction in modern marine sediments. Sedimentology 9, 169-202.

Bentor, Y.K. and Kastner, M. (1965) Notes on the mineralogy and origin of glauconite. J. Sed. Petrol. 35, 155-166.

Birch, G.F., Willis, J.P., and Rickard, R.S. (1976) An electron microprobe study of glauconites from the continental margin off the west coast of S. Africa. Marine Geol. 22, 271-284.

Borchert, H. and Braun, H. (1963) Zum Chemismus von drei Glaukonittypen. Chem. Erde 23, 82-90.

Boyer, P.S., Guinness, E.A., Lynch-Blosse, M.A., and Stolzman, R.A. (1977) Greensand fecal pellets from New Jersey. J. Sed. Petrol. 47, 267-280.

Brindley, G.W. and Brown, G. (1980) Crystal Structures of Clay Minerals and Their X-ray Identification. Mineralogical Society, Monograph No. 5, London, U.K.

Buckley, H.A., Bevan, J.C., Brown, K.M., Johnson, L.R., and Farmer, V.C. (1968) Glauconite and celadonite: two separate mineral species. Mineral. Mag. 42, 373-378.

Burst, J.F. (1958a) "Glauconite" pellets: their mineral nature and application to stratigraphic interpretation. Bull. Am. Assoc. Petrol. Geol. 42, 310-327.

Burst, J.F. (1958b) Mineral heterogeneity in "glauconite" pellets. Am. Mineral. 43, 481-497.

Cimbalnikova, A. (1971) Chemical variability and structural heterogeneity of glauconites. Am. Mineral. 56, 1385-1392.

Ehlmann, A.J., Hulings, N.C., and Glover, E.D. (1963) Stages of glauconite formation in modern foraminiferal sediments. J. Sed. Petrol. 33, 87-96.

Foster, M.D. (1969) Studies of celadonite and glauconite. U.S. Geol. Surv. Prof. Paper 614-F, 1-17.

Galliher, E.W. (1935) Geology of glauconite. Bull. Am. Assoc. Petrol. Geol. 19, 1569-1601.

Giresse, P. and Odin, G.S. (1973) Nature mineralogique et origine des glauconies du plateau continental du Gabon et du Congo. Sedimentology 20, 457-488.

Glocker, E.F. (1847) Generum et Specierum Mineralium secundum Ordines Naturales digestorum Synopsis. Halle, p. 193.

Hendricks, S.B. and Ross, C.S. (1941) Chemical composition and genesis of glauconite and celadonite. Am. Mineral. 26, 683-709.

Hower, J. (1961) Some factors concerning the nature and origin of glauconite. Am. Mineral. 46, 313-334.

Keller, W.D. (1958) Glauconitic mica in the Morrison formation of Colorado. Clays & Clay Minerals 5, 120-128.

Lamboy, M. and Odin, G.S. (1975) Nouveaux aspects concernant les glauconies du plateau continental Nord-Ouest espagnol. Rev. geogr. phys. Geol. dyn. XVII, 99-120.

Manghnani, M.H. and Hower, J. (1964) Glauconites: cation exchange capacities and infrared spectra. Part I. The cation exchange capacity of glauconite. Am. Mineral. 49, 586-498.

McRae, S.C. (1972) Glauconite. Earth-Sci. Rev. 8, 397-440.

Millot, G. (197) Geology of Clays. Springer-Verlag, New York.

Odin, G.S. (1975) De glauconiarum constitutione, origine, aetateque. Unpublished Ph.D. thesis, Univ. Paris, France.

Odin, G.S. and Matter, A. (1981) De glauconiarum origine. Sedimentology 28, 611-641.

Odom, I.E. (1976) Microstructure, mineralogy and chemistry of Cambrian glauconite pellets and glauconite, central U.S.A. Clays & Clay Minerals 24, 232-238.

Odom, I.E. and Culver, H.S. (1980) Computer-based match procedure for the identification of clay minerals in rocks from energy dispersive X-ray spectra. Program: 17th Ann. Mtg. Clay Minerals Soc., Waco, Texas.

Parry, W.T. and Reeves, C.C. (1966) Lacustrine glauconitic mica from pluvial Lake Mound, Lynn and Terry Counties, Texas. Am. Mineral. 51, 229-235.

Polevaya, N.I., Murina, G.A., and Kozakov, G.A. (1961) Glauconite in absolute dating. Ann. N.Y. Acad. Sci. 91, 298-310.

Porrenga, D.H. (1966) Clay minerals in Recent sediments of the Niger delta. Geol. Mijnb. 44, 400-403.

Porrenga, D.H. (1968) Nonmarine glauconitic illite in the Lower Oligocene of Aardeburg, Belgium. Clay Minerals 7, 421-430.

Pratt, W.L. (1963) Glauconite from the sea floor Southern California. In: R.L. Miller, ed., Essays in Marine Geology in Honor of K.O. Emery, pp. 97-119. Univ. of California Press, Los Angeles, California.

Reynolds, R.C. and Hower, J. (1970) The nature of interlayering in mixed-layer illite-montmorillonite. Clays & Clay Minerals 18, 25-36.

Rolf, R.M., Kimball, C.W., and Odom, I.E. (1977) Mössbauer characteristics of Cambrian glauconite, central U.S.A. Clays & Clay Minerals 25, 131-137.

Seed, D.P. (1965) The formation of vermicular pellets in New Zealand glauconites. Am. Mineral. 50, 1097-1106.

Smulikowski, K. (1954) The problem of glauconite. Arch. Mineral. 18, 21-120.

Takahashi, J. (1939) Synopsis of glauconitization. In: P.D. Trask, ed., Recent Marine Sediments, Am. Assoc. Petrol. Geol., Tulsa, Oklahoma, pp. 503-512.

Tapper, M. and Fanning, D.S. (1968) Glauconite pellets: similar X-ray patterns from individual pellets of lobate and vermiform morphology. Clays & Clay Minerals 16, 275-383.

Thompson, G.R. and Hower, J. (1975) The mineralogy of glauconite. Clays & Clay Minerals 23, 289-300.

Toler, L.G. and Hower, J. (1969) Determination of mixed layering in glauconites by index of refraction. Am. Mineral. 44, 1314-1318.

Triplehorn, D.M. (1966) Morphology, internal structure and origin of glauconite pellets. Sedimentology 6, 247-266.

Twenhofel, W.H. (1936) The greensands of Wisconsin. Econ. Geol. 31, 472-487.

Tyler, S.A. and Bailey, S.W. (1961) Secondary glauconite in the Biwabic iron-formation of Minnesota. Econ. Geol. 56, 1033-1044.

Velde, B. and Odin, G.S. (1975) Further information related to the origin of glauconite. Clays & Clay Minerals 23, 376-381.

Warshaw, C.M. (1957) The mineralogy of glauconite. Ph.D. dissertation, Penn. State Univ., State College, Pennsylvania.

Weaver, C.E. and Pollard, L.D. (1975) The Chemistry of Clay Minerals. Elsevier Sci. Publ. Co., New York.

Wermund, E.G. (1961) Glauconite in early Tertiary sediments of Gulf Coastal Province. Bull. Am. Assoc. Petrol. Geol. 45, 1667-1696.

Wise, W.S. and Eugster, H.P. (1964) Celadonite -- synthesis, thermal stability, and occurrence. Am. Mineral. 49, 1031-1083.

Zumpe, H.H. (1971) Microstructure in Cenomanian glauconite from the Isle of Wight, England. Mineral. Mag. 38, 215-224.

APPENDIX
X-RAY POWDER PATTERNS of MICAS

DIOCTAHEDRAL TRUE MICAS
TRIOCTAHEDRAL TRUE MICAS
BRITTLE MICAS

REFERENCES

Ankinovich, S.G., Ankinovich, E.A., Rozhdestvenskaya, I.V., and Frank-Kamenetskii, V.A. (1972) Chernykhite, a new barium-vanadium mica from northwestern Karatau. Zap vses. mineral. Obshch. 101, 451-458 (in Russian).

Aoki, K. and Shimoda, N. (1965) Margarite from the Shin-Kiura mine, Oita Prefecture. Mineral. J. (Japan) 7, 87-93.

Bailey, S.W. (1980) Structures of layer silicates. Ch. 1 in Crystal Structures of Clay Minerals and their X-ray Identification, Mineral. Soc. Monograph 5, 1-123.

Chatterjee, N.D. (1970) Synthesis and upper stability of paragonite. Contrib. Mineral. Petrol. 27, 244-257.

Eugster, H.P. and Wones, D.R. (1962) Stability relations of the ferruginous biotite, annite. J. Petrol. 3, 82-125.

Forman, S.A., Kodama, H., and Maxwell, J.A. (1967) The trioctahedral brittle micas. Am. Mineral. 52, 1122-1128.

Frondel, C. and Ito, J. (1966) Hendricksite, a new species of mica. Am. Mineral. 51, 1107-1123.

Gottesmann, B. (1962) Über einige Lithium-Glimmer aus Zinnwald und Altenberg in Sachsen. Geologie 11, 1164-1176.

Harada, K., Honda, M., Nagashima, K., and Kanisawa, S. (1976) Masutomilite, manganese analogue of zinnwaldite, with special reference to masutomilite-zinnwaldite series. Mineral. J. (Japan) 8, 95-109.

Higashi, S. (1982) A new ammonium dioctahedral mica. Mineral J. (Japan) 11, 138-146.

Juster, T.C. (1984) Very low-grade metamorphism of pelites associated with coal, northeastern Pennsylvania. M.S. thesis. Univ. Wisconsin-Madison.

Keusen, H.R. and Peters, Tj. (1980) Preiswerkite, an Al-rich trioctahedral sodium mica from the Geisspfad ultramafic complex (Penninic Alps). Am. Mineral. 65, 1134-1137.

LaLonde, R.E. (1963) X-ray diffraction data for taeniolite. Am. Mineral. 48, 204-205.

MaCauley, J.W. and Newnham, R.E. (1973) Structure refinement of a barium mica. Z. Kristallogr. 137, 360-367.

Pattiaratchi, D.B., Saari, E., and Sahama, Th.G. (1967) Anadite, a new barium iron silicate from Wilagedera, North Western Province, Ceylon. Mineral. Mag. 36, 1-4.

Robert, J.-L. and Maury, R.C. (1979) Natural occurrence of a (Fe,Mn,Mg) tetrasilicic potassium mica. Contrib. Mineral. Petrol. 68, 117-123.

Shimoda, S. (1970) A hydromuscovite from the Shakanai mine, Akita Prefecture, Japan. Clays & Clay Minerals 18, 269-274.

Smith, J.V. and Yoder, H.S. (1956) Experimental and theoretical studies of the mica polymorphs. Mineral. Mag. 31, 209-235.

Warshaw, C.M. (1957) The mineralogy of glauconite. Ph.D. thesis, Pennsylvania State University, State College, Pennsylvania.

Wise, W.S. and Eugster, H.P. (1964) Celadonite: synthesis, thermal stability and occurrence. Am. Mineral. 49, 1031-1083.

Yoder, H.S. and Eugster, H.P. (1954) Phlogopite synthesis and stability range. Geochim. Cosmochim. Acta 6, 157-185.

_____ and _____ (1955) Synthetic and natural muscovites. Geochim. Cosmochim. Acta 8, 225-280.

DIOCTAHEDRAL TRUE MICAS

Polytypic forms of synthetic muscovite

A. 1M

hkℓ	I	d(Å)
001	>100	10.077
002	37	5.036
020	90	4.488
111	27	4.349
021	16	4.115
11$\bar{2}$	60	3.660
003⎤ 022⎦	>100	3.356
112	50	3.073
11$\bar{3}$	6	2.929
023	16	2.689
13$\bar{0}$	50	2.582
131	90	2.565
200	22	2.550
131	11	2.450
13$\bar{2}$	4	2.405
11$\bar{4}$	12	2.380
040	8	2.246
220	7	2.219
041	4	2.191
13$\bar{3}$	20	2.156
202	6	2.109
005	32	2.013
133	7	1.957
13$\bar{4}$	4	1.900
116	18	1.668
151	12	1.653
204	12	1.635
135	4	1.514
060	33	1.499

B. 2M₁

hkℓ	I	d(Å)
002	>100	10.014
004	55	5.021
110	55	4.478
11$\bar{1}$	65	4.458
021	14	4.391
111	21	4.296
11$\bar{2}$	14	4.109
022	14	4.109
112	12	3.973
11$\bar{3}$	37	3.889
023	32	3.735
11$\bar{4}$	44	3.500
006⎤ 024⎦	>100	3.351
114	47	3.208
025	47	2.999
115	35	2.871
11$\bar{6}$	22	2.803
131	50	2.589
13$\bar{1}$	45	2.580
116	90	2.562
202	20	2.514
008	19	2.458
13$\bar{3}$	12	2.446
202	10	2.396
20$\bar{4}$	24	2.380
133		
22$\bar{1}$	12	2.247
040⎤ 24$\bar{1}$⎦		2.236
041	5	2.236
221	5	2.201
22$\bar{3}$	7	2.184
222	10	2.149
043		
135	23	2.132
044	6	2.051
0,0,10	75	2.010
137	14	1.975
13$\bar{9}$	6	1.736
150⎤ 24$\bar{1}$⎦	6	1.699
2,0,$\bar{10}$⎤ 3$\bar{1}4$⎦	12	1.670
313	17	1.653
33$\bar{1}$	7	1.602
060	40	1.499

Polytypic forms of muscovite

C. 2M₂

hkℓ	I	d(Å)
002	55	10.25
004	42	5.06
110;11$\bar{1}$	60	4.49
202	15	4.37
111	5	4.31
112	5	3.966
11$\bar{3}$	21	3.921
20$\bar{4}$	43	3.681
11$\bar{4}$	33	3.520
006	58	3.348
114	28	3.211
204	40	3.066
20$\bar{6}$	15	2.946
115	22	2.869
11$\bar{6}$	18	2.812
021	90	2.583
008	10	2.513
31$\bar{4}$	16	2.450
317;402	15	2.426
023	14	2.402
312	4	2.285
313	8	2.252
22$\bar{1}$;40$\bar{7}$	7	2.210
221;400	7	2.186
025;223	20	2.082
317;402	15	2.055
026	30	2.006
0,0,10	3	1.750
22$\bar{8}$;1,1,1$\bar{1}$	10	1.717
1,3,$\bar{10}$	15	1.694
029;131;22$\bar{7}$	15	1.668
318;2,0,1$\bar{2}$	8	1.633
133;42$\bar{5}$	8	1.614
3,1,$\bar{11}$;51$\bar{6}$	8	1.585
13$\bar{5}$;0,2,10;423	5	1.569
319	30	1.500
602;33$\bar{1}$;42$\bar{8}$	4	1.481
2,0,12;331;600		

D. 3T

hkℓ	I	d(Å)
0003	>100	9.969
0006	53	4.991
10$\bar{1}$0	19	4.492
1011	19	4.460
10$\bar{1}$4	10	3.873
10$\bar{1}$5	8	3.596
0009	>100	3.331
10$\bar{1}$7	10	3.110
10$\bar{1}$8	16	2.884
11$\bar{2}$1	15	2.589
11$\bar{2}$2	27	2.564
0,0,0,12	11	2.499
11$\bar{2}$4	7	2.457
1125	8	2.384
20$\bar{2}$0	5	2.254
11$\bar{2}$7	4	2.222
20$\bar{2}$3	4	2.197
11$\bar{2}$8	12	2.136
2026	3	2.056
0,0,0,15	47	1.999
1,1,$\bar{2}$,10	7	1.966
1,1,$\bar{2}$,11	2	1.885
1,1,$\bar{2}$,14	10	1.654
2,1,$\bar{3}$,5	3	1.638
2,1,$\bar{3}$,6	3	1.614
2,1,$\bar{3}$,8	2	1.551
1,1,$\bar{2}$,16	6	1.521
3,0,$\bar{3}$,0	11	1.502

A. 1M $a = 5.208$, $b = 8.995$, $c = 10.275$ Å, $\beta = 101.58°$
Synthetic specimen (Yoder and Eugster, 1955).

B. 2M₁ $a = 5.189$, $b = 8.995$, $c = 20.097$ Å, $\beta = 95.18°$
Synthetic specimen (Yoder and Eugster, 1955).

C. 2M₂ $a = 9.017$, $b = 5.210$, $c = 20.437$ Å, $\beta = 100.4°$
Hydromuscovite specimen from Shakanai mine, Japan.
(Indexing modified from Shimoda, 1970.)

D. 3T $a = 5.203$, $c = 29.988$ Å, $\beta = 120°$
Specimen from Sultan Basin, Washington, U.S.A.
(Yoder and Eugster, 1955.)

Polytypic forms of synthetic paragonite (Chatterjee, 1970)

	A. 1M			B. 2M$_1$		
hkl	I(obs.)	d(obs.)	d(calc.)	hkl	I(obs.)	d(obs.)	d(calc.)
001	80	9.665 Å	9.6334 Å	002	31	9.633 Å	9.6226 Å
002	50	4.822	4.8167	004	21	4.814	4.8113
020	35	4.442	4.4423	110	37	4.437	4.4416
11$\bar{1}$	20	4.236	4.2283	11$\bar{1}$	42	4.392	4.3925
021	3	4.029	4.0340	111	4	4.270	4.2660
112	18	3.497	3.4930	11$\bar{2}$	7	4.140	4.1391
11$\bar{2}$	100	3.212	3.2111	022	4	4.040	4.0352
003	31	3.058	3.0547	11$\bar{3}$	7	3.763	3.7713
112	6	2.778	2.7785	023	6	3.654	3.6536
113	8	2.602	2.6024	114	10	3.368	3.3777
023	12	2.557	2.5581	024	21	3.267	3.2649
13$\bar{1}$	19	2.523	2.5215	006	100	3.207	3.2075
131	9	2.426	2.4262	114	19	3.163	3.1603
201	3	2.364	2.3666	11$\bar{5}$	3	3.020	3.0097
13$\bar{2}$	3	2.336	2.3357	025	16	2.910	2.9097
041	3	2.164	2.1643	115	16	2.822	2.8174
024	4	2.116	2.1172	11$\bar{6}$	10	2.680	2.6867
13$\bar{3}$	6	2.083	2.0812	131	22	2.557	2.5561
005	25	1.9279	1.9267	13$\bar{1}$ \} 20$\bar{2}$	48	2.528	{ 2.5304 / 2.5261
115	3	1.6851	1.6846	202	19	2.427	2.4312
15$\bar{1}$	3	1.6653	1.6667	13$\bar{3}$	20	2.416	2.4145
204	3	1.6268	1.6265	133	11	2.350	2.3512
006	5	1.6045	1.6056	027	6	2.335	2.3382
060	10	1.4816	1.4808	13$\bar{5}$	10	2.172	2.1738
20$\bar{6}$	4	1.4623	1.4626	222	2	2.133	2.1330
				043	12	2.100	2.1000
				22$\bar{4}$	6	2.073	2.0695
				044	3	2.018	2.0176
				0,0,10	31	1.9248	1.9245
				137	2	1.8413	1.8408
				046	4	1.8275	1.8268
				15$\bar{1}$	7	1.6771	1.6772
				24$\bar{4}$	8	1.6114	1.6110
				060	31	1.4814	1.4816
				331	17	1.4685	1.4687

A. 1M $a = 5.139$, $b = 8.885$, $c = 9.750$ Å, $\beta = 98.87°$

B. 2M$_1$ $a = 5.143$, $b = 8.890$, $c = 19.302$ Å, $\beta = 94.41°$

A. Glauconite-1M (Warshaw, 1957)			B. Celadonite-1M (Wise and Eugster, 1964)				C. Illite-1M		
hkl	I	d(obs.)	hkl	I	d(obs.)	d(calc.)	hkl	I	d(Å)
001	100	10.1 Å	001	47	9.97 Å	9.94 Å	001	35	10.7
002	*		020	85	4.53	4.53	002	30	5.0
11$\bar{1}$	80	4.98	11$\bar{1}$	42	4.35	4.36	020	100	4.43
021	20	4.53	021	37	4.14	4.123	11$\bar{1}$	30	4.33
112	10	4.35	112	80	3.635	3.638	021	14	4.11
003,022	40	4.12	022	60	3.351	3.349	112	40	3.66
11$\bar{2}$	60	3.63	003	70	3.318^1	3.314	003	35	3.31
112	40	3.33	112	80	3.087	3.081	112	40	3.06
11$\bar{3}$	5	3.09	11$\bar{3}$	10	2.90 B	2.901	11$\bar{3}$	16	2.931
023	10	2.89	023	75	2.678	2.675	023	12	2.675
		2.67	130,13$\bar{1}$	70	2.604	2.605	13$\bar{1}$	85	2.560
130,13$\bar{1}$	100	2.587	200	100	2.580	2.581	131	16	2.445
200			132,20$\bar{1}$	75	2.402	2.404	11$\bar{4}$	25	2.386
132,20$\bar{1}$	60	2.396	040,22$\bar{1}$	18	2.264	2.265	040	12	2.242
040,22$\bar{1}$	20	2.263	041	25	2.209	2.208	220	10	2.216
220,041	10	2.213	13$\bar{3}$		2.148	{ 2.150 / 2.125	13$\bar{3}$	12 B	2.151
13$\bar{3}$,202	20	2.154	005	31			202	8	2.093
005	20 B	1.994	22$\bar{4}$ } 31$\bar{1}$,24$\bar{1}$	15	2.092	1.988	005	8	2.023
22$\bar{4}$	5	1.817	151	60	1.65 B	1.665		12	1.989
31$\bar{1}$,24$\bar{1}$	10	1.715	060		1.509	1.510	13$\bar{3}$	8	1.957
240,31$\bar{2}$	30 B	1.66					13$\bar{4}$	3	1.903
310,24$\bar{1}$							-	3	1.82
060,33$\bar{1}$	60	1.511					-	10	1.688
330	10	1.495					11$\bar{6}$	12	1.663
260,400	30	1.307					204	12	1.631
170,350	10	1.258					-	4 B	1.585
420							-	4 B	1.536
							060	40	1.496
							-	4 B	1.372
							-	6 B	1.341
							-	12	1.292

A. *obtained only with oriented sample.
 $a = 5.234$, $b = 9.066$, $c = 10.16$ Å, $\beta = 100.5°$
 Specimen from Vidono Fm., Anzoategui, Venezuela.

B. ^{1}The exact location of these two peaks is difficult to obtain. B=broad peak.
 $a = 5.23$, $b = 9.06$, $c = 10.13$ Å, $\beta = 100°55'$
 Specimen from Wind River area, Washington.

C. $a = 5.18$, $b = 8.98$, $c = 10.32$Å, $\beta \cong 101.83°$
 Specimen from Hungary, data modified from PDF card 29-1496.

A. Tobelitic muscovite-$2M_1$				B. Tobelite-$1M$ (Higashi, 1982)				A. Chernykhite-$2M_1$ (Ankinovich et al., 1973)							B. Roscoelite-$1M$		
hkl	I	d(obs.)	d(calc.)	hkl	I	d(Å)		hkl	I	d(Å)	hkl	I	d(Å)		hkl	I	d(obs)	d(calc)
002	80	10.179 Å	10.048 Å	001	100	10.44		002	15	10.0	138	4	1.858		001	80	10.0 Å	9.996 Å
004	15	5.019	5.024	002	70	5.12		004	6	4.99	0,2,$\overline{10}$	10	1.836		002	10	4.99	4.998
110	100	4.46	{4.491 / 4.457}	020	70	4.486		020	3	4.59	226	4	1.806		020	85	4.546	4.554
11$\overline{1}$				11$\overline{1}$	30	4.360		110	15	4.56	138	1	1.763		11$\overline{1}$	50	4.387	4.384
021	10}B	4.35	{4.397 / 4.313}	021	5	4.131		11$\overline{1}$	10	4.37		10	1.735		021	30	4.146	4.144
111				112	30	3.685		111	1	4.26		4	1.681		112	80	3.660	3.653
022	10	4.099	4.111	003	60	3.408		022	4	4.17	312	15	1.671		022	20}B	3.350	3.366
11$\overline{3}$	15	3.874	3.872	112	35	3.103		11$\overline{3}$	7	4.01	153	60	1.660		003	80	3.326	3.332
023	15	3.710	3.739	11$\overline{3}$	8	2.969		113	20	3.92		4	1.633		112	75	3.103	3.109
114	20	3.499	3.485	023	10	2.720		023	20	3.77		4	1.619		11$\overline{3}$	30	2.917	2.913
024	40	3.344	{3.354 / 3.349}	13$\overline{1}$,004	45	2.566		113	1	3.61		4	1.580		023	50	2.688	2.689
006				202	8	2.487		11$\overline{4}$	30	3.52		2	1.574		201	20}B	2.627	{2.633 / 2.621}
114	20	3.203	3.227	131	15	2.452		024	25	3.38		4	1.561		130			
025	20	3.005	2.999	13$\overline{2}$,11$\overline{4}$	15	2.402		006	100	3.33		1	1.553		13$\overline{1}$	100	2.597	2.595
115	15}B	2.88	2.886	201	10	2.374		114	40	3.23	060	50	1.530		200			
11$\overline{6}$	10	2.79	2.791	040	5	2.243		115	2	3.14		6	1.520		004	5	2.496	2.499
131	40}B	2.590	{2.592 / 2.591}	220,132	20	2.167		025	50	3.01		4	1.516		132	60	2.415	{2.416 / 2.406}
116				133	10	2.048		115	40	2.887		1	1.496		20$\overline{1}$			
20$\overline{2}$			2.590	20$\overline{4}$	10	2.020		11$\overline{6}$	40	2.802		6	1.465		22$\overline{1}$			
131	80}B	2.571	2.564	131	10	1.9746		130	5	2.639		1	1.456		040	20	2.278	{2.280 / 2.277}
13$\overline{3}$	35	2.465	2.460	13$\overline{4}$	5	1.9155		202	70	2.607		6	1.443		203			
202			2.456	224	5	1.8436		132	20	2.584		3	1.425		132	10	2.219	{2.237 / 2.260 / 2.220}
204	45	2.390	2.390	13$\overline{5}$	15	1.6894		117	20	2.513		4	1.393		041			
13$\overline{3}$			2.387	151	10	1.6515		13$\overline{3}$	20	2.492		25	1.356		13$\overline{3}$	20 B	2.159	{2.160 / 2.147}
22$\overline{1}$	20	2.244	{2.251 / 2.246}	060	20	1.4977		20$\overline{4}$	30	2.420		15	1.347		202			
042	5	2.19	{2.198 / 2.194}					040	15	2.291		15	1.326		221	5	2.131	2.127
220				A. $a = 5.201$, $b = 9.012$, $c = 20.177$Å				118	20	2.269		15	1.323		00$\overline{5}$			
221	35	2.135	{2.144 / 2.139}	$\beta = 95.12°$				13$\overline{5}$	4	2.255		20	1.305		20$\overline{4}$	40	1.998	{1.999 / 1.994}
206				Pattern modified from Juster (1984).				221	5	2.243		10	1.275		133	5	1.972	1.972
135				Specimen from NE Pennsylvania.				028	2	2.186		2	1.266		13$\overline{5}$	40}B	1.666	{1.666 / 1.654}
0,0,10				Irrational 00l values suggest				043	10	2.170		2	1.257		006	10		1.654
13$\overline{7}$	15}B	2.02	2.010	interestratified phengite-tobelite.				135	40	2.160		15	1.245		204	40	1.654	1.604
206	15	1.96	{1.971 / 1.966}					118	15	2.109		4	1.225		243	10	1.603	1.596
31$\overline{2}$			1.697	B. $a = 5.219$, $b = 8.986$, $c = 10.447$ Å,				044	8	2.086		4	1.218		152	10	1.596	
313	15}	1.70	1.680	$\beta = 101.31°$				22$\overline{5}$	4	2.056		4	1.210		33$\overline{1}$	80}B	1.5227	{1.5230 / 1.5224}
2,0,$\overline{10}$			1.661	Specimen from Tobe, Ehime Pref.,				029	60	1.996		2	1.204		135			
139	15		1.657	Japan.				139	15	1.972		2	1.200		060	60	1.5187	1.5181
24$\overline{7}$			1.640					226	4	1.957		2	1.193		007	5	1.4287	1.4280
153	15	1.63	1.638					119	3	1.929		1	1.129		136			{1.3500
154			1.627					137	5	1.893					155	30	1.346	1.3448}
060	60	1.502	{1.502 / 1.502}															
33$\overline{1}$								A. $a = 5.29$, $b = 9.182$, $c = 20.023$ Å, $\beta = 95.68°$										
2,2,1$\overline{1}$	5}B	1.37	{1.366 / 1.356}					Specimen from Northwest Karatau, Kazakhstan, USSR.										
2,1,$\overline{10}$																		
1,3,1$\overline{3}$	5	1.34	{1.352 / 1.344}					B. $a = 5.282$, $b = 9.109$, $c = 10.172$ Å, $\beta = 100.67°$										
2,0,12								Specimen from Stockslager Mine, Calif.										
0,4,12								Pattern by Bailey (unpublished), CrKα, 114.6 mm camera, Wilson mount, Int.										
260	20	1.296	1.299					est. visually, Indices assigned by comparison with single crystal intensities.										
26$\overline{2}$			1.296															
400			1.295															
17$\overline{1}$	15	1.249	1.249															
406			1.246															

TRIOCTAHEDRAL TRUE MICAS
Polytypic forms of phlogopite

A. 1M			B. 3T			C. 2M₁				A. Synthetic annite (Eugster and Wones, 1962)					B. Hendrickxite (Frondel and Ito, 1966)		
hkℓ	I	d(Å)	hkℓ	I	d(Å)	hkℓ	I	d(Å)		1M setting hkℓ	3T setting hkℓ	I	d(obs.)		hkℓ	I	d(obs.)
001	>100	9.97	0003	>100	10.129	002	>100	10.129		001	003	100	10.264 Å		001	100	10.20 Å
002	6	5.02	0006	23	5.022	004	18	5.056		002	006	3	5.070		002	36	5.094
020	7	4.588	10$\bar{1}$0	8	4.596	110	19	4.612		020	100	4	4.644		003	60	3.398
110	5	4.553	10$\bar{1}$1	8	4.546	021	7			111	104	3	3.975		130,20$\bar{1}$	7	2.682
111	3	3.923	10$\bar{1}$4	7	3.941	112	5	4.515		11$\bar{2}$	105	3	3.711		13$\bar{1}$,200	13	2.652
11$\bar{2}$	7	3.654	10$\bar{1}$5	45	3.663	021	7	4.079		003	009	80	3.380		004	35	2.546
022	20	3.389	10$\bar{1}$6	45	3.393	023	18	3.814		112	107	15	3.179		201,13$\bar{2}$	9	2.462
003	>100	3.348	0009	>100	3.354	11$\bar{4}$	33	3.540		?	?	10	3.106		202,13$\bar{3}$	9	2.202
112	10	3.144	10$\bar{1}$7	25	3.148	006	>100	3.362		023	109	70	2.734		005	8	2.034
11$\bar{3}$	7	2.916	10$\bar{1}$8	21	2.917	114	38	3.283		13$\bar{1}$	112	70	2.654		20$\bar{4}$,133	9	2.016
023	3	2.708	10$\bar{1}$9	12	2.710	11$\bar{5}$	9	3.156		200,20$\bar{2}$,131	0,0,12;114	15	2.532		006	17	1.696
130	7	2.642	11$\bar{2}$1	10	2.643	025	40	3.040		004,20$\bar{2}$,131	115	40	2.465		060	10	1.554
20$\bar{1}$			11$\bar{2}$2	30	2.618	115	8	2.926		201,13$\bar{2}$	118	20	2.199		205	3	1.495
20$\bar{2}$	30	2.614	0,0,0,12			11$\bar{6}$	22	2.818		202,13$\bar{3}$	1,1,10	10	2.018		007	6	1.454
13$\bar{1}$			1,0,$\bar{1}$,10			131	22	2.651		20$\bar{4}$,133	1,1,11	5	1.932		027,136	6	1.378
004			11$\bar{2}$4			116	>100	2.624		203,13$\bar{4}$	0,0,1$\bar{8}$;1,1,14	20	1.692		117,206	3	1.327
113	15	2.515	11$\bar{2}$5	17	2.431	008	28	2.522		006,204,13$\bar{5}$	217	1	1.635		008	5	1.272
13$\bar{1}$			202$\bar{1}$	3	2.296	133	40	2.439		152,24$\bar{3}$	0,0,1$\bar{8}$;1,1,16	40	1.556			5	1.002
20$\bar{2}$			11$\bar{2}$7	2	2.259	117	16	2.361		060,33$\bar{1}$,206,135	300;1,1,16	40	1.538			5	0.974
			20$\bar{2}$3	2	2.243					061,332	303	10	1.485				
13$\bar{2}$	15	2.430	11$\bar{2}$8	21	2.171	220	9	2.304		062,331	306	5					
20$\bar{1}$			0,0,0,15	>100	2.011	040	9	2.270									
22$\bar{1}$	3	2.292	1,1,$\bar{2}$,10	12	1.996	135	45	2.180		A. 1M a = 5.39, b = 9.30, c = 10.30 Å, β = 100.1°							
220			1,1,$\bar{2}$,11	4	1.908	224	21	2.039		3T a = b = 5.39, c = 30.43 Å, γ = 120°							
132	?	2.26	1,1,$\bar{2}$,13	3	1.748	0,0,10	66	2.017		Indexing modified from Eugster and Wones (1962).							
20$\bar{3}$			1,1,$\bar{2}$,14	35	1.675	137	20	2.000									
041	?	2.24	3030	25	1.536	135	5	1.914		B. 1M a = 5.37, b = 9.32, c = 10.30 Å, β = 99°							
13$\bar{3}$			3033	27	1.516	137	3	1.751		Specimen from Franklin, New Jersey.							
202	15	2.170				139	5	1.737									
005	30	2.012				227	47	1.677									
20$\bar{4}$	7	1.995				153											
133						060	50	1.538									
13$\bar{4}$	2	1.910				330											
20$\bar{3}$						062	10	1.521									
20$\bar{5}$	?	1.748															
13$\bar{5}$																	
204	15	1.674															
060	15	1.535															
061																	
330	27	1.517															

A. 1M a = 5.312, b = 9.20, c = 10.204 Å, β = 99.82°

B. 3T a = 5.317, c = 30.168 Å
For similar composition specimens, it is likely that the 1M and 3T patterns would be indistinguishable. (Modified from Smith and Yoder, 1956.)

C. a = 5.347, b = 9.227, c = 20.252Å, β = 95.02°
Specimen from Burgess, Ontario (U.S.N.M. 106758). (Yoder and Eugster, 1954.)

Ferrian biotite-1M (lepidomelane)

hkℓ	I	d(obs)	d(calc)	hkℓ	I	d(obs)	d(calc)
001	100	10.02 Å	10.014 Å	243̄	10	1.617	{1.618 / 1.617
020	30	4.58	{4.619 / 4.566	152̄			
110	25	3.930	3.934	153	5	1.582	{1.583 / 1.583
111	35	3.669	3.663	242̄			
112̄	35	3.399	3.395	331̄	60}B	1.539	{1.540 / 1.540
022	80	3.334	3.338	060			
003	35	3.142	3.143	206̄	40	1.534	{1.536 / 1.535
112	30	2.919	2.915	135̄			
113̄	10	2.702	2.706	332̄			
023				330	15	1.522	{1.522 / 1.522
131̄	60	2.626	{2.626 / 2.626	061			
200				136̄	15	1.472	{1.473 / 1.473
004	15	2.504	2.503	205̄			
132̄	55	2.438	{2.438 / 2.437	007	15	1.431	1.431
201				334̄			
040				332̄	2	1.399	{1.398 / 1.398
221̄	5	2.305	{2.309 / 2.303	063̄			
220				207	50	1.361	{1.361 / 1.360
203̄				136			
132	5 B	2.265	{2.283 / 2.266 / 2.265	401̄	5}B	1.331	{1.332 / 1.332
041̄				261̄			
222̄	2	2.209	{2.251 / 2.208	402̄	25}B	1.329	{1.329 / 1.328
135̄				260			
202	30	2.175	{2.175 / 2.174	335̄			
005	25}B	2.002	{2.003 / 1.995	064	30	1.312	{1.312 / 1.311
204̄	15}B	1.993	1.994	333			
133̄				403̄	10	1.302	{1.302 / 1.302
134̄	10	1.906	{1.908 / 1.908	261			
203				263̄	15	1.273	{1.274 / 1.273
135̄				401			
204	50	1.671	{1.672 / 1.671	226	5	1.260	1.259
006̄			1.669	008	15	1.253	1.252
313̄	2	1.647	1.648				

a = 5.335, b = 9.239, c = 10.172 Å, β = 100.11°
Specimen from Silver Crater Mine, Bancroft, Ontario, Canada.
Pattern by Bailey (unpublished), CrKα, 114.6 mm Gandolfi camera, Wilson mount,
int. est. visually, indices assigned by comparison with single crystal intensities.

Manganoan phlogopite-1M (manganophyllite)

hkℓ	I	d(obs)	d(calc)	hkℓ	I	d(obs)	d(calc)
001	90	10.12 Å	10.111 Å	204	45	1.679	{1.679 / 1.679
002	2	5.05	5.056	135̄			
020	15}B	4.59	4.602	311	5	1.6444	{1.645 / 1.645
110	15}B	4.56	4.550	313̄			
111	15	3.933	3.934	152	5	1.613	{1.614 / 1.614
112̄	30	3.664	3.666	243			
022	20	3.402	3.403	242	5	1.582	{1.581 / 1.581
003	100	3.371	3.370	153̄			
112	35	3.153	3.155	135	30	1.543	{1.543 / 1.543
113̄	20	2.928	2.926	206̄			
023	10	2.717	2.719	060	80	1.534	{1.534 / 1.534
201̄	5	2.647	{2.647 / 2.647	331̄			
130̄				061	15	1.516	{1.517 / 1.517
131̄	85	2.618	{2.617 / 2.617	330			
200				332̄			
004	10}B	2.526	2.528	205̄	15	1.481	{1.481 / 1.481
202̄	10}B		{2.508 / 2.507	136			
131		2.509		007	10	1.445	1.444
201	35	2.434	{2.434 / 2.434	225		1.410	1.410
132̄				063̄	2	1.397	{1.396 / 1.396
221̄	5	2.293	2.294	332̄			
132	5	2.266	{2.265 / 2.265	334̄			
203̄				136	90	1.369	{1.368 / 1.368
041	2	2.242	2.244	207			
202	35	2.174	{2.176 / 2.176	260	20	1.324	{1.323 / 1.323
133̄				402̄			
005	15	2.021	2.022	064	25	1.311	{1.311 / 1.311
133	15	1.997	1.999	333			
204̄				335̄			
203	5	1.912	{1.913 / 1.913	261	15	1.298	{1.298 / 1.298
134̄				403̄			
134	2	1.755	{1.753 / 1.753				
205̄							

a = 5.314, b = 9.204, c = 10.265 Å, β = 99.93°
Specimen from Långban, Sweden (2.7 wt. % MnO).
Pattern by Bailey (unpublished), CrKα, 114.6 mm Gandolfi camera, Wilson mount,
int. est. visually, indices assigned by comparison with single crystal intensities.

Lepidolites (Bailey, 1980)

A. 1M

hkℓ	I	d(obs.)	d(calc.)	hkℓ	I	d(obs.)	d(calc.)
001	40	9.97 A	9.959 A	33$\bar{1}$			{1.504
002	35	4.981	4.979	060	50	1.501	1.501
020	50	4.498	4.502	061			1.484
11$\bar{1}$	20	4.329	4.430	116	10	1.480	1.480
021	20	4.107	4.103	13$\bar{6}$			1.460
111	20	3.847	3.845	154	3	1.461	{1.445
11$\bar{2}$	70	3.615	3.617	22$\bar{6}$	3	1.444	1.445
022			{3.340	007			{1.423
003	70	3.329	3.320	117	3	1.421	1.422
112	70	3.086	3.085	225	10	1.378	1.380
113	35	2.895	2.890	207	10	1.353	1.353
023	30	2.669	2.672	136	20	1.339	1.342
20$\bar{1}$			2.599	046			1.336
130	80	2.594	{2.599	261			{1.300
13$\bar{1}$	100	2.565	2.565	40$\bar{2}$	35 B	1.297	1.299
11$\bar{3}$			{2.480	137			1.296
20$\bar{2}$	20	2.479	2.469	227			1.295
131	30	2.452	2.452	260	2	1.276	1.276
13$\bar{2}$	40	2.389	2.390	40$\bar{3}$	5	1.267	1.269
040			2.251	261			1.246
22$\bar{1}$	25	2.247	{2.251	26$\bar{3}$	15	1.246	{1.245
132	7	2.217	2.216	008			1.242
041	10	2.193	2.196	401	3	1.242	1.242
13$\bar{3}$	30	2.141	2.140	06$\bar{5}$			{1.199
202	20 B	2.125	2.126	172	10	1.197	1.198
221	10	2.102	2.103	42$\bar{1}$			1.198
042	2	2.051	2.051	334			1.195
005	40	1.994	1.992				
133	20	1.955	1.956				
22$\bar{4}$	10	1.809	1.809				
20$\bar{5}$	5	1.734	1.733				
13$\bar{5}$	12	1.715	1.717				
15$\bar{2}$	30	1.657	1.654				
24$\bar{1}$			{1.639				
31$\bar{3}$	10	1.637	1.635				
24$\bar{3}$			1.619				
152	10	1.619	{1.586				
242	20 B	1.582	1.579				
20$\bar{6}$	10	1.546	1.546				
314	2	1.524	1.526				
31$\bar{4}$			1.524				

A. a = 5.212, b = 9.005, c = 10.129 Å, β = 100.5°. Specimen from Auburn, Maine. 114.6 mm camera, Wilson mount, CuKα, int. est. visually, indices assigned by comparison with single crystal intensities.

B. 2M$_2$

hkℓ	I	d(obs.)	d(calc.)	hkℓ	I	d(obs.)	d(calc.)
002	70	10.01 A	10.000 A	133			{1.637
004	40	5.002	5.000	42$\bar{5}$	10	1.633	1.634
11$\bar{1}$	70	4.464	4.465	51$\bar{6}$			{1.609
202	10	4.338	4.342	3,1,11	5	1.611	1.600
11$\bar{3}$	20	3.883	3.882	42$\bar{3}$			{1.590
20$\bar{4}$	40	3.638	3.641	13$\bar{5}$	5 B	1.584	1.586
114	20	3.495	3.493	513			{1.572
11$\bar{4}$	70	3.338	3.333	319	5	1.572	1.572
006	60	3.205	3.208	13$\bar{5}$			1.549
114	40	3.119	3.123	427			{1.544
11$\bar{5}$	5			424	3	1.545	1.542
204	40	3.075	3.073	514			{1.522
20$\bar{6}$	25	2.910	2.913	1,1,12	2	1.521	1.521
115	30	2.872	2.868	33$\bar{1}$			1.503
116	30	2.794	2.793	60$\bar{2}$	60	1.502	1.502
021			{2.581	600			{1.481
310	100 B	2.578	2.575	2,0,12	2	1.478	1.480
116			2.574	2,2,11			1.476
008	8	2.502	2.500	2,2,10			{1.434
31$\bar{4}$			2.439	2,0,14	5	1.432	1.432
023	40 B	2.429	2.425	0,0,14			1.429
31$\bar{2}$			2.409	1,1,13			{1.419
31$\bar{5}$			2.326	3,1,13	5 B	1.417	1.415
024	3	2.317	2.309	0,2,12			1.404
31$\bar{3}$			2.291	1,1,14	6	1.397	{1.396
22$\bar{1}$	3	2.293	2.253	3,1,11			1.392
402	10	2.253	2.251	4,0,10	2	1.375	1.375
221	4	2.211	2.211	2,2,11	4	1.355	1.357
025	6	2.182	2.182	0,2,13			{1.324
317	10 B	2.070	2.070	2,2,13	2 B	1.324	1.323
026			{2.052	6,0,10			{1.300
315	10	2.041	2.033	041	30 B	1.298	1.299
408	40	2.001	2.000	62$\bar{1}$			1.297
0,0,10	5	1.821	1.820	339			1.294
3,1,10			1.705				
51$\bar{2}$			1.704				
42$\bar{1}$	10	1.701	1.703				
227			1.702				
131			{1.697				
029	5 B	1.689	1.692				
318			1.690				
424	5	1.674	{1.675				
			1.667				

B. a = 9.016, b = 5.206, c = 20.292 Å, β = 99.7°. Specimen from Hebron, Maine. 114.6 mm camera, Wilson mount, CuKα, int. est. visually, indices assigned by comparison with single crystal intensities.

Lepidolites (Bailey, 1980) continued

C. 3T

hk.ℓ	I	d(obs.)	d(calc.)	hk.ℓ	I	d(obs.)	d(calc.)
00.3	30	9.93 A	9.924 A	21.8	10	1.548	1.549
00.6	20	4.967	4.962	20.14			1.546
10.1	50	4.449	4.455	21.9	5	1.513	1.514
10.2	3	4.308	4.313	11.16			1.513
10.3	30	4.095	4.103	30.0	30	1.502	1.502
10.4	30	3.854	3.854	10.19			1.485
10.5	30	3.592	3.593	30.3	5 B	1.482	1.480
10.6		3.331	3.336	11.17	2	1.453	1.453
00.9	100 B	3.310	3.308	20.16	2	1.436	1.435
10.7	30	3.093	3.093	00.21			1.418
10.8	30	2.871	2.869	10.20	2	1.415	1.413
10.9	10	2.664	2.666	20.17	1	1.382	1.383
11.1̄	80	2.591	2.592	10.21	1	1.352	1.352
11.2	80	2.562	2.563	11.19			1.342
10.10	5	2.481	2.484	20.18	5	1.332	1.333
00.12			2.481	21.14			1.329
11.4	20	2.456	2.456	22.1			1.300
11.5	25	2.386	2.384	10.22	15 B	1.298	1.296
20.0			2.253	22.2̄			1.296
20.1	10	2.248	2.246	20.19			1.287
20.2	5	2.229	2.228	30.12	10	1.285	1.285
20.3	10	2.197	2.197	22.5	8	1.271	1.271
20.4	5	2.157	2.156	31.2̄			1.245
11.8	20	2.133	2.132	22.7	12 B	1.243	1.244
20.5	2	2.106	2.107	20.20			1.242
20.6	8	2.051	2.051	31.5			1.323
00.15	20	1.985	1.985	21.17	3	1.221	1.221
11.10	15	1.959	1.959	11.22			1.201
11.11	2	1.875	1.876	20.21	10	1.199	1.200
2,0,9	5	1.860	1.862	30.15			1.198
11.12	2	1.796	1.795				
20.11	3	1.731	1.731				
11.13	8	1.720	1.719				
21.1	5	1.699	1.700				
21.2	2	1.678	1.678				
21.3	2	1.660	1.660				
21.4		1.645	1.646				
11.14	20 B	1.639	1.637				
21.5	15	1.611	1.611				
21.6	8	1.580	1.581				

C. $a = b = 5.203$, $c = 29.771$ A, $\gamma = 120°$
Specimen from Kalgoorlie, Australia.
114.6 mm camera, Wilson mount, CuKα, int. est. visually, indices assigned by comparison with single crystal intensities.

Zinnwaldite-1M

hkℓ	I	d(obs)	d(calc)	hkℓ	I	d(obs)	d(calc)
001	100	9.90 A	9.893 A	044	5	1.674	1.674
002	5	4.92	4.947	151̄			1.674
020	30	4.54	4.550	135̄	50	1.655	1.656
110	25 B	4.49	4.497	204	50	1.640	1.640
111̄	10	4.365	4.372	225̄	5	1.620	1.623
021	10	4.131	4.134	311			1.621
111	20	3.863	3.865	243̄	5 B	1.600	1.600
112̄	40	3.641	3.637	152			1.592
022	60	3.349	3.349	242	5 B	1.557	1.556
003	50	3.295	3.298	026	5	1.550	1.550
112	50	3.089	3.086	206̄			1.527
113̄	30	2.895	2.895	331̄	45	1.518	1.519
023	20	2.667	2.670	060	40	1.511	1.511
201̄	15	2.624	2.626	135			1.499
130̄			2.617	061	5 B	1.499	1.499
131̄			2.591	330			1.472
200	60	2.588	2.586	116	5	1.472	1.472
202̄	5	2.490	2.493	136̄	10	1.459	1.459
113			2.474	226̄			1.447
004	20	2.473	2.473	205̄	10	1.447	1.446
131			2.472	117̄			1.416
132̄	30	2.411	2.410	007	10	1.415	1.413
201	25	2.391	2.394	063	30	1.379	1.378
114̄	15	2.333	2.334	225			1.378
040̄	15	2.274	2.275	207̄	35	1.352	1.352
221̄			2.275	155̄			1.339
220	15	2.248	2.249	136	60 B	1.338	1.339
041̄	15	2.216	2.217	046̄			1.335
222̄	15	2.185	2.186	401̄			1.314
133̄	25	2.153	2.152	261̄	10	1.313	1.313
202	20	2.131	2.133	402̄			1.313
042	2	2.066	2.067	260	5	1.308	1.308
223̄	2	2.019	2.018	335̄	10	1.302	1.302
005	30	1.977	1.979	137̄			1.296
133	20	1.958	1.961	262̄	15	1.295	1.296
134̄			1.889	064			1.293
043̄	2	1.871	1.873	403̄			1.289
203	2	1.818	1.818	333	15	1.289	1.288
224̄	2	1.735	1.737	261	5	1.281	1.281
205̄	5	1.731	1.731	263̄	2	1.259	1.258
223			1.718	226̄	35	1.238	1.237
134	15	1.717	1.717	008			1.237
150							

$a = 5.267$, $b = 9.101$, $c = 10.073$ A, $\beta = 100.51°$
Specimen from Zinnwald, Erzgebirge, Germany, U. Wis. Museum spec. #707/43141.
Pattern by Bailey (unpublished), CuKα, 114.6 mm Gandolfi camera, Wilson mount, int. est. visually, indices assigned by comparison with single crystal intensities.

A. Siderophyllite-1M			B. Masutomilite-1M			Ephesite-2M$_1$					
hkℓ	I	d(Å)	hkℓ	I	d(Å)	hkℓ	I	d(obs)	d(calc)	hkℓ	I	d(obs)	d(calc)

hkℓ	I	d(Å)	hkℓ	I	d(Å)	hkℓ	I	d(obs)	d(calc)	hkℓ	I	d(obs)	d(calc)
001	100	9.99	001	70	10.10	002	60	9.63 A	9.618 A	066	30	1.3422	1.3425
020,110	100	4.57	002	30	4.99	004	20	4.81	4.809	337			1.3424
111	20	3.89	020	25	4.54	110	75	4.40	4.421	335			1.3420
11$\bar{2}$	50	3.65	11$\bar{1}$	14	4.37	11$\bar{1}$			4.383	1,3,1$\bar{3}$	35	1.3060	1.3063
022	90	3.36	021	7	4.13	021	2	4.28	4.322	2,0,1$\bar{2}$			1.3058
		3.36	111	12	3.86	111			4.238	402			1.2786
003	90	3.27	11$\bar{2}$	45	3.64	022	5	4.04	4.027	2,2,1$\bar{3}$	20	1.2768	1.2758
112	50	3.13	112	65	3.35	11$\bar{3}$	10	3.776	3.778	26$\bar{2}$			1.2752
11$\bar{3}$	40	2.90	022	100	3.09	023	10	3.651	3.647	400			1.2748
200,13$\bar{1}$,20$\bar{1}$,130	100	2.62	003	60	2.903	113	10	3.509	3.515	339			1.2593
20$\bar{2}$,131,004,113	20	2.50	112	35	2.679	11$\bar{4}$	10	3.392	3.387	337			1.2593
201,13$\bar{2}$	80	2.432	11$\bar{3}$	20	2.607	006	60	3.213	3.206	1,1,1$\bar{5}$	10	1.2589	1.2589
040,22$\bar{1}$	10	2.301	023	30	2.589	114	10	3.136	3.137	2,0,14			1.2578
041	30	2.245	201,130	45	2.484	11$\bar{5}$	5	3.019	3.019	1,3,1$\bar{3}$			1.2570
20$\bar{2}$,13$\bar{3}$,420,221	70	2.158	200,13$\bar{1}$	17	2.401	025	15	2.910	2.906	264	5	1.2500	1.2498
20$\bar{4}$,133,005	70	1.979	20$\bar{2}$,131	25	2.270	115	15	2.799	2.797	40$\bar{2}$			1.2494
203,13$\bar{4}$	10	1.900	201	10	2.144	11$\bar{6}$	15	2.694	2.696	42$\bar{1}$			1.2294
025,22$\bar{4}$	<10	1.818	040,220,22$\bar{1}$	19	1.989	13$\bar{1}$	40	2.550	2.550	17$\bar{1}$	10	1.2290	1.2289
205,134	10	1.738	202,13$\bar{3}$	45	1.964	200			2.550	42$\bar{2}$			1.2286
204,13$\bar{5}$,241,15$\bar{2}$	80 B	1.666	005,204	12	1.737	13$\bar{1}$	85 B	2.521	2.521	264	10	1.221	1.2225
152,24$\bar{3}$	10	1.618	133	15	1.717	20$\bar{3}$,223			2.521	40$\bar{6}$			1.2225
242,15$\bar{3}$	10	1.576	20$\bar{5}$,223	6	1.653	13$\bar{5}$			2.413	353			1.2220
060,33$\bar{1}$,314	90	1.542	31$\bar{1}$,24$\bar{1}$	5	1.646	202	80	2.412	2.412	3,3,1$\bar{1}$	20	1.1719	1.1722
206,135,061,33$\bar{7}$+	20	1.523	13$\bar{5}$				20$\bar{4}$			2.360	0,6,10			1.1721
205,136	10	1.467	204,241	18	1.646	133	30	2.339	2.340	339			1.1718
007,117	<10	1.419	152,243	5	1.592	02$\bar{7}$			2.336	44$\bar{1}$			1.1083
063,332,33$\bar{4}$	10	1.393	15$\bar{3}$,224,242	4	1.552	13$\bar{5}$	30	2.177	2.175	44$\bar{2}$	5	1.1079	1.1077
207,136,027,046	30 B	1.351	060,33$\bar{1}$,135	6	1.512	20$\bar{4}$			2.174	2,6,10			1.0876
260,40$\bar{2}$	40	1.325	116	1	1.480	20$\bar{6}$	45	2.088	2.088	408			1.0871
064,31$\bar{5}$,333,206+	30 B	1.301	007,117	2	1.419	135	30 B		2.088	3,3,1$\bar{3}$	20 B	1.0870	1.0868
401,263,422,42$\bar{1}$+	10 B	1.275	063,33$\bar{4}$	1	1.371	0,0,10		1.928	1.924	0,6,12			1.0867
008,351,35$\bar{3}$	20	1.243	207	2	1.353	13$\bar{7}$	20	1.919	1.915	2,6,12			1.0864
065,334,336,047+	10 B	1.211	046,136	2	1.341	20$\bar{6}$			1.914	4,0,1$\bar{2}$			1.0864
			064,335	.1	1.297	20$\bar{8}$	20	1.833	1.832	2,0,18	15	1.0438	1.0439
						13$\bar{7}$			1.832	1,3,17			1.0438
						208	25	1.6761	1.6773	1,3,15	10	1.0182	1.0184
						2,0,1$\bar{0}$	40	1.6054	1.6766	0,6,14			1.0182
						139			1.6057	3,3,1$\bar{3}$			1.0065
						33$\bar{1}$			1.6053	4,0,14	10	1.0062	1.0064
						060	100	1.4788	1.4783	2,6,12			1.0061
						2,0,1$\bar{2}$			1.4780	4,0,1$\bar{6}$	10	0.9787	0.9789
						1,3,11	10	1.4152	1.4148	2,6,12			0.9787
						0,0,14	5	1.3738	1.3740	46$\bar{0}$			0.9654
										53$\bar{3}$,1,3,1$\bar{9}$	10	0.9653	0.9654
										2,0,18	20	0.9560	0.9560
													0.9558

A. a = 5.328, b = 9.228, c = 10.096 Å, β = 100.16°
Specimen from Altenberg, Saxony.
Pattern modified from Gottesmann (1962).

B. a = 5.252, b = 9.065, c = 10.107 Å, β = 100.49°
Specimen from Tanakamiyama, Shiga Prefecture, Japan.
Indexing modified from Harada et al. (1976) by comparison with single crystal intensities.

a = 5.120, b = 8.870, c = 19.312 Å, β = 95.09°
Specimen from Postmasburg, South Africa, USNM #117225.
Pattern by Bailey (unpublished), 114.6 mm camera, Wilson mount, CrKα, int. est. visually, indices assigned by comparison with single crystal intensities.

Preiswerkite-1Md

hkl	I	d(obs)	d(calc)	hkl	I	d(obs)	d(calc)
001	80	9.65 Å	9.649 Å	402̄	20	1.302	{1.301 / 1.301}
002	5	4.82	4.824	260			1.279
02-1̄	30 B	4.498	{4.525 / 4.472}	333̄			{1.279 / 1.279}
11-1̄				335̄			1.275
003	85	3.216	3.216	064	15 B	1.276	1.274
200	100	2.573	{2.572 / 2.571}	403			1.254
13̄1				261			1.246
202̄	25	2.458	{2.458 / 2.457}	17-1̄	5} B	1.254	1.246
131				401	5}	1.246	1.245
201	25	2.382	{2.382 / 2.381}	263̄			1.189
13̄2				35-1̄			1.188
203̄	20	2.210	{2.209 / 2.209}	334	10	1.188	1.188
132				336			1.170
13̄3̄	35	2.119	{2.119 / 2.118}	065			1.170
202				405	2	1.170	1.104
204̄	15 } B	1.939	{1.940 / 1.939}	263			1.104
133	20 }	1.931	1.930	406̄	10	1.103	1.059
005	15	1.853	1.854	264			1.059
203				404	15	1.059	1.059
134̄				266			1.024
15-1̄	5 } B	1.704	{1.707 / 1.700}	138			1.024
24-1̄	5 }			209̄	5 } B	1.024	1.018
205̄		1.695	{1.695 / 1.695}	208̄	5 }	1.017	1.018
134	20	1.622	{1.622 / 1.621}	139̄			1.017
135̄				336̄			0.9919
006	2	1.607	1.608	338̄			0.9916
33̄1̄	80	1.508	{1.509 / 1.508}	067	2	0.9918	0.9854
060				405̄			0.9853
206	40	1.488	{1.487 / 1.487}	267̄	15	0.9853	0.9851
135̄				531̄			0.9736
205̄	2	1.428	{1.427 / 1.426}	462			0.9735
136̄	2	1.379	1.378	191̄	15	0.9735	0.9733
007				530			0.9606
332				463̄			0.9606
334̄	20	1.365	{1.366 / 1.366}	192̄			0.9604
063̄				534	15	0.9604	
207̄	20	1.315	{1.316 / 1.316}	461			
136				192			

a = 5.227, b = 9.049, c = 9.804 Å, β = 100.23°.
Specimen from Geisspfad ultramafic complex, Pennine Alps. Pattern by Bailey (unpublished), 114.6 mm Gandolfi camera, CrKα, int. est. visually. Keusen & Peters (1980) have described a 2M₁ polytype from this locality, but their powder pattern is impure.

A. Montdorite-1M (Robert and Maury, 1979)   B. Taeniolite-1M (LaLonde, 1963)

hkl	I	d(Å)	hkl	I	d(obs)
001	43	9.97	001	74	10.04 Å
002	20	5.01	002	44	5.01
111̄	8	3.93	020	7	4.48*
112̄	12	3.66	111	7	3.87
022	35	3.41	112̄	17	3.60*
003	100	3.339	003	100	3.34
112	26	3.142	112	26	3.12
113̄	21	2.906	113	25	2.88
023	34	2.710	023	12	2.68
20̄1,130	5	2.649	131	12	2.60
200,131̄	20	2.617	200	13	2.575
113	39	2.519	004	9	2.50*
201,132̄	16	2.429	201,132̄	17	2.398
220	10	2.277	221̄	2	2.25
222̄	5	2.204	220	2	2.23
202,133̄	25	2.172	202,133̄	11	2.145
221	14	2.154	005	35	1.999
005	26	2.003	134̄	4	1.977*
204,133	17	1.992	135,006	17	1.655*
006			331̄	7	1.521
204,135̄	31	1.669	060	5	1.507
31̄4̄	10	1.543	116	3	1.491*
060,206̄	19	1.533	007	4	1.423
			136,207	11	1.350
			117,206	6	1.30

A. a = 5.310, b = 9.20, c = 10.175 Å, β = 99.9°. Specimen from La Bourboule, France.

B. *Intensity may be influenced by presence of dickite. a = 5.27, b = 9.13, c = 10.25, Å, β = 100°. Specimen from Magnet Cove, Ark. Single crystal study of a different specimen from this locality indicates the 3T structure.

BRITTLE MICAS

A. Clintonite-1M

hkℓ	I	d(obs.)	d(calc)
001	5	9.68 A	9.67 A
020	3	4.50	4.52
111	1/2	3.82	3.82
1̄12	1/2	3.55	3.56
022	1/2	3.30	3.30
003	7	3.21	3.22
112	1	3.05	3.04
1̄13	2	2.83	2.83
201̄	1/2	2.59	{2.59, 2.59}
130			
200	10	2.56	{2.56, 2.57}
1̄31			
202̄	5	2.45	{2.45, 2.45}
131			
201	4	2.37	{2.37, 2.38}
132̄			
203̄	4	2.20	{2.20, 2.20}
041̄			
132			
202	7	2.11	{2.11, 2.12, 2.10}
1̄33			
221̄			
005	4	1.931	1.931
043	2	1.851	1.849
150̄	2	1.705	{1.704, 1.702}
3̄11			
151	*	1.661	{1.661, 1.661}
242̄			
135	*	1.622	1.621
006	1/2	1.609	1.609
152̄	*	1.578	1.577
2̄43			
242	*	1.539	1.541
060	6	1.505	{1.506, 1.506}
3̄14			
330			
135̄	5	1.485	{1.486, 1.485}
3̄32			
063̄	3	1.363	{1.364, 1.362}
334̄			

*line is barely visible.

A. a = 5.20₄, b = 9.02₆, c = 9.81₂ A, β = 100.33°
Specimen from Zlatoust, S. Urals, Russia. (Forman et al., 1967).

B. Margarite-2M₁

hkℓ	I	d(obs.)
002	10	9.56 A
004	8	4.77
110	10	4.42
11̄1	10	4.37
021	4	4.28
111	4	4.23
022	6	4.01
11̄3	6	3.771
113	8	3.636
023̄	6	3.381
11̄4	8	3.224
024	8	3.180
006	100	3.123
114̄	14	3.001
115̄	4	2.883
025	8	2.775
115	10	2.687
11̄6	6	2.549
200	10	2.540
131̄	10	2.517
20̄2̄	25	2.508
131,116	18	2.407
202	12	2.399
133̄	8	2.386
207̄,133+	6	2.343
008	6	2.199
040	4	2.187
041,221	8	2.168
20̄4̄	8	2.161
135,22̄3̄	8	2.118
222,206̄	8	2.077
043,135+	12	1.947
136	6	1.908
0,0,10;2̄06	35	1.903
137,241̄	18	1.665
139̄,150+	10	1.596
2,0,10̄	10	1.593
139	10	1.486
33̄7̄	12	1.475
060	16	1.466

B. a = 5.118, b = 8.799, c = 19.159 A, β = 95.15°
Specimen from Shin-kiura mine, Oita Pref., Japan. (Aoki and Shimoda, 1965).

C. Anandite-2M₁

hkℓ	I	d(obs.)	d(calc.)
002	60	9.92 A	9.97 A
004	85	4.995	4.98
022	10	4.27	4.27
113̄	20	3.700	3.696
024	40	3.430	3.430
006	100	3.320	3.322
113̄	25	3.165	3.160
115	25	2.929	2.926
130	50	2.716	{2.722, 2.718, 2.713}
026			
13̄1̄			
117̄	45	2.681	2.681
131	15	2.521	2.522
008	80	2.490	2.492
222	20	2.240	2.242
0,0,10	2	1.991	{1.993, 1.989}
226̄	35		
0,0,12;24̄5̄	35	1.660	1.661

C. a = 5.415, b = 9.458, c = 20.013 A, β = 95.1° (Indexing modified from Pattiaratchi et al., 1967). Specimen from Wilagedera, Sri Lanka.

Anandite-2Or (Bailey, 1980)

hkℓ	I	d(obs.)	d(calc.)	hkℓ	I	d(obs.)	d(calc.)
002	20	9.91	9.926	155	10	1.633	{1.635, 1.633}
004	20	4.960	4.963	245			
022	5	4.279	4.283	1,3,10	10	1.607	{1.607, 1.606}
113	20	3.847	3.851	2,0,10			
024̄	40	3.431	3.431	060	25	1.582	{1.582, 1.579}
114̄	35	3.309	{3.426, 3.309}	330			
006	35	3.309	3.309	316	20 B	1.573	1.574
115	40	3.039	3.043	062̄	15 B	1.562	{1.563, 1.562, 1.559}
026̄				0,2,12̄			
131̄	100	2.710	{2.714, 2.713, 2.712, 2.707}	332			
116̄				064			
201				1,3,11̄	20 B	1.507	{1.508, 1.507, 1.506, 1.504}
132̄	40	2.635	{2.640, 2.634}	2,0,11̄			
202				334			
133̄	2	2.527	{2.530, 2.526}	1,1,13	20	1.453	1.453
203				0,0,14			
008	30	2.481	2.482	1,3,12	15	1.418	{1.418, 1.416, 1.415}
117	20	2.432	2.433	2,0,12̄			
134̄	10	2.395	2.398	261			
204	5 B	2.372	2.394	401	5	1.365	{1.366, 1.363}
221̄	5	2.348	2.374	0,2,14̄			
042̄	5 B	2.307	{2.351, 2.308, 2.304}	0,4,12	35 B	1.357	{1.359, 1.357, 1.356, 1.353}
222				2,2,12̄			
135	5	2.254	2.254	262			
205	2	2.230	2.251	402			
223	5	2.228	2.230	068			
028				1,3,13	20 B	1.332	{1.334, 1.334, 1.333, 1.332}
118	5	2.197	2.198	2,0,13̄			
044̄	7	2.140	{2.199, 2.141, 2.137}	338			
224̄				2,2,13̄	15	1.282	1.283
136	40	2.108	2.107	0,0,16	25	1.242	1.241
206							
225	2	2.030	2.034				
0,0,10	20	1.984	1.985				
137	5	1.969	{1.970, 1.968}				
207							
139̄	30	1.717	{1.718, 1.716}				
209							
154							
1,1,11	5 B	1.685	{1.687, 1.687, 1.685}				
244							
314							
0,0,12	25	1.654	1.683				

a = 5.465, b = 9.494, c = 19.853 A. Specimen from Wilagedera, Sri Lanka. 114.6 mm camera, Wilson mount, FeKα. Int. est. visually, indices assigned by comparison with single crystal intensities.

A. Bityite-2M₁

hkl	I	d(obs.)	d(calc.)
002	22	9.49 A	9.47 A
1̄11	22	4.31	
111	2	4.15	4.15
022	<2	3.94	3.95
1̄13	8	3.72	3.71
023	2	3.58	3.58
113	2	3.46	3.45
1̄14	2	3.332	3.327
024	<2	3.207	3.200
006	51	3.161	3.158
114	2	3.081	3.079
025	2	2.862	2.855
115	2	2.755	2.748
1̄31	20	2.496	2.497
131	100	2.469	{2.468 / 2.365
1̄33	35	2.363	2.363
202			{2.297
027	10	2.296	2.296
2̄04			{2.170
2̄21	3	2.164	2.165
220			{2.134
1̄35	5	2.129	2.132
204			{2.050
2̄06	83	2.046	2.047
135	21	1.894	1.895
0,0,10			{1.883
045			1.881
1̄37	12	1.882	1.879
2̄06			{1.801
208	10	1.798	1.798
137			{1.648
1̄39	17	1.648	1.646
208			{1.579
2,0,10	37 B	1.578	1.577
139			{1.450
1̄,3,11			1.448
2,0,10	60 B	1.448	1.448
3̄31			1.447
060			{1.315
335	8	1.315	1.315
066			{1.285
1̄,3,13	10	1.286	1.284
2,0,12			{1.249
400	7	1.250	1.249
2̄62			{1.224
2̄64	<2	1.235	1.224
402			{1.199
4̄06	<2	1.199	1.199
264			{1.197
1̄73			{1.187
0,0,16	<2	1.185	1.184
352			1.183
0,6,10			{1.150
339	3 B	1.149	1.149

B. Kinoshitalite-1M

hkl	I	d(obs.)	d(calc.)
001	3	10.0 A	10.00 A
002	3	4.99	5.00
020	6	4.61	4.62
1̄11	9	4.41	4.41
021	32	4.20	4.19
111	15	3.92	3.93
1̄12	22	3.66	3.66
022	34	3.39	3.39
003	4	3.33	3.34
112	29	3.144	3.141
1̄13	34	2.914	2.912
023	18	2.693	2.704
130	20	2.648	{2.654
2̄01			2.653
1̄31			{2.624
131	100	2.625	2.623
004	12	2.509	2.502
132	6	2.438	2.436
2̄04			{2.350
1̄14	6	2.350	2.350
040	10	2.308	2.308
041	12	2.251	2.249
220			{2.210
2̄22	8	2.210	2.206
1̄33			{2.173
133	25	2.171	2.172
202			{2.096
042	13	2.097	2.096
2̄04			{1.993
133	31	1.994	1.993
2̄05			{1.745
206	10	1.743	1.745
134			{1.718
3̄12	7	1.719	1.718
151	5	1.699	1.699
(?)241			{1.675
135	25 B	1.670	1.671
204			{1.670
311	19	1.647	1.646
2̄43			{1.616
152	8	1.616	1.616
060			{1.539
3̄31	33 B	1.537	1.538
206			{1.534
135			1.534

A. $a = 5.015$, $b = 8.682$, $c = 19.024$ A, $\beta = 95.14°$
Specimen from Mt. Bity, Madagascar, Harvard Museum spec. #87680.
Pattern by S. Guggenheim (unpublished), 114.6 mm Gandolfi camera, mono. CuKα with calibration ring, int. est. visually. Observed d-values are average of two photographs.

B. $a = 5.327$, $b = 9.233$, $c = 10.165$ A, $\beta = 100.09°$
Specimen from Nodatamagawa, Iwate Pref., Japan. Pattern by S. Guggenheim (unpublished), 114.6 mm Gandolfi camera, mono. CuKα with calibration ring, int. est. visually, some preferred orientation present. Observed d-values are average of two photographs.

Synthetic Ba-Li fluormica-1M

hkl	I	d(obs.)	d(calc.)
001	>100	9.9 A	9.94 A
002	85	5.0	4.97
021	13	4.19	4.18
1̄11	77	3.93	3.92
1̄12	62	3.65	3.64
022	54	3.39	3.38
003	17	3.312	3.312
112	54	3.127	3.126
1̄13	56	2.899	2.897
023	44	2.685	2.689
2̄01	60	2.645	2.648
1̄31	100	2.614	{2.617 / 2.617
200			{2.502
113	10	2.503	2.502
1̄32			{2.428
132	38	2.427	2.428
2̄01			{2.296
221	29	2.298	2.296
220	21	2.281	2.275
041	15	2.241	2.243
2̄22	15	2.199	2.200
1̄33			{2.164
202	87	2.164	2.164
221	15	2.150	2.148
042	12	2.088	2.089
2̄23	12	2.023	2.026
(?)005			{1.987
2̄04	48	1.984	1.984
133			{1.984
2̄05			{1.737
134	15	1.734	1.736
204			{1.662
135	60	1.662	1.662
3̄13			{1.643
311	27	1.641	1.642
2̄43			{1.611
152	27	1.612	1.611
242			{1.577
1̄53	25	1.577	1.577
3̄31	67	1.534	{1.535 / 1.535
060			{1.526
206	44 B	1.531	1.526
135	13	1.502	1.504
045			

$a = 5.317$, $b = 9.209$, $c = 10.094$ A, $\beta = 100.13°$
Composition $Ba_{0.97}(Mg_{2.23}Li_{0.77})(Si_{2.84}Al_{1.10})O_{9.90}F_{2.08}$.
Synthesis described by McCauley and Newnham (1973).
Pattern by S. Guggenheim (unpublished), 114.6 mm Gandolfi camera, mono. CuKα with calibration ring, int. measured by optical densitometer. Observed d-values are average of two photographs.